Ergebnisse der Mathematik und ihrer Grenzgebiete

Volume 54

3. Folge

A Series of Modern Surveys in Mathematics

For other titles published in this series, go to
www.springer.com/series/728

Michel Talagrand

Mean Field Models for Spin Glasses

Volume I: Basic Examples

Springer

Michel Talagrand
Université Paris 6
Institut de mathématiques
UMR 7586 CNRS
Place Jussieu 4
75252 Paris Cedex 05
France
michel@talagrand.net

This volume is the first part of a treatise on Spin Glasses in the series Ergebnisse der Mathematik und ihrer Grenzgebiete. The second part is Vol. 55 of the Ergebnisse series. The first edition of the treatise appeared as Vol. 46 of the same series (978-3-540-00356-4).

ISBN 978-3-642-15201-6 e-ISBN 978-3-642-15202-3
DOI 10.1007/978-3-642-15202-3
Springer Heidelberg Dordrecht London New York

Ergebnisse der Mathematik und ihrer Grenzgebiete. 3. Folge / A Series of Modern Surveys in Mathematics ISSN 0071-1136

Mathematics Subject Classification (2010): Primary: 82D30, 82B44. Secondary: 82C32, 60G15, 60K40

Cover design: VTEX, Vilnius

Printed on acid-free paper

Springer is part of Springer Science+Business Media (www.springer.com)

To Wansoo T. Rhee, for so many reasons.

Contents

Introduction

Let us denote by $\mathbb{S}_N$ the sphere of $\mathbb{R}^N$ of center 0 and radius $\sqrt{N}$, and by μ_N the uniform measure on $\mathbb{S}_N$. For $i, k \geq 1$, consider independent standard Gaussian random variables (r.v.s) $g_{i,k}$ and the subset U_k of $\mathbb{R}^N$ given by

$$U_k = \Big\{(x_1, \ldots, x_N) \in \mathbb{R}^N \ ; \ \sum_{i \leq N} g_{i,k} x_i \geq 0\Big\} .$$

The direction of the vector $(g_{i,k})_{i \leq N}$ is random (with uniform distribution over all possible directions) so that U_k is simply a half-space through the origin of random direction. (It might not be obvious now why we use Gaussian r.v.s to define a space of random direction, but this will become gradually clear.) Consider the set $\mathbb{S}_N \cap_{k \leq M} U_k$, the intersection of $\mathbb{S}_N$ with many such half-spaces. Denoting by E mathematical expectation, it should be obvious that

$$\mathsf{E}\bigg(\mu_N\Big(\mathbb{S}_N \bigcap_{k \leq M} U_k\Big)\bigg) = 2^{-M} , \tag{0.1}$$

because every point of $\mathbb{S}_N$ has a probability 2^{-M} to belong to all the sets U_k, $k \leq M$. This however is not really interesting. The fascinating fact is that when N is large and $M/N \simeq \alpha$, if $\alpha > 2$ the set $\mathbb{S}_N \cap_{k \leq M} U_k$ is typically empty (a classical result), while if $\alpha < 2$, with probability very close to 1, we have

$$\frac{1}{N} \log \mu_N\Big(\mathbb{S}_N \bigcap_{k \leq M} U_k\Big) \simeq \mathrm{RS}(\alpha) . \tag{0.2}$$

Here,

$$\mathrm{RS}(\alpha) = \min_{0<q<1} \bigg(\alpha \,\mathsf{E} \log \mathcal{N}\Big(\frac{z\sqrt{q}}{\sqrt{1-q}}\Big) + \frac{1}{2}\frac{q}{1-q} + \frac{1}{2}\log(1-q)\bigg) ,$$

where $\mathcal{N}(x)$ denotes the probability that a standard Gaussian r.v. g is $\geq x$, and where $\log x$ denotes (as everywhere through the book) the natural logarithm of x. Of course you should rush to require medical attention if this formula seems transparent to you. We simply give it now to demonstrate that we deal with a situation whose depth cannot be guessed beforehand. The wonderful fact (0.2) was not discovered by a mathematician, but by a physicist, E. Gardner. More generally theoretical physicists have discovered wonderful new areas of mathematics, which they have explored by their methods. This book is an attempt to correct this anomaly by exploring these areas using mathematical methods, and an attempt to bring these marvelous questions to the attention of the mathematical community. This is a book of *mathematics.* No knowledge of physics or statistical mechanics whatsoever is required or probably even useful to read it. If you read enough of this volume and the next, then in Volume II you will be able to understand (0.2).

More specifically, this is a book of probability theory (mostly). Attempting first a description at a "philosophical" level, a fundamental problem is as follows. Consider a large finite collection $(X_k)_{k\leq M}$ of random variables. What can we say about the largest of them? More generally, what can we say about the "few largest" of them? When the variables X_k are probabilistically independent, everything is rather easy. This is no longer the case when the variables are correlated. Even when the variables are identically distributed, the answer depends very much on their correlation structure. What are the correlation structures of interest? Most of the familiar correlation structures in Probability are low-dimensional, or even "one-dimensional". This is because they model random phenomena indexed by time, or, equivalently, by the real line, a one-dimensional object. In contrast with these familiar situations, the correlation structures considered here will be "high-dimensional" – in a sense that will soon become clear – and will create new and truly remarkable phenomena. This is a direction of probability theory that has not yet received the attention it deserves.

A natural idea to study the few largest elements of a given realization of the family $(X_k)_{k\leq M}$ is to assign weights to these elements, giving large weights to the large elements. Ideas from statistical mechanics suggest that, considering a parameter β, weights proportional to $\exp \beta X_k$ are particularly appropriate. That is, the (random) weight of the k-th element is

$$\frac{\exp \beta X_k}{\sum_{i\leq M} \exp \beta X_i} . \tag{0.3}$$

These weights define a random probability measure on the index set. Under an appropriate normalization, one can expect that this probability measure will be essentially supported by the indices k for which X_k is approximately a certain value $x(\beta)$. This is because the number of variables X_k close to a given large value x should decrease as x increases, while the corresponding weights increase, so that an optimal compromise should be reached at a certain level. The number $x(\beta)$ will increase with β. Thus we have a kind of "scanner" that enables us to look at the values of the family $(X_k)_{k\leq M}$ close to the (large) number $x(\beta)$, and this scanner is tuned with the parameter β.

We must stress an essential point. We are interested in what happens for a *typical* realization of the family (X_k). This can be very different (and much harder to understand) than what happens in average of all realizations. To understand the difference between typical and average, consider the situation of the Spinland State Lottery. It sells 10^{23} tickets at a unit price of one spin each. One ticket wins the single prize of 10^{23} spins. The average gain of a ticket is 1 spin, but the typical gain is zero. The average value is very different from the typical value because there is a large contribution coming from a set of very small probability. This is exactly the difference between (0.1) and (0.2). If $M/N \simeq \alpha < 2$, in average, $\mu_N(\mathbb{S}_N \cap_{k\le M} U_k) = 2^{-\alpha N}$, but typically $N^{-1}\log \mu_N(\mathbb{S}_N \cap_{k\le M} U_k) \simeq \mathrm{RS}(\alpha)$.

In an apparently unrelated direction, let us consider a physical system that can be in a (finite) number of possible configurations. In each configuration, the system has a given energy. It is maintained by the outside world at a given temperature, and is subject to thermal fluctuations. If we observe the system after it has been left undisturbed for a long time, what is the probability to observe it in a given configuration?

The system we will mostly consider is $\Sigma_N = \{-1,1\}^N$, where N is a (large) integer. A *configuration* $\boldsymbol{\sigma} = \{\sigma_1,\ldots,\sigma_N\}$ is an element of Σ_N. It tells us the values of the N "spins" σ_i, each of which can take the values ± 1. When in the configuration $\boldsymbol{\sigma}$, the system has an energy $H_N(\boldsymbol{\sigma})$. Thus H_N is simply a real-valued function on Σ_N. It is called the Hamiltonian of the system. We consider a parameter β (that physically represents the inverse of the temperature). We weigh each configuration proportionally to its so-called Boltzmann factor $\exp(-\beta H_N(\boldsymbol{\sigma}))$. This defines Gibbs' measure, a probability measure on Σ_N given by

$$G_N(\{\boldsymbol{\sigma}\}) = \frac{\exp(-\beta H_N(\boldsymbol{\sigma}))}{Z_N} \tag{0.4}$$

where the normalizing factor Z_N is given by

$$Z_N = Z_N(\beta) = \sum_{\boldsymbol{\sigma}} \exp(-\beta H_N(\boldsymbol{\sigma}))\,. \tag{0.5}$$

The summation is of course over $\boldsymbol{\sigma}$ in Σ_N. The factor Z_N is called the *partition function.* Statistical mechanics asserts that Gibbs' measure represents the probability of observing a configuration $\boldsymbol{\sigma}$ after the system has reached equilibrium with an infinite heat bath at temperature $1/\beta$. (Thus the expression "high temperature" will mean "β small" while the expression "low temperature" will mean "β large".) Of course the reader might wonder why (0.4) is the "correct definition". This is explained in physics books such as [102], [126], [125], and is not of real concern to us. That this definition is very fruitful will soon become self-evident.

The reason for the minus sign in Boltzmann's factor $\exp(-\beta H_N(\boldsymbol{\sigma}))$ is that the system favors low (and not high) energy configurations. It should be stressed that the (considerable...) depth of the innocuous looking definition (0.4) stems from the normalizing factor Z_N. This factor, the partition function, is the sum of many terms of widely different orders of magnitude, and it is unclear how to estimate it. The (few) large values become more important as β increases, and predominate over the more numerous smaller values. Thus the problem of understanding Gibbs' measure gets typically harder for large β (low temperature) than for small β (high temperature).

At this stage, **the reader has already learned *all* the statistical mechanics** (s)he needs to know to read this work.

The energy levels $H_N(\boldsymbol{\sigma})$ are closely related to the "interactions" between the spins. When we try to model a situation of "disordered interactions" these energy levels will become random variables, or, equivalently, the Hamiltonian, and hence Gibbs' measure, will become random. There are two levels of randomness (a probabilist's paradise). The "disorder", that is, the randomness of the Hamiltonian H_N, is given with our sample system. It does not evolve with the thermal fluctuations. It is frozen, or "quenched" as the physicists say. The word "glass" of the expression "spin glasses" conveys (among many others) this idea of frozen disorder.

Probably the reader has met with skepticism the statement that no further knowledge of statistical mechanics is required to read this book. She might think that this could be formally true, but that nonetheless it would be very helpful for her intuition to understand some of the classical models of statistical mechanics. This is not the case. When one studies systems at "high temperature" the fundamental mental picture is that of the model with random Hamiltonian $H_N(\boldsymbol{\sigma}) = -\sum_{i\le N} h_i\sigma_i$ where h_i are i.i.d. Gaussian random variables (that are not necessarily centered). This particular model is completely trivial because there is no interaction between the sites, so it reduces to a collection of N models consisting each of one single spin, and each acting on their own. (All the work is of course to show that this is in some sense the way things happen in more complicated models.) When one studies systems at "low temperature", matters are more complicated, but this is a completely new subject, and simply nothing of what had rigorously been proved before is of much help.

In modeling disordered interactions between the spins, the problem is to understand Gibbs' measure for a typical realization of the disorder. As we explained, this is closely related to the problem of understanding the large values among a typical realization of the family $(-H_N(\boldsymbol{\sigma}))$. This family is correlated. One reason for the choice of the index set Σ_N is that it is suitable to create extremely interesting correlation structures with simple formulas.

At the beginning of the already long story of spin glasses are "real" spin glasses, alloys with strange magnetic properties, which are of considerable interest, both experimentally and theoretically. It is believed that their remarkable properties arise from a kind of disorder among the interactions of magnetic impurities. To explain (at least qualitatively) the behavior of real spin glasses, theoretical physicists have invented a number of models. They fall into two broad categories: the "realistic" models, where the interacting atoms are located at the vertices of a lattice, and where the strength of the interaction between two atoms decreases when their distance increases; and the "mean-field" models, where the geometric location of the atoms in space is forgotten, and where each atom interacts with all the others. The mean-field models are of special interest to mathematicians because they are very basic mathematical objects and yet create extremely intricate structures. (As for the "realistic" models, they appear to be intractable at the moment.) Moreover, some physicists believe that these structures occur in a wide range of situations. The breadth, and the ambition, of these physicists' work can in particular be admired in the book "Spin Glass Theory and Beyond" by M. Mézard, G. Parisi, M.A. Virasoro, and in the book "Field Theory, Disorder and Simulation" by G. Parisi. The methods used by the physicists are probably best described here as a combination of heuristic arguments and numerical simulation. They are probably reliable, but they have no claim to rigor, and it is often not even clear how to give a precise mathematical formulation to some of the central predictions. The recent book [102] by M. Mézard and A. Montanari is much more friendly to the mathematically minded reader. It covers a wide range of topics, and succeeds well at conveying the depth of the physicists' ideas.

It was rather paradoxical for a mathematician like the author to see simple, basic mathematical objects being studied by the methods of theoretical physics. It was also very surprising, given the obvious importance of what the physicists have done, and the breadth of the paths they have opened, that mathematicians had not succeeded yet in proving any of their conjectures.

Despite considerable efforts in recent years, the program of giving a sound mathematical basis to the physicists' work is still in its infancy. We already have tried to make the case that in essence this program represents a new direction of probability theory. It is hence not surprising that, as of today, one has not yet been able to find anywhere in mathematics an already developed set of tools that would bear on these questions. Most of the methods used in this book belong in spirit to the area loosely known as "high-dimensional probability", but they are developed here from first principles. In fact, for much of the book, the most advanced tool that is not proved in complete detail is Hölder's inequality. The book is long because it attempts to fulfill several goals (that will be described below) but reading the first two chapters should be sufficient to get a very good idea of what spin glasses are about, as far as rigorous results are concerned.

The author believes that the present area has a tremendous long-term potential to yield incredibly beautiful results. There is of course no way of telling when progress will be made on the really difficult questions, but to provide an immediate incitement to seriously learn this topic, the author has stated as research problems a number of interesting questions (the solution of which would likely deserve to be published) that he believes are within the reach of the already established methods, but that he purposely did not, and will not, try to solve. (On the other hand, there is ample warning about the potentially truly difficult problems.)

This book, together with a forthcoming second volume, forms a second edition of our previous work [157],"Spin Glasses, a Challenge for Mathematicians". One of the goals in writing [157] was to increase the chance of significant progress by making sure that no stone was left unturned. This strategy greatly helped the author to obtain the solution of what was arguably at the time the most important problem about mean-field spin glasses, the validity of the "Parisi Formula". This advance occurred a few weeks before [157] appeared, and therefore could not be included there. Explaining this result in the appropriate context is a main motivation for this second edition, which also provides an opportunity to reorganize and rewrite with considerably more details all the material of the first edition.

The programs of conferences on spin glasses include many topics that are not touched here. This book is not an encyclopedia, but represents the coherent development of a line of thought. The author feels that the real challenge is the study of spin systems, and, among those, considers only pure mean-field models from the "statics" point of view. A popular topic is the study of "dynamics". In principle this topic also bears on mean-field models for spin glasses, but in practice it is as of today entirely disjoint from what we do here.

This work is divided in two volumes, that total a rather large number of pages. How is a reader supposed to attack this? The beginning of an answer is that many of the chapters are largely independent of each other, so that in practice these two volumes contain several "sub-books" that can be read somewhat independently of each other. For example, there is the "perceptron book" (Chapters 2, 3, 8, 9). On the other hand, we must stress that we progressively learn how to handle technical details. Unless the reader is already an expert, we highly recommend that she studies most of the first four chapters before attempting to read anything else in detail.

We now proceed to a more detailed description of the contents of the present volume. In Chapter 1 we study in great detail the Sherrington-Kirkpatrick model (SK), the "original" spin glass, at sufficiently high temperature. This model serves as an introduction to the basic ideas and methods. In the remainder of the present volume we introduce six more models. In this manner we try to demonstrate that the theory of spin glasses does not deal only with such and such very specific model, but that the basic phenomena occur again and again, as a kind of new law of nature (or at least of probability theory). We present enough material to provide a solid understanding of these models, but without including any of the really difficult results. In Chapters 2 and 3, we study the so-called "perceptron capacity model". This model is fundamental in the theory of neural networks, but the underlying mathematical problem is the rather attractive question of computing the "proportion" of the discrete cube (resp. the unit sphere) that is typically contained in the intersection of many random half-spaces, the question to which (0.2) answers in a special case. Despite the fact that the case of the cube and of the sphere are formally similar, the case of the sphere is substantially easier, because one can use there fundamental tools from convexity theory. In Chapter 4 we study the Hopfield model, using an approach invented by A. Bovier and V. Gayrard, that relies on the same tools from convexity as Chapter 3. This approach is substantially simpler than the approach first invented by the author, although it yields less complete results, and in particular does not seem to be able to produce either the correct rates of convergence or even to control a region of parameters of the correct shape. Chapter 5 introduces a new model, based on V-statistics. It is connected to the Perceptron model of Chapter 2, but with a remarkable twist. The last two chapters present models that are much more different from the previous ones than those are from each other. They require somewhat different methods, but illustrate well the great generality of the underlying phenomena. Chapter 6 studies a common generalization of the diluted SK model, and of the K-sat problem (a fundamental question of computer science). It is essentially different from the models of the previous chapters, since it is a model with "finite connectivity", that is, a spin interacts in average only with a number of spins that remains bounded as the size of the system increases (so we can kiss goodbye to the Central Limit Theorem). Chapter 7 is motivated by the random assignment problem. It is the least understood of all the models presented here, but must be included because of all the challenges it provides. An appendix recalls many basic facts of probability theory.

Let us now give a preview of the contents of the forthcoming Volume II. We shall first develop advanced results about the high-temperature behavior of some of the models that we introduce in the present volume. This work is heartfully dedicated to all the physicists who think that the expressions "high-temperature" and "advanced results" are contradictory. We shall demonstrate the depth of the theory even in this supposedly easier situation, and we shall present some of its most spectacular results. We shall return to the Perceptron model, to prove the celebrated "Gardner formula" that gives the proportion of the discrete cube (resp. the sphere, of which (0.2) is a special case) that lies in the intersection of many random half spaces. We shall return to the Hopfield model to present the approach through the cavity method that yields the correct rates of convergence, as well as a region of parameters of the correct shape. And we shall return to the SK model to study in depth the high-temperature phase in the absence of an external field.

In the rest of Volume II, we shall present low-temperature results. Besides the celebrated Ghirlanda-Guerra identities that hold very generally, essentially nothing is known outside the case of the SK model and some of its obvious generalizations, such as the p-spin interaction model for p even. For these models we shall present the basic ideas that underline the proof of the validity of the Parisi formula, as well as the complete proof itself. We shall bring attention to the apparently deep mysteries that remain, even for the SK model, the problem of ultrametricity and the problem of chaos. A final chapter will be devoted to the case of the p-spin interaction model, for p odd, for which the validity of the Parisi formula will be proved in a large region of parameters using mostly the cavity method.

At some point I must apologize for the countless typos, inaccuracies, or downright mistakes that this book is bound to contain. I have corrected many of each from the first edition, but doubtlessly I have missed some and created others. This is unavoidable. I am greatly indebted to Sourav Chatterjee, Albert Hanen and Marc Yor for laboring through this entire volume and suggesting literally hundreds of corrections and improvements. Their input was really invaluable, both at the technical level, and by the moral support it provided to the author. Special thanks are also due to Tim Austin, David Fremlin and Fréderique Watbled. Of course, all remaining mistakes are my sole responsibility.

This book owes its very existence to Gilles Godefroy. While Director of the Jussieu Institute of Mathematics he went out of his way to secure what has been in practice unlimited typing support for the author. Without such support this work would not even have been started.

While writing this book (and, more generally, while devoting a large part of my life to mathematics) it was very helpful to hold a research position without any other duties whatsoever. So it is only appropriate that I express here a life-time of gratitude to three colleagues, who, at crucial junctures, went far beyond their mere duties to give me a chance to get or to keep this position: Jean Braconnier, Jean-Pierre Kahane, Paul-André Meyer.

It is customary for authors, at the end of an introduction, to warmly thank their spouse for having granted them the peaceful time needed to complete their work. I find that these thanks are far too universal and overly enthusiastic to be believable. Yet, I must say simply that I have been privileged with a life-time of unconditional support. Be jealous, reader, for I yet have to hear the words I dread the most: "Now is not the time to work".

1. The Sherrington-Kirkpatrick Model

1.1 Introduction

Consider a large population of individuals (or atoms) that we label from 1 to N. Let us assume that each individual knows all the others. The feelings of the individual i towards the individual j are measured by a number g_{ij} that can be positive, or, unfortunately, negative. Let us assume symmetry, $g_{ij} = g_{ji}$, so only the numbers $(g_{ij})_{i<j}$ are relevant. We are trying to model a situation where these feelings are random. We are not trying to make realistic assumptions, but rather to find the simplest possible model; so let us assume that the numbers $(g_{ij})_{i<j}$ are independent random variables. (Throughout the book, the word "independent" should always be understood in the probabilistic sense.) Since we are aiming for simplicity, let us also assume that these random variables (r.v.s) are standard Gaussian. This is the place to point out that Gaussian r.v.s will often be denoted by lower case letters.

A very important feature of this model (called frustration, in physics) is that even if $g_{ij} > 0$ and $g_{jk} > 0$ (that is, i and j are friends, and j and k are friends), then i and k are just as likely to be enemies as they are to be friends. The interactions (g_{ij}) describe a very complex social situation.

Let us now think that we fix a typical realization of the numbers (g_{ij}). Here and elsewhere we say that an event is "typical" if (for large N) it occurs with probability close to 1. For example, the situation where nearly half of the r.v.s g_{ij} are > 0 is typical, but the situation where all of them are < 0 is certainly not typical. Let us choose the goal of separating the population in two classes, putting, as much as possible, friends together and enemies apart. It should be obvious that at best this can be done very imperfectly: some friends will be separated and some enemies will cohabit. To introduce a quantitative way to measure how well we have succeeded, it is convenient to assign to each individual i a number $\sigma_i \in \{-1, 1\}$, thereby defining two classes of individuals. Possibly the simplest measure of how well these two classes unite friends and separate enemies is the quantity

$$\sum_{i<j} g_{ij}\sigma_i\sigma_j \ . \tag{1.1}$$

M. Talagrand, *Mean Field Models for Spin Glasses*, Ergebnisse der Mathematik und ihrer Grenzgebiete. 3. Folge / A Series of Modern Surveys in Mathematics 54, DOI 10.1007/978-3-642-15202-3_1, © Springer-Verlag Berlin Heidelberg 2011

Trying to make this large invites making the quantities $g_{ij}\sigma_i\sigma_j$ positive, and thus invites in turn taking σ_i and σ_j of the same sign when $g_{ij} > 0$, and of opposite signs when $g_{ij} < 0$.

Despite the simplicity of the expression (1.1), the optimization problem of finding the maximum of this quantity (for a typical realization of the g_{ij}) over the configuration $\boldsymbol{\sigma} = (\sigma_1, \ldots, \sigma_N)$ appears to be of extreme difficulty, and little is rigorously known about it. Equivalently, one can look for a minimum of the function

$$-\sum_{i<j} g_{ij}\sigma_i\sigma_j .$$

Finding the minimum value of a function of the configurations is called in physics a zero-temperature problem, because at zero temperature a system is always found in its configuration of lowest energy. To a zero-temperature problem is often associated a version of the problem "with a temperature", here the problem corresponding to the Hamiltonian

$$H_N(\boldsymbol{\sigma}) = -\frac{1}{\sqrt{N}}\sum_{i<j} g_{ij}\sigma_i\sigma_j . \tag{1.2}$$

That is, we think of the quantity (1.2) as being the energy of the configuration $\boldsymbol{\sigma}$. The purpose of the normalization factor $N^{-1/2}$ will be apparent after (1.9) below. The energy level of a given configuration depends on the (g_{ij}), and this randomness models the "disorder" of the situation.

The minus signs in the Boltzmann factor $\exp(-\beta H_N(\boldsymbol{\sigma}))$ that arise from the physical requirement to favor configurations of low energy are a nuisance for mathematics. This nuisance is greatly decreased if we think that the object of interest is $(-H_N)$, i.e. that the minus sign is a part of the Hamiltonian. We will use this strategy throughout the book. Keeping with this convention, we write formula (1.2) as

$$-H_N(\boldsymbol{\sigma}) = \frac{1}{\sqrt{N}}\sum_{i<j} g_{ij}\sigma_i\sigma_j . \tag{1.3}$$

One goal is to understand the system governed by the Hamiltonian (1.3) at a given (typical) realization of the disorder (i.e. the r.v.s g_{ij}), or, equivalently, at a given realization of the (random) Hamiltonian H_N. To understand better this Hamiltonian, we observe that the energies $H_N(\boldsymbol{\sigma})$ are centered Gaussian r.v.s. The energies of two different configurations are however not independent. In fact, for two configurations $\boldsymbol{\sigma}^1$ and $\boldsymbol{\sigma}^2$, we have

$$\begin{aligned}\mathsf{E}(H_N(\boldsymbol{\sigma}^1)H_N(\boldsymbol{\sigma}^2)) &= \frac{1}{N}\sum_{i<j}\sigma_i^1\sigma_j^1\sigma_i^2\sigma_j^2 \\ &= \frac{N}{2}\left(\frac{1}{N}\sum_{i\le N}\sigma_i^1\sigma_i^2\right)^2 - \frac{1}{2} .\end{aligned} \tag{1.4}$$

Here we see the first occurrence of a quantity which plays an essential part in the sequel, namely

$$R_{1,2} = R_{1,2}(\boldsymbol{\sigma}^1, \boldsymbol{\sigma}^2) = \frac{1}{N} \sum_{i \le N} \sigma_i^1 \sigma_i^2 \,. \tag{1.5}$$

This quantity is called the *overlap* of the configurations $\boldsymbol{\sigma}^1, \boldsymbol{\sigma}^2$, because the closer it is to 1, the closer they are to each other. It depends on $\boldsymbol{\sigma}^1$ and $\boldsymbol{\sigma}^2$, even though the compact notation keeps the dependence implicit. The words "which plays an essential part in the sequel" have of course to be understood as "now is the time to learn and remember this definition, that will be used again and again". We can rewrite (1.4) as

$$\mathsf{E}(H_N(\boldsymbol{\sigma}^1) H_N(\boldsymbol{\sigma}^2)) = \frac{N}{2} R_{1,2}^2 - \frac{1}{2} \,. \tag{1.6}$$

Let us denote by $d(\boldsymbol{\sigma}^1, \boldsymbol{\sigma}^2)$ the Hamming distance of $\boldsymbol{\sigma}^1, \boldsymbol{\sigma}^2$, that is the proportion of coordinates where $\boldsymbol{\sigma}^1, \boldsymbol{\sigma}^2$ differ,

$$d(\boldsymbol{\sigma}^1, \boldsymbol{\sigma}^2) = \frac{1}{N} \mathrm{card}\{i \le N \,;\, \sigma_i^1 \ne \sigma_i^2\} \,. \tag{1.7}$$

Then

$$R_{1,2} = 1 - 2d(\boldsymbol{\sigma}^1, \boldsymbol{\sigma}^2) \,, \tag{1.8}$$

and this shows that $R_{1,2}$, and hence the correlation of the family $(H_N(\boldsymbol{\sigma}))$, is closely related to the structure of the metric space (Σ_N, d), where $\Sigma_N = \{-1, 1\}^N$. This structure is very rich, and this explains why the simple expression (1.3) suffices to create a complex situation. Let us also note that (1.4) implies

$$\mathsf{E} H_N^2(\boldsymbol{\sigma}) = \frac{N-1}{2} \,. \tag{1.9}$$

Here is the place to point out that to lighten notation we write $\mathsf{E} H_N^2$ rather than $\mathsf{E}(H_N^2)$, a quantity that should not be confused with $(\mathsf{E} H_N)^2$. The reader should remember this when she will meet expressions such as $\mathsf{E}|X - Y|^2$.

We can explain to the reader having some basic knowledge of Gaussian r.v.s the reason behind the factor $\sqrt{N}$ in (1.3). The 2^N Gaussian r.v.s $-H_N(\boldsymbol{\sigma})$ are not too much correlated; each one is of "size about $\sqrt{N}$". Their maximum should be of size about $\sqrt{N}\sqrt{\log 2^N}$, i.e. about N, see Lemma A.3.1. If one keeps in mind the physical picture that $H_N(\boldsymbol{\sigma})$ is the energy of the configuration $\boldsymbol{\sigma}$, a configuration of a N-spin system, it makes a lot of sense that as N becomes large the "average energy per spin" $H_N(\boldsymbol{\sigma})/N$ remains in a sense bounded independently of N. With the choice (1.3), some of the terms $\exp(-\beta H_N(\boldsymbol{\sigma}))$ will be (on a logarithmic scale) of the same order as the entire sum $Z_N(\beta) = \sum_{\boldsymbol{\sigma}} \exp(-\beta H_N(\boldsymbol{\sigma}))$, a challenging situation.

In the heuristic considerations leading to (1.3), we have made the assumption that any two individuals interact. This is certainly not the case if the individuals are atoms in a macroscopic sample of matter. Thus (1.3) is not, by any means, a realistic model for disordered interactions among atoms. A more realistic model would locate the atoms at the vertices of a lattice (e.g. $\mathbb{Z}^2$) and would assume that the strength of the interaction between two atoms decreases as their distance increases. The problem is that these models, while more interesting from the point of view of physics, are also extremely difficult. Even if one makes the simplifying assumption that an atom interacts only with its nearest neighbors, they are so difficult that, at the time of this writing, no consensus has been reached among physicists regarding their probable behavior. It is to bypass this difficulty that Sherrington and Kirkpatrick introduced the Hamiltonian (1.3), where the geometric location of the atoms is forgotten and where they all interact with each other. Such a simplification is called a "mean-field approximation", and the corresponding models are called "mean-field models".

The Hamiltonian (1.3) presents a very special symmetry. It is invariant under the transformation $\boldsymbol{\sigma} \mapsto -\boldsymbol{\sigma}$, and so is the corresponding Gibbs measure (0.4). This special situation is somewhat misleading. In order not to get hypnotized by special features, we will consider the version of (1.3) "with an external field", i.e. where the Hamiltonian is

$$-H_N(\boldsymbol{\sigma}) = \frac{1}{\sqrt{N}} \sum_{i<j} g_{ij}\sigma_i\sigma_j + h \sum_{i \le N} \sigma_i \,. \tag{1.10}$$

The reader might observe that the sentence "the Hamiltonian is" preceding (1.10) is not strictly true, since this formula actually give the value of $-H_N$ rather than H_N. It seems however harmless to allow such minor slips of language. The last term in (1.10) represents the action of an "external field", that is a magnetic field created by an apparatus outside the sample of matter we study. The external field favors the + spins over the − spins when $h > 0$. With the Hamiltonian (1.10), the Boltzmann factor $\exp(-\beta H_N(\boldsymbol{\sigma}))$ becomes

$$\exp\left(\frac{\beta}{\sqrt{N}} \sum_{i<j} g_{ij}\sigma_i\sigma_j + \beta h \sum_{i \le N} \sigma_i\right) . \tag{1.11}$$

The coefficient βh of $\sum_{i \le N} \sigma_i$ makes perfect sense to a physicist. However, when one looks at the mathematical structure of (1.11), one sees that the two terms $N^{-1/2} \sum_{i<j} g_{ij}\sigma_i\sigma_j$ and $\sum_{i \le N} \sigma_i$ appear to be of different natures. Therefore, it would be more convenient to have unrelated coefficients in front of these terms. For example, it is more cumbersome to take derivatives in β when using the factors (1.11) than when using the factors

$$\exp\left(\frac{\beta}{\sqrt{N}} \sum_{i<j} g_{ij}\sigma_i\sigma_j + h \sum_{i \le N} \sigma_i\right) .$$

Thus for the sake of mathematical clarity, it is better to abandon the physical point of view of having a "main parameter" β. Rather, we will think of the Hamiltonian as depending upon parameters $\beta, h, \ldots$ That is, we will write

$$- H_N(\boldsymbol{\sigma}) = -H_N(\boldsymbol{\sigma}, \beta, h) = \frac{\beta}{\sqrt{N}} \sum_{i<j} g_{ij}\sigma_i\sigma_j + h \sum_{i\leq N} \sigma_i \tag{1.12}$$

for the Hamiltonian of the Sherrington-Kirkpatrick model. The Boltzmann factor $\exp(-\beta H_N(\boldsymbol{\sigma}))$ then becomes

$$\exp(-H_N(\boldsymbol{\sigma}, \beta, h))$$

or, with simpler notation, when β and h are implicit it becomes

$$\exp(-H_N(\boldsymbol{\sigma})) \, . \tag{1.13}$$

Once one has survived the initial surprise of not seeing the customary term β in (1.13), this notation works out appropriately. The formulas familiar to the physicists can be recovered by replacing our term h by βh.

1.2 Notations and Simple Facts

The purpose of this book is to study "spin systems". The main parameter on which these depend is the number N of points of the system. To lighten notation, the number N remains often implicit in the notation, such as in (1.16) below. For all but the most trivial situations, certain exact computations seem impossibly difficult at a given value of N. Rather, we will obtain "approximate results" that become asymptotically exact as $N \to \infty$. As far as possible, we have tried to do quantitative work, that is to obtain optimal rates of convergence as $N \to \infty$.

Let us recall that throughout the book we write $\Sigma_N = \{-1,1\}^N$. Given a Hamiltonian H_N on $\Sigma_N = \{-1,1\}^N$, that is, a family of numbers $(H_N(\boldsymbol{\sigma}))_{\boldsymbol{\sigma}\in\Sigma_N}$, $\boldsymbol{\sigma} \in \Sigma_N$, we define its *partition function* Z_N by

$$Z_N = \sum_{\boldsymbol{\sigma}} \exp(-H_N(\boldsymbol{\sigma})) \, . \tag{1.14}$$

(Thus, despite its name, the partition function is a number, not a function.) Let us repeat that we are interested in understanding what happens for N very large. It is very difficult then to study Z_N, as there are so many terms, all random, in the sum. Throughout the book, we keep the letter Z to denote a partition function.

The Gibbs measure G_N on Σ_N with Hamiltonian H_N is defined by

$$G_N(\{\boldsymbol{\sigma}\}) = \frac{\exp(-H_N(\boldsymbol{\sigma}))}{Z_N} \, . \tag{1.15}$$

Exercise 1.2.1. Characterize the probability measures on Σ_N that arise as the Gibbs measure of a certain Hamiltonian. If the answer is not obvious to you, start with the case $N = 1$.

Given a function f on Σ_N, we denote by $\langle f \rangle$ its average for G_N,

$$\langle f \rangle = \int f(\boldsymbol{\sigma}) \mathrm{d}G_N(\boldsymbol{\sigma}) = \frac{1}{Z_N} \sum_{\boldsymbol{\sigma}} f(\boldsymbol{\sigma}) \exp(-H_N(\boldsymbol{\sigma})) \ . \tag{1.16}$$

Given a function f on $\Sigma_N^n = (\Sigma_N)^n$, we denote

$$\begin{aligned} \langle f \rangle &= \int f(\boldsymbol{\sigma}^1, \ldots, \boldsymbol{\sigma}^n) \mathrm{d}G_N(\boldsymbol{\sigma}^1) \cdots \mathrm{d}G_N(\boldsymbol{\sigma}^n) \\ &= \frac{1}{Z_N^n} \sum_{\boldsymbol{\sigma}^1, \ldots, \boldsymbol{\sigma}^n} f(\boldsymbol{\sigma}^1, \ldots, \boldsymbol{\sigma}^n) \exp\Big(-\sum_{\ell \le n} H_N(\boldsymbol{\sigma}^\ell)\Big) \ . \end{aligned} \tag{1.17}$$

The notations (1.16) and (1.17) are in agreement. For example, if, say, a function $f(\boldsymbol{\sigma}^1, \boldsymbol{\sigma}^2)$ on Σ_N^2 depends only on $\boldsymbol{\sigma}^1$, we can also view it as a function on Σ_N; and whether we compute $\langle f \rangle$ using (1.16) or (1.17), we get the same result.

The formula (1.17) means simply that we integrate f on the space $(\Sigma_N^n, G_N^{\otimes n})$. The configurations $\boldsymbol{\sigma}^1, \boldsymbol{\sigma}^2, \ldots$ belonging to the different copies of Σ_N involved there are called *replicas*. In probabilistic terms, the sequence $(\boldsymbol{\sigma}^\ell)_{\ell \ge 1}$ is simply an i.i.d. sequence distributed like the Gibbs measure. Replicas will play a fundamental role. In physics, they are called "real replicas", to distinguish them from the n replicas of the celebrated "replica method", where "n is an integer tending to 0". (There is no need yet to consult your analyst if the meaning of this last expression is unclear to you.) Throughout the book we denote replicas by upper indices. Again, this simply means that these configurations are integrated independently with respect to Gibbs' measure.

Replicas can be used in particular for "linearization" i.e. replacing a product of brackets $\langle \cdot \rangle$ by a single bracket. In probabilistic terms, this is simply the identity $\mathsf{E}X\mathsf{E}Y = \mathsf{E}XY$ when X and Y are independent r.v.s. Thus (with slightly informal but convenient notation) we have, for a function f on Σ_N,

$$\langle f \rangle^2 = \langle f(\boldsymbol{\sigma}^1) f(\boldsymbol{\sigma}^2) \rangle \ . \tag{1.18}$$

The partition function Z_N is exponentially large. It is better studied on a logarithmic scale through the quantity $N^{-1} \log Z_N$. This quantity is random; we denote by p_N its expectation

$$p_N = \frac{1}{N} \mathsf{E} \log Z_N \ . \tag{1.19}$$

Here, E denotes expectation over the "disorder", i.e. the randomness of the Hamiltonian. (Hence in the case of the Hamiltonian (1.12), this means expectation with respect to the r.v.s g_{ij}.) One has to prove in principle that the expectation exists. A sketch of proof is that the integrability of the function $\log Z_N$ can be obtained on the set $\{Z_N \geq 1\}$ by using that $\log x \leq x$ and that $\mathsf{E}Z_N < \infty$ and on the set $\{Z_N \leq 1\}$ by using that $Z_N \geq \exp(-H_N(\boldsymbol{\sigma}_0))$, where $\boldsymbol{\sigma}_0$ is any given point of Σ_N. This argument is "semi-trivial" in the sense that there is "a lot of room", and that it contains nothing fancy or clever. We have claimed in the introduction that this is a fully rigorous work. It seems however better to lighten the exposition in the beginning of this work by not proving a number of "semi-trivial" facts as above, and a great many statements will be given without a complete formal proof. Of course triviality is in the eye of the beholder, but it seems that either the reader is trained enough in analysis to complete the proofs of these facts without much effort (in the unlikely event that she feels this is really necessary), or else that she better take these facts for granted, since in any case *they are quite beside the main issues* we try to tackle. We fear that too much technicality at this early stage could discourage readers before they feel the beauty of the topic and are therefore better prepared to accept the unavoidable pain of technical work (which will be necessary soon enough). This policy of skipping some "details" will be used only at the beginning of this work, when dealing with "simple situations". In contrast, when we will later be dealing with more complicated situations, we will prove everything in complete detail.

The number p_N of (1.19) is of fundamental importance, and we first try to explain in words why. There will be many informal explanations such as this, in which the statements are a sometimes imprecise and ambiguous description of what happens, and are *usually by no means obvious.* Later (not necessarily in the same section) will come formal statements and complete proofs. If you find these informal descriptions confusing, please just skip them, and stick to the formal statements.

In some sense, as $N \to \infty$, the number p_N captures much important information about the r.v. $N^{-1} \log Z_N$. This is because (in all the cases of interest), this number p_N stays bounded below and above independently of N, while (under rather general conditions) the fluctuations of the r.v. $N^{-1} \log Z_N$ become small as $N \to \infty$ (its variance is about $1/N$). In physics' terminology, the random quantity $N^{-1} \log Z_N$ is "self-averaging". At a crude first level of approximation, one can therefore think of the r.v. $N^{-1} \log Z_N$ as being constant and equal to p_N. For the SK model, this will be proved on page 18.

Let us demonstrate another way that p_N encompasses much information about the system. For example, consider $p_N = p_N(\beta, h)$ obtained in the case of the Hamiltonian (1.12). Then we have

$$\begin{aligned}\frac{\partial}{\partial h}\frac{1}{N}\log Z_N &= \frac{1}{N}\frac{1}{Z_N}\frac{\partial Z_N}{\partial h}\\ &= \frac{1}{N}\frac{1}{Z_N}\sum_{\boldsymbol{\sigma}}\frac{\partial(-H_N(\boldsymbol{\sigma}))}{\partial h}\exp(-H_N(\boldsymbol{\sigma}))\end{aligned}$$

$$= \frac{1}{N} \frac{1}{Z_N} \sum_{\boldsymbol{\sigma}} \Bigl(\sum_{i \le N} \sigma_i \Bigr) \exp(-H_N(\boldsymbol{\sigma}))$$

$$= \frac{1}{N} \Bigl\langle \sum_{i \le N} \sigma_i \Bigr\rangle , \tag{1.20}$$

and *therefore*, taking expectation,

$$\frac{\partial p_N}{\partial h} = \frac{1}{N} \mathsf{E} \Bigl\langle \sum_{i \le N} \sigma_i \Bigr\rangle . \tag{1.21}$$

The *therefore* involves the interchange of a derivative and an expectation, which is in principle a non-trivial fact. Keeping in mind that Z_N is a rather simple function, a finite sum of very simple functions, we certainly do not expect any difficulty there or in similar cases. We have provided in Proposition A.2.1 a simple result that is sufficient for our needs, although we will not check this every time. In the present case the interchange is made legitimate by the fact that the quantities (1.20) are bounded by 1, so that (A.1) holds. Let us stress the main point. The interchange of limits is done here **at a given value of N**. In contrast, any statement involving limits as $N \to \infty$ (and first of all the existence of such limits) is typically much more delicate.

Let us note that

$$\frac{1}{N} \mathsf{E} \Bigl\langle \sum_{i \le N} \sigma_i \Bigr\rangle = \mathsf{E} \langle \sigma_1 \rangle ,$$

which follows from the fact that $\mathsf{E}\langle \sigma_i \rangle$ does not depend on i by symmetry. This argument will often be used. It is called "symmetry between sites". (A site is simply an $i \le N$, the name stemming from the physical idea that it is the site of a small magnet.) Therefore

$$\frac{\partial p_N}{\partial h} = \mathsf{E} \langle \sigma_1 \rangle , \tag{1.22}$$

the "average magnetization."

Since the quantity p_N encompasses much information, its exact computation cannot be trivial, even in the limit $N \to \infty$ (the existence of which is absolutely not obvious). As a first step one can try to get lower and upper bounds. A very useful fact for the purpose of finding bounds is Jensen's inequality, that asserts that for a convex function φ, one has

$$\varphi(\mathsf{E} X) \le \mathsf{E} \varphi(X) . \tag{1.23}$$

This inequality will be used a great many times (which means, as already pointed out, that it would be helpful to learn it now). For concave functions the inequality goes the other way, and the concavity of the log implies that

$$p_N = \frac{1}{N} \mathsf{E} \log Z_N \le \frac{1}{N} \log \mathsf{E} Z_N . \tag{1.24}$$

The right-hand side of (1.24) is not hard to compute, but the bound (1.24) is not really useful, as the inequality is hardly ever an equality.

Exercise 1.2.2. Construct a sequence Z_N of r.v.s with $Z_N \geq 1$ such that $\lim_{N\to\infty} N^{-1}\mathsf{E}\log Z_N = 0$ but $\lim_{N\to\infty} N^{-1}\log \mathsf{E}Z_N = 1$.

Throughout the book we denote by $\mathrm{ch}(x)$, $\mathrm{sh}(x)$ and $\mathrm{th}(x)$ the hyperbolic cosine, sine and tangent of x, and we write $\mathrm{ch}x, \mathrm{sh}x, \mathrm{th}x$ when no confusion is possible.

Exercise 1.2.3. Use (A.6) to prove that for the Hamiltonian (1.12) we have

$$\frac{1}{N}\log \mathsf{E}Z_N = \frac{\beta^2}{4}\Big(1-\frac{1}{N}\Big) + \log 2 + \log \mathrm{ch}(h)\,. \tag{1.25}$$

If follows from Jensen's inequality and the convexity of the exponential function that for a random variable X we have $\mathsf{E}\exp X \geq \exp \mathsf{E}X$. Using this for the uniform probability over Σ_N we get

$$2^{-N}\sum_{\boldsymbol{\sigma}}\exp(-H_N(\boldsymbol{\sigma})) \geq \exp\Big(2^{-N}\sum_{\boldsymbol{\sigma}} -H_N(\boldsymbol{\sigma})\Big)\,,$$

and taking logarithm and expectation this proves that $p_N \geq \log 2$. Therefore, combining with (1.24) and (1.25) we have (in the case of the Hamiltonian (1.12)), and lightening notation by writing $\mathrm{ch}h$ rather than $\mathrm{ch}(h)$,

$$\log 2 \leq p_N \leq \frac{\beta^2}{4}(1-\frac{1}{N}) + \log 2 + \log \mathrm{ch}h\,. \tag{1.26}$$

This rather crude bound will be much improved later. Let us also point out that the computation of p_N for every $\beta > 0$ provides the solution of the "zero-temperature problem" of finding

$$\frac{1}{N}\mathsf{E}\max_{\boldsymbol{\sigma}}(-H_N(\boldsymbol{\sigma}))\,. \tag{1.27}$$

Indeed,

$$\exp\big(\beta\max_{\boldsymbol{\sigma}}(-H_N(\boldsymbol{\sigma}))\big) \leq \sum_{\boldsymbol{\sigma}}\exp(-\beta H_N(\boldsymbol{\sigma})) \leq 2^N\exp\big(\beta\max_{\boldsymbol{\sigma}}(-H_N(\boldsymbol{\sigma}))\big)$$

so that, taking logarithm and expectation we have

$$\begin{aligned}\frac{\beta}{N}\mathsf{E}\max_{\boldsymbol{\sigma}}(-H_N(\boldsymbol{\sigma})) \leq p_N(\beta) &:= \frac{1}{N}\mathsf{E}\log\sum_{\boldsymbol{\sigma}}\exp(-\beta H_N(\boldsymbol{\sigma}))\\ &\leq \log 2 + \frac{\beta}{N}\mathsf{E}\max_{\boldsymbol{\sigma}}(-H_N(\boldsymbol{\sigma}))\end{aligned}$$

and thus

$$0 \le \frac{p_N(\beta)}{\beta} - \frac{1}{N}\mathsf{E}\max_{\boldsymbol{\sigma}}(-H_N(\boldsymbol{\sigma})) \le \frac{\log 2}{\beta} . \tag{1.28}$$

Of course the computation of $p_N(\beta)$ for large β (even in the limit $N \to \infty$) is very difficult but it is not quite as hopeless as a direct evaluation of $\mathsf{E}\max_{\boldsymbol{\sigma}}(-H_N(\boldsymbol{\sigma}))$.

For the many models we will consider in this book, the computation of p_N will be a central objective. We will be able to perform this computation in many cases at "high temperature", but the computation at "low temperature" remains a formidable challenge.

We now pause for a while and introduce a different and simpler Hamiltonian. It is not really obvious that this Hamiltonian is relevant to the study of the SK model, and that this is indeed the case is a truly remarkable feature. We consider an i.i.d. sequence $(z_i)_{i \le N}$ of standard Gaussian r.v.s. Consider the Hamiltonian

$$-H_N(\boldsymbol{\sigma}) = \sum_{i \le N} \sigma_i(\beta z_i \sqrt{q} + h) , \tag{1.29}$$

where q is a parameter, that will be adjusted in due time. The sequence $(\beta z_i \sqrt{q} + h)$ is simply an i.i.d. sequence of Gaussian r.v.s. (that are not centered if $h \neq 0$), so the random Hamiltonian (1.29) is rather canonical. It is also rather trivial, because there is no interaction between sites: the Hamiltonian is the sum of the terms $\sigma_i(\beta z_i \sqrt{q} + h)$, each of which depends only on the spin at one site. Let us first observe that if we are given numbers $a_i(1)$ and $a_i(-1)$ we have the identity

$$\sum_{\boldsymbol{\sigma}} \prod_{i \le N} a_i(\sigma_i) = \prod_{i \le N} (a_i(1) + a_i(-1)) . \tag{1.30}$$

Using this relation for

$$a_i(\sigma) = \exp(\sigma(\beta z_i \sqrt{q} + h)) \tag{1.31}$$

we obtain

$$\begin{aligned} Z_N &= \sum_{\sigma_i = \pm 1} \exp\Big(\sum_{i \le N} \sigma_i(\beta z_i \sqrt{q} + h)\Big) \\ &= \prod_{i \le N} (\exp(\beta z_i \sqrt{q} + h) + \exp(-(\beta z_i \sqrt{q} + h)) \\ &= 2^N \prod_{i \le N} \mathrm{ch}(\beta z_i \sqrt{q} + h) , \end{aligned} \tag{1.32}$$

where we recall that $\mathrm{ch}(x)$ denotes the hyperbolic cosine of x, so that

$$p_N = \log 2 + \mathsf{E} \log \mathrm{ch}(\beta z \sqrt{q} + h) \tag{1.33}$$

where z is a standard Gaussian r.v.

Consider now functions f_i on $\{-1,1\}$ and the function

$$f(\boldsymbol{\sigma}) = \prod_{i \le N} f_i(\sigma_i) \,.$$

Then, using (1.30) yields

$$\sum_{\boldsymbol{\sigma}} f(\boldsymbol{\sigma}) \prod_{i \le N} a_i(\sigma_i) = \sum_{\boldsymbol{\sigma}} \prod_{i \le N} f_i(\sigma_i) a_i(\sigma_i) = \prod_{i \le N} (f_i(1)a_i(1) + f_i(-1)a_i(-1)) \,.$$

Combining with (1.32) we get

$$\langle f(\boldsymbol{\sigma}) \rangle = \prod \langle f_i \rangle_i \,, \tag{1.34}$$

where

$$\langle f_i \rangle_i = \frac{f_i(1)a_i(1) + f_i(-1)a_i(-1)}{a_i(1) + a_i(-1)} \,. \tag{1.35}$$

This shows that Gibbs' measure is a product measure. It is determined by the averages $\langle \sigma_i \rangle$ because a probability measure μ on $\{-1,1\}$ is determined by $\int x \mathrm{d}\mu(x)$. To compute the average $\langle \sigma_i \rangle$, we use the case where $f(\boldsymbol{\sigma}) = \sigma_i$ and (1.34), (1.35), (1.31) to obtain

$$\langle \sigma_i \rangle = \frac{\exp(\beta z_i \sqrt{q} + h) - \exp(-(\beta z_i \sqrt{q} + h))}{\exp(\beta z_i \sqrt{q} + h) + \exp(-(\beta z_i \sqrt{q} + h))} \,,$$

and thus

$$\langle \sigma_i \rangle = \mathrm{th}(\beta z_i \sqrt{q} + h) \,, \tag{1.36}$$

where $\mathrm{th} x$ denotes the hyperbolic tangent of x. Moreover the quantities (1.36) are probabilistically independent.

In words, we can reduce the study of the system with Hamiltonian (1.29) to the study of the system with one single spin σ_i taking the possible values $\sigma_i = \pm 1$, and with Hamiltonian $H(\sigma_i) = -\sigma_i(\beta z_i \sqrt{q} + h)$.

Exercise 1.2.4. Given a number a, compute the averages $\langle \exp a\sigma_i \rangle$ and $\langle \exp a\sigma_i^1 \sigma_i^2 \rangle$ for the Hamiltonian (1.29). Of course as usual, the upper indexes denote different replicas, so $\langle \exp a\sigma_i^1 \sigma_i^2 \rangle$ is a "double integral". As in the case of (1.36), this reduces to the case of a system with one spin, and it is surely a good idea to master these before trying to understand systems with N spins. If you need a hint, look at (1.107) below.

Exercise 1.2.5. Show that if a Hamiltonian H on Σ_N decomposes as $H(\boldsymbol{\sigma}) = H_1(\boldsymbol{\sigma}) + H_2(\boldsymbol{\sigma})$ where $H_1(\boldsymbol{\sigma})$ depends only on the values of σ_i for $i \in I \subset \{1, \ldots, N\}$, while $H_1(\boldsymbol{\sigma})$ depends only on the values of σ_i for i in the complement of I, then Gibbs' measure is a product measure in a natural way. Prove the converse of this statement.

For Hamiltonians that are more complicated than (1.29), and in particular when different sites interact, the Gibbs measure will not be a product measure. Remarkably, however, it will often nearly be a product if one looks only at a "finite number of spins". That is, given any integer n (that does not depend on N), as $N \to \infty$, the law of the Gibbs measure under the map $\boldsymbol{\sigma} \mapsto (\sigma_1, \ldots, \sigma_n)$ becomes nearly a (random) product measure. Moreover, the r.v.s $(\langle \sigma_i \rangle)_{i \le n}$ become nearly independent. It will be proved in this work that this is true at high temperature for many models.

If one thinks about it, this is the simplest possible structure, the default situation. It is of course impossible to interest a physicist in such a situation. What else could happen, will he tell you. What else, indeed, but finding proofs is quite another matter.

Despite the triviality of the situation (1.29), an (amazingly successful) intuition of F. Guerra is that it will help to compare this situation with that of the SK model. This will be explained in the next section. This comparison goes quite far. In particular it will turn out that (when β is not too large) for each n the sequence $(\langle \sigma_i \rangle)_{i \le n}$ will asymptotically have the same law as the sequence $(\mathrm{th}(\beta z_i \sqrt{q} + h))_{i \le n}$, where z_i are i.i.d. standard Gaussian r.v.s and where the number q depends on β and h only. This should be compared to (1.36).

1.3 Gaussian Interpolation and the Smart Path Method

To study a difficult situation one can compare it to a simpler one, by finding a path between them and controlling derivatives along this path. This is an old idea. In practice we are given the difficult situation, and the key to the effectiveness of the method is to find the correct simple situation to which it should be compared. This can be done only after the problem is well understood. To insist upon the fact that the choice of the path is the real issue, we call this method the *smart path method*. (More precisely, the real issue is in the choice of the "easy end of the path". Once this has been chosen, the choice of the path itself will be rather canonical, except for its "orientation". We make the convention that the "smart path" moves *from* the "easy end" *to* the "hard end") The smart path method, under various forms, will be the main tool throughout the book.

In the present section, we introduce this method in the case of Gaussian processes. We obtain a general result of fundamental importance, Theorem 1.3.4 below, as well as two spectacular applications to the SK model. At the same time, we introduce the reader to some typical calculations.

Consider an integer M and an infinitely differentiable function F on $\mathbb{R}^M$ (such that all its partial derivatives are of "moderate growth" in the sense of (A.18)). Consider two centered jointly Gaussian families $\mathbf{u} = (u_i)_{i \le M}$, $\mathbf{v} = (v_i)_{i \le M}$. How do we compare

$$\mathsf{E}F(\mathbf{u}) \quad \text{and} \quad \mathsf{E}F(\mathbf{v})\ ? \tag{1.37}$$

Of course the quantity $\mathsf{E}F(\mathbf{u})$ is determined by the distribution of $\mathbf{u}$, and it might help to make sense of the formula (1.40) below to remember that this distribution is determined by its covariance matrix, i.e. by the quantities $\mathsf{E}u_i u_j$ (a fundamental property of Gaussian distributions). There is a canonical method to compare the quantities (1.37) (going back to [137]). Since we are comparing a function of the law of $\mathbf{u}$ with a function of the law of $\mathbf{v}$, we can assume without loss of generality that the families $\mathbf{u}$ and $\mathbf{v}$ are independent. We consider $\mathbf{u}(t) = (u_i(t))_{i\le M}$ where

$$u_i(t) = \sqrt{t}u_i + \sqrt{1-t}v_i\ , \tag{1.38}$$

so that $\mathbf{u} = \mathbf{u}(1)$ and $\mathbf{v} = \mathbf{u}(0)$, and we consider the function

$$\varphi(t) = \mathsf{E}F(\mathbf{u}(t))\ . \tag{1.39}$$

The following lemma relies on the Gaussian integration by parts formula (A.17), one of the most constantly used tools in this work.

Lemma 1.3.1. *For $0 < t < 1$ we have*

$$\varphi'(t) = \frac{1}{2}\sum_{i,j}(\mathsf{E}u_i u_j - \mathsf{E}v_i v_j)\mathsf{E}\frac{\partial^2 F}{\partial x_i \partial x_j}(\mathbf{u}(t))\ . \tag{1.40}$$

Proof. Let

$$u_i'(t) = \frac{\mathrm{d}}{\mathrm{d}t}u_i(t) = \frac{1}{2\sqrt{t}}u_i - \frac{1}{2\sqrt{1-t}}v_i$$

so that

$$\varphi'(t) = \mathsf{E}\sum_{i\le M} u_i'(t)\frac{\partial F}{\partial x_i}(\mathbf{u}(t))\ . \tag{1.41}$$

Now

$$\mathsf{E}u_i'(t)u_j(t) = \frac{1}{2}(\mathsf{E}u_i u_j - \mathsf{E}v_i v_j)$$

so the Gaussian integration by parts formula (A.17) yields (1.40). □

Of course (and this is nearly the last time in this chapter that we worry about this kind of problem) there is some extra work to do to give a complete ε-δ proof of this statement, and in particular to deduce (1.41) from (1.39) using Proposition A.2.1. The details of the argument are given in Section A.2.

Since Lemma 1.3.1 relies on Gaussian integration by parts, the reader might have already formed the question of what happens when one deals with non-Gaussian situations, such as when one replaces the r.v.s g_{ij} of (1.12) by, say, independent Bernoulli r.v.s (i.e. random signs), or by more general r.v.s. Generally speaking, the question of what happens in a probabilistic situation when one replaces Gaussian r.v.s by random signs can lead to very difficult (and interesting) problems, but in the case of the SK model, it is largely a purely technical question. While progressing through our various models, we will gradually learn how to address such technical problems. It will then become obvious that most of the results of the present chapter remain true in the Bernoulli case.

Even though the purpose of this work is to study spin glasses rather than to develop abstract mathematics, it might help to make a short digression about what is really going on in Lemma 1.3.1. The joint law of the Gaussian family $\mathbf{u}$ is determined by the matrix of the covariances $a_{ij} = \mathsf{E}u_i u_j$. This matrix is symmetric, $a_{ij} = a_{ji}$, so it is completely determined by the triangular array $\mathbf{a} = (a_{ij})_{1 \le i \le j \le n}$ and we can think of the quantity $\mathsf{E}F(\mathbf{u})$ as a function $\Psi(\mathbf{a})$. The domain of definition of Ψ is a convex cone with non-empty interior (since $\Psi(\mathbf{a})$ is defined if and only if the symmetric matrix $(a_{ij})_{i,j \le n}$ is positive definite), so it (often) makes sense to think of the derivatives $\partial\Psi/\partial a_{ij}$. The fundamental formula is as follows.

Proposition 1.3.2. *If $i < j$ we have*

$$\frac{\partial\Psi}{\partial a_{ij}}(\mathbf{a}) = \mathsf{E}\,\frac{\partial^2 F}{\partial x_i \partial x_j}(\mathbf{u}) \,, \tag{1.42}$$

while

$$\frac{\partial\Psi}{\partial a_{ii}}(\mathbf{a}) = \frac{1}{2}\mathsf{E}\frac{\partial^2 F}{\partial x_i^2}(\mathbf{u}) \,. \tag{1.43}$$

Let us first explain why this implies Lemma 1.3.1. If one thinks of a Gaussian family as determined by its matrix of covariance, the magic formula (1.40) is just the canonical interpolation in $\mathbb{R}^{n(n+1)/2}$ between the points $(a_{ij}) = (\mathsf{E}u_i u_j)$ and $(b_{ij}) := (\mathsf{E}v_i v_j)$, since

$$a_{ij}(t) := \mathsf{E}u_i(t)u_i(t) = t\mathsf{E}u_i u_j + (1-t)\mathsf{E}v_i v_j = t a_{ij} + (1-t) b_{ij} \,.$$

Therefore Lemma 1.3.1 follows from (1.42) and the chain rule, as is obvious if we observe that $\varphi(t) = \Psi(\mathbf{a}(t))$ where $\mathbf{a}(t) = (a_{ij}(t))_{1 \le i \le j \le n}$ and if we reformulate (1.40) as

$$\begin{aligned}\varphi'(t) = &\sum_{1 \le i < j \le n} (\mathsf{E}u_i u_j - \mathsf{E}v_i v_j)\mathsf{E}\,\frac{\partial^2 F}{\partial x_i \partial x_j}(\mathbf{u}(t)) \\ &+ \sum_{1 \le i \le n} \frac{1}{2}(\mathsf{E}\,u_i^2 - \mathsf{E}\,v_i^2)\mathsf{E}\,\frac{\partial^2 F}{\partial x_i^2}(\mathbf{u}(t)) \,.\end{aligned}$$

Proof of Proposition 1.3.2. Considering the triangular array

$$\mathbf{b} = (b_{ij})_{1 \le i \le j \le n} = (\mathsf{E}v_i v_j)_{1 \le i \le j \le n}$$

and integrating (1.40) between 0 and 1 we get

$$\begin{aligned}\Psi(\mathbf{a}) - \Psi(\mathbf{b}) &= \mathsf{E}F(\mathbf{u}) - \mathsf{E}F(\mathbf{v}) = \varphi(1) - \varphi(0) \\ &= \sum_{i,j} \frac{1}{2}(a_{ij} - b_{ij}) \int_0^1 \mathsf{E}\, \frac{\partial^2 F}{\partial x_i \partial x_j}(\mathbf{u}(t)) \mathrm{d}t \\ &= \sum_{1 \le i < j \le n} (a_{ij} - b_{ij}) \int_0^1 \mathsf{E}\, \frac{\partial^2 F}{\partial x_i \partial x_j}(\mathbf{u}(t)) \mathrm{d}t \\ &\quad + \sum_{i \le n} \frac{1}{2}(a_{ii} - b_{ii}) \int_0^1 \mathsf{E}\, \frac{\partial^2 F}{\partial x_i^2}(\mathbf{u}(t)) \mathrm{d}t\,. \end{aligned} \tag{1.44}$$

Now, as $\mathbf{b}$ gets close to $\mathbf{a}$ the integral

$$\int_0^1 \mathsf{E}\, \frac{\partial^2 F}{\partial x_i \partial x_j}(\mathbf{u}(t)) \mathrm{d}t$$

tends to

$$\mathsf{E}\, \frac{\partial^2 F}{\partial x_i \partial x_j}(\mathbf{u})\,,$$

because uniformly in t the distribution of $\mathbf{u}(t)$ gets close to the distribution of $\mathbf{u}$. Therefore

$$\begin{aligned}\Psi(\mathbf{b}) - \Psi(\mathbf{a}) = &\sum_{1 \le i < j \le n} (b_{ij} - a_{ij}) \mathsf{E}\, \frac{\partial^2 F}{\partial x_i \partial x_j}(\mathbf{u}) + \sum_{i \le n} \frac{1}{2}(b_{ii} - a_{ii}) \mathsf{E}\, \frac{\partial^2 F}{\partial x_i^2}(\mathbf{u})\,, \\ &+ \|\mathbf{b} - \mathbf{a}\| o(\|\mathbf{b} - \mathbf{a}\|)\,, \end{aligned} \tag{1.45}$$

where $\|\cdot\|$ denotes the Euclidean norm and $o(x)$ a quantity that goes to 0 with x. This concludes the proof. □

This ends our mathematical digression. To illustrate right away the power of the smart path method, let us prove a classical result (extensions of which will be useful in Volume II).

Proposition 1.3.3. (Slepian's lemma) *Assume that*

$$\forall i \ne j, \qquad \frac{\partial^2 F}{\partial x_i \partial x_j} \ge 0$$

and

$$\forall i \le M,\ \mathsf{E}u_i^2 = \mathsf{E}v_i^2;\ \forall i \ne j,\ \mathsf{E}u_i u_j \ge \mathsf{E}v_i v_j\,.$$

Then

$$\mathsf{E}F(\mathbf{u}) \ge \mathsf{E}F(\mathbf{v})\,.$$

Proof. It is obvious from (1.40) that $\varphi'(t) \ge 0$. □

The following is a fundamental property.

Theorem 1.3.4. *Consider a Lipschitz function F on $\mathbb{R}^M$, of Lipschitz constant $\le A$. That is, we assume that, given $\mathbf{x}, \mathbf{y}$ in $\mathbb{R}^M$, we have*

$$|F(\mathbf{x}) - F(\mathbf{y})| \le Ad(\mathbf{x}, \mathbf{y}) , \tag{1.46}$$

where d denotes the Euclidean distance on $\mathbb{R}^M$. If $g_1, \ldots, g_M$ are independent standard Gaussian r.v.s, and if $\mathbf{g} = (g_1, \ldots, g_M)$, then for each $t > 0$ we have

$$\mathsf{P}(|F(\mathbf{g}) - \mathsf{E}F(\mathbf{g})| \ge t) \le 2\exp\Big(-\frac{t^2}{4A^2}\Big) . \tag{1.47}$$

The remarkable part of this statement is that (1.47) *does not depend on* M. It is a typical occurrence of the "concentration of measure phenomenon".

When F is differentiable (and this will be the case for all the applications we will consider in this work) (1.46) is equivalent to the following

$$\forall \mathbf{x} \in \mathbb{R}^M , \ \|\nabla F(\mathbf{x})\| \le A , \tag{1.48}$$

where ∇F denotes the gradient of F, and $\|\mathbf{x}\|$ the Euclidean norm of the vector $\mathbf{x}$.

Proof. Let us first assume that F is infinitely differentiable (a condition that is satisfied in all the cases where we use this result). Given a parameter s, we would like to bound

$$\mathsf{E}\exp\big(s(F(\mathbf{g}) - \mathsf{E}F(\mathbf{g}))\big) . \tag{1.49}$$

At the typographic level, a formula as above is on the heavy side, and we will often omit the outer brackets when this creates no ambiguity, i.e. we will write $\exp s(F(\mathbf{g}) - \mathsf{E}F(\mathbf{g}))$. To bound the quantity (1.49) it is easier (using a fundamental idea of probability called symmetrization) to control first $\mathsf{E}\exp s(F(\mathbf{g}) - F(\mathbf{g}'))$ where $\mathbf{g}'$ is an independent copy of $\mathbf{g}$. (If you have never seen this, observe as a motivation that $F(\mathbf{g}) - F(\mathbf{g}')$ is the difference between two independent copies of the r.v. $F(\mathbf{g}) - \mathsf{E}F(\mathbf{g})$.)

Given s, we consider the function G on $\mathbb{R}^{2M}$ given by

$$G((y_i)_{i\le 2M}) = \exp s(F((y_i)_{i\le M}) - F((y_{i+M})_{i\le M})) .$$

We consider a family $\mathbf{u} = (u_i)_{i\le 2M}$ of independent standard Gaussian r.v.s and we would like to bound $\mathsf{E}G(\mathbf{u})$. The idea is to compare this situation with the much simpler quantity $\mathsf{E}G(\mathbf{v})$ where $v_i = v_{i+M}$ when $i \le M$ (so that $G(\mathbf{v}) = 1$ and hence $\mathsf{E}G(\mathbf{v}) = 1$). So let us consider a family $(v_i)_{i\le 2M}$ such that the r.v.s $(v_i)_{i\le M}$ are independent standard Gaussian, independent of the sequence $(u_i)_{i\le 2M}$, and such that $v_i = v_{i-M}$ if $i \ge M+1$, and therefore $v_i = v_{i+M}$ if $i \le M$. We note that

$$\mathsf{E}u_iu_j - \mathsf{E}v_iv_j = 0$$

unless $j = i + M$ or $i = j + M$ in which case

$$\mathsf{E}u_i u_j - \mathsf{E}v_i v_j = -1 \ .$$

We consider $\mathbf{u}(t)$ as in (1.38), and $\varphi(t) = \mathsf{E}G(\mathbf{u}(t))$. Using (1.40) for G rather than F, we get

$$\varphi'(t) = -\mathsf{E}\sum_{i\le M} \frac{\partial^2 G}{\partial y_i \partial y_{i+M}}(\mathbf{u}(t)) \ . \tag{1.50}$$

The reason why it is legitimate to use (1.40) is that "$\exp sF$ is of moderate growth" (as defined on page 440) since F is Lipschitz. We compute

$$\frac{\partial^2 G}{\partial y_i\, \partial y_{i+M}}(\mathbf{y}) = -s^2 \frac{\partial F}{\partial x_i}((y_i)_{i\le M}) \frac{\partial F}{\partial x_i}((y_{i+M})_{i\le M})\, G(\mathbf{y}) \ . \tag{1.51}$$

Now, (1.48) implies

$$\forall \mathbf{x} \in \mathbb{R}^M \ , \ \sum_{i\le M} \left(\frac{\partial F}{\partial x_i}(\mathbf{x})\right)^2 \le A^2 \ ,$$

and (1.50), (1.51) and the Cauchy-Schwarz inequality shows that

$$\varphi'(t) \le s^2 A^2 \varphi(t) \ . \tag{1.52}$$

As pointed out the choice of the family $(v_i)_{i\le 2M}$ ensures that $\varphi(0) = \mathsf{E}G(\mathbf{v}) = 1$, so that (1.52) implies that $\varphi(1) \le \exp s^2 A^2$, i.e.

$$\mathsf{E}\exp s(F(u_1,\dots,u_M) - F(u_{M+1},\dots,u_{2M})) \le \exp s^2 A^2 \ .$$

We use Jensen's inequality (1.23) for the convex function $\exp(-sx)$ while taking expectation in $u_{M+1},\dots,u_{2M}$, so that

$$\begin{aligned} &\mathsf{E}\exp s(F(u_1,\dots,u_M) - \mathsf{E}F(u_{M+1},\dots,u_{2M})) \\ &\le \mathsf{E}\exp s(F(u_1,\dots,u_M) - F(u_{M+1},\dots,u_{2M})) \le \exp s^2 A^2 \ . \end{aligned}$$

Going back to the notation $\mathbf{g}$ of Theorem 1.3.4, and since

$$\mathsf{E}F(u_{M+1},\cdots,u_{2M}) = \mathsf{E}F(\mathbf{g})$$

we have

$$\mathsf{E}\exp s(F(\mathbf{g}) - \mathsf{E}F(\mathbf{g})) \le \exp s^2 A^2 \ .$$

Using Markov's inequality (A.7) we get that for $s, t > 0$

$$\mathsf{P}(F(\mathbf{g}) - \mathsf{E}F(\mathbf{g}) \ge t) \le \exp(s^2 A^2 - st)$$

and, taking $s = t/(2A^2)$,

$$\mathsf{P}(F(\mathbf{g}) - \mathsf{E}F(\mathbf{g}) \ge t) \le \exp\Big(-\frac{t^2}{4A^2}\Big) \ .$$

Applying the same inequality to $-F$ completes the proof when F is infinitely differentiable (or even twice continuously differentiable). The general case (that is not needed in this book) reduces to this special case by convolution with a smooth function. □

The importance of Theorem 1.3.4 goes well beyond spin glasses, but it seems appropriate to state a special case that we will use many times.

Proposition 1.3.5. *Consider a finite set S and for $s \in S$ consider a vector $\mathbf{a}(s) \in \mathbb{R}^M$ and a number $w_s > 0$. Consider the function F on $\mathbb{R}^M$ given by*

$$F(\mathbf{x}) = \log \sum_{s \in S} w_s \exp \mathbf{x} \cdot \mathbf{a}(s) .$$

Then F has a Lipschitz constant $\le A = \max_{s \in S} \|\mathbf{a}(s)\|$.

Consequently if $g_1, \ldots, g_M$ are independent standard Gaussian r.v.s, and if $\mathbf{g} = (g_1, \ldots, g_M)$, then for each $t > 0$ we have

$$\mathsf{P}(|F(\mathbf{g}) - \mathsf{E}F(\mathbf{g})| \ge t) \le 2 \exp\left(-\frac{t^2}{4 \max_{s \in S} \|\mathbf{a}(s)\|^2}\right) . \tag{1.53}$$

Proof. The gradient $\boldsymbol{\nabla} F(\mathbf{x})$ of F at $\mathbf{x}$ is given by

$$\boldsymbol{\nabla} F(\mathbf{x}) = \frac{\sum_{s \in S} w_s \mathbf{a}(s) \exp \mathbf{x} \cdot \mathbf{a}(s)}{\sum_{s \in S} w_s \exp \mathbf{x} \cdot \mathbf{a}(s)} ,$$

so that $\|\boldsymbol{\nabla} F(\mathbf{x})\| \le \max_{s \in S} \|\mathbf{a}(s)\|$, and we conclude from (1.47) using the equivalence of (1.46) and (1.48). □

As a first example of application, let us consider the case where $M = N(N-1)/2$, $S = \Sigma_N$, and, for $s = \boldsymbol{\sigma} \in S$, $w_{\boldsymbol{\sigma}} = \exp h \sum_{i \le N} \sigma_i$ and

$$\mathbf{a}(\boldsymbol{\sigma}) = \left(\frac{\beta}{\sqrt{N}} \sigma_i \sigma_j\right)_{1 \le i < j \le N} .$$

Therefore

$$\|\mathbf{a}(\boldsymbol{\sigma})\| = \frac{\beta}{\sqrt{N}} \left(\frac{N(N-1)}{2}\right)^{1/2} \le \beta \sqrt{\frac{N}{2}} .$$

It follows from (1.53) that the partition function Z_N of the Hamiltonian (1.12) satisfies

$$\mathsf{P}(|\log Z_N - \mathsf{E} \log Z_N| \ge t) \le 2 \exp\Big(-\frac{t^2}{2\beta^2 N}\Big) . \tag{1.54}$$

If $U_N = N^{-1} \log Z_N$, we can rewrite this as

$$\mathsf{P}(|U_N - \mathsf{E}U_N| \ge t) \le 2 \exp\Big(-\frac{t^2 N}{2\beta^2}\Big) .$$

The right hand side starts to become small for t about $N^{-1/2}$, i.e. it is unlikely that U_N will differ from its expectation by more than a quantity of order $N^{-1/2}$. In words, the fluctuations of the quantity $U_N = N^{-1}\log Z_N$ are typically of order at most $N^{-1/2}$, while the quantity itself is of order 1. This quantity is "self-averaging", a fundamental fact, as was first mentioned on page 7.

Let us try now to use (1.40) to compare two Gaussian Hamiltonians. This technique is absolutely fundamental. It will be first used to make precise the intuition of F. Guerra mentioned on page 12, but at this stage we try to obtain a result that can also be used in other situations. We take $M = 2^N = \mathrm{card}\Sigma_N$. We consider two jointly Gaussian families $\mathbf{u} = (u_{\boldsymbol{\sigma}})$ and $\mathbf{v} = (v_{\boldsymbol{\sigma}})$ $(\boldsymbol{\sigma} \in \Sigma_N)$, which we assume to be independent from each other. We recall the notation

$$u_{\boldsymbol{\sigma}}(t) = \sqrt{t}u_{\boldsymbol{\sigma}} + \sqrt{1-t}v_{\boldsymbol{\sigma}} \quad ; \quad \mathbf{u}(t) = (u_{\boldsymbol{\sigma}}(t))_{\boldsymbol{\sigma}} \ ,$$

and we set

$$U(\boldsymbol{\sigma},\boldsymbol{\tau}) = \frac{1}{2}(\mathsf{E}u_{\boldsymbol{\sigma}}u_{\boldsymbol{\tau}} - \mathsf{E}v_{\boldsymbol{\sigma}}v_{\boldsymbol{\tau}}) \ . \tag{1.55}$$

Then (1.40) asserts that for a (well-behaved) function F on $\mathbb{R}^M$, if $\varphi(t) = \mathsf{E}F(\mathbf{u}(t))$ we have

$$\varphi'(t) = \mathsf{E}\sum_{\boldsymbol{\sigma},\boldsymbol{\tau}} U(\boldsymbol{\sigma},\boldsymbol{\tau}) \frac{\partial^2 F}{\partial x_{\boldsymbol{\sigma}}\partial x_{\boldsymbol{\tau}}}(\mathbf{u}(t)) \ . \tag{1.56}$$

Let us assume that we are given numbers $w_{\boldsymbol{\sigma}} > 0$. For $\mathbf{x} = (x_{\boldsymbol{\sigma}}) \in \mathbb{R}^M$ let us define

$$F(\mathbf{x}) = \frac{1}{N}\log Z(\mathbf{x}) \ ; \ Z(\mathbf{x}) = \sum_{\boldsymbol{\sigma}} w_{\boldsymbol{\sigma}} \exp x_{\boldsymbol{\sigma}} \ . \tag{1.57}$$

Thus, if $\boldsymbol{\sigma} \neq \boldsymbol{\tau}$ we have

$$\frac{\partial^2 F}{\partial x_{\boldsymbol{\sigma}}\partial x_{\boldsymbol{\tau}}}(\mathbf{x}) = -\frac{1}{N}\frac{w_{\boldsymbol{\sigma}}w_{\boldsymbol{\tau}}\exp(x_{\boldsymbol{\sigma}}+x_{\boldsymbol{\tau}})}{Z(\mathbf{x})^2} \ ,$$

while if $\boldsymbol{\sigma} = \boldsymbol{\tau}$ we have

$$\frac{\partial^2 F}{\partial x_{\boldsymbol{\sigma}}^2}(\mathbf{x}) = \frac{1}{N}\left(\frac{w_{\boldsymbol{\sigma}}\exp x_{\boldsymbol{\sigma}}}{Z(\mathbf{x})} - \frac{w_{\boldsymbol{\sigma}}^2\exp 2x_{\boldsymbol{\sigma}}}{Z(\mathbf{x})^2}\right) \ .$$

Exercise 1.3.6. Prove that the function F and its partial derivatives of order 1 satisfy the "moderate growth condition" (A.18). (Hint: Use a simple bound from below on $Z(\mathbf{x})$, such as $Z(\mathbf{x}) \geq w_{\boldsymbol{\tau}}\exp x_{\boldsymbol{\tau}}$ for a given $\boldsymbol{\tau}$ in Σ_N.)

This exercise shows that it is legitimate to use (1.56) to compute the derivative of $\varphi(t) = \mathsf{E}F(\mathbf{u}(t))$, which is therefore

$$\varphi'(t) = \frac{1}{N}\mathsf{E}\Big(\frac{1}{Z(\mathbf{u}(t))}\sum_{\boldsymbol{\sigma}} U(\boldsymbol{\sigma},\boldsymbol{\sigma}) w_{\boldsymbol{\sigma}} \exp u_{\boldsymbol{\sigma}}(t)$$
$$- \frac{1}{Z(\mathbf{u}(t))^2}\sum_{\boldsymbol{\sigma},\boldsymbol{\tau}} U(\boldsymbol{\sigma},\boldsymbol{\tau}) w_{\boldsymbol{\sigma}} w_{\boldsymbol{\tau}} \exp(u_{\boldsymbol{\sigma}}(t) + u_{\boldsymbol{\tau}}(t))\Big). \quad (1.58)$$

Let us now denote by

$$\langle\cdot\rangle_t$$

an average for the Gibbs measure with Hamiltonian

$$-H_t(\boldsymbol{\sigma}) = u_{\boldsymbol{\sigma}}(t) + \log w_{\boldsymbol{\sigma}} = \sqrt{t}u_{\boldsymbol{\sigma}} + \sqrt{1-t}v_{\boldsymbol{\sigma}} + \log w_{\boldsymbol{\sigma}} . \quad (1.59)$$

Any function f on Σ_N satisfies the formula

$$\langle f\rangle_t = \frac{\sum_{\boldsymbol{\sigma}} f(\boldsymbol{\sigma})\exp(-H_t(\boldsymbol{\sigma}))}{\sum_{\boldsymbol{\sigma}}\exp(-H_t(\boldsymbol{\sigma}))} = \frac{\sum_{\boldsymbol{\sigma}} w_{\boldsymbol{\sigma}} f(\boldsymbol{\sigma})\exp u_{\boldsymbol{\sigma}}(t)}{\sum_{\boldsymbol{\sigma}} w_{\boldsymbol{\sigma}}\exp u_{\boldsymbol{\sigma}}(t)} .$$

The notation $\langle\cdot\rangle_t$ will be used *many times* in the sequel. It would be nice to remember now that the index t simply refers to the value of the interpolating parameter. This will be the case whenever we use an interpolating Hamiltonian. If you forget the meaning of a particular notation, you might try to look for it in the glossary or the index, that attempt to list for many of the typical notations the page where it is defined.

Thus (1.58) simply means that

$$\varphi'(t) = \frac{1}{N}(\mathsf{E}\langle U(\boldsymbol{\sigma},\boldsymbol{\sigma})\rangle_t - \mathsf{E}\langle U(\boldsymbol{\sigma}^1,\boldsymbol{\sigma}^2)\rangle_t) . \quad (1.60)$$

In the last term the bracket is a double integral for Gibbs' measure, and the variables are denoted $\boldsymbol{\sigma}^1$ and $\boldsymbol{\sigma}^2$ rather than $\boldsymbol{\sigma}$ and $\boldsymbol{\tau}$.

The very general formula (1.60) applies to the interpolation between any two Gaussian Hamiltonians, and is rather fundamental in the study of such Hamiltonians.

We should observe for further use that (1.60) even holds if the quantities $w_{\boldsymbol{\sigma}}$ are random, provided their randomness is independent of the randomness of $u_{\boldsymbol{\sigma}}$ and $v_{\boldsymbol{\sigma}}$. This is seen by proving (1.60) at $w_{\boldsymbol{\sigma}}$ given, and taking a further expectation in the randomness of these quantities. (When doing this, we permute expectation in the r.v.s $w_{\boldsymbol{\sigma}}$ and differentiation in t. Using Proposition A.2.1 this is permitted by the fact that the quantity (1.60) is uniformly bounded over all choices of $(w_{\boldsymbol{\sigma}})$ by (1.65) below.)

The consideration of Hamiltonians such as (1.29) shows that it is natural to consider "random external fields". That is, we consider an i.i.d. sequence $(h_i)_{i\le N}$ of random variables, having the same distribution as a given r.v. h (with moments of all orders). We assume that this sequence is independent of all the other r.v.s. Rather than the Hamiltonian (1.12) we consider instead the more general Hamiltonian

$$-H_N(\boldsymbol{\sigma}) = \frac{\beta}{\sqrt{N}} \sum_{i<j} g_{ij}\sigma_i\sigma_j + \sum_{i\le N} h_i\sigma_i \,. \tag{1.61}$$

There is presently nothing to change either to the notation or the proofs to consider this more general case, so this will be our setting. Whenever extra work would be needed to handle this case, we will come back to the case where h_i is non-random.

Since there are now two sources of randomness in the disorder, namely the g_{ij} and the h_i, this is the place to mention that throughout the book, and unless it is explicitly specified otherwise, as is absolutely standard, the notation E will stand for expectation over *all* these sources of randomness. When we have two (or more) independent sources of randomness like here, and we want to take expectation only on, say, the r.v.s g_{ij}, we will say just that, or (as probabilists often do) that we take expectation conditionally on the r.v.s h_i, or given the r.v.s h_i.

To compare (following Guerra) the Hamiltonian (1.61) with the simpler Hamiltonian (1.29) we use (1.60) in the case

$$u_{\boldsymbol{\sigma}} = \frac{\beta}{\sqrt{N}} \sum_{i<j\le N} g_{ij}\sigma_i\sigma_j \; ; \; v_{\boldsymbol{\sigma}} = \beta \sum_{i\le N} z_i\sqrt{q}\sigma_i \; ; \; w_{\boldsymbol{\sigma}} = \exp\Big(\sum_{i\le N} h_i\sigma_i\Big) \,, \tag{1.62}$$

where $0 \le q \le 1$ is a parameter. Recalling the fundamental notation (1.5), relation (1.6) implies

$$\mathsf{E}u_{\boldsymbol{\sigma}^1}u_{\boldsymbol{\sigma}^2} = \frac{\beta^2}{2}\left(NR_{1,2}^2 - 1\right) \tag{1.63}$$

and

$$\mathsf{E}v_{\boldsymbol{\sigma}^1}v_{\boldsymbol{\sigma}^2} = N\beta^2 q R_{1,2} \,.$$

Recalling (1.55), we have

$$\frac{1}{N}U(\boldsymbol{\sigma}^1,\boldsymbol{\sigma}^2) = \frac{\beta^2}{4}\left(R_{1,2}^2 - \frac{1}{N}\right) - \frac{\beta^2}{2} q R_{1,2} \,, \tag{1.64}$$

and since $R_{1,2}(\boldsymbol{\sigma},\boldsymbol{\sigma}) = 1$, we get

$$\begin{aligned}\varphi'(t) &= \frac{\beta^2}{4}(1 - 2q - \mathsf{E}\langle R_{1,2}^2\rangle_t + 2q\mathsf{E}\langle R_{1,2}\rangle_t) \\ &= -\frac{\beta^2}{4}\mathsf{E}\langle (R_{1,2}-q)^2\rangle_t + \frac{\beta^2}{4}(1-q)^2 \,.\end{aligned} \tag{1.65}$$

A miracle has occurred. The difficult term is negative, so that

$$\varphi'(t) \le \frac{\beta^2}{4}(1-q)^2 \,. \tag{1.66}$$

Needless to say, such a miracle will not occur for many models of interest, so we better enjoy it while we can. The relation $\varphi(1) = \varphi(0) + \int_0^1 \varphi'(t)dt$ implies

$$\varphi(1) \le \varphi(0) + \frac{\beta^2}{4}(1-q)^2 . \tag{1.67}$$

When considering an interpolating Hamiltonian H_t we will always lighten notation by writing H_0 rather than $H_{t=0}$. Recalling the choice of $v_{\boldsymbol{\sigma}}$ in (1.62) it follows from (1.59) that

$$-H_0(\boldsymbol{\sigma}) = \sum_{i \le N} \sigma_i(\beta z_i \sqrt{q} + h_i) , \tag{1.68}$$

and, as in (1.33), we obtain

$$\varphi(0) = \log 2 + \mathsf{E} \log \operatorname{ch}(\beta z \sqrt{q} + h) , \tag{1.69}$$

where of course the expectation is over the randomness of z and h.

Let us now consider the partition function of the Hamiltonian (1.61),

$$Z_N(\beta, h) = \sum_{\boldsymbol{\sigma}} \exp\Big(\frac{\beta}{\sqrt{N}} \sum_{i \le N} g_{ij}\sigma_i\sigma_j + \sum_{i \le N} h_i \sigma_i\Big) . \tag{1.70}$$

Here we have chosen convenient but technically incorrect notation. The notation (1.70) is incorrect, since $Z_N(\beta, h)$ depends on the actual realization of the r.v.s h_i, not only on h. Speaking of incorrect notation, we will go one step further and write

$$p_N(\beta, h) := \frac{1}{N} \mathsf{E} \log Z_N(\beta, h) . \tag{1.71}$$

The expectation in the right hand side is over all sources of randomness, in this case the r.v.s h_i, and (despite the notation) the quantity $p_N(\beta, h)$ is a **number** depending only on β and the law of h. If $\mathcal{L}(h)$ denotes the law of h, it would probably be more appropriate to write $p_N(\beta, \mathcal{L}(h))$ rather than $p_N(\beta, h)$. The simpler notation $p_N(\beta, h)$ is motivated by the fact that the most important case (at least in the sense that it is as hard as the general case) is the case where h is constant. If this notation disturbs you, please assume everywhere that h is constant and you will not lose much.

Thus with these notations we have

$$p_N(\beta, h) = \varphi(1) . \tag{1.72}$$

In the statement of the next theorem E stands as usual for expectation in all sources of randomness, here the r.v.s z and h. This theorem is a consequence of (1.72), (1.67) and (1.69).

Theorem 1.3.7. (Guerra's replica-symmetric bound). *For any choice of β, h and q we have*

$$p_N(\beta, h) \le \log 2 + \mathsf{E} \log \operatorname{ch}(\beta z \sqrt{q} + h) + \frac{\beta^2}{4}(1-q)^2 . \tag{1.73}$$

Again, despite the notation, the quantity $p_N(\beta, h)$ is a number. The expression "replica-symmetric" is physics' terminology. Its meaning might gradually become clear. The choice $q = 0$, h constant essentially recovers (1.25).

It is now obvious what is the best choice of q: the choice that minimizes the right-hand side of (1.73), i.e.

$$0 = \mathsf{E}\,\frac{\beta z}{2\sqrt{q}}\mathrm{th}(\beta z\sqrt{q}+h) - \frac{\beta^2}{2}(1-q) = \frac{\beta^2}{2}\left(\mathsf{E}\,\frac{1}{\mathrm{ch}^2(\beta z\sqrt{q}+h)} - (1-q)\right),$$

using Gaussian integration by parts (A.14). Since $\mathrm{ch}^{-2}(x) = 1 - \mathrm{th}^2(x)$, this means that we have the **absolutely fundamental** relation

$$q = \mathsf{E}\mathrm{th}^2(\beta z\sqrt{q}+h)\,. \tag{1.74}$$

Of course at this stage this equation looks rather mysterious. The mystery will gradually recede, in particular in (1.105) below. The reader might wonder at this stage why we do not give a special name, such as q^*, to the fundamental quantity defined by (1.74), to distinguish it from the generic value of q. The reason is simply that in the long range it is desirable that the simplest name goes to the most used quantity, and the case where q is not the solution of (1.74) is only of some limited interest.

It will be convenient to know that the equation (1.74) has a unique solution.

Proposition 1.3.8. (Latala-Guerra) *The function*

$$\Psi(x) = \mathsf{E}\,\frac{\mathrm{th}^2(z\sqrt{x}+h)}{x}$$

is strictly decreasing on $\mathbb{R}^+$ *and vanishes as* $x \to \infty$. *Consequently if* $\mathsf{E}h^2 > 0$ *there is a unique solution to (1.74).*

The difficult part of the statement is the proof that the function Ψ is strictly decreasing. In that case, since $\lim_{x\to 0^+} x\Psi(x) = \mathsf{E}\mathrm{th}^2 h > 0$, we have $\lim_{x\to 0^+}\Psi(x) = \infty$, and since $\lim_{x\to\infty}\Psi(x) = 0$ there is a unique solution to the equation $\Psi(x) = 1/\beta^2$ and hence (1.74). But $\mathsf{E}\mathrm{th}^2 h = 0$ only when $h = 0$ a.e. (in which case when $\beta > 1$ there are 2 solutions to (1.74), one of which being 0).

Proposition 1.3.8 is nice but not really of importance. The proof is very beautiful, but rather tricky, and the tricky ideas are not used anywhere else. To avoid distraction, we postpone this proof until Section A.14. At this stage we give the proof only in the case where $\beta < 1$, because the ideas of this simple argument will be used again and again. Given a (smooth) function f the function $\psi(x) = \mathsf{E}f(\beta z\sqrt{x}+h)$ satisfies

$$\psi'(x) = \beta\mathsf{E}\,\frac{z}{2\sqrt{x}}f'(\beta z\sqrt{x}+h) = \frac{\beta^2}{2}\mathsf{E}f''(\beta z\sqrt{x}+h)\,, \tag{1.75}$$

using Gaussian integration by parts (A.14). We use this for the function $f(y) = \mathrm{th}^2 y$, that satisfies

$$f'(y) = 2\frac{\mathrm{th} y}{\mathrm{ch}^2 y} \quad ; \quad f''(y) = 2\frac{1 - 2\mathrm{sh}^2 y}{\mathrm{ch}^4 y} \leq 2 \,.$$

Thus, if $\beta < 1$, we deduce from (1.75) that the function $\psi(q) = \mathsf{E}\mathrm{th}^2(\beta z\sqrt{q} + h)$ satisfies $\psi'(q) < 1$. This function maps the unit interval into itself, so that it has a unique fixed point.

Let us denote by $\mathrm{SK}(\beta, h)$ the right-hand side of (1.73) when q is as in (1.74). As in the case of $p_N(\beta, h)$ this is a number depending only on β and the law of h. Thus (1.73) implies that

$$p_N(\beta, h) \leq \mathrm{SK}(\beta, h) \,. \tag{1.76}$$

We can hope that when q satisfies (1.74) there is near equality in (1.76), so that the right hand-side of (1.76) is not simply a good bound for $p_N(\beta, h)$, but essentially the value of this quantity as $N \to \infty$. Moreover, we have a clear road to prove this, namely (see (1.65)) to show that $\int_0^1 \mathsf{E}\langle (R_{1,2} - q)^2 \rangle_t dt$ is small. We will pursue this idea in Section 1.4, where we will prove that this is indeed the case when β is not too large. The case of large β (low temperature) is much more delicate, but will be approached in Volume II through a much more elaborated version of the same ideas.

Theorem 1.3.9. (Guerra-Toninelli [75]) *For all values of β, h, the sequence $(Np_N(\beta, h))_{N \geq 1}$ is superadditive, that is, for integers N_1 and N_2 we have*

$$p_{N_1+N_2}(\beta, h) \geq \frac{N_1}{N_1 + N_2} p_{N_1}(\beta, h) + \frac{N_2}{N_1 + N_2} p_{N_2}(\beta, h). \tag{1.77}$$

Consequently, the limit

$$p(\beta, h) = \lim_{N \to \infty} p_N(\beta, h) \tag{1.78}$$

exists.

Of course this does not tell us what is the value of $p(\beta, h)$, although we know by (1.76) that $p(\beta, h) \leq \mathrm{SK}(\beta, h)$.

Proof. Let $N = N_1 + N_2$. The idea is to compare the SK Hamiltonian of size N with two non-interacting SK Hamiltonians of sizes N_1 and N_2. Consider $u_{\boldsymbol{\sigma}}$ as in (1.62) and

$$v_{\boldsymbol{\sigma}} = \frac{\beta}{\sqrt{N_1}} \sum_{i < j \leq N_1} g'_{ij} \sigma_i \sigma_j + \frac{\beta}{\sqrt{N_2}} \sum_{N_1 < i < j \leq N} g'_{ij} \sigma_i \sigma_j \,,$$

where g'_{ij} are i.i.d. standard Gaussian r.v.s independent of the r.v.s g_{ij}. Considering φ as in (1.39) with $F(\mathbf{x})$ as in (1.57) and $w_{\boldsymbol{\sigma}} = \exp \sum_{i \le N} h_i \sigma_i$, we have

$$\begin{aligned}\varphi(0) &= \frac{N_1}{N} p_{N_1}(\beta, h) + \frac{N_2}{N} p_{N_2}(\beta, h) \\ \varphi(1) &= p_N(\beta, h) .\end{aligned}$$

Let us recall yet another time the fundamental notation (1.9),

$$R_{1,2} = N^{-1} \sum_{i \le N} \sigma_i^1 \sigma_i^2 ,$$

and let us define similarly

$$R' = N_1^{-1} \sum_{i \le N_1} \sigma_i^1 \sigma_i^2 \ ; \ R'' = N_2^{-1} \sum_{N_1 < i \le N} \sigma_i^1 \sigma_i^2 ,$$

so that

$$R_{1,2} = \frac{N_1}{N} R' + \frac{N_2}{N} R'' .$$

The convexity of the function $x \mapsto x^2$ implies

$$R_{1,2}^2 \le \frac{N_1}{N} R'^2 + \frac{N_2}{N} R''^2 . \tag{1.79}$$

Rather than (1.64), a few lines of elementary algebra now yield

$$\frac{1}{N} U(\boldsymbol{\sigma}^1, \boldsymbol{\sigma}^2) = \frac{\beta^2}{4} \left(R_{1,2}^2 - \frac{N_1}{N} R'^2 - \frac{N_2}{N} R''^2 + \frac{1}{N} \right) . \tag{1.80}$$

When $\boldsymbol{\sigma}^1 = \boldsymbol{\sigma}^2$ we have $R_{1,2} = R' = R'' = 1$, so that (1.60) entails

$$\varphi'(t) = -\frac{\beta^2}{4} \mathsf{E} \Big\langle R_{1,2}^2 - \frac{N_1}{N} R'^2 - \frac{N_2}{N} R''^2 \Big\rangle_t \ge 0$$

by (1.79). The fact that $\lim_{N \to \infty} r_N / N$ exists for a superadditive sequence (r_N) is classical. It is called "Fekete's lemma" and is even mentioned (with a reference to the original paper) in Wikipedia. □

Exercise 1.3.10. Carry out the proof of (1.80).

Generally speaking it seems plausible that "all limits exist". Some information can be gained using an elementary fact known as Griffiths' lemma in statistical mechanics. This is, if a sequence φ_N of convex (differentiable) functions converges pointwise in an interval to a (necessarily convex) function φ, then $\lim_{N \to \infty} \varphi'_N(x) = \varphi'(x)$ at every point x for which $\varphi'(x)$ exists (which is everywhere outside a countable set of possible exceptional values).

If Griffiths' lemma does not seem obvious to you, please do not worry, for the time being this is only a side story, the real point of it being a pretense to introduce Lemma 1.3.11 below, a step in our learning of Gaussian integration by parts. Later on, in Volume II, we will use quantitative versions of Griffiths' lemma with complete proofs.

It is a special case of Hölder's inequality that the function

$$\beta \mapsto \log \int f^\beta \mathrm{d}\mu$$

is convex (whenever $f > 0$) for any probability measure μ. Indeed, this means that for $0 < a < 1$ and $\beta_1, \beta_2 > 0$ we have

$$\int f^{a\beta_1+(1-a)\beta_2} \mathrm{d}\mu \le \left(\int f^{\beta_1} \mathrm{d}\mu\right)^a \left(\int f^{\beta_2} \mathrm{d}\mu\right)^{1-a},$$

and setting $U = f^{\beta_1}$ and $V = f^{\beta_2}$ this is the inequality

$$\int U^a V^{1-a} \mathrm{d}\mu \le \left(\int U \mathrm{d}\mu\right)^a \left(\int V \mathrm{d}\mu\right)^{1-a}.$$

Consequently (thinking of the sum in (1.70) as an integral) the function

$$\beta \mapsto \Phi(\beta) = \frac{1}{N} \log Z_N(\beta, h) \tag{1.81}$$

is a convex random function (a fact that will turn out to be essential much later). An alternative proof of the convexity of the function Φ, more in line with the ideas of statistical mechanics, is as follows. As in (1.62) we define

$$u_{\boldsymbol{\sigma}} = \frac{1}{\sqrt{N}} \sum_{i<j\le N} g_{ij}\sigma_i\sigma_j . \tag{1.82}$$

Let $w_{\boldsymbol{\sigma}} = \exp \sum_{i\le N} h_i\sigma_i$, so that $Z_N := Z_N(\beta,h) = \sum_{\boldsymbol{\sigma}} w_{\boldsymbol{\sigma}} \exp \beta u_{\boldsymbol{\sigma}}$ and

$$N\Phi(\beta) = \log \sum_{\boldsymbol{\sigma}} w_{\boldsymbol{\sigma}} \exp \beta u_{\boldsymbol{\sigma}} .$$

Thus

$$N\Phi'(\beta) = \frac{1}{Z_N} \sum_{\boldsymbol{\sigma}} w_{\boldsymbol{\sigma}} u_{\boldsymbol{\sigma}} \exp \beta u_{\boldsymbol{\sigma}} \quad (= \langle u_{\boldsymbol{\sigma}} \rangle) \tag{1.83}$$

and

$$\begin{aligned} N\Phi''(\beta) &= \frac{1}{Z_N} \sum_{\boldsymbol{\sigma}} w_{\boldsymbol{\sigma}} u_{\boldsymbol{\sigma}}^2 \exp \beta u_{\boldsymbol{\sigma}} - \left(\frac{1}{Z_N} \sum_{\boldsymbol{\sigma}} w_{\boldsymbol{\sigma}} u_{\boldsymbol{\sigma}} \exp \beta u_{\boldsymbol{\sigma}}\right)^2 \\ &= \langle u_{\boldsymbol{\sigma}}^2 \rangle - \langle u_{\boldsymbol{\sigma}} \rangle^2 \ge 0 , \end{aligned}$$

where the last inequality is simply the Cauchy-Schwarz inequality used in the probability space (Σ_N, G_N).

In particular $p_N(\beta, h)$ is a convex function of β. By Theorem 1.3.9, $p(\beta, h) = \lim_{N\to\infty} p_N(\beta, h)$ exists. The function $\beta \mapsto p(\beta, h)$ is convex and therefore is differentiable at every point outside a possible countable set of exceptional values. Now, we have the following important formula.

Lemma 1.3.11. *For any value of β we have*

$$\frac{\partial}{\partial\beta} p_N(\beta, h) = \frac{\beta}{2}(1 - \mathsf{E}\langle R_{1,2}^2 \rangle) . \tag{1.84}$$

Thus Griffiths' lemma proves that, given h, $\lim_{N\to\infty} \mathsf{E}\langle R_{1,2}^2 \rangle$ exists for each value of β where the map $\beta \mapsto p(\beta, h)$ is differentiable.

It is however typically much more difficult to prove that no such exceptional values exist. We will be able to prove it after considerable work in Volume II.

Proof of Lemma 1.3.11. We recall (1.82). Defining

$$R(\boldsymbol{\sigma}, \boldsymbol{\tau}) = N^{-1} \sum_{i\le N} \sigma_i \tau_i ,$$

we can rewrite (1.63) (where we take $\beta = 1$) as

$$\mathsf{E} u_{\boldsymbol{\sigma}} u_{\boldsymbol{\tau}} = \frac{1}{2}(N R(\boldsymbol{\sigma}, \boldsymbol{\tau})^2 - 1) .$$

Let again $w_{\boldsymbol{\sigma}} = \exp \sum_{i\le N} h_i \sigma_i$, so that taking expectation in (1.83) yields

$$\begin{aligned} \frac{\partial}{\partial\beta} p_N(\beta, h) &= \frac{1}{N} \mathsf{E} \frac{\sum_{\boldsymbol{\sigma}} w_{\boldsymbol{\sigma}} u_{\boldsymbol{\sigma}} \exp \beta u_{\boldsymbol{\sigma}}}{\sum_{\boldsymbol{\tau}} w_{\boldsymbol{\tau}} \exp \beta u_{\boldsymbol{\tau}}} \\ &= \frac{1}{N} \sum_{\boldsymbol{\sigma}} \mathsf{E} u_{\boldsymbol{\sigma}} \frac{w_{\boldsymbol{\sigma}} \exp \beta u_{\boldsymbol{\sigma}}}{\sum_{\boldsymbol{\tau}} w_{\boldsymbol{\tau}} \exp \beta u_{\boldsymbol{\tau}}} . \end{aligned} \tag{1.85}$$

To compute

$$\mathsf{E} u_{\boldsymbol{\sigma}} \frac{w_{\boldsymbol{\sigma}} \exp \beta u_{\boldsymbol{\sigma}}}{\sum_{\boldsymbol{\tau}} w_{\boldsymbol{\tau}} \exp \beta u_{\boldsymbol{\tau}}} ,$$

we first think of the quantities $w_{\boldsymbol{\tau}}$ as being fixed numbers, with $w_{\boldsymbol{\tau}} > 0$. We then apply the Gaussian integration by parts formula (A.17) to the jointly Gaussian family $(u_{\boldsymbol{\tau}})_{\boldsymbol{\tau}}$ and the function

$$F_{\boldsymbol{\sigma}}(\mathbf{x}) = \frac{w_{\boldsymbol{\sigma}} \exp \beta x_{\boldsymbol{\sigma}}}{\sum_{\boldsymbol{\tau}} w_{\boldsymbol{\tau}} \exp \beta x_{\boldsymbol{\tau}}} ,$$

to get

$$\frac{1}{N}\mathsf{E} u_{\boldsymbol{\sigma}} \frac{w_{\boldsymbol{\sigma}} \exp \beta u_{\boldsymbol{\sigma}}}{\sum_{\boldsymbol{\tau}} w_{\boldsymbol{\tau}} \exp \beta u_{\boldsymbol{\tau}}} = \frac{\beta}{2}\Big(1 - \frac{1}{N}\Big)\mathsf{E}\frac{w_{\boldsymbol{\sigma}} \exp \beta u_{\boldsymbol{\sigma}}}{\sum_{\boldsymbol{\tau}} w_{\boldsymbol{\tau}} \exp \beta u_{\boldsymbol{\tau}}}$$
$$- \frac{\beta}{2}\sum_{\boldsymbol{\tau}} \mathsf{E}\frac{w_{\boldsymbol{\sigma}} w_{\boldsymbol{\tau}} (R(\boldsymbol{\sigma},\boldsymbol{\tau})^2 - 1/N) \exp \beta(u_{\boldsymbol{\sigma}} + u_{\boldsymbol{\tau}})}{(\sum_{\boldsymbol{\tau}} w_{\boldsymbol{\tau}} \exp \beta u_{\boldsymbol{\tau}})^2} .$$

Taking a further expectation in the randomness of the r.v.s h_i, this equality remains true if $w_{\boldsymbol{\sigma}} = \exp \sum_{i \le N} h_i \sigma_i$, and using this formula in (1.85) we get the result. □

Although the reader might not have noticed it, in the proof of (1.84) we have done something remarkable, and it is well worth to spell it out. Looking at the definition of the Hamiltonian, it would be quite natural to think of the quantity

$$\frac{1}{N} \log Z_N(\beta, h)$$

as a function of the $N(N-1)/2$ r.v.s g_{ij}. Instead, we have been thinking of it as a function of the much larger family of the 2^N r.v.s $u_{\boldsymbol{\sigma}}$. Such a shift in point of view will be commonplace in many instances where we will use the Gaussian integration by parts formula (A.17). The use of this formula can greatly simplify if one uses a clever choice for the Gaussian family of r.v.s.

Exercise 1.3.12. To make clear the point of the previous remark, derive formula (1.84) by considering Z_N as a function of the r.v.s (g_{ij}).

Of course, after having proved (1.60), no great inventiveness was required to think of basing the integration by parts on the family $(u_{\boldsymbol{\sigma}})$, in particular in view of the following (that reveals that the only purpose of the direct proof of Lemma 1.3.11 we gave was to have the reader think a bit more about Gaussian integration by parts (A.17)).

Exercise 1.3.13. Show that (1.84) can in fact be deduced from (1.60). Hint: use $u_{\boldsymbol{\sigma}}$ as in (1.62) but take now $v_{\boldsymbol{\sigma}} = 0$.

As the next exercise shows, the formula (1.84) is not an accident, but a first occurrence of a general principle that we will use a great many times later. In the long range the reader would do well to really master this result.

Exercise 1.3.14. Consider a jointly Gaussian family of r.v.s $(H_N(\boldsymbol{\sigma}))_{\boldsymbol{\sigma} \in \Sigma_N}$ and another family $(H'_N(\boldsymbol{\sigma}))_{\boldsymbol{\sigma} \in \Sigma_N}$ of r.v.s. These two families are assumed to be independent of each other. Let

$$p_N(\beta) = \frac{1}{N}\mathsf{E} \log \sum_{\boldsymbol{\sigma}} \exp(-\beta H_N(\boldsymbol{\sigma}) - H'_N(\boldsymbol{\sigma})) .$$

Prove that

$$\frac{\mathrm{d}}{\mathrm{d}\beta} p_N(\beta) = \frac{\beta}{N}(\mathsf{E}\langle U(\boldsymbol{\sigma},\boldsymbol{\sigma})\rangle - \mathsf{E}\langle U(\boldsymbol{\sigma}^1,\boldsymbol{\sigma}^2)\rangle) ,$$

where $U(\boldsymbol{\sigma}^1, \boldsymbol{\sigma}^2) = \mathsf{E} H_N(\boldsymbol{\sigma}^1) H_N(\boldsymbol{\sigma}^2)$, and where the bracket is an average for the Gibbs measure with Hamiltonian $\beta H_N + H'_N$. Prove that this formula generalizes (1.84).

Research Problem 1.3.15. (Level 2) Prove that $\lim_{N\to\infty} \mathsf{E}\langle R_{1,2}^3 \rangle$ exists for all β.

In fact, it does not even seem to have been shown that, given h, there exist values of β beyond the Almeida-Thouless line of Section 1.9 where this limit exists.

To help the newcomer to the area, the research problems are ranked level 1 to level 3. The solution to level 1 problems should be suitable for publication, but the author feels that they can be successfully attacked by the methods explained in this book, or simple extensions of these methods. To put it another way, the author feels that he would be rather likely to solve them in (expected) finite time if he tried (which he won't). Level 2 problems are more likely to require ingredients substantially beyond what is found in the book. On the other hand these problems do not touch what seem to be the central issues of spin glass theory, and there is no particular reason to think that they are very hard. Simply, they have not been tried. Level 3 problems seem to touch essential issues, and there is currently no way of telling how difficult they might be. It goes without saying that this classification is based on the author's current understanding, and comes with no warranty whatsoever. (In particular, problems labeled level 2 as the above might well turn out to be level 3.)

1.4 Latala's Argument

It will turn out in many models that at high temperature "the overlap is essentially constant". That is, there exists a number q, depending only on the system, such that if one picks two configurations $\boldsymbol{\sigma}^1$ and $\boldsymbol{\sigma}^2$ independently according to Gibbs' measure, one observes that typically

$$R_{1,2} \simeq q \,. \tag{1.86}$$

The symbol $\simeq$ stands of course for approximate equality. It will often be used in our informal explanations, and in each case its precise meaning will soon become apparent. It is not surprising in the least that a behavior such as (1.86) occurs. If we remember that $NR_{1,2} = \sum_{i\le N} \sigma_i^1 \sigma_i^2$, and if we expect to have at least some kind of "weak independence" between the sites, then (1.86) should hold by the law of large numbers. The reader might have also observed that a condition of the type (1.86) is precisely what is required to nullify the dangerous term $\mathsf{E}\langle (R_{1,2} - q)^2 \rangle_t$ in (1.65).

What is not intuitive is that (1.86) has very strong consequences. In particular it implies that at given typical disorder, a few spins are nearly independent under Gibbs measure, as is shown in Theorem 1.4.15 below. (The expression "a few spins" means that we consider a fixed number of spins, and then take N very large.) For many of the models we will study the proof of (1.86) will be a major goal, and the key step in the computation of p_N.

In this section we present a beautiful (unpublished!!) argument of R. Latala that probably provides the fastest way to prove (1.86) for the SK model at high enough temperature (i.e. β small enough). This argument is however not easy to generalize in some directions, and we will learn a more versatile method in Section 1.6.

From now on we lighten notation by writing $\nu(f)$ for $\mathsf{E}\langle f\rangle$. In this section the Gibbs measure is relative to the Hamiltonian (1.61), that is

$$-H_N(\boldsymbol{\sigma}) = \frac{\beta}{\sqrt{N}}\sum_{i<j} g_{ij}\sigma_i\sigma_j + \sum_{i\le N} h_i\sigma_i \,.$$

The next theorem provides a precise version of (1.86), in the form of a strong exponential inequality.

Theorem 1.4.1. *Assume $\beta < 1/2$. Then for $2s < 1-4\beta^2$ we have*

$$\nu\big(\exp sN(R_{1,2}-q)^2\big) \le \frac{1}{\sqrt{1-2s-4\beta^2}} \,, \tag{1.87}$$

where q is the unique solution of (1.74), i.e. $q = \mathsf{E}\mathrm{th}^2(\beta z\sqrt{q}+h)$.

Of course here to lighten notation we write $\exp sN(R_{1,2}-q)^2$ rather than $\exp(sN(R_{1,2}-q)^2)$. Since $\exp x \ge x^k/k!$ for $x \ge 0$ and $k \ge 1$, this shows that

$$\frac{1}{k!}\nu((sN)^k(R_{1,2}-q)^{2k}) \le \frac{1}{\sqrt{1-2s-4\beta^2}} \,,$$

so that, since $k! \le k^k$,

$$\nu((R_{1,2}-q)^{2k}) \le \frac{1}{\sqrt{1-2s-4\beta^2}}\Big(\frac{ks}{N}\Big)^k \,,$$

and in particular

$$\nu\big((R_{1,2}-q)^{2k}\big) \le \Big(\frac{Kk}{N}\Big)^k \,, \tag{1.88}$$

where K does not depend on N or k.

The important relationship between growth of moments and exponential integrability is detailed in Section A.6. This relation is explained there for probabilities. It is perfectly correct to think of ν (and of its avatar ν_t defined below) as **being the expectation for a certain probability**. This can be made formal. We do not explain this since it requires an extra level of abstraction that does not seem very fruitful.

An important special case of (1.88) is:

$$\nu\big((R_{1,2} - q)^2\big) \le K/N\,. \tag{1.89}$$

Equation (1.88) is the first of **very** many that involve an unspecified constant K. There are several reasons why it is desirable to use such constants. A clean explicit value might be hard to get, or, like here, it might be irrelevant and rather distracting. When using such constants, *it is understood throughout the book that their value might not be the same at each occurrence.* The use of the word "constant" to describe K is because this number is never, ever permitted to depend on N. On the other hand, it is typically permitted to depend on β and h. Of course we will try to be more specific when the need arises. An unspecified constant that does not depend on any parameter (a so-called universal constant) will be denoted by L,and the value of this quantity might also not be the same at each occurrence (as e.g. in the relation $L = L + 2$). Of course, K_0, L_1, etc. denote specific quantities. These conventions will be used *throughout the book* and it surely would help to remember them from now on.

It is a very non-trivial question to determine the supremum of the values of β for which one can control $\nu(\exp sN(R_{1,2} - q)^2)$ for some $s > 0$, or the supremum of the values of β for which (1.89) holds. (It is believable that these are the same.) The method of proof of Theorem 1.4.1 does not allow one to reach this value, so we do not attempt to push the method to its limit, but rather to give a clean statement. There is nothing magic about the condition $\beta < 1/2$, which is an artifact of the method of proof. In Volume II, we will prove that actually (1.88) holds in a much larger region.

We now turn to a general principle of fundamental importance. We go back to the general case of Gaussian families $(u_{\boldsymbol{\sigma}})$ and $(v_{\boldsymbol{\sigma}})$, and for $\boldsymbol{\sigma} \in \Sigma_N$ we consider a number $w_{\boldsymbol{\sigma}} > 0$. We recall that we denote by

$$\langle \cdot \rangle_t$$

an average for the Gibbs measure with Hamiltonian (1.59), that is,

$$-H_t(\boldsymbol{\sigma}) = \sqrt{t}u_{\boldsymbol{\sigma}} + \sqrt{1-t}v_{\boldsymbol{\sigma}} + \log w_{\boldsymbol{\sigma}} = u_{\boldsymbol{\sigma}}(t) + \log w_{\boldsymbol{\sigma}}\,.$$

Then, for a function f on $\Sigma_N^n (= (\Sigma_N)^n)$ we have

$$\langle f \rangle_t = Z(\mathbf{u}(t))^{-n} \sum_{\boldsymbol{\sigma}^1,\ldots,\boldsymbol{\sigma}^n} f(\boldsymbol{\sigma}^1,\ldots,\boldsymbol{\sigma}^n) w_{\boldsymbol{\sigma}^1} \cdots w_{\boldsymbol{\sigma}^n} \exp\Big(\sum_{\ell \le n} u_{\boldsymbol{\sigma}^\ell}(t)\Big)\,,$$

where $Z(\mathbf{u}(t)) = \sum_{\boldsymbol{\sigma}} w_{\boldsymbol{\sigma}} \exp u_{\boldsymbol{\sigma}}(t)$. We write

$$\nu_t(f) = \mathsf{E}\langle f \rangle_t\;;\; \nu_t'(f) = \frac{\mathrm{d}}{\mathrm{d}t}(\nu_t(f))\,.$$

The general principle stated in Lemma 1.4.2 below provides an explicit formula for $\nu_t'(f)$. It is in a sense a straightforward application of Lemma 1.3.1.

However, since Lemma 1.3.1 requires computing the second partial derivatives of the function F, when this function is complicated, (e.g. is a quotient of 2 factors) we must face the unavoidable fact that this will produce formulas that are not as simple as we might wish. We should be well prepared for this, as we all know that computing derivatives can lead to complicated expressions.

We recall the function of two configurations $U(\boldsymbol{\sigma},\boldsymbol{\tau})$ given by (1.55), that is, $U(\boldsymbol{\sigma},\boldsymbol{\tau}) = 1/2(\mathsf{E}u_{\boldsymbol{\sigma}}u_{\boldsymbol{\tau}} - \mathsf{E}v_{\boldsymbol{\sigma}}v_{\boldsymbol{\tau}})$. Thus, in the formula below, the quantity $U(\boldsymbol{\sigma}^\ell,\boldsymbol{\sigma}^{\ell'})$ is

$$U(\boldsymbol{\sigma}^\ell,\boldsymbol{\sigma}^{\ell'}) = \frac{1}{2}(\mathsf{E}u_{\boldsymbol{\sigma}^\ell}u_{\boldsymbol{\sigma}^{\ell'}} - \mathsf{E}v_{\boldsymbol{\sigma}^\ell}v_{\boldsymbol{\sigma}^{\ell'}})\,.$$

We also point out that in this formula, to lighten notation, f stands for $f(\boldsymbol{\sigma}^1,\ldots,\boldsymbol{\sigma}^n)$.

Lemma 1.4.2. *If f is a function on $\Sigma_N^n (= (\Sigma_N)^n)$, then*

$$\begin{aligned}\nu_t'(f) = &\sum_{1\le \ell,\ell'\le n} \nu_t(U(\boldsymbol{\sigma}^\ell,\boldsymbol{\sigma}^{\ell'})f) - 2n\sum_{\ell\le n}\nu_t(U(\boldsymbol{\sigma}^\ell,\boldsymbol{\sigma}^{n+1})f)\\ &- n\nu_t(U(\boldsymbol{\sigma}^{n+1},\boldsymbol{\sigma}^{n+1})f) + n(n+1)\nu_t(U(\boldsymbol{\sigma}^{n+1},\boldsymbol{\sigma}^{n+2})f)\,. \quad (1.90)\end{aligned}$$

This formula looks scary the first time one sees it, but one should observe that the right-hand side is a linear combination of terms of the same nature, each of the type

$$\nu_t(U(\boldsymbol{\sigma}^\ell,\boldsymbol{\sigma}^{\ell'})f) = \mathsf{E}\langle U(\boldsymbol{\sigma}^\ell,\boldsymbol{\sigma}^{\ell'})f(\boldsymbol{\sigma}^1,\ldots,\boldsymbol{\sigma}^n)\rangle_t\,.$$

The complication is purely algebraic (as it should be). One can observe that even though f depends only on n replicas, (1.90) involves two new independent replicas $\boldsymbol{\sigma}^{n+1}$ and $\boldsymbol{\sigma}^{n+2}$.

We will use countless times a principle called *symmetry between replicas*, a name not to be confused with the expression "replica-symmetric". This principle asserts e.g. that $\nu(f(\boldsymbol{\sigma}^1)U(\boldsymbol{\sigma}^1,\boldsymbol{\sigma}^2)) = \nu(f(\boldsymbol{\sigma}^1)U(\boldsymbol{\sigma}^1,\boldsymbol{\sigma}^3))$. The reason for this is simply that the sequence $(\boldsymbol{\sigma}^\ell)$ is an i.i.d. sequence under Gibbs' measure, so that for any permutation π of the replica indices, and any function $f(\boldsymbol{\sigma}^1,\ldots,\boldsymbol{\sigma}^n)$, one has $\langle f(\boldsymbol{\sigma}^1,\ldots,\boldsymbol{\sigma}^n)\rangle = \langle f(\boldsymbol{\sigma}^{\pi(1)},\ldots,\boldsymbol{\sigma}^{\pi(n)})\rangle$, and hence taking expectation,

$$\nu(f(\boldsymbol{\sigma}^1,\ldots,\boldsymbol{\sigma}^n)) = \nu(f(\boldsymbol{\sigma}^{\pi(1)},\ldots,\boldsymbol{\sigma}^{\pi(n)}))\,.$$

In particular, if f is a function on Σ_N^n, then the value of $\nu(U(\boldsymbol{\sigma}^\ell,\boldsymbol{\sigma}^r)f)$ does not depend on the value of r if $r\ge n+1$. Similarly, if $\ell>\ell'>n$, the value of $\nu(U(\boldsymbol{\sigma}^\ell,\boldsymbol{\sigma}^{\ell'})f)$ does not depend on ℓ or ℓ'.

Exercise 1.4.3. a) Let us take for f the function on Σ_N^n that is constant equal to 1. Then $\langle f\rangle_t = 1$, so that $\nu_t(f) = 1$ for each t and hence $\nu_t'(f) = 0$. Prove that in that case the right-hand side of (1.90) is 0.

b) A function f on Σ_N^n can also be seen as a function on $\Sigma_N^{n'}$ for $n' > n$. Prove that the right-hand sides of (1.90) computed for n and n' coincide (the extra terms in the case of n' cancel out).

Proof of Lemma 1.4.2. Consider as before $\mathbf{x} = (x_{\boldsymbol{\sigma}})$ and let

$$Z(\mathbf{x}) = \sum_{\boldsymbol{\sigma}} w_{\boldsymbol{\sigma}} \exp x_{\boldsymbol{\sigma}}$$

$$F(\mathbf{x}) = Z(\mathbf{x})^{-n} \sum_{\boldsymbol{\sigma}^1,\ldots,\boldsymbol{\sigma}^n} w_{\boldsymbol{\sigma}^1} \cdots w_{\boldsymbol{\sigma}^n} f(\boldsymbol{\sigma}^1,\ldots,\boldsymbol{\sigma}^n) \exp\Big(\sum_{\ell\le n} x_{\boldsymbol{\sigma}^\ell}\Big) .$$

We recall the formula (1.56):

$$\varphi'(t) = \mathsf{E} \sum_{\boldsymbol{\sigma},\boldsymbol{\tau}} U(\boldsymbol{\sigma},\boldsymbol{\tau}) \frac{\partial^2 F}{\partial x_{\boldsymbol{\sigma}} x_{\boldsymbol{\tau}}}(\mathbf{u}(t)) ,$$

that we will apply to this function, carefully collecting the terms. Let us set

$$F_1(\mathbf{x}) = \sum_{\boldsymbol{\sigma}^1,\ldots,\boldsymbol{\sigma}^n} w_{\boldsymbol{\sigma}^1} \cdots w_{\boldsymbol{\sigma}^n} f(\boldsymbol{\sigma}^1,\ldots,\boldsymbol{\sigma}^n) \exp\Big(\sum_{\ell\le n} x_{\boldsymbol{\sigma}^\ell}\Big) ,$$

so that

$$F(\mathbf{x}) = Z(\mathbf{x})^{-n} F_1(\mathbf{x}) ,$$

and therefore

$$\frac{\partial F}{\partial x_{\boldsymbol{\sigma}}}(\mathbf{x}) = Z(\mathbf{x})^{-n} \frac{\partial F_1}{\partial x_{\boldsymbol{\sigma}}}(\mathbf{x}) - n Z^{-n-1}(\mathbf{x}) \frac{\partial Z}{\partial x_{\boldsymbol{\sigma}}}(\mathbf{x}) F_1(\mathbf{x}) .$$

Consequently,

$$\begin{aligned}
\frac{\partial^2 F}{\partial x_{\boldsymbol{\sigma}} \partial x_{\boldsymbol{\tau}}}(\mathbf{x}) &= Z(\mathbf{x})^{-n} \frac{\partial^2 F_1}{\partial x_{\boldsymbol{\sigma}} \partial x_{\boldsymbol{\tau}}}(\mathbf{x}) \\
&\quad - n Z(\mathbf{x})^{-n-1} \Big(\frac{\partial Z}{\partial x_{\boldsymbol{\sigma}}}(\mathbf{x}) \frac{\partial F_1}{\partial x_{\boldsymbol{\tau}}}(\mathbf{x}) + \frac{\partial Z}{\partial x_{\boldsymbol{\tau}}}(\mathbf{x}) \frac{\partial F_1}{\partial x_{\boldsymbol{\sigma}}}(\mathbf{x}) \Big) \\
&\quad - n Z(\mathbf{x})^{-n-1} \frac{\partial^2 Z}{\partial x_{\boldsymbol{\sigma}} \partial x_{\boldsymbol{\tau}}}(\mathbf{x}) F_1(\mathbf{x}) \\
&\quad + n(n+1) Z(\mathbf{x})^{-n-2} \frac{\partial Z}{\partial x_{\boldsymbol{\sigma}}}(\mathbf{x}) \frac{\partial Z}{\partial x_{\boldsymbol{\tau}}}(\mathbf{x}) F_1(\mathbf{x}) .
\end{aligned} \tag{1.91}$$

Each of the four terms of (1.91) corresponds to a term in (1.90). We will explain this in detail for the first and the last terms. We observe first that

$$\frac{\partial Z}{\partial x_{\boldsymbol{\sigma}}}(\mathbf{x}) = w_{\boldsymbol{\sigma}} \exp(x_{\boldsymbol{\sigma}}) ,$$

so that the last term of (1.91) is

$$C(\boldsymbol{\sigma},\boldsymbol{\tau},\mathbf{x}) := n(n+1)Z(\mathbf{x})^{-n-2}w_{\boldsymbol{\sigma}}w_{\boldsymbol{\tau}}\exp(x_{\boldsymbol{\sigma}}+x_{\boldsymbol{\tau}})F_1(\mathbf{x}) \,.$$

Consequently,

$$\begin{aligned}&\sum_{\boldsymbol{\sigma},\boldsymbol{\tau}} U(\boldsymbol{\sigma},\boldsymbol{\tau})C(\boldsymbol{\sigma},\boldsymbol{\tau},\mathbf{u}(t)) \\ &= n(n+1)Z(\mathbf{u}(t))^{-n-2}\sum_{\boldsymbol{\sigma},\boldsymbol{\tau}} U(\boldsymbol{\sigma},\boldsymbol{\tau})w_{\boldsymbol{\sigma}}w_{\boldsymbol{\tau}}\exp(u_{\boldsymbol{\sigma}}(t)+u_{\boldsymbol{\tau}}(t))F_1(\mathbf{u}(t)) \,.\end{aligned} \tag{1.92}$$

Recalling the value of $F_1(\mathbf{x})$, we obtain

$$F_1(\mathbf{u}(t)) = \sum_{\boldsymbol{\sigma}^1,\ldots,\boldsymbol{\sigma}^n} f(\boldsymbol{\sigma}^1,\ldots,\boldsymbol{\sigma}^n)\Big(\prod_{1\le\ell\le n} w_{\boldsymbol{\sigma}^\ell}\Big)\exp\Big(\sum_{1\le\ell\le n} u_{\boldsymbol{\sigma}^\ell}(t)\Big) \,,$$

and using this in the second line below we find

$$\begin{aligned}&\sum_{\boldsymbol{\sigma},\boldsymbol{\tau}} U(\boldsymbol{\sigma},\boldsymbol{\tau})w_{\boldsymbol{\sigma}}w_{\boldsymbol{\tau}}\exp(u_{\boldsymbol{\sigma}}(t)+u_{\boldsymbol{\tau}}(t))F_1(\mathbf{u}(t)) \\ &= \sum_{\boldsymbol{\sigma}^{n+1},\boldsymbol{\sigma}^{n+2}} U(\boldsymbol{\sigma}^{n+1},\boldsymbol{\sigma}^{n+2})w_{\boldsymbol{\sigma}^{n+1}}w_{\boldsymbol{\sigma}^{n+2}}\exp(u_{\boldsymbol{\sigma}^{n+1}}(t)+u_{\boldsymbol{\sigma}^{n+2}}(t))F_1(\mathbf{u}(t)) \\ &= \sum_{\boldsymbol{\sigma}^1,\ldots,\boldsymbol{\sigma}^{n+2}} U(\boldsymbol{\sigma}^{n+1},\boldsymbol{\sigma}^{n+2})f(\boldsymbol{\sigma}^1,\ldots,\boldsymbol{\sigma}^n)\Big(\prod_{1\le\ell\le n+2} w_{\boldsymbol{\sigma}^\ell}\Big)\exp\sum_{1\le\ell\le n+2} u_{\boldsymbol{\sigma}^\ell}(t) \,.\end{aligned}$$

It is of course in this computation that the new independent replicas occur. Combining with (1.92), we get, by definition of $\langle\cdot\rangle_t$,

$$\sum_{\boldsymbol{\sigma},\boldsymbol{\tau}} U(\boldsymbol{\sigma},\boldsymbol{\tau})C(\boldsymbol{\sigma},\boldsymbol{\tau},\mathbf{u}(t)) = n(n+1)\langle U(\boldsymbol{\sigma}^{n+1},\boldsymbol{\sigma}^{n+2})f(\boldsymbol{\sigma}^1,\ldots,\boldsymbol{\sigma}^n)\rangle_t \,,$$

so that $\mathsf{E}\sum_{\boldsymbol{\sigma},\boldsymbol{\tau}} U(\boldsymbol{\sigma},\boldsymbol{\tau})C(\boldsymbol{\sigma},\boldsymbol{\tau},\mathbf{u}(t))$ is indeed the last term of (1.90).

Let us now treat in detail the contribution of the first term of (1.91). We have

$$\frac{\partial^2 F_1}{\partial x_{\boldsymbol{\sigma}}\partial x_{\boldsymbol{\tau}}}(\mathbf{x}) = \sum_{\ell,\ell'\le n} C_{\ell,\ell'}(\boldsymbol{\sigma},\boldsymbol{\tau},\mathbf{x}) \,,$$

where

$$\begin{aligned}&C_{\ell,\ell'}(\boldsymbol{\sigma},\boldsymbol{\tau},\mathbf{x}) \\ &= \sum_{\boldsymbol{\sigma}^1,\ldots,\boldsymbol{\sigma}^n} \mathbf{1}_{\{\boldsymbol{\sigma}^\ell=\boldsymbol{\sigma}\}}\mathbf{1}_{\{\boldsymbol{\sigma}^{\ell'}=\boldsymbol{\tau}\}}w_{\boldsymbol{\sigma}^1}\cdots w_{\boldsymbol{\sigma}^n}f(\boldsymbol{\sigma}^1,\ldots,\boldsymbol{\sigma}^n)\exp\Big(\sum_{\ell_1\le n} x_{\boldsymbol{\sigma}^{\ell_1}}\Big) \,,\end{aligned}$$

and where $\mathbf{1}_{\{\boldsymbol{\sigma}^\ell=\boldsymbol{\sigma}\}} = 1$ if $\boldsymbol{\sigma}^\ell = \boldsymbol{\sigma}$ and is 0 otherwise. Therefore

$$Z(\mathbf{u}(t))^{-n}\sum_{\boldsymbol{\sigma},\boldsymbol{\tau}} U(\boldsymbol{\sigma},\boldsymbol{\tau})C_{\ell,\ell'}(\boldsymbol{\sigma},\boldsymbol{\tau},\mathbf{u}(t)) \tag{1.93}$$

$$= Z(\mathbf{u}(t))^{-n}\sum_{\boldsymbol{\sigma}^1,\dots,\boldsymbol{\sigma}^n} U(\boldsymbol{\sigma}^\ell,\boldsymbol{\sigma}^{\ell'})w_{\boldsymbol{\sigma}^1}\cdots w_{\boldsymbol{\sigma}^n} f(\boldsymbol{\sigma}^1,\dots,\boldsymbol{\sigma}^n)\exp\Bigl(\sum_{\ell_1\le n} u_{\boldsymbol{\sigma}^{\ell_1}}(t)\Bigr)$$

$$= \langle U(\boldsymbol{\sigma}^\ell,\boldsymbol{\sigma}^{\ell'})f(\boldsymbol{\sigma}^1,\dots,\boldsymbol{\sigma}^n)\rangle_t\ , \tag{1.94}$$

and the contribution of the second term of (1.91) is indeed the second term of (1.90). The case of the other terms is similar. □

Exercise 1.4.4. In the proof of Lemma 1.4.2 write in full detail the contribution of the other terms of (1.91).

The reader is urged to complete this exercise, and to meditate the proof of Lemma 1.4.2 until she fully understands it. The algebraic mechanism at work in (1.90) will occur on several occasions (since Gibbs' measures are intrinsically given by a ratio of two quantities). More generally, calculations of a similar nature will be needed again and again.

It will often be the case that $U(\boldsymbol{\sigma},\boldsymbol{\sigma})$ is a number that does not depend on $\boldsymbol{\sigma}$, in which case the third sum in (1.90) cancels the diagonal of the first one, and (1.90) simplifies to

$$\nu_t'(f) = 2\Biggl(\sum_{1\le\ell<\ell'\le n}\nu_t(U(\boldsymbol{\sigma}^\ell,\boldsymbol{\sigma}^{\ell'})f) - n\sum_{\ell\le n}\nu_t(U(\boldsymbol{\sigma}^\ell,\boldsymbol{\sigma}^{n+1})f) + \frac{n(n+1)}{2}\nu_t(U(\boldsymbol{\sigma}^{n+1},\boldsymbol{\sigma}^{n+2})f)\Biggr)\ . \tag{1.95}$$

What we have done in Lemma 1.4.2 is very general. We now go back to the study of the Hamiltonian (1.61) and as in (1.62) we define

$$u_{\boldsymbol{\sigma}} = \frac{\beta}{\sqrt{N}}\sum_{i<j\le N} g_{ij}\sigma_i\sigma_j\ ;\ v_{\boldsymbol{\sigma}} = \beta\sum_{i\le N} z_i\sqrt{q}\sigma_i\ ;\ w_{\boldsymbol{\sigma}} = \exp\Bigl(\sum_{i\le N} h_i\sigma_i\Bigr)\ .$$

Then (1.90) still holds true despite the fact that the numbers $w_{\boldsymbol{\sigma}}$ are now random. This is seen by first using (1.90) at a given realization of the r.v.s h_i, and then taking a further expectation in the randomness of these. Let us next compute in the present setting the quantities $U(\boldsymbol{\sigma}^\ell,\boldsymbol{\sigma}^{\ell'})$. Let us define

$$R_{\ell,\ell'} = \frac{1}{N}\sum_{i\le N}\sigma_i^\ell\sigma_i^{\ell'}\ . \tag{1.96}$$

This notation will be used **in the entire book** a countless number of times. We will also use countless times that by symmetry between replicas, we have e.g. that $\nu(R_{1,2}) = \nu(R_{1,3})$ or $\nu(R_{1,2}R_{2,3}) = \nu(R_{1,2}R_{1,3})$. On the other hand, if a function f depends only on $\boldsymbol{\sigma}^1$ and $\boldsymbol{\sigma}^2$, it is true that $\nu((R_{1,3}-q)^2 f) = \nu((R_{1,4}-q)^2 f)$, but not in general that $\nu((R_{1,2}-q)^2 f) = \nu((R_{1,3}-q)^2 f)$.

As in (1.64) we have

$$\frac{1}{N}U(\boldsymbol{\sigma}^\ell, \boldsymbol{\sigma}^{\ell'}) = \frac{\beta^2}{4}\left(R_{\ell,\ell'}^2 - \frac{1}{N}\right) - \frac{\beta^2}{2} q R_{\ell,\ell'} \,.$$

Using (1.95) for $n=2$, and completing the squares we get

$$\nu_t'(f) = \frac{N\beta^2}{2}\Big(\nu_t((R_{1,2}-q)^2 f) - 2\sum_{\ell\le 2}\nu_t((R_{\ell,3}-q)^2 f) + 3\nu_t\big((R_{3,4}-q)^2 f\big)\Big) \,. \tag{1.97}$$

Up to Corollary 1.4.7 below, the results are true for every value of q, not only the solution of (1.74).

Lemma 1.4.5. *Consider any number $\lambda > 0$. Then*

$$\nu_t\big((R_{3,4}-q)^2 \exp \lambda N (R_{1,2}-q)^2\big) \le \nu_t\big((R_{1,2}-q)^2 \exp \lambda N (R_{1,2}-q)^2\big) \,. \tag{1.98}$$

Proof. First, we observe a general form of Hölder's inequality,

$$\nu_t(f_1 f_2) \le \nu_t(f_1^{\tau_1})^{1/\tau_1} \nu_t(f_2^{\tau_2})^{1/\tau_2} \,, \tag{1.99}$$

for $f_1, f_2 \ge 0$, $1/\tau_1 + 1/\tau_2 = 1$. This is obtained by using Hölder's inequality for the probability ν_t(or by using it successively for $\langle\cdot\rangle_t$ and then for E). Using (1.99) with $\tau_1 = k+1$, $\tau_2 = (k+1)/k$ we deduce that, using symmetry between replicas in the second line,

$$\begin{aligned}
&\nu_t\big((R_{3,4}-q)^2 (R_{1,2}-q)^{2k}\big) \\
&\le \nu_t\big((R_{3,4}-q)^{2k+2}\big)^{1/k+1} \nu_t\big((R_{1,2}-q)^{2k+2}\big)^{k/k+1} \\
&= \nu_t\big((R_{1,2}-q)^{2k+2}\big) \,.
\end{aligned}$$

To prove (1.98), we simply expand $\exp \lambda N (R_{1,2}-q)^2$ as a power series of $(R_{1,2}-q)^2$ and we apply the preceding inequality to each term, i.e. we write

$$\begin{aligned}
\nu_t\big((R_{3,4}-q)^2 \exp \lambda N (R_{1,2}-q)^2\big) &= \sum_{k\ge 0} \frac{(N\lambda)^k}{k!} \nu_t\big((R_{3,4}-q)^2 (R_{1,2}-q)^{2k}\big) \\
&\le \sum_{k\ge 0} \frac{(N\lambda)^k}{k!} \nu_t\big((R_{1,2}-q)^{2k+2}\big) \\
&= \nu_t\big((R_{1,2}-q)^2 \exp \lambda N (R_{1,2}-q)^2\big) \,. \quad \square
\end{aligned}$$

Combining with (1.97) we get:

Corollary 1.4.6. *If* $\lambda > 0$ *then*

$$\nu_t'\big(\exp \lambda N(R_{1,2} - q)^2\big) \le 2N\beta^2 \nu_t\big((R_{1,2} - q)^2 \exp \lambda N(R_{1,2} - q)^2\big) . \quad (1.100)$$

Corollary 1.4.7. *For* $t < \lambda/2\beta^2$ *we have*

$$\frac{\mathrm{d}}{\mathrm{d}t}\Big(\nu_t\Big(\exp\big((\lambda - 2t\beta^2)N(R_{1,2} - q)^2\big)\Big)\Big) \le 0 ,$$

or in other words, the function

$$t \mapsto \nu_t\Big(\exp\big((\lambda - 2t\beta^2)N(R_{1,2} - q)^2\big)\Big) \quad (1.101)$$

is non-increasing.

Proof. In the function (1.101) there are two sources of dependence in t, through ν_t and through the term $-2t\beta^2$, so that

$$\begin{aligned}
&\frac{\mathrm{d}}{\mathrm{d}t}\Big(\nu_t\Big(\exp\big((\lambda - 2t\beta^2)N(R_{1,2} - q)^2\big)\Big)\Big) \\
&= \nu_t'\Big(\exp\big((\lambda - 2t\beta^2)N(R_{1,2} - q)^2\big)\Big) \\
&- 2N\beta^2\nu_t\Big((R_{1,2} - q)^2 \exp\big((\lambda - 2t\beta^2)N(R_{1,2} - q)^2\big)\Big) ,
\end{aligned}$$

and we use (1.100). □

Proposition 1.4.8. *When* q *is the solution of (1.74), for* $\lambda < 1/2$ *we have*

$$\nu_0\big(\exp \lambda N(R_{1,2} - q)^2\big) \le \frac{1}{\sqrt{1 - 2\lambda}} . \quad (1.102)$$

Whenever, like here, we state a result without proof or reference, the reason is always that (unless it is an obvious corollary of what precedes) the proof can be found later in the same section, but that we prefer to demonstrate its use before giving this proof.

At this point we may try to formulate in words the idea underlying the proof of Theorem 1.4.1: it is to transfer the excellent control (1.102) of $R_{1,2} - q$ for ν_0 to ν_1 using Lemma 1.4.2.

Proof of Theorem 1.4.1. Taking $\lambda = s + 2\beta^2 < 1/2$ we deduce from (1.102) and Corollary 1.4.7 that for all $0 \le t \le 1$,

$$\nu_t\Big(\exp\big((s + 2(1 - t)\beta^2)N(R_{1,2} - q)^2\big)\Big) \le \frac{1}{\sqrt{1 - 2s - 4\beta^2}} ,$$

because this is true for $t = 0$ and because the left-hand side is a non-increasing function of t. Since $s + 2(1 - t)\beta^2 \ge s$ this shows that for each t (and in particular for $t = 1$) we have

$$\nu_t\big(\exp sN(R_{1,2}-q)^2\big) \le \frac{1}{\sqrt{1-2s-4\beta^2}} . \qquad \square$$

As a consequence (1.88) holds uniformly in t and in $\beta \le \beta_0 < 1/2$, i.e.

$$\nu_t\big((R_{1,2}-q)^{2k}\big) \le (Kk/N)^k , \tag{1.103}$$

where K does not depend on t or β.

Exercise 1.4.9. Prove that if $\varepsilon = \pm 1$ we have

$$\nu_0(\mathbf{1}_{\{\sigma_i^1=\sigma_i^2=\varepsilon\}}) = \frac{1+q}{4}$$

and

$$\nu_0(\mathbf{1}_{\{\sigma_i^1=-\sigma_i^2=\varepsilon\}}) = \frac{1-q}{4} .$$

These relations will never be used, so do not worry if you can't solve this exercise. Its purpose is to help learning to manipulate simple objects. Some hints might be contained in the proof of (1.104) below.

Proof of Proposition 1.4.8. Let us first recall that ν_0 is associated to the Hamiltonian (1.68), so that for ν_0 there is no correlation between sites, so this is a (nice) exercise in Calculus. Let $Y_i = \beta z_i\sqrt{q} + h_i$, so (1.68) means that $-H_0(\boldsymbol{\sigma}) = \sum_{i\le N}\sigma_i Y_i$. Recalling that $\langle\cdot\rangle_0$ denotes an average for the Gibbs measure with Hamiltonian H_0, we get that $\langle\sigma_i\rangle_0 = \mathrm{th}Y_i$ and, since $q = \mathsf{E}\mathrm{th}^2 Y_i$ by (1.74) we have

$$\nu_0(\sigma_i^1\sigma_i^2) = \mathsf{E}\langle\sigma_i^1\sigma_i^2\rangle_0 = \mathsf{E}\langle\sigma_i\rangle_0^2 = \mathsf{E}\mathrm{th}^2 Y_i = q . \tag{1.104}$$

At this point the probabilistically oriented reader should think of the sequence $(\sigma_i^1\sigma_i^2)_{1\le i\le N}$ as (under ν_0) an i.i.d. sequence of $\{-1,1\}$-valued r.v.s of expectation q, for which all kinds of estimates are classical. Nonetheless we give a simple self-contained proof. The main step of this proof is to show that for every u we have

$$\nu_0\big(\exp Nu(R_{1,2}-q)\big) \le \exp\frac{Nu^2}{2} . \tag{1.105}$$

Since (1.105) holds for every value of u it holds when u is a Gaussian r.v. with $\mathsf{E}u^2 = 2\lambda/N$, independent of all the other sources of randomness. Taking expectation in u in (1.105) and using (A.11) yields (1.102).

To prove (1.105), we first evaluate

$$\begin{aligned}\nu_0(\exp Nu(R_{1,2}-q)) &= \nu_0\Big(\exp u\sum_{i\le N}(\sigma_i^1\sigma_i^2-q)\Big)\\ &= \prod_{i\le N}\nu_0\left(\exp u(\sigma_i^1\sigma_i^2-q)\right) ,\end{aligned} \tag{1.106}$$

by independence between the sites. Using that when $|\varepsilon| = 1$ we have $\exp \varepsilon x = \mathrm{ch} x + \mathrm{sh} \varepsilon x = \mathrm{ch} x + \varepsilon\, \mathrm{sh} x$, we obtain

$$\exp u\sigma_i^1\sigma_i^2 = \mathrm{ch} u + \sigma_i^1\sigma_i^2\, \mathrm{sh} u \ ,$$

and thus

$$\nu_0(\exp u\sigma_i^1\sigma_i^2) = \mathrm{ch} u + \nu_0(\sigma_i^1\sigma_i^2)\, \mathrm{sh} u \ . \tag{1.107}$$

Therefore (1.104) implies

$$\nu_0\left(\exp u(\sigma_i^1\sigma_i^2)\right) = \mathrm{ch} u + q\, \mathrm{sh} u \ ,$$

and consequently

$$\nu_0\left(\exp u(\sigma_i^1\sigma_i^2 - q)\right) = \exp(-qu)\left(\mathrm{ch} u + q\, \mathrm{sh} u\right) \ .$$

Now, for $q \geq 0$ and all u we have

$$(\mathrm{ch} u + q\mathrm{sh} u)\exp(-qu) \leq \exp \frac{u^2}{2} \ .$$

Indeed the function

$$f(u) = \log(\mathrm{ch} u + q\mathrm{sh} u) - qu$$

satisfies $f(0) = 0$,

$$f'(u) = \frac{\mathrm{sh} u + q\mathrm{ch} u}{\mathrm{ch} u + q\mathrm{sh} u} - q \ ; \ f''(u) = 1 - \left(\frac{\mathrm{sh} u + q\mathrm{ch} u}{\mathrm{ch} u + q\mathrm{sh} u}\right)^2$$

so that $f'(0) = 0$ and $f''(u) \leq 1$, and therefore $f(u) \leq u^2/2$. Thus

$$\nu_0\left(\exp u(\sigma_i^1\sigma_i^2 - q)\right) \leq \exp \frac{u^2}{2}$$

and (1.106) yields (1.105). This completes the proof. □

Let us recall that we denote by $\mathrm{SK}(\beta, h)$ the right-hand side of (1.73) when q is as in (1.74). As in the case of $p_N(\beta, h)$ this is a number depending only on β and the law of h.

Theorem 1.4.10. *If $\beta < 1/2$ then*

$$|p_N(\beta, h) - \mathrm{SK}(\beta, h)| \leq \frac{K}{N} \ , \tag{1.108}$$

where K does not depend on N.

Thus, when $\beta < 1/2$, (1.73) is a near equality, and in particular $p(\beta, h) = \lim_{N\to\infty} p_N(\beta, h) = \mathrm{SK}(\beta, h)$. Of course, this immediately raises the question as for which values of (β, h) this equality remains true. This is a difficult question that will be investigated later. It suffices to say now that, given h, the equality fails for large enough β, but this statement itself is far from being obvious.

We have observed that, as a consequence of Hölder's inequality, the function $\beta \mapsto p_N(\beta, h)$ is convex. It then follows from (1.108) that, when $\beta < 1/2$, the function $\beta \mapsto \mathrm{SK}(\beta, h)$ is also convex. Yet, this is not really obvious on the definition of this function. It should not be very difficult to find a calculus proof of this fact, but what is needed is to understand *really* why this is the case. Much later, we will be able to give a complicated analytic expression (the Parisi formula) for $\lim_{N\to\infty} p_N(\beta, h)$, which is valid for any value of β, and it is still not known how to prove by a direct argument that this analytical expression is a convex function of β.

In a statement such as (1.108) the constant K can in principle depend on β and h. It will however be shown in the proof that for $\beta \le \beta_0 < 1/2$, it can be chosen so that it does not depend on β or h.

Proof of Theorem 1.4.10. We have proved in (1.103) that if $\beta \le \beta_0 < 1/2$ then $\nu_t((R_{1,2}-q)^2) \le K/N$, where K depends on β_0 only. Now (1.65) implies

$$\left|\varphi'(t) - \frac{\beta^2}{4}(1-q)^2\right| \le \frac{K}{N} ,$$

where K depends on β_0 only. Thus

$$\left|\varphi(1) - \varphi(0) - \frac{\beta^2}{4}(1-q)^2\right| \le \frac{K}{N} ,$$

and

$$\varphi(1) = p_N(\beta, h) \ ; \ \ \varphi(0) = \log 2 + \mathsf{E} \log \mathrm{ch}(\beta\sqrt{q} + h) . \qquad \square$$

Theorem 1.4.10 controls the expected value (= first moment) of the quantity $N^{-1} \log Z_N(\beta, h) - \mathrm{SK}(\beta, h)$. In Theorem 1.4.11 below we will be able to accurately compute the higher moments of this quantity. Of course this requires a bit more work. This result will not be used in the sequel, so it can in principle be skipped at first reading. However we must mention that one of the goals of the proof is to further acquaint the reader with the mechanisms of integration by parts.

Let us denote by $a(k)$ the k-th moment of a standard Gaussian r.v. (so that $a(0) = 1 = a(2)$, $a(1) = 0$ and, by integration by parts, $a(k) = \mathsf{E} g g^{k-1} = (k-1)a(k-2)$). Consider q as in (1.74) and $Y = \beta z \sqrt{q} + h$. Let

$$b = \mathsf{E}(\log \mathrm{ch} Y)^2 - (\mathsf{E} \log \mathrm{ch} Y)^2 - \frac{\beta^2 q^2}{2} .$$

Theorem 1.4.11. *Assume that the r.v. h is Gaussian (not necessarily centered). Then if $\beta < 1/2$, for each $k \geq 1$ we have*

$$\left|\mathsf{E}\left(\frac{1}{N}\log Z_N(\beta,h) - \mathrm{SK}(\beta,h)\right)^k - \frac{1}{N^{k/2}}a(k)b^{k/2}\right| \leq \frac{K}{N^{(k+1)/2}}\,, \tag{1.109}$$

where K does not depend on N.

Let us recall that, to lighten notation, we write

$$\mathsf{E}\left(\frac{1}{N}\log Z_N(\beta,h) - \mathrm{SK}(\beta,h)\right)^k$$

instead of

$$\mathsf{E}\left(\left(\frac{1}{N}\log Z_N(\beta,h) - \mathrm{SK}(\beta,h)\right)^k\right).$$

A similar convention will be used whenever there is no ambiguity.

The case $k = 1$ of (1.109) recovers (1.108). We can view (1.109) as a "quantitative central limit theorem". With accuracy about $N^{-1/2}$, the k-th moment of the r.v.

$$\sqrt{N}\left(\frac{1}{N}\log Z_N(\beta,h) - \mathrm{SK}(\beta,h)\right) \tag{1.110}$$

is about that of $\sqrt{b}z$ where z is a standard Gaussian r.v. In particular the r.v. (1.110) has in the limit the same law as the r.v. $\sqrt{b}z$.

When dealing with central limit theorems, it will be convenient to denote by $O(k)$ any quantity A such that $|A| \leq KN^{-k/2}$ where K does not depend on N (of course K will depend on k). This is very different from the "standard" meaning of this notation (that we will never use). Thus we can write (1.109) as

$$\mathsf{E}\left(\frac{1}{N}\log Z_N(\beta,h) - \mathrm{SK}(\beta,h)\right)^k = \frac{1}{N^{k/2}}a(k)b^{k/2} + O(k+1)\,. \tag{1.111}$$

Let us also note that

$$O(k)O(\ell) = O(k+\ell)\;;\; O(2k)^{1/2} = O(k)\,. \tag{1.112}$$

Lemma 1.4.12. *If the r.v. h is Gaussian (not necessarily centered) then for any value of β we have*

$$\mathsf{E}\left|\frac{1}{N}\log Z_N(\beta,h) - p_N(\beta,h)\right|^k = O(k)\,. \tag{1.113}$$

Moreover the constant K implicit in the notation $O(k)$ remains bounded as both β and the variance of h remain bounded.

The hypothesis that h is Gaussian can be considerably weakened; but here is not the place for such refinements.

Proof. Let us write $h_i = cy_i + d$, where y_i are i.i.d. centered Gaussian. The key to the proof is that for $t > 0$ we have

$$\mathsf{P}(|\log Z_N(\beta,h) - Np_N(\beta,h)| \geq t) \leq 2\exp\left(-\frac{t^2}{4c^2N + 2\beta^2(N-1)}\right). \tag{1.114}$$

This can be proved by an obvious adaptation of the proof of (1.54). Equivalently, to explain the same argument in a different way, let us think of the quantity $\log Z_N(\beta,h)$ as a function F of the variables $(g_{ij})_{i<j}$ and $(y_i)_{i\leq N}$. It is obvious (by writing the value of the derivative) that

$$\left|\frac{\partial F}{\partial g_{ij}}\right| \leq \frac{\beta}{\sqrt{N}} \; ; \quad \left|\frac{\partial F}{\partial y_i}\right| \leq c\,.$$

Thus the gradient ∇F of F satisfies $\|\nabla F\|^2 \leq c^2N + \beta^2(N-1)/2$ and (1.47) implies (1.114). Then (1.114) yields

$$\mathsf{P}(|X| \geq t) \leq 2\exp\left(-\frac{Nt^2}{A^2}\right), \tag{1.115}$$

where

$$X = \frac{1}{N}\log Z_N(\beta,h) - p_N(\beta,h)$$

and $A^2 = 4c^2 + 2\beta^2$. Now (1.113) follows from (1.115) by computing moments through the formula $\mathsf{E}Y^k = \int_0^\infty kt^{k-1}\mathsf{P}(Y \geq t)dt$ for $Y \geq 0$, as is explained in detail in Section A.6. □

Proof of Theorem 1.4.11. We use again the path (1.59). Let

$$A(t) = \frac{1}{N}\log\sum_{\boldsymbol{\sigma}}\exp(-H_t(\boldsymbol{\sigma})) \tag{1.116}$$

$$\mathrm{SK}(t) = \log 2 + \mathsf{E}\log\mathrm{ch}Y + \frac{\beta^2 t}{4}(1-q)^2$$

$$V(t) = A(t) - \mathrm{SK}(t)$$

$$b(t) = \mathsf{E}(\log\mathrm{ch}Y)^2 - (\mathsf{E}\log\mathrm{ch}Y)^2 - \frac{\beta^2q^2t}{2}\,.$$

Thus, the quantities $\mathsf{E}A(t)$, $\mathrm{SK}(t)$ and $b(t)$ correspond along the interpolation respectively to the quantities $p_N(\beta,h)$, $\mathrm{SK}(\beta,h)$ and b. In this proof we write $O(k)$ for a quantity A such that $|A| \leq KN^{-k/2}$ where K does not depend on N *or* of the interpolation parameter t.

Let us write explicitly the interpolating Hamiltonian (1.59) using (1.62):

$$-H_t(\boldsymbol{\sigma}) = \frac{\beta\sqrt{t}}{\sqrt{N}}\sum_{i<j\leq N} g_{ij}\sigma_i\sigma_j + \sum_{i\leq N}(h_i + \beta\sqrt{1-t}z_i\sqrt{q})\sigma_i\,. \tag{1.117}$$

It is of the type (1.61), but we have replaced β by $\beta\sqrt{t} \le \beta$ and the r.v. h by $h + \beta\sqrt{1-t}z\sqrt{q}$, where z is a standard Gaussian r.v. independent of h. Thus (1.113) implies

$$\mathsf{E}V(t)^k = O(k) . \tag{1.118}$$

We will prove by induction over $k \ge 1$ that

$$\mathsf{E}V(t)^k = \frac{1}{N^{k/2}} a(k) b(t)^{k/2} + O(k+1) . \tag{1.119}$$

For $t = 1$ this is (1.111). To start the induction, we observe that by Theorem 1.4.10 and (1.117), this is true for $k = 1$. For the induction step, let us fix k and assume that (1.119) has been proved for all $k' \le k - 1$. Let us define

$$\psi(t) = \mathsf{E}V(t)^k .$$

The basic idea is to prove that

$$\psi'(t) = \frac{k(k-1)}{2N} b'(t) \mathsf{E}V(t)^{k-2} + O(k+1) . \tag{1.120}$$

The induction hypothesis then yields

$$\begin{aligned} \psi'(t) &= \frac{k(k-1)}{2N^{k/2}} b'(t) a(k-2) b(t)^{k/2-1} + O(k+1) \\ &= \frac{1}{N^{k/2}} a(k) \Big(\frac{k}{2} b'(t) b(t)^{k/2-1} \Big) + O(k+1) . \end{aligned} \tag{1.121}$$

Assume now that we can prove that

$$\psi(0) = \frac{1}{N^{k/2}} a(k) b(0)^{k/2} + O(k+1) . \tag{1.122}$$

Then by integration of (1.121) we get

$$\psi(t) = \frac{1}{N^{k/2}} a(k) b(t)^{k/2} + O(k+1) , \tag{1.123}$$

which is (1.119). We now start the real proof, the first step of which is to compute $\psi'(t)$. For a given number a we consider

$$\varphi(t, a) = \mathsf{E}(A(t) - a)^k , \tag{1.124}$$

and we compute $\partial\varphi(t,a)/\partial t$ using (1.40). This is done by a suitable extension of (1.60). Keeping the notation of this formula, as well as the notation (1.60), consider the function $W(x) = (x-a)^k$ and for $\mathbf{x} = (x_{\boldsymbol{\sigma}})$, consider the function

$$F(\mathbf{x}) = W\Big(\frac{1}{N} \log Z(\mathbf{x}) \Big) .$$

Thus

$$\frac{\partial F}{\partial x_{\boldsymbol{\tau}}}(\mathbf{x}) = \frac{1}{N}\frac{w_{\boldsymbol{\tau}} \exp x_{\boldsymbol{\tau}}}{Z(\mathbf{x})} W'\left(\frac{1}{N}\log Z(\mathbf{x})\right).$$

If $x_{\boldsymbol{\sigma}} \neq x_{\boldsymbol{\tau}}$ we then have

$$\begin{aligned}\frac{\partial^2 F}{\partial x_{\boldsymbol{\sigma}} \partial x_{\boldsymbol{\tau}}}(\mathbf{x}) = & -\frac{1}{N}\frac{w_{\boldsymbol{\sigma}} w_{\boldsymbol{\tau}} \exp(x_{\boldsymbol{\sigma}} + x_{\boldsymbol{\tau}})}{Z(\mathbf{x})^2} W'\left(\frac{1}{N}\log Z(\mathbf{x})\right) \\ & + \frac{1}{N^2}\frac{w_{\boldsymbol{\sigma}} w_{\boldsymbol{\tau}} \exp(x_{\boldsymbol{\sigma}} + x_{\boldsymbol{\tau}})}{Z(\mathbf{x})^2} W''\left(\frac{1}{N}\log Z(\mathbf{x})\right),\end{aligned}$$

while

$$\begin{aligned}\frac{\partial^2 F}{\partial x_{\boldsymbol{\sigma}}^2}(\mathbf{x}) = & \frac{1}{N}\left(\frac{w_{\boldsymbol{\sigma}} \exp x_{\boldsymbol{\sigma}}}{Z(\mathbf{x})} - \frac{w_{\boldsymbol{\sigma}}^2 \exp 2x_{\boldsymbol{\sigma}}}{Z(\mathbf{x})^2}\right) W'\left(\frac{1}{N}\log Z(\mathbf{x})\right) \\ & + \frac{1}{N^2}\frac{w_{\boldsymbol{\sigma}}^2 \exp 2x_{\boldsymbol{\sigma}}}{Z(\mathbf{x})^2} W''\left(\frac{1}{N}\log Z(\mathbf{x})\right).\end{aligned}$$

Therefore, proceeding as in the proof of (1.60), we conclude that the function

$$\varphi(t,a) = \mathsf{E}W\left(\frac{1}{N}\log Z(\mathbf{u}(t))\right) = \mathsf{E}W(A(t))$$

satisfies

$$\begin{aligned}\frac{\partial \varphi}{\partial t}(t,a) = & \frac{1}{N}\mathsf{E}\big((\langle U(\boldsymbol{\sigma},\boldsymbol{\sigma})\rangle_t - \langle U(\boldsymbol{\sigma}^1,\boldsymbol{\sigma}^2)\rangle_t) W'(A(t))\big) \\ & + \frac{1}{N^2}\mathsf{E}\big(\langle U(\boldsymbol{\sigma}^1,\boldsymbol{\sigma}^2)\rangle_t W''(A(t))\big),\end{aligned}$$

and replacing W by its value this is

$$\begin{aligned}\frac{\partial \varphi}{\partial t}(t,a) = & \frac{k}{N}\mathsf{E}\big((\langle U(\boldsymbol{\sigma},\boldsymbol{\sigma})\rangle_t - \langle U(\boldsymbol{\sigma}^1,\boldsymbol{\sigma}^2)\rangle_t)(A(t)-a)^{k-1}\big) \\ & + \frac{k(k-1)}{N^2}\mathsf{E}\big(\langle U(\boldsymbol{\sigma}^1,\boldsymbol{\sigma}^2)\rangle_t (A(t)-a)^{k-2}\big).\end{aligned} \tag{1.125}$$

This is a generalization of (1.60), that corresponds to the case $k = 1$.

There is an alternate way to explain the structure of the formula (1.125) (but the proof is identical). It is to say that straightforward (i.e. applying only the most basic rules of Calculus) differentiation of (1.116) yields

$$A'(t) = \frac{1}{N}\frac{\sum_{\boldsymbol{\sigma}} -H_t'(\boldsymbol{\sigma})\exp(-H_t(\boldsymbol{\sigma}))}{\sum_{\boldsymbol{\sigma}} \exp(-H_t(\boldsymbol{\sigma}))} = \frac{1}{N}\langle -H_t'(\boldsymbol{\sigma})\rangle_t ,$$

where

$$-H_t'(\boldsymbol{\sigma}) := \frac{\mathrm{d}}{\mathrm{d}t}(-H_t(\boldsymbol{\sigma})) = \frac{1}{2\sqrt{t}}u_{\boldsymbol{\sigma}} - \frac{1}{2\sqrt{1-t}}v_{\boldsymbol{\sigma}} ,$$

so that

$$\frac{\partial\varphi}{\partial t}(t,a) = k\mathsf{E}(A'(t)(A(t)-a)^{k-1}) = \frac{k}{N}\mathsf{E}\big(\langle -H'_t(\boldsymbol{\sigma})\rangle_t(A(t)-a)^{k-1}\big) \ . \quad (1.126)$$

One then integrates by parts, while using the key relation $U(\boldsymbol{\sigma},\boldsymbol{\tau}) = \mathsf{E}H'_t(\boldsymbol{\sigma})H_t(\boldsymbol{\tau})$. (Of course making this statement precise amounts basically to reproducing the previous calculation.) The dependence of the bracket $\langle\cdot\rangle_t$ on the Hamiltonian creates the first term in (1.125) (we have actually already done this computation), while the dependence of $A(t)$ on this Hamiltonian creates the second term.

This method of explanation is convenient to guide the reader (once she has gained some experience) through the many computations (that will soon become routine) involving Gaussian integration by parts, without reproducing the computations in detail (which would be unbearable). For this reason we will gradually shift (in particular in the next chapters) to this convenient method of giving a high-level description of these computations. Unfortunately, there is no miracle, and to gain the experience that will make these formulas transparent to the reader, she has to work through a few of them in complete detail, and doing in detail the integration by parts in (1.126) is an excellent start.

Using (1.64) and completing the squares in (1.125) yields

$$\begin{aligned}\frac{\partial\varphi}{\partial t}(t,a) &= -\frac{\beta^2 k}{4}\mathsf{E}\big(\langle(R_{1,2}-q)^2\rangle_t(A(t)-a)^{k-1}\big)\\ &+\frac{\beta^2}{4}k(1-q)^2\mathsf{E}(A(t)-a)^{k-1}\\ &+\frac{\beta^2}{4}\frac{k(k-1)}{N}\mathsf{E}\big(\langle(R_{1,2}-q)^2\rangle_t(A(t)-a)^{k-2}\big)\\ &-\frac{\beta^2}{4}\frac{k(k-1)}{N}q^2\mathsf{E}(A(t)-a)^{k-2}-\frac{\beta^2}{4N^2}k(k-1)\mathsf{E}(A(t)-a)^{k-2} \ .\end{aligned}$$

Now, since

$$\frac{\mathrm{d}}{\mathrm{d}t}\varphi(t,\mathrm{SK}(t)) = \frac{\partial\varphi}{\partial t}(t,\mathrm{SK}(t)) + \mathrm{SK}'(t)\frac{\partial\varphi}{\partial a}(t,\mathrm{SK}(t)) \ ,$$

and

$$\frac{\partial\varphi}{\partial a}(t,a) = -k\mathsf{E}(A(t)-a)^{k-1} \ ; \ \mathrm{SK}'(t) = \frac{\beta^2}{4}(1-q)^2 \ ,$$

one gets

$$\psi'(t) = \mathrm{I}+\mathrm{II}$$

where

$$\begin{aligned}\mathrm{I} &= -\frac{\beta^2 q^2}{4}\frac{k(k-1)}{N}\mathsf{E}V(t)^{k-2}\\ \mathrm{II} &= -\frac{\beta^2 k}{4}\mathsf{E}\big(\langle(R_{1,2}-q)^2\rangle_t V(t)^{k-1}\big)+\frac{\beta^2}{4}\frac{k(k-1)}{N}\mathsf{E}\big(\langle(R_{1,2}-q)^2\rangle_t V(t)^{k-2}\big)\\ &-\frac{\beta^2}{4N^2}k(k-1)\mathsf{E}V(t)^{k-2} \ .\end{aligned}$$

We claim that

$$\mathrm{II} = O(k+1) .$$

To see this we note that by (1.118) (used for $2(k-1)$ rather than k) we have $\mathsf{E}(V(t)^{2(k-1)}) = O(2(k-1))$ and we write, using (1.103),

$$\begin{aligned}\mathsf{E}\big(\langle(R_{1,2}-q)^2\rangle_t V(t)^{k-1}\big) &\le (\mathsf{E}\langle(R_{1,2}-q)^4\rangle_t)^{1/2}\mathsf{E}\big(V(t)^{2(k-1)}\big)^{1/2} \\ &= O(2)O(k-1) = O(k+1) .\end{aligned}$$

The case of the other terms is similar. Thus, we have proved that $\psi'(t) = \mathrm{I} + O(k+1)$, and since $b'(t) = -\beta^2 q^2/2$ we have also proved (1.120). To complete the induction it remains only to prove (1.122). With obvious notation,

$$V(0) = \frac{1}{N}\sum_{i\le N}(\log \mathrm{ch} Y_i - \mathsf{E}\log \mathrm{ch} Y) .$$

The r.v.s $X_i = \log \mathrm{ch} Y_i - \mathsf{E}\log \mathrm{ch} Y$ form an i.i.d. sequence of centered variables, so the statement in that case is simply (a suitable quantitative version of) the central limit theorem. We observe that by (1.118), for each k, we have $\mathsf{E}V(0)^k = O(k)$. (Of course the use of Lemma 1.4.12 here is an overkill.) To evaluate $\mathsf{E}V(0)^k$ we use symmetry to write

$$\mathsf{E}V(0)^k = \mathsf{E}\big(X_N V(0)^{k-1}\big) = \mathsf{E}\bigg(X_N\left(\frac{X_N}{N}+B\right)^{k-1}\bigg)$$

where $B = N^{-1}\sum_{i\le N-1} X_i$. We observe that since $B = V(0) - X_N/N$, for each k we have $\mathsf{E}B^k = O(k)$. We expand the term $(X_N/N+B)^{k-1}$ and since $\mathsf{E}X_N = 0$ we get the relation

$$\mathsf{E}V(0)^k = \frac{k-1}{N}\mathsf{E}X_N^2\mathsf{E}B^{k-2} + O(k+1) .$$

Using again that $B = V(0) - X_N/N$ and since $\mathsf{E}X_N^2 = b(0)$ we then obtain

$$\mathsf{E}V(0)^k = \frac{k-1}{N}b(0)\mathsf{E}V(0)^{k-2} + O(k+1) ,$$

from which the claim follows by induction. □

Here is one more exercise to help the reader think about interpolation between two Gaussian Hamiltonians $u_{\boldsymbol{\sigma}}$ and $v_{\boldsymbol{\sigma}}$.

Exercise 1.4.13. Consider a (reasonable) function $W(y_1,\ldots,y_{m+1})$ of $m+1$ variables. Consider m functions $f_1,\ldots,f_m$ on Σ_N^n. Compute the derivative of

$$\varphi(t) = W(\nu_t(f_1),\ldots,\nu_t(f_m), N^{-1}\log Z(\mathbf{u}(t))) ,$$

where the notation is as usual.

Our next result makes apparent that the (crucial) property $\nu((R_{1,2} - q)^2) \le K/N$ implies some independence between the sites.

Proposition 1.4.14. *For any p and any q with $0 \le q \le 1$ we have*

$$\mathsf{E}(\langle\sigma_1 \cdots \sigma_p\rangle - \langle\sigma_1\rangle \cdots \langle\sigma_p\rangle)^2 \le K(p)\nu\big((R_{1,2} - q)^2\big) , \tag{1.127}$$

where $K(p)$ depends on p only.

This statement is clearly of importance: it means that when the right-hand side is small "the spins decorrelate". (When $p = 2$, the quantity $\langle\sigma_1\sigma_2\rangle - \langle\sigma_1\rangle\langle\sigma_2\rangle$ is the covariance of the spins σ_1 and σ_2, seen as r.v.s on the probability space (Σ_N, G_N). The physicists call this quantity the truncated correlation.) Equation (1.127) is true for any value of q, but we will show in Proposition 1.9.5 below that essentially the only value of q for which the quantity $\nu((R_{1,2} - q)^2)$ might be small is the solution of (1.74).

We denote by $\cdot$ the dot product in $\mathbb{R}^N$, so that e.g. $R_{1,2} = \boldsymbol{\sigma}^1 \cdot \boldsymbol{\sigma}^2/N$.

A notable feature of the proof of Proposition 1.4.14 is that the *only* feature of the model it uses is symmetry between sites, so this proposition can be applied to many of the models we will study.

Proof of Proposition 1.4.14. Throughout the proof $K(p)$ denotes a number depending on p only, that need not be the same at each occurrence. The proof goes by induction on p, and the case $p = 1$ is obvious. For the induction from $p - 1$ to p it suffices to prove that

$$\mathsf{E}(\langle\sigma_1 \cdots \sigma_p\rangle - \langle\sigma_1 \cdots \sigma_{p-1}\rangle\langle\sigma_p\rangle)^2 \le K(p)\nu\big((R_{1,2} - q)^2\big) . \tag{1.128}$$

Let $\dot\sigma_i = \sigma_i - \langle\sigma_i\rangle$ and $\dot{\boldsymbol{\sigma}} = (\dot\sigma_i)_{i \le N}$. Therefore

$$\langle\sigma_1 \cdots \sigma_p\rangle - \langle\sigma_1 \cdots \sigma_{p-1}\rangle\langle\sigma_p\rangle = \langle\sigma_1\sigma_2 \cdots \sigma_{p-1}\dot\sigma_p\rangle .$$

Using replicas, we have

$$\begin{aligned}(\langle\sigma_1 \cdots \sigma_p\rangle - \langle\sigma_1 \cdots \sigma_{p-1}\rangle\langle\sigma_p\rangle)^2 &= \langle\sigma_1\sigma_2 \cdots \sigma_{p-1}\dot\sigma_p\rangle^2 \\ &= \langle\sigma_1^1\sigma_1^2 \cdots \sigma_{p-1}^1\sigma_{p-1}^2\dot\sigma_p^1\dot\sigma_p^2\rangle ,\end{aligned}$$

so that

$$\mathsf{E}(\langle\sigma_1 \cdots \sigma_p\rangle - \langle\sigma_1 \cdots \sigma_{p-1}\rangle\langle\sigma_p\rangle)^2 = \nu(\sigma_1^1\sigma_1^2 \cdots \sigma_{p-1}^1\sigma_{p-1}^2\dot\sigma_p^1\dot\sigma_p^2) . \tag{1.129}$$

Using symmetry between sites,

$$\begin{aligned}&N(N-1)\cdots(N-p+1)\nu(\sigma_1^1\sigma_1^2 \cdots \sigma_{p-1}^1\sigma_{p-1}^2\dot\sigma_p^1\dot\sigma_p^2) \\ &= \sum_{i_1,\dots,i_p \text{ all different}} \nu(\sigma_{i_1}^1\sigma_{i_1}^2\sigma_{i_2}^1\sigma_{i_2}^2 \cdots \sigma_{i_{p-1}}^1\sigma_{i_{p-1}}^2\dot\sigma_{i_p}^1\dot\sigma_{i_p}^2) \\ &\le \sum_{\text{all } i_1,\dots,i_p} \nu(\sigma_{i_1}^1\sigma_{i_1}^2\sigma_{i_2}^1\sigma_{i_2}^2 \cdots \sigma_{i_{p-1}}^1\sigma_{i_{p-1}}^2\dot\sigma_{i_p}^1\dot\sigma_{i_p}^2) \\ &= N^p\nu\left(R_{1,2}^{p-1}\frac{\dot{\boldsymbol{\sigma}}^1 \cdot \dot{\boldsymbol{\sigma}}^2}{N}\right) = N^p\nu\left((R_{1,2}^{p-1} - q^{p-1})\frac{\dot{\boldsymbol{\sigma}}^1 \cdot \dot{\boldsymbol{\sigma}}^2}{N}\right) ,\end{aligned} \tag{1.130}$$

where the inequality follows from the fact that since

$$\langle \sigma^1_{i_1}\sigma^2_{i_1}\sigma^1_{i_2}\sigma^2_{i_2}\cdots\sigma^1_{i_{p-1}}\sigma^2_{i_{p-1}}\dot\sigma^1_{i_p}\dot\sigma^2_{i_p}\rangle = \langle \sigma_{i_1}\sigma_{i_2}\cdots\sigma_{i_{p-1}}\dot\sigma_{i_p}\rangle^2$$

all terms are ≥ 0, and where the last equality uses that $\langle \dot{\boldsymbol\sigma}^1\cdot\dot{\boldsymbol\sigma}^2\rangle = 0$. Of course here $\dot{\boldsymbol\sigma}^1\cdot\dot{\boldsymbol\sigma}^2 = \sum_{i\leq N}\dot\sigma^1_i\dot\sigma^2_i$, and the vector notation is simply for convenience. Using the inequality $|x^{p-1}-y^{p-1}|\leq(p-1)|x-y|$ for $|x|,|y|\leq 1$ and the Cauchy-Schwarz inequality we obtain

$$\nu\left((R^{p-1}_{1,2}-q^{p-1})\frac{\dot{\boldsymbol\sigma}^1\cdot\dot{\boldsymbol\sigma}^2}{N}\right)\leq(p-1)\nu\left(|R_{1,2}-q|\left|\frac{\dot{\boldsymbol\sigma}^1\cdot\dot{\boldsymbol\sigma}^2}{N}\right|\right) \tag{1.131}$$
$$\leq(p-1)\nu\big((R_{1,2}-q)^2\big)^{1/2}\nu\left(\left(\frac{\dot{\boldsymbol\sigma}^1\cdot\dot{\boldsymbol\sigma}^2}{N}\right)^2\right)^{1/2}.$$

Now we have

$$\left\langle\left(\frac{\dot{\boldsymbol\sigma}^1\cdot\dot{\boldsymbol\sigma}^2}{N}\right)^2\right\rangle=\left\langle\left(\frac{(\boldsymbol\sigma^1-\langle\boldsymbol\sigma^1\rangle)\cdot(\boldsymbol\sigma^2-\langle\boldsymbol\sigma^2\rangle)}{N}\right)^2\right\rangle$$
$$=\left\langle\left(\frac{(\boldsymbol\sigma^1-\langle\boldsymbol\sigma^3\rangle)\cdot(\boldsymbol\sigma^2-\langle\boldsymbol\sigma^4\rangle)}{N}\right)^2\right\rangle.$$

To bound the right-hand side, we move the averages in $\boldsymbol\sigma^3$ and $\boldsymbol\sigma^4$ outside the square (and we note that the function $x\mapsto x^2$ is convex). Jensen's inequality (1.23) therefore asserts that

$$\left\langle\left(\frac{(\boldsymbol\sigma^1-\langle\boldsymbol\sigma^3\rangle)\cdot(\boldsymbol\sigma^2-\langle\boldsymbol\sigma^4\rangle)}{N}\right)^2\right\rangle\leq\left\langle\left(\frac{(\boldsymbol\sigma^1-\boldsymbol\sigma^3)\cdot(\boldsymbol\sigma^2-\boldsymbol\sigma^4)}{N}\right)^2\right\rangle.$$

Finally we write

$$\left\langle\left(\frac{(\boldsymbol\sigma^1-\boldsymbol\sigma^3)\cdot(\boldsymbol\sigma^2-\boldsymbol\sigma^4)}{N}\right)^2\right\rangle=\langle(R_{1,2}-R_{1,4}-R_{3,2}+R_{3,4})^2\rangle$$
$$\leq 4\langle(R_{1,2}-q)^2\rangle,$$

using that $(\sum_{i\leq 4}x_i)^2\leq 4\sum_{i\leq 4}x_i^2$. Combining the three previous inequalities and taking expectation and square root we reach

$$\nu\left(\left(\frac{\dot{\boldsymbol\sigma}^1\cdot\dot{\boldsymbol\sigma}^2}{N}\right)^2\right)^{1/2}\leq 2\nu\big((R_{1,2}-q)^2\big)^{1/2}.$$

Combining with (1.129), (1.130) and (1.131) we then get

$$N(N-1)\cdots(N-p+1)\mathsf{E}(\langle\sigma_1\cdots\sigma_p\rangle-\langle\sigma_1\cdots\sigma_{p-1}\rangle\langle\sigma_p\rangle)^2$$
$$\leq 2(p-1)N^p\nu\big((R_{1,2}-q)^2\big),$$

and this finishes the proof since

$$\sup_{N \geq p} \frac{N^p}{N(N-1)\cdots(N-p+1)} < \infty\,. \qquad \square$$

As a consequence, when one looks only at a given number of spins, one fixes $\beta < 1/2$ and lets $N \to \infty$, Gibbs' measure is asymptotically a product measure. To see this, we first observe that combining (1.127) and (1.89) implies

$$\mathsf{E}(\langle \sigma_1 \cdots \sigma_p \rangle - \langle \sigma_1 \rangle \cdots \langle \sigma_p \rangle)^2 \leq \frac{K(p)}{N}\,. \tag{1.132}$$

Next, given $\eta_1, \ldots, \eta_n \in \{-1, 1\}$, consider the set

$$A = \{\boldsymbol{\sigma} \in \Sigma_N;\ \forall i \leq n,\ \sigma_i = \eta_i\}\,, \tag{1.133}$$

where the dependence on $\eta_1, \ldots, \eta_n$ is kept implicit. Then, denoting by $\mathbf{1}_A$ the function such that $\mathbf{1}_A(\boldsymbol{\sigma}) = 1$ if $\boldsymbol{\sigma} \in A$ and $\mathbf{1}_A(\boldsymbol{\sigma}) = 0$ otherwise, we have

$$\mathbf{1}_A(\boldsymbol{\sigma}) = 2^{-n} \prod_{i \leq n} (1 + \sigma_i \eta_i) = 2^{-n} \sum_{I \subset \{1, \ldots, n\}} \sigma_I \eta_I\,,$$

where $\sigma_I = \prod_{i \in I} \sigma_i$ and $\eta_I = \prod_{i \in I} \eta_i$. Thus, using (1.132),

$$\begin{aligned} G_N(A) = \langle \mathbf{1}_A \rangle = 2^{-n} \sum_{I \subset \{1, \ldots, n\}} \eta_I \langle \sigma_I \rangle &\simeq 2^{-n} \sum_{I \subset \{1, \ldots, n\}} \eta_I \prod_{i \in I} \langle \sigma_i \rangle \\ &= 2^{-n} \prod_{i \leq n} (1 + \eta_i \langle \sigma_i \rangle) \\ &= \mu_n(\{\boldsymbol{\eta}\})\,, \end{aligned} \tag{1.134}$$

where $\boldsymbol{\eta} = (\eta_1, \ldots, \eta_n)$ and μ_n is the product probability on $\{-1, 1\}^n$ with density $\prod_{i \leq n} (1 + \eta_i \langle \sigma_i \rangle)$ with respect to the uniform measure. (Let us observe that μ_n is the only probability measure on $\{0, 1\}^n$ such that for each i the average of σ_i for μ_n is equal to $\langle \sigma_i \rangle$.)

Formally, we have the following.

Theorem 1.4.15. *Assume $\beta < 1/2$. Denote by $G_{N,n}$ the law of $(\sigma_1, \ldots, \sigma_n)$ under G_N, and consider μ_n as above, the probability on $\{-1, 1\}^n$ with density $\prod_{i \leq n} (1 + \eta_i \langle \sigma_i \rangle)$ with respect to the uniform measure. Then*

$$\mathsf{E} \| G_{N,n} - \mu_n \|^2 \leq \frac{K(n)}{N}\,,$$

where $\| \cdot \|$ denotes the total variation distance.

Thus, to understand well the random measure $G_{N,n}$ it remains only to understand the random sequence $(\langle \sigma_i \rangle)_{i \leq n}$. This will be achieved in Theorem 1.7.1 below.

Proof. By definition of the total variation distance (and (A.79)), it holds

$$\|G_{N,n} - \mu_n\| = \sum_{\boldsymbol{\eta}} |G_{N,n}(\{\boldsymbol{\eta}\}) - \mu_n(\{\boldsymbol{\eta}\})| ,$$

where the summation is over $\boldsymbol{\eta}$ in $\{-1,1\}^n$. Since there are 2^N terms in the summation, using the Cauchy-Schwarz inequality as in $(\sum_{i\in I} a_i)^2 \leq \mathrm{card} I \sum_{i\in I} a_i^2$ and taking expectation we get

$$\mathsf{E}\|G_{N,n} - \mu_n\|^2 \leq 2^n \sum_{\boldsymbol{\eta}} \mathsf{E}(G_{N,n}(\{\boldsymbol{\eta}\}) - \mu_n(\{\boldsymbol{\eta}\}))^2 .$$

Now $G_{N,n}(\{\boldsymbol{\eta}\}) = G_N(A)$ where A is given by (1.133), and the result follows by making formal the computation (1.134). Namely, we write

$$\begin{aligned}(G_N(A) - \mu_n(\{\boldsymbol{\eta}\}))^2 &= \Big(2^{-n} \sum_{I\subset\{1,\dots,n\}} \eta_I(\langle\sigma_I\rangle - \prod_{i\in I}\langle\sigma_i\rangle)\Big)^2 \\ &\leq 2^{-n} \sum_{I\subset\{1,\dots,n\}} (\langle\sigma_I\rangle - \prod_{i\in I}\langle\sigma_i\rangle)^2 ,\end{aligned}$$

so that, taking expectation and using (1.132),

$$\mathsf{E}(G_N(A) - \mu_n(\{\boldsymbol{\eta}\}))^2 \leq \frac{K(n)}{N} .$$ □

This result raises all kinds of open problems. Here is an obvious question.

Research Problem 1.4.16. How fast can $n(N)$ grow so that $\|G_{N,n(N)} - \mu_{n(N)}\| \to 0$?

Of course, it will be easy to prove that one can take $n(N) \to \infty$, but finding the best rate might be hard. One might also conjecture the following.

Conjecture 1.4.17. When $\beta > 0$ we have

$$\lim_{N\to\infty} \mathsf{E} \inf \|G_N - \mu\| = 2 ,$$

where the infimum is computed over all the classes of measures μ that are product measures.

Conjecture 1.4.18. When $\beta < 1/2$ we have

$$\lim_{N\to\infty} \mathsf{E} d(G_N, \mu) = 0 ,$$

where μ is the product measure on Σ_N such that for each $i \leq N$ we have $\int \sigma_i \mathrm{d}\mu(\boldsymbol{\sigma}) = \langle\sigma_i\rangle$, and where now d denotes the transportation-cost distance (see Section A.11) associated with the Hamming distance (1.7).

A solution of the previous conjectures would not yield much information. Are there more fruitful questions to be asked concerning the global structure of Gibbs' measure?

1.5 A Kind of Central Limit Theorem

This short section brings forward a fundamental fact, a kind of random central limit theorem (CLT). The usual CLT asserts (roughly speaking) that a sum $\sum a_i X_i$ is nearly Gaussian provided the r.v.s X_i are independent and none of the terms of the sum is large compared to the sum itself. The situation here is different: the terms X_i are not really independent, but we do not look at all sums $\sum a_i X_i$, only at sums where the coefficients a_i are random.

More specifically consider a probability measure $\mu = \mu_N$ on $\mathbb{R}^N$ (a Gibbs measure is the case of interest). Assume that for two numbers q and ρ we have, for large N,

$$\int \left(\frac{\mathbf{x}^1 \cdot \mathbf{x}^2}{N} - q \right)^2 \mathrm{d}\mu(\mathbf{x}^1) \mathrm{d}\mu(\mathbf{x}^2) \simeq 0 \tag{1.135}$$

$$\int \left(\frac{\|\mathbf{x}\|^2}{N} - \rho \right)^2 \mathrm{d}\mu(\mathbf{x}) \simeq 0 \,. \tag{1.136}$$

Consider the case where μ is Gibbs' measure for the SK model. Then (1.135) means that $\langle (R_{1,2}-q)^2 \rangle \simeq 0$, while (1.136) is automatically satisfied for $\rho = 1$, because μ is supported by Σ_N and for $\boldsymbol{\sigma} \in \Sigma_N$ we have $\|\boldsymbol{\sigma}\|^2 = \sum_{i \le N} \sigma_i^2 = N$. (We will later consider systems where the individual spins can take values in $\mathbb{R}$, and (1.136) will become relevant.) Let

$$\mathbf{b} = \int \mathbf{x} \mathrm{d}\mu(\mathbf{x}) = \left(\int x_i \mathrm{d}\mu(\mathbf{x}) \right)_{i \le N}$$

be the barycenter of μ. The fundamental fact is as follows. Consider independent Gaussian standard r.v.s g_i and $\mathbf{g} = (g_i)_{i \le N}$. Then for a typical value of $\mathbf{g}$ (i.e. unless we have been unlucky enough to pick $\mathbf{g}$ in a small exceptional set), we have

The image of μ under the map $\mathbf{x} \mapsto \mathbf{g} \cdot \mathbf{x}/\sqrt{N}$ is nearly a Gaussian measure of mean $\mathbf{g} \cdot \mathbf{b}/\sqrt{N}$ and of variance $\rho - q$. (1.137)

The reader should not worry about the informality of the statement which is designed only to create the correct intuition. We shall never need a formal statement, but certain constructions we shall use are based on the intuition provided by (1.137). The reason why (1.135) and (1.136) imply (1.137) is very simple. Let us consider a bounded, continuous function f, and the two r.v.s

$$U = \int f \left(\frac{\mathbf{g} \cdot \mathbf{x}}{\sqrt{N}} \right) \mathrm{d}\mu(\mathbf{x}) \quad \text{and} \quad V = \mathsf{E}_\xi f \left(\frac{\mathbf{g} \cdot \mathbf{b}}{\sqrt{N}} + \xi \sqrt{\rho - q} \right),$$

where ξ is a standard Gaussian r.v. independent of $\mathbf{g}$ and where **throughout the book we denote by E_ξ expectation in the r.v.s ξ only**, that is, for all the other r.v.s given.

We will show that, given the function f, with probability close to 1 we have $U \simeq V$ i.e.

$$\int f\left(\frac{\mathbf{g}\cdot\mathbf{x}}{\sqrt{N}}\right) d\mu(\mathbf{x}) \simeq \mathsf{E}_\xi f\left(\frac{\mathbf{g}\cdot\mathbf{b}}{\sqrt{N}} + \xi\sqrt{\rho-q}\right) . \tag{1.138}$$

Therefore, given a finite set $\mathcal{F}$ of functions, with probability close to 1 (i.e. "for a typical value of $\mathbf{g}$"), (1.138) occurs simultaneously for each f in $\mathcal{F}$, which is what we meant by (1.137).

To prove that $U \simeq V$, we compute

$$\mathsf{E}(U-V)^2 = \mathsf{E}U^2 + \mathsf{E}V^2 - 2\mathsf{E}UV ,$$

and we show that $\mathsf{E}U^2 \simeq \mathsf{E}V^2 \simeq \mathsf{E}UV$. Now,

$$\begin{aligned}\mathsf{E}U^2 &= \mathsf{E}\int f\left(\frac{\mathbf{g}\cdot\mathbf{x}^1}{\sqrt{N}}\right) f\left(\frac{\mathbf{g}\cdot\mathbf{x}^2}{\sqrt{N}}\right) d\mu(\mathbf{x}^1)d\mu(\mathbf{x}^2) \\ &= \int \mathsf{E}f\left(\frac{\mathbf{g}\cdot\mathbf{x}^1}{\sqrt{N}}\right) f\left(\frac{\mathbf{g}\cdot\mathbf{x}^2}{\sqrt{N}}\right) d\mu(\mathbf{x}^1)d\mu(\mathbf{x}^2) .\end{aligned}$$

For $\ell = 1, 2$, let $g^\ell = \mathbf{g}\cdot\mathbf{x}^\ell/\sqrt{N}$. These two Gaussian r.v.s are such that

$$\mathsf{E}(g^\ell)^2 = \frac{\|\mathbf{x}^\ell\|^2}{N} ; \qquad \mathsf{E}(g^1g^2) = \frac{\mathbf{x}^1\cdot\mathbf{x}^2}{N} .$$

Using (1.135) and (1.136) we see that, generically (i.e. for most of the points $\mathbf{x}^1, \mathbf{x}^2$) we have $\mathsf{E}(g^\ell)^2 \simeq \rho$ and $\mathsf{E}(g^1g^2) \simeq q$. Since the distribution of a finite jointly Gaussian family (g_p) is determined by the quantities $\mathsf{E}g_pg_{p'}$, the pair (g^1, g^2) has nearly the distribution of the pair $(z\sqrt{q} + \xi^1\sqrt{\rho-q}, z\sqrt{q} + \xi^2\sqrt{\rho-q})$ where z, ξ^1 and ξ^2 are independent standard Gaussian r.v.s. Hence

$$\begin{aligned}\mathsf{E}f\left(\frac{\mathbf{g}\cdot\mathbf{x}^1}{\sqrt{N}}\right) f\left(\frac{\mathbf{g}\cdot\mathbf{x}^2}{\sqrt{N}}\right) &\simeq \mathsf{E}f(z\sqrt{q} + \xi^1\sqrt{\rho-q})f(z\sqrt{q} + \xi^2\sqrt{\rho-q}) \\ &= \mathsf{E}(\mathsf{E}_\xi f(z\sqrt{q} + \xi\sqrt{\rho-q}))^2 ,\end{aligned}$$

the last equality not being critical here, but preparing for future formulas. This implies that

$$\mathsf{E}U^2 \simeq \mathsf{E}(\mathsf{E}_\xi f(z\sqrt{q} + \xi\sqrt{\rho-q}))^2 . \tag{1.139}$$

The same argument proves that $\mathsf{E}UV$ and $\mathsf{E}V^2$ are also nearly equal to the right-hand side of (1.139), so that $\mathsf{E}(U-V)^2 \simeq 0$, completing the argument.

In practice, we will need estimates for quantities such as

$$\mathsf{E}W\left(\int f\left(\frac{\mathbf{g}\cdot\mathbf{x}^1}{\sqrt{N}}, \ldots, \frac{\mathbf{g}\cdot\mathbf{x}^n}{\sqrt{N}}\right) d\mu(\mathbf{x}^1)\cdots d\mu(\mathbf{x}^n)\right) , \tag{1.140}$$

where W is a real-valued function and f is now a function of n variables. We will compare such a quantity with

$$\mathsf{E}W\big(\mathsf{E}_\xi f(z\sqrt{q}+\xi^1\sqrt{\rho-q},\dots,z\sqrt{q}+\xi^n\sqrt{\rho-q})\big)\,, \tag{1.141}$$

using the standard path to obtain quantitative estimates.

The variables ξ will be "invisible" in the present chapter because they will occur in terms such as

$$\mathsf{E}_\xi \exp a(z\sqrt{q}+\xi\sqrt{\rho-q}) = \exp\frac{a^2(\rho-q)}{2}\exp az\sqrt{q}\,. \tag{1.142}$$

They will however be essential in subsequent chapters.

1.6 The Cavity Method

As pointed out, the arguments of Theorems 1.3.7 and 1.3.9 are very special; but even Latala's argument is not easy to extend to other models. The purpose of this section is to develop an other method, the cavity method, which we will be able to use for many models that do not share the special features of the SK model. Moreover, even in the case of the SK model, the cavity method is essential to obtain certain types of information, as we will demonstrate in the rest of this chapter.

Originally, the cavity method is simply induction over N. To reduce the system to a smaller system, one removes a spin, creating a "cavity". The basic step is to bring forward the dependence of the Hamiltonian on the last spin by writing

$$\begin{aligned}-H_N(\boldsymbol{\sigma}) &= \frac{\beta}{\sqrt{N}}\sum_{i<j\le N} g_{ij}\sigma_i\sigma_j + \sum_{i\le N} h_i\sigma_i \\ &= -H_{N-1}(\boldsymbol{\sigma}) + \sigma_N\left(\frac{\beta}{\sqrt{N}}\sum_{i<N} g_i\sigma_i + h_N\right)\,,\end{aligned} \tag{1.143}$$

where $g_i = g_{iN}$ and

$$-H_{N-1}(\boldsymbol{\sigma}) = \frac{\beta}{\sqrt{N}}\sum_{i<j\le N-1} g_{ij}\sigma_i\sigma_j + \sum_{i\le N-1} h_i\sigma_i\,. \tag{1.144}$$

Thus, if we write $\boldsymbol{\rho} = (\sigma_1,\dots,\sigma_{N-1})$, we see that $-H_{N-1}(\boldsymbol{\sigma}) = -H_{N-1}(\boldsymbol{\rho})$ depends on $\boldsymbol{\rho}$ only. It is the Hamiltonian of a $(N-1)$-spin system (for a different value of β). Let us denote by $\langle\cdot\rangle_-$ an average for this Hamiltonian. Then we have the following absolutely fundamental identity.

Proposition 1.6.1. *For a function f on Σ_N, it holds that*

$$\langle f\rangle = \frac{\Big\langle \mathrm{Av}\Big(f(\boldsymbol{\sigma})\exp\sigma_N\big(\frac{\beta}{\sqrt{N}}\sum_{i<N} g_i\sigma_i + h_N\big)\Big)\Big\rangle_-}{\Big\langle \mathrm{Av}\Big(\exp\sigma_N\big(\frac{\beta}{\sqrt{N}}\sum_{i<N} g_i\sigma_i + h_N\big)\Big)\Big\rangle_-}\,. \tag{1.145}$$

Here, Av means average over $\sigma_N = \pm 1$; the result of this averaging is a function of $\boldsymbol{\rho}$ only, which is then integrated with respect to $\langle\cdot\rangle_-$. Of course the denominator is simply

$$\left\langle \mathrm{ch}\left(\frac{\beta}{\sqrt{N}}\sum_{i<N} g_i\sigma_i + h_N\right)\right\rangle_-$$

and it can only help that it is always ≥ 1.

Proof. There is no magic. One replaces each of the two brackets on the right-hand side of (1.145) by their definition; each of these brackets is a fraction. The denominators are the same and cancel out. What remains is

$$\frac{\sum_{\boldsymbol{\rho}} \mathrm{Av} f(\boldsymbol{\sigma}) \exp\left(\sigma_N\left(\frac{\beta}{\sqrt{N}}\sum_{i<N} g_i\sigma_i + h_N\right) - H_{N-1}(\boldsymbol{\sigma})\right)}{\sum_{\boldsymbol{\rho}} \mathrm{Av} \exp\left(\sigma_N\left(\frac{\beta}{\sqrt{N}}\sum_{i<N} g_i\sigma_i + h_N\right) - H_{N-1}(\boldsymbol{\sigma})\right)} ,$$

where, to lighten the formula, we have written $\mathrm{Av}U$ rather than $\mathrm{Av}(U)$ both in numerator and denominator. Recalling (1.144) this is

$$\frac{\sum_{\boldsymbol{\rho}} \mathrm{Av} f(\boldsymbol{\sigma}) \exp(-H_N(\boldsymbol{\sigma}))}{\sum_{\boldsymbol{\rho}} \mathrm{Av} \exp(-H_N(\boldsymbol{\sigma}))} .$$

Multiplying both numerator and denominator by 2 and recalling the meaning of Av we see that this quantity is $\langle f\rangle$. □

Let us assume now that the $(N-1)$-spin system with Hamiltonian $-H_{N-1}$ behaves well (i.e. satisfies (1.89)). Then, according to the intuition developed in Section 1.5, we expect that the maps $(\sigma_1, \ldots, \sigma_{N-1}) \mapsto (\beta/\sqrt{N})\sum_{i<N} g_i\sigma_i$ behaves, under Gibbs' measure like a Gaussian r.v. To compute the right-hand side of (1.145), we will follow the intuition of comparing a quantity as in (1.140) to a quantity as in (1.141) (remembering that we can forget about the variables ξ because of (1.142))). For this, we will replace in (1.145) the quantity $(\beta/\sqrt{N})\sum_{i<N} g_i\sigma_i$ by $\beta z\sqrt{q}$ (where $0 \leq q \leq 1$ will be chosen later), that is, we will consider the Hamiltonian

$$-H_{N-1}(\boldsymbol{\rho}) + \sigma_N(\beta z\sqrt{q} + h_N) . \tag{1.146}$$

This Hamiltonian is the Hamiltonian of an N-spin system, but in which the last spin is "decoupled" from the first $(N-1)$-spins. It turns out to be easier to compare the N-spin system to this decoupled system rather than to the $(N-1)$-spin system. We will interpolate between the Hamiltonians (1.143) and (1.146) using

$$-H_t(\boldsymbol{\sigma}) = -H_{N-1}(\boldsymbol{\rho}) + \sigma_N\left(\sqrt{t}\frac{\beta}{\sqrt{N}}\sum_{i<N} g_i\sigma_i + \sqrt{1-t}\beta z\sqrt{q} + h_N\right) . \tag{1.147}$$

We denote by $\langle\cdot\rangle_t$ an average for the corresponding Gibbs measure, and $\nu_t(\cdot) = \mathsf{E}\langle\cdot\rangle_t$. The notations are identical to those of the previous sections, for a different interpolating Hamiltonian. This should not create confusion since from now on, at least in the present chapter, ν_t will always refer to the interpolating Hamiltonian (1.147). The Hamiltonian defined by (1.59) and (1.62) was designed to compare the Hamiltonian of the SK model with a situation where all the spins are independent of each other. In contrast, the Hamiltonian (1.147) is designed to compare the SK model to a situation where the last spin is independent of the first $N-1$ spins.

We write, for $\boldsymbol{\sigma}^\ell, \boldsymbol{\sigma}^{\ell'} \in \Sigma_N$

$$R^-_{\ell,\ell'} = \frac{1}{N}\sum_{i<N} \sigma_i^\ell \sigma_i^{\ell'} . \tag{1.148}$$

To lighten notation, when considering replicas $\boldsymbol{\sigma}^1, \boldsymbol{\sigma}^2, \ldots$ we write

$$\varepsilon_\ell = \sigma_N^\ell .$$

With this notation we have

$$R_{\ell,\ell'} = R^-_{\ell,\ell'} + \frac{\varepsilon_\ell \varepsilon_{\ell'}}{N} . \tag{1.149}$$

The fact that ν_0 decouples the last spin is expressed by the following, where

$$Y = \beta z \sqrt{q} + h .$$

Lemma 1.6.2. *For any function f^- on Σ^n_{N-1} and any set $I \subset \{1, \ldots, n\}$ we have*

$$\nu_0\Big(f^- \prod_{i\in I} \varepsilon_i\Big) = \nu_0\Big(\prod_{i\in I} \varepsilon_i\Big)\nu_0(f^-) = \mathsf{E}(\mathrm{th} Y)^{\mathrm{card} I} \nu_0(f^-) . \tag{1.150}$$

Proof. Since when $t=0$ the Hamiltonian H_t is the sum of a term depending only on the first $N-1$ spins and of a term depending only on the last spin it should be obvious that

$$\Big\langle f^- \prod_{i\in I} \varepsilon_i \Big\rangle_0 = \langle f^- \rangle_0 \Big\langle \prod_{i\in I} \varepsilon_i \Big\rangle_0 = \langle f^- \rangle_0 (\mathrm{th} Y)^{\mathrm{card} I}$$

and the result follows taking expectation, since the randomnesses of Y and H_{N-1} are independent. □

Lemma 1.6.2 in particular computes $\nu_0(f)$ when f depends only on the last spin, with formulas such as $\nu_0(\varepsilon_1\varepsilon_2\varepsilon_3) = \mathsf{E}\mathrm{th}^3 Y$.

The fundamental tool is as follows, where we recall that $\varepsilon_\ell = \sigma_N^\ell$.

Lemma 1.6.3. *Consider a function f on $\Sigma_N^n = (\Sigma_N)^n$; then for $0 < t < 1$ we have*

$$\nu_t'(f) := \frac{\mathrm{d}}{\mathrm{d}t}\nu_t(f) = \beta^2 \sum_{1\le \ell<\ell'\le n} \nu_t(f\varepsilon_\ell\varepsilon_{\ell'}(R^-_{\ell,\ell'} - q))$$
$$- \beta^2 n \sum_{\ell\le n} \nu_t(f\varepsilon_\ell\varepsilon_{n+1}(R^-_{\ell,n+1} - q))$$
$$+ \beta^2 \frac{n(n+1)}{2} \nu_t(f\varepsilon_{n+1}\varepsilon_{n+2}(R^-_{n+1,n+2} - q)) \quad (1.151)$$

and also

$$\nu_t'(f) = \beta^2 \sum_{1\le \ell<\ell'\le n} \nu_t(f\varepsilon_\ell\varepsilon_{\ell'}(R_{\ell,\ell'} - q))$$
$$- \beta^2 n \sum_{\ell\le n} \nu_t(f\varepsilon_\ell\varepsilon_{n+1}(R_{\ell,n+1} - q))$$
$$+ \beta^2 \frac{n(n+1)}{2} \nu_t(f\varepsilon_{n+1}\varepsilon_{n+2}(R_{n+1,n+2} - q)) \,. \quad (1.152)$$

This fundamental formula looks very complicated the first time one sees it, although the shock should certainly be milder once one has seen (1.95). A second look reveals that fortunately as in (1.95) the complication is only algebraic. Counting terms with their order of multiplicity, the right-hand side of (1.151) is the sum of $2n^2$ simple terms of the type $\pm\beta^2\nu_t(f\varepsilon_\ell\varepsilon_{\ell'}(R^-_{\ell,\ell'} - q))$.

Proof. The formula (1.151) is the special case of formula (1.95) where

$$u_{\boldsymbol{\sigma}} = \frac{\beta}{\sqrt{N}}\sigma_N \sum_{i<N} g_i\sigma_i \,; \qquad v_{\boldsymbol{\sigma}} = \beta\sigma_N z\sqrt{q} \quad (1.153)$$

$$w_{\boldsymbol{\sigma}} = \exp\left(-H_{N-1}(\boldsymbol{\rho}) + h_N\sigma_N\right) \,,$$

so that (1.55) implies:

$$U(\boldsymbol{\sigma}^\ell, \boldsymbol{\sigma}^{\ell'}) = \varepsilon_\ell\varepsilon_{\ell'}\frac{\beta^2}{2}(R^-_{\ell,\ell'} - q) \,.$$

Finally (1.152) follows from (1.151) and (1.149), as the extra terms cancel out since $\varepsilon_\ell^2 = 1$. □

The reader has observed that the choice (1.153) is fundamentally different from the choice (1.62). In words, in (1.153) we decouple the last spin from the others, rather than "decoupling all the spins at the same time" as in (1.62).

Since the formula (1.151) is the fundamental tool of the cavity method, we would like to help the reader overcome his expected dislike of this formula by explaining why, if one leaves aside the algebra, it is very simple. It helps to think of $R^-_{\ell,\ell'} - q$ as a small quantity. Then all the terms of the right-hand side of (1.151) are small, and thus, $\nu(f) = \nu_1(f) \sim \nu_0(f)$. This is very helpful when f depends only on the last spin, e.g. $f(\boldsymbol{\sigma}) = \varepsilon_1\varepsilon_2$ because in that case we can calculate $\nu_0(f)$ using Lemma 1.6.2. That same lemma lets us also simplify the terms $\nu_0(f\varepsilon_\ell\varepsilon_{\ell'}(R^-_{\ell,\ell'} - q))$, at least when f does not depend on the last spin. We will get then very interesting information simply by writing that $\nu(f) \sim \nu_0(f) + \nu'_0(f)$.

For pedagogical reasons, we now derive some of the results of Section 1.4 through the cavity method.

Lemma 1.6.4. *For a function $f \geq 0$ on Σ_N^n, we have*

$$\nu_t(f) \leq \exp(4n^2\beta^2)\nu(f) . \tag{1.154}$$

Proof. Here of course as usual $\nu(f) = \nu_1(f)$. Since $|R^-_{\ell,\ell'}| \leq 1$ and $q \in [0,1]$ we have $|R^-_{\ell,\ell'} - q| \leq 2$, so that (1.151) yields

$$|\nu'_t(f)| \leq 4n^2\beta^2\nu_t(f) \tag{1.155}$$

and we integrate. □

Proposition 1.6.5. *Consider a function f on Σ_N^n, and $\tau_1, \tau_2 > 0$ with $1/\tau_1 + 1/\tau_2 = 1$. Then we have*

$$|\nu(f) - \nu_0(f)| \leq 2n^2\beta^2 \exp(4n^2\beta^2)\nu(|f|^{\tau_1})^{1/\tau_1}\nu(|R_{1,2} - q|^{\tau_2})^{1/\tau_2} . \tag{1.156}$$

Proof. We have

$$|\nu(f) - \nu_0(f)| = \left|\int_0^1 \nu'_t(f)\,\mathrm{d}t\right| \leq \sup_{0<t<1} |\nu'_t(f)| .$$

Now, Hölder's inequality for ν_t implies

$$|\nu_t(f\varepsilon_\ell\varepsilon_{\ell'}(R_{\ell,\ell'} - q))| \leq \nu_t(|f||R_{\ell,\ell'} - q|) \leq \nu_t(|f|^{\tau_1})^{1/\tau_1}\nu_t(|R_{\ell,\ell'} - q|^{\tau_2})^{1/\tau_2}$$

and thus by (1.152) (and since $n(n+1)/2 + n^2 + n(n-1)/2 = 2n^2$),

$$|\nu'_t(f)| \leq 2n^2\beta^2\nu_t(|f|^{\tau_1})^{1/\tau_1}\nu_t(|R_{1,2} - q|^{\tau_2})^{1/\tau_2} .$$

We then use (1.154) for $|f|^{\tau_1}$ and $|R_{1,2} - q|^{\tau_2}$. □

Proposition 1.6.6. *There exists $\beta_0 > 0$ such that if $\beta \leq \beta_0$ then*

$$\nu\big((R_{1,2} - q)^2\big) \leq \frac{2}{N} ,$$

where q is the solution of (1.74).

The larger the value of β_0, the harder it is to prove the result. It seems difficult by the cavity method to reach the value $\beta_0 = 1/2$ that we obtained with Latala's argument in (1.87) and (1.88).

Proof. Recalling that $\varepsilon_\ell = \sigma_N^\ell$, we use symmetry among sites to write

$$\nu\big((R_{1,2}-q)^2\big) = \frac{1}{N}\sum_{i\le N}\nu((\sigma_i^1\sigma_i^2 - q)(R_{1,2}-q)) = \nu(f) \tag{1.157}$$

where

$$f = (\varepsilon_1\varepsilon_2 - q)(R_{1,2}-q) \ .$$

The simple idea underlying (1.157) is simply "to bring out as much as possible of the dependence on the last spin". This is very natural, since the cavity method brings forward the influence of this last spin. It is nonetheless extremely effective.

Using (1.149), and since $\varepsilon_\ell^2 = 1$, we have

$$f = (\varepsilon_1\varepsilon_2 - q)(R_{1,2}-q) = \frac{1}{N}(1-\varepsilon_1\varepsilon_2 q) + (\varepsilon_1\varepsilon_2 - q)(R_{1,2}^- - q) \ .$$

The key point is that Lemma 1.6.2 implies

$$\begin{aligned}\nu_0((\varepsilon_1\varepsilon_2 - q)(R_{1,2}^- - q)) &= \nu_0(\varepsilon_1\varepsilon_2 - q)\nu_0(R_{1,2}^- - q)\\ &= (\mathsf{E}\,\mathrm{th}^2 Y - q)\nu_0(R_{1,2}^- - q)\\ &= 0\end{aligned}$$

because $\nu_0(\varepsilon_1\varepsilon_2) = \mathsf{E}\mathrm{th}^2 Y$ using Lemma 1.6.2 again and since (1.74) means that $q = \mathsf{E}\,\mathrm{th}^2 Y$. Furthermore,

$$\nu_0(f) = \frac{1}{N}\nu_0(1-\varepsilon_1\varepsilon_2 q) = \frac{1}{N}(1-q\mathsf{E}\,\mathrm{th}^2 Y) = \frac{1}{N}(1-q^2) \ . \tag{1.158}$$

We now use (1.156) with $\tau_1 = \tau_2 = 2$ and $n = 2$. Since $|\varepsilon_1\varepsilon_2 - q| \le 2$, we get

$$|\nu(f) - \nu_0(f)| \le \big(16\,\beta^2\exp(16\beta^2)\big)\,\nu\big((R_{1,2}-q)^2\big)$$

and using (1.157) and (1.158),

$$\nu\big((R_{1,2}-q)^2\big) \le \frac{1}{N} + \big(16\,\beta^2\exp(16\beta^2)\big)\,\nu\big((R_{1,2}-q)^2\big) \ .$$

Thus, if β_0 is chosen so that

$$16\,\beta_0^2\exp 16\beta_0^2 \le \frac{1}{2}$$

we obtain

$$\nu\big((R_{1,2}-q)^2\big) \le \frac{1}{N} + \frac{1}{2}\nu\big((R_{1,2}-q)^2\big) \ , \tag{1.159}$$

and thus

$$\nu\big((R_{1,2}-q)^2\big) \le \frac{2}{N} \ . \qquad \square$$

In essence the previous proof is a kind of contraction argument, as is shown by (1.159). When β is small "the operation of adding one spin improves the behavior of $(R_{1,2} - q)^2$". A great many of the arguments we will use to control various models under a "high-temperature" condition will be of the same nature, although they will typically require more work.

An elementary inductive argument will allows us to control the higher moments of $(R_{1,2} - q)^2$.

Proposition 1.6.7. *There exists $\beta_0 > 0$ such that for $\beta \le \beta_0$ and any $k \ge 1$ we have*

$$\nu((R_{1,2} - q)^{2k}) \le \left(\frac{64k}{N}\right)^k . \tag{1.160}$$

In Section A.6 we explain a general principle relating growth of moments and exponential integrability. This principle shows that (1.160) implies that for a certain constant L we have

$$\nu\left(\exp \frac{N}{L}(R_{1,2} - q)^2\right) \le 2 ,$$

a statement similar to (1.102).

Proof of Proposition 1.6.7. For $1 \le n \le N$, let

$$A_n = \frac{1}{N} \sum_{n \le i \le N} (\sigma_i^1 \sigma_i^2 - q) ,$$

so that $R_{1,2} - q = A_1$. We will prove by induction over k that (provided $\beta \le \beta_0$) we have

$$\forall n \le N,\ \nu\left(A_n^{2k}\right) \le \left(\frac{64k}{N}\right)^k . \tag{1.161}$$

This tricky induction hypothesis should not mislead the reader into thinking that the argument is difficult. It is actually very robust, and the purpose of the tricky hypothesis is simply to avoid a few lines of unappetizing computations. To perform the induction from k to $k+1$, we observe that we can assume $n < N$, since if $n = N$ (1.161) is always true because $|A_N| \le 2/N$. Symmetry between sites implies

$$\nu\left(A_n^{2k+2}\right) = \frac{1}{N} \sum_{n \le i \le N} \nu\left((\sigma_i^1 \sigma_i^2 - q) A_n^{2k+1}\right) = \frac{N-n+1}{N} \nu(f) \le |\nu(f)| , \tag{1.162}$$

where

$$f = (\varepsilon_1 \varepsilon_2 - q) A_n^{2k+1} .$$

It follows that

$$\nu\left(A_n^{2k+2}\right) \le |\nu_0(f)| + \sup_t |\nu_t'(f)| . \tag{1.163}$$

We first evaluate $\nu_0(f)$. Let

$$A' = \frac{1}{N} \sum_{n \le i \le N-1} (\sigma_i^1 \sigma_i^2 - q) .$$

Lemma 1.6.2 implies

$$\nu_0\big((\varepsilon_1\varepsilon_2 - q)A'^{2k+1}\big) = 0 ,$$

so that since $|\varepsilon_1\varepsilon_2 - q| \le 2$,

$$\begin{aligned} |\nu_0(f)| &= \big|\nu_0\big((\varepsilon_1\varepsilon_2 - q)A_n^{2k+1}\big) - \nu_0\big((\varepsilon_1\varepsilon_2 - q)A'^{2k+1}\big)\big| \\ &\le 2\nu_0(|A_n^{2k+1} - A'^{2k+1}|) . \end{aligned} \tag{1.164}$$

We use the inequality

$$|x^{2k+1} - y^{2k+1}| \le (2k+1)|x-y|(x^{2k} + y^{2k})$$

for $x = A_n$ and $y = A'$. Since $|x - y| \le 2/N$ we deduce from (1.164) that

$$|\nu_0(f)| \le \frac{4(2k+1)}{N}(\nu_0(A'^{2k}) + \nu_0(A_n^{2k})) . \tag{1.165}$$

Assuming $\beta_0 \le 1/8$, we obtain from (1.154) that

$$\nu_t(f^*) \le 2\nu(f^*) , \tag{1.166}$$

whenever $f^* \ge 0$ is a function on Σ_N^2. Then (1.165) implies

$$|\nu_0(f)| \le \frac{8(2k+1)}{N}(\nu(A'^{2k}) + \nu(A_n^{2k})) .$$

We now observe that A' and A_{n+1} are equal in distribution under ν because $n < N$. Thus the induction hypothesis yields

$$|\nu_0(f)| \le \frac{16(2k+1)}{N}\left(\frac{64k}{N}\right)^k \le \frac{1}{2}\left(\frac{64(k+1)}{N}\right)^{k+1} . \tag{1.167}$$

Next, we compute $\nu_t'(f)$ using (1.152) with $n = 2$. There are 8 terms (counting them with their order of multiplicity), and in each of them we bound $\varepsilon_1\varepsilon_2 - q$ by 2. We apply Hölder's inequality with $\tau_1 = (2k+2)/(2k+1)$ and $\tau_2 = 2k+2$ in the first line and (1.166) in the second line to get

$$\begin{aligned} |\nu_t'(f)| &\le 16\beta^2 \nu_t\big(A_n^{2k+2}\big)^{1/\tau_1} \nu_t\big((R_{1,2} - q)^{2k+2}\big)^{1/\tau_2} \\ &\le 32\beta^2 \nu\big(A_n^{2k+2}\big)^{1/\tau_1} \nu\big((R_{1,2} - q)^{2k+2}\big)^{1/\tau_2} \\ &\le 32\beta^2 \big(\nu(A_n^{2k+2}) + \nu((R_{1,2} - q)^{2k+2})\big) , \end{aligned}$$

using that $xy \le x^{\tau_1} + y^{\tau_2}$. (A nice feature of using Hölder's inequality is that "it separates replicas", and in the end we only need to consider two replicas.) Combining with (1.163) and (1.167), when $32\beta_0^2 \le 1/4$ we get

$$\nu\big(A_n^{2k+2}\big) \le \frac{1}{2}\left(\frac{64(k+1)}{N}\right)^{k+1} + \frac{1}{4}\big(\nu(A_n^{2k+2}) + \nu((R_{1,2}-q)^{2k+2})\big) \ . \quad (1.168)$$

Since $A_1 = R_{1,2} - q$, when $n = 1$ the previous inequality implies:

$$\nu\big((R_{1,2}-q)^{2k+2}\big) = \nu\big(A_1^{2k+2}\big) \le \left(\frac{64(k+1)}{N}\right)^{k+1} . \quad (1.169)$$

Using (1.169) in (1.168) yields that for the other values of n as well we have

$$\nu\big(A_n^{2k+2}\big) \le \left(\frac{64(k+1)}{N}\right)^{k+1} . \qquad \square$$

The following provides another method to estimate $p_N(\beta, h)$.

Proposition 1.6.8. *For any choices of β, h, q (with $q \in [0,1]$) we have*

$$|(N+1)p_{N+1}(\beta,h) - Np_N(\beta,h) - A(\beta,h,q)| \le K\left(\frac{1}{N} + \nu(|R_{1,2}-q|)\right) , \quad (1.170)$$

where

$$A(\beta,h,q) = \log 2 + \frac{\beta^2}{4}(1-q)^2 + \mathsf{E}\log \mathrm{ch}(\beta z\sqrt{q} + h)$$

and where K depends only on β and h.

As a consequence of this formula, by summation, and with obvious notation we get

$$|p_N(\beta,h) - A(\beta,h,q)| \le K\left(\frac{\log N}{N} + \frac{1}{N}\sum_{M \le N} \nu_M(|R_{1,2}-q|)\right) . \quad (1.171)$$

The proof of Proposition 1.6.8 does not use Guerra's interpolation (i.e the interpolation of Theorem 1.3.7), but rather an explicit formula ((1.176) below) that is the most interesting part of this approach. This method is precious in situations where we do not wish to (or cannot) use interpolation. Several such situations will occur later. Another positive feature of (1.170) is that it is valid for any value of β, h and q. The way this provides information is that the average $N^{-1}\sum_{M \le N} \nu_M(|R_{1,2}-q|)$ cannot be small unless the left-hand side of (1.171) is small. Let us also remark that combining (1.171) with (1.73) shows that this average can be small only if $A(\beta, h, q)$ is close to its infimum in q, i.e. only if q is a near solution of (1.74).

On the other hand, the best we can do about the right-hand side of (1.170) is to write (when $\beta \le \beta_0$)

$$\nu_N(|R_{1,2} - q|) \le \nu_N\big((R_{1,2} - q)^2\big)^{1/2} \le \frac{K}{\sqrt{N}} ,$$

so that (1.170) does not recover (1.108), since it gives only a rate $K/\sqrt{N}$ instead of K/N. It is possible however to prove a better version of (1.170), where one replaces the error term $\nu(|R_{1,2} - q|)$ by $\nu((R_{1,2} - q)^2)$. This essentially requires replacing the "order 1 Taylor expansions" by "order 2 Taylor expansions", a technique that will become familiar later.

Proof. In this proof we will consider an $(N + 1)$-spin system for different values of β, so we write the Hamiltonian as $H_{N+1,\beta}$ to make clear which value of β is actually used. Consider the number β_+ given by $\beta_+ = \beta\sqrt{1 + 1/N}$, so that

$$\frac{\beta_+}{\sqrt{N+1}} = \frac{\beta}{\sqrt{N}} .$$

We write the Hamiltonian $-H_{N+1,\beta_+}(\sigma_1, \dots, \sigma_{N+1})$ of an $(N + 1)$-spin system with parameter β_+ rather than β, and we gather the terms containing the last spin as in (1.143), so that

$$-H_{N+1,\beta_+}(\sigma_1, \dots, \sigma_{N+1}) = -H_N(\sigma_1, \dots, \sigma_N) + \sigma_{N+1}\Big(\frac{\beta}{\sqrt{N}} \sum_{i \le N} g_i\sigma_i + h_{N+1}\Big)$$

where $g_i = g_{i,N+1}$. With obvious notation, the identity

$$Z_{N+1}(\beta_+) = 2Z_N(\beta)\Big\langle \mathrm{ch}\Big(\frac{\beta}{\sqrt{N}} \sum_{i \le N} g_i\sigma_i + h_{N+1}\Big)\Big\rangle \tag{1.172}$$

holds, where of course $Z_N(\beta) = \sum_{\boldsymbol{\sigma}} \exp(-H_N(\boldsymbol{\sigma}))$. This is obvious if one replaces the bracket by its value and one writes that $2\mathrm{ch}(x) = \exp x + \exp(-x)$. Hence taking logarithm and expectation we obtain

$$\begin{aligned}(N + 1)p_{N+1}(\beta_+, h) &= Np_N(\beta, h) + \log 2 \\ &\quad + \mathsf{E}\log\Big\langle \mathrm{ch}\Big(\frac{\beta}{\sqrt{N}} \sum_{i \le N} g_i\sigma_i + h_{N+1}\Big)\Big\rangle . \end{aligned} \tag{1.173}$$

On the left-hand side we have $p_{N+1}(\beta_+, h)$ rather than $p_{N+1}(\beta, h)$, and we proceed to relate these two quantities. Consider a new independent sequence g'_{ij} of standard Gaussian r.v.s. Then

$$\begin{aligned} - H_{N+1,\beta_+}(\sigma_1, \dots, \sigma_{N+1}) &\stackrel{\mathcal{D}}{=} -H_{N+1,\beta}(\sigma_1, \dots, \sigma_{N+1}) \\ &\quad + \frac{\beta}{\sqrt{N(N+1)}} \sum_{i<j\le N+1} g'_{ij}\sigma_i\sigma_j , \end{aligned} \tag{1.174}$$

where $\mathcal{D}$ means equality in distribution. This is because

$$\frac{1}{\sqrt{N+1}} g_{ij} + \frac{1}{\sqrt{N(N+1)}} g'_{ij} \stackrel{\mathcal{D}}{=} \frac{1}{\sqrt{N}} g_{ij}$$

since $1/N = 1/(N+1) + 1/(N(N+1))$. We then observe the identity

$$Z_{N+1}(\beta_+) \stackrel{\mathcal{D}}{=} Z_{N+1}(\beta) \Big\langle \exp \frac{\beta}{\sqrt{N(N+1)}} \sum_{i<j\le N+1} g'_{ij}\sigma_i\sigma_j \Big\rangle' ,$$

where $\langle\cdot\rangle'$ denotes an average for the Gibbs measure with Hamiltonian $-H_{N+1,\beta}$. (Please note that this $\langle\cdot\rangle'$ does *not* indicate a derivative of any kind, but rather a shortage of available symbols.) This is proved as in (1.172). Thus, taking logarithm and expectation,

$$\begin{aligned}(N+1)p_{N+1}(\beta_+, h) &= (N+1)p_{N+1}(\beta, h) \\ &\quad + \mathsf{E}\log\Big\langle \exp \frac{\beta}{\sqrt{N(N+1)}} \sum_{i<j\le N+1} g'_{ij}\sigma_i\sigma_j \Big\rangle' . \qquad (1.175)\end{aligned}$$

Comparing (1.175) and (1.173) we get

$$\begin{aligned}&(N+1)p_{N+1}(\beta, h) - Np_N(\beta, h) \\ &= \log 2 + \mathsf{E}\log\Big\langle \mathrm{ch}\Big(\frac{\beta}{\sqrt{N}} \sum_{i\le N} g_i\sigma_i + h_{N+1}\Big)\Big\rangle \\ &- \mathsf{E}\log\Big\langle \exp \frac{\beta}{\sqrt{N(N+1)}} \sum_{i<j\le N+1} g'_{ij}\sigma_i\sigma_j \Big\rangle' . \qquad (1.176)\end{aligned}$$

To prove (1.170) we will calculate the last two terms of (1.176). The next exercise provides motivation for the result. The second part of the exercise is rather challenging, and should be all the more profitable.

Exercise 1.6.9. Convince yourself, using the arguments of Section 1.5, that one should have

$$\mathsf{E}\log\Big\langle \mathrm{ch}\Big(\frac{\beta}{\sqrt{N}} \sum_{i\le N} g_i\sigma_i + h_{N+1}\Big)\Big\rangle \simeq \frac{\beta^2}{2}(1-q) + \mathsf{E}\log \mathrm{ch}(\beta z\sqrt{q} + h) .$$

Then extend the arguments of Section 1.5 to get convinced that one should have

$$\mathsf{E}\log\Big\langle \exp \frac{\beta}{\sqrt{N(N+1)}} \sum_{i<j\le N+1} g'_{ij}\sigma_i\sigma_j \Big\rangle' \simeq \frac{\beta^2}{4}(1-q^2) .$$

The rigorous computation of the last two terms of (1.176) uses suitable interpolations. This takes about three pages. In case the reader finds the detail of the arguments tedious, she can simply skip them, they are not important for the sequel. Let us consider a (well behaved) function f and for $\boldsymbol{\sigma} \in \Sigma_N$, let us consider a number $w_{\boldsymbol{\sigma}} > 0$. For $\mathbf{x} = (x_{\boldsymbol{\sigma}})$, let us consider the function

$$F(\mathbf{x}) = \log \sum_{\boldsymbol{\sigma}} w_{\boldsymbol{\sigma}} f(x_{\boldsymbol{\sigma}}) .$$

Let us consider two independent jointly Gaussian families $(u_{\boldsymbol{\sigma}})$ and $(v_{\boldsymbol{\sigma}})$, and define as usual $u_{\boldsymbol{\sigma}}(t) = \sqrt{t} u_{\boldsymbol{\sigma}} + \sqrt{1-t} v_{\boldsymbol{\sigma}}$, $\mathbf{u}(t) = (u_{\boldsymbol{\sigma}}(t))_{\boldsymbol{\sigma}}$ and

$$U(\boldsymbol{\sigma}^1, \boldsymbol{\sigma}^2) := \frac{1}{2}(\mathsf{E} u_{\boldsymbol{\sigma}^1} u_{\boldsymbol{\sigma}^2} - \mathsf{E} v_{\boldsymbol{\sigma}^1} v_{\boldsymbol{\sigma}^2}) .$$

Then, a computation very similar to that leading to (1.58) shows that

$$\begin{aligned} \frac{\mathrm{d}}{\mathrm{d}t} \mathsf{E} F(\mathbf{u}(t)) = & \ \mathsf{E} \frac{1}{D(\mathbf{u}(t))} \sum_{\boldsymbol{\sigma}} w_{\boldsymbol{\sigma}} U(\boldsymbol{\sigma}, \boldsymbol{\sigma}) f''(u_{\boldsymbol{\sigma}}(t)) \\ & - \mathsf{E} \frac{1}{D(\mathbf{u}(t))^2} \sum_{\boldsymbol{\sigma}^1, \boldsymbol{\sigma}^2} w_{\boldsymbol{\sigma}^1} w_{\boldsymbol{\sigma}^2} U(\boldsymbol{\sigma}^1, \boldsymbol{\sigma}^2) f'(u_{\boldsymbol{\sigma}^1}(t)) f'(u_{\boldsymbol{\sigma}^2}(t)) , \end{aligned} \tag{1.177}$$

where $D(\mathbf{u}(t)) = \sum_{\boldsymbol{\sigma}} w_{\boldsymbol{\sigma}} f(u_{\boldsymbol{\sigma}}(t))$. Let us now consider the average $\langle \cdot \rangle$ for the Gibbs measure with Hamiltonian $-H(\boldsymbol{\sigma}) = \log w_{\boldsymbol{\sigma}}$. Then (1.177) simply means that

$$\begin{aligned} \frac{\mathrm{d}}{\mathrm{d}t} \mathsf{E} \log \langle f(u_{\boldsymbol{\sigma}}(t)) \rangle = & \ \mathsf{E} \frac{\langle U(\boldsymbol{\sigma}, \boldsymbol{\sigma}) f''(u_{\boldsymbol{\sigma}}(t)) \rangle}{\langle f(u_{\boldsymbol{\sigma}}(t)) \rangle} \\ & - \mathsf{E} \frac{\langle U(\boldsymbol{\sigma}^1, \boldsymbol{\sigma}^2) f'(u_{\boldsymbol{\sigma}^1}(t)) f'(u_{\boldsymbol{\sigma}^2}(t)) \rangle}{\langle f(u_{\boldsymbol{\sigma}}(t)) \rangle^2} . \end{aligned} \tag{1.178}$$

Let us consider the case where

$$u_{\boldsymbol{\sigma}} = \frac{\beta}{\sqrt{N}} \sum_{i \le N} g_i \sigma_i \quad ; \quad v_{\boldsymbol{\sigma}} = \beta z \sqrt{q}$$

so that

$$U(\boldsymbol{\sigma}^1, \boldsymbol{\sigma}^2) = \frac{1}{2}(\mathsf{E} u_{\boldsymbol{\sigma}^1} u_{\boldsymbol{\sigma}^2} - \mathsf{E} v_{\boldsymbol{\sigma}^1} v_{\boldsymbol{\sigma}^2}) = \frac{\beta^2}{2}(R_{1,2} - q) .$$

Let us now define

$$\varphi(t) = \mathsf{E} \log \langle \mathrm{ch}(u_{\boldsymbol{\sigma}}(t) + h_{N+1}) \rangle ,$$

where the bracket means that the function $\boldsymbol{\sigma} \mapsto \mathrm{ch}(u_{\boldsymbol{\sigma}}(t) + h_{N+1})$ is averaged for the Gibbs measure.

Let us apply the formula (1.178) to the case where $f(x) = \mathrm{ch}(x + h_{N+1})$ and $w_{\boldsymbol{\sigma}} = \exp(-H_N(\boldsymbol{\sigma}))$, at a given realization of h_{N+1} and H_N. (These quantities are random, but their randomness is independent of the randomnesses of $u_{\boldsymbol{\sigma}}$ and $v_{\boldsymbol{\sigma}}$.) Let us finally take a further expectation (this time in the randomness of h_{N+1} and H_N). We then get

$$\varphi'(t) = \mathsf{E}\frac{\langle U(\boldsymbol{\sigma},\boldsymbol{\sigma})\mathrm{ch}(u_{\boldsymbol{\sigma}}(t)+h_{N+1})\rangle}{\langle \mathrm{ch}(u_{\boldsymbol{\sigma}}(t)+h_{N+1})\rangle} - \mathsf{E}\frac{\langle U(\boldsymbol{\sigma}^1,\boldsymbol{\sigma}^2)\mathrm{sh}(u_{\boldsymbol{\sigma}^1}(t)+h_{N+1})\mathrm{sh}(u_{\boldsymbol{\sigma}^2}(t)+h_{N+1})\rangle}{\langle \mathrm{ch}(u_{\boldsymbol{\sigma}}(t)+h_{N+1})\rangle^2} .$$

Since $U(\boldsymbol{\sigma},\boldsymbol{\sigma}) = \beta^2(1-q)/2$ the first term is $\beta^2(1-q)/2$, and since $\mathrm{ch}(u_{\boldsymbol{\sigma}}(t)+h_{N+1}) \ge 1$ we obtain

$$\left|\varphi'(t) - \frac{\beta^2}{2}(1-q)\right| \le \frac{\beta^2}{2}\mathsf{E}\langle |R_{1,2}-q||\mathrm{sh}(u_{\boldsymbol{\sigma}^1}(t)+h_{N+1})\mathrm{sh}(u_{\boldsymbol{\sigma}^2}(t)+h_{N+1})|\rangle . \tag{1.179}$$

Taking first expectations in the r.v.s g_i, z and h_{N+1} (which are independent of the randomness of $\langle\cdot\rangle$), we get

$$\left|\varphi'(t) - \frac{\beta^2}{2}(1-q)\right| \le K\nu(|R_{1,2}-q|) , \tag{1.180}$$

where K does not depend on N. Therefore, we have

$$\begin{aligned}&\left|\mathsf{E}\log\left\langle \mathrm{ch}\left(\frac{\beta}{\sqrt{N}}\sum_{i\le N} g_i\sigma_i + h_{N+1}\right)\right\rangle - \frac{\beta^2}{2}(1-q) - \mathsf{E}\log\mathrm{ch}(\beta z\sqrt{q}+h)\right| \\ &= \left|\varphi(1)-\varphi(0) - \frac{\beta^2}{2}(1-q)\right| \\ &\le \sup_t\left|\varphi'(t) - \frac{\beta^2}{2}(1-q)\right| \le K\nu(|R_{1,2}-q|) . \end{aligned} \tag{1.181}$$

We use a similar procedure to evaluate the last term of (1.176). We will use (1.178) for the function $f(x) = \exp x$ and $N+1$ instead of N. We denote now by $\boldsymbol{\tau} = (\sigma_1,\ldots,\sigma_{N+1}) \in \Sigma_{N+1}$ the generic element of Σ_{N+1}, and we consider the case where

$$u_{\boldsymbol{\tau}} = \frac{\beta}{\sqrt{N(N+1)}}\sum_{i<j\le N+1} g'_{ij}\sigma_i\sigma_j \; ; \quad v_{\boldsymbol{\tau}} = \frac{\beta}{\sqrt{2}}qz .$$

Let us set

$$R'_{1,2} = R'_{1,2}(\boldsymbol{\tau}^1,\boldsymbol{\tau}^2) = \frac{1}{N}\sum_{i\le N+1}\sigma_i^1\sigma_i^2 .$$

Using (1.63) for $N+1$ rather than N we get

$$\begin{aligned}\mathsf{E}u_{\boldsymbol{\tau}^1}u_{\boldsymbol{\tau}^2} &= \frac{N+1}{N}\frac{\beta^2}{2}\left(\left(\frac{1}{N+1}\sum_{i\le N+1}\sigma_i^1\sigma_i^2\right)^2 - \frac{1}{N+1}\right) \\ &= \frac{\beta^2}{2}\left(\frac{N}{N+1}{R'_{1,2}}^2 - \frac{1}{N}\right) ,\end{aligned}$$

and thus

$$U(\boldsymbol{\tau}^1, \boldsymbol{\tau}^2) = \frac{1}{2}(\mathsf{E}u_{\boldsymbol{\tau}^1}u_{\boldsymbol{\tau}^2} - \mathsf{E}v_{\boldsymbol{\tau}^1}v_{\boldsymbol{\tau}^2}) = \frac{\beta^2}{4}\left(\frac{N}{N+1}{R'_{1,2}}^2 - q^2 - \frac{1}{N}\right) .$$

We choose $w_{\boldsymbol{\tau}} = \exp(-H_{N+1,\beta}(\boldsymbol{\tau}))$, and we define

$$\varphi(t) = \mathsf{E}\log\langle\exp(u_{\boldsymbol{\tau}}(t))\rangle' .$$

Using (1.178) we find that

$$\varphi'(t) = \mathsf{E}\frac{\langle U(\boldsymbol{\tau}, \boldsymbol{\tau})\exp(u_{\boldsymbol{\tau}}(t))\rangle'}{\langle\exp u_{\boldsymbol{\tau}}(t)\rangle'} - \mathsf{E}\frac{\langle U(\boldsymbol{\tau}^1, \boldsymbol{\tau}^2)\exp(u_{\boldsymbol{\tau}^1}(t) + u_{\boldsymbol{\tau}^2}(t))\rangle'}{\langle\exp u_{\boldsymbol{\tau}}(t)\rangle'^2} ,$$

and since $R'(\boldsymbol{\tau}, \boldsymbol{\tau}) = (N+1)/N$ we have

$$U(\boldsymbol{\tau}, \boldsymbol{\tau}) = \frac{\beta^2}{4}(1 - q^2)$$

and thus

$$\left|\varphi'(t) - \frac{\beta^2}{4}(1-q^2)\right| \le \mathsf{E}\frac{\langle|{R'_{1,2}}^2 - q^2|\exp(u_{\boldsymbol{\tau}^1}(t) + u_{\boldsymbol{\tau}^2}(t))\rangle'}{\langle\exp u_{\boldsymbol{\tau}}(t)\rangle'^2} .$$

Now $R'_{1,2} \le (N+1)/N \le 2$ and $|q| \le 1$ so that

$$|{R'_{1,2}}^2 - q^2| \le 3|R'_{1,2} - q| \le 3|R_{1,2} - q| + \frac{2}{N}$$

and therefore

$$\left|\varphi'(t) - \frac{\beta^2}{4}(1-q^2)\right| \le \frac{K}{N} + 3\mathsf{E}\frac{\langle|R_{1,2} - q|\exp(u_{\boldsymbol{\tau}^1}(t) + u_{\boldsymbol{\tau}^2}(t))\rangle'}{\langle\exp u_{\boldsymbol{\tau}}(t)\rangle'^2} . \tag{1.182}$$

To finish the proof it suffices to bound this last term by $K\nu(|R_{1,2} - q|)$, since then (1.182) gives, since $\varphi(0) = 0$,

$$\left|\mathsf{E}\log\left\langle\exp\frac{\beta}{\sqrt{N(N+1)}}\sum_{1\le i<j\le N+1} g'_{ij}\sigma_i\sigma_j\right\rangle' - \frac{\beta^2}{4}(1-q^2)\right|$$
$$= \left|\varphi(1) - \varphi(0) - \frac{\beta^2}{4}(1-q^2)\right| \le \frac{K}{N} + K\nu(|R_{1,2} - q|) ,$$

and combining with (1.181) and (1.176) finishes the proof.

To bound the last term of (1.182), we consider the function

$$\psi(t) = 3\mathsf{E}\,\frac{\langle|R_{1,2} - q|\exp(u_{\boldsymbol{\tau}^1}(t) + u_{\boldsymbol{\tau}^2}(t))\rangle'}{\langle\exp u_{\boldsymbol{\tau}}(t)\rangle'^2} .$$

Let us set $w_{\boldsymbol{\tau}} = \exp(-H_{N+1,\beta}(\boldsymbol{\tau}))$, and consider the Hamiltonian $-H_t(\boldsymbol{\tau}) = \log w_{\boldsymbol{\tau}} + u_{\boldsymbol{\tau}}(t)$. Denoting by $\langle\cdot\rangle_t$ an average for this Hamiltonian, we have

$$\psi(t) = 3\mathsf{E}\langle |R_{1,2} - q| \rangle_t \ .$$

We compute $\psi'(t)$ by Lemma 1.4.2, used for $N+1$ rather than N. The exact expression thus obtained is not important. What matters here is that using the bound $|U(\boldsymbol{\tau}^1, \boldsymbol{\tau}^2)| \le K$, we find an inequality

$$|\psi'(t)| \le K\psi(t) \ ,$$

and by integration this shows that $\psi(t) \le K\psi(1)$. Denoting by $\langle \cdot \rangle_+$ an average for the Gibbs measure with Hamiltonian H_{N+1,β_+} and using (1.174) we then observe the identity

$$\psi(1) = 3\mathsf{E}\ \langle |R_{1,2} - q| \rangle_1 = 3\mathsf{E}\langle |R_{1,2} - q| \rangle_+ \ .$$

Next, using the cavity method for the Hamiltonian $-H_{N+1,\beta_+}$ we obtain

$$\begin{aligned} &\mathsf{E}\langle |R_{1,2} - q| \rangle_+ \\ &= \mathsf{E}\frac{\left\langle \mathrm{Av}|R_{1,2} - q| \exp \sum_{\ell \le 2} \sigma_{N+1}^\ell \left(\frac{\beta}{\sqrt{N}} \sum_{i \le N} g_{i,N+1}\sigma_i^\ell + h_{N+1} \right) \right\rangle}{\left\langle \mathrm{Av} \exp \sigma_{N+1} \left(\frac{\beta}{\sqrt{N}} \sum_{i \le N} g_{i,N+1}\sigma_i + h_{N+1} \right) \right\rangle^2} \\ &\le \mathsf{E}\left\langle |R_{1,2} - q| \mathrm{Av} \exp \sum_{\ell \le 2} \sigma_{N+1}^\ell \left(\frac{\beta}{\sqrt{N}} \sum_{i \le N} g_{i,N+1}\sigma_i^\ell + h_{N+1} \right) \right\rangle \ , \end{aligned}$$

where Av means average over $\sigma_{N+1}^\ell = \pm 1$, and since $\mathrm{Av}\exp(\sigma_{N+1}x) = \mathrm{ch}x \ge 1$. Taking expectation in the r.v.s $g_{i,N+1}$ and h_{N+1} we conclude that this is $\le K\nu(|R_{1,2} - q|)$. □

1.7 Gibbs' Measure; the TAP Equations

In this section we assume that the external field term of the Hamiltonian is $h\sum_{i \le N} \sigma_i$, although it probably would not require much more effort to handle the case of a term $\sum_{i \le N} h_i \sigma_i$.

We have shown in Theorem 1.4.15 that when we fix a number n of spins, and we look at the behavior of these n spins under Gibbs' measure, it is nearly determined by the random sequence $(\langle \sigma_i \rangle)_{i \le n}$. What is the behavior of this sequence? Here, again, the situation is as simple as possible: the sequence $(\langle \sigma_i \rangle)_{i \le n}$ is asymptotically independently identically distributed. Moreover we can provide a precise rate for this.

Theorem 1.7.1. *Given $\beta < 1/2$, and an integer n, we can find independent standard Gaussian r.v.s $(z_i)_{i \le n}$ such that*

$$\mathsf{E}\sum_{i \le n} \left(\langle \sigma_i \rangle - \mathrm{th}(\beta z_i \sqrt{q} + h) \right)^2 \le \frac{K}{N} \ , \qquad (1.183)$$

where q is the solution of (1.74) i.e. $q = \mathsf{E}\mathrm{th}^2(\beta z \sqrt{q} + h)$, and where K does not depend on N (but will of course depend on n).

Exercise 1.7.2. Assume that (1.132) and Theorem 1.7.1 hold. Prove that $\nu\big((R_{1,2}-q)^2\big) \le K/N$. (Hint: replace $R_{1,2}$ by its value, expand and use symmetry between sites.)

Research Problem 1.7.3. Find approximation results when $n = n(N) \to \infty$. (The level of the problem might depend upon how much you ask for.)

We recall the notation $\boldsymbol{\rho} = (\sigma_1, \ldots, \sigma_{N-1})$, and we consider the Hamiltonian

$$- H_{N-1}(\boldsymbol{\rho}) = \frac{\beta}{\sqrt{N}} \sum_{i<j\le N-1} g_{ij}\sigma_i\sigma_j + h \sum_{i\le N-1} \sigma_i \,. \tag{1.184}$$

This is the Hamiltonian of the SK model of an $(N-1)$-spin system, but the value of β has been changed into β_- such that

$$\frac{\beta_-}{\sqrt{N-1}} = \frac{\beta}{\sqrt{N}} \,.$$

Let us note that $|\beta - \beta_-| \le K/N$. We denote by $\langle\cdot\rangle_-$ an average for the corresponding Gibbs' measure. We recall that we write $g_i = g_{iN}$ for $i < N$. The following fact essentially allows us to compute $\langle\sigma_N\rangle$ as a function of the $(N-1)$-spin system.

Lemma 1.7.4. *For $\beta < 1/2$ we have*

$$\mathsf{E}\Big(\langle\sigma_N\rangle - \operatorname{th}\Big(\frac{\beta}{\sqrt{N}}\sum_{i<N} g_i\langle\sigma_i\rangle_- + h\Big)\Big)^2 \le \frac{K}{N} \tag{1.185}$$

$$\mathsf{E}(\langle\sigma_1\rangle - \langle\sigma_1\rangle_-)^2 \le \frac{K}{N} \,. \tag{1.186}$$

We will prove this at the end of the section as a consequence of a general principle (Theorem 1.7.11 below), but we can explain right now why (1.185) is true. The cavity method (i.e. (1.145)) implies

$$\langle\sigma_N\rangle = \frac{\big\langle \operatorname{sh}\big(\frac{\beta}{\sqrt{N}}\sum_{i\le N-1} g_i\sigma_i + h\big)\big\rangle_-}{\big\langle \operatorname{ch}\big(\frac{\beta}{\sqrt{N}}\sum_{i\le N-1} g_i\sigma_i + h\big)\big\rangle_-} \,.$$

As we have seen in (1.137), under $\langle\cdot\rangle_-$, the cavity field $\frac{1}{\sqrt{N}}\sum_{i\le N-1} g_i\sigma_i$ is approximately Gaussian with mean $\frac{1}{\sqrt{N}}\sum_{i\le N-1} g_i\langle\sigma_i\rangle_-$ and variance $1-q$; and if z is a Gaussian r.v. with expectation μ and arbitrary variance, one has

$$\frac{\mathsf{E}\operatorname{sh} z}{\mathsf{E}\operatorname{ch} z} = \operatorname{th}\mu \,.$$

Relation (1.185) is rather fundamental. Not only is it the key ingredient to Theorem 1.7.1, but it is also at the root of the Thouless-Anderson-Palmer (TAP) equations that are stated in (1.192) below.

Replacing β by $\beta_- = \beta\sqrt{1-1/N}$ slightly changes q into q_- such that $q_- = \mathsf{E}\mathrm{th}^2(\beta_- z\sqrt{q_-}+h)$. The following should not come as a surprise. We recall that L denotes a universal constant.

Lemma 1.7.5. *For $\beta < 1/2$ we have*

$$|q - q_-| \le \frac{L}{N} . \tag{1.187}$$

Proof. This proof is straightforward and uninteresting. If we define $F(\beta, q) = \mathsf{E}\mathrm{th}^2(\beta z\sqrt{q}+h)$ and define $q(\beta)$ by

$$q(\beta) = F(\beta, q(\beta))$$

then

$$q'(\beta) = \frac{\frac{\partial F}{\partial \beta}(\beta, q(\beta))}{1 - \frac{\partial F}{\partial q}(\beta, q(\beta))} .$$

Now if $f(x) = \mathrm{th}^2(x)$, we have $f''(x) = (2 - 4\,\mathrm{sh}^2(x))/\mathrm{ch}^4(x) \le 2$. Computation using Gaussian integration by parts shows that $\frac{\partial F}{\partial \beta}(\beta, q) = \beta q \mathsf{E} f''(\beta z\sqrt{q}+h)$ remains bounded and that $\frac{\partial F}{\partial q}(\beta, q) = (\beta^2/2)\mathsf{E} f''(\beta z\sqrt{q}+h) \le 1/4$. Therefore $q'(\beta)$ remains bounded for $\beta \le 1/2$. □

Lemma 1.7.6. *We can find a standard Gaussian r.v. z, depending only on the r.v.s $(g_{ij})_{i<j\le N}$, which is independent of the r.v.s $(g_{ij})_{i<j\le N-1}$, and such that*

$$\mathsf{E}\big(\langle\sigma_N\rangle - \mathrm{th}(\beta z\sqrt{q}+h)\big)^2 \le \frac{K}{N} .$$

It is important to read carefully the previous statement. It does not say (and this is not true) that z depends only on the r.v.s $(g_{iN})_{i<N}$. One would certainly wish in this result to have the constant K remain bounded as $0 \le h \le h_0$; unfortunately our argument does not yield this (there is a kind of discontinuity as $h \to 0$).

Proof. We can and do assume $h \ne 0$, for otherwise $q = 0$ and $\langle\sigma_N\rangle \equiv 0$, so there is nothing to prove. Looking at (1.185) the basic idea is simply that one should have $z\sqrt{q} \simeq N^{-1/2}\sum_{i\le N-1} g_i\langle\sigma_i\rangle_-$. However some renormalization is necessary to ensure that $\mathsf{E}z^2 = 1$, so that we define

$$z = \frac{1}{A}\sum_{i\le N-1} g_i\langle\sigma_i\rangle_- ,$$

where $A^2 = \sum_{i\le N-1}\langle\sigma_i\rangle_-^2$ and $g_i = g_{iN}$. Thus z depends only upon the r.v.s $(g_{ij})_{i<j\le N}$. Conditionally upon the r.v.s $(g_{ij})_{i<j\le N-1}$, the r.v. z is standard Gaussian, because these r.v.s determine the numbers $\langle\sigma_i\rangle_-$ and are independent of the r.v.s g_i. Therefore (as surprising as this might be the first time one thinks about this), the r.v. z is independent of the r.v.s $(g_{ij})_{i<j\le N-1}$.

Combining (1.185) and the inequality $|\operatorname{th} x - \operatorname{th} y| \le |x-y|$, it remains only to prove that

$$\mathsf{E}\Big(z\sqrt{q} - \frac{1}{\sqrt{N}}\sum_{i\le N-1} g_i\langle\sigma_i\rangle_-\Big)^2 \le \frac{K(h)}{N}\,. \tag{1.188}$$

Taking first the expectation in $(g_i)_{i\le N-1}$, we obtain

$$\begin{aligned}\mathsf{E}\Big(z\sqrt{q} - \frac{1}{\sqrt{N}}\sum_{i\le N-1} g_i\langle\sigma_i\rangle_-\Big)^2 &= \mathsf{E}\Big(\sum_{i\le N-1} g_i\langle\sigma_i\rangle_-\Big(\frac{\sqrt{q}}{A} - \frac{1}{\sqrt{N}}\Big)\Big)^2\\ &= \mathsf{E}\sum_{i\le N-1}\langle\sigma_i\rangle_-^2\Big(\frac{\sqrt{q}}{A} - \frac{1}{\sqrt{N}}\Big)^2\\ &= \mathsf{E}A^2\Big(\frac{\sqrt{q}}{A} - \frac{1}{\sqrt{N}}\Big)^2 = \mathsf{E}\Big(\sqrt{q} - \frac{A}{\sqrt{N}}\Big)^2\,.\end{aligned}$$

Now,

$$\Big(\sqrt{q} - \frac{A}{\sqrt{N}}\Big)^2 = \frac{(q - A^2/N)^2}{(\sqrt{q} + A/\sqrt{N})^2} \le \frac{1}{q}\Big(q - \frac{A^2}{N}\Big)^2\,.$$

Finally, since $A^2 = \sum_{i\le N-1}\langle\sigma_i\rangle_-^2 = \sum_{i\le N-1}\langle\sigma_i^1\sigma_i^2\rangle_- = N\langle R^-_{1,2}\rangle_-$, we get

$$\mathsf{E}\Big(q - \frac{A^2}{N}\Big)^2 = \mathsf{E}(q - \langle R^-_{1,2}\rangle_-)^2 = \mathsf{E}\langle R^-_{1,2} - q\rangle_-^2 \le \mathsf{E}\langle (R^-_{1,2} - q)^2\rangle_-\,. \tag{1.189}$$

Using (1.89) for the $(N-1)$-spin system yields that $\mathsf{E}\langle (R^-_{1,2} - q_-)^2\rangle_- \le K/N$ and (1.187) then implies that $\mathsf{E}\langle (R^-_{1,2} - q)^2\rangle_- \le K/N$. □

Proof of Theorem 1.7.1. The proof goes by induction over n. When we use the cavity method, we replace β by β_-, that depends on N, so we cannot "use (1.183) for β_- instead of β". Since $\beta_- \le \beta$, this difficulty disappears if one formulates the induction hypothesis as follows:

Given n and $\beta_0 < 1/2$, there exists a number $K(n, \beta_0)$ such that for $\beta \le \beta_0$ and any N one can find r.v.s $(z_i)_{i\le n}$, depending only on the r.v.s $(g_{ij})_{1\le i<j\le N}$ such that

$$\sum_{i\le n}\mathsf{E}\big(\langle\sigma_i\rangle - \operatorname{th}(\beta z_i\sqrt{q} + h)\big)^2 \le \frac{K(n,\beta_0)}{N}\,. \tag{1.190}$$

The reader notices that we assume that the r.v.s $(z_i)_{i\le n}$ are functions of the variables $(g_{ij})_{i<j\le N}$ as part of the induction hypothesis. That this induction hypothesis is true for $n = 1$ follows from Lemma 1.7.6, exchanging the sites 1 and N. For the induction step from n to $n+1$, we apply the induction

hypothesis to the $(N-1)$-spin system with Hamiltonian H_{N-1} given by (1.184). This amounts to replacing β by $\beta_- \leq \beta$. We then get from (1.190) that

$$\sum_{i \leq n} \mathsf{E}\big(\langle\sigma_i\rangle_- - \mathrm{th}(\beta_-\, z_i\sqrt{q_-} + h)\big)^2 \leq \frac{K(n,\beta_0)}{N-1}\,, \tag{1.191}$$

where the variables $(z_i)_{i\leq n}$ are i.i.d. standard Gaussian and depend only on $(g_{ij})_{i<j\leq N-1}$. We observe that, since $|\sqrt{a}-\sqrt{b}| \leq \sqrt{|a-b|}$ we have

$$(\sqrt{q_-} - \sqrt{q})^2 \leq |q_- - q| \leq \frac{L}{N}$$

by (1.187), so that

$$(\beta_-\sqrt{q_-} - \beta\sqrt{q})^2 \leq \frac{L}{N}$$

and, since $|\mathrm{th}x - \mathrm{th}y| \leq |x-y|$, this implies

$$\mathsf{E}\big((\mathrm{th}(\beta_-\, z_i\sqrt{q_-} + h) - \mathrm{th}(\beta\, z_i\sqrt{q} + h)\big)^2 \leq \mathsf{E}\big(z_i(\beta_-\sqrt{q_-} - \beta\sqrt{q})\big)^2 \leq \frac{L}{N}\,.$$

Combining with (1.186) and (1.191) we obtain

$$\sum_{i \leq n} \mathsf{E}\big(\langle\sigma_i\rangle - \mathrm{th}(\beta\, z_i\sqrt{q} + h)\big)^2 \leq \frac{K}{N}\,,$$

where K depends only on β_0, h and n. We now appeal to Lemma 1.7.6. The r.v. z is standard Gaussian and probabilistically independent of the r.v.s $(z_i)_{i\leq n}$ because these are functions of the r.v.s $(g_{ij})_{i<j\leq N-1}$ and z is independent of these r.v.s. Moreover, setting $z_N = z$, we have

$$\sum_{i \in \{1,\cdots,n,N\}} \mathsf{E}\big(\langle\sigma_i\rangle - \mathrm{th}(\beta\, z_i\sqrt{q} + h)\big)^2 \leq \frac{K}{N}\,.$$

Exchanging the sites N and $n+1$ concludes the proof. □

We now turn to the Thouless-Anderson-Palmer (TAP) equations [160]. These equations, at a given disorder (hopefully) determine the numbers $\langle\sigma_i\rangle$ (the mean magnetization at site i). They can be stated as

$$\langle\sigma_i\rangle \approx \mathrm{th}\bigg(\frac{\beta}{\sqrt{N}} \sum_{j\neq i} g_{ij}\langle\sigma_j\rangle + h - \beta^2(1-q)\langle\sigma_i\rangle\bigg)\,. \tag{1.192}$$

The physicists have no qualms writing exact equalities in (1.192), but it is certainly not obvious that these equations hold simultaneously for every i, even approximately. This will be a consequence of the next result, which, as Lemma 1.7.4, depends on the general principle of Theorem 1.7.11 below.

Theorem 1.7.7. *For $\beta < 1/2$, any h, any integer $k \geq 1$ we have*

$$\mathsf{E}\Big(\langle\sigma_N\rangle - \mathrm{th}\Big(\frac{\beta}{\sqrt{N}}\sum_{i\leq N-1} g_i\langle\sigma_i\rangle + h - \beta^2(1-q)\langle\sigma_N\rangle\Big)\Big)^{2k} \leq \frac{K}{N^k}\,, \quad (1.193)$$

where K depends on β and k but not on N.

In most of the statements of the rest of this section, the constant K is as above, it might depend on β and will certainly depend on k. Even though we will not mention this every time, if we fix $\beta_0 < 1/2$, one can check that for $\beta \leq \beta_0$ the constant K depends on k only.

There is an obvious relationship between (1.185) and (1.193). We have introduced a kind of correction term in (1.193), but now all the quantities that appear are defined in terms of the N-spin system. A big difference however is that (in order to control all spins at the same time) we need to control higher moments and that this requires new ideas compared to Section 1.6.

Corollary 1.7.8. *For any $\beta < 1/2$, any h, and any $\varepsilon > 0$ we have*

$$\mathsf{E}\max_{i\leq N}\Big|\langle\sigma_i\rangle - \mathrm{th}\Big(\frac{\beta}{\sqrt{N}}\sum_{j\neq i} g_{ij}\langle\sigma_j\rangle + h - \beta^2(1-q)\langle\sigma_i\rangle\Big)\Big| \leq \frac{K(\beta,\varepsilon)}{N^{1/2-\varepsilon}}\,. \quad (1.194)$$

Proof. Let

$$\Delta_i = \langle\sigma_i\rangle - \mathrm{th}\Big(\frac{\beta}{\sqrt{N}}\sum_{j\neq i} g_{ij}\langle\sigma_j\rangle + h - \beta^2(1-q)\langle\sigma_i\rangle\Big)\,, \quad (1.195)$$

so that by (1.193) and symmetry between sites we have $\mathsf{E}\Delta_i^{2k} \leq K(\beta,k)N^{-k}$ and

$$\mathsf{E}\Big(\max_{i\leq N}|\Delta_i|\Big)^{2k} \leq \sum_{i\leq N}\mathsf{E}\Delta_i^{2k} \leq \frac{K(\beta,k)}{N^{k-1}}$$

so

$$\mathsf{E}\max_{i\leq N}|\Delta_i| \leq \frac{K(\beta,k)}{N^{1/2-1/2k}}\,,$$

and taking k with $2k \geq 1/\varepsilon$ concludes the proof. □

Research Problem 1.7.9. (Level 1^+) Is it true that for some K that does not depend on N one has

$$\mathsf{E}\exp\frac{N\Delta_N^2}{K} \leq 2\ ?$$

Research Problem 1.7.10. (Level 1^+) Is it true that the r.v. $\sqrt{N}\Delta_N$ converges in law to a Gaussian limit?

These are good problems. Even though the SK model is well under control for $\beta < 1/2$, matters seem rather complicated here; that is, until one finds a good way to look at them.

We turn to the general principle on which much of the section relies. Let us consider a standard Gaussian r.v. ξ. Let us remind the reader that throughout the book we denote by E_ξ expectation in the r.v. ξ only, that is, when all other r.v.s are given.

Theorem 1.7.11. *Assume $\beta < 1/2$. Consider a function U on $\mathbb{R}$, which is infinitely differentiable. Assume that for all numbers ℓ and k, for any Gaussian r.v. z, we have*

$$\mathsf{E}|U^{(\ell)}(z)|^k < \infty . \tag{1.196}$$

Consider independent standard Gaussian r.v.s y_i and ξ, which are independent of the randomness of $\langle\cdot\rangle$. Then, using the notation $\dot{\sigma}_i = \sigma_i - \langle\sigma_i\rangle$, for each k we have

$$\mathsf{E}\left(\left\langle U\left(\frac{1}{\sqrt{N}}\sum_{i\le N} y_i\dot{\sigma}_i\right)\right\rangle - \mathsf{E}_\xi U(\xi\sqrt{1-q})\right)^{2k} \le \frac{K}{N^k}, \tag{1.197}$$

where of course q is the solution of (1.74), and the constant K depends on k, U, β, but not on N.

According to (1.137) we expect that the r.v. $\boldsymbol{\sigma} \mapsto N^{-1/2}\sum_{i\le N} y_i\dot{\sigma}_i$ should be approximately Gaussian of variance $1-q$ under $\langle\cdot\rangle$, so that we should have $\langle U(N^{-1/2}\sum_{i\le N} y_i\dot{\sigma}_i)\rangle \simeq \mathsf{E}_\xi U(\xi\sqrt{1-q})$, and (1.197) makes this statement precise.

As we shall see, the proof of (1.197) is made possible by special symmetries. It would be useful to know other statements, such as the following (which, unfortunately is probably not true).

Research Problem 1.7.12. (Level 1+). Under the preceding conditions, and recalling that E_ξ denotes expectation in ξ only, is it true that

$$\mathsf{E}\left(\left\langle U\left(\frac{1}{\sqrt{N}}\sum_{i\le N} y_i\sigma_i\right)\right\rangle - \mathsf{E}_\xi U\left(\frac{1}{\sqrt{N}}\sum_{i\le N} y_i\langle\sigma_i\rangle + \xi\sqrt{1-q}\right)\right)^{2k} \le \frac{K}{N^k}\ ? \tag{1.198}$$

The subsequent proofs use many times the following observation. If two random quantities A and B (depending on N) satisfy $\mathsf{E}A^{2k} \le K/N^k$ and $\mathsf{E}B^{2k} \le K/N^k$, then $\mathsf{E}(A+B)^{2k} \le K/N^k$ (for a different constant K). This follows from the inequality $(A+B)^{2k} \le 2^{2k}(A^{2k}+B^{2k})$. (The reader observes that in fact the previous inequality also holds with the factor 2^{2k-1} rather than 2^{2k}. Our policy however is to often write crude but sufficient inequalities.)

Proof of Theorem 1.7.11. Consider the function

$$V(x) = U(x) - \mathsf{E}_\xi U(\xi\sqrt{1-q})$$

so that

$$\mathsf{E}_\xi V(\xi\sqrt{1-q}) = 0\,. \tag{1.199}$$

Using replicas, and defining $\dot{S}^\ell = N^{-1/2}\sum_{i\le N} y_i \dot{\sigma}_i^\ell$, the left-hand side of (1.197) is

$$\mathsf{E}\left\langle V(\dot{S}^1)\right\rangle^{2k} = \mathsf{E}\left\langle \prod_{\ell\le 2k} V(\dot{S}^\ell)\right\rangle\,. \tag{1.200}$$

Consider independent standard Gaussian r.v. $(\xi^\ell)_{\ell\le 2k}$ and the function

$$\varphi(t) = \mathsf{E}\left\langle \prod_{\ell\le 2k} V(\sqrt{t}\dot{S}^\ell + \sqrt{1-t}\xi^\ell\sqrt{1-q})\right\rangle\,,$$

so that the quantity (1.200) is $\varphi(1)$. To prove the theorem, it suffices to prove that $\varphi^{(r)}(0) = 0$ for $r < 2k$ and that $|\varphi^{(2k)}(t)| \le KN^{-k}$.

For $\mathbf{x} = (x_\ell)_{\ell\le 2k}$, let us consider the function $F(\mathbf{x})$ given by $F(\mathbf{x}) = \prod_{\ell\le 2k} V(x_\ell)$. Let us define $\mathbf{X}_t = (X_\ell)_{\ell\le 2k}$ for $X_\ell = \sqrt{t}\dot{S}^\ell + \sqrt{1-t}\xi^\ell\sqrt{1-q}$. With this notation we have $\varphi(t) = \mathsf{E}\langle F(\mathbf{X}_t)\rangle$. We observe that

$$\varphi(t) = \mathsf{E}\langle F(\mathbf{X}_t)\rangle = \mathsf{E}\langle \mathsf{E}_0 F(\mathbf{X}_t)\rangle\,,$$

where E_0 denotes expectation in the r.v.s y_i and ξ^ℓ only.

Let

$$T_{\ell,\ell} = \mathsf{E}_0(\dot{S}^\ell)^2 - \mathsf{E}_0(\xi^\ell\sqrt{1-q})^2 = \frac{1}{N}\sum_{i\le N}(\dot{\sigma}_i^\ell)^2 - (1-q)$$

and, for $\ell \ne \ell'$, let

$$T_{\ell,\ell'} = \mathsf{E}_0\dot{S}^\ell\dot{S}^{\ell'} - \mathsf{E}_0\xi^\ell\xi^{\ell'}(1-q) = \frac{1}{N}\sum_{i\le N}\dot{\sigma}_i^\ell\dot{\sigma}_i^{\ell'}\,.$$

We will prove that these quantities satisfy

$$\forall r\,,\quad \mathsf{E}\langle T_{\ell,\ell'}^{2r}\rangle \le \frac{K}{N^r}\,. \tag{1.201}$$

Let us explain this in the case $\ell = \ell'$, the case $\ell \ne \ell'$ being similar. We observe that, since $(\dot{\sigma}_i^\ell)^2 = (\sigma_i^\ell - \langle\sigma_i\rangle)^2 = 1 - 2\sigma_i^\ell\langle\sigma_i\rangle + \langle\sigma_i\rangle^2$,

$$T_{\ell,\ell} = \frac{1}{N}\sum_{i\le N}(\dot{\sigma}_i^\ell)^2 - (1-q) = -2\left(\frac{1}{N}\sum_{i\le N}\sigma_i^\ell\langle\sigma_i\rangle - q\right) + \frac{1}{N}\sum_{i\le N}\langle\sigma_i\rangle^2 - q\,.$$

To control the first term on the right-hand side we write

$$\left\langle\left(\frac{1}{N}\sum_{i\le N}\sigma_i^\ell\langle\sigma_i\rangle-q\right)^{2r}\right\rangle=\left\langle\left(\frac{1}{N}\sum_{i\le N}\sigma_i^1\langle\sigma_i^2\rangle-q\right)^{2r}\right\rangle\le\langle(R_{1,2}-q)^{2r}\rangle\,,$$

by using Jensen's inequality (1.23) to average in $\boldsymbol{\sigma}^2$ outside the power $2r$ rather than inside. Then (1.88) implies $\mathsf{E}\langle(R_{1,2}-q)^{2r}\rangle\le KN^{-r}$. We control the other term similarly.

To compute the derivatives of $\varphi(t)$ we apply iteratively (1.40) to the function $t\mapsto \mathsf{E}_0F(\mathbf{X}_t)$ (given the randomness of $\langle\cdot\rangle$). We observe that for $\mathbf{s}=(s_\ell)_{\ell\le 2k}$ the corresponding partial derivative $F^{(\mathbf{s})}$ of F is given by

$$F^{(\mathbf{s})}(\mathbf{x})=\prod_{\ell\le 2k}V^{(s_\ell)}(x_\ell)\,. \tag{1.202}$$

Consider a list $\ell_1,\ell_1',\ell_2,\ell_2',\dots,\ell_r'$ of integers $\le 2k$, and the sequence $\mathbf{s}=(s_\ell)_{\ell\le 2k}$ that is obtained from this list as follows: for each $\ell\le 2k$, s_ℓ counts the number of times ℓ occurs in the list. Then it follows from (1.40) and induction on r that $\varphi^{(r)}(t)$ is given by

$$\varphi^{(r)}(t)=2^{-r}\sum_{\ell_1,\ell_1',\dots,\ell_r,\ell_r'}\mathsf{E}\langle T_{\ell_1,\ell_1'}\cdots T_{\ell_r,\ell_r'}F^{(\mathbf{s})}(\mathbf{X}_t)\rangle\,, \tag{1.203}$$

where the summation is over all choices of $\ell_1,\ell_1',\dots,\ell_r'$. If we combine (1.196), (1.202) and (1.201) with Hölder's inequality we see that $|\varphi^{(r)}(t)|\le KN^{-r/2}$ (as usual, "each factor $T_{\ell,\ell'}$ contributes as $N^{-1/2}$"). Let us now examine $\varphi^{(r)}(0)$, with the aim of proving that this is 0 unless $r\ge 2k$. Since the randomness of ξ^ℓ is independent of the randomness of $\langle\cdot\rangle$, $\varphi^{(r)}(0)$ is of the type

$$\varphi^{(r)}(0)=2^{-r}\sum_{\ell_1,\ell_1',\dots,\ell_r,\ell_r'}\mathsf{E}F^{(\mathbf{s})}((\xi^\ell\sqrt{1-q})_{\ell\le 2k})\mathsf{E}\langle T_{\ell_1,\ell_1'}\cdots T_{\ell_r,\ell_r'}\rangle\,. \tag{1.204}$$

Using independence and (1.202) we note first that

$$\mathsf{E}F^{(\mathbf{s})}((\xi^\ell\sqrt{1-q})_{\ell\le 2k})=\prod_{\ell\le 2k}\mathsf{E}V^{(s_\ell)}(\xi\sqrt{1-q})\,.$$

Combining with (1.199) we obtain

$$\mathsf{E}F^{(\mathbf{s})}((\xi^\ell\sqrt{1-q})_{\ell\le 2k})=0\ \text{ unless }\ \forall\ell\le 2k\,,\quad s_\ell\ge 1\,.$$

This implies that when we consider a non-zero term in the sum (1.204), each number $\ell\le 2k$ occurs at least one time in the list $\ell_1,\ell_1',\ell_2,\ell_2',\dots,\ell_r,\ell_r'$. Let us assume this is the case. We also observe that for $\ell\ne\ell'$ the averages of $T_{\ell,\ell'}$ over $\boldsymbol{\sigma}^\ell$ and over $\boldsymbol{\sigma}^{\ell'}$ are both zero. It follows that when

$$\langle T_{\ell_1,\ell_1'}\cdots T_{\ell_r,\ell_r'}\rangle\ne 0 \tag{1.205}$$

no number $\ell \le 2k$ can occur exactly once in the list $\ell_1, \ell'_1, \ell_2, \ell'_2, \dots, \ell'_r$. Since we assume that each of these numbers occurs at least once in this list, it must occur at least twice (for otherwise averaging $T_{\ell_1,\ell'_1} \cdots T_{\ell_r,\ell'_r}$ in $\boldsymbol{\sigma}^\ell$ would already be 0). Since the length of the list is $2r$ we must have $2r \ge 4k$ i.e. $r \ge 2k$. Therefore $r \ge 2k$ whenever $\varphi^{(r)}(0) \ne 0$. □

Corollary 1.7.13. *Given $\beta \le \beta_0 < 1/2$, h, $\varepsilon = \pm 1$ and $k \ge 1$ we have*

$$\mathsf{E}\Big(\Big\langle \exp \frac{\varepsilon\beta}{\sqrt{N}} \sum_{i\le N} y_i\sigma_i \Big\rangle - \exp\frac{\beta^2}{2}(1-q) \exp \frac{\varepsilon\beta}{\sqrt{N}} \sum_{i \le N} y_i \langle\sigma_i\rangle\Big)^{2k} \le \frac{K}{N^k} \tag{1.206}$$

and

$$\mathsf{E}\Big(\Big\langle \frac{1}{\sqrt{N}} \sum_{i\le N} y_i \dot{\sigma}_i \exp \frac{\varepsilon\beta}{\sqrt{N}} \sum_{i \le N} y_i\sigma_i \Big\rangle$$
$$-\varepsilon\beta(1-q)\exp\frac{\beta^2}{2}(1-q) \exp\frac{\varepsilon\beta}{\sqrt{N}} \sum_{i\le N} y_i\langle\sigma_i\rangle\Big)^{2k} \le \frac{K}{N^k} \tag{1.207}$$

where K does not depend on N and, given k and h, K stays bounded with $\beta \le \beta_0$.

Proof. To prove (1.206) we use (1.197) with $U(x) = \exp \varepsilon\beta x$ to get

$$\mathsf{E}A^{4k} \le \frac{K}{N^{2k}}$$

where

$$A = \Big\langle \exp \frac{\varepsilon\beta}{\sqrt{N}} \sum_{i\le N} y_i \dot{\sigma}_i \Big\rangle - \exp\frac{\beta^2}{2}(1-q)\;.$$

Now, if $B = \exp \varepsilon\beta N^{-1/2} \sum_{i\le N} y_i\langle\sigma_i\rangle$, (A.6) entails that $\mathsf{E}B^{4k} \le K^k$, and therefore

$$\mathsf{E}(AB)^{2k} \le \big(\mathsf{E}A^{4k}\mathsf{E}B^{4k}\big)^{1/2} \le \frac{K}{N^k}\;.$$

This proves (1.206).

We proceed similarly for (1.207), using now $U(x) = x\exp\varepsilon\beta x$, noting that then (using Gaussian integration by parts and (A.6))

$$\mathsf{E}_\xi U(\xi\sqrt{1-q}) = \varepsilon\beta(1-q)\exp(\beta^2(1-q)/2)\;.$$

The reason why K remains bounded for $\beta \le \beta_0$ is simply that all the estimates are uniform over that range. □

Lemma 1.7.14. *If $|A'| \le B'$ and $B \ge 1$ we have*

$$\left|\frac{A'}{B'} - \frac{A}{B}\right| \le |A - A'| + |B - B'|\;.$$

Proof. We write

$$\begin{aligned}\frac{A'}{B'} - \frac{A}{B} &= \frac{A'}{B'} - \frac{A'}{B} + \frac{A'}{B} - \frac{A}{B} \\ &= \frac{A'}{B'}\left(\frac{B - B'}{B}\right) + \frac{1}{B}(A' - A)\,,\end{aligned}$$

and the result is then obvious. □

Corollary 1.7.15. *Let*

$$\mathcal{E} = \exp\left(\frac{\varepsilon\beta}{\sqrt{N}}\sum_{i\le N} y_i\sigma_i + \varepsilon h\right). \tag{1.208}$$

Recalling that Av *denotes average over* $\varepsilon = \pm 1$, *we have*

$$\mathsf{E}\left(\frac{\langle \mathrm{Av}\varepsilon\mathcal{E}\rangle}{\langle \mathrm{Av}\mathcal{E}\rangle} - \mathrm{th}\left(\frac{\beta}{\sqrt{N}}\sum_{i\le N} y_i\langle\sigma_i\rangle + h\right)\right)^{2k} \le \frac{K}{N^k} \tag{1.209}$$

$$\mathsf{E}\left(\frac{1}{\sqrt{N}}\sum_{i\le N} y_i\frac{\langle\sigma_i\mathrm{Av}\mathcal{E}\rangle}{\langle \mathrm{Av}\mathcal{E}\rangle} - \beta(1-q)\frac{\langle \mathrm{Av}\varepsilon\mathcal{E}\rangle}{\langle \mathrm{Av}\mathcal{E}\rangle} - \frac{1}{\sqrt{N}}\sum_{i\le N} y_i\langle\sigma_i\rangle\right)^{2k} \le \frac{K}{N^k}\,. \tag{1.210}$$

Proof. Defining

$$A(\varepsilon) = \left\langle \exp\frac{\varepsilon\beta}{\sqrt{N}}\sum_{i\le N} y_i\sigma_i\right\rangle - \exp\frac{\beta^2}{2}(1-q)\exp\frac{\varepsilon\beta}{\sqrt{N}}\sum_{i\le N} y_i\langle\sigma_i\rangle\,,$$

we deduce from (1.206) that

$$\mathsf{E}\left(\frac{1}{2}A(1)\exp h \pm \frac{1}{2}A(-1)\exp(-h)\right)^{2k} \le \frac{K}{N^k}\,,$$

i.e.

$$\mathsf{E}\left(\langle \mathrm{Av}\varepsilon\mathcal{E}\rangle - \mathrm{sh}\left(\frac{\beta}{\sqrt{N}}\sum_{i\le N} y_i\langle\sigma_i\rangle + h\right)\exp\frac{\beta^2}{2}(1-q)\right)^{2k} \le \frac{K}{N^k}$$

and

$$\mathsf{E}\left(\langle \mathrm{Av}\mathcal{E}\rangle - \mathrm{ch}\left(\frac{\beta}{\sqrt{N}}\sum_{i\le N} y_i\langle\sigma_i\rangle + h\right)\exp\frac{\beta^2}{2}(1-q)\right)^{2k} \le \frac{K}{N^k}\,, \tag{1.211}$$

from which (1.209) follows using Lemma 1.7.14. From (1.207) we obtain by the same method

$$\mathsf{E}\left(\left\langle\frac{1}{\sqrt{N}}\sum_{i\le N} y_i\dot{\sigma}_i\mathrm{Av}\mathcal{E}\right\rangle - \beta(1-q)\exp\frac{\beta^2}{2}(1-q)\mathrm{sh}\left(\frac{\beta}{\sqrt{N}}\sum_{i\le N} y_i\langle\sigma_i\rangle + h\right)\right)^{2k}$$

$$\leq \frac{K}{N^k} .$$

Combining with (1.211) and Lemma 1.7.14 we get

$$\mathsf{E}\left(\frac{\langle \frac{1}{\sqrt{N}} \sum_{i \leq N} y_i \dot{\sigma}_i \mathrm{Av}\mathcal{E} \rangle}{\langle \mathrm{Av}\mathcal{E} \rangle} - \beta(1-q)\mathrm{th}\left(\frac{\beta}{\sqrt{N}} \sum_{i \leq N} y_i \langle \sigma_i \rangle + h\right)\right)^{2k} \leq \frac{K}{N^k} .$$

Since

$$\frac{\langle \frac{1}{\sqrt{N}} \sum_{i \leq N} y_i \dot{\sigma}_i \mathrm{Av}\mathcal{E} \rangle}{\langle \mathrm{Av}\mathcal{E} \rangle} = \frac{1}{\sqrt{N}} \sum_{i \leq N} y_i \frac{\langle \sigma_i \mathrm{Av}\mathcal{E} \rangle}{\langle \mathrm{Av}\mathcal{E} \rangle} - \frac{1}{\sqrt{N}} \sum_{i \leq N} y_i \langle \sigma_i \rangle ,$$

combining with (1.209) proves (1.210). □

Proof of Theorem 1.7.7. The Hamiltonian (1.184) is the Hamiltonian of an $(N-1)$-spin system with parameter

$$\beta_- = \beta \sqrt{\frac{N-1}{N}} \leq \beta .$$

The cavity method yields

$$\langle \sigma_N \rangle = \frac{\langle \mathrm{Av}\varepsilon\mathcal{E} \rangle_-}{\langle \mathrm{Av}\mathcal{E} \rangle_-} ,$$

where, recalling that $g_i = g_{iN}$,

$$\mathcal{E} = \exp\left(\frac{\varepsilon\beta_-}{\sqrt{N-1}} \sum_{i \leq N-1} g_i \sigma_i + \varepsilon h\right) = \exp\left(\frac{\varepsilon\beta}{\sqrt{N}} \sum_{i \leq N-1} g_i \sigma_i + \varepsilon h\right) .$$

We then apply (1.209) to the $(N-1)$-spin system with Hamiltonian (1.184), and to the sequence $y_i = g_i$ to get

$$\mathsf{E}\left(\langle \sigma_N \rangle - \mathrm{th}\left(\frac{\beta}{\sqrt{N}} \sum_{i \leq N-1} g_i \langle \sigma_i \rangle_- + h\right)\right)^{2k} \leq \frac{K}{N^k} \tag{1.212}$$

and in particular (1.185). Similarly we obtain from (1.210) that

$$\mathsf{E}\left(\frac{1}{\sqrt{N-1}} \sum_{i \leq N-1} g_i \langle \sigma_i \rangle - \beta_-(1-q_-)\langle \sigma_N \rangle - \frac{1}{\sqrt{N-1}} \sum_{i \leq N-1} g_i \langle \sigma_i \rangle_-\right)^{2k} \leq \frac{K}{N^k},$$

and, since $\beta_- = \beta\sqrt{N-1}/\sqrt{N}$, multiplying by β_-^{2k} and observing that $|\beta_-^2 - \beta^2| \leq 1/N$ and that $|q - q_-| \leq K/N$ by (1.187), we get

$$\mathsf{E}\left(\frac{\beta}{\sqrt{N}} \sum_{i \leq N-1} g_i \langle \sigma_i \rangle - \beta^2(1-q)\langle \sigma_N \rangle - \frac{\beta}{\sqrt{N}} \sum_{i \leq N-1} g_i \langle \sigma_i \rangle_-\right)^{2k} \leq \frac{K}{N^k} .$$

Therefore if

$$A = \frac{\beta}{\sqrt{N}} \sum_{i \le N-1} g_i \langle \sigma_i \rangle_- + h$$

$$B = \frac{\beta}{\sqrt{N}} \sum_{i \le N-1} g_i \langle \sigma_i \rangle - \beta^2 (1-q) \langle \sigma_N \rangle + h \ ,$$

we have

$$\mathsf{E}(\mathrm{th} A - \mathrm{th} B)^{2k} \le \mathsf{E}(A-B)^{2k} \le \frac{K}{N^k} \ ,$$

and combining with (1.212) this yields (1.193). □

Proof of Lemma 1.7.4. Since (1.185) follows from (1.212), it remains only to prove (1.186). Recalling (1.208), it suffices to prove that

$$\mathsf{E} \frac{\langle \dot{\sigma}_1 \mathrm{Av} \mathcal{E} \rangle^2}{\langle \mathrm{Av} \mathcal{E} \rangle^2} \le \frac{K}{N} \ . \tag{1.213}$$

Indeed, we have

$$\frac{\langle \dot{\sigma}_1 \mathrm{Av} \mathcal{E} \rangle}{\langle \mathrm{Av} \mathcal{E} \rangle} = \frac{\langle \sigma_1 \mathrm{Av} \mathcal{E} \rangle}{\langle \mathrm{Av} \mathcal{E} \rangle} - \langle \sigma_1 \rangle \ .$$

Using (1.213) for the $(N-1)$-spin system, and noticing that then by the cavity method the right-hand side is $\langle \sigma_1 \rangle - \langle \sigma_1 \rangle_-$, we obtain (1.186). Thus it suffices to prove that

$$\mathsf{E} \langle \dot{\sigma}_1 \mathrm{Av} \mathcal{E} \rangle^2 \le \frac{K}{N} \ .$$

Let us define

$$\mathcal{E}_\ell = \exp\Big(\frac{\varepsilon_\ell \beta}{\sqrt{N}} \sum_{i \le N} y_i \sigma_i^\ell + \varepsilon_\ell h \Big) \ ,$$

so that using replicas

$$\langle \dot{\sigma}_1 \mathrm{Av} \mathcal{E} \rangle^2 = \langle \dot{\sigma}_1^1 \dot{\sigma}_1^2 \mathrm{Av} \mathcal{E}_1 \mathrm{Av} \mathcal{E}_2 \rangle = \langle \dot{\sigma}_1^1 \dot{\sigma}_1^2 \mathrm{Av} \mathcal{E}_1 \mathcal{E}_2 \rangle \ ,$$

where from now on Av means average over $\varepsilon_1, \varepsilon_2 = \pm 1$.

Using symmetry between sites, and taking first expectation in the r.v.s y_i (that are independent of the randomness of the bracket) we get

$$\mathsf{E} \langle \dot{\sigma}_1^1 \dot{\sigma}_1^2 \mathrm{Av} \mathcal{E}_1 \mathcal{E}_2 \rangle = \mathsf{E} \Big\langle \frac{1}{N} \sum_{i \le N} \dot{\sigma}_i^1 \dot{\sigma}_i^2 \mathrm{Av} \mathcal{E}_1 \mathcal{E}_2 \Big\rangle = \mathsf{E} \Big\langle \frac{1}{N} \sum_{i \le N} \dot{\sigma}_i^1 \dot{\sigma}_i^2 \mathrm{Av} \mathsf{E}\, \mathcal{E}_1 \mathcal{E}_2 \Big\rangle \ .$$

Now, using (A.6),

$$\begin{aligned} \mathsf{E}\, \mathcal{E}_1 \mathcal{E}_2 &= \mathsf{E} \exp\Big(\frac{\varepsilon_1 \beta}{\sqrt{N}} \sum_{i \le N} y_i \sigma_i^1 + \frac{\varepsilon_2 \beta}{\sqrt{N}} \sum_{i \le N} y_i \sigma_i^2 + \varepsilon_1 h + \varepsilon_2 h \Big) \\ &= \exp(\beta^2 + \beta^2 \varepsilon_1 \varepsilon_2 R_{1,2} + \varepsilon_1 h + \varepsilon_2 h) \ . \end{aligned}$$

We observe that if $\varepsilon = \pm 1$ we have $\exp \varepsilon x = \mathrm{ch} x + \varepsilon \mathrm{sh} x$. Writing the above quantity as a product of four exponentials, to each of which we apply this formula, we get

$$\mathrm{Av}\mathsf{E}\, \mathcal{E}_1 \mathcal{E}_2 = \exp \beta^2 \big(\mathrm{ch}(\beta^2 R_{1,2}) \mathrm{ch}^2 h + \mathrm{sh}(\beta^2 R_{1,2}) \mathrm{sh}^2 h\big) .$$

Thus it suffices to show that

$$\left|\mathsf{E}\Big\langle \frac{1}{N}\sum_{i\le N} \dot\sigma_i^1 \dot\sigma_i^2 f(R_{1,2})\Big\rangle\right| = \left|\mathsf{E}\Big\langle \frac{1}{N}\sum_{i\le N} \dot\sigma_i^1 \dot\sigma_i^2 (f(R_{1,2}) - f(q))\Big\rangle\right| \le \frac{K}{N} ,$$

where either $f(x) = \mathrm{ch}(\beta^2 x)$ or $f(x) = \mathrm{sh}(\beta^2 x)$, and where the equality follows from the fact that $\langle \dot\sigma_i^1 \dot\sigma_i^2 \rangle = 0$. Since $\beta \le 1$ we have $|f(x) - f(q)| \le L|x - q|$ and thus

$$\left|\mathsf{E}\Big\langle \frac{1}{N}\sum_{i\le N} \dot\sigma_i^1 \dot\sigma_i^2 (f(R_{1,2}) - f(q))\Big\rangle\right| \le L\mathsf{E}\Big\langle \Big|\frac{1}{N}\sum_{i\le N} \dot\sigma_i^1 \dot\sigma_i^2\Big| |R_{1,2} - q|\Big\rangle .$$

Now, each of the factors on the right "contributes as $1/\sqrt{N}$". This is seen by using the Cauchy-Schwarz inequality, (1.89) and the fact that by Jensen's inequality (1.23) and (1.89) again we have

$$\mathsf{E}\Big\langle \Big(N^{-1}\sum_{i\le N} \dot\sigma_i^1 \dot\sigma_i^2\Big)^2\Big\rangle \le \frac{K}{N} . \qquad \square$$

1.8 Second Moment Computations and the Almeida-Thouless Line

In this section, q is always the solution of (1.74). Theorem 1.4.1 shows that $\nu((R_{1,2}-q)^2) \le K/N$ for $\beta < 1/2$, so we expect that $\lim_{N\to\infty} N\nu((R_{1,2}-q)^2)$ exists, and we would like to compute it. The present section develops the machinery to do this. Our computations will be proven to hold true for $\beta < 1/2$, but an interesting side story is that it will be obvious that the result of these calculations can be correct only when $\beta^2 \mathsf{E}\mathrm{ch}^{-4}(\beta z\sqrt{q} + h) < 1$. It is conjectured that this is exactly the region where this is the case. When h is non-random, the line

$$\beta^2 \mathsf{E} \frac{1}{\mathrm{ch}^4(\beta z\sqrt{q} + h)} = 1 \tag{1.214}$$

in the (β, h) plane is called the Almeida-Thouless (AT) line. In the SK model, it is the (conjectured) boundary between the "high-temperature" region (where the replica-symmetric solution is correct) and the "low-temperature" region (where the situation is much more complicated).

The basic tool is as follows, where ν_t is as in Section 1.6, and where we recall that $R^-_{1,2} = N^{-1}\sum_{i<N} \sigma_i^1 \sigma_i^2$.

Proposition 1.8.1. *Consider a function f on n replicas. Then, if $\tau_1, \tau_2 > 0$ and $1/\tau_1 + 1/\tau_2 = 1$ we have*

$$|\nu(f) - \nu_0(f)| \le K(n,\beta)\nu(|f|^{\tau_1})^{1/\tau_1}\nu(|R_{1,2}^- - q|^{\tau_2})^{1/\tau_2} \quad (1.215)$$

$$|\nu(f) - \nu_0(f) - \nu_0'(f)| \le K(n,\beta)\nu(|f|^{\tau_1})^{1/\tau_1}\nu(|R_{1,2}^- - q|^{2\tau_2})^{1/\tau_2}. \quad (1.216)$$

Of course $K(n,\beta)$ does not depend on N.

One should think of $|R_{1,2}^- - q|$ as being small (about $1/\sqrt{N}$). The difference between the right-hand sides of (1.215) and (1.216) is that we have in the latter an exponent $2\tau_2$ rather than τ_2. Higher-order expansions yield smaller error terms.

Proof. To prove (1.215) we simply bound $\nu_t'(f)$ using (1.151), (1.154) and Hölder's inequality. To prove (1.216) we compute $\nu_t''(f)$ by iteration of (1.151), observing that the new differentiation "brings a new factor $(R_{\ell,\ell'}^- - q)$ in each term", we bound $|\nu_t''(f)|$ as previously, and we use that

$$|\nu(f) - \nu_0(f) - \nu_0'(f)| \le \sup_{0\le t\le 1} |\nu_t''(f)| . \qquad \square$$

When $\beta < 1/2$ we know that $\nu(|R_{1,2}^- - q|^k)$ is small; but, as we see later, there is some point in making the computation for each value of β. There are two aspects in the computation; one is to get the correct error terms, which is very simple; the other is to perform the algebra, and this runs into (algebraic!) complications. Before we start the computation itself, we explain its mechanism (which will be used again and again). This will occupy the next page and a half.

To lighten notation in the argument we denote by $\mathcal{R}$ any quantity such that

$$|\mathcal{R}| \le K\left(\frac{1}{N^{3/2}} + \nu(|R_{1,2}^- - q|^3)\right) , \quad (1.217)$$

where K does not depend on N. Using the inequality $xy \le x^{3/2} + y^3$ for $x, y \ge 0$ we observe first that

$$\frac{1}{N}\nu(|R_{1,2}^- - q|) = \mathcal{R} . \quad (1.218)$$

We start the computation of $\nu\big((R_{1,2} - q)^2\big)$ as usual, recalling the notation $\varepsilon_\ell = \sigma_N^\ell$ and writing $f = (\varepsilon_1\varepsilon_2 - q)(R_{1,2} - q)$, $f^\sim = (\varepsilon_1\varepsilon_2 - q)(R_{1,2}^- - q)$, so that

$$\begin{aligned}\nu\big((R_{1,2} - q)^2\big) &= \nu(f)\\ &= \nu(f^\sim) + \frac{1}{N}\nu(1 - \varepsilon_1\varepsilon_2 q) .\end{aligned} \quad (1.219)$$

Using (1.215) with $\tau_1 = \infty$, $\tau_2 = 1$ and (1.218) we obtain

$$\frac{1}{N}\nu(1 - \varepsilon_1\varepsilon_2 q) = \frac{1}{N}\nu_0(1 - \varepsilon_1\varepsilon_2 q) + \mathcal{R} . \tag{1.220}$$

We know that $\nu_0(1 - \varepsilon_1\varepsilon_2 q) = 1 - q^2$ using Lemma 1.6.2.

Next, we apply (1.216) to $f^{\sim}$ with $\tau_1 = 3$, $\tau_2 = 3/2$, to get

$$\nu(f^{\sim}) = \nu_0'(f^{\sim}) + \mathcal{R} ,$$

because $\nu_0(f^{\sim}) = 0$ by Lemma 1.6.2 and (1.74). Therefore we have

$$\nu\big((R_{1,2} - q)^2\big) = \frac{1 - q^2}{N} + \nu_0'((\varepsilon_1\varepsilon_2 - q)(R^-_{1,2} - q)) + \mathcal{R} . \tag{1.221}$$

As is shown by (1.151), the quantity $\nu_0'(f^{\sim})$ is a sum of terms of the type

$$\pm\beta^2\nu_0\big(\varepsilon_\ell\varepsilon_{\ell'}(\varepsilon_1\varepsilon_2 - q)(R^-_{\ell,\ell'} - q)(R^-_{1,2} - q)\big) .$$

Using Lemma 1.6.2, such a term is of the form

$$\pm\, b(\ell, \ell')\nu_0((R^-_{\ell,\ell'} - q)(R^-_{1,2} - q)) \tag{1.222}$$

where

$$b(\ell, \ell') = \beta^2\nu_0(\varepsilon_\ell\varepsilon_{\ell'}(\varepsilon_1\varepsilon_2 - q)) .$$

Next, we apply (1.215) to $f = (R^-_{\ell,\ell'} - q)(R^-_{1,2} - q)$, this time with $\tau_1 = 3/2$ and $\tau_2 = 3$ to get (after a further use of Hölder's inequality)

$$\nu_0((R^-_{\ell,\ell'} - q)(R^-_{1,2} - q)) = \nu((R^-_{\ell,\ell'} - q)(R^-_{1,2} - q)) + \mathcal{R} .$$

Using the formula $R^-_{\ell,\ell'} = R_{\ell,\ell'} - \varepsilon_\ell\varepsilon_{\ell'}/N$, we obtain (using (1.218))

$$\nu((R^-_{\ell,\ell'} - q)(R^-_{1,2} - q)) = \nu((R_{\ell,\ell'} - q)(R_{1,2} - q)) + \mathcal{R} . \tag{1.223}$$

Because of the symmetry between replicas the quantity $\nu((R_{\ell,\ell'} - q)(R_{1,2} - q))$ can take only 3 values, namely

$$U = \nu\big((R_{1,2} - q)^2\big) ; \tag{1.224}$$

$$V = \nu((R_{1,2} - q)(R_{1,3} - q)) ; \tag{1.225}$$

$$W = \nu((R_{1,2} - q)(R_{3,4} - q)) . \tag{1.226}$$

Thus from (1.221) we have obtained the relation

$$U = \frac{1}{N}(1 - q^2) + c_1 U + c_2 V + c_3 W + \mathcal{R} , \tag{1.227}$$

for certain numbers c_1, c_2, c_3. We repeat this work for V and W; specifically, we write

$$V = \nu(f^\sim) + \frac{1}{N}\nu((\varepsilon_1\varepsilon_2 - q)\varepsilon_1\varepsilon_3) \tag{1.228}$$
$$W = \nu((\varepsilon_1\varepsilon_2 - q)(R^-_{3,4} - q)) + \frac{1}{N}\nu((\varepsilon_1\varepsilon_2 - q)\varepsilon_2\varepsilon_4)\ ,$$

where now $f^\sim = (\varepsilon_1\varepsilon_2 - q)(R^-_{1,3} - q)$ and we proceed as above. In this manner, we get a system of 3 linear equations in U, V, W, the solution of which yields the values of these quantities (at least in the case $\beta < 1/2$, where we know that $|\mathcal{R}| \le KN^{-3/2}$).

Having finished to sketch the method of proof, we now turn to the computation of the actual coefficients in (1.227). It is convenient to consider the quantity

$$\widehat{q} = \nu_0(\varepsilon_1\varepsilon_2\varepsilon_3\varepsilon_4) = \mathsf{E}\mathrm{th}^4 Y = \mathsf{E}\mathrm{th}^4(\beta z\sqrt{q} + h)\ . \tag{1.229}$$

Let (using Lemma 1.6.2)

$$b(2) = \beta^2\nu_0(\varepsilon_1\varepsilon_2(\varepsilon_1\varepsilon_2 - q)) = \beta^2\nu_0(1 - \varepsilon_1\varepsilon_2 q) = \beta^2(1 - q^2)$$
$$b(1) = \beta^2\nu_0(\varepsilon_1\varepsilon_3(\varepsilon_1\varepsilon_2 - q)) = \beta^2\nu_0(\varepsilon_2\varepsilon_3 - \varepsilon_1\varepsilon_3 q) = \beta^2(q - q^2)$$
$$b(0) = \beta^2\nu_0(\varepsilon_3\varepsilon_4(\varepsilon_1\varepsilon_2 - q)) = \beta^2\nu_0(\varepsilon_1\varepsilon_2\varepsilon_3\varepsilon_4 - q\varepsilon_3\varepsilon_4) = \beta^2(\widehat{q} - q^2)\ .$$

For two integers x, y we define

$$b(\ell, \ell'; x, y) = b(\mathrm{card}(\{\ell, \ell'\} \cap \{x, y\}))\ .$$

Lemma 1.8.2. *Consider a function f^- on Σ^n_{N-1} and two integers $x, y \le n$, $x \ne y$. Then*

$$\begin{aligned}\nu_0'((\varepsilon_x\varepsilon_y - q)f^-) = &\sum_{1\le\ell<\ell'\le n} b(\ell, \ell'; x, y)\nu_0(f^-(R^-_{\ell,\ell'} - q))\\ &- n\sum_{\ell\le n} b(\ell, n+1; x, y)\nu_0(f^-(R^-_{\ell,n+1} - q))\\ &+ \frac{n(n+1)}{2} b(0)\nu_0(f^-(R^-_{n+1,n+2} - q))\ .\end{aligned} \tag{1.230}$$

This is of course an immediate consequence of (1.151), Lemma 1.6.2, and the definition of $b(\ell, \ell'; x, y)$. The reason why we bring this formula forward is that it contains the *entire* algebraic structure of our calculations. In particular these calculations will hold for other models provided (1.230) is true (possibly with different values of $b(0), b(1)$ and $b(2)$). Let us also note that $b(0) = b(n+1, n+2; x, y)$.

Using (1.230) with $f^- = R^-_{1,2} - q$ and $n = 2$ yields

$$\begin{aligned}\nu_0'((\varepsilon_1\varepsilon_2 - q)(R^-_{1,2} - q)) = &\, b(2)\nu_0\big((R^-_{1,2} - q)^2\big)\\ &- 2b(1)\sum_{\ell\le 2}\nu_0((R^-_{1,2} - q)(R^-_{\ell,3} - q))\\ &+ 3b(0)\nu_0((R^-_{1,2} - q)(R^-_{3,4} - q))\ ,\end{aligned}$$

so that going back to (1.221) and recalling the definitions (1.224) to (1.226) and (1.223) we can fill the coefficients in (1.227):

$$U = \frac{1-q^2}{N} + b(2)U - 4b(1)V + 3b(0)W + \mathcal{R}\,. \tag{1.231}$$

To treat the situation (1.228) we use (1.230) with $n = 3$ and $f^- = R^-_{1,3} - q$. One needs to be patient in counting how many terms of each type there are; one gets the relation

$$V = \frac{q-q^2}{N} + b(1)U + (b(2) - 2b(1) - 3b(0))V + (6b(0) - 3b(1))W + \mathcal{R} \tag{1.232}$$

and similarly

$$W = \frac{\widehat{q}-q^2}{N} + b(0)U + (4b(1) - 8b(0))V + (b(2) - 8b(1) + 10b(0))W + \mathcal{R}\,. \tag{1.233}$$

Of course, this is not as simple as one might wish. This brings forward the matrix

$$\begin{pmatrix} b(2) & -4b(1) & 3b(0) \\ b(1) & b(2) - 2b(1) - 3b(0) & 6b(0) - 3b(1) \\ b(0) & 4b(1) - 8b(0) & b(2) - 8b(1) + 10b(0) \end{pmatrix}. \tag{1.234}$$

Rather amazingly, the transpose of this matrix has eigenvectors $(1, -2, 1)$ and $(1, -4, 3)$ with eigenvalues respectively

$$b(2) - 2b(1) + b(0) = \beta^2(1 - 2q + \widehat{q}) \tag{1.235}$$

$$b(2) - 4b(1) + 3b(0) = \beta^2(1 - 4q + 3\widehat{q})\,. \tag{1.236}$$

The second eigenvalue has multiplicity 2, but this multiplicity appears in the form of a two-dimensional Jordan block so that the corresponding eigenspace has dimension 1. The amazing point is of course that the eigenvectors do not depend on the specific values of $b(0), b(1), b(2)$. Not surprisingly the quantities (1.235) and (1.236) will occur in many formulas.

Using eigenvectors is certainly superior to brute force in solving a system of linear equations, so one should start the computation of U, V, W by computing first $U - 2V + W$. There is more however to (1.230) than the matrix (1.234). This will become much more apparent later in Section 1.10. The author cannot help feeling that there is some simple underlying algebraic structure, probably in the form of an operator between two rather large spaces.

Research Problem 1.8.3. (Level 2) Clarify the algebraic structure underlying (1.230).

Even without solving this problem, the idea of eigenvectors gives the feeling that matters will simplify considerably if one considers well-chosen combinations of (1.230) for various values of x and y, such as the following, which brings out the value (1.235).

Lemma 1.8.4. *Consider a function f^- on Σ_{N-1}^n. Then*

$$\begin{aligned}
&\nu_0'((\varepsilon_1-\varepsilon_2)(\varepsilon_3-\varepsilon_4)f^-) \\
&=(b(2)-2b(1)+b(0))\nu_0(f^-(R_{13}^- - R_{14}^- - R_{23}^- + R_{24}^-)) \\
&=\beta^2(1-2q+\widehat{q})\nu_0(f^-(R_{13}^- - R_{14}^- - R_{23}^- + R_{24}^-)) \,. \qquad (1.237)
\end{aligned}$$

Proof. The magic here lies in the cancellation of most of the terms in the sums $\sum_{1\le\ell<\ell'\le n}$ and $\sum_{\ell\le n}$ coming from (1.230). We use (1.230) four times for $x=1,2$ and $y=3,4$ and we compute

$$c(\ell,\ell')=b(\ell,\ell';1,3)-b(\ell,\ell';1,4)-b(\ell,\ell';2,3)+b(\ell,\ell';2,4)\,.$$

We see that this is zero except in the following cases:

$$c(1,3)=c(2,4)=-c(1,4)=-c(2,3)=b(2)-2b(1)+b(0)\,. \qquad \square$$

"Rectangular sums" such as $R_{1,3}-R_{1,4}-R_{2,3}+R_{2,4}$ or $R_{1,3}^- - R_{1,4}^- - R_{2,3}^- + R_{2,4}^-$ will occur frequently.

Now that we have convinced the reader that the error terms in our computation are actually of the type (1.217) we will for simplicity assume that $\beta<1/2$, in which case the error terms are $O(3)$, where we recall that $O(k)$ means a quantity A such that $|A|\le KN^{-k/2}$ where K does not depend on N.

We will continue the computation of U,V,W later, but to immediately make the point that (1.237) simplifies the algebra we prove the following, where we recall that "$\cdot$" denotes the dot product in $\mathbb{R}^N$, so that $\boldsymbol{\sigma}^1\cdot\boldsymbol{\sigma}^2 = NR_{1,2}$. It is worth making the effort to fully understand the mechanism of the next result, which is a prototype for many of the later calculations.

Proposition 1.8.5. *If $\beta<1/2$ we have*

$$\nu\Big(\Big(\frac{(\boldsymbol{\sigma}^1-\boldsymbol{\sigma}^2)\cdot(\boldsymbol{\sigma}^3-\boldsymbol{\sigma}^4)}{N}\Big)^2\Big)=\frac{4(1-2q+\widehat{q})}{N(1-\beta^2(1-2q+\widehat{q}))}+O(3)\,. \qquad (1.238)$$

Proof. Let $a_i=(\sigma_i^1-\sigma_i^2)(\sigma_i^3-\sigma_i^4)$, so that

$$\frac{(\boldsymbol{\sigma}^1-\boldsymbol{\sigma}^2)\cdot(\boldsymbol{\sigma}^3-\boldsymbol{\sigma}^4)}{N}=R_{1,3}-R_{1,4}-R_{2,3}+R_{2,4}=\frac{1}{N}\sum_{i\le N}a_i\,. \qquad (1.239)$$

Therefore, if $f=R_{1,3}-R_{1,4}-R_{2,3}+R_{2,4}$, we have

$$\begin{aligned}
\nu\Big(\Big(\frac{(\boldsymbol{\sigma}^1-\boldsymbol{\sigma}^2)\cdot(\boldsymbol{\sigma}^3-\boldsymbol{\sigma}^4)}{N}\Big)^2\Big)&=\nu((R_{1,3}-R_{1,4}-R_{2,3}+R_{2,4})^2) \\
&=\frac{1}{N}\nu\Big(\sum_{i\le N}a_i f\Big)=\nu(a_N f) \\
&=\nu((\varepsilon_1-\varepsilon_2)(\varepsilon_3-\varepsilon_4)f)\,. \qquad (1.240)
\end{aligned}$$

Moreover

$$\nu((\varepsilon_1-\varepsilon_2)(\varepsilon_3-\varepsilon_4)f)=\nu((\varepsilon_1-\varepsilon_2)(\varepsilon_3-\varepsilon_4)f^-)+\frac{1}{N}\nu(((\varepsilon_1-\varepsilon_2)(\varepsilon_3-\varepsilon_4))^2)\ , \tag{1.241}$$

where $f^- = R^-_{1,3} - R^-_{1,4} - R^-_{2,3} + R^-_{2,4}$. First we observe that

$$\nu_0(((\varepsilon_1-\varepsilon_2)(\varepsilon_3-\varepsilon_4))^2)=4\nu_0((1-\varepsilon_1\varepsilon_2)(1-\varepsilon_3\varepsilon_4))=4(1-2q+\widehat{q})\ . \tag{1.242}$$

We use (1.216) for $f^* = (\varepsilon_1-\varepsilon_2)(\varepsilon_3-\varepsilon_4)f^-$ with $\tau_1 = 3$ and $\tau_2 = 3/2$ to obtain

$$|\nu(f^*)-\nu_0(f^*)-\nu_0'(f^*)|\le K\nu(|R^-_{1,2}-q|^3)=O(3)\ .$$

Next, by Lemma (1.6.2) we have $\nu_0(f^*) = \nu_0((\varepsilon_1-\varepsilon_2)(\varepsilon_3-\varepsilon_4)f^-) = 0$, and (1.237) implies

$$\begin{aligned}\nu_0'(f^*)&=\nu_0'((\varepsilon_1-\varepsilon_2)(\varepsilon_3-\varepsilon_4)f^-)\\&=\beta^2(1-2q+\widehat{q})\nu_0\left((R^-_{1,3}-R^-_{1,4}-R^-_{2,3}+R^-_{2,4})^2\right)\ .\end{aligned}$$

Next, we observe that $\nu((R^-_{1,2}-q)^4)=O(4)$, so that $\nu((R^-_{1,3}-R^-_{1,4}-R^-_{2,3}+R^-_{2,4})^4)=O(4)$ and we apply (1.216) with e.g $\tau_1=\tau_2=2$ to obtain

$$\nu_0\left((R^-_{1,3}-R^-_{1,4}-R^-_{2,3}+R^-_{2,4})^2\right)=\nu\left((R^-_{1,3}-R^-_{1,4}-R^-_{2,3}+R^-_{2,4})^2\right)+O(3)\ .$$

We then use the relation $R^-_{\ell,\ell'} = R_{\ell,\ell'} - \varepsilon_\ell\varepsilon_{\ell'}/N$ and expansion to get

$$\nu\left((R^-_{1,3}-R^-_{1,4}-R^-_{2,3}+R^-_{2,4})^2\right)=\nu\left((R_{1,3}-R_{1,4}-R_{2,3}+R_{2,4})^2\right)+O(3)\ .$$

Finally we have reached the relation

$$\begin{aligned}\nu(f^*)&=\nu((\varepsilon_1-\varepsilon_2)(\varepsilon_3-\varepsilon_4)f^-)\\&=\beta^2(1-2q+\widehat{q})\nu\left((R_{1,3}-R_{1,4}-R_{2,3}+R_{2,4})^2\right)+O(3)\end{aligned}$$

and thus combining with (1.240) and (1.241) we get

$$(1-\beta^2(1-2q+\widehat{q}))\nu\left((R_{1,3}-R_{1,4}-R_{2,3}+R_{2,4})^2\right)=\frac{4}{N}(1-2q+\widehat{q})+O(3)\ .$$

Since for $\beta\le 1/2$ we have $\beta^2(1-2q+\widehat{q})<1/4<1$ the result follows. □

Since error terms are always handled by the same method, this will not be detailed any more.

One can note the nice (accidental?) expression

$$1-2q+\widehat{q}=\mathsf{E}\,\frac{1}{\mathrm{ch}^4 Y}\ .$$

We have proved (1.238) for $\beta < 1/2$, but we may wonder for which values of β it might hold. Since the left hand side of (1.238) is ≥ 0, this relation cannot hold unless

$$\beta^2(1-2q+\widehat{q}) = \beta^2 \mathsf{E}\frac{1}{\mathrm{ch}^4 Y} < 1 \ , \tag{1.243}$$

i.e. unless we are on the high-temperature side of the AT line (1.214), a point to which we will return in the next section.

Corollary 1.8.6. *If $\beta < 1/2$ we have*

$$\mathsf{E}\big((\langle\sigma_1\sigma_2\rangle - \langle\sigma_1\rangle\langle\sigma_2\rangle)^2\big) = \frac{\beta^2(1-2q+\widehat{q})^2}{N(1-\beta^2(1-2q+\widehat{q}))} + O(3) \ . \tag{1.244}$$

Proof. Recalling (1.239) and using symmetry between sites,

$$\begin{aligned}\nu\Big(\Big(\frac{(\boldsymbol{\sigma}^1-\boldsymbol{\sigma}^2)\cdot(\boldsymbol{\sigma}^3-\boldsymbol{\sigma}^4)}{N}\Big)^2\Big) &= \nu\Big(\Big(\frac{1}{N}\sum_{i\le N} a_i\Big)^2\Big) \\ &= \frac{1}{N}\nu(a_N^2) + \frac{N-1}{N}\nu(a_1a_2) \ .\end{aligned}$$

Now, (1.242) and (1.215) imply

$$\nu(a_N^2) = 4(1-2q+\widehat{q}) + O(1)$$

and

$$\begin{aligned}\nu(a_1a_2) &= \mathsf{E}\langle(\sigma_1^1-\sigma_1^2)(\sigma_1^3-\sigma_1^4)(\sigma_2^1-\sigma_2^2)(\sigma_2^3-\sigma_2^4)\rangle \\ &= \mathsf{E}\langle(\sigma_1^1-\sigma_1^2)(\sigma_2^1-\sigma_2^2)\rangle^2 \\ &= 4\mathsf{E}(\langle\sigma_1\sigma_2\rangle - \langle\sigma_1\rangle\langle\sigma_2\rangle)^2 \ .\end{aligned}$$

The result then follows from (1.238). □

After these parentheses, we can get back to the computation of U, V and W. If we were interested only in these values, the shortest route would certainly be to solve the equations (1.231), (1.232), (1.233). We choose a less direct approach, that will be much easier than the brute force method to generalize to higher moments in Section 1.10. The computation is very pretty and natural, but, as we have already discovered, the result of this computation will be a bit complicated. It is given at the end of this section. Rather than scaring away the reader with these formulas, we take the gentler road of gradually discovering how they come into existence.

Pursuing the idea that the computation simplifies if we "use the correct basis" we introduce the quantities

$$T_{\ell,\ell'} = \frac{(\boldsymbol{\sigma}^\ell - \mathbf{b})\cdot(\boldsymbol{\sigma}^{\ell'}-\mathbf{b})}{N} \ ; T_\ell = \frac{(\boldsymbol{\sigma}^\ell-\mathbf{b})\cdot\mathbf{b}}{N} \ ; T = \frac{\mathbf{b}\cdot\mathbf{b}}{N} - q \ , \tag{1.245}$$

where as usual $\mathbf{b} = \langle\boldsymbol{\sigma}\rangle = (\langle\sigma_i\rangle)_{i\le N}$. Using the notation $\dot{\sigma}_i = \sigma_i - \langle\sigma_i\rangle$, we can also write (1.245) as

$$T_{\ell,\ell'} = \frac{1}{N}\sum_{i\le N} \dot{\sigma}_i^\ell \dot{\sigma}_i^{\ell'} \; ; \; T_\ell = \frac{1}{N}\sum_{i\le N} \dot{\sigma}_i^\ell \langle \sigma_i \rangle \; ; \; T = \frac{1}{N}\sum_{i\le N} \langle \sigma_i \rangle^2 - q \, .$$

These quantities will be proved to be "independent" in a certain sense. They will allow to recover the quantities $R_{\ell,\ell'} - q$ by the formula

$$R_{\ell,\ell'} - q = T_{\ell,\ell'} + T_\ell + T_{\ell'} + T \, . \tag{1.246}$$

Proposition 1.8.7. *If $\beta < 1/2$ we have*

$$\nu(T_{1,2}^2) = A^2 + O(3) \tag{1.247}$$

where

$$A^2 = \frac{1 - 2q + \widehat{q}}{N(1 - \beta^2(1 - 2q + \widehat{q}))} \, . \tag{1.248}$$

Proof. A basic problem is that $T_{1,2}$ is a function of $\boldsymbol{\sigma}^1$ and $\boldsymbol{\sigma}^2$, but that it depends on the disorder through $\mathbf{b}$, so that one cannot directly use results such as Lemma 1.6.3 for $f = T_{1,2}$. There is a basic technique to go around this difficulty. It will be used again and again in the rest of this section, and in Section 1.10. It is basically to "replace each occurrence of $\mathbf{b}$ by $\boldsymbol{\sigma}^\ell$ for a value of ℓ that has never been used before". For example we have

$$\begin{aligned}\langle T_{1,2}^2 \rangle &= \left\langle \frac{(\boldsymbol{\sigma}^1 - \mathbf{b}) \cdot (\boldsymbol{\sigma}^2 - \mathbf{b})}{N} \frac{(\boldsymbol{\sigma}^1 - \mathbf{b}) \cdot (\boldsymbol{\sigma}^2 - \mathbf{b})}{N} \right\rangle \\ &= \left\langle \frac{(\boldsymbol{\sigma}^1 - \boldsymbol{\sigma}^3) \cdot (\boldsymbol{\sigma}^2 - \boldsymbol{\sigma}^4)}{N} \frac{(\boldsymbol{\sigma}^1 - \boldsymbol{\sigma}^5) \cdot (\boldsymbol{\sigma}^2 - \boldsymbol{\sigma}^6)}{N} \right\rangle \, .\end{aligned}$$

To understand this formula we keep in mind that $\boldsymbol{\sigma}^\ell$ are averaged independently for Gibbs' measure, and that for any given vectors $\mathbf{x}, \mathbf{y} \in \mathbb{R}^N$ we have the formula $\langle \mathbf{x} \cdot (\mathbf{y} + \boldsymbol{\sigma}) \rangle = \mathbf{x} \cdot (\mathbf{y} + \mathbf{b})$. Applying this four times, to the integrations in $\boldsymbol{\sigma}^\ell$ for $\ell = 3, 4, 5, 6$ proves the above equality. Therefore we have

$$\nu(T_{1,2}^2) = \nu\left(\frac{(\boldsymbol{\sigma}^1 - \boldsymbol{\sigma}^3) \cdot (\boldsymbol{\sigma}^2 - \boldsymbol{\sigma}^4)}{N} \frac{(\boldsymbol{\sigma}^1 - \boldsymbol{\sigma}^5) \cdot (\boldsymbol{\sigma}^2 - \boldsymbol{\sigma}^6)}{N} \right) ,$$

or to match better with the notation of Proposition 1.8.5, and using symmetry between replicas,

$$\nu(T_{1,2}^2) = \nu\left(\frac{(\boldsymbol{\sigma}^1 - \boldsymbol{\sigma}^2) \cdot (\boldsymbol{\sigma}^3 - \boldsymbol{\sigma}^4)}{N} \frac{(\boldsymbol{\sigma}^1 - \boldsymbol{\sigma}^5) \cdot (\boldsymbol{\sigma}^3 - \boldsymbol{\sigma}^6)}{N} \right) .$$

Thus, using again (1.239) and symmetry among sites,

$$\nu(T_{1,2}^2) = \nu((\varepsilon_1 - \varepsilon_2)(\varepsilon_3 - \varepsilon_4) f) \, ,$$

where $f = R_{1,3} - R_{1,6} - R_{5,3} + R_{5,6}$, so that

$$\nu(T_{1,2}^2) = \frac{1}{N}\nu((\varepsilon_1-\varepsilon_2)(\varepsilon_3-\varepsilon_4)(\varepsilon_1-\varepsilon_5)(\varepsilon_3-\varepsilon_6)) + \nu((\varepsilon_1-\varepsilon_2)(\varepsilon_3-\varepsilon_4)f^-)\,,$$

for $f^- = R^-_{1,3} - R^-_{1,6} - R^-_{5,3} + R^-_{5,6}$. We observe that

$$\begin{aligned}
&\langle(\varepsilon_1-\varepsilon_2)(\varepsilon_3-\varepsilon_4)(\varepsilon_1-\varepsilon_5)(\varepsilon_3-\varepsilon_6)\rangle_0 \\
&= \langle(\varepsilon_1-\varepsilon_2)(\varepsilon_1-\varepsilon_5)\rangle_0^2 = \langle 1-\varepsilon_1\varepsilon_5-\varepsilon_1\varepsilon_2+\varepsilon_2\varepsilon_5\rangle_0^2 \\
&= (1-\mathrm{th}^2 Y)^2 = 1 - 2\,\mathrm{th}^2 Y + \mathrm{th}^4 Y
\end{aligned} \tag{1.249}$$

so that

$$\nu_0\big((\varepsilon_1-\varepsilon_2)(\varepsilon_3-\varepsilon_4)(\varepsilon_1-\varepsilon_5)(\varepsilon_3-\varepsilon_6)\big) = 1 - 2\,q + \widehat{q}\,.$$

One then proceeds exactly as in Proposition 1.8.5 to prove (1.248). □

Proposition 1.8.8. *If $\ell < \ell'$ and $(\ell,\ell') \neq (1,2)$ we have*

$$\nu(T_{1,2}T_{\ell,\ell'}) = 0\,. \tag{1.250}$$

For any ℓ we have

$$\nu(T_{1,2}T_\ell) = 0\,. \tag{1.251}$$

Finally we have

$$\nu(T_{1,2}T) = 0\,. \tag{1.252}$$

Proof. For example, if $1 \notin \{\ell,\ell'\}$ we have that $\langle T_{1,2}T_{\ell,\ell'}\rangle = 0$ by integrating in $\boldsymbol{\sigma}^1$. □

The following is in the spirit of Lemma 1.8.4. It is simpler than (1.230), yet it allows more computations than (1.237).

Lemma 1.8.9. *Consider a function f^- on Σ_{N-1}^n. Then*

$$\begin{aligned}
&\nu_0'((\varepsilon_1-\varepsilon_2)\varepsilon_3 f^-) = (b(2)-b(1))\nu_0(f^-(R^-_{1,3}-R^-_{2,3})) \\
&+ (b(1)-b(0))\Bigg(\sum_{4\le\ell\le n}\nu_0(f^-(R^-_{1,\ell}-R^-_{2,\ell})) - n\nu_0(f^-(R^-_{1,n+1}-R^-_{2,n+1}))\Bigg)\,.
\end{aligned} \tag{1.253}$$

Moreover, when f does not depend on the third replica we have

$$\begin{aligned}
&\nu_0'((\varepsilon_1-\varepsilon_2)\varepsilon_3 f^-) \\
&= (b(2)-4b(1)+3b(0))\nu_0(f^-(R^-_{1,3}-R^-_{2,3})) \\
&+ (b(1)-b(0))\sum_{4\le\ell\le n}\nu_0(f^-(R^-_{1,\ell}-R^-_{2,\ell}-R^-_{1,n+1}+R^-_{2,n+1})) \\
&= \beta^2(1-4q+3\widehat{q})\nu_0(f^-(R^-_{1,3}-R^-_{2,3})) \\
&+ \beta^2(q-\widehat{q})\sum_{4\le\ell\le n}\nu_0(f^-(R^-_{1,\ell}-R^-_{2,\ell}-R^-_{1,n+1}+R^-_{2,n+1}))\,.
\end{aligned} \tag{1.254}$$

Proof. We use again (1.230). For $\ell < \ell'$ we compute

$$c(\ell, \ell') := b(\ell, \ell'; 1, 3) - b(\ell, \ell'; 2, 3) .$$

This is 0 if $\ell \geq 3$ or $\ell = 1$, $\ell' = 2$. Moreover

$$\begin{aligned} c(1,3) &= -c(2,3) = b(2) - b(1) \\ c(1,\ell') &= -c(2,\ell') = b(1) - b(0) \quad \text{if } \ell' \geq 4. \end{aligned}$$

This proves (1.253). To prove (1.254) when f does not depend on the third replica we simply notice that then

$$\nu_0(f^-(R^-_{1,3} - R^-_{2,3})) = \nu_0(f^-(R^-_{1,n+1} - R^-_{2,n+1})) ,$$

and we move the corresponding terms from (1.254) around. □

Proposition 1.8.10. *If $\beta < 1/2$ we have*

$$\nu(T_1^2) = B^2 + O(3) \tag{1.255}$$

where

$$B^2 = \frac{1}{N} \frac{q - \widehat{q}}{(1 - \beta^2(1 - 2q + \widehat{q}))(1 - \beta^2(1 - 4q + 3\widehat{q}))} . \tag{1.256}$$

Moreover,

$$\nu(T_1 T_2) = \nu(T_1 T) = 0 . \tag{1.257}$$

Proof. We start with the observation that

$$\begin{aligned} \langle T_1^2 \rangle &= \left\langle \frac{(\boldsymbol{\sigma}^1 - \mathbf{b}) \cdot \mathbf{b}}{N} \frac{(\boldsymbol{\sigma}^1 - \mathbf{b}) \cdot \mathbf{b}}{N} \right\rangle \\ &= \left\langle \frac{(\boldsymbol{\sigma}^1 - \boldsymbol{\sigma}^2) \cdot \boldsymbol{\sigma}^3}{N} \frac{(\boldsymbol{\sigma}^1 - \boldsymbol{\sigma}^4) \cdot \boldsymbol{\sigma}^5}{N} \right\rangle \\ &= \langle (R_{1,3} - R_{2,3})(R_{1,5} - R_{4,5}) \rangle . \end{aligned} \tag{1.258}$$

We then write as usual

$$\begin{aligned} \nu(T_1^2) &= \nu((\varepsilon_1 - \varepsilon_2)\varepsilon_3(R_{1,5} - R_{4,5})) \\ &= \frac{1}{N}\nu((\varepsilon_1 - \varepsilon_2)\varepsilon_3(\varepsilon_1 - \varepsilon_4)\varepsilon_5) + \nu((\varepsilon_1 - \varepsilon_2)\varepsilon_3(R^-_{1,5} - R^-_{4,5})) . \end{aligned}$$

We have

$$\nu_0((\varepsilon_1 - \varepsilon_2)\varepsilon_3(\varepsilon_1 - \varepsilon_4)\varepsilon_5) = \nu_0((1 - \varepsilon_1\varepsilon_4 - \varepsilon_2\varepsilon_1 + \varepsilon_2\varepsilon_4)\varepsilon_3\varepsilon_5) = q - \widehat{q} .$$

Using (1.254) for $n = 5$, $f^- = R^-_{1,5} - R^-_{4,5}$, our usual scheme of proof yields

$$\nu(T_1^2) = \frac{1}{N}(q - \widehat{q}) + \beta^2(1 - 4q + 3\widehat{q})\nu((R_{1,3} - R_{2,3})(R_{1,5} - R_{4,5})) \quad (1.259)$$
$$+ \beta^2(q - \widehat{q}) \sum_{\ell=4,5} \nu\big((R_{1,5} - R_{4,5})(R_{1,\ell} - R_{2,\ell} - R_{1,6} + R_{2,6})\big) + O(3) .$$

From (1.258) we deduce that

$$\nu((R_{1,3} - R_{2,3})(R_{1,5} - R_{4,5})) = \nu(T_1^2) .$$

To evaluate the last term in (1.259) we write, using (1.246)

$$R_{1,5} - R_{4,5} = T_{1,5} - T_{4,5} + T_1 - T_4$$
$$R_{1,\ell} - R_{2,\ell} - R_{1,6} + R_{2,6} = T_{1,\ell} - T_{2,\ell} - T_{1,6} + T_{2,6} .$$

Proposition 1.8.8 and (1.247) imply

$$\sum_{\ell=4,5} \nu\big((R_{1,5} - R_{4,5})(R_{1,\ell} - R_{2,\ell} - R_{1,6} + R_{2,6})\big) = \nu(T_{1,5}^2) = A^2 + O(3) ,$$

so that (1.259) means that

$$(1 - \beta^2(1 - 4q + 3\widehat{q}))\nu(T_1^2) = \frac{1}{N}(q - \widehat{q}) + \beta^2(q - \widehat{q})A^2 + O(3) . \quad (1.260)$$

Since $\widehat{q} \leq q$ as is obvious from the definition of $\widehat{q}$, we have $1 - 4q + 3\widehat{q} \leq 1$, so that $\beta^2(1 - 4q + 3\widehat{q}) < 1$ and (1.260) implies (1.256). The rest is obvious as in Proposition 1.8.8. □

We can then take the last step.

Proposition 1.8.11. *If $\beta < 1/2$ we have*

$$\nu(T^2) = C^2 + O(3) \quad (1.261)$$

where

$$(1-\beta^2(1-4q+3\widehat{q}))C^2 = \frac{\widehat{q} - q^2}{N} + \beta^2(\widehat{q}-q^2)A^2 + 2\beta^2(2q+q^2-3\widehat{q})B^2 . \quad (1.262)$$

Proof. We know exactly how to proceed. We write

$$\begin{aligned}\nu(T^2) &= \nu((R_{1,2} - q)(R_{3,4} - q)) \\ &= \nu((\varepsilon_1\varepsilon_2 - q)(R_{3,4} - q)) \\ &= \frac{1}{N}\nu((\varepsilon_1\varepsilon_2 - q)\varepsilon_3\varepsilon_4) + \nu((\varepsilon_1\varepsilon_2 - q)(R_{3,4}^- - q)) .\end{aligned}$$

We observe that

$$\nu_0((\varepsilon_1\varepsilon_2 - q)\varepsilon_3\varepsilon_4) = \widehat{q} - q^2 ,$$

and we use (1.230) with $n = 4$ and $f^- = R^-_{3,4} - q$ to get, writing $b(\ell,\ell') = b(\ell,\ell';1,2)$, that

$$\nu(T^2) = \frac{1}{N}(\widehat{q} - q^2) + \sum_{1\le\ell<\ell'\le 4} b(\ell,\ell')\nu((R_{\ell,\ell'} - q)(R_{3,4} - q))$$
$$- 4\sum_{\ell\le 4} b(\ell,5)\nu((R_{\ell,5} - q)(R_{3,4} - q))$$
$$+ 10b(5,6)\nu((R_{5,6} - q)(R_{3,4} - q)) + O(3)\,. \quad (1.263)$$

Using (1.246) and Propositions 1.8.7, 1.8.8 and 1.8.10 we know that

$$\nu\big((R_{\ell,\ell'} - q)^2\big) = A^2 + 2\,B^2 + \nu(T^2) + O(3)$$
$$\nu((R_{\ell,\ell'} - q)(R_{\ell_1,\ell_2} - q)) = B^2 + \nu(T^2) + O(3)$$

if $\mathrm{card}(\{\ell,\ell'\}\cap\{\ell_1,\ell_2\}) = 1$, while if $\{\ell,\ell'\}\cap\{\ell_1,\ell_2\} = \emptyset$, then

$$\nu((R_{\ell,\ell'} - q)(R_{\ell_1,\ell_2} - q)) = \nu(T^2)\,.$$

We substitute these expressions in the right-hand side of (1.263) and we collect the terms. The coefficient of $\nu(T^2)$ is

$$\sum_{1\le\ell<\ell'\le 4} b(\ell,\ell') - 4\sum_{\ell\le 4} b(\ell,5) + 10\,b(5,6) = \beta^2\big(1 - q^2 + 4\,q(1-q) + (\widehat{q} - q^2)$$
$$- 4(2\,q(1-q) + 2(\widehat{q} - q^2)) + 10(\widehat{q} - q^2)\big)$$
$$= \beta^2(1 - 4\,q + 3\,\widehat{q})\,.$$

The coefficient of A^2 is $\beta^2(\widehat{q} - q^2)$, and the coefficient of B^2 is

$$\sum_{\ell=1,2,\ell'=3,4} b(\ell,\ell') + 2b(3,4) - 4(b(3,5) + b(4,5))$$
$$= \beta^2\big(4(q - q^2) + 2(\widehat{q} - q^2) - 8(\widehat{q} - q^2)\big) = 2\,\beta^2(2\,q + q^2 - 3\,\widehat{q})\,.$$

We then get

$$\nu(T^2) = \frac{1}{N}(\widehat{q} - q^2) + \beta^2(1 - 4\,q + 3\,\widehat{q})\,\nu(T^2) + \beta^2(\widehat{q} - q^2)\,A^2$$
$$+ 2\,\beta^2(2\,q + q^2 - 3\,\widehat{q})\,B^2 + O(3) \quad (1.264)$$

and this implies the result. □

Using (1.246) again, we have proved the following, where A, B, C are given respectively by (1.248), (1.256) and (1.262).

Theorem 1.8.12. *For $\beta < 1/2$ we have*

$$\nu\big((R_{1,2} - q)^2\big) = A^2 + 2B^2 + C^2 + O(3) \quad (1.265)$$
$$\nu((R_{1,2} - q)(R_{1,3} - q)) = B^2 + C^2 + O(3)$$
$$\nu((R_{1,2} - q)(R_{3,4} - q)) = C^2 + O(3)\,.$$

1.9 Beyond the AT Line

We recall that q is the solution of (1.74) and that $\widehat{q} = \mathsf{E}\mathrm{th}^4 Y = \mathsf{E}\mathrm{th}^4(\beta z\sqrt{q} + h)$. We should mention for the specialist that we will (much) later prove that "beyond the AT line", that is, when

$$\beta^2(1 - 2q + \widehat{q}) = \beta^2 \mathsf{E}\frac{1}{\mathrm{ch}^4 Y} > 1 \tag{1.266}$$

the left-hand side of (1.171) is bounded below independently of N, and that, consequently, for some number δ that does not depend on N, we have

$$\frac{1}{N}\sum_{M \le N} \nu_M(|R_{1,2} - q|) > \delta > 0 , \tag{1.267}$$

where the index M refers to an M-spin system. This fact, however, relies on an extension of Theorem 1.3.7, and, like this theorem, uses very special properties of the SK model.

Research Problem 1.9.1. (Level 2) Prove that beyond the AT line we have in fact for each N that

$$\nu(|R_{1,2} - q|) > \delta . \tag{1.268}$$

As we will explain later, we know with considerable work how to deduce (1.268) from (1.267) in many cases, for example when in the term $\sum_{i \le N} h_i \sigma_i$ of the Hamiltonian the r.v.s h_i are i.i.d. Gaussian with a non-zero variance; but we do not know how to prove (1.268) when $h_i = h \ne 0$.

In contrast with the previous arguments, the results of the present section rely on a very general method, which has the potential to be used for a great many models, and that provides results for every N. This method simply analyzes what goes wrong in the proof of (1.238) when (1.266) occurs. The main result is as follows.

Proposition 1.9.2. *Under (1.266), there exists a number $\delta > 0$, that does not depend on N, such that for N large enough, we have*

$$\nu\bigl(|R_{1,2} - q|^3\bigr) \ge \delta \nu\bigl((R_{1,2} - q)^2\bigr) \ge \frac{\delta^2}{N} . \tag{1.269}$$

This is not as nice as (1.268), but this shows something remarkable: the set where $|R_{1,2} - q| \ge \delta/2$ is not exponentially small (in contrast with what happens in (1.87)). To see this we write, since $|R_{1,2} - q| \le 2$,

$$|R_{1,2} - q|^3 \le \frac{\delta}{2}(R_{1,2} - q)^2 + 8 \cdot \mathbf{1}_{\{|R_{1,2} - q| \ge \delta/2\}} , \tag{1.270}$$

where $\mathbf{1}_{\{|R_{1,2} - q| \ge \delta/2\}}$ is the function of $\boldsymbol{\sigma}^1$ and $\boldsymbol{\sigma}^2$ that is 1 when $|R_{1,2} - q| \ge \delta/2$ and is 0 otherwise. Using the first part of (1.269) in the first inequality and (1.270) in the second one, we obtain

$$\delta\nu\big((R_{1,2}-q)^2\big) \le \nu\big(|R_{1,2}-q|^3\big) \le \frac{\delta}{2}\nu\big((R_{1,2}-q)^2\big) + 8\nu\big(\mathbf{1}_{\{|R_{1,2}-q|\ge\delta/2\}}\big).$$

Hence, using the second part of (1.269) in the second inequality,

$$\nu\big(\mathbf{1}_{\{|R_{1,2}-q|\ge\delta/2\}}\big) \ge \frac{\delta}{16}\nu\big((R_{1,2}-q)^2\big) \ge \frac{\delta^2}{16N}\,. \tag{1.271}$$

Lemma 1.9.3. *For each values of β and h, we have*

$$\beta^2(1-4q+3\widehat{q}) < 1\,. \tag{1.272}$$

Proof. Consider the function

$$\Phi(x) = \mathsf{E}\,\mathrm{th}^2(\beta z\sqrt{x}+h)\,. \tag{1.273}$$

Then Proposition A.14.1 shows that $\Phi(x)/x$ is decreasing, so that

$$x\,\Phi'(x) - \Phi(x) < 0\,,$$

and since $q = \Phi(q)$, we have $\Phi'(q) < 1$. Now, using Gaussian integration by parts, and writing as usual $Y = \beta z\sqrt{q}+h$,

$$\begin{aligned}\Phi'(q) &= \beta\,\mathsf{E}\left(\frac{z}{\sqrt{q}}\frac{\mathrm{th}\,Y}{\mathrm{ch}^2\,Y}\right) = \beta^2\,\mathsf{E}\left(\frac{1}{\mathrm{ch}^4\,Y} - 2\frac{\mathrm{sh}^2\,Y}{\mathrm{ch}^4\,Y}\right)\\ &= \beta^2\,\mathsf{E}\left(\frac{3}{\mathrm{ch}^4\,Y} - \frac{2}{\mathrm{ch}^2\,Y}\right) = \beta^2(3(1-2\,q+\widehat{q}) - 2(1-q))\\ &= \beta^2(1-4\,q+3\,\widehat{q})\,,\end{aligned} \tag{1.274}$$

and this finishes the proof. □

Lemma 1.9.4. *We recall the quantities A^2, B^2, C^2 of Section 1.8. Then, under (1.266) we have*

$$\nu\left(\left(\frac{(\boldsymbol{\sigma}^1-\boldsymbol{\sigma}^2)(\boldsymbol{\sigma}^3-\boldsymbol{\sigma}^4)}{N}\right)^2\right) = \frac{4}{N}\frac{(1-2q+\widehat{q})}{(1-\beta^2(1-2q+\widehat{q}))} + \mathcal{R} \tag{1.275}$$

$$\nu((R_{1,2}-q)^2) = A^2 + 2B^2 + C^2 + \mathcal{R} \tag{1.276}$$

where

$$|\mathcal{R}| \le K(\beta,h)\left(\frac{1}{N^{3/2}} + \nu(|R_{1,2}-q|^3)\right)\,. \tag{1.277}$$

Proof. As explained at the beginning of Section 1.8 all the computations there are done modulo an error term as in (1.277); and (1.266) and (1.272) show that we are permitted to divide by $(1-\beta^2(1-2q+\widehat{q}))$ and $(1-\beta^2(1-4q+\widehat{q}))$, so that (1.275) and (1.276) are what we actually proved in (1.238) and (1.265) respectively. □

Proof of Proposition 1.9.2. We deduce from (1.275) that

$$\mathcal{R} \geq -\frac{4}{N}\frac{(1-2q+\widehat{q})}{(1-\beta^2(1-2q+\widehat{q}))} \geq \frac{1}{KN}$$

because $1-2q+\widehat{q} = \mathsf{E}\mathrm{ch}^{-4}Y > 0$ and the denominator is < 0 by (1.266). It follows from (1.277) that for N large enough

$$\nu(|R_{1,2}-q|^3) \geq \frac{1}{KN} . \tag{1.278}$$

Now since $A^2, B^2, C^2 \leq K/N$, (1.276) shows that, using (1.278) in the third following inequality,

$$\nu((R_{1,2}-q)^2) \leq \frac{K}{N} + \mathcal{R} \leq K\left(\frac{1}{N} + \nu(|R_{1,2}-q|^3)\right) \leq K\nu(|R_{1,2}-q|^3) ,$$

and this proves that there exists δ, that does not depend on N, such that $\nu(|R_{1,2}-q|^3) \geq \delta\nu((R_{1,2}-q)^2)$. Moreover, using (1.278), and since $|R_{1,2}-q| \leq 2$ we have $\nu((R_{1,2}-q)^2) \geq \nu(|R_{1,2}-q|^3)/2 \geq 1/(KN)$. □

We might think that the unpleasant behavior (1.271) arises from the fact that $\nu(|R_{1,2}-x|) \simeq 0$ for some $x \neq q$. This is not the case.

Proposition 1.9.5. *a) There exists K depending on β and h only such that for all $x \geq 0$ we have*

$$|x-q| \leq K\left(\nu(|R_{1,2}-x|) + \frac{1}{N}\right) . \tag{1.279}$$

b) Under (1.266) there exists a number δ' such that for N large enough

$$\forall x \geq 0 , \ \nu\big(\mathbf{1}_{\{|R_{1,2}-x| \geq \delta'\}}\big) \geq \frac{\delta'}{N} . \tag{1.280}$$

Proof. We use (1.215), but where ν_t is defined using x rather than q, to get

$$|\nu(\varepsilon_1\varepsilon_2) - \nu_0(\varepsilon_1\varepsilon_2)| \leq K\nu(|R^-_{1,2}-x|) . \tag{1.281}$$

We have $\nu_0(\varepsilon_1\varepsilon_2) = \Phi(x)$, where Φ is given by (1.273), and $\nu(\varepsilon_1\varepsilon_2) = \nu(R_{1,2})$ by symmetry among sites, and therefore (1.281) implies

$$|\Phi(x) - \nu(R_{1,2})| \leq K\left(\nu(|R_{1,2}-x|) + \frac{1}{N}\right) . \tag{1.282}$$

Jensen's inequality entails

$$|\nu(R_{1,2}) - x| \leq \nu(|R_{1,2}-x|) ,$$

so (1.282) yields

$$|\Phi(x) - x| \le |\Phi(x) - \nu(R_{1,2})| + |\nu(R_{1,2}) - x| \le K\left(\nu(|R_{1,2} - x|) + \frac{1}{N}\right) .$$

Now, the function $\Phi(x)$ satisfies $\Phi(q) = q$ and, as seen in the proof of Lemma 1.9.3 we have $\Phi'(q) < 1$, so that $|x - \Phi(x)| \ge K^{-1}|x - q|$ when $|x - q|$ is small. Since Proposition A.14.1 shows that $\Phi(x)/x$ is decreasing it follows that $|x - \Phi(x)| \ne 0$ for $x \ne q$, and the previous inequality holds for all $x \ge 0$ and this proves (1.279).

To prove (1.280), we observe that if $|x - q| \le \delta/4$ then by (1.271) we have

$$\nu\big(\mathbf{1}_{\{|R_{1,2}-x|\ge\delta/4\}}\big) \ge \nu\big(\mathbf{1}_{\{|R_{1,2}-q|\ge\delta/2\}}\big) \ge \frac{\delta^2}{16N} ,$$

so it is enough to consider the case $|x - q| \ge \delta/4$. But then by (1.279) it holds

$$\frac{\delta}{4} \le K\left(\nu(|R_{1,2} - x|) + \frac{1}{N}\right) ,$$

so that for N large enough we get $\nu(|R_{1,2} - x|) \ge \delta/(8K) := 1/K_0$ and thus since $|R_{1,2} - x| \le 2$ we obtain

$$\frac{1}{K_0} \le \nu(|R_{1,2} - x|) \le 2\nu\big(\mathbf{1}_{\{|R_{1,2}-x|\ge 1/(2K_0)\}}\big) + \frac{1}{2K_0} .$$

Consequently,

$$\nu\big(\mathbf{1}_{\{|R_{1,2}-x|\ge 1/2K_0\}}\big) \ge \frac{1}{4K_0} . \qquad \square$$

In the rest of this section we show that (1.280) has consequences with a nice physical interpretation (although the underlying mathematics is elementary large deviation theory).

For this we consider the Hamiltonian

$$-H_{N,\lambda}(\boldsymbol{\sigma}^1, \boldsymbol{\sigma}^2) = -H_N(\boldsymbol{\sigma}^1) - H_N(\boldsymbol{\sigma}^2) + \lambda N R_{1,2} . \tag{1.283}$$

This is the Hamiltonian of a system made from two copies of (Σ_N, G_N) that *interact* through the term $\lambda N R_{1,2}$. We define

$$Z_{N,\lambda} = \sum_{\boldsymbol{\sigma}^1,\boldsymbol{\sigma}^2} \exp(-H_{N,\lambda}(\boldsymbol{\sigma}^1, \boldsymbol{\sigma}^2)) \tag{1.284}$$

$$\psi_N(\lambda) = \frac{1}{N}\mathsf{E}\log Z_{N,\lambda} - \frac{1}{N}\mathsf{E}\log Z_{N,0} \tag{1.285}$$

so that the identity

$$\psi_N(\lambda) = \frac{1}{N}\mathsf{E}\log\langle\exp \lambda N R_{1,2}\rangle \tag{1.286}$$

holds, where $\langle\cdot\rangle$ denotes an average for the Gibbs measure with Hamiltonian H_N. This quantity is natural to consider to study the fluctuations of $R_{1,2}$. We denote by $\langle\cdot\rangle_\lambda$ an average for the Gibbs measure with Hamiltonian (1.283); thus

$$\psi'_N(\lambda) = \mathsf{E}\,\frac{\langle R_{1,2}\exp\lambda N R_{1,2}\rangle}{\langle\exp\lambda N R_{1,2}\rangle} = \mathsf{E}\langle R_{1,2}\rangle_\lambda\,.$$

We also observe that ψ_N is a convex function of λ, as is obvious from (1.286) and Hölder's inequality. (One can also compute $\psi''_N(\lambda) = N(\mathsf{E}\langle R_{1,2}^2\rangle_\lambda - \mathsf{E}\langle R_{1,2}\rangle_\lambda^2) \geq 0$.)

Theorem 1.9.6. $\psi(\lambda) = \lim_{N\to\infty}\psi_N(\lambda)$ *exists for all* β, h *and (under (1.266)) is* **not** *differentiable at* $\lambda = 0$.

The important part of Theorem 1.9.6 is the non-differentiability of the function ψ. We shall prove the following, of which Theorem 1.9.6 is an immediate consequence once we know the existence of the limit $\lim_{N\to\infty}\psi_N(\lambda)$. The existence of this limit is only a side story in Theorem 1.9.6. It requires significant work, so we refer the reader to [76] for a proof.

Proposition 1.9.7. *Assume (1.266), and consider* δ' *as in (1.280). Then for any* $\lambda > 0$ *we have* $\psi'_N(\lambda) - \psi'_N(-\lambda) > \delta'/2$ *provided* N *is large enough.*

To deduce Theorem 1.9.6, consider the subset U of $\mathbb{R}$ such that $\psi'(\pm\lambda)$ exists for $\lambda \in U$. Since ψ is convex, the complement of U is at most countable. Griffiths' lemma (see page 25) asserts that $\lim_{N\to\infty}\psi'_N(\pm\lambda) = \psi'(\pm\lambda)$ for λ in U. By Proposition 1.9.7 for any $\lambda \in U$, $\lambda > 0$, we have $\psi'(\lambda) - \psi'(-\lambda) \geq \delta'/2$. Now, since ψ is convex, the limit $\lim_{\lambda\to 0_+,\lambda\in U}\psi'(\lambda)$ is the right-derivative $\psi'_+(0)$ and similarly, while the limit $\lambda \to 0_-$ is the left derivative $\psi'_-(0)$. Therefore $\psi'_+(0) - \psi'_-(0) > \delta'/2$ and ψ is not differentiable at 0.

In words, an arbitrarily small change of λ around 0 produces a change in $\mathsf{E}\langle R_{1,2}\rangle_\lambda$ of at least $\delta'/2$, a striking instability.

Proof of Proposition 1.9.7. Let $x_N = \mathsf{E}\langle R_{1,2}\rangle = \psi'_N(0)$. Using (1.280) we see that at least one of the following occurs

$$\nu\big(\mathbf{1}_{\{R_{1,2}\geq x_N+\delta'\}}\big) \geq \frac{\delta'}{2N} \tag{1.287}$$

$$\nu\big(\mathbf{1}_{\{R_{1,2}\leq x_N-\delta'\}}\big) \geq \frac{\delta'}{2N}\,. \tag{1.288}$$

We assume (1.287); the proof in the case (1.288) is similar. We have

$$\langle\exp\lambda N R_{1,2}\rangle \geq \exp\lambda N(x_N+\delta')\big\langle\mathbf{1}_{\{R_{1,2}\geq x_N+\delta'\}}\big\rangle$$

so that

$$\frac{1}{N}\log\langle\exp\lambda N R_{1,2}\rangle \geq \lambda(x_N+\delta') + \frac{1}{N}\log\big\langle\mathbf{1}_{\{R_{1,2}\geq x_N+\delta'\}}\big\rangle\,.$$

The r.v. $X = \langle \mathbf{1}_{\{R_{1,2} \geq x_N + \delta'\}} \rangle$ satisfies $\mathsf{E}X \geq \delta'/2N$ by (1.287), so that, since $X \leq 1$, we have

$$\mathsf{P}\left(X \geq \frac{\delta'}{4N}\right) \geq \frac{\delta'}{4N} .$$

(Note that $\mathsf{E}X \leq \varepsilon + \mathsf{P}(X \geq \varepsilon)$ for each ε and take $\varepsilon = \delta'/(4N)$.) Thus

$$\mathsf{P}\left(\frac{1}{N}\log\langle\exp\lambda N R_{1,2}\rangle \geq \lambda(x_N + \delta') + \frac{1}{N}\log\left(\frac{\delta'}{4N}\right)\right) \geq \frac{\delta'}{4N} . \tag{1.289}$$

On the other hand the r.v.

$$F = \frac{1}{N}\log\langle\exp\lambda N R_{1,2}\rangle = \frac{1}{N}\log Z_{N,\lambda} - \frac{1}{N} Z_{N,0}$$

satisfies

$$\forall x > 0\,,\ \mathsf{P}(|F - \mathsf{E}F| \geq x) \leq 2\exp\left(-\frac{Nx^2}{K}\right)$$

by Proposition 1.3.5 (as used in the proof of (1.54)). Taking $x = \lambda\delta'/4$ we get

$$\mathsf{P}\left(\left|\frac{1}{N}\log\langle\exp\lambda N R_{1,2}\rangle - \psi_N(\lambda)\right| \geq \frac{\lambda\delta'}{4}\right) \leq 2\exp\left(-\frac{N}{K}\right) ,$$

and in particular

$$\mathsf{P}\left(\frac{1}{N}\log\langle\exp\lambda N R_{1,2}\rangle \geq \psi_N(\lambda) + \frac{\lambda\delta'}{4}\right) \leq 2\exp\left(-\frac{N}{K}\right) .$$

Comparing with (1.289) implies that for N large enough we have

$$\psi_N(\lambda) + \frac{\lambda\delta'}{4} \geq \lambda(x_N + \delta') + \frac{1}{N}\log\left(\frac{\delta'}{4N}\right) ,$$

and in particular

$$\psi_N(\lambda) \geq \lambda\left(x_N + \frac{\delta'}{2}\right)$$

and therefore, since $\psi_N(0) = 0$ and ψ_N is convex,

$$\psi'_N(\lambda) \geq \frac{\psi_N(\lambda) - \psi_N(0)}{\lambda} \geq x_N + \frac{\delta'}{2} = \psi'_N(0) + \frac{\delta'}{2} \geq \psi'_N(-\lambda) + \frac{\delta'}{2} . \qquad \square$$

1.10 Central Limit Theorem for the Overlaps

In this section we continue the work of Section 1.8 but now for higher moments. We show that the quantities $\sqrt{N}T_{\ell,\ell'}$, $\sqrt{N}T_\ell$, $\sqrt{N}T$ of (1.245) behave asymptotically as $N \to \infty$ like an independent family of Gaussian r.v.s. As a consequence, we show that as $N \to \infty$, for the typical disorder, a given family of variables

$$(N^{1/2}(R_{\ell,\ell'} - \langle R_{1,2} \rangle))_{1 \le \ell < \ell' \le n}$$

becomes nearly Gaussian on the probability space (Σ_N^n, G_N^n) with an explicit covariance structure that is independent of the disorder. Moreover, the r.v.s $N^{1/2}(\langle R_{1,2} \rangle - q)$ are asymptotically Gaussian.

Since all the important ideas were laid down in Section 1.8, the point is really to write things down properly. This requires significant work. This work is not related to any further material in this volume, so the reader who may not enjoy these calculations should simply skip the rest of this section.

We recall that the quantities A, B, C have been defined in (1.248), (1.256), (1.262), and we use the notation $a(k) = \mathsf{E} g^k$ where g is a standard Gaussian r.v.

Theorem 1.10.1. *Assume that $\beta < 1/2$. Fix an integer n. For $1 \le \ell < \ell' \le n$ consider integers $k(\ell, \ell')$, and for $1 \le \ell \le n$ consider integers $k(\ell)$. Set*

$$k_1 = \sum_{1 \le \ell < \ell' \le n} k(\ell, \ell') \; ; \; k_2 = \sum_{1 \le \ell \le n} k(\ell) \, ,$$

consider a further integer k_3 and finally set $k = k_1 + k_2 + k_3$. Then

$$\nu\Big(\prod_{1 \le \ell < \ell' \le n} T_{\ell,\ell'}^{k(\ell,\ell')} \prod_{1 \le \ell \le n} T_\ell^{k(\ell)} T^{k_3} \Big) \tag{1.290}$$
$$= \prod_{1 \le \ell < \ell' \le n} a(k(\ell,\ell')) \prod_{1 \le \ell \le n} a(k(\ell)) \, a(k_3) A^{k_1} B^{k_2} C^{k_3} + O(k+1) \, .$$

Here, as usual, $O(k+1)$ denotes a quantity W with $|W| \le K N^{-(k+1)/2}$, where K does not depend on N (but will depend on the integers $k(\ell,\ell'), k(\ell), k_3$). The left-hand side and the first term in the right-hand side of (1.290) are both of order $N^{-k/2}$. The product

$$\prod_{1 \le \ell < \ell' \le n} T_{\ell,\ell'}^{k(\ell,\ell')} \prod_{1 \le \ell \le n} T_\ell^{k(\ell)} T^{k_3}$$

is simply any (finite) product of quantities of the type $T_{\ell,\ell'}, T_\ell, T$, and the rôle of the integer n is simply to record "on how many replicas this product depends", which is needed to apply the cavity method.

One can reformulate (1.290) as follows. Consider independent Gaussian r.v.s $U_{\ell,\ell'}, U_\ell, U$ and assume

$$\mathsf{E} U_{\ell,\ell'}^2 = NA^2 \;\; ; \;\; \mathsf{E} U_\ell^2 = NB^2 \;\; ; \;\; \mathsf{E} U^2 = NC^2 \, .$$

The point of this definition is that the quantities NA^2, NB^2, and NC^2 do not depend on N. Then (1.290) means

$$N^{k/2}\nu\Big(\prod_{1\le\ell<\ell'\le n} T_{\ell,\ell'}^{k(\ell,\ell')} \prod_{1\le\ell\le n} T_\ell^{k(\ell)}\, T^{k_3}\Big)$$
$$= \mathsf{E}\Big(\prod_{1\le\ell<\ell'\le n} U_{\ell,\ell'}^{k(\ell,\ell')} \prod_{1\le\ell\le n} U_\ell^{k(\ell)}\, U^{k_3}\Big) + O(1)\ , \qquad (1.291)$$

where, in agreement with the notation for $O(k)$ (see above) $O(1)$ denotes a quantity W such that $|W| \le K/\sqrt{N}$.

We now explain why this statement contains the fact that the r.v.s $(N^{1/2}(R_{\ell,\ell'} - \langle R_{1,2}\rangle))_{1\le\ell<\ell'\le n}$ are asymptotically Gaussian under Gibbs' measure. We still consider numbers $k(\ell,\ell')$ for $1\le\ell<\ell'\le n$ and numbers $k(\ell)$ for $1\le\ell\le n$, and let $k = \sum_{1\le\ell<\ell'\le n} k(\ell,\ell') + \sum_{1\le\ell\le n} k(\ell)$. First we show that the quantity

$$V = N^{k/2}\Big\langle \prod_{1\le\ell<\ell'\le n} T_{\ell,\ell'}^{k(\ell,\ell')} \prod_{1\le\ell\le n} T_\ell^{k(\ell)}\Big\rangle \qquad (1.292)$$

is essentially non-random. Indeed, we use replicas to express V^2 as

$$V^2 = N^k\Big\langle \prod_{1\le\ell<\ell'\le n} T_{\ell,\ell'}^{k(\ell,\ell')} \prod_{1\le\ell\le n} T_\ell^{k(\ell)} \prod_{1\le\ell<\ell'\le n} T_{\ell+n,\ell'+n}^{k(\ell,\ell')} \prod_{1\le\ell\le n} T_{\ell+n}^{k(\ell)}\Big\rangle\ ,$$

and we apply (1.291) to compute $\mathsf{E}V^2$ and $\mathsf{E}V$; we thus obtain:

$$\mathsf{E}V^2 = (\mathsf{E}V)^2 + O(1)\ , \qquad (1.293)$$

because the r.v.s $U_{\ell,\ell'}, U_\ell$ for $\ell,\ell' \le n$ are independent from the r.v.s $U_{\ell,\ell'}, U_\ell$ for $n+1 \le \ell,\ell'$. Consequently,

$$\mathsf{E}(V - \mathsf{E}V)^2 = O(1)\ . \qquad (1.294)$$

Let $\overline{q} = \langle R_{1,2}\rangle$, and observe that $T = \overline{q} - q$, so that by (1.246) we obtain

$$R_{\ell,\ell'} - \overline{q} = T_{\ell,\ell'} + T_\ell + T_{\ell'}\ . \qquad (1.295)$$

When a product $\prod_{\ell,\ell'}(R_{\ell,\ell'} - \overline{q})^{k(\ell,\ell')}$ contains k factors, the quantity

$$W = N^{k/2}\Big\langle\prod_{\ell,\ell'}(R_{\ell,\ell'} - \overline{q})^{k(\ell,\ell')}\Big\rangle$$

satisfies

$$\mathsf{E}(W - \mathsf{E}W)^2 = O(1)\ ,$$

because we may use (1.295) and expand this quantity as a sum of terms of the type (1.292). Consider the r.v.s $g_{\ell,\ell'} = U_{\ell,\ell'} + U_\ell + U_{\ell'}$. Expanding and using (1.291) we see that

$$\mathsf{E}W = \mathsf{E}\prod_{\ell,\ell'} g_{\ell,\ell'}^{k(\ell,\ell')} + O(1)\ .$$

The above facts make precise the statement that for the typical disorder, the quantities $(\sqrt{N}(R_{\ell,\ell'} - \overline{q}))_{1\le \ell<\ell'\le n}$ are asymptotically Gaussian on the space (Σ_N^n, G_N^n). Indeed, for the typical disorder,

$$N^{k/2}\Big\langle \prod_{\ell,\ell'} (R_{\ell,\ell'} - \overline{q})^{k(\ell,\ell')} \Big\rangle \simeq \mathsf{E} \prod_{\ell,\ell'} g_{\ell,\ell'}^{k(\ell,\ell')} .$$

The Gaussian family $(g_{\ell,\ell'})$ may also be described by the following properties: $\mathsf{E} g^2_{\ell,\ell'} = N(A^2 + 2B^2)$, $\mathsf{E} g_{\ell,\ell'} g_{\ell_1,\ell_2} = NB^2$ if $\mathrm{card}(\{\ell,\ell'\} \cap \{\ell_1,\ell_2\}) = 1$ and $\mathsf{E} g_{\ell,\ell'} g_{\ell_1,\ell_2} = 0$ if $\{\ell,\ell'\}$ and $\{\ell_1,\ell_2\}$ are disjoint.

We now prepare for the proof of Theorem 1.10.1. In the next two pages, until we start the proof itself, the letter k denotes any integer. We start the work with the easy part, which is the control of the error terms. By (1.88), for each k we have $\nu\big((R_{1,2} - q)^{2k}\big) \le K/N^k$, and thus

$$\nu\big(|R_{1,2} - q|^k\big) \le \frac{K}{N^{k/2}} .$$

Consequently,

$$\nu\big(|R^-_{1,2} - q|^k\big) \le \frac{K}{N^{k/2}} , \tag{1.296}$$

and this entails a similar bound for any quantity W that is a linear combination of the quantities $R^-_{\ell,\ell'} - q$, e.g. $W = R^-_{1,3} - R^-_{2,3}$.

Let us say that a function f *is of order* k if it is the product of k such quantities $W_1, \ldots, W_k$ and of a function W_0 with $|W_0| \le 4$. The reason for the condition $|W_0| \le 4$ is simply that typical choices for W_0 will be $W_0 = 1$ and $W_0 = (\varepsilon_1 - \varepsilon_2)(\varepsilon_3 - \varepsilon_4)$ (and this latter choice satisfies $|W_0| \le 4$). Thus a typical example of a function of order 1 is $(R^-_{\ell,\ell'} - q)(\varepsilon_1 - \varepsilon_2)(\varepsilon_3 - \varepsilon_4)$.

As a consequence of (1.296) and Hölder's inequality, if f is a function of order k then we have

$$\nu(f^2)^{1/2} = O(k) . \tag{1.297}$$

In words, each factor W_ℓ for $\ell = 1, \ldots, k$ contributes as $N^{-1/2}$ while the factor W_0 contributes as a constant. Consequently, (1.215) and (1.216) used for $\tau_1 = \tau_2 = 2$ imply that a function f of order k satisfies

$$\nu(f) = \nu_0(f) + O(k+1) \tag{1.298}$$

$$\nu(f) = \nu_0(f) + \nu_0'(f) + O(k+2) . \tag{1.299}$$

In order to avoid repetition, we will spell out the exact property we will use in the proof of Theorem 1.10.1. The notation is as in Lemma 1.8.2.

Lemma 1.10.2. *Consider integers $x, y \le n$ as well as a function f^- on Σ_N^n, which is the product of k terms of the type $R^-_{\ell,\ell'} - q$. Then the identity*

$$\begin{aligned}\nu((\varepsilon_x\varepsilon_y - q)f^-) = &\sum_{1\le\ell<\ell'\le n} b(\ell,\ell';x,y)\nu(f^-(R^-_{\ell,\ell'} - q))\\ &- n\sum_{\ell\le n} b(\ell,n+1;x,y)\nu(f^-(R^-_{\ell,n+1} - q))\\ &+ \frac{n(n+1)}{2} b(0)\nu(f^-(R^-_{n+1,n+2} - q)) + O(k+2) \quad (1.300)\end{aligned}$$

holds.

Proof. We use (1.299) for $f = f^-(\varepsilon_x\varepsilon_y - q)$, so that $\nu_0(f) = 0$ and we use (1.230) to compute $\nu'_0(f)$. We then use (1.298) with $k+1$ instead of k to see that $\nu_0(f^-(R^-_{\ell,\ell'} - q)) = \nu(f^-(R^-_{\ell,\ell'} - q)) + O(k+2)$. □

Of course Lemma 1.10.2 remains true when f is a product of k terms which are linear combinations of terms of the type $R^-_{\ell,\ell'} - q$.

Corollary 1.10.3. *Consider a function f^- on Σ^n_N, that is the product of k terms of the type $R^-_{\ell,\ell'} - q$. Then we have*

$$\begin{aligned}&\nu((\varepsilon_1-\varepsilon_2)(\varepsilon_3-\varepsilon_4)f^-)\\ &= (b(2) - 2b(1) + b(0))\nu((R^-_{1,3} - R^-_{1,4} - R^-_{2,3} + R^-_{2,4})f^-)\\ &+ O(k+2)\ . \qquad (1.301)\end{aligned}$$

Moreover, whenever f^- does not depend on the third replica $\boldsymbol{\sigma}^3$ we also have

$$\begin{aligned}\nu((\varepsilon_1-\varepsilon_2)\varepsilon_3 f^-) = &(b(2) - 4b(1) + 3b(0))\nu((R^-_{1,3} - R^-_{2,3})f^-)\\ &+ (b(1) - b(0))\sum_{4\le\ell\le n}\nu((R^-_{1,\ell} - R^-_{2,\ell} - R^-_{1,n+1} + R^-_{2,n+1})f^-)\\ &+ O(k+2)\ . \qquad (1.302)\end{aligned}$$

Proof. It suffices to reproduce the calculations of Lemmas 1.8.4 and 1.8.9, using (1.300) in place of (1.230). □

A fundamental idea in the proof of Theorem 1.10.1 is that we should attack first a term $T_{\ell,\ell'}$ (if there is any), using symmetry among sites to write the left-hand side of (1.290) as $\nu((\varepsilon_1-\varepsilon_2)(\varepsilon_3-\varepsilon_4)f)$ for a suitable f. The goal is then to use (1.301); for this, one has to understand the influence of the dependence of f on the last coordinate. This requires the knowledge of (1.290), but where k_1 has been decreased, opening the door to induction. If no term $T_{\ell,\ell'}$ is present, one instead attacks a term T_ℓ (if there is any). We will then have to use the more complicated formula (1.302) rather than (1.301), but this is compensated by the fact that, by the previous step, we already know (1.290) when $k_1 > 0$, so that the values of many of the terms resulting of the use of (1.302) are already known. Finally, if there is no term $T_{\ell,\ell'}$ or T_ℓ in the left-hand side of (1.290), we are forced to use the formidable formula (1.300) itself, but we will be saved by the fact that most of the resulting terms have already been computed in the previous steps. If one thinks about it, this is exactly the way we have proceeded in Section 1.8.

We now start the proof of Theorem 1.10.1, the notation of which we use; in particular:

$$k = k_1 + k_2 + k_3 \ .$$

Proposition 1.10.4. *We have*

$$\nu\Big(\prod_{1\le\ell<\ell'\le n} T_{\ell,\ell'}^{k(\ell,\ell')} \prod_{1\le\ell\le n} T_\ell^{k(\ell)} T^{k_3}\Big) \tag{1.303}$$

$$= \prod_{1\le\ell<\ell'\le n} a(k(\ell,\ell'))A^{k_1}\nu\Big(\prod_{1\le\ell\le n} T_\ell^{k(\ell)} T^{k_3}\Big) + O(k+1) \ .$$

Proof. This is true for $k_1 = 0$. The proof goes by induction on k_1. We assume $k_1 > 0$, and, without loss of generality, we assume $k(1,2) \ge 1$. Before starting any computation, we must address the fact that $T_{\ell,\ell'}$ and T_ℓ depend on the disorder, and we must express the right-hand side of (1.303) as the average of a non-random function on $\Sigma_N^{n'}$ for a certain n'. It eases notation to label properly the terms with which we are working, and to enumerate them as a sequence. For $v \le k_1$ we consider two integers $\ell(v)$ and $\ell'(v)$ in such a way that each pair (ℓ,ℓ') for $1\le\ell<\ell'\le n$ is equal to the pair $(\ell(v),\ell'(v))$ for exactly $k(\ell,\ell')$ values of $v \le k_1$, and that

$$(\ell(v),\ell'(v)) = (1,2) \Leftrightarrow v \le k(1,2) \ .$$

It then holds that

$$\prod_{\ell,\ell'} T_{\ell,\ell'}^{k(\ell,\ell')} = \prod_{v\le k_1} T_{\ell(v),\ell'(v)} \ .$$

For $k_1 < v \le k_1 + k_2$ we consider an integer $\ell(v)$ such that for each $\ell \le n$, we have $\ell = \ell(v)$ for exactly $k(\ell)$ values of v. It then holds that

$$\prod_\ell T_\ell^{k(\ell)} = \prod_{k_1<v\le k_1+k_2} T_{\ell(v)} \ . \tag{1.304}$$

Now we shall use on a massive scale the technique described on page 88 of "replacing each copy of $\mathbf{b}$ by $\boldsymbol\sigma^\ell$ for a new value of ℓ". For this purpose for $1 \le v \le k$ we consider for two integers $j(v), j'(v) > n$, such that these are all distinct as v varies. For $v \le k_1$ we set

$$U(v) = \frac{(\boldsymbol\sigma^{\ell(v)} - \boldsymbol\sigma^{j(v)})\cdot(\boldsymbol\sigma^{\ell'(v)} - \boldsymbol\sigma^{j'(v)})}{N}$$
$$= R_{\ell(v),\ell'(v)} - R_{\ell(v),j'(v)} - R_{j(v),\ell'(v)} + R_{j(v),j'(v)} \ ;$$

for $k_1 < v \le k_1 + k_2$ we set

$$U(v) = \frac{(\boldsymbol\sigma^{\ell(v)} - \boldsymbol\sigma^{j(v)})\cdot\boldsymbol\sigma^{j'(v)}}{N} = R_{\ell(v),j'(v)} - R_{j(v),j'(v)} \ ;$$

and for $k_1 + k_2 < v \le k = k_1 + k_2 + k_3$ we set

$$U(v) = \frac{\boldsymbol{\sigma}^{j(v)} \cdot \boldsymbol{\sigma}^{j'(v)}}{N} - q = R_{j(v),j'(v)} - q \,.$$

As we just assumed, the integers $j(v), j'(v)$ are all distinct for $v \le k$. Averaging first in all the $\boldsymbol{\sigma}^\ell$'s for ℓ any one of these integers we see that

$$\Big\langle \prod_{1\le \ell<\ell'\le n} T_{\ell,\ell'}^{k(\ell,\ell')} \prod_{\ell\le n} T_\ell^{k(\ell)} T^{k_3} \Big\rangle = \Big\langle \prod_{v\le k} U(v) \Big\rangle \tag{1.305}$$

and $\prod_{v\le k} U(v)$ is now independent of the disorder. This quantity can be considered as a function on $\Sigma_N^{n'}$, where n' is any integer larger than all the numbers $j(v)$ and $j'(v)$.

Let us define

$$\varepsilon(v) = (\varepsilon_{\ell(v)} - \varepsilon_{j(v)})(\varepsilon_{\ell'(v)} - \varepsilon_{j'(v)}) \tag{1.306}$$

for $v \le k_1$;

$$\varepsilon(v) = (\varepsilon_{\ell(v)} - \varepsilon_{j(v)})\varepsilon_{j'(v)} \tag{1.307}$$

for $k_1 < v \le k_1 + k_2$; and finally, for $k_1 + k_2 < v \le k + k_1 + k_2 + k_3$, let

$$\varepsilon(v) = \varepsilon_{j(v)}\varepsilon_{j'(v)} \,. \tag{1.308}$$

Using (1.305) and symmetry between sites in the second line,

$$\begin{aligned} V := \nu\Big(\prod_{1\le \ell<\ell'\le n} T_{\ell,\ell'}^{k(\ell,\ell')} \prod_{\ell\le n} T_\ell^{k(\ell)} T^{k_3}\Big) &= \nu\Big(\prod_{v\le k} U(v)\Big) \\ &= \nu\Big(\varepsilon(1) \prod_{2\le v\le k} U(v)\Big) \,. \end{aligned} \tag{1.309}$$

Now we want to bring out the dependence of $\prod_{2\le v\le k} U(v)$ on the last spin. We define $U^-(v) = U(v) - \varepsilon(v)/N$, and we expand the product

$$\begin{aligned} \prod_{2\le v\le k} U(v) &= \prod_{2\le v\le k} \Big(\frac{\varepsilon(v)}{N} + U^-(v)\Big) \\ &= \prod_{2\le v\le k} U^-(v) + \sum_{2\le u\le k} \frac{\varepsilon(u)}{N} \prod_{v\ne u} U^-(v) + \mathcal{S} \end{aligned}$$

where the notation $\prod_{v\ne u} U^-(v)$ means that the product is over $2 \le v \le k$, $v \ne u$, and where $\mathcal{S}$ is the sum of all the other terms. Each term W of $\mathcal{S}$ is the product of $k-1$ factors, each of which being either $\varepsilon(v)/N$ or $U^-(v)$, and at least 2 of these factors are of the type $\varepsilon(v)/N$.

If we do think of each factor $U^-(u)$ as "contributing like $1/\sqrt{N}$" by (1.296) and each factor $\varepsilon(u)/N$ as "contributing like $1/N$"(an argument which is made rigorous by using Hölder's inequality as in the proof of (1.297)) we see that $\nu(|W|) = O(k+1)$. Therefore we have

$$\nu\Bigl(\varepsilon(1)\prod_{2\le v\le k} U(v)\Bigr) = \nu\Bigl(\varepsilon(1)\prod_{2\le v\le k} U^-(v)\Bigr) + \mathrm{I} + O(k+1) \tag{1.310}$$

where

$$\mathrm{I} = \frac{1}{N}\sum_{2\le u\le k} \nu\Bigl(\varepsilon(1)\varepsilon(u)\prod_{v\ne u} U^-(v)\Bigr) . \tag{1.311}$$

We first study the term I. Since the product $\prod_{v\ne u} U^-(v)$ contains $k-2$ factors, the function $\varepsilon(1)\varepsilon(u)\prod_{v\ne u} U^-(v)$ is a function of order $k-2$ as defined four lines above (1.297); and then (1.298) entails

$$\frac{1}{N}\nu\Bigl(\varepsilon(1)\varepsilon(u)\prod_{v\ne u} U^-(v)\Bigr) = \frac{1}{N}\nu_0\Bigl(\varepsilon(1)\varepsilon(u)\prod_{v\ne u} U^-(v)\Bigr) + O(k+1) .$$

Now Lemma 1.6.2 implies

$$\nu_0\Bigl(\varepsilon(1)\varepsilon(u)\prod_{v\ne u} U^-(v)\Bigr) = \nu_0(\varepsilon(1)\varepsilon(u))\nu_0\Bigl(\prod_{v\ne u} U^-(v)\Bigr) .$$

For $u \le k(1,2)$ we have

$$\begin{aligned}\nu_0(\varepsilon(1)\varepsilon(u)) &= \nu_0((\varepsilon_1-\varepsilon_{j(1)})(\varepsilon_2-\varepsilon_{j'(1)})(\varepsilon_1-\varepsilon_{j(u)})(\varepsilon_2-\varepsilon_{j'(u)}))\\ &= 1-2q+\widehat{q} ,\end{aligned}$$

as is seen by computation: we expand the product and compute each of the 16 different terms. One of the terms is 1. Each of the other terms is either of the type $\pm\nu_0(\varepsilon_j\varepsilon_{j'})$ for $j\ne j'$, and hence equal to $\pm q$, or of the type $\pm\nu_0(\varepsilon_{j_1}\varepsilon_{j_2}\varepsilon_{j_3}\varepsilon_{j_4})$ where the indexes j_1, j_2, j_3, j_4 are all different and hence the term is $\pm\widehat{q}$.

For $u > k(1,2)$ we now show that $\nu_0(\varepsilon(1)\varepsilon(u)) = 0$. First one checks on the various cases (1.306) to (1.308) that $\varepsilon(u)$ does not depend on both ε_1 and ε_2. If, say, it does not depend on ε_1, then since $\varepsilon(u)$ does not depend on $\varepsilon_{j(1)}$ (because the integers $j(v), j(v')$ are all distinct and $> n$) the factor $\varepsilon_1 - \varepsilon_{j(1)}$ in $\varepsilon(1)$ ensures that $\langle\varepsilon(1)\varepsilon(u)\rangle = 0$. This proves that, recalling the notation I of (1.311),

$$\begin{aligned}\mathrm{I} &= \frac{1}{N}(1-2q+\widehat{q})\sum_{2\le u\le k(1,2)} \nu_0\Bigl(\prod_{v\ne u} U^-(v)\Bigr) + O(k+1)\\ &= \frac{1}{N}(1-2q+\widehat{q})\sum_{2\le u\le k(1,2)} \nu\Bigl(\prod_{v\ne u} U^-(v)\Bigr) + O(k+1) ,\end{aligned}$$

using again (1.298). Moreover, when $u \le k(1,2)$, by symmetry we have

$$\nu\Big(\prod_{v\neq u} U^-(v)\Big) = \nu\Big(\prod_{3\leq v\leq k} U^-(v)\Big) .$$

Thus we obtain

$$\mathrm{I} = \frac{k(1,2)-1}{N}(1-2q+\widehat{q})\nu\Big(\prod_{3\leq v\leq k} U^-(v)\Big) + O(k+1) . \tag{1.312}$$

Next, we use (1.301) (with indices $1, j(1), 2, j'(1)$ rather than $1, 2, 3, 4$, and with $k-1$ rather than k) to see that

$$\begin{aligned}
&\nu\Big(\varepsilon(1)\prod_{2\leq v\leq k} U^-(v)\Big) \\
&= \beta^2(1-2q+\widehat{q})\nu\Big((R^-_{1,2} - R^-_{1,j'(1)} - R^-_{j(1),2} + R^-_{j(1),j'(1)})\prod_{2\leq v\leq k} U^-(v)\Big) \\
&+ O(k+1) \\
&= \beta^2(1-2q+\widehat{q})\nu\Big(\prod_{1\leq v\leq k} U^-(v)\Big) + O(k+1) .
\end{aligned}$$

Combining with (1.310) and (1.309) we reach the equality

$$\begin{aligned}
&\nu\Big(\varepsilon(1)\prod_{2\leq v\leq k} U(v)\Big) \\
&= \beta^2(1-2q+\widehat{q})\nu\Big(\prod_{1\leq v\leq k} U^-(v)\Big) \\
&+ \frac{k(1,2)-1}{N}(1-2q+\widehat{q})\nu\Big(\prod_{3\leq v\leq k} U^-(v)\Big) + O(k+1) .
\end{aligned}$$

We claim that on the right-hand side we may replace each term $U^-(v)$ by $U(v)$ up to an error of $O(k+1)$. To see this we simply use the relation $U^-(v) = U(v) - \varepsilon(v)/N$ and we expand the products. All the terms except the one where all factors are $U(v)$ are $O(k+1)$, as follows from (1.297). Recalling (1.309) we have proved that

$$(1-\beta^2(1-2q+\widehat{q}))V = \frac{k(1,2)-1}{N}(1-2q+\widehat{q})\nu\Big(\prod_{3\leq v\leq k} U(v)\Big) + O(k+1) . \tag{1.313}$$

The proof is finished if $k(1,2) = 1$, since $a(1) = 0$. If $k(1,2) \geq 2$, we have

$$\nu\Big(\prod_{3\leq v\leq k} U(v)\Big) = \nu\Big(\prod_{1\leq \ell<\ell'\leq n} T_{\ell,\ell'}^{k'(\ell,\ell')} \prod_{1\leq \ell\leq n} T_\ell^{k(\ell)} T^{k_3}\Big) ,$$

where $k'(\ell,\ell') = k(\ell,\ell')$ unless $\ell = 1$, $\ell' = 2$, in which case $k'(1,2) = k(1,2) - 2$. This term is of the same type as the left-hand side of (1.303), but with $k_1 - 2$ instead of k_1. We can therefore apply the induction hypothesis to get

$$\nu\Big(\prod\nolimits_{1\le\ell<\ell'\le n} T_{\ell,\ell'}^{k'(\ell,\ell')} \prod\nolimits_{1\le\ell\le n} T_\ell^{k(\ell)}\, T^{k_3}\Big)$$
$$= \prod\nolimits_{1\le\ell<\ell'\le n} a(k'(\ell,\ell'))A^{k_1-2}\nu\Big(\prod\nolimits_{1\le\ell\le n} T_\ell^{k(\ell)}\, T^{k_3}\Big) + O(k+1) \ .$$

Combining with (1.313) and using the value of A we get

$$V = (k(1,2)-1) \prod_{1\le\ell<\ell'\le n} a(k'(\ell,\ell'))A^{k_1}\nu\Big(\prod_{1\le\ell\le n} T_\ell^{k(\ell)}\, T^{k_3}\Big) + O(k+1) \ .$$

Using that $a(k(1,2)) = (k(1,2)-1)a(k'(1,2))$, this completes the induction and the proof of Proposition 1.10.4. □

Proposition 1.10.5. *With the notation of Theorem 1.10.1 we have*

$$\nu\Big(\prod_{1\le\ell<\ell'\le n} T_{\ell,\ell'}^{k(\ell,\ell')} \prod_{1\le\ell\le n} T_\ell^{k(\ell)}\, T^{k_3}\Big) \tag{1.314}$$
$$= \prod_{1\le\ell<\ell'\le n} a(k(\ell,\ell')) \prod_{1\le\ell\le n} a(k(\ell))A^{k_1}B^{k_2}\nu(T^{k_3}) + O(k+1) \ .$$

Proof. We already know from Proposition 1.10.4 that we can assume that $k_1 = 0$. So we fix $k_1 = 0$ and we prove Proposition 1.10.5 by induction over k_2. Thus assume $k_2 > 0$ and also without loss of generality that $k(1) > 0$. We keep the notation of Proposition 1.10.4. Recalling (1.304) we assume

$$\ell(v) = 1 \Leftrightarrow v \le k(1) \ .$$

Using (1.305) we write, using symmetry between sites,

$$V := \nu\Big(\prod_{1\le\ell\le n} T_\ell^{k(\ell)}\, T^{k_3}\Big) = \nu\Big(\prod_{v\le k} U(v)\Big)$$
$$= \nu\Big(\varepsilon(1)\prod_{2\le v\le k} U(v)\Big) \tag{1.315}$$

and (1.310) remains valid. For $v \le k(1)$ we have

$$\begin{aligned}\nu_0(\varepsilon(1)\varepsilon(v)) &= \nu_0((\varepsilon_1 - \varepsilon_{j(1)})\varepsilon_{j'(1)}(\varepsilon_1 - \varepsilon_{j(v)})\varepsilon_{j'(v)}) \\ &= q - \widehat{q}\end{aligned}$$

and for $v > k(1)$ we have $\nu_0(\varepsilon(1)\varepsilon(v)) = 0$ because $\varepsilon(v)$ does not depend on either ε_1 or $\varepsilon_{j(1)}$. Thus, instead of (1.312) we now have (recalling that the term I has been defined in (1.311))

$$\mathrm{I} = \frac{k(1)-1}{N}(q-\widehat{q})\nu\Big(\prod_{3\le v\le k} U^-(v)\Big) + O(k+1)\,. \tag{1.316}$$

We use (1.302) (with the indices $1, j(1), j'(1)$ rather than $1,2,3$ and with $k-1$ rather than k) to obtain

$$\begin{aligned}
&\nu\Big(\varepsilon(1)\prod_{2\le v\le k} U^-(v)\Big)\\
&= \beta^2(1-4q+3\widehat{q})\nu\Big((R^-_{1,j'(1)} - R^-_{j(1),j'(1)})\prod_{2\le v\le k} U^-(v)\Big)\\
&+\mathrm{II} + O(k+1)
\end{aligned} \tag{1.317}$$

for

$$\mathrm{II} = \beta^2(q-\widehat{q})\sum_{\ell}\nu\Big((R^-_{1,\ell} - R^-_{j(1),\ell} - R^-_{1,n'+1} + R^-_{j(1),n'+1})\prod_{2\le v\le k} U^-(v)\Big)\,, \tag{1.318}$$

where n' is an integer larger than all indices $j(v), j'(v)$, and where the summation is over $2 \le \ell \le n'$, $\ell \ne j(1)$, $\ell \ne j'(1)$.

Compared to the proof of Proposition 1.10.4 the new (and non-trivial) part of the argument is to establish the relation

$$\mathrm{II} = (k(1)-1)\beta^2(q-\widehat{q})\nu\Big(\prod_{3\le v\le k} T_{\ell(v)}\,T^2_{1,n'+1}\Big) + O(k+1) \tag{1.319}$$

and we explain first how to conclude once (1.319) has been established. As usual in (1.316) and (1.317) we can replace $U^-(v)$ by $U(v)$ and $R^-_{\ell,\ell'}$ by $R_{\ell,\ell'}$ with an error $O(k+1)$, so that

$$\mathrm{I} = \frac{k(1)-1}{N}(q-\widehat{q})\nu\Big(\prod_{3\le v\le k} U(v)\Big) + O(k+1)\,,$$

and also

$$\begin{aligned}
\nu\Big((R^-_{1,j'(1)} - R^-_{j(1),j'(1)})\prod_{2\le v\le k} U^-(v)\Big) &= \nu\Big(\prod_{1\le v\le k} U^-(v)\Big)\\
&= \nu\Big(\prod_{1\le v\le k} U(v)\Big) + O(k+1)\\
&= V + O(k+1)\,.
\end{aligned}$$

Combining with (1.310) and (1.317) we get

$$\begin{aligned}
(1-\beta^2(1-4q+3\widehat{q}))V &= (k(1)-1)(q-\widehat{q})\Big(\frac{1}{N}\nu\Big(\prod_{3\le v\le k} U(v)\Big)\\
&+\beta^2\nu\Big(\prod_{3\le v\le k} T_{\ell(v)}\,T^2_{1,n+1}\Big)\Big) + O(k+1)\,.
\end{aligned} \tag{1.320}$$

This completes the proof if $k(1) = 1$ since $a(1) = 0$. If $k(1) \geq 2$, we have

$$\nu\Big(\prod_{3\leq v\leq k} U(v)\Big) = \nu\Big(\prod_{\ell\leq n} T_\ell^{k'(\ell)}\, T^{k_3}\Big)\,, \tag{1.321}$$

where $k'(\ell) = k(\ell)$ for $\ell > 1$ and $k'(1) = k(1) - 2$ if $\ell = 1$. Thus the induction hypothesis implies

$$\frac{1}{N}\nu\Big(\prod_{1\leq \ell\leq n} T_\ell^{k'(\ell)}\, T^{k_3}\Big) = \frac{1}{N}\prod_{1\leq \ell\leq n} a(k'(\ell)) B^{k_2-2} \nu(T^{k_3}) + O(k+1)\,. \tag{1.322}$$

We can use Proposition 1.10.4 and the induction hypothesis to compute the term

$$\nu\Big(\prod_{3\leq v\leq k} T_{\ell(v)}\, T_{1,n'+1}^2\Big)$$

because this term contains only $k_2 - 2$ factors T_ℓ, and we find

$$\nu\Big(\prod_{3\leq v\leq k} T_{\ell(v)}\, T_{1,n'+1}^2\Big) = A^2 \prod_{1\leq \ell\leq n} a(k'(\ell)) B^{k_2-2} \nu(T^{k_3}) + O(k+1)\,. \tag{1.323}$$

Combining (1.320) to (1.323) we get

$$\begin{aligned}&(1-\beta^2(1-4q+3\widehat{q}))V\\ &= (q-\widehat{q})\Big(\frac{1}{N}+\beta^2 A^2\Big)\prod_{1\leq \ell\leq n} a(k'(\ell)) B^{k_2-2}\nu(T^{k_3}) + O(k+1)\,.\end{aligned}$$

Using the relation

$$(1-\beta^2(1-4q+3\widehat{q}))B^2 = (q-\widehat{q})\Big(\frac{1}{N}+\beta^2 A^2\Big)\,,$$

and that $(k(1)-1)a(k(1)-2) = a(k(1))$ then completes the induction.

Now we turn to the proof of (1.319). As usual we have

$$\begin{aligned}\mathrm{II} = \beta^2(q-\widehat{q})\sum_\ell \nu\Big((R_{1,\ell} - R_{j(1),\ell} - R_{1,n'+1} + R_{j(1),n'+1})\prod_{2\leq v\leq k} U(v)\Big)\\ + O(k+1)\,,\end{aligned}$$

where the summation is as in (1.318). Moreover (1.246) implies

$$\begin{aligned}\mathrm{II} = \beta^2(q-\widehat{q})\sum_\ell \nu\Big((T_{1,\ell} - T_{j(1),\ell} - T_{1,n'+1} + T_{j(1),n'+1})\prod_{2\leq v\leq k} U(v)\Big)\\ + O(k+1)\,,\end{aligned} \tag{1.324}$$

and, for $2 \le v \le k_2$

$$\begin{aligned} U(v) &= R_{\ell(v),j'(v)} - R_{j(v),j'(v)} \\ &= T_{\ell(v),j'(v)} - T_{j(v),j'(v)} + T_{\ell(v)} - T_{j(v)} , \end{aligned}$$

while, for $k_2 < v \le k_3$

$$U(v) = R_{j(v),j'(v)} - q = T_{j(v),j'(v)} + T_{j(v)} + T_{j'(v)} + T .$$

This looks complicated, but we shall prove that when we expand the product most of the terms are $O(k+1)$. We know from Proposition 1.10.4 that in order for a term not to be $O(k+1)$, each factor $T_{\ell,\ell'}$ must occur at an *even* power because $a(k) = 0$ for odd k. In order for the terms $T_{1,\ell}$ (or $T_{j(1),\ell}$, or $T_{1,n'+1}$ or $T_{j(1),n'+1}$) to occur at an even power in the expansion, one has to pick the same term again in one of the factors $U(v)$ for $v \ge 2$. Since all the integers $j(v), j'(v)$ are $\le n'$, this is impossible for the terms $T_{1,n'+1}$ and $T_{j(1),n'+1}$.

Can this happen for the term $T_{j(1),\ell}$? We can never have $\{j(1), \ell\} = \{j(v), j'(v)\}$ for $v \ge 2$ because the integers $j(v), j'(v)$ are all distinct. We can never have $\{j(1), \ell\} = \{\ell(v), j'(v)\}$ because $j(1) \notin \{\ell(v), j'(v)\}$ since $j(1) > n$, $\ell(v) \le n$ and $j(1) \ne j'(v)$, so this cannot happen either for this term $T_{j(1),\ell}$.

Can it happen then for the term $T_{1,\ell}$? Since $j(v), j'(v) \ge n$, we can never have $\{1, \ell\} = \{j(v), j'(v)\}$. Since $j'(v) > n$, we have $\{1, \ell\} = \{\ell(v), j'(v)\}$ exactly when $\ell(v) = 1$ and $\ell = j'(v)$. Since $2 \le v \le k$, there are exactly $k(1) - 1$ possibilities for v, namely $v = 2, \ldots, k(1)$. For each of these values of v, there is exactly one possibility for ℓ, namely $\ell = j'(v)$.

So, only for the terms $T_{1,\ell}$ where $\ell \in \{j'(2), \ldots, j'(k(1))\}$ can we pick another copy of this term in the product $\prod_{2 \le v \le k} U(v)$, and this term is found in $U(u)$ for the unique $2 \le u \le k(1)$ for which $\ell = j'(u)$. Therefore in that case we have

$$\nu\Big((T_{1,\ell} - T_{j(1),\ell} - T_{1,n'+1} + T_{j(1),n'+1}) \prod_{2 \le v \le k} U(v) \Big) = \nu\Big(T_{1,\ell}^2 \prod_{v \ne u} U(v) \Big) .$$

Moreover, since $\ell = j'(u)$ we then have $\ell > n$, and since $\ell(v) \le n$ and all the numbers $j(v)$ and $j'(v)$ are distinct, ℓ does not belong to any of the sets $\{\ell(v), j(v), j'(v)\}$ for $v \ne u$, so that, by symmetry between replicas,

$$\nu\Big(T_{1,\ell}^2 \prod_{v \ne u} U(v) \Big) = \nu\Big(T_{1,n'+1}^2 \prod_{3 \le v \le k} U(v) \Big) = \nu\Big(T_{1,n'+1}^2 \prod_{3 \le v \le k} T_{\ell(v)} \Big) ,$$

and since there are exactly $k(1) - 1$ such contributions, this completes the proof of (1.319), hence of Proposition 1.10.5. □

Proof of Theorem 1.10.1. We prove by induction over k that

$$\nu(T^k) = a(k)C^k + O(k+1)$$

where C is as in (1.262). This suffices according to Propositions 1.10.4 and 1.10.5. We write

$$\nu(T^k) = \nu\Big(\prod_{1\le v\le k}(R_{2v-1,2v}-q)\Big) = \nu\Big((\varepsilon_1\varepsilon_2 - q)\prod_{2\le v\le k}(R_{2v-1,2v}-q)\Big) .$$

We proceed as before, using now (1.230) for $n = 2k$ to obtain

$$\nu(T^k) = \mathrm{I} + \mathrm{II} + O(k+1) , \tag{1.325}$$

where, defining $b(\ell,\ell') = b(\mathrm{card}\{\ell,\ell'\}\cap\{1,2\})$, we have

$$\mathrm{I} = \frac{k-1}{N}(\widehat{q}-q^2)\nu(T^{k-2}) \tag{1.326}$$

$$\begin{aligned}\mathrm{II} = &\sum_{1\le\ell<\ell'\le n} b(\ell,\ell')\nu((R_{\ell,\ell'}-q)f)\\ &- n\sum_{\ell\le n} b(\ell,n+1)\nu((R_{\ell,n+1}-q)f)\\ &+ \frac{n(n+1)}{2}b(n+1,n+2)\nu((R_{n+1,n+2}-q)f)\end{aligned} \tag{1.327}$$

for $f = \prod_{2\le v\le k}(R_{2v-1,2v}-q)$. The key computation is the relation

$$\begin{aligned}\mathrm{II} = &\beta^2(1-4q+3\widehat{q})\nu(T^k)\\ &+ (k-1)\big(\beta^2(\widehat{q}-q^2)\nu(T_{1,2}^2 T^{k-2}) + 2\beta^2(2q+q^2-3\widehat{q})\nu(T_1^2\, T^{k-2})\big) .\end{aligned} \tag{1.328}$$

Once this has been proved one can compute the last two terms using the induction hypothesis and Propositions 1.10.4 and 1.10.5, namely

$$\nu(T_{1,2}^2\, T^{k-2}) = A^2 a(k-2)C^{k-2} + O(k+1)$$

and

$$\nu(T_1^2\, T^{k-2}) = B^2 a(k-2)C^{k-2} + O(k+1) .$$

Combining this value of II with (1.325) and (1.326), and using (1.262) one then completes the induction.

It would be nice to have a one-line argument to prove (1.328); maybe such an argument exists if one finds the correct approach, which probably means that one has to solve Research Problem 1.8.3. For the time being, one carefully collects the terms of (1.327). Here are the details of this computation (a more general version of which will be given in Volume II). In order to compute $\nu((R_{\ell,\ell'}-q)f)$ we can replace each factor $R_{2v-1,2v}-q$ of f by T whenever $\{2v-1,2v\}\cap\{\ell,\ell'\}=\emptyset$. Thus we see first that

$$\nu((R_{n+1,n+2}-q)f) = \nu(T^{k-1}(R_{n+1,n+2}-q)) = \nu(T^k) .$$

If w is the unique integer $\leq k$ such that $\ell \in \{2w-1, 2w\}$, then for $w = 1$ (and since f does not contain the factor $R_{1,2}-q$) we have $\nu((R_{\ell,n+1}-q)f) = \nu(T^k)$, whereas for $w \geq 2$ we have

$$\nu((R_{\ell,n+1} - q)f) = \nu\big((R_{\ell,n+1} - q)(R_{2w-1,2w} - q)T^{k-2}\big) ,$$

as is seen simply by averaging first in $\boldsymbol{\sigma}^{2v-1}$ and $\boldsymbol{\sigma}^{2v}$ for $v \neq w$. To compute this term, we use (1.246) to write

$$R_{\ell,n+1} - q = T_{\ell,n+1} + T_\ell + T_{n+1} + T$$

$$R_{2w-1,2w} - q = T_{2w-1,2w} + T_{2w-1} + T_{2w} + T ,$$

and we expand the product of these quantities. Since $a(1) = 0$, the induction hypothesis shows that

$$\nu((R_{\ell,n+1} - q)f) = \nu(T^k) + \nu(T_\ell^2\, T^{k-2}) + O(k+1) .$$

To compute $\nu((R_{\ell,\ell'} - q)f)$ for $1 \leq \ell < \ell' \leq n$, we first consider the case where for some $1 \leq w \leq n$, we have $\ell = 2w - 1$ and $\ell' = 2w$. If $w \geq 2$ we have

$$\nu((R_{\ell,\ell'} - q)f) = \nu((R_{\ell,\ell'} - q)^2 T^{k-2}) .$$

Using again (1.246), the induction hypothesis, and the fact that $a(1) = 0$, we get

$$\begin{aligned}\nu((R_{\ell,\ell'} - q)^2\, T^{k-2}) &= \nu(T^k) + \nu(T_{\ell,\ell'}^2 T^{k-2}) + \nu(T_\ell^2\, T^{k-2}) \\ &\quad + \nu(T_{\ell'}^2\, T^{k-2}) + O(k+1) .\end{aligned}$$

If $w = 1$, we have instead $\nu((R_{\ell,\ell'} - q)f) = \nu((R_{1,2} - q)f) = \nu(T^k)$.

Next we consider the case where $\ell \in \{2w-1, 2w\}$, $\ell' \in \{2w'-1, 2w'\}$ for some $1 \leq w < w'$.

- If $w \geq 2$ we have

$$\nu((R_{\ell,\ell'} - q)f) = \nu\big((R_{\ell,\ell'} - q)(R_{2w-1,2w} - q)(R_{2w'-1,2w'} - q)T^{k-3}\big) ,$$

and proceeding as before we get

$$\nu((R_{\ell,\ell'} - q)f) = \nu(T^k) + \nu(T_\ell^2\, T^{k-2}) + \nu(T_{\ell'}^2\, T^{k-2}) + O(k+1) .$$

- If $w = 1$, we find instead

$$\begin{aligned}\nu((R_{\ell,\ell'} - q)f) &= \nu\big((R_{\ell,\ell'} - q)(R_{2w'-1,2w'} - q)T^{k-2} f\big) \\ &= \nu(T^k) + \nu(T_{\ell'}^2\, T^{k-2}) + O(k+1) .\end{aligned}$$

It remains to gather these terms as in the right-hand side of (1.327). The coefficient of $\nu(T^k)$ is

$$\sum_{\ell<\ell'\le n} b(\ell,\ell') - n\sum_{\ell\le n} b(\ell,n+1) + \frac{n(n+1)}{2} b(n+1,n+2)$$
$$= \beta^2\Big(1-q^2+2(n-2)(q-q^2)+\frac{(n-2)(n-1)}{2}(\widehat{q}-q^2)$$
$$-2n(q-q^2)-(n-2)n(\widehat{q}-q^2)+\frac{n(n+1)}{2}(\widehat{q}-q^2)\Big)$$
$$= \beta^2((1-q^2)-4(q-q^2)+3(\widehat{q}-q^2)) = \beta^2(1-4q+3\widehat{q}) .$$

We observe that $\nu(T^2_{\ell,\ell'}T^{k-2}) = \nu(T^2_{1,2}T^{k-2})$. The coefficient of $\nu(T^2_{1,2}T^{k-2})$ is

$$\sum_{2\le w\le k} b(2w-1,2w) = (k-1)\beta^2(\widehat{q}-q^2) .$$

The coefficient of $\nu(T_1^2 T^{k-2})$ is

$$\beta^2(2(n-2)(q-q^2)+2\frac{(n-3)(n-2)}{2}(\widehat{q}-q^2)-(n-2)n(\widehat{q}-q^2))$$
$$= 2\beta^2(k-1)(2q+q^2-3\widehat{q}) ,$$

since $n-2 = 2(k-1)$.

This completes the proof of (1.328) and of Theorem 1.10.1. □

1.11 Non Gaussian Behavior: Hanen's Theorem

After reading the previous section one could form the impression that every simple quantity defined in terms of a few spins will have asymptotic Gaussian behavior when properly normalized. This however is not quite true. In this section we prove the following remarkable result of A. Hanen, where, as usual, $Y = \beta z\sqrt{q}+h$, q is the root of the equation (1.74), $a(k) = \mathsf{E}g^k$ for $k \in \mathbb{N}$ and a standard Gaussian r.v. g and $\widehat{q} = \mathsf{E}\mathrm{th}^4 Y$.

Theorem 1.11.1. *(A. Hanen [79]) If $\beta < 1/2$, for each k we have*

$$\mathsf{E}(\langle\sigma_1\sigma_2\rangle - \langle\sigma_1\rangle\langle\sigma_2\rangle)^k = a(k)\left(\frac{\beta^2}{N(1-\beta^2(1-2q+\widehat{q}))}\right)^{k/2}\left(\mathsf{E}\frac{1}{\mathrm{ch}^{2k}Y}\right)^2$$
$$+ O(k+1) . \tag{1.329}$$

Of course $O(k+1)$ denotes a quantity U with $|U| \le K/N^{(k+1)/2}$ where K does not depend on N. Since the right-hand side is not of the type $a(k)D^k$, this is not a Gaussian behavior. In fact, the meaning of (1.329) is that in the limit $N\to\infty$ we have

$$\sqrt{N}(\langle\sigma_1\sigma_2\rangle - \langle\sigma_1\rangle\langle\sigma_2\rangle) \stackrel{\mathcal{D}}{=} \frac{\beta}{\sqrt{1-\beta^2(1-2q+\widehat{q})}} g \frac{1}{\mathrm{ch}^2 Y_1}\frac{1}{\mathrm{ch}^2 Y_2} , \tag{1.330}$$

where $\mathcal{D}$ means equality in distribution, g, Y_1 and Y_2 are independent, g is standard Gaussian and Y_1 and Y_2 are independent copies of Y.

Research Problem 1.11.2. A decomposition such as (1.330) can hardly be accidental. Rather, it is likely to arise from some underlying structure. Find it.

In some sense the proof of Theorem 1.11.1 is not very difficult. It relies on the cavity method and Taylor's expansions. On the other hand, it is among the deepest of this volume, and the reader should be comfortable with arguments such as those used in the proof of Theorem 1.7.11 before attempting to follow all the details of the proof.

To start this proof we note that

$$\mathsf{E}(\langle\sigma_1\sigma_2\rangle - \langle\sigma_1\rangle\langle\sigma_2\rangle)^k = \mathsf{E}(\langle\sigma_N\sigma_{N-1}\rangle - \langle\sigma_N\rangle\langle\sigma_{N-1}\rangle)^k \tag{1.331}$$

and that

$$\begin{aligned}\langle\sigma_N\sigma_{N-1}\rangle - \langle\sigma_N\rangle\langle\sigma_{N-1}\rangle &= \frac{1}{2}\langle(\sigma_N^1 - \sigma_N^2)(\sigma_{N-1}^1 - \sigma_{N-1}^2)\rangle \\ &= \frac{1}{2}\langle(\varepsilon_1 - \varepsilon_2)(\sigma_{N-1}^1 - \sigma_{N-1}^2)\rangle \end{aligned} \tag{1.332}$$

where as usual $\varepsilon_\ell = \sigma_N^\ell$. Using replicas, we then have

$$\langle(\varepsilon_1-\varepsilon_2)(\sigma_{N-1}^1-\sigma_{N-1}^2)\rangle^k = \langle(\varepsilon_1-\varepsilon_2)(\varepsilon_3-\varepsilon_4)\cdots(\varepsilon_{2k-1}-\varepsilon_{2k})f^-\rangle\,, \tag{1.333}$$

where $f^- = (\sigma_{N-1}^1 - \sigma_{N-1}^2)\cdots(\sigma_{N-1}^{2k-1} - \sigma_{N-1}^{2k})$.

For $v \geq 1$, let us set $\eta_v = \varepsilon_{2v-1} - \varepsilon_{2v}$, and for a set V of integers let us set

$$\eta_V = \prod_{v\in V}\eta_v\,.$$

Let us also set

$$V^* = \bigcup_{v\in V}\{2v-1, 2v\}\,,$$

so that η_V depends only on the variables ε_ℓ for $\ell \in V^*$. From (1.331) to (1.333) we see the relevance of studying quantities such as $\nu(\eta_V f^-)$ where f^- is a function on Σ_{N-1}^n. These quantities will be studied by making a Taylor expansion of the functions $t \mapsto \nu_t(\eta_V f^-)$ at $t = 0$, where ν_t refers to the interpolating Hamiltonian (1.147). We denote by $\nu_t^{(m)}(f)$ the m-th derivative of the function $t \mapsto \nu_t(f)$, and we first learn how to control these derivatives.

Lemma 1.11.3. *If f is a function on Σ_N^n we have*

$$|\nu_t^{(m)}(f)| \leq \frac{K(m,n)}{N^{m/2}}\nu(f^2)^{1/2}\,. \tag{1.334}$$

Proof. This is because "each derivative brings out a factor $R^-_{\ell,\ell'} - q$ that contributes as $N^{-1/2}$." More formally, by (1.151), counting each term with its order of multiplicity, $\nu_t'(f)$ is the sum of $2n^2$ terms of the type

$$\pm\beta^2\nu_t(\varepsilon_\ell\varepsilon_{\ell'}(R^-_{\ell,\ell'} - q)f) ,$$

where $\ell, \ell' \le n+2$, so that by iteration $\nu_t^{(m)}(f)$ is the sum of at most

$$2n^2(2(n+2)^2)\cdots(2(n+2(m-1))^2)$$

terms of the type

$$\pm\beta^{2m}\nu_t\Big(\prod_{r\le m}\varepsilon_{\ell_r}\varepsilon_{\ell'_r}(R^-_{\ell_r,\ell'_r} - q)f\Big)$$

and we bound each term through Hölder's inequality and (1.103). □

For a set J of integers, we define

$$\varepsilon_J = \prod_{\ell\in J}\varepsilon_\ell .$$

A basic idea is that the quantity $\nu_0(\eta_V\varepsilon_J)$ has a great tendency to be zero, because each factor $\eta_v = \varepsilon_{2v-1} - \varepsilon_{2v}$ gives it a chance. And taking the product by ε_J cannot destroy all these chances if $\mathrm{card}J < \mathrm{card}V$, as is made formal in the next lemma.

Lemma 1.11.4. *Assume that* $\mathrm{card}(V^*\cap J) < \mathrm{card}V$, *and consider a function* $\widehat{f}$ *of* $(\varepsilon_\ell)_{\ell\notin V^*}$. *Then*

$$\nu_0(\eta_V\varepsilon_J\widehat{f}) = 0 .$$

Proof. Recalling the definition of V^* and since $\mathrm{card}(V^*\cap J) < \mathrm{card}V$, there exists v in V such that $\{2v-1, 2v\}\cap J = \emptyset$. Defining $V' = V\setminus\{v\}$ we get

$$\eta_V\varepsilon_J\widehat{f} = \eta_v\eta_{V'}\varepsilon_J\widehat{f} ,$$

where $\eta_{V'}\varepsilon_J\widehat{f}$ depends only on ε_ℓ for $\ell \ne \{2v-1, 2v\}$. Thus

$$\langle\eta_V\varepsilon_J\widehat{f}\rangle_0 = \langle\eta_v\rangle_0\langle\eta_{V'}\varepsilon_J\widehat{f}\rangle_0 = 0$$

because $\langle\eta_v\rangle_0 = 0$. □

As a consequence of Lemma 1.11.4 terms η_V create a great tendency for certain derivatives to vanish.

Lemma 1.11.5. *Consider two integers* r, s *and sets* V *and* J *with* $\mathrm{card}V = s$ *and* $\mathrm{card}(V^*\cap J) \le r$. *Consider a function* $\widehat{f}$ *of* $(\varepsilon_\ell)_{\ell\notin V^*}$ *and a function* f^- *on* Σ^n_{N-1}. *Then for* $2m + r < s$ *we have*

$$\nu_0^{(m)}(\eta_V\varepsilon_J\widehat{f}f^-) = 0 .$$

Proof. Lemma 1.6.2 implies the following fundamental equality:

$$\nu_0(\eta_V \varepsilon_J \widehat{f} f^-) = \nu_0(\eta_V \varepsilon_J \widehat{f})\nu_0(f^-) .$$

Therefore for $m = 0$ the conclusion follows from Lemma 1.11.4 since $\nu_0^{(0)} = \nu_0$. The proof is then by induction over m. We simply observe that $\nu_0'(\eta_V \varepsilon_J \widehat{f} f^-)$ is a sum of $2n^2$ terms

$$\pm\beta^2\nu_0(\eta_V \varepsilon_J \varepsilon_\ell \varepsilon_{\ell'} \widehat{f}(R^-_{\ell,\ell'} - q) f^-)$$

and that $\varepsilon_J \varepsilon_\ell \varepsilon_{\ell'} = \varepsilon_{J'}$ with $J' \subset J \cup \{\ell, \ell'\}$, so that

$$\mathrm{card}(V^* \cap J') \le 2 + \mathrm{card}(V^* \cap J) \le 2 + r .$$

Moreover

$$2(m-1) + 2 + r = 2m + r < s .$$

The induction hypothesis then yields

$$\nu_0^{(m-1)}(\eta_V \varepsilon_J \varepsilon_\ell \varepsilon_{\ell'} \widehat{f}(R^-_{\ell,\ell'} - q) f^-) = 0 ,$$

and this concludes the proof. □

The next corollary takes advantage of the fact that many derivatives vanish through Taylor's formula.

Corollary 1.11.6. *Consider sets V and J with* $\mathrm{card} V = s$ *and* $\mathrm{card}(V^* \cap J) \le r$. *Consider a function $\widehat{f}$ of $(\varepsilon_\ell)_{\ell \notin V^*}$ and a function f^- on Σ^n_{N-1}. Assume that $\eta_J, \varepsilon_J, \widehat{f}$ are functions on Σ^n_N (that is, they depend only on ε_ℓ for $\ell \le n$). Then*

$$|\nu(\eta_V \varepsilon_J \widehat{f} f^-)| \le \frac{K(s,n)}{N^{a/2}} \nu((\widehat{f} f^-)^2)^{1/2} ,$$

where

$$a = \frac{s-r+1}{2} \text{ if } s-r \text{ is odd; } a = \frac{s-r}{2} \text{ if } s-r \text{ is even.} \tag{1.335}$$

Proof. Consider the largest integer m with $2m < s - r$ so

$$m = \frac{s-r-1}{2} \text{ if } s-r \text{ is odd; } m = \frac{s-r}{2} - 1 \text{ if } s-r \text{ is even.}$$

Thus $a = m + 1$. Moreover Lemma 1.11.5 implies that $\nu_0^{(m')}(\eta_V \varepsilon_J \widehat{f} f^-) = 0$ whenever $m' \le m$. Taylor's formula and Lemma 1.11.3 then yield

$$|\nu(\eta_V \varepsilon_J \widehat{f} f^-)| \le \sup_{|t| \le 1} |\nu_t^{(m+1)}(\eta_V \varepsilon_J \widehat{f} f^-)| \le \frac{K(s,n)}{N^{(m+1)/2}} \nu((\widehat{f} f^-)^2)^{1/2} .$$

The reason why $K(s,n)$ depends only on s and n is simply that $m \le s$. □

Corollary 1.11.7. *Consider a number q' (that may depend on N) with $|q - q'| \le L/N$. Consider a set V with* card$V = s$, *and for $u \le m$ consider integers $\ell(u) < \ell'(u)$. Then*

$$\nu\Big(\eta_V \prod_{u \le m} (R_{\ell(u),\ell'(u)} - q')\Big) = O(b + m)$$

where $b = (s+1)/2$ if s is odd and $b = s/2$ if s is even. Moreover these estimates are uniform over $\beta \le \beta_0 < 1/2$.

Proof. Let us write

$$R_{\ell(u),\ell'(u)} - q' = R^-_{\ell(u),\ell'(u)} - q' + \frac{\varepsilon_{\ell(u)}\varepsilon_{\ell(u')}}{N}$$

and expand the product $\prod_{u \le m}(R_{\ell(u),\ell'(u)} - q')$ according to this decomposition. We find

$$\nu\Big(\eta_V \prod_{u \le m} (R_{\ell(u),\ell'(u)} - q')\Big) = \sum_{I \subset \{1,\dots,m\}} W_I$$

where

$$W_I = \nu\Big(\eta_V \prod_{u \le m} C_u\Big)$$

with

$$C_u = \frac{\varepsilon_{\ell(u)}\varepsilon_{\ell'(u)}}{N} \;\text{ if } u \in I;\;\; C_u = R^-_{\ell(u),\ell'(u)} - q' \;\text{ if } u \notin I.$$

Let $r' :=$ card$I$, so that

$$\prod_{u \in I} C_u = \frac{1}{N^{r'}} \varepsilon_J$$

where card$J \le 2r'$. Let

$$f^- := \prod_{u \notin I} C_u = \prod_{u \notin I} (R^-_{\ell(u),\ell'(u)} - q') ,$$

so that

$$W_I = \frac{1}{N^{r'}} \nu(\eta_V \varepsilon_J f^-) .$$

We may use Corollary 1.11.6 with $r = 2r'$, $\widehat{f} = 1$ to obtain

$$|\nu(\eta_V \varepsilon_J f^-)| \le \frac{K(s,n)}{N^{a/2}} \nu((f^-)^2)^{1/2} ,$$

where $a = (s+1)/2 - r'$ if s is odd and $a = s/2 - r'$ if s is even so that $a = b - r'$.

Also, by (1.103) we have $\nu_t((R^-_{1,2}-q')^{2k}) \le KN^{-k}$ and Hölder's inequality implies

$$\nu((f^-)^2)^{1/2} \le \frac{K(m)}{N^{(m-r')/2}} .$$

Therefore

$$|W_I| \le \frac{1}{N^{r'}} \frac{K(s,n)}{N^{b/2-r'/2}} \frac{K(m)}{N^{(m-r')/2}} = \frac{K(s,n,m)}{N^{(b+m)/2}} .$$

The uniformity of these estimates over $\beta \le \beta_0 < 1/2$ should be obvious. □

Lemma 1.11.8. *Consider a set V with* card$V = 2m$, *a function $\widehat{f}$ of $(\varepsilon_\ell)_{\ell \notin V^*}$, and a function f^- on Σ^n_{N-1}. Assume that η_V and $\widehat{f}$ are functions on Σ^n_N. Then*

$$\nu_0^{(m)}(\eta_V \widehat{f} f^-) = \beta^2 \sum_{\ell < \ell' \le n} \nu_0^{(m-1)}(\eta_V \varepsilon_\ell \varepsilon_{\ell'} (R_{\ell,\ell'} - q) \widehat{f} f^-) . \tag{1.336}$$

Proof. From (1.151) we know that $\nu'_t(\eta_V \widehat{f} f^-)$ is the sum of $2n^2$ terms of the type

$$\pm \beta^2 \nu_t(\eta_V \varepsilon_\ell \varepsilon_{\ell'} (R^-_{\ell,\ell'} - q) \widehat{f} f^-) .$$

Now it follows from Lemma 1.11.5 used for $s = 2m$ and $\nu_0^{(m-1)}$ rather than $\nu_0^{(m)}$ that

$$\nu_0^{(m-1)}(\eta_V \varepsilon_\ell \varepsilon_{\ell'} (R^-_{\ell,\ell'} - q) \widehat{f} f^-) = 0$$

unless $\ell, \ell' \in V^*$. Looking again at (1.151) we observe that the only terms for which this occurs are the terms

$$\beta^2 \nu_t(\eta_V \varepsilon_\ell \varepsilon_{\ell'} (R^-_{\ell,\ell'} - q) \widehat{f} f^-) \text{ for } \ell < \ell' \le n . \qquad \square$$

The next result is the heart of the matter. Given a set V with card$V = 2m$, we denote by $\mathcal{I}$ a partition of V in sets J with card$J = 2$. When $J = \{u, v\}$ we consider the "rectangular sums"

$$U^-_J = R^-_{2u-1,2v-1} - R^-_{2u-1,2v} - R^-_{2u,2v-1} + R^-_{2u,2v} ,$$

and

$$U_J = R_{2u-1,2v-1} - R_{2u-1,2v} - R_{2u,2v-1} + R_{2u,2v} .$$

Theorem 1.11.9. *Consider a set V with* card$V = 2m$, *a function $\widehat{f}$ of $(\varepsilon_\ell)_{\ell \notin V^*}$ and a function f^- on Σ^n_{N-1}. Then*

$$\nu_0^{(m)}(\eta_V \widehat{f} f^-) = \beta^{2m} m! \sum_{\mathcal{I}} \mathsf{E}\Big(\langle \widehat{f} \rangle_0 \frac{1}{\mathrm{ch}^{4m} Y}\Big) \nu_0\Big(f^- \prod_{J \in \mathcal{I}} U^-_J\Big) , \tag{1.337}$$

where the summation is over the possible choices of the partition $\mathcal{I}$ of V.

When $m = 1$ and $\widehat{f} = 1$, this is (1.237).

Proof. We may assume that n is large enough so that η_V and $\widehat{f}$ are functions on Σ_N^n. Iteration of (1.336) and use of Lemma 1.6.2 show that

$$\nu_0^{(m)}(\eta_V \widehat{f} f^-) = \beta^{2m} \sum \nu_0(\eta_V \varepsilon_{\ell_1}\varepsilon_{\ell'_1} \cdots \varepsilon_{\ell_m}\varepsilon_{\ell'_m} \widehat{f}) \nu_0\Big(f^- \prod_{r \le m} (R^-_{\ell_r,\ell'_r} - q)\Big) , \tag{1.338}$$

where the summation is over all choices of $1 \le \ell_1 < \ell'_1 \le n, \ldots, 1 \le \ell_m < \ell'_m \le n$. Now, as shown in the proof of Lemma 1.11.4,

$$\nu_0(\eta_V \varepsilon_{\ell_1}\varepsilon_{\ell'_1} \cdots \varepsilon_{\ell_m}\varepsilon_{\ell'_m} \widehat{f}) = 0 \tag{1.339}$$

unless each of the sets $\{2v-1, 2v\}$ for $v \in V$ contains at least one of the points ℓ_r or ℓ'_r ($r \le m$). There are $2m$ such sets and $2m$ such points; hence each set must contain exactly one point. When this is the case let us define

$$J_r = \{v_r, v_{r'}\} \text{ where } \ell_r \in \{2v_r - 1, 2v_r\} \,;\; \ell'_r \in \{2v'_r - 1, 2v'_r\} \,.$$

Then $\{J_1, \ldots, J_m\}$ forms a partition of V. Moreover,

$$\langle \eta_V \varepsilon_{\ell_1}\varepsilon_{\ell'_1} \cdots \varepsilon_{\ell_m}\varepsilon_{\ell'_m} \widehat{f} \rangle_0 = \langle \widehat{f} \rangle_0 \prod_{r \le m} \langle \eta_{v_r} \varepsilon_{\ell_r} \rangle_0 \prod_{r \le m} \langle \eta_{v'_r} \varepsilon_{\ell'_r} \rangle_0$$

and

$$\langle \eta_{v_r} \varepsilon_{\ell_r} \rangle_0 = \langle (\varepsilon_{2v_r-1} - \varepsilon_{2v_r}) \varepsilon_{\ell_r} \rangle_0 = \begin{cases} 1 - \mathrm{th}^2 Y = 1/\mathrm{ch}^2 Y & \text{if } \ell_r = 2v_r - 1 \\ -(1 - \mathrm{th}^2 Y) = -1/\mathrm{ch}^2 Y & \text{if } \ell_r = 2v_r \end{cases}$$

and similarly for $\langle \eta_{v'_r} \varepsilon_{\ell'_r} \rangle_0$. Let us then define

$$\tau_r = 1 \text{ if } \ell_r = 2v_r - 1 \,;\; \tau_r = -1 \text{ if } \ell_r = 2v_r \,,$$

and τ'_r similarly. Then the quantity (1.338) is

$$\beta^{2m} \mathsf{E}\Big(\langle \widehat{f} \rangle_0 \frac{1}{\mathrm{ch}^{4m} Y}\Big) \sum \Big(\prod_{r \le m} \tau_r \tau'_r\Big) \nu_0\Big(f^- \prod_{r \le m} (R^-_{\ell_r,\ell'_r} - q)\Big) , \tag{1.340}$$

where the summation is over all the choices of the partition $\{J_1, \ldots, J_m\}$ of V in sets of two elements, and all choices of ℓ_r and ℓ'_r as above. Given the set J_r, there are two possible choices for ℓ_r (namely $\ell_r = 2v_r - 1$ and $\ell_r = 2v_r$) and similarly there are two possible choices for ℓ'_r. Thus, given the sets $J_1, \ldots, J_r$, there are 2^{2m} choices for the indices ℓ_r and ℓ'_r, $r \le m$. In the next step, we add the 2^{2m} terms in the right-hand side of (1.340) for which the sets $J_1, \ldots, J_m$ take given values. We claim that this gives a combined term of the form

$$\beta^{2m}\mathsf{E}\Big(\langle\widehat{f}\rangle_0\frac{1}{\mathrm{ch}^{4m}Y}\Big)\nu_0\Big(f^-\prod_{r\le m}U_{J_r}^-\Big)\,.$$

To understand this formula, one simply performs the computation when $m = 1$ and one observes that "there is factorization over the different values of $r \le m$". If we keep in mind the fact that there are $m!$ choices of the sequence $J_1, \dots, J_m$ for which $\{J_1, \dots, J_m\}$ forms a given partition $\mathcal{I}$ of V in sets of 2 elements, we have proved (1.337). □

We are now ready to start the real computation. We recall the notation A^2 of (1.248).

Proposition 1.11.10. *Consider a set V with* $\mathrm{card}V = 2m$*, and a partition $\mathcal{I}$ of V in sets with 2 elements. Consider a function $\widehat{f}$ of $(\varepsilon_\ell)_{\ell\notin V^*}$. Then*

$$\nu\Big(\eta_V\widehat{f}\prod_{J\in\mathcal{I}}U_J^-\Big)=\mathsf{E}\Big(\langle\widehat{f}\rangle_0\frac{1}{\mathrm{ch}^{4m}Y}\Big)(4\beta^2A^2)^m+O(2m+1)\,. \tag{1.341}$$

Proof. First, since $\nu((\prod_{J\in\mathcal{I}}U_J^-)^2)^{1/2}=O(m)$ (because there are m factors in the product, each counting as $1/\sqrt{N}$), it follows from (1.334), used for $m+1$ rather than m that $\nu_t^{(m+1)}(\eta_V\widehat{f}\prod_{J\in\mathcal{I}}U_J^-)=O(2m+1)$ (uniformly in t). Next, it follows from Lemma 1.11.5 (used for $r=0$ and $s=2m$) that for $p<m$ we have $\nu_t^{(p)}(\eta_V\widehat{f}\prod_{J\in\mathcal{I}}U_J^-)=0$. Combining these facts with Taylor's formula, we obtain:

$$\nu\Big(\eta_V\widehat{f}\prod_{J\in\mathcal{I}}U_J^-\Big)=\frac{1}{m!}\nu_0^{(m)}\Big(\eta_V\widehat{f}\prod_{J\in\mathcal{I}}U_J^-\Big)+O(2m+1)\,. \tag{1.342}$$

From (1.337) we get

$$\nu_0^{(m)}\Big(\eta_V\widehat{f}\prod_{J\in\mathcal{I}}U_J^-\Big)=\beta^{2m}m!\mathsf{E}\Big(\langle\widehat{f}\rangle_0\frac{1}{\mathrm{ch}^{4m}Y}\Big)\sum_{\mathcal{I}'}\nu_0\Big(\prod_{J\in\mathcal{I}}U_J^-\prod_{J'\in\mathcal{I}'}U_{J'}^-\Big)\,, \tag{1.343}$$

where the summation is over all partitions $\mathcal{I}'$ of V in sets with 2 elements. Both $\mathcal{I}$ and $\mathcal{I}'$ have m elements. In the previous section we have explained in detail why

$$\nu_0\Big(\prod_{J\in\mathcal{I}}U_J^-\prod_{J'\in\mathcal{I}'}U_{J'}^-\Big)=\nu\Big(\prod_{J\in\mathcal{I}}U_J\prod_{J'\in\mathcal{I}'}U_{J'}\Big)+O(2m+1)\,. \tag{1.344}$$

Now, if $J=\{v,v'\}$ we obtain, recalling the notation $T_{\ell,\ell'}$ of (1.245),

$$\begin{aligned}U_J&=R_{2v-1,2v'-1}-R_{2v-1,2v'}-R_{2v,2v'-1}+R_{2v,2v'}\\&=T_{2v-1,2v'-1}-T_{2v-1,2v'}-T_{2v,2v'-1}+T_{2v,2v'}\,.\end{aligned} \tag{1.345}$$

In this manner each term U_J is decomposed as the sum of 4 terms $\pm T_{\ell,\ell'}$, so that the right-hand side of (1.343) can be computed through Theorem 1.10.1 (using only the much easier case where $k_2 = k_3 = 0$). The fundamental fact is that if a term $T_{\ell,\ell'}$ occurs both in the decompositions of U_J and $U_{J'}$ then we must have $J = J'$ because ℓ and ℓ' determine J by the formula

$$J = \left\{ \left[\frac{\ell+1}{2}\right], \left[\frac{\ell'+1}{2}\right] \right\},$$

i.e. J is the two-point set whose elements are the integer parts of $(\ell+1)/2$ and $(\ell'+1)/2$ respectively. In order for the quantity

$$\nu\Big(\prod_{J\in\mathcal{I}} U_J \prod_{J'\in\mathcal{I}'} U_{J'}\Big)$$

not to be $O(2m+1)$, the following must occur: given any $J_0 \in \mathcal{I}$, at least one of the terms $T_{\ell,\ell'}$ of the decomposition (1.345) of U_{J_0} must occur in the decomposition of another U_J, $J \in \mathcal{I} \cup \mathcal{I}'$, $J \neq J_0$. (This is because $a(1) = 0$.) The only possibility is that $J_0 \in \mathcal{I}'$. Since this must hold for any choice of J_0, we must have $\mathcal{I}' = \mathcal{I}$, and thus (1.343) implies

$$\nu_0^{(m)}\Big(\eta_V \widehat{f} \prod_{J\in\mathcal{I}} U_J^-\Big) = \beta^{2m} m! \mathsf{E}\Big(\langle \widehat{f}\rangle_0 \frac{1}{\mathrm{ch}^{4m} Y}\Big)\, \nu\Big(\prod_{J\in\mathcal{I}} U_J^2\Big) + O(2m+1)\,. \tag{1.346}$$

Expanding U_J^2 using (1.345), and using Theorem 1.10.1 then shows that

$$\nu\Big(\prod_{J\in\mathcal{I}} U_J^2\Big) = (4A^2)^m + O(2m+1)\,,$$

and combining with (1.342), (1.343) and (1.346) completes the proof. □

Proposition 1.11.11. *Consider a set V with* $\mathrm{card}V = 2p$ *and a partition $\mathcal{I}$ of V in sets with two elements. Then*

$$\nu\Big(\eta_V \prod_{J\in\mathcal{I}} U_J\Big) = \Big(4\Big(\frac{1}{N} + \beta^2 A^2\Big)\Big)^p \mathsf{E}\frac{1}{\mathrm{ch}^{4p} Y} + O(2p+1)\,. \tag{1.347}$$

Proof. We observe the relation

$$U_J = U_J^- + \frac{\eta_J}{N}$$

so that

$$\prod_{J\in\mathcal{I}} U_J = \prod_{J\in\mathcal{I}} \Big(U_J^- + \frac{\eta_J}{N}\Big)\,.$$

We shall prove (1.347) by expanding the product and using (1.346) for each term. We have

$$\prod_{J\in\mathcal{I}}\left(U_J^- + \frac{\eta_J}{N}\right) = \sum_{\mathcal{I}'}\left(\prod_{J\notin\mathcal{I}'}\frac{\eta_J}{N}\right)\left(\prod_{J\in\mathcal{I}'}U_J^-\right), \tag{1.348}$$

where the sum is over all subsets $\mathcal{I}'$ of $\mathcal{I}$. Consider such a subset with card$\mathcal{I}' = m \le p = $ card$\mathcal{I}$. Let $V' = \bigcup\{J\,;\, J\in\mathcal{I}'\}$ and observe that

$$\eta_V = \eta_{V'}\prod_{J\notin\mathcal{I}'}\eta_J$$

so that

$$\eta_V \prod_{J\notin\mathcal{I}'}\eta_J \prod_{J\in\mathcal{I}'}U_J^- = \eta_{V'}\prod_{J\notin\mathcal{I}'}\eta_J^2\prod_{J\in\mathcal{I}'}U_J^- \,.$$

We can then use (1.341) with V' instead of V, $\mathcal{I}'$ instead of $\mathcal{I}$, $m = $ card$\mathcal{I}'$ and $\widehat{f} = \prod_{J\notin\mathcal{I}'}\eta_J^2$. We observe that

$$\langle\widehat{f}\rangle_0 = \prod_{J\notin\mathcal{I}'}\langle\eta_J^2\rangle_0$$

and that if $J = \{v, v'\}$,

$$\begin{aligned}\langle\eta_J^2\rangle_0 &= \langle(\varepsilon_{2v-1}-\varepsilon_{2v})^2(\varepsilon_{2v'-1}-\varepsilon_{2v'})^2\rangle_0\\ &= \langle(\varepsilon_{2v-1}-\varepsilon_{2v})^2\rangle_0^2 = (2-2\mathrm{th}^2Y)^2 = \frac{4}{\mathrm{ch}^4Y}\,.\end{aligned}$$

Therefore (1.341) proves that

$$\begin{aligned}\nu\left(\eta_V\left(\prod_{J\notin\mathcal{I}'}\frac{\eta_J}{N}\right)\left(\prod_{J'\in\mathcal{I}'}U_{J'}^-\right)\right) &= \frac{1}{N^{p-m}}\mathsf{E}\left(\left(\frac{4}{\mathrm{ch}^4Y}\right)^{p-m}\frac{1}{\mathrm{ch}^{4m}Y}\right)(4\beta^2A^2)^m\\ &\quad+\frac{1}{N^{p-m}}O(2m+1)\\ &= \frac{1}{N^{p-m}}\mathsf{E}\left(\frac{4^p}{\mathrm{ch}^{4p}Y}\right)(\beta^2A^2)^m + O(2p+1)\,,\end{aligned}$$

and performing the summation in (1.348) completes the proof. □

Proof of Theorem 1.11.1. Combining (1.331), (1.332) and (1.333) we obtain

$$\nu\big((\langle\sigma_1\sigma_2\rangle - \langle\sigma_1\rangle\langle\sigma_2\rangle)^k\big) = 2^{-k}\nu(\eta_V f^-) \tag{1.349}$$

where $V = \{1,\ldots,k\}$ and

$$f^- = (\sigma_{N-1}^1 - \sigma_{N-1}^2)\cdots(\sigma_{N-1}^{2k-1} - \sigma_{N-1}^{2k})\,. \tag{1.350}$$

Using Taylor's formula and (1.334) proves that

$$\nu(\eta_V f^-) = \sum_{m\le k}\frac{1}{m!}\nu_0^{(m)}(\eta_V f^-) + O(k+1)\,. \tag{1.351}$$

Let us denote $p = (k+1)/2$ when k is odd and $p = k/2$ when k is even. We claim that

$$\nu_0^{(m)}(\eta_V f^-) = O(p+m) \,. \tag{1.352}$$

To prove this we recall that by (1.151), and Lemma 1.6.2, the quantity $\nu_0^{(m)}(\eta_V f^-)$ is the sum of terms of the type

$$\pm\beta^{2m}\nu_0\Big(\eta_V \varepsilon_{\ell_1}\varepsilon_{\ell'_1}\cdots\varepsilon_{\ell_m}\varepsilon_{\ell'_m} f^- \prod_{r\le m}(R^-_{\ell_r,\ell'_r}-q)\Big)$$

$$= \pm\beta^{2m}\nu_0(\eta_V \varepsilon_{\ell_1}\varepsilon_{\ell'_1}\cdots\varepsilon_{\ell_m}\varepsilon_{\ell'_m})\nu_0\Big(f^- \prod_{r\le m}(R^-_{\ell_r,\ell'_r}-q)\Big) \,.$$

Thus it suffices to prove that

$$\nu_0\Big(f^- \prod_{r\le m}(R^-_{\ell_r,\ell'_r}-q)\Big) = O(m+p) \,. \tag{1.353}$$

This will be shown by using Corollary 1.11.7 for the $(N-1)$-spin system with Hamiltonian (1.144). First, we observe that if $\langle\cdot\rangle_-$ denotes an average for the Gibbs measure with Hamiltonian (1.144) then for a function f' on Σ^n_{N-1} we have $\langle f'\rangle_0 = \langle f'\rangle_-$. So, if $\nu_-(\cdot) = \mathsf{E}\langle\cdot\rangle_-$, (1.353) shall follow from

$$\nu_-\Big(f^- \prod_{r\le m}(R^-_{\ell_r,\ell'_r}-q)\Big) = O(m+p) \,. \tag{1.354}$$

Since the overlaps for the $(N-1)$-spin system are given by

$$R^\sim_{\ell,\ell'} = \frac{1}{N-1}\sum_{i\le N-1}\sigma_i^\ell\sigma_i^{\ell'} = \frac{N}{N-1}R^-_{\ell,\ell'} \,, \tag{1.355}$$

it suffices to prove (1.354) to show that

$$\nu_-\Big(f^- \prod_{r\le m}\Big(R^\sim_{\ell_r,\ell'_r} - \frac{N}{N-1}q\Big)\Big) = O(m+p) \,. \tag{1.356}$$

The $(N-1)$-spin system with Hamiltonian (1.144) has parameter

$$\beta_- = \sqrt{\frac{N-1}{N}}\beta \le \beta \,, \tag{1.357}$$

and the corresponding value q_- satisfies $|q-q_-| \le L/N$ by (1.187), so that

$$\Big|\frac{N}{N-1}q - q_-\Big| \le \frac{L}{N-1} \,.$$

Recalling the value (1.350) of f^- we see that indeed (1.356) follows from Corollary 1.11.7, because the estimate in that corollary is uniform over $\beta \le \beta_0 < 1/2$ (and thus the fact that β_- in (1.357) depends on N is irrelevant).

Thus we have proved (1.352), and combining with (1.351) we get

$$\nu(\eta_V f^-) = \sum_{m \le k-p} \frac{1}{m!} \nu_0^{(m)}(\eta_V f^-) + O(k+1) . \tag{1.358}$$

When k is odd, we have $k - p = (k-1)/2$, and for $m \le k - p$ we have $2m < k$. It then follows from Lemma 1.11.5 (used for $r = 0$ and $s = k$) that $\nu_0^{(m)}(\eta_V f^-) = 0$ for $m \le k - p$. In that case $\nu(\eta_V f^-) = O(k+1)$, and since $a(k) = 0$ when k is odd, we have proved (1.329) in that case.

So we assume that k is even, $k = 2p$. It then follows from Lemma 1.11.5 (used for $r = 0$ and $s = 2p$) that $\nu_0^{(m)}(\eta_V f^-) = 0$ for $m < p$. Therefore from (1.358) we obtain, using (1.337),

$$\begin{aligned} \nu(\eta_V f^-) &= \frac{1}{p!} \nu_0^{(p)}(\eta_V f^-) + O(k+1) \\ &= \beta^{2p} \sum_{\mathcal{I}} \mathsf{E}\left(\frac{1}{\mathrm{ch}^{2k} Y}\right) \nu_0\left(f^- \prod_{J \in \mathcal{I}} U_J^-\right) + O(k+1) , \end{aligned} \tag{1.359}$$

where the summation is over all partitions $\mathcal{I}$ of V in sets of two elements. Now we use (1.347) for the $(N-1)$-spin system to see that, using (1.355) and defining A_- in the obvious manner:

$$\nu_0\left(f^- \prod_{J \in \mathcal{I}} U_J^-\right) = \left(4\left(\frac{1}{N-1} + \beta^2 A_-^2\right)\right)^p \mathsf{E}\frac{1}{\mathrm{ch}^{4p} Y_-} + O(2p+1)$$

where $Y_- = \beta_- z\sqrt{q_-} + h$. It is a very simple matter to check that

$$\left(4\left(\frac{1}{N-1} + \beta^2 A_-^2\right)\right)^p \mathsf{E}\frac{1}{\mathrm{ch}^{4p} Y_-} = \left(4\left(\frac{1}{N} + \beta^2 A^2\right)\right)^p \mathsf{E}\frac{1}{\mathrm{ch}^{4p} Y} + O(2p+1)$$

and thus

$$\nu_0\left(f^- \prod_{J \in \mathcal{I}} U_J^-\right) = \left(4\left(\frac{1}{N} + \beta^2 A^2\right)\right)^p \mathsf{E}\frac{1}{\mathrm{ch}^{4p} Y} + O(2p+1) .$$

Each choice of $\mathcal{I}$ gives the same contribution. To count the number of partitions $\mathcal{I}$, we observe that if $1 \in J$, and $\mathrm{card} J = 2$, J is determined by its other element so there are $2p-1$ choices for J. In this manner induction over p shows that

$$\mathrm{card}\mathcal{I} = (2p-1)(2p-3)\cdots = a(2p) = a(k) .$$

Therefore

$$\nu(\eta_V f^-) = a(k)\beta^k \left(\mathsf{E}\,\frac{1}{\mathrm{ch}^{2k} Y}\right)^2 \left(4\left(\frac{1}{N} + \beta^2 A^2\right)\right)^{k/2} + O(k+1)\,,$$

which completes the proof recalling (1.349) and since

$$\frac{1}{N} + \beta^2 A^2 = \frac{1}{N}\frac{1}{1-\beta^2(1-2q+\widehat{q})}\,. \qquad \square$$

Having succeeded to make this computation one can of course ask all kinds of questions.

Research Problem 1.11.12. (Level 1) Compute

$$\lim_{N\to\infty} N^{k/2}\mathsf{E}(\langle\sigma_1\sigma_2\sigma_3\rangle - \langle\sigma_1\rangle\langle\sigma_2\rangle\langle\sigma_3\rangle)^k\,.$$

Research Problem 1.11.13. (Level 1^-) Recall the notation $\dot{\sigma}_i = \sigma_i - \langle\sigma_i\rangle$. Consider a number t, and i.i.d. standard Gaussian r.v.s g_i, independent of the randomness of H_N. Compute

$$\lim_{N\to\infty} N^{k/2}\mathsf{E}\left(\left\langle \exp t\sum_{i\le N}\frac{g_i\dot{\sigma}_i}{\sqrt{N}}\right\rangle - \exp\frac{t^2}{2}(1-q)\right)^k\,.$$

(Hint: read very carefully the proof of Theorem 1.7.11.)

1.12 The SK Model with d-component Spins

A model where spins take only the values ± 1 could be an oversimplification. It is more physical to consider spins as vectors in $\mathbb{R}^3$ or $\mathbb{R}^d$. This is what we will do in this section. The corresponding model is of obvious interest. It has been investigated in less detail than the standard SK model, so many questions remain unanswered. On the one hand, this is somewhat specialized material, and it is not directly related to the rest of this volume. On the other hand, this is as simple a situation as one might wish to describe a "replica-symmetric solution" beyond the case of the ordinary SK model.

In the SK model with d-component spins, the individual spin σ_i is a vector $(\sigma_{i,1},\ldots,\sigma_{i,d})$ of $\mathbb{R}^d$. We will denote by $(\cdot,\cdot)$ the dot product in $\mathbb{R}^d$, so that

$$(\sigma_i,\sigma_j) = \sum_{u\le d}\sigma_{i,u}\sigma_{j,u}\,.$$

The Hamiltonian is given by

$$-H_N = \frac{\beta}{\sqrt{N}} \sum_{1\le i<j\le N} g_{ij}(\sigma_i, \sigma_j) \tag{1.360}$$

where, of course, $(g_{ij})_{i<j}$ are independent standard normal r.v.s. We may rewrite (1.360) as

$$-H_N = \frac{\beta}{\sqrt{N}} \sum_{u\le d} \sum_{i<j} g_{ij}\sigma_{i,u}\sigma_{j,u} , \tag{1.361}$$

a formula that is reminiscent of a Hamiltonian depending on d configurations $\boldsymbol{\sigma}_u = (\sigma_{1,u}, \ldots, \sigma_{N,u})$ for $u \le d$, each of which has the same energy as in the SK model. A first difference is that now $\sigma_{i,u}$ varies in $\mathbb{R}$ rather than in $\{-1,1\}$. A deeper difference is that $\boldsymbol{\sigma}_1, \ldots, \boldsymbol{\sigma}_d$ are not independent configurations but are interacting. (This does not show in the Hamiltonian itself, the interaction takes place through the measure μ below.) In order to compare with the SK model, and to accommodate the case $\sigma_{i,u} = \pm 1$, we will assume that

$$\forall i , \quad \sum_{u\le d} \sigma_{i,u}^2 \le d , \tag{1.362}$$

or, in words, that σ_i belongs to the Euclidean ball B_d centered at 0, of radius $\sqrt{d}$. Thus the configuration space is now

$$S_N = B_d^N .$$

We consider a probability measure μ on B_d. We will define Gibbs' measure as the probability measure on $S_N = B_d^N$ of density proportional to $\exp(-H_N)$ with respect to $\mu^{\otimes N}$. The case $d = 1$ is already of interest. This case is simply the generalization of the standard SK model where the individual spin σ_i is permitted to be any number in the interval $[-1,1]$. When moreover μ is supported by $\{-1,1\}$, and for $\varepsilon \in \{-1,1\}$ has a density proportional to $\exp \varepsilon h$ with respect to the uniform measure on $\{-1,1\}$, we recover the case of the standard SK model with non-random external field. (Thus it might be correct to think of μ as determining a kind of "external field" and to expect that the behavior of the model will be very sensitive to the value of μ.) Also of special interest is the case where $d = 2$, μ is supported by $\{-1,1\}^2$, and for $(\varepsilon_1, \varepsilon_2) \in \{-1,1\}^2$ has a density proportional to

$$\exp(\varepsilon_1 h + \varepsilon_2 h + \lambda \varepsilon_1 \varepsilon_2)$$

with respect to the uniform measure on $\{-1,1\}^2$. This is the case of "two coupled copies of the SK model" considered in Section 1.9. This case is of fundamental importance. It seems connected to some of the deepest remaining mysteries of the low temperature phase of the SK model. For large values of β, this case of "two coupled copies of the SK model" is far from being completely understood at the time of this writing. One major reason for this is that it is not clear how to use arguments in the line of the arguments of Theorem 1.3.7. The main difficulty is that some of the terms one obtains when trying to use Guerra's interpolation have the wrong sign, a topic to which we will return later.

Let us define

$$Z_N = Z_N(\beta, \mu) = \int \exp(-H_N) \mathrm{d}\mu(\sigma_1) \cdots \mathrm{d}\mu(\sigma_N) , \tag{1.363}$$

where H_N is the Hamiltonian (1.360). (Let us note that in the case where $d = 1$ and μ is supported by $\{-1, 1\}$ this differs from our previous definition of Z_N because we replace a sum over configurations by an average over configurations.) Let us write

$$p_N(\beta, \mu) = \frac{1}{N} \mathsf{E} \log Z_N(\beta, \mu) . \tag{1.364}$$

One of our objectives is the computation of $\lim_{N\to\infty} p_N(\beta, \mu)$. It will be achieved when β is small enough. This computation has applications to the theory of "large deviations". For example, in the case of "two coupled copies of the SK model", computing $\lim_{N\to\infty} p_N(\beta, \mu)$ amounts to computing

$$\lim_{N\to\infty} \frac{1}{N} \mathsf{E} \log\langle \exp \lambda N R_{1,2} \rangle , \tag{1.365}$$

where now the bracket is an average for the Gibbs measure of the usual SK model. "Concentration of measure" (as in Theorem 1.3.4) shows that $N^{-1} \log\langle \exp \lambda N R_{1,2}\rangle$ fluctuates little with the disorder. Thus computing (1.365) amounts to computing the value of $\langle \exp \lambda N R_{1,2}\rangle$ for the typical disorder. Since we can do this for every λ this is very much the same as computing $N^{-1} \log G_N^{\otimes 2}(\{R_{1,2} \geq q + a\})$ and $N^{-1} \log G_N^{\otimes 2}(\{R_{1,2} \leq q - a\})$ for $a > 0$ and a suitable median value q. In summary, the result of (1.365) can be transfered in a result about the "large deviations of $R_{1,2}$" for the typical disorder. See [151] and [162] for more on this.

We will be able to compute $\lim_{N\to\infty} p_N(\beta, \mu)$ under the condition $L\beta d \leq 1$, where as usual L is a universal constant. Despite what one might think at first, the quality of this result does not decrease as d becomes large. It controls "the same proportion of the high-temperature region independently of d". Indeed, if μ gives mass 1/2 to the two points $(\pm\sqrt{d}, 0, \ldots, 0)$, the corresponding model is "a clone" of the usual SK model at temperature βd. The problem of computing $\lim_{N\to\infty} p_N(\beta, \mu)$ is much more difficult (and unsolved) if βd is large.

The SK model with d-component spins offers new features compared with the standard SK model. One of these is that if μ is "spread out" then one can understand the system up to values of β much larger than $1/d$. For example, if μ is uniform on $\{-1, 1\}^d$, the model simply consists in d replicas of the SK model with $h = 0$, and we understand it for $\beta < 1/2$, independently of the value of d. Comparable results will be proved later in Volume II when μ is the uniform measure on the boundary of B_d.

The good behavior of the SK model at small $\beta < 1/2$ is largely expressed by (1.89), i.e. the fact that $\nu((R_{1,2} - q)^2) \le L/N$. The situation is more complicated here. Consider $1 \le u,\ v \le d$, and set

$$R^{u,v} = \frac{1}{N} \sum_{i \le N} \sigma_{i,u} \sigma_{i,v} \,. \tag{1.366}$$

This is a function of a single configuration $(\sigma_1, \ldots, \sigma_N) \in S_N$, where $\sigma_i = (\sigma_{i,u})_{u \le d}$. Consider now two configurations $(\sigma_1^1, \ldots, \sigma_N^1)$ and $(\sigma_1^2, \ldots, \sigma_N^2)$. Consider the following function of these two configurations

$$R_{1,2}^{u,v} = \frac{1}{N} \sum_{i \le N} \sigma_{i,u}^1 \sigma_{i,v}^2 \,. \tag{1.367}$$

In the present context, similar to (1.89), we have the following.

Theorem 1.12.1. *If $L\beta d \le 1$, we can find numbers $(q_{u,v})$, $(\rho_{u,v})$ such that*

$$\sum_{u,v \le d} \nu\big((R^{u,v} - \rho_{u,v})^2\big) \le \frac{K(d)}{N} \,, \tag{1.368}$$

$$\sum_{u,v \le d} \nu\big((R_{1,2}^{u,v} - q_{u,v})^2\big) \le \frac{K(d)}{N} \,. \tag{1.369}$$

Here $K(d)$ depends on d only; $\nu(\cdot) = \mathsf{E}\langle\cdot\rangle$, $\langle\cdot\rangle$ is an average for Gibbs' measure, over one configuration in (1.368), and over two configurations in (1.369). To get a first feeling for these conditions, consider the case $d = 1$. Then (1.369) is the usual assertion that $R_{1,2} \simeq q$, but (1.368) is a new feature which means that $N^{-1} \sum_{i \le N} \sigma_i^2 \simeq \rho$. This of course was automatic with $\rho = 1$ when we required that the individual spins be ± 1.

In order to give a proper description of these numbers $(q_{u,v})_{1 \le u,v \le d}$, $(\rho_{u,v})_{1 \le u,v \le d}$, let us consider (see Appendix page 439) for each symmetric positive definite matrix $(q_{u,v})_{1 \le u,v \le d}$, a centered jointly Gaussian family $(Y_u)_{u \le d}$ with covariance

$$\mathsf{E} Y_u Y_v = \beta^2 q_{u,v} \,. \tag{1.370}$$

For each family of real numbers $(\rho_{u,v})_{1 \le u,v \le d}$ and for $x = (x_1, \ldots, x_d)$ in $\mathbb{R}^d$, let us set further

$$\mathcal{E} = \mathcal{E}(x) = \exp\Big(\sum_{u \le d} x_u Y_u + \frac{\beta^2}{2} \sum_{u,v \le d} x_u x_v (\rho_{u,v} - q_{u,v})\Big) \,. \tag{1.371}$$

Theorem 1.12.2. *Assuming that $L\beta d \le 1$, and setting $Z = \int \mathcal{E}(x)\,\mathrm{d}\mu(x)$, the following equations have a unique solution, and (1.368) and (1.369) hold for these numbers:*

$$q_{u,v} = \mathsf{E}\left(\frac{1}{Z^2}\int x_u \mathcal{E}(x)\mathrm{d}\mu(x)\int x_v \mathcal{E}(x)\mathrm{d}\mu(x)\right) , \tag{1.372}$$

$$\rho_{u,v} = \mathsf{E}\left(\frac{1}{Z}\int x_u x_v \mathcal{E}(x)\mathrm{d}\mu(x)\right) . \tag{1.373}$$

Of course the above theorem subsumes Theorem 1.12.1, and moreover the proof of Theorem 1.12.1 requires the relations (1.372) and (1.373). But for pedagogical reasons we will prove first Theorem 1.12.1 and only then obtain the information above.

Theorem 1.12.3. *If $\beta L d \le 1$, then*

$$\lim_{N\to\infty} p_N(\beta,\mu) = -\frac{\beta^2}{4}\sum_{u,v\le d}(\rho_{u,v}^2 - q_{u,v}^2) + \mathsf{E}\log\int \mathcal{E}(x)\mathrm{d}\mu(x) \tag{1.374}$$

where $\mathcal{E}(x)$ is given by (1.371).

Our argument gives a rate of convergence in $N^{-1/2}$, but it is almost certain that a little bit more work would yield the usual rate in $1/N$.

We will later explain why the solutions to (1.372) and (1.373) exist and are unique for $L\beta d \le 1$, but let us accept this for the time being and turn to the fun part, the search for the "smart path". We will compare the system with a version of it where the last spin is "decoupled".

We consider the "configurations of the $(N-1)$-spin system"

$$\boldsymbol{\rho}_u = (\sigma_{1,u}, \ldots, \sigma_{N-1,u}) .$$

(One will distinguish between the configuration $\boldsymbol{\rho}_u$ and the numbers $\rho_{u,v}$.) We define

$$g(\boldsymbol{\rho}_u) = \frac{\beta}{\sqrt{N}}\sum_{i\le N-1} g_{iN}\sigma_{i,u}$$

$$g_t(\boldsymbol{\rho}_u) = \sqrt{t}g(\boldsymbol{\rho}_u) + \sqrt{1-t}Y_u .$$

We consider the Hamiltonian

$$\begin{aligned} -H_{N,t}(\sigma_1,\ldots,\sigma_N) &= \frac{\beta}{\sqrt{N}}\sum_{u\le d}\sum_{i<j\le N-1} g_{ij}\sigma_{i,u}\sigma_{j,u} + \sum_{u\le d}\sigma_{N,u}g_t(\boldsymbol{\rho}_u) \\ &\quad + \frac{\beta^2}{2}(1-t)\sum_{u,v\le d}\sigma_{N,u}\sigma_{N,v}(\rho_{u,v} - q_{u,v}) . \end{aligned} \tag{1.375}$$

The last term is the new feature compared to the standard case.

For a function f on S_N^n we write

$$\nu_t(f) = \mathsf{E}\langle f\rangle_t ,$$

where $\langle\cdot\rangle_t$ denotes integration with respect to (the nth power of) the Gibbs measure relative to the Hamiltonian (1.375). A function f on S_N^n depends on configurations $(\sigma_1^1,\ldots,\sigma_N^1),(\sigma_1^2,\ldots,\sigma_N^2),\ldots,(\sigma_1^n,\ldots,\sigma_N^n)$. We define

$$R_{\ell,\ell'}^{-,u,v} = \frac{1}{N}\sum_{i\le N-1}\sigma_{i,u}^{\ell}\sigma_{i,v}^{\ell'}$$

for $\ell,\ell'\le n$ and $u,v\le d$. As usual, we write $\nu_t'(f)=\frac{\mathrm{d}}{\mathrm{d}t}\nu_t(f)$. We define

$$q_{u,v}(\ell,\ell') = q_{u,v} \quad \text{if} \quad \ell\ne\ell'$$

$$q_{u,v}(\ell,\ell) = \rho_{u,v} \;.$$

Proposition 1.12.4. *We have*

$$\begin{aligned}
\nu_t'(f) &= \frac{\beta^2}{2}\sum_{\ell,\ell'\le n,u,v\le d}\nu_t\big(f\varepsilon_u^\ell\varepsilon_v^{\ell'}(R_{\ell,\ell'}^{-,u,v}-q_{u,v}(\ell,\ell'))\big)\\
&\quad - n\beta^2\sum_{\ell\le n,u,v\le d}\nu_t\big(f\varepsilon_u^\ell\varepsilon_v^{n+1}(R_{\ell,n+1}^{-,u,v}-q_{u,v}(\ell,n+1))\big)\\
&\quad - n\frac{\beta^2}{2}\sum_{u,v\le d}\nu_t\big(f\varepsilon_u^{n+1}\varepsilon_v^{n+1}(R_{n+1,n+1}^{-,u,v}-q_{u,v}(n+1,n+1))\big) \qquad (1.376)\\
&\quad + \frac{n(n+1)}{2}\beta^2\sum_{u,v\le d}\nu_t\big(f\varepsilon_u^{n+1}\varepsilon_v^{n+2}(R_{n+1,n+2}^{-,u,v}-q_{u,v}(n+1,n+2))\big)\;.
\end{aligned}$$

Here we have set $\varepsilon_u^\ell=\sigma_{N,u}^\ell$. First let us explain why (1.376) coincides with (1.151) in the case of the standard SK model. In such a case we have $d=1$, $\varepsilon_1^\ell=\varepsilon_\ell$, $R_{\ell,\ell'}^{-,1,1}=R_{\ell,\ell'}^-$, $q_{1,1}=q$, $\rho_{1,1}=1$ (because $x^2=1$ if $x\in\{-1,1\}$ cf. (1.373)). Let us also point out that $R_{\ell,\ell}^-=(N-1)/N$, so that the contribution of the case $\ell'=\ell$ in the first sum of the right-hand side of (1.376) cancels out with the contribution of the third sum. We finally observe that $\sum_{\ell\ne\ell'}=2\sum_{\ell<\ell'}$.

Proof. Of course this formula is yet another avatar of (1.90), the new feature being the last term of the Hamiltonian (1.375), which creates extra terms. We leave to the reader the minimal work of deducing (1.376) from (1.90). A direct proof goes as follows. We use straightforward differentiation (i.e. use of rules of Calculus) in the definition of $\nu_t(f)$ to obtain

$$\nu_t'(f) = -\frac{\beta^2}{2}\sum_{\ell\le n}\sum_{u,v\le d}\nu_t(f\varepsilon_u^\ell\varepsilon_v^\ell(\rho_{u,v}-q_{u,v}))$$

$$+\frac{n\beta^2}{2}\sum_{u,v\le d}\nu_t(f\varepsilon_u^{n+1}\varepsilon_v^{n+1}(\rho_{u,v}-q_{u,v}))$$
$$+\frac{1}{2}\sum_{\ell\le n}\sum_{u\le d}\nu_t\left(f\varepsilon_u^\ell\left(\frac{1}{\sqrt{t}}g(\boldsymbol{\rho}_u^\ell)-\frac{1}{\sqrt{1-t}}Y_u\right)\right)$$
$$-\frac{n}{2}\sum_{u\le d}\nu_t\left(f\varepsilon_u^{n+1}\left(\frac{1}{\sqrt{t}}g(\boldsymbol{\rho}_u^{n+1})-\frac{1}{\sqrt{1-t}}Y_u\right)\right)\,. \tag{1.377}$$

The first two terms are produced by the last term of the Hamiltonian (1.375), and the last 2 terms by the dependence of $g_t(\boldsymbol{\rho})$ on t. One then performs Gaussian integration by parts in the last two terms of (1.377), which yields an expression similar to (1.376), except that one has $q_{u,v}$ rather than $q_{u,v}(\ell,\ell')$ everywhere. Combining this with the first two terms on the right-hand side of (1.377) yields (1.376). □

The proof of Theorem 1.12.1 will follow the scheme of that of Proposition 1.6.6, but getting a dependence on d of the correct order requires some caution.

Corollary 1.12.5. *If $n=2$, we have*

$$|\nu_t'(f)|\le L\beta^2 d\,\nu_t^{1/2}(f^2)\left(\nu_t\left(\sum_{u,v\le d}(R_{1,1}^{-,u,v}-\rho_{u,v})^2\right)^{1/2}\right.$$
$$\left.+\,\nu_t\left(\sum_{u,v\le d}(R_{1,2}^{-,u,v}-q_{u,v})^2\right)^{1/2}\right) \tag{1.378}$$

and also

$$|\nu_t'(f)|\le L\beta^2 d^2\nu_t(|f|)\,. \tag{1.379}$$

Here and throughout the book we lighten notation by writing $\nu_t(f)^{1/2}$ rather than $(\nu_t(f))^{1/2}$, etc. The quantity $\nu_t(f)^{1/2}$ cannot be confused with the quantity $\nu_t((f)^{1/2})$ simply because we will never, ever, consider this latter quantity.

Proof. We write

$$\sum_{u,v\le d}\nu_t(f\varepsilon_u^\ell\varepsilon_v^{\ell'}(R_{\ell,\ell'}^{-,u,v}-q_{u,v}(\ell,\ell')))=\nu_t\left(\sum_{u,v\le d}f\varepsilon_u^\ell\varepsilon_v^{\ell'}(R_{\ell,\ell'}^{-,u,v}-q_{u,v}(\ell,\ell'))\right)\,.$$

Next, we observe that since we are assuming that for each i we have

$$\sum_{u\le d}\sigma_{i,u}^2\le d\,, \tag{1.380}$$

taking $i=N$, for each ℓ we have

$$\sum_{u\le d}(\varepsilon_u^\ell)^2\le d\,. \tag{1.381}$$

Now, by the Cauchy-Schwarz inequality, and using (1.380), we have

$$\begin{aligned}
&\Big|\sum_{u,v\le d} \varepsilon_u^\ell \varepsilon_v^{\ell'} (R_{\ell,\ell'}^{-,u,v} - q_{u,v}(\ell,\ell'))\Big| \\
&\le \Big(\sum_{u,v\le d} (\varepsilon_u^\ell \varepsilon_v^{\ell'})^2\Big)^{1/2} \Big(\sum_{u,v\le d} (R_{\ell,\ell'}^{-,u,v} - q_{u,v}(\ell,\ell'))^2\Big)^{1/2} \\
&\le d\Big(\sum_{u,v\le d} (R_{\ell,\ell'}^{-,u,v} - q_{u,v}(\ell,\ell'))^2\Big)^{1/2} ,
\end{aligned}$$

so that use of the Cauchy-Schwarz inequality for ν_t shows that

$$\begin{aligned}
&\Big|\nu_t\Big(\sum_{u,v\le d} f\varepsilon_u^\ell \varepsilon_v^{\ell'} (R_{\ell,\ell'}^{-,u,v} - q_{u,v}(\ell,\ell'))\Big)\Big| \\
&\le d\,\nu_t^{1/2}(f^2)\nu_t\Big(\sum_{u,v\le d} (R_{\ell,\ell'}^{-,u,v} - q_{u,v}(\ell,\ell'))^2\Big)^{1/2} .
\end{aligned}$$

The right-hand side takes only two possible values, depending on whether $\ell = \ell'$ or not. This yields (1.378).

To deduce (1.379) from (1.376), it suffices to show that, for each ℓ, ℓ' we have

$$\Big|\sum_{u,v\le d} \varepsilon_u^\ell \varepsilon_v^{\ell'} (R_{\ell,\ell'}^{-,u,v} - q_{u,v}(\ell,\ell'))\Big| \le 2d^2 .$$

Using (1.381) and the Cauchy-Schwarz inequality, it suffices to prove that

$$\sum_{u,v\le d} (R_{\ell,\ell'}^{-,u,v} - q_{u,v}(\ell,\ell'))^2 \le 4d^2 ,$$

which follows from

$$\sum_{u,v\le d} (R_{\ell,\ell'}^{-,u,v})^2 \le d^2 \; ; \; \sum_{u,v\le d} q_{u,v}(\ell,\ell')^2 \le d^2 . \tag{1.382}$$

To prove this we observe first that

$$\begin{aligned}
\sum_{u,v\le d} (R_{\ell,\ell'}^{-,u,v})^2 &= \sum_{u,v\le d} \Big(\frac{1}{N}\sum_{i\le N-1} \sigma_{i,u}^\ell \sigma_{i,v}^{\ell'}\Big)^2 \\
&\le \frac{1}{N} \sum_{u,v\le d}\sum_{i\le N-1} (\sigma_{i,u}^\ell \sigma_{i,v}^{\ell'})^2 \le d^2
\end{aligned}$$

by (1.380). Next, we observe by (1.372) that

$$q_{u,v}^2 \le \mathsf{E}\Big(\frac{1}{Z^4}\Big(\int x_u \mathcal{E}(x)\mathrm{d}\mu(x)\Big)^2 \Big(\int x_v \mathcal{E}(x)\mathrm{d}\mu(x)\Big)^2\Big) .$$

The Cauchy-Schwarz inequality implies

$$\left(\int x_u \mathcal{E}(x) \mathrm{d}\mu(x)\right)^2 \leq Z \int x_u^2 \mathcal{E}(x) \mathrm{d}\mu(x)$$

so that

$$\sum_{u,v} q_{u,v}^2 \leq \mathsf{E}\left(\frac{1}{Z^2} \int \sum_u x_u^2 \mathcal{E}(x) \mathrm{d}\mu(x) \int \sum_v x_v^2 \mathcal{E}(x) \mathrm{d}\mu(x)\right) \leq d^2$$

since $\sum_{u \leq d} x_u^2 \leq d$ for x in the support of μ. The inequality $\sum_{u,v} \rho_{u,v}^2 \leq d^2$ is similar. □

Proof of Theorem 1.12.1. In this proof we assume the existence of numbers $q_{u,v}$, $\rho_{u,v}$ satisfying (1.372) and (1.373). This existence will be proved later. Symmetry between sites implies

$$A := \nu\left(\sum_{u,v \leq d} (R_{1,2}^{u,v} - q_{u,v})^2\right) = \nu(f) \ , \tag{1.383}$$

where

$$f = \sum_{u,v \leq d} (\varepsilon_u^1 \varepsilon_v^2 - q_{u,v})(R_{1,2}^{u,v} - q_{u,v}) \ .$$

Using (1.381) and (1.382) we obtain

$$\sum_{u,v \leq d} (\varepsilon_u^1 \varepsilon_v^2 - q_{u,v})^2 \leq 2 \sum_{u,v \leq d} (\varepsilon_u^1 \varepsilon_v^2)^2 + 2 \sum_{u,v \leq d} q_{u,v}^2 \leq 4d^2 \ ,$$

and the Cauchy-Schwarz inequality entails

$$f^2 \leq 4d^2 \sum_{u,v \leq d} (R_{1,2}^{u,v} - q_{u,v})^2 \ . \tag{1.384}$$

Next, as in the case of the ordinary SK model, (1.372) implies that for a function f^- on S_{N-1}^n, we have

$$\nu_0((\varepsilon_u^1 \varepsilon_v^2 - q_{u,v}) f^-) = 0$$

and thus, as in the case of the ordinary SK model,

$$|\nu_0(f)| \leq \frac{K(d)}{N} \ .$$

If $\beta d \leq 1$, (1.379) implies that $\nu_t(f) \leq L \nu_1(f)$ whenever $f \geq 0$. Combining this with (1.378), and the usual relation

$$\nu(f) \leq \nu_0(f) + \sup_{0<t<1} |\nu_t'(f)| \ ,$$

we get that

$$\nu(f) \le \frac{K(d)}{N} + L\beta^2 d\,\nu(f^2)^{1/2}\biggl(\nu\Bigl(\sum_{u,v\le d}(R_{1,1}^{-,u,v}-\rho_{u,v})^2\Bigr)^{1/2} + \nu\Bigl(\sum_{u,v\le d}(R_{1,2}^{-,u,v}-q_{u,v})^2\Bigr)^{1/2}\biggr) .$$

Using (1.383), (1.384) and the fact that replacing $R_{1,2}^{-,u,v}$ by $R_{1,2}^{u,v}$ or $R_{1,1}^{-,u,v}$ by $R^{u,v}$ creates an error term of at most $K(d)/N$, we get the relation

$$A \le \frac{K(d)}{N} + L\beta^2 d^2 A^{1/2}(B^{1/2}+A^{1/2}) , \tag{1.385}$$

where A is defined in (1.383) and

$$B = \nu\Bigl(\sum_{u,v\le d}(R^{u,v}-\rho_{u,v})^2\Bigr) .$$

The same argument (using now (1.373) rather than (1.372)) yields the relation

$$B \le \frac{K(d)}{N} + L\beta^2 d^2 B^{1/2}(B^{1/2}+A^{1/2}) .$$

Combining with (1.385) we get

$$A+B \le \frac{K(d)}{N} + L_0\beta^2 d^2(A+B) ,$$

so that if $L_0\beta^2 d^2 \le 1/2$ this implies that $A+B \le K(d)/N$. □

The above arguments prove Theorems 1.12.1, except that it remains to show the existence of solutions to the equations (1.372) and (1.373). It seems to be a general fact that "the proof of the existence at high temperature of solutions to the replica-symmetric equations is implicitly part of the proof of the validity of the replica-symmetric solution". What we mean here is that an argument proving the existence of a solution to (1.372) and (1.373) can be extracted from the smart path method as used in the above proof of Theorem 1.12.1. The same phenomenon will occur in many places.

Consider a positive definite symmetric matrix $Q = (q_{u,v})_{u,v\le d}$, and a symmetric matrix $Q' = (\rho_{u,v})_{u,v\le d}$. Consider a centered jointly Gaussian family $(Y_u)_{u\le d}$ as in (1.370). Consider the matrices $T(Q,Q')$ and $T'(Q,Q')$ given by the right-hand sides of (1.372) and (1.373) respectively. The proof of the existence of a solution to (1.372) and (1.373) consists in showing that if we provide the set of pairs of matrices (Q,Q') as above with Euclidean distance (when seen as a subset of $(\mathbb{R}^{d^2})^2$), the map $(Q,Q') \mapsto (T(Q,Q'),T'(Q,Q'))$ is a contraction provided $L\beta d \le 1$. (Thus it admits a unique fixed point.) To see this, considering another pair $(\widehat{Q}, \widehat{Q}')$ of matrices, we move from the pair (Q, Q') to the pair $(\widehat{Q}, \widehat{Q}')$ using the path $t \mapsto (Q(t), Q'(t))$, where

$$Q(t) = (tq_{u,v} + (1-t)\widehat{q}_{u,v})_{u,v\le d} \tag{1.386}$$

$$Q'(t) = (t\rho_{u,v} + (1-t)\widehat{\rho}_{u,v})_{u,v\le d} \; .$$

As already observed on page 14 this is very closely related to the smart path used in the proof of Theorem 1.12.1, since (with obvious notation) the Gaussian process $Y_u(t)$ associated to $Q(t)$ is given by

$$Y_u(t) = \sqrt{t}Y_u + \sqrt{1-t}\widehat{Y}_u$$

where Y_u, $\widehat{Y}_u$ are assumed to be independent. This is simply because

$$\mathsf{E}\, Y_u(t)Y_v(t) = t\mathsf{E}\, Y_uY_v + (1-t)\mathsf{E}\, \widehat{Y}_u\widehat{Y}_v \; .$$

All we have to do is to compute the derivative of the map $t \mapsto (Q(t), Q'(t))$ and to exhibit a convenient upper bound for the modulus of this derivative, depending on the distance between the pairs (Q, Q') and $(\widehat{Q}, \widehat{Q}')$, i.e. on

$$\left(\sum_{u,v}(q_{u,v} - \widehat{q}_{u,v})^2 + (\rho_{u,v} - \widehat{\rho}_{u,v})^2\right)^{1/2} .$$

The estimates required are very similar to those of Corollary 1.12.5 and the details are better left to the reader.

Proof of Theorem 1.12.2. We just proved the existence of solutions to the equations (1.372) and (1.373). The uniqueness follows from Theorem 1.12.1. □

We begin our preparations for the proof of Theorem 1.12.3. It seems very likely that one could use interpolation as in (1.108) or adapt the proof of (1.170). We sketch yet another approach, which is rather instructive in a different way. We start with the relation

$$\begin{aligned}\frac{\partial p_N}{\partial \beta}(\beta, \mu) &= \frac{1}{N^{3/2}}\sum_{i<j}\sum_{u\le d}\mathsf{E}\,(g_{ij}\langle\sigma_{i,u}\sigma_{j,u}\rangle) \\ &= \frac{\beta}{N^2}\sum_{i<j}\sum_{u,v\le d}(\nu(\sigma_{i,u}\sigma_{j,u}\sigma_{i,v}\sigma_{j,v}) - \nu(\sigma^1_{i,u}\sigma^1_{j,u}\sigma^2_{i,v}\sigma^2_{j,v})),\end{aligned}$$

where the first equality is by straightforward differentiation, and the second one by integration by parts. Thus, since e.g.

$$\frac{1}{N^2}\sum_{i<j}\sigma^1_{i,u}\sigma^1_{j,u}\sigma^2_{i,v}\sigma^2_{j,v} = \frac{1}{2}(R^{u,v}_{1,2})^2 - \frac{1}{2N^2}\sum_{i\le N}(\sigma^1_{i,u}\sigma^2_{i,v})^2 \; ,$$

we obtain

$$\left|\frac{\partial p_N}{\partial \beta}(\beta,\mu) - \frac{\beta}{2}\sum_{u,v}\nu((R^{u,v})^2 - (R^{u,v}_{1,2})^2)\right|$$

$$\leq \frac{\beta}{N^2}\sum_{u,v}\left|\sum_{i\leq N}\nu((\sigma_{i,u}\sigma_{i,v})^2 - (\sigma^1_{i,u}\sigma^2_{i,v})^2)\right| \leq \frac{K(d)}{N}$$

and thus, by Theorem 1.12.1,

$$\left|\frac{\partial p_N}{\partial \beta}(\beta,\mu) - \frac{\beta}{2}\sum_{u,v}(\rho^2_{u,v} - q^2_{u,v})\right| \leq \frac{K(d)}{\sqrt{N}}\;.$$

Therefore, (and since the result is obvious for $\beta = 0$) all we have to check is that the derivative of

$$-\frac{\beta^2}{4}\sum_{u,v\leq d}(\rho^2_{u,v} - q^2_{u,v}) + \mathsf{E}\log\int \mathcal{E}(x)\mathrm{d}\mu(x) \tag{1.387}$$

with respect to β is $\beta\sum_{u,v\leq d}(\rho^2_{u,v} - q^2_{u,v})/2$. The crucial fact is as follows.

Lemma 1.12.6. *The relations (1.372) and (1.373) mean that the partial derivatives of the quantity (1.387) with respect to* $q_{u,v}$ *and* $\rho_{u,v}$ *are zero.*

The reader will soon observe that each time we succeed in computing the limiting value of p_N for a certain model, we find this limit as a function F of certain parameters (here β, μ, $(q_{u,v})$ and $(\rho_{u,v})$). Some of these parameters are intrinsic to the model (here β and μ) while others are "free" (here $(q_{u,v})$ and $(\rho_{u,v})$). It seems to be a general fact that the "free parameters" are determined by the fact that the partial derivatives of the function F with respect to these are 0.

Research Problem 1.12.7. Such a phenomenon as just described above cannot be accidental. Understand the underlying structure.

Proof of Lemma 1.12.6. The case of the derivative with respect to $\rho_{u,v}$ is completely straightforward, so we explain only the case of the derivative with respect to $q_{u,v}$. We recall the definition (1.371) of $\mathcal{E}(x)$:

$$\mathcal{E}(x) = \exp\left(\sum_{u'\leq d} x_{u'}Y_{u'} + \frac{\beta^2}{2}\sum_{u',v'\leq d} x_{u'}x_{v'}(\rho_{u',v'} - q_{u',v'})\right),$$

where the r.v.s $Y_{u'}$ are jointly Gaussian and satisfy $\mathsf{E}Y_{u'}Y_{v'} = \beta^2 q_{u',v'}$. Let us now consider another jointly Gaussian family $W_{u'}$ and let $a_{u',v'} = \mathsf{E}W_{u'}W_{v'}$. Let us define

$$\mathcal{E}^*(x) = \exp\Bigl(\sum_{u' \le d} x_{u'} W_{u'} + \frac{\beta^2}{2} \sum_{u',v' \le d} x_{u'} x_{v'} (\rho_{u',v'} - q_{u',v'})\Bigr) ,$$

which we think of as a function of the families $(q_{u',v'})$ and $(a_{u',v'})$ (the quantities $(\rho_{u',v'})$ being fixed once and for all). The purpose of this is to distinguish the two different manners in which $\mathcal{E}(x)$ depends on $q_{u,v}$. Thus we have

$$\frac{\partial}{\partial q_{u,v}} \mathsf{E} \log \int \mathcal{E}(x) \mathrm{d}\mu(x) = \mathrm{I} + \mathrm{II} , \tag{1.388}$$

where

$$\mathrm{I} = \frac{\partial}{\partial q_{u,v}} \mathsf{E} \log \int \mathcal{E}^*(x) \mathrm{d}\mu(x) , \tag{1.389}$$

and

$$\mathrm{II} = \beta^2 \frac{\partial}{\partial a_{u,v}} \mathsf{E} \log \int \mathcal{E}^*(x) \mathrm{d}\mu(x) . \tag{1.390}$$

In both these relations, $\mathcal{E}^*(x)$ is computed at the values $a_{u',v'} = \beta^2 q_{u',v'}$. To perform the computation, on has to keep in mind that

$$\sum_{u',v' \le d} x_{u'} x_{v'} (\rho_{u',v'} - q_{u',v'}) = 2 \sum_{1 \le u' < v' \le d} x_{u'} x_{v'} (\rho_{u',v'} - q_{u',v'}) + \sum_{u' \le d} x_{u'}^2 (\rho_{u',u'} - q_{u',u'}) .$$

For simplicity we will consider only the case $u < v$. The case $u = v$ is entirely similar. Recalling the notation $Z = \int \mathcal{E}(x) \mathrm{d}\mu(x)$, it should be obvious that

$$\mathrm{I} = -\beta^2 \mathsf{E}\Bigl(\frac{1}{Z} \int x_u x_v \mathcal{E}(x) \mathrm{d}\mu(x)\Bigr) . \tag{1.391}$$

To compute the term II we consider the function

$$G(y_1, \ldots, y_d) = \log \int \exp\Bigl(\sum_{u' \le d} x_{u'} y_{u'} + \frac{\beta^2}{2} \sum_{u',v' \le d} x_{u'} x_{v'} (\rho_{u',v'} - q_{u',v'})\Bigr) \mathrm{d}\mu(x) ,$$

so that

$$\mathsf{E} \log \int \mathcal{E}^*(x) \mathrm{d}\mu(x) = \mathsf{E} G(W_1, \ldots, W_d) ,$$

and to compute the term II we simply appeal to Proposition 1.3.2. Since we assume $u \neq v$ we obtain

$$\frac{\partial}{\partial a_{u,v}} \mathsf{E} \log \int \mathcal{E}^*(x) \mathrm{d}\mu(x) = \mathsf{E} \frac{\partial^2 G}{\partial y_u \partial y_v} (W_1, \ldots, W_d) ,$$

and when this is computed at the values $a_{u',v'} = \beta^2 q_{u',v'}$ this is

$$\mathsf{E}\Big(\frac{1}{Z}\int x_u x_v \mathcal{E}(x)\mathrm{d}\mu(x)\Big) - \mathsf{E}\Big(\frac{1}{Z^2}\int x_u \mathcal{E}(x)\mathrm{d}\mu(x)\int x_v \mathcal{E}(x)\mathrm{d}\mu(x)\Big)\,.$$

Recalling (1.388) this yields the formula

$$\frac{\partial}{\partial q_{u,v}}\mathsf{E}\log\int \mathcal{E}(x)\mathrm{d}\mu(x) = -\beta^2\mathsf{E}\Big(\frac{1}{Z^2}\int x_u \mathcal{E}(x)\mathrm{d}\mu(x)\int x_v \mathcal{E}(x)\mathrm{d}\mu(x)\Big)\,,$$

from which the conclusion readily follows. □

Proof of Theorem 1.12.3. It follows from Lemma 1.12.6 that to differentiate in β the quantity (1.387) we can pretend that $q_{u,v}$ and $\rho_{u,v}$ do not depend on β. To explain why this is the case in a situation allowing for simpler notation, this is simple consequence of the chain rule,

$$\frac{\mathrm{d}}{\mathrm{d}\beta}F(\beta,p(\beta),q(\beta)) = \frac{\partial F}{\partial \beta} + p'(\beta)\frac{\partial F}{\partial p} + q'(\beta)\frac{\partial F}{\partial q}\,, \tag{1.392}$$

so that $\mathrm{d}F(\beta,p(\beta),q(\beta))/\mathrm{d}\beta = \partial F/\partial\beta$ when the last two partial derivatives of (1.392) are 0. Thus it suffices to prove that

$$\frac{\partial}{\partial\beta}\mathsf{E}\log\int \mathcal{E}(x)\mathrm{d}\mu(x) = \beta\sum_{u,v}(\rho_{u,v}^2 - q_{u,v}^2)\,. \tag{1.393}$$

Consider a jointly Gaussian family $(X_u)_{u\le d}$ such that $\mathsf{E}X_uX_v = q_{u,v}$, and which, like $q_{u,v}$, we may pretend does not depend on β. We may choose these so that $Y_u = \beta X_u$ and now

$$\frac{\partial}{\partial\beta}\mathcal{E}(x) = \Big(\sum_{u\le d} x_u X_u + \beta\sum_{u,v\le d} x_u x_u(\rho_{u,v} - q_{u,v})\Big)\mathcal{E}(x)\,.$$

Therefore, using (1.373) in the third line,

$$\begin{aligned}
&\mathsf{E}\frac{\partial}{\partial\beta}\log\int\mathcal{E}(x)\mathrm{d}\mu(x)\\
&= \mathsf{E}\frac{\int(\sum_{u\le d}x_uX_u + \beta\sum_{u,v\le d}x_ux_u(\rho_{u,v}-q_{u,v}))\mathcal{E}(x)\mathrm{d}\mu(x)}{\int\mathcal{E}(x)\mathrm{d}\mu(x)}\\
&= \beta\sum_{u,v\le d}(\rho_{u,v}^2 - \rho_{u,v}q_{u,v}) + \mathsf{E}\frac{\int\sum_{u\le d}x_uX_u\mathcal{E}(x)\mathrm{d}\mu(x)}{\int\mathcal{E}(x)\mathrm{d}\mu(x)}\,.
\end{aligned}$$

Using Gaussian integration by parts and (1.372) and (1.373) we then reach that

$$\begin{aligned}
\mathsf{E}\frac{\int\sum_{u\le d}x_uX_u\mathcal{E}(x)\mathrm{d}\mu(x)}{\int\mathcal{E}(x)\mathrm{d}\mu(x)} &= \beta\sum_{u,v\le d}q_{u,v}\mathsf{E}\frac{\int x_ux_v\mathcal{E}(x)\mathrm{d}\mu(x)}{\int\mathcal{E}(x)\mathrm{d}\mu(x)}\\
&\quad - \beta\sum_{u,v\le d}q_{u,v}\mathsf{E}\frac{\int x_u\mathcal{E}(x)\mathrm{d}\mu(x)\int x_v\mathcal{E}(x)\mathrm{d}\mu(x)}{(\int\mathcal{E}(x)\mathrm{d}\mu(x))^2}\\
&= \beta\sum_{u,v}(q_{u,v}\rho_{u,v} - q_{u,v}^2)\,,
\end{aligned}$$

and this completes the proof of (1.393). □

Exercise 1.12.8. Find another proof of (1.393) using Proposition 1.3.2 as in Lemma 1.12.6.

One should comment that the above method of taking the derivative in β is rather similar in spirit to the method of (1.108); but unlike the proof of (1.105) it does not use the "right path", and as a penalty one would have to work to get the correct rate of convergence K/N instead of obtaining it for free.

Exercise 1.12.9. Write down a complete proof of Theorem 1.12.3 using interpolation in the spirit of (1.108).

Research Problem 1.12.10. (Level 1) In this problem ν refers to the Hamiltonian H_N of (1.12). Consider a number λ and the following random function on Σ_N

$$\varphi(\boldsymbol{\sigma}) = \frac{1}{N} \log \sum_{\boldsymbol{\tau}} \exp(\lambda N R(\boldsymbol{\sigma}, \boldsymbol{\tau}) - H_N(\boldsymbol{\tau})) . \tag{1.394}$$

Develop the tools to be able to compute (when β is small enough) the quantity $\nu(\varphi(\boldsymbol{\sigma}))$. Compute also $\nu(\varphi(\boldsymbol{\sigma})^2)$.

The relationship with the material of the present section is that by Jensen's inequality we have

$$\begin{aligned}\langle \varphi(\boldsymbol{\sigma}) \rangle \le & \frac{1}{N} \log \sum_{\boldsymbol{\sigma},\boldsymbol{\tau}} \exp(\lambda N R(\boldsymbol{\sigma}, \boldsymbol{\tau}) - H_N(\boldsymbol{\sigma}) - H_N(\boldsymbol{\tau})) \\ & - \frac{1}{N} \log \sum_{\boldsymbol{\sigma}} \exp(-H_N(\boldsymbol{\sigma})) ,\end{aligned}$$

and that the expected value of this quantity can be computed using (1.374) by a suitable choice of μ in a 2-component spin model.

A possible solution to Problem 1.12.10 involves developing the cavity method in a slightly different setting than we have done so far. Carrying out the details should be a very good exercise for the truly interested reader.

Research Problem 1.12.11. (Level 2). With the notation above, is it true that at *any* temperature for large N one has

$$\nu(\varphi^2) \simeq \nu(\varphi)^2 \ ? \tag{1.395}$$

Quantities similar to those above are considered in physics, see e.g. [43]. The physicists find it natural to assert that the quantity (1.394) is "self-averaging", which means here that it is essentially independent of the disorder and the value of $\boldsymbol{\sigma}$ (when weighted with the Gibbs measure), which is the meaning of (1.395).

1.13 The Physicist's Replica Method

Physicists have discovered their results about the SK model using the "replica method", a method that has certainly contributed to arouse the interest of mathematicians in spin glasses. In this section, we largely follow the paper [81], where the authors attempt as far as possible to make the replica method rigorous. We start with the following, where we consider only the case of non-random external field.

Theorem 1.13.1. *Consider an integer $n \geq 1$. Then*

$$\lim_{N\to\infty} \frac{1}{N} \log \mathsf{E}\, Z_N^n(\beta, h) = n \log 2 + \max_q \left(\frac{n\beta^2}{4}(1-q)^2 - \frac{n^2\beta^2}{4} q^2 + \log \mathsf{E} \operatorname{ch}^n(\beta z\sqrt{q} + h) \right) \tag{1.396}$$

where z is standard normal.

We do not know if the arguments we will present extend to the case of random external field, but (1.396) remains true in that case, and even if $n \geq 1$ is not an integer. This is proved in [159]. The proof uses a fundamental principle called the Ghirlanda-Guerra identities that we will present in Section 12.5 when we start to concentrate on low-temperature results. In some sense this general argument is much more interesting than the specific arguments of the present section, which, however beautiful, look more like tricks than general principles.

To prove (1.396) we write

$$Z_N^n = \sum_{\boldsymbol{\sigma}} \exp \sum_{\ell \leq n} \left(\frac{\beta}{\sqrt{N}} \sum_{i<j} g_{ij} \sigma_i^\ell \sigma_j^\ell + h \sum_{i \leq N} \sigma_i^\ell \right).$$

Now we have

$$\begin{aligned} \mathsf{E}\left(\sum_{i<j} g_{ij} \sum_{\ell \leq n} \sigma_i^\ell \sigma_j^\ell \right)^2 &= \sum_{i<j} \left(\sum_{\ell \leq n} \sigma_i^\ell \sigma_j^\ell \right)^2 \\ &= \sum_{\ell,\ell'} \sum_{i<j} \sigma_i^\ell \sigma_i^{\ell'} \sigma_j^\ell \sigma_j^{\ell'} \\ &= \frac{1}{2} \sum_{\ell,\ell'} \left(\left(\sum_{i \leq N} \sigma_i^\ell \sigma_i^{\ell'} \right)^2 - N \right) \\ &= \frac{1}{2}(nN^2 - n^2 N) + \sum_{1 \leq \ell < \ell' \leq n} (\boldsymbol{\sigma}^\ell \cdot \boldsymbol{\sigma}^{\ell'})^2 , \end{aligned} \tag{1.397}$$

so that combining (1.397) with (A.6) we get

$$\mathsf{E}\, Z_N^n = \exp\Big(\frac{\beta^2}{4} n(N-n)\Big)$$
$$\times \sum_{\boldsymbol{\sigma}} \exp\Big(\frac{\beta^2}{2N} \sum_{1 \le \ell < \ell' \le n} (\boldsymbol{\sigma}^\ell \cdot \boldsymbol{\sigma}^{\ell'})^2 + h \sum_{\ell \le n, i \le N} \sigma_i^\ell\Big) , \quad (1.398)$$

where $\sum_{\boldsymbol{\sigma}}$ means that the summation is over $(\boldsymbol{\sigma}^1, \ldots, \boldsymbol{\sigma}^n)$ in Σ_N^n. Consider now $\mathbf{g} = (g_{\ell,\ell'})_{1 \le \ell < \ell' \le n}$ where $(g_{\ell,\ell'})$ are i.i.d. Gaussian r.v.s with $\mathsf{E}\, g_{\ell,\ell'}^2 = 1/N$. (Despite the similarity in notation these r.v.s play a very different rôle than the interaction r.v.s (g_{ij}).) It follows from (A.6) that

$$\sum_{\boldsymbol{\sigma}} \exp\Big(\frac{\beta^2}{2N} \sum_{1 \le \ell < \ell' \le n} (\boldsymbol{\sigma}^\ell \cdot \boldsymbol{\sigma}^{\ell'})^2 + h \sum_{\ell \le n, i \le N} \sigma_i^\ell\Big)$$
$$= \mathsf{E} \sum_{\boldsymbol{\sigma}} \exp\Big(\beta \sum_{1 \le \ell < \ell' \le n} g_{\ell,\ell'} \boldsymbol{\sigma}^\ell \cdot \boldsymbol{\sigma}^{\ell'} + h \sum_{\ell \le n, i \le N} \sigma_i^\ell\Big)$$
$$= \mathsf{E} \sum_{\boldsymbol{\sigma}} \prod_{i \le N} \exp\Big(\beta \sum_{1 \le \ell < \ell' \le n} g_{\ell,\ell'} \sigma_i^\ell \sigma_i^{\ell'} + h \sum_{\ell \le n} \sigma_i^\ell\Big)$$
$$= \mathsf{E} \Bigg(\sum_{\varepsilon_1, \ldots, \varepsilon_n = \pm 1} \exp\Big(\beta \sum_{1 \le \ell < \ell' \le n} g_{\ell,\ell'} \varepsilon_\ell \varepsilon_{\ell'} + h \sum_{\ell \le n} \varepsilon_\ell\Big)\Bigg)^N$$
$$= \mathsf{E} \exp N A(\mathbf{g}) ,$$

where

$$A(\mathbf{g}) = \log \sum_{\varepsilon_1, \ldots, \varepsilon_n = \pm 1} \exp\Big(\beta \sum_{1 \le \ell < \ell' \le n} g_{\ell,\ell'} \varepsilon_\ell \varepsilon_{\ell'} + h \sum_{\ell \le n} \varepsilon_\ell\Big) .$$

Now,

$$\mathsf{E} \exp N A(\mathbf{g}) = \Big(\frac{N}{2\pi}\Big)^{n(n-1)/4} \int \exp N\Big(A(\mathbf{g}) - \frac{1}{2} \sum_{1 \le \ell < \ell' \le n} g_{\ell,\ell'}^2\Big) \mathrm{d}\mathbf{g}$$

where the integral is taken with respect to Lebesgue's measure $\mathrm{d}\mathbf{g}$ on $\mathbb{R}^{n(n-1)/2}$. Since $|A(\mathbf{g})| \le K(\|\mathbf{g}\| + 1)$, it is elementary to show that

$$\lim_{N \to \infty} \frac{1}{N} \log \mathsf{E} \exp N A(\mathbf{g}) = \max_{\mathbf{g}} \Big(A(\mathbf{g}) - \frac{1}{2} \sum_{\ell < \ell'} g_{\ell,\ell'}^2\Big)$$

and from (1.398) we get

$$\lim_{N \to \infty} \frac{1}{N} \log \mathsf{E}\, Z_N^n = \frac{\beta^2 n}{4} + \max_{\mathbf{g}} \Big(A(\mathbf{g}) - \frac{1}{2} \sum_{\ell < \ell'} g_{\ell,\ell'}^2\Big) . \quad (1.399)$$

Consider the function

$$B(\mathbf{g}) = A(\mathbf{g}) - \frac{1}{2} \sum_{1 \le \ell < \ell' \le n} g_{\ell,\ell'}^2 \,.$$

We say that $\mathbf{g}$ is a *maximizer* of B if B attains its maximum at $\mathbf{g}$. The following is based on an idea of [81] (attributed to Elliot Lieb) and further elaboration of this idea by D. Panchenko.

Proposition 1.13.2. *a) If $h > 0$ and $\mathbf{g}$ is a maximizer of B, there exists a number $a \ge 0$ with $g_{\ell,\ell'} = a$ for each $1 \le \ell < \ell' \le n$.*
b) If $h = 0$ and $\mathbf{g}$ is a maximizer of B, there exists a number $a \ge 0$ and a subset I of $\{1, \ldots, n\}$ such that $g_{\ell,\ell'} = a$ if $\ell, \ell' \in I$ or $\ell, \ell' \notin I$ and $g_{\ell,\ell'} = -a$ in the other cases.

Let us denote by $\mathbf{a}$ the sequence such that $a_{\ell,\ell'} = a$ for $1 \le \ell < \ell' \le n$. In the case b), we have $B(\mathbf{g}) = B(\mathbf{a})$ as is shown by the transformation $\varepsilon'_\ell = \varepsilon_\ell$ if $\ell \in I$ and $\varepsilon'_\ell = -\varepsilon_\ell$ if $\ell \notin I$. Therefore the maximizer cannot be unique for symmetry reasons. In the case a), this symmetry is broken by the external field.

Corollary 1.13.3. *To compute $\max_{\mathbf{g}} B(\mathbf{g})$ it suffices to maximize over the sequences $\mathbf{g}$ where all coordinates are equal.*

We start the proof of Proposition 1.13.2. The proof is pretty but is unrelated to any other argument in this work. It occupies the next two and a half pages. The fun argument starts again on page 145.

Lemma 1.13.4. *Consider numbers a_1, a_2, g. Then*

$$\begin{aligned} &\mathrm{ch}a_1\mathrm{ch}a_2\mathrm{ch}g + \mathrm{sh}a_1\mathrm{sh}a_2\mathrm{sh}g \\ &\le (\mathrm{ch}^2a_1\mathrm{ch}|g| + \mathrm{sh}^2a_1\mathrm{sh}|g|)^{1/2}(\mathrm{ch}^2a_2\mathrm{ch}|g| + \mathrm{sh}^2a_2\mathrm{sh}|g|)^{1/2} \,. \end{aligned} \tag{1.400}$$

Moreover, if there is equality in (1.400) and if $g \neq 0$, we have $a_1 = a_2$ if $g > 0$ and $a_1 = -a_2$ if $g < 0$.

Proof. For numbers $c_1, c_2, u \ge 0$ and s_1, s_2, v, we write, using the Cauchy-Schwarz inequality in the second line,

$$c_1c_2u + s_1s_2v \le c_1c_2u + |s_1||s_2||v| \tag{1.401}$$
$$\le (c_1^2u + s_1^2|v|)^{1/2}(c_2^2u + s_2^2|v|)^{1/2} \,, \tag{1.402}$$

and we use this for

$$c_j = \mathrm{ch}a_j \ ; \ u = \mathrm{ch}g \ ; \ s_j = \mathrm{sh}a_j \ ; \ v = \mathrm{sh}g \,. \tag{1.403}$$

Then if $g \neq 0$ (so that $|v| = |\mathrm{sh}g| \neq 0$) there can be equality in (1.402) only if for some λ we have

$$(c_1, |s_1|) = \lambda(c_2, |s_2|)$$

i.e. we have $|\text{th}a_1| = |\text{th}a_2|$ and $|a_1| = |a_2|$. If we moreover have equality in (1.401) we have $\text{sh}a_1\text{sh}a_2\text{sh}g = s_1 s_2 v \geq 0$. The result follows. □

Lemma 1.13.5. *Given* $\mathbf{g}$, *consider the sequences* $\mathbf{g}'$ *(resp.* $\mathbf{g}''$*) obtained from* $\mathbf{g}$ *by replacing* $g_{1,2}$ *by* $|g_{1,2}|$ *and* $g_{2,\ell}$ *by* $g_{1,\ell}$ *(resp.* $g_{1,\ell}$ *by* $g_{2,\ell}$*) for* $3 \leq \ell \leq n$. *Now, if* $\mathbf{g}$ *is a maximizer, then both* $\mathbf{g}'$ *and* $\mathbf{g}''$ *are maximizers. Moreover if* $g_{1,2} > 0$ *we have* $g_{1,\ell} = g_{2,\ell}$ *for* $\ell \geq 3$, *while if* $g_{1,2} < 0$ *we have* $g_{1,\ell} = -g_{2,\ell}$ *for* $\ell \geq 3$.

Proof. We will prove that

$$A(\mathbf{g}) \leq \frac{1}{2}A(\mathbf{g}') + \frac{1}{2}A(\mathbf{g}'') . \tag{1.404}$$

Since

$$\sum_{\ell<\ell'} g_{\ell,\ell'}^2 = \frac{1}{2}\left(\sum_{\ell<\ell'}(g'_{\ell,\ell'})^2 + \sum_{\ell<\ell'}(g''_{\ell,\ell'})^2\right) ,$$

this implies

$$B(\mathbf{g}) \leq \frac{1}{2}(B(\mathbf{g}') + B(\mathbf{g}'')) \tag{1.405}$$

so that both $\mathbf{g}'$ and $\mathbf{g}''$ are maximizers. Moreover, since $\mathbf{g}$ is a maximizer, we have $B(\mathbf{g}) = B(\mathbf{g}') = B(\mathbf{g}'')$ so in fact

$$A(\mathbf{g}) = \frac{1}{2}A(\mathbf{g}') + \frac{1}{2}A(\mathbf{g}'') . \tag{1.406}$$

Let us introduce the notation

$$\alpha = (\varepsilon_1, \ldots, \varepsilon_n) \; ; \; A_j(\alpha) = \beta \sum_{3\leq\ell\leq n} g_{j,\ell}\varepsilon_\ell + h \quad \text{for } j = 1, 2$$

$$w(\alpha) = \exp\left(\beta \sum_{3\leq\ell<\ell'\leq n} g_{\ell,\ell'}\varepsilon_\ell\varepsilon_{\ell'} + h \sum_{3\leq\ell\leq n} \varepsilon_\ell\right) .$$

Then, using (1.400) in the last line, we have

$$\begin{aligned}
\exp A(\mathbf{g}) &= \sum_{\varepsilon_1,\ldots,\varepsilon_n=\pm1} \exp\left(\beta \sum_{1\leq\ell<\ell'\leq n} g_{\ell,\ell'}\varepsilon_\ell\varepsilon_{\ell'} + h \sum_{3\leq\ell\leq n} \varepsilon_\ell\right) \\
&= \sum_\alpha w(\alpha) \sum_{\varepsilon_1,\varepsilon_2=\pm1} \exp(A_1(\alpha)\varepsilon_1 + A_2(\alpha)\varepsilon_2 + \beta g_{1,2}\varepsilon_1\varepsilon_2) \\
&= 4\sum_\alpha w(\alpha)(\text{ch}A_1(\alpha)\text{ch}A_2(\alpha)\text{ch}\beta g_{1,2} + \text{sh}A_1(\alpha)\text{sh}A_2(\alpha)\text{sh}\beta g_{1,2}) \\
&\leq 4\sum_\alpha w(\alpha)B_1(\alpha)^{1/2}B_2(\alpha)^{1/2}
\end{aligned} \tag{1.407}$$

where

$$B_j(\alpha) = \mathrm{ch}^2 A_j(\alpha)\mathrm{ch}\beta|g_{1,2}| + \mathrm{sh}^2 A_j(\alpha)\mathrm{sh}\beta|g_{1,2}| \, .$$

The Cauchy-Schwarz inequality implies

$$\begin{aligned} 4\sum_\alpha w(\alpha)B_1(\alpha)^{1/2}B_2(\alpha)^{1/2} &\le \Big(4\sum_\alpha w(\alpha)B_1(\alpha)\Big)^{1/2}\Big(4\sum_\alpha w(\alpha)B_2(\alpha)\Big)^{1/2} \\ &= (\exp A(\mathbf{g}')\exp A(\mathbf{g}''))^{1/2} \, , \end{aligned}$$

where the equality follows from the computation performed in the first three lines of (1.407). Combining with (1.407) proves (1.404).

In order to have (1.406) we must have equality in (1.407). Since each quantity $w(\alpha)$ is > 0, for each α we must have

$$\mathrm{ch}A_1(\alpha)\mathrm{ch}A_2(\alpha)\mathrm{ch}\beta g_{1,2} + \mathrm{sh}A_1(\alpha)\mathrm{sh}A_2(\alpha)\mathrm{sh}\beta g_{1,2} = B_1(\alpha)^{1/2}B_2(\alpha)^{1/2} \, .$$

If $g_{1,2} > 0$ Lemma 1.13.4 shows that $A_1(\alpha) = A_2(\alpha)$ for each α, and thus $g_{1,\ell} = g_{2,\ell}$ for each $\ell \ge 3$. If $g_{1,2} < 0$ Lemma 1.13.4 shows that $A_1(\alpha) = -A_2(\alpha)$ for each α, so that ($h = 0$ and) $g_{1,\ell} = -g_{2,\ell}$ for each $\ell \ge 3$. □

Proof of Proposition 1.13.2. Consider a maximizer $\mathbf{g}$. There is nothing to prove if $\mathbf{g} = \mathbf{0}$, so we assume that this is not the case. In a first step we prove that $|g_{\ell,\ell'}|$ does not depend on ℓ, ℓ'. Assuming $g_{2,3} \ne 0$, we prove that $|g_{1,2}| = |g_{2,3}|$; this clearly suffices. By Lemma 1.13.5, $\mathbf{g}''$ is a maximizer, and by definition $g''_{1,3} = g_{2,3} \ne 0$. Since $g''_{1,3} \ne 0$, and $\mathbf{g}''$ is a maximizer, Lemma 1.13.5 shows that $|g''_{1,\ell}| = |g''_{3,\ell}|$ for $\ell \notin \{1,3\}$, and in particular $|g''_{1,2}| = |g''_{2,3}|$, i.e. $|g_{1,2}| = |g_{2,3}|$.

Next, consider a subset $I \subset \{1,\dots,n\}$ with the property that

$$\ell < \ell' \, , \ \ell,\ell' \in I \ \Rightarrow \ g_{\ell,\ell'} > 0 \, .$$

If no such set exists, $g_{\ell,\ell'} < 0$ for each ℓ,ℓ' and we are done. Otherwise consider I as large as possible. Without loss of generality, assume that $I = \{1,\dots,m\}$, and note that $m \ge 1$. If $m = n$ we are done. Otherwise, consider first $\ell > m$. We observe by Lemma 1.13.5 that if $\ell_1 < \ell_2 \le m$ we have $g_{\ell_1,\ell} = g_{\ell_2,\ell}$, and since we have assumed that I is as large as possible, we have $g_{\ell_1,\ell} < 0$. Next consider $\ell_1 < m < \ell < \ell'$. Then as we have just seen both $g_{\ell_1,\ell}$ and $g_{\ell_1,\ell'}$ are < 0 so that Lemma 1.13.5 shows that $g_{\ell,\ell'} > 0$. Therefore, for a certain number $a \ge 0$ we have, for $\ell < \ell'$

$$\begin{aligned} g_{\ell,\ell'} &= a \ \text{ if } \ell < \ell' \le m \text{ or } m < \ell < \ell' \\ g_{\ell,\ell'} &= -a \ \text{ if } \ell \le m < \ell' \, . \end{aligned}$$

This proves b). To prove a) we observe that when $h > 0$ we have shown that in fact $g_{\ell,\ell'} \ge 0$ when $\mathbf{g}$ is a maximizer. □

We go back to the main computation. By Corollary 1.13.3, in equation (1.399) we can restrict the max to the case where for a certain number q we have $g_{\ell,\ell'} = \beta q$ for each $1 \le \ell < \ell' \le n$. Then

$$\begin{aligned}
&\sum_{\varepsilon_1,\ldots,\varepsilon_n=\pm1} \exp\Big(\beta^2 q \sum_{\ell<\ell'} \varepsilon_\ell \varepsilon_{\ell'} + h \sum_{\ell\le n} \varepsilon_\ell\Big) \\
&= \sum_{\varepsilon_1,\ldots,\varepsilon_n=\pm1} \exp\bigg(\frac{\beta^2 q}{2}\Big(\sum_{\ell\le n} \varepsilon_\ell\Big)^2 - \frac{n\beta^2 q}{2} + h \sum_{\ell\le n} \varepsilon_\ell\bigg) \\
&= \sum_{\varepsilon_1,\ldots,\varepsilon_n=\pm1} \mathsf{E}\exp\Big((\beta z\sqrt{q} + h)\sum_{\ell\le n}\varepsilon_\ell - \frac{n\beta^2 q}{2}\Big) \\
&= \exp\Big(-\frac{n\beta^2 q}{2}\Big)\mathsf{E}(2\,\mathrm{ch}(\beta z\sqrt{q}+h))^n \,,
\end{aligned}$$

where z is a standard Gaussian r.v. and where the summations are over $\varepsilon_1, \ldots, \varepsilon_n = \pm 1$. Thus

$$\begin{aligned}
&\max_{\mathbf{g}}\Big(A(\mathbf{g}) - \frac{1}{2}\sum_{\ell<\ell'} g_{\ell,\ell'}^2\Big) \\
&= \max_q\Big(-\frac{n\beta^2 q}{2} + n\log 2 + \log \mathsf{E}\,\mathrm{ch}^n(\beta z\sqrt{q}+h) - \frac{n(n-1)}{4}\beta^2 q^2\Big) \,.
\end{aligned}$$

Combining this with (1.399) proves (1.396). □

It is a simple computation (using of course Gaussian integration by parts) to see that the maximum in (1.396) is obtained for a value q_n such that

$$q_n = \frac{\mathsf{E}\,(\mathrm{ch}^n Y \mathrm{th}^2 Y)}{\mathsf{E}\,\mathrm{ch}^n Y} \tag{1.408}$$

where $Y = \beta z\sqrt{q_n} + h$.

Let us also observe that

$$\lim_{t\to 0_+} \frac{1}{t}\log \mathsf{E}\, Z_N^t = \mathsf{E}\log Z_N \,, \tag{1.409}$$

as follows from the fact that $Z_N^t \simeq 1 + t\log Z_N$ for small t.

Now we take a deep breath. We pretend that Theorem 1.13.1 is true not only for n integer, but for any number $n > 0$. We rewrite (1.409) as

$$\frac{1}{N}\mathsf{E}\log Z_N = \lim_{n\to 0}\frac{1}{Nn}\log \mathsf{E}\, Z_N^n \,. \tag{1.410}$$

Let us moreover pretend that we can exchange the limits $N \to \infty$ and $n \to 0$. Then presumably $q = \lim_{n\to 0} q_n$ exists, and (1.396) yields

$$\lim_{N\to\infty}\frac{1}{N}\mathsf{E}\log Z_N = \frac{\beta^2}{4}(1-q)^2 + \log 2 + \mathsf{E}\log \mathrm{ch} Y \,, \tag{1.411}$$

where $Y = \beta z\sqrt{q} + h$ and (1.408) becomes $q = \mathsf{E}\,\mathrm{th}^2 Y$.

When trying to justify this procedure one is tempted to think about analytic continuation. However the information contained in Theorem 1.13.1 about the large values of n seems to be completely irrelevant to the problem at hand. To get convinced of this, one can consider the case where $h = 0$ and $\beta < 1$; then it is not difficult to get convinced that $\lim_{n\to\infty} q_n = 1$ (because for n large only the large values of $\mathrm{ch}Y$ become relevant for the computation of $\mathsf{E}\mathrm{ch}^n Y$, and for these values $\mathrm{th}Y$ gets close to one) and this is hard to relate to the fact that $q = 0$.

It is not difficult a posteriori to justify the previous method. The function $\psi_N(t) = N^{-1}\log \mathsf{E}Z_N^t$ is convex (by Hölder's inequality) and for β small enough its limit $\psi(t)$ as $N \to \infty$ exists and is differentiable at zero. (This can be shown by generalizing (1.108) for any $t \neq 0$ using essentially the same method). Therefore $\psi'(0) = \lim_{N\to\infty} \psi'_N(0)$, which means exactly that the exchange of the limits $N \to \infty$ and $n \to 0$ in (1.410) is justified; but of course this has very limited interest since the computation of $\psi(t)$ is not any easier than that of the limit in (1.411).

Moreover the nice formula (1.411) is wrong for large β (low-temperature). The book [105], following ground-breaking work of G. Parisi, attempts to explain how one should (from a physicist's point of view) modify at low temperature the computation (1.396) when $n < 1$. (This is particularly challenging because the number of variables $g_{\ell,\ell'}$, which is $n(n-1)/2$, is negative in that case...) As a mathematician, the author does not feel qualified to try to explain these ideas or even to comment on them.

Hundreds of papers have been written relying on the replica method; the authors of these papers seem to have little doubt that this method always gives the correct answer. Its proponents hope that at some point it will be made rigorous. At the present time however it is difficult, at least for this author, to see in it more than a way to guess the correct formulas. Certainly the predictive power of the method is impressive. The future will tell whether this is the case because its experts are guided by a set of intuitions that is correct at a still deeper level, or whether the power comes from the method itself.

1.14 Notes and Comments

The SK model has a rather interesting history. The paper of Sherrington and Kirkpatrick [136] that introduces the model is called "Solvable model of a spin glass". The authors felt that $\lim_{N\to\infty} p_N(\beta, h) = \mathrm{SK}(\beta, h)$ for all values of β and h. They however already noticed that something must be wrong with this result, and this was confirmed soon after [5]. Whatever one may think of the methods of this branch of theoretical physics (and I do not really know what I think myself about them), their reliability is not guaranteed. One can find a description of these methods in the book of Mézard, Parisi, Virasoro [105], but the only part of the book I feel I understand is the introduction (on which Section 1.1 relies heavily). Two later (possibly more accessible) books about spin glasses written by physicists are [59] and [112]. The recent book by M. Mézard and A. Montanari [102] is much more accessible to a mathematically minded reader. It covers a wide range of topics, and remarkably succeeds at conveying the breath and depth of the physical ideas.

The first rigorous results on the SK model concern only the case $h = 0$. They are proved by Aizenman, Lebowitz and Ruelle in [4] using a "cluster expansion technique", which is a common tool in physics. Their methods seem to apply only to the case $h = 0$. At about the same time, Fröhlich and Zegarlinski [61] prove (as a consequence of a more general approach that is also based on a cluster expansion technique) that the spin correlations vanish if $\beta \leq \beta_0$, even if $h \neq 0$. In fact they prove that

$$\mathsf{E}(\langle\sigma_1\sigma_2\rangle - \langle\sigma_1\rangle\langle\sigma_2\rangle)^2 \leq \frac{L}{N}\,. \tag{1.412}$$

A later paper by Comets and Neveu [49] provides more elegant proofs of several of the main results of [4] using stochastic calculus. Their method unfortunately does not appear to extend beyond the case $h = 0$. They prove a central limit theorem for the overlap $R_{1,2}$.

Theorem 1.3.4 is a special occurrence of the general phenomenon of concentration of measure. This phenomenon was first discovered by P. Lévy, and its importance was brought to light largely through the efforts of V.D. Milman [106]. It is arguably one of the truly great ideas of probability theory. More references, and applications to probability theory can be found in [139] and [140]. In the fundamental case of Gaussian measure, the optimal result is already obtained in [86], and Theorem 1.3.4 is a weak consequence of this result. Interestingly, it took almost 20 years after the paper [86] before results similar to (1.54) were obtained in the theory of disordered systems, by Pastur and Shcherbina [120], using martingale difference sequences. A very nice exposition of most of what is known about concentration of measure can be found in the book of M. Ledoux [93]

It was not immediately understood that, while the case $\beta < 1$, $h = 0$ of the SK model is not very difficult, the case $h \neq 0$ is an entirely different matter. The first rigorous attempt at justifying the mysterious expression in the right-hand side of (1.73) is apparently that of Pastur and Shcherbina [120]. They prove that this formula holds in the domain where

$$\lim_{N\to\infty} \mathrm{Var}\,(\langle R_{1,2}\rangle) = \lim_{N\to\infty} \mathsf{E}(\langle R_{1,2}\rangle - \mathsf{E}\,\langle R_{1,2}\rangle)^2 = 0\,, \tag{1.413}$$

but they do *not* prove that (1.413) is true for small β. Their proof required them to add a strange perturbation term to the Hamiltonian. The result was later clarified by Shcherbina [127], who used the Hamiltonian (1.61) with h_i Gaussian. Using arguments somewhat similar to those of the Ghirlanda-Guerra identities, (which we will study in Volume II) she proved that (1.413) is equivalent (over a certain domain) to

$$\lim_{N\to\infty} \mathsf{E}\langle (R_{1,2} - \langle R_{1,2}\rangle)^2\rangle = 0 \,. \tag{1.414}$$

She did not prove (1.414). She was apparently unaware that (1.414) is proved in [61] for small β. Since the paper [127] was not published, I was not aware of it and rediscovered its results in Section 4 of [141] with essentially the same proof. I also gave a very simple proof of (1.412) for small β. Discovering this simple proof was an absolute disaster, because I wasted considerable energy trying to use the same principle in other situations, which invariably led to difficult proofs of suboptimal results. I will not describe in detail the contents of [141] or my other papers because this now does not seem so interesting any more. I hope that the proofs presented here are much cleaner than those of these previous papers.

In a later paper Shcherbina [128] proved that $\lim_{N\to\infty} p_N(\beta,h) = \mathrm{SK}(\beta,h)$ in a remarkably large region containing in particular all values $\beta < 1$. The ideas of this paper are not really transparent to me. A later version [129] is more accessible, but I became aware of its existence too late to have the energy to analyze it. It would be interesting to decide if this approach succeeds because of a special trick, or if it contains the germ of a powerful method. One should however point out that her use of relations similar to the Ghirlanda-Guerra identities seems to preclude obtaining the correct rates of convergence.

I proved in [149] an expansion somewhat similar to (1.151), using a more complicated method that does not seem to extend to the model to be considered in Chapter 2. This paper proves weaker versions of many of the results of Section 1.6 and Section 1.8 to Section 1.10. The existence of the limits of quantities such as $N^{k/2}\mathsf{E}\langle A\rangle$, where A is the product of k terms of the type $R_{\ell,\ell'}$ is proved by a recursion method very similar to the one used here, but the limit is not computed explicitly.

I do not know who first used the "smart path method". The proof of Proposition 1.3.3 is due to J.P. Kahane [87] and that of Theorem 1.3.4 is due to G. Pisier [124]. I had known these papers since they appeared, but it took a very, very long time to realize that it was the route to take in the cavity method. The smart path method was first used in this context in [147], and then systematically in [158]. Interestingly, Guerra and Toninelli arrived independently at the very similar idea of interpolating between Hamiltonians as in Section 1.3. Proposition 1.3.2 must have been known for a very long time, at least as far back as [137].

The reader might wonder about the purpose of (1.152), since we nearly always use (1.151) instead. One use is that, using symmetry between sites, we can get a nice expression for $\nu_1'(f)$. This idea will be used in Volume II. We do not use it here, because, besides controlling the quantities $R_{1,2}$, it requires controlling $R_{1,2,3,4} = N^{-1}\sum_{i\le N}\sigma_i^1\sigma_i^2\sigma_i^3\sigma_i^4$. To give a specific example, if $f = R_{1,2} - q$, we get from (1.152) that

$$\begin{aligned}\nu_1'((\varepsilon_1\varepsilon_2 - q)f) &= \beta^2\nu((1-\varepsilon_1\varepsilon_2 q)(R_{1,2}-q)f)\\ &\quad - 4\beta^2\nu((\varepsilon_2\varepsilon_3 - q\varepsilon_1\varepsilon_3)(R_{1,3}-q)f)\\ &\quad + 3\beta^2\nu((\varepsilon_1\varepsilon_2\varepsilon_3\varepsilon_4 - q\varepsilon_3\varepsilon_4)(R_{3,4}-q)f)\\ &= \beta^2\nu((1-R_{1,2}q)(R_{1,2}-q)f)\\ &\quad - 4\beta^2\nu((R_{2,3}-qR_{1,3})(R_{1,3}-q)f)\\ &\quad + 3\beta^2\nu((R_{1,2,3,4}-qR_{3,4})(R_{3,4}-q)f)\,.\end{aligned}$$

If we know that $\nu(|R_{1,2}-q|^3)^{1/3} \le L/\sqrt{N}$ and $\nu(|R_{1,2,3,4}-\widehat{q}|^3)^{1/3} \le L/\sqrt{N}$, we get

$$\begin{aligned}\nu_1'((\varepsilon_1\varepsilon_2 - q)f) &= \beta^2(1-q^2)\nu((R_{1,2}-q)^2)\\ &\quad - 4\beta^2(q-q^2)\nu((R_{1,2}-q)(R_{2,3}-q))\\ &\quad + 3\beta^2(\widehat{q}-q^2)\nu((R_{1,2}-q)(R_{3,4}-q)) + O(3)\end{aligned}$$

a relation that we may combine with

$$\nu((R_{1,2}-q)^2) = \nu((\varepsilon_1\varepsilon_2 - q)f) = \nu_0((\varepsilon_1\varepsilon_2 - q)f) + \nu_1'((\varepsilon_1\varepsilon_2 - q)f) + O(3)\,.$$

In this way we have fewer error terms to control in the course of proving the central limit theorems presented here. The drawback is that one must prove first that $\nu((R_{1,2,3,4}-\widehat{q})^{2n}) \le K/N^n$ (which is not very difficult).

Two months after the present Chapter was widely circulated at the time of [157] (in a version that already contained the central limit theorems of Section 1.10), the paper [74] came out, offering very similar results, together with a CLT for $N^{-1}\log Z_N(\beta,h)$, of which Theorem 1.4.11 is a quantitative improvement.

I am grateful to M. Mézard for having explained to me the idea of coupling two copies of the SK model, and the discontinuity this should produce beyond the A-T line. This led to Theorem 1.9.6.

Guerra's bound of (1.73) is proved in [71] where Proposition 1.3.8 can also be found. (This lemma was also proved independently by R. Latala in an unpublished paper.)

The present work should make self-apparent the amount of energy already spent in trying to reach a mathematical understanding of mean field models related to spin glasses. It is unfortunate that some of the most precise results about the SK model rely on very specific properties of this model. However fascinating, the SK model is a rather specific object, and as such its importance can be questioned. I feel that the appeal of the "theory" of spin glasses does not lie in any particular model, but rather in the apparent generality of the phenomenon it predicts. About this, we still understand very little, despite all the examples that will be given in forthcoming chapters.

2. The Perceptron Model

2.1 Introduction

The name of this chapter comes from the theory of neural networks. An accessible introduction to neural networks is provided in [83], but what these are is not relevant to our purpose, which is to study the underlying mathematics. Roughly speaking, the basic problem is as follows. What "proportion" of $\Sigma_N = \{-1,1\}^N$ is left when one intersects this set with many random half-spaces? A natural definition for a random half-space is a set $\{\mathbf{x} \in \mathbb{R}^N \, ; \, \mathbf{x} \cdot \mathbf{v} \ge 0\}$ where the random vector $\mathbf{v}$ is uniform over the unit sphere of $\mathbb{R}^N$. More conveniently one can consider the set $\{\mathbf{x} \in \mathbb{R}^N ; \mathbf{x} \cdot \mathbf{g} \ge 0\}$, where $\mathbf{g}$ is a standard Gaussian vector, i.e. $\mathbf{g} = (g_i)_{i \le N}$, where g_i are independent standard Gaussian r.v.s. This is equivalent because the vector $\mathbf{g}/\|\mathbf{g}\|$ is uniformly distributed on the unit sphere of $\mathbb{R}^N$. Consider now M such Gaussian vectors $\mathbf{g}_k = (g_{i,k})_{i \le N}$, $k \le M$, all independent, the half-spaces

$$U_k = \{\mathbf{x} \, ; \, \mathbf{x} \cdot \mathbf{g}_k \ge 0\} = \Big\{\mathbf{x} \, , \, \sum_{i \le N} g_{i,k} x_i \ge 0\Big\} \, ,$$

and the set

$$\Sigma_N \cap \bigcap_{k \le M} U_k \, . \tag{2.1}$$

A given point of Σ_N has exactly a 50% chance to belong to U_k, so that

$$\mathsf{E}\,\mathrm{card}\Big(\Sigma_N \cap \bigcap_{k \le M} U_k\Big) = 2^{N-M} \, . \tag{2.2}$$

The case of interest is when N becomes large and M is proportional to N, $M/N \to \alpha > 0$. A consequence of (2.2) is that if $\alpha > 1$ the set (2.1) is typically empty when N is large, because the expected value of its cardinality is $\ll 1$. When $\alpha < 1$, what is interesting is *not* however the expected value (2.2) of the cardinality of the set (2.1), but rather the typical value of this cardinality, which is likely to be smaller. Our ultimate goal is the computation of this typical value, which we will achieve only for α small enough.

A similar problem was considered in (0.2) where Σ_N is replaced by the sphere $\mathbb{S}_N$ of center 0 and radius $\sqrt{N}$. The situation with Σ_N is usually

M. Talagrand, *Mean Field Models for Spin Glasses*, Ergebnisse der Mathematik und ihrer Grenzgebiete. 3. Folge / A Series of Modern Surveys in Mathematics 54, DOI 10.1007/978-3-642-15202-3_2,

called the *binary perceptron*, while the situation with $\mathbb{S}_N$ is usually called the *spherical perceptron*. The spherical perceptron will motivate the next chapter. We will return to both the binary and the spherical perceptron in Volume II, in Chapter 8 and Chapter 9 respectively. Both the spherical and the binary perceptron admit another popular version, where the Gaussian r.v.s $g_{i,j}$ are replaced by independent Bernoulli r.v.s (i.e. independent random signs), and we will also study these. Thus we will eventually investigate a total of four related but different models. It is not very difficult to replace the Gaussian r.v.s by random signs; but it is very much harder to study the case of Σ_N than the case of the sphere.

Research Problem 2.1.1. (Level 3!) Prove that there exists a number α^* and a function $\varphi : [0, \alpha^*) \to \mathbb{R}$ with the following properties:

1 - If $\alpha > \alpha^*$, then as $N \to \infty$ and $M/N \to \alpha$ the probability that the set (2.1) is not empty is at most $\exp(-N/K(\alpha))$.

2 - If $\alpha < \alpha^*$, $N \to \infty$ and $M/N \to \alpha$, then

$$\frac{1}{N} \log \operatorname{card}\Big(\Sigma_N \cap \bigcap_{k \le M} U_k\Big) \to \varphi(\alpha) \tag{2.3}$$

in probability. Compute α^* and φ.

This problem is a typical example of a situation where one expects "regularity" as $N \to \infty$, but where it is unclear how to even start doing anything relevant. In Volume II, we will prove (2.3) when α is small enough, and we will compute $\varphi(\alpha)$ in that case. (We expect that the case of larger α is much more difficult.) As a corollary, we will prove that there exists a number $\alpha_0 < 1$ such that if $M = \lfloor \alpha N \rfloor$, $\alpha > \alpha_0$, then the set (2.1) is typically empty for N large, despite the fact that the expected value of its cardinality is $2^{N-M} \gg 1$.

One way to approach the (very difficult) problem mentioned above is to introduce a version "with a temperature". We observe that if $x \ge 0$ we have $\lim_{\beta \to \infty} \exp(-\beta x) = 0$ if $x > 0$ and $= 1$ if $x = 0$. Using this for $x = \sum_{k \le M} \mathbf{1}_{\{\boldsymbol{\sigma} \notin U_k\}}$ where $\boldsymbol{\sigma} \in \Sigma_N$ implies

$$\operatorname{card}\Big(\Sigma_N \cap \bigcap_{k \le M} U_k\Big) = \lim_{\beta \to \infty} \sum_{\boldsymbol{\sigma} \in \Sigma_N} \exp\Big(-\beta \sum_{k \le M} \mathbf{1}_{\{\boldsymbol{\sigma} \notin U_k\}}\Big) , \tag{2.4}$$

so that to study (2.3) it should be relevant to use the Hamiltonian

$$- H_{N,M}(\boldsymbol{\sigma}) = -\beta \sum_{k \le M} \mathbf{1}_{\{\boldsymbol{\sigma} \notin U_k\}} . \tag{2.5}$$

If one can compute the corresponding partition function (and succeed in exchanging the limits $N \to \infty$ and $\beta \to \infty$), one will then prove (2.3).

More generally, we will consider Hamiltonians of the type

$$-H_{N,M}(\boldsymbol{\sigma}) = \sum_{k \le M} u\Big(\frac{1}{\sqrt{N}} \sum_{i \le N} g_{i,k}\sigma_i\Big)\,, \tag{2.6}$$

where u is a function, and where $(g_{i,k})$ are independent standard normal r.v.s. Of course the Hamiltonian depends on u, but the dependence is kept implicit. The role of the factor $N^{-1/2}$ is to make the quantity $N^{-1/2}\sum_{i \le N} g_{i,k}\sigma_i$ typically of order 1. There is no parameter β in the right-hand side of (2.6), since this parameter can be thought of as being included in the function u.

Since it is difficult to prove anything at all without using integration by parts we will always assume that u is differentiable. But if we want the Hamiltonian (2.6) to be a fair approximation of the Hamiltonian (2.5), we will have to accept that u' takes very large values. Then, in the formulas where u' occurs, we will have to show that somehow these large values cancel out. There is no magic way to do this, one has to work hard and prove delicate estimates (as we will do in Chapter 8). Another source of difficulty is that we want to approximate the Hamiltonian (2.5) for large values of β. That makes it difficult to bound from below a number of quantities that occur naturally as denominators in our computations.

On the other hand, there is a kind of beautiful "algebraic" structure connected to the Hamiltonian (2.6), which is uncorrelated to the analytical problems described above. We feel that it is appropriate, in a first stage, to bring this structure forward, and to set aside the analytical problems (to which we will return later). Thus, in this chapter we will assume a very strong condition on u, namely that for a certain constant D we have

$$\forall \ell\,,\ 0 \le \ell \le 3\,,\ \ |u^{(\ell)}| \le D\,. \tag{2.7}$$

Given values of N and M we will try to "describe the system generated by the Hamiltonian (2.6)" within error terms that become small for N large. We will be able to do this when the ratio $\alpha = M/N$ is small enough, $\alpha \le \alpha(D)$. The notation $\alpha = M/N$ will be used through this chapter and until Chapter 4.

Let us now try to give an overview of what will happen, without getting into details. We recall the notation $R_{\ell,\ell'} = N^{-1}\sum_{i \le N} \sigma_i^\ell \sigma_i^{\ell'}$. As is the case for the SK model, we expect that in the high-temperature regime we have

$$R_{1,2} \simeq q \tag{2.8}$$

for a certain number q depending on the system. Let us define

$$S_k = \frac{1}{\sqrt{N}} \sum_{i \le N} g_{i,k}\sigma_i \ \ ; \ \ S_k^\ell = \frac{1}{\sqrt{N}} \sum_{i \le N} g_{i,k}\sigma_i^\ell\,. \tag{2.9}$$

After one works some length of time with the system, one gets the irresistible feeling that (in the high-temperature regime) "the quantities S_k behave like individual spins", and (2.8) has to be complemented by the relation

$$\frac{1}{N}\sum_{k\le M} u'(S_k^1)u'(S_k^2) \simeq r \tag{2.10}$$

where r is another number attached to the system. Probably the reader would expect a normalization factor M rather than N in (2.10), but since we should think of M/N as $M/N \to \alpha > 0$, this is really the same. Also, the occurrence of u' will soon become clear.

We will use the cavity method twice. In Section 2.2 we "remove one spin" as in Chapter 1. This lets us guess what is the correct expression of q as a function of r. In Section 2.3, we then use the "cavity in M", comparing the system with the similar system where M has been replaced by $M-1$. This lets us guess what the expression of r should be as a function of q. The two relations between r and q that are obtained in this manner are called the "replica-symmetric equations" in physics. We prove in Section 2.4 that these equations do have a solution, and that (2.8) and (2.10) hold for these values of q and r. For N large and M/N small, we will then (approximately) compute the value of

$$p_{N,M}(u) = \frac{1}{N}\mathsf{E}\log\sum_{\boldsymbol{\sigma}}\exp(-H_{M,N}(\boldsymbol{\sigma}))\ , \tag{2.11}$$

(for the Hamiltonian defined by (2.6)) by an interpolation method motivated by the idea that the quantities S_k "behave like individual spins".

2.2 The Smart Path

It would certainly help to understand how the Hamiltonian (2.6) depends on the last spin. Let us write

$$S_k^0 = \frac{1}{\sqrt{N}}\sum_{i\le N-1} g_{i,k}\sigma_i\ ,$$

so that $S_k = S_k^0 + N^{-1/2}g_{N,k}\sigma_N$ and if u is differentiable,

$$\sum_{k\le M} u(S_k) = \sum_{k\le M} u(S_k^0) + \sigma_N\sum_{k\le M}\frac{g_{N,k}}{\sqrt{N}}u'(S_k^0) + \frac{\sigma_N^2}{2}\sum_{k\le M}\frac{g_{N,k}^2}{N}u''(S_k^0) + \cdots \tag{2.12}$$

The terms $\cdots$ are of lower order. We observe that $\sigma_N^2 = 1$. (This will no longer be the case in Chapter 3, when we will consider spins taking all possible values, so that σ_N^2 will no longer be constant.) We also observe that the r.v.s $g_{N,k}$ are independent. So it is reasonable according to the law of large numbers to expect that the third term on the right-hand side should behave like a constant and not influence the Hamiltonian. By the central limit theorem, one should expect the second term on the right-hand side of (2.12) to behave like

$\sigma_N Y$, where Y is a Gaussian r.v. independent of all the other r.v.s (Of course at some point we will have to guess what is the right choice for $r = \mathsf{E}Y^2$, but the time will come when this guess will be obvious.) Thus we expect that

$$\sum_{k\le M} u(S_k) \simeq \sum_{k\le M} u(S_k^0) + \sigma_N Y + \text{constant} . \tag{2.13}$$

Rather than using power expansions (which are impractical when we do not have a good control on higher derivatives) it is more fruitful to find a suitable interpolation between the left and the right-hand sides of (2.13). The first idea that comes to mind is to use the Hamiltonian

$$\sum_{k\le M} u\left(S_k^0 + \sqrt{\frac{t}{N}} g_{N,k}\sigma_N\right) + \sigma_N\sqrt{1-t}Y . \tag{2.14}$$

This is effective and was used in [157]. However, the variance of the Gaussian r.v. $S_k^0 + \sqrt{t/N} g_{N,k}\sigma_N$ depends on t; when differentiating, this creates terms that we will avoid by being more clever. Let us consider the quantity

$$\begin{aligned} S_{k,t} = S_{k,t}(\boldsymbol{\sigma}, \xi_k) &= S_k^0 + \sqrt{\frac{t}{N}} g_{N,k}\sigma_N + \sqrt{\frac{1-t}{N}}\xi_k \\ &= \frac{1}{\sqrt{N}}\sum_{i<N} g_{i,k}\sigma_i + \sqrt{\frac{t}{N}} g_{N,k}\sigma_N + \sqrt{\frac{1-t}{N}}\xi_k . \end{aligned} \tag{2.15}$$

In this expression, we should think of $(\xi_k)_{k\le M}$ not just as random constants ensuring that the variance of $S_{k,t}$ is constant but also as "new spins". That is, let $\boldsymbol{\xi} = (\xi_k)_{k\le M} \in \mathbb{R}^M$, and consider the Hamiltonian

$$-H_{N,M,t}(\boldsymbol{\sigma}, \boldsymbol{\xi}) = \sum_{k\le M} u(S_{k,t}) + \sigma_N\sqrt{1-t}Y . \tag{2.16}$$

The configurations are now points $(\boldsymbol{\sigma}, \boldsymbol{\xi})$ in $\Sigma_N \times \mathbb{R}^M$. Let us denote by γ the canonical Gaussian measure on $\mathbb{R}^M$. We define Gibbs' measure on $\Sigma_N \times \mathbb{R}^M$ by the formula

$$\langle f\rangle_t = \frac{1}{Z}\sum_{\boldsymbol{\sigma}}\int f(\boldsymbol{\sigma}, \boldsymbol{\xi})\exp(-H_{N,M,t}(\boldsymbol{\sigma}, \boldsymbol{\xi}))\mathrm{d}\gamma(\boldsymbol{\xi}) ,$$

where f is a function on $\Sigma_N \times \mathbb{R}^M$ and where Z is the normalizing factor,

$$Z = \sum_{\boldsymbol{\sigma}}\int \exp(-H_{N,M,t}(\boldsymbol{\sigma}, \boldsymbol{\xi}))\mathrm{d}\gamma(\boldsymbol{\xi}) .$$

More generally for a function f on $(\Sigma_N \times \mathbb{R}^M)^n = \Sigma_N^n \times \mathbb{R}^{Mn}$, we define

$$\langle f \rangle_t = \frac{1}{Z^n} \sum_{\boldsymbol{\sigma}^1,\ldots,\boldsymbol{\sigma}^n} \int \cdots \int f(\boldsymbol{\sigma}^1,\ldots,\boldsymbol{\sigma}^n,\boldsymbol{\xi}^1,\ldots,\boldsymbol{\xi}^n)$$

$$\times \exp\Bigl(-\sum_{\ell\le n} H^\ell_{N,M,t}\Bigr) \mathrm{d}\gamma(\boldsymbol{\xi}^1)\cdots\mathrm{d}\gamma(\boldsymbol{\xi}^n)\,, \tag{2.17}$$

where Z is as above and

$$H^\ell_{N,M,t} = H_{N,M,t}(\boldsymbol{\sigma}^\ell,\boldsymbol{\xi}^\ell)\,. \tag{2.18}$$

Integration of $\boldsymbol{\xi}$ with respect to γ means simply that we think of $(\xi_k)_{k\le M}$ as independent Gaussian r.v.s and we take expectation. We recall **the convention that E_ξ denotes expectation with respect to all r.v.s labeled** ξ (be it with subscripts or superscripts). We thus rewrite (2.17) as

$$\langle f \rangle_t = \frac{1}{Z^n}\mathsf{E}_\xi \sum_{\boldsymbol{\sigma}^1,\ldots,\boldsymbol{\sigma}^n} f(\boldsymbol{\sigma}^1,\ldots,\boldsymbol{\sigma}^n,\boldsymbol{\xi}^1,\ldots,\boldsymbol{\xi}^n)\exp\Bigl(-\sum_{\ell\le n} H^\ell_{N,M,t}\Bigr)\ ; \tag{2.19}$$

$$Z = \mathsf{E}_\xi \sum_{\boldsymbol{\sigma}} \exp(-H_{N,M,t}(\boldsymbol{\sigma},\boldsymbol{\xi}))\,.$$

In these formulas, $\boldsymbol{\xi}^\ell = (\xi^\ell_k)_{k\le M}$, ξ^ℓ_k are independent Gaussian r.v.s. One should think of $\boldsymbol{\xi}^\ell$ as being a "replica" of $\boldsymbol{\xi}$. In this setting, replicas are simply independent copies.

Exercise 2.2.1. Prove that when f depends on $\boldsymbol{\sigma}^1,\ldots,\boldsymbol{\sigma}^n$, but not on $\boldsymbol{\xi}^1,\ldots,\boldsymbol{\xi}^n$, then $\langle f\rangle_t$ in (2.19) is exactly the average of f with respect to the Hamiltonian

$$-H = \sum_{k\le M} u_t\Bigl(\frac{1}{\sqrt{N}}\sum_{i\le N-1} g_{i,k}\sigma_i + \sqrt{\frac{t}{N}}g_{N,k}\sigma_N\Bigr) + \sigma_N\sqrt{1-t}Y\,,$$

where u_t is defined by

$$\exp u_t(x) = \mathsf{E}\exp u\Bigl(x + \sqrt{\frac{1-t}{N}}\xi\Bigr)\,, \tag{2.20}$$

for ξ a standard normal r.v.

The reader might wonder whether it is really worth the effort to introduce this present setting simply in order to avoid an extra term in Proposition 2.2.3 below, a term with which it is not so difficult to deal anyway. The point is that the mechanism of "introducing new spins" is fundamental and must be used in Section 2.3, so we might as well learn it now.

Consistently with our notation, if f is a function on $\Sigma_N^n \times \mathbb{R}^{Mn}$, we define

$$\nu_t(f) = \mathsf{E}\langle f\rangle_t\,;\ \nu_t'(f) = \frac{\mathrm{d}}{\mathrm{d}t}\nu_t(f)\,, \tag{2.21}$$

where $\langle f\rangle_t$ is given by (2.19).

We also write $\nu(f) = \nu_1(f)$. When f does not depend on the r.v.s $\boldsymbol{\xi}^\ell$, then $\nu(f) = \mathsf{E}\langle f\rangle$, where $\langle\cdot\rangle$ refers to Gibbs' measure with Hamiltonian (2.6). As in Chapter 1, we write $\varepsilon_\ell = \sigma_N^\ell$, and we recall the r.v. Y of (2.16).

Lemma 2.2.2. *Given a function f^- on Σ_{N-1}^n, and a subset I of $\{1,\dots,n\}$, we have*

$$\nu_0\Bigl(f^-\prod_{\ell\in I}\varepsilon_\ell\Bigr) = \mathsf{E}\bigl((\mathrm{th}Y)^{\mathrm{card}I}\bigr)\nu_0(f^-) = \nu_0\Bigl(\prod_{\ell\in I}\varepsilon_\ell\Bigr)\nu_0(f^-)\,.$$

This lemma holds whatever the value of $r = \mathsf{E}Y^2$. The proof is identical to that of Lemma 1.6.2. The Hamiltonian $H_{N,M,0}$ decouples the last spin from the first $N-1$ spins, which is what it is designed to do.

We now turn to the computation of $\nu_t'(f)$. Throughout the chapter, we write $\alpha = M/N$. Implicitly, we think of N and M as being large but fixed. The model then depends on the parameters N and α (and of course of u). We recall the definition (2.15) of $S_{k,t}$, and consistently with the notation (2.18) we write

$$S_{k,t}^\ell = \frac{1}{\sqrt{N}}\sum_{i<N} g_{i,k}\sigma_i^\ell + \sqrt{\frac{t}{N}}g_{N,k}\varepsilon_\ell + \sqrt{\frac{1-t}{N}}\xi_k^\ell\,. \tag{2.22}$$

Proposition 2.2.3. *Assume that u is twice differentiable and let $r = \mathsf{E}Y^2$. Then for a function f on Σ_N^n, we have*

$$\nu_t'(f) = \mathrm{I} + \mathrm{II} \tag{2.23}$$

$$\begin{aligned}\mathrm{I} = \alpha&\sum_{1\le\ell<\ell'\le n}\nu_t\bigl(\varepsilon_\ell\varepsilon_{\ell'}u'(S_{M,t}^\ell)u'(S_{M,t}^{\ell'})f\bigr)\\ &-\alpha n\sum_{\ell\le n}\nu_t\bigl(\varepsilon_\ell\varepsilon_{n+1}u'(S_{M,t}^\ell)u'(S_{M,t}^{n+1})f\bigr)\\ &+\alpha\frac{n(n+1)}{2}\nu_t\bigl(\varepsilon_{n+1}\varepsilon_{n+2}u'(S_{M,t}^{n+1})u'(S_{M,t}^{n+2})f\bigr)\,.\end{aligned} \tag{2.24}$$

$$\begin{aligned}\mathrm{II} = -r\biggl(&\sum_{1\le\ell<\ell'\le n}\nu_t(\varepsilon_\ell\varepsilon_{\ell'}f) - n\sum_{\ell\le n}\nu_t(\varepsilon_\ell\varepsilon_{n+1}f)\\ &+\frac{n(n+1)}{2}\nu_t(\varepsilon_{n+1}\varepsilon_{n+2}f)\biggr)\,.\end{aligned} \tag{2.25}$$

The proposition resembles Lemma 1.6.3, so it should not be so scary anymore. As in Lemma 1.6.3, the complication is algebraic, and each of the terms I and II is made up of simple pieces. Moreover both terms have similar structures. This formula will turn out to be much easier to use than one might think at first. In particular one should observe that by symmetry, and since $\alpha = M/N$, in the expression for I we can replace the term $\alpha u'(S^\ell_{M,t})u'(S^{\ell'}_{M,t})$ by

$$\frac{1}{N}\sum_{k\le M} u'(S^\ell_{k,t})u'(S^{\ell'}_{k,t})\ ,$$

so that if (2.10) is indeed correct, the terms I and II should have a good will to cancel each other out.

Proof. We could make this computation appear as a consequence of (1.40), but for the rest of the book we will change policy, and proceed directly, i.e. we write the value of the derivative and we integrate by parts. It is immediate from (2.19) that

$$\frac{\mathrm{d}}{\mathrm{d}t}\langle f\rangle_t = \sum_{\ell\le n}\left\langle \frac{\mathrm{d}}{\mathrm{d}t}(-H^\ell_{N,M,t})f\right\rangle_t - n\left\langle \frac{\mathrm{d}}{\mathrm{d}t}(-H^{n+1}_{N,M,t})f\right\rangle_t\ , \tag{2.26}$$

and, writing g_k for $g_{N,k}$,

$$\frac{\mathrm{d}}{\mathrm{d}t}(-H^\ell_{N,M,t}) = \sum_{k\le M}\frac{1}{2\sqrt{N}}\left(\frac{g_k\varepsilon_\ell}{\sqrt{t}} - \frac{\xi^\ell_k}{\sqrt{1-t}}\right)u'(S^\ell_{k,t}) - \frac{\varepsilon_\ell Y}{2\sqrt{1-t}}\ . \tag{2.27}$$

We observe the symmetry for $k \le M$. All the values of k bring the same contribution. There are M of them, and $M/\sqrt{N} = \alpha\sqrt{N}$, so that

$$\nu_t'(f) = \mathrm{III} + \mathrm{IV} + \mathrm{V}$$

$$\mathrm{III} = \frac{\alpha}{2}\sqrt{\frac{N}{t}}\left(\sum_{\ell\le n}\nu_t\big(g_M\varepsilon_\ell u'(S^\ell_{M,t})f\big) - n\nu_t\big(g_M\varepsilon_{n+1}u'(S^{n+1}_{M,t})f\big)\right) \tag{2.28}$$

$$\mathrm{IV} = -\frac{\alpha}{2}\sqrt{\frac{N}{1-t}}\left(\sum_{\ell\le n}\nu_t\big(\xi^\ell_M u'(S^\ell_{M,t})f\big) - n\nu_t\big(\xi^{n+1}_M u'(S^{n+1}_{M,t})f\big)\right)$$

$$\mathrm{V} = -\frac{1}{2}\frac{1}{\sqrt{1-t}}\left(\sum_{\ell\le n}\nu_t(\varepsilon_\ell Yf) - n\nu_t(\varepsilon_{n+1}Yf)\right)\ .$$

It remains to integrate by parts in these formulas to get the result. The easiest case is that of the term IV, because "different replicas use independent copies of ξ". We write the explicit formula for $\langle \xi^\ell_M u'(S^\ell_{M,t})f\rangle_t$, that is

$$\begin{aligned}&\langle \xi^\ell_M u'(S^\ell_{M,t})f\rangle_t\\ &= \frac{1}{Z^n}\mathsf{E}_\xi\left(\xi^\ell_M\sum_{\boldsymbol{\sigma}^1,\dots,\boldsymbol{\sigma}^n} u'(S^\ell_{M,t})f(\boldsymbol{\sigma}^1,\dots,\boldsymbol{\sigma}^n)\exp\Big(-\sum_{\ell\le n}H^\ell_{M,N,t}\Big)\right)\ ,\end{aligned}$$

and we see that we only have to integrate by parts in the numerator. The dependence on ξ_M^ℓ is through $u'(S_{M,t}^\ell)$ and through the term $u(S_{M,t}^\ell)$ in the Hamiltonian and moreover

$$\frac{\partial S_{M,t}^\ell}{\partial \xi_M^\ell} = \sqrt{\frac{1-t}{N}}\,, \tag{2.29}$$

so that

$$\langle \xi_M^\ell u'(S_{M,t}^\ell) f\rangle_t = \sqrt{\frac{1-t}{N}}\big\langle (u''(S_{M,t}^\ell) + u'^2(S_{M,t}^\ell))f\big\rangle_t\,,$$

and therefore

$$\mathrm{IV} = -\frac{\alpha}{2}\Big(\sum_{\ell\le n}\nu_t\big(((u''(S_{M,t}^\ell)+u'^2(S_{M,t}^\ell))f\big)-n\nu_t\big((u''(S_{M,t}^{n+1})+u'^2(S_{M,t}^{n+1}))f\big)\Big).$$

The second easiest case is that of V, because we have done the same computation (implicitly at least) in Chapter 1; since $\mathsf{E}Y^2 = r$, we have V = II. Of course, the reader who does not find this formula obvious should simply write

$$\nu_t(\varepsilon_\ell Y f) = \mathsf{E}Y\langle \varepsilon_\ell f\rangle_t\,,$$

and carry out the integration by parts, writing the explicit formula for $\langle \varepsilon_\ell f\rangle_t$. To compute the term III, there is no miracle. We write

$$\nu_t(g_M\varepsilon_\ell u'(S_{M,t}^\ell)f) = \mathsf{E}g_M\langle \varepsilon_\ell u'(S_{M,t}^\ell)f\rangle_t$$

and we use the integration by parts formula $\mathsf{E}(g_M F(g_M)) = \mathsf{E}F'(g_M)$ when seeing $\langle \varepsilon_\ell u'(S_{M,t}^\ell)f\rangle_t$ as a function of g_M. The dependence on g_M is through the quantities $S_{M,v}^\ell$, and

$$\frac{\partial S_{M,v}^\ell}{\partial g_M} = \varepsilon_\ell\sqrt{\frac{t}{N}}\,.$$

Writing the (cumbersome) explicit formula for $\langle \varepsilon_\ell u'(S_{M,t}^\ell)f\rangle_t$, we get that

$$\frac{\partial}{\partial g_M}\langle \varepsilon_\ell u'(S_{M,t}^\ell)f\rangle_t = \sqrt{\frac{t}{N}}\Big(\langle u''(S_{M,t}^\ell)f\rangle_t$$
$$+\sum_{\ell'\le n}\langle \varepsilon_\ell\varepsilon_{\ell'}u'(S_{M,t}^\ell)u'(S_{M,t}^{\ell'})f\rangle_t - n\langle \varepsilon_\ell\varepsilon_{n+1}u'(S_{M,t}^\ell)u'(S_{M,t}^{n+1})f\rangle_t\Big)\,.$$

The first term arises from the dependence of the factor $u'(S_{M,t}^\ell)$ on g_M and the other terms from the dependence of the Hamiltonian on g_M. Consequently we obtain

$$\nu_t(\varepsilon_\ell u'(S_{M,t}^\ell)f) = \sqrt{\frac{t}{N}}\Big(\nu_t(u''(S_{M,t}^\ell)f)$$
$$+\sum_{\ell'\le n}\nu_t(\varepsilon_\ell\varepsilon_{\ell'}u'(S_{M,t}^\ell)u'(S_{M,t}^{\ell'})f) - n\nu_t(\varepsilon_\ell\varepsilon_{n+1}u'(S_{M,t}^\ell)u'(S_{M,t}^{n+1})f)\Big)\,.$$

Similarly we have

$$\begin{aligned}\frac{\partial}{\partial g_M}\langle \varepsilon_{n+1} u'(S^{n+1}_{M,t}) f\rangle_t = \sqrt{\frac{t}{N}}\Bigg(&\langle u''(S^{n+1}_{M,t}) f\rangle_t \\ &+ \sum_{\ell' \le n+1} \big\langle \varepsilon_{\ell'}\varepsilon_{n+1} u'(S^{\ell'}_{M,t}) u'(S^{n+1}_{M,t}) f\big\rangle_t \\ &- (n+1)\big\langle \varepsilon_{n+1}\varepsilon_{n+2} u'(S^{n+1}_{M,t}) u'(S^{n+2}_{M,t}) f\big\rangle_t\Bigg) ,\end{aligned}$$

and consequently

$$\begin{aligned}\nu_t(\varepsilon_{n+1} u'(S^{n+1}_{M,t}) f) = \sqrt{\frac{t}{N}}\Bigg(&\nu_t(u''(S^{n+1}_{M,t}) f) \\ &+ \sum_{\ell' \le n+1} \nu_t(\varepsilon_{\ell'}\varepsilon_{n+1} u'(S^{\ell'}_{M,t}) u'(S^{n+1}_{M,t}) f) \\ &- (n+1)\nu_t(\varepsilon_{n+1}\varepsilon_{n+2} u'(S^{n+1}_{M,t}) u'(S^{n+2}_{M,t}) f)\Bigg) .\end{aligned}$$

Regrouping the terms, we see that $\mathrm{III} + \mathrm{IV} = \mathrm{I}$. □

Exercise 2.2.4. Suppose that we had not been as sleek as we were, and that instead of (2.15) and (2.22) we had defined

$$S_{k,t} = S_{k,t}(\boldsymbol{\sigma}) = S^0_k + \sqrt{\frac{t}{N}} g_{N,k}\sigma_N = \frac{1}{\sqrt{N}} \sum_{i<N} g_{i,k}\sigma_i + \sqrt{\frac{t}{N}} g_{N,k}\sigma_N$$

and

$$S^\ell_{k,t} = \frac{1}{\sqrt{N}} \sum_{i<N} g_{i,k}\sigma^\ell_i + \sqrt{\frac{t}{N}} g_{N,k}\sigma^\ell_N .$$

Prove that then in the formula (2.23) we would get the extra term

$$\mathrm{VI} = \frac{\alpha}{2}\Bigg(\sum_{\ell\le n} \nu_t\Big(\big(u'(S^\ell_{M,t})^2 + u''(S^\ell_{M,t})\big) f\Big) - n\nu_t\Big(\big(u'(S^{n+1}_{M,t})^2 + u''(S^{n+1}_{M,t})\big) f\Big)\Bigg) .$$

2.3 Cavity in M

To pursue the idea that the terms I and II in (2.23) should nearly cancel out each other, the first thing to do is to try to make sense of the term I, and to understand the influence of the quantities $u'(S^\ell_{M,t})$. The quantities $S^\ell_{M,t}$ also occur in the Hamiltonian, and we should make this dependence explicit. For this we introduce a new Hamiltonian

$$-H_{N,M-1,t}(\boldsymbol{\sigma},\boldsymbol{\xi}) = \sum_{k\le M-1} u(S_{k,t}(\boldsymbol{\sigma},\xi_k)) + \sigma_N\sqrt{1-t}Y\ , \tag{2.30}$$

where the dependence on $\boldsymbol{\xi}$ is stressed to point out that it will be handled as in the case of the Hamiltonian (2.16), that is, an average $\langle\cdot\rangle_{t,\sim}$ with respect to this Hamiltonian will be computed with the formula (2.31) below. Let us first notice that, even though the right-hand side of (2.30) does not depend on ξ_M, we denote for simplicity of notation the Hamiltonian as a function of $\boldsymbol{\sigma}$ and $\boldsymbol{\xi}$. If f is a function on $\Sigma_N^n \times \mathbb{R}^{Mn}$, we then define

$$\langle f\rangle_{t,\sim} = \frac{1}{Z_\sim^n}\mathsf{E}_\xi \sum_{\boldsymbol{\sigma}^1,\ldots,\boldsymbol{\sigma}^n} f(\boldsymbol{\sigma}^1,\ldots,\boldsymbol{\sigma}^n,\boldsymbol{\xi}^1,\ldots,\boldsymbol{\xi}^n)\exp\Big(-\sum_{\ell\le n} H_{N,M-1,t}^\ell\Big)\ , \tag{2.31}$$

where

$$Z_\sim = \mathsf{E}_\xi \sum_{\boldsymbol{\sigma}} \exp(-H_{N,M-1,t}(\boldsymbol{\sigma},\boldsymbol{\xi}))\ ,$$

and where $H_{N,M-1,t}^\ell = H_{N,M-1,t}(\boldsymbol{\sigma}^\ell,\boldsymbol{\xi}^\ell)$. There of course E_ξ includes expectation in the r.v.s ξ_M^ℓ, even though the Hamiltonian does not depend on those. Since $-H_{N,M,t}^\ell = -H_{N,M-1,t}^\ell + u(S_{M,t}^\ell)$, the identity

$$\begin{aligned} Z = \mathsf{E}_\xi \sum_{\boldsymbol{\sigma}} \exp(-H_{N,M,t}^1) &= \mathsf{E}_\xi \sum_{\boldsymbol{\sigma}} \exp u(S_{M,t}^1)\exp(-H_{N,M-1,t}^1) \\ &= Z_\sim \langle \exp u(S_{M,t}^1)\rangle_{t,\sim} \end{aligned}$$

holds, and, similarly,

$$\begin{aligned} &\mathsf{E}_\xi \sum_{\boldsymbol{\sigma}^1,\ldots,\boldsymbol{\sigma}^n} f(\boldsymbol{\sigma}^1,\ldots,\boldsymbol{\sigma}^n,\boldsymbol{\xi}^1,\ldots,\boldsymbol{\xi}^n)\exp\Big(-\sum_{\ell\le n} H_{N,M,t}^\ell\Big) \\ &= Z_\sim^n \Big\langle f\exp\sum_{\ell\le n} u(S_{M,t}^\ell)\Big\rangle_{t,\sim}\ . \end{aligned}$$

Combining these two formulas with (2.31) yields that if f is a function on $\Sigma_N^n \times \mathbb{R}^{Mn}$, we have

$$\langle f\rangle_t = \frac{\langle f\exp(\sum_{\ell\le n} u(S_{M,t}^\ell))\rangle_{t,\sim}}{\langle \exp u(S_{M,t}^1)\rangle_{t,\sim}^n}\ . \tag{2.32}$$

Our best guess now is that the quantities $S_{M,t}^\ell$, when seen as functions of the system with Hamiltonian (2.30), will have a jointly Gaussian behavior under Gibbs' measure, with pairwise correlation q, allowing us to approximately compute the right-hand side of (2.32) in Proposition 2.3.5 below. This again will be shown by interpolation. Let us consider a new parameter $0\le q\le 1$ and standard Gaussian r.v.s (ξ^ℓ) and z that are independent of all the other r.v.s already considered. (The reader will not confuse the r.v.s ξ^ℓ with the r.v.s ξ^ℓ_M.) Let us set

$$\theta^\ell = z\sqrt{q} + \xi^\ell\sqrt{1-q} \,. \tag{2.33}$$

Thus these r.v.s share the common randomness z and are independent given that randomness. For $0 \le v \le 1$ we define

$$S^\ell_v = \sqrt{v}S^\ell_{M,t} + \sqrt{1-v}\theta^\ell \,. \tag{2.34}$$

The dependence on t is kept implicit; when using S^ℓ_v we think of t (and M) as being fixed.

Let us pursue the idea that in (2.31), E_ξ denotes **expectation in all r.v.s labeled** ξ *including the variables* ξ^ℓ and let us further define with this convention

$$\nu_{t,v}(f) = \mathsf{E}\frac{\langle f\exp(\sum_{\ell\le n} u(S^\ell_v))\rangle_{t,\sim}}{\langle \exp u(S^1_v)\rangle^n_{t,\sim}} \,. \tag{2.35}$$

Using (2.32) yields

$$\nu_{t,1}(f) = \nu_t(f) \,.$$

The idea of (2.35) is of course that in certain cases $\nu_{t,0}(f)$ should be much easier to evaluate than $\nu_t(f) = \nu_{t,1}(f)$ and that these quantities should be close to each other if q is appropriately chosen. Before we go into the details however, we would like to explain the pretty idea that is hidden behind this construction. The idea is simply that we consider ξ "as a new spin". To explain this, consider a spin system where the space of configurations is the collection of all triplets $(\boldsymbol{\sigma}, \boldsymbol{\xi}, \xi)$ for $\boldsymbol{\sigma} \in \Sigma_N$, $\boldsymbol{\xi} \in \mathbb{R}^M$ and $\xi \in \mathbb{R}$. Consider the Hamiltonian

$$-H(\boldsymbol{\sigma}, \boldsymbol{\xi}, \xi) = -H_{N,M-1,t}(\boldsymbol{\sigma}, \boldsymbol{\xi}) + u(S_v) \,,$$

where $S_v = \sqrt{v}S_{M,t} + \sqrt{1-v}\theta$, for $\theta = z\sqrt{q} + \sqrt{1-q}\xi$. Then, for a function f of $\boldsymbol{\sigma}^1, \dots, \boldsymbol{\sigma}^n, \boldsymbol{\xi}^1, \dots, \boldsymbol{\xi}^n$ and $\xi^1, \dots, \xi^n$ we can define a quantity $\langle f\rangle_{t,v}$ by a formula similar to (2.19) and (2.31). As in (2.32), we have

$$\langle f\rangle_{t,v} = \frac{\langle f\exp(\sum_{\ell\le n} u(S^\ell_v))\rangle_{t,\sim}}{\langle \exp u(S^1_v)\rangle^n_{t,\sim}} \,,$$

so that in fact $\nu_{t,v} = \mathsf{E}\langle\cdot\rangle_{t,v}$. Let us observe that the r.v. θ depends also on z, but this r.v. is not considered as a "new spin", but rather as "new randomness".

The present idea of considering ξ as a new spin is essential. As we mentioned on page 156, the idea of considering $\xi_1, \dots, \xi_M$ as new spins was not essential, but since it is the same idea, we decided to make the minimal extra effort to use the setting of (2.19).

First, we reveal the magic of the computation of $\nu_{t,0}$.

Lemma 2.3.1. *Consider $0 \le q \le 1$ and define*

$$\widehat{r} = \mathsf{E}\left(\frac{\mathsf{E}_\xi u'(\theta)\exp u(\theta)}{\mathsf{E}_\xi \exp u(\theta)}\right)^2 , \tag{2.36}$$

where $\theta = z\sqrt{q} + \xi\sqrt{1-q}$ for independent standard Gaussian r.v.s z and ξ and where E_ξ denotes expectation in ξ only. Consider a function f on Σ_N^n. This function might depend on the variables ξ_k^ℓ for $k < M$ and $\ell \le n$, but it does not depend on the randomness of the variables ξ_M^ℓ or ξ^ℓ. Then

$$\nu_{t,0}(f) = \mathsf{E}\langle f\rangle_{t,\sim} , \tag{2.37}$$

and

$$\nu_{t,0}(u'(S_0^1)u'(S_0^2)f) = \widehat{r}\mathsf{E}\langle f\rangle_{t,\sim} . \tag{2.38}$$

In particular we have $\nu_{t,0}(u'(S_0^1)u'(S_0^2)f) = \widehat{r}\nu_{t,0}(f)$. If such an equality is nearly true for $v = 1$ rather than for $v = 0$, we are in good shape to use Proposition 2.2.3.

Proof. First we have

$$\Big\langle f \exp\sum_{\ell\le n} u(\theta^\ell)\Big\rangle_{t,\sim} = \langle f\rangle_{t,\sim}\mathsf{E}_\xi \exp\sum_{\ell\le n} u(\theta^\ell) . \tag{2.39}$$

This follows from the formula (2.31). The quantities θ^ℓ do not depend on the spins $\boldsymbol{\sigma}$, and their randomness "in the variables labeled ξ" is independent of the randomness of the other terms. Now, independence implies

$$\mathsf{E}_\xi \exp\sum_{\ell\le n} u(\theta^\ell) = (\mathsf{E}_\xi \exp u(\theta))^n .$$

Moreover $\langle \exp u(\theta)\rangle_{t,\sim} = \mathsf{E}_\xi \exp u(\theta)$, as (an obvious) special case of (2.39). This proves (2.37).

To prove (2.38), proceeding in a similar manner and using now that

$$\mathsf{E}_\xi\Big(u'(\theta^1)u'(\theta^2)\exp\sum_{\ell\le n} u(\theta^\ell)\Big) = \big(\mathsf{E}_\xi u'(\theta)\exp u(\theta)\big)^2\big(\mathsf{E}_\xi \exp u(\theta)\big)^{n-2} ,$$

we get

$$\begin{aligned}\nu_{t,0}(u'(S_0^1)u'(S_0^2)f) &= \mathsf{E}\frac{\big\langle f u'(\theta^1)u'(\theta^2)\exp\sum_{\ell\le n} u(\theta^\ell)\big\rangle_{t,\sim}}{\langle \exp u(\theta)\rangle_{t,\sim}^n} \\ &= \widehat{r}\mathsf{E}\langle f\rangle_{t,\sim} ,\end{aligned}$$

and this finishes the proof. □

We now turn to the proof that $\nu_{t,0}$ and $\nu_{t,1}$ are close. We recall that D is the constant of (2.7).

Lemma 2.3.2. *Consider a function f on Σ_N^n. This function depend on the variables ξ_k^ℓ for $k < M$ and $\ell \le n$, but it does not depend on the randomness of the variables $z, g_{i,M}, \xi_M^\ell$ or ξ^ℓ. Then if $B_v \equiv 1$ or $B_v = u'(S_v^1)u'(S_v^2)$, whenever $1/\tau_1 + 1/\tau_2 = 1$ we have*

$$\left|\frac{\mathrm{d}}{\mathrm{d}v}\nu_{t,v}(B_v f)\right| \le K(n,D)\left(\nu_{t,v}(|f|^{\tau_1})^{1/\tau_1}\nu_{t,v}(|R_{1,2}-q|^{\tau_2})^{1/\tau_2} + \frac{1}{N}\nu_{t,v}(|f|)\right). \tag{2.40}$$

Here $K(n,D)$ depends on n and D only.

Therefore the left-hand side is small if we can find q such that $R_{1,2} \simeq q$. The reason why we write a derivative in the left-hand side rather than a partial derivative is that when considering $\nu_{t,v}$ we always think of t as fixed.

Proof. The core of the proof is to compute $\mathrm{d}(\nu_{t,v}(B_v f))/\mathrm{d}v$ by differentiation and integration by parts, after which the bound (2.40) basically follows from Hölder's inequality. It turns out that if one looks at things the right way, there is a relatively simple expression for $\mathrm{d}(\nu_{t,v}(B_v f))/\mathrm{d}v$. We will not reveal this magic formula now. Our immediate concern is to explain in great detail the mechanism of integration by parts, that will occur again and again, and for this we decided to use a completely pedestrian approach, writing only absolutely explicit formulas.

First, we compute $\mathrm{d}(\nu_{t,v}(B_v f))/\mathrm{d}v$ by straightforward differentiation of the formula (2.35). In the case where $B_v = u'(S_v^1)u'(S_v^2)$, setting

$$S_v^{\ell\prime} = \frac{1}{2\sqrt{v}}S_{M,t}^\ell - \frac{1}{2\sqrt{1-v}}\theta^\ell ,$$

we find

$$\begin{aligned}\frac{\mathrm{d}}{\mathrm{d}v}(\nu_{t,v}(B_v f)) &= \nu_{t,v}\big(f S_v^{1\prime}u''(S_v^1)u'(S_v^2)\big) + \nu_{t,v}\big(f S_v^{2\prime}u'(S_v^1)u''(S_v^2)\big)\\ &\quad + \sum_{\ell\le n}\nu_{t,v}\big(f S_v^{\ell\prime}u'(S_v^\ell)u'(S_v^1)u'(S_v^2)\big)\\ &\quad - (n+1)\nu_{t,v}\big(f S_v^{n+1\prime}u'(S_v^{n+1})u'(S_v^1)u'(S_v^2)\big).\end{aligned} \tag{2.41}$$

Of course the first term occurs because of the factor $u'(S_v^1)$ in B_v, the second term because of the factor $u'(S_v^2)$ and the other terms because of the dependence of the Hamiltonian on v. The rest of the proof consists in integrating by parts. In some sense it is a straight forward application of the Gaussian integration by parts formula (A.17). However, since we are dealing with complicated expressions, it will take several pages to fill in all the details. The notation is complicated, and this obscures the basic simplicity of the argument. Probably the ambitious reader should try to compute everything on her own in simple case, and look at our presentation only if she gets stuck.

Even though we have written the previous formula in a compact form using $\nu_{t,v}$, to integrate by parts we have to spell out the dependence of the

Hamiltonian on the variables S_v^ℓ by using the formula (2.35). For example, the first term in the right-hand side of (2.41) is

$$\mathsf{E}\frac{\langle f S_v^{1\prime} u''(S_v^1)u'(S_v^2)\exp(\sum_{\ell\le n} u(S_v^\ell))\rangle_{t,\sim}}{\langle \exp u(S_v^1)\rangle_{t,\sim}^n} . \tag{2.42}$$

To keep the formulas manageable, let us write

$$w = w(\boldsymbol{\sigma}^1,\ldots,\boldsymbol{\sigma}^n,\boldsymbol{\xi}^1,\ldots,\boldsymbol{\xi}^n) = \exp\Bigl(-\sum_{\ell\le n} H_{N,M-1,t}^\ell\Bigr)$$

and let us define

$$w_*^\ell = w_*(\boldsymbol{\sigma}^\ell,\boldsymbol{\xi}^\ell) = \exp(-H_{N,M-1,t}^\ell) .$$

These quantities are probabilistically independent of the randomness of the variables S_v^ℓ (which is why we introduced the Hamiltonian $H_{N,M-1,t}$ in the first place).

The quantity (2.42) is then equal to

$$\mathsf{E}\frac{\mathsf{E}_\xi \sum_{\boldsymbol{\sigma}^1,\ldots,\boldsymbol{\sigma}^n} w S_v^{1\prime} C}{Z^n} , \tag{2.43}$$

where

$$Z = \mathsf{E}_\xi \sum_{\boldsymbol{\sigma}^1} w_*^1 \exp u(S_v^1) ,$$

and where

$$C = f u''(S_v^1)u'(S_v^2)\exp\Bigl(\sum_{\ell\le n} u(S_v^\ell)\Bigr) .$$

Let us now make an observation that will be used **many times**. The r.v. Z is independent of all the r.v.s labeled ξ, so that

$$\frac{\mathsf{E}_\xi \sum_{\boldsymbol{\sigma}^1,\ldots,\boldsymbol{\sigma}^n} w\, S_v^{1\prime} C}{Z^n} = \mathsf{E}_\xi \frac{\sum_{\boldsymbol{\sigma}^1,\ldots,\boldsymbol{\sigma}^n} w\, S_v^{1\prime} C}{Z^n} ,$$

and thus the quantity (2.43) is then equal to

$$\mathsf{E}\mathsf{E}_\xi \sum_{\boldsymbol{\sigma}^1,\ldots,\boldsymbol{\sigma}^n} w\, S_v^{1\prime}\frac{C}{Z^n} = \mathsf{E} \sum_{\boldsymbol{\sigma}^1,\ldots,\boldsymbol{\sigma}^n} w\, S_v^{1\prime}\frac{C}{Z^n} . \tag{2.44}$$

Let us now denote by E_0 integration in the randomness of $g_{i,M}$, ξ_M^ℓ, z and ξ^ℓ, given all the other sources of randomness. Therefore, since the quantities w do not depend on any of the variables $g_{i,M}$, ξ_k^ℓ, z or ξ^ℓ, the quantity (2.44) equals

$$\mathsf{E} \sum_{\boldsymbol{\sigma}^1,\ldots,\boldsymbol{\sigma}^n} w\, \mathsf{E}_0 S_v^{1\prime}\frac{C}{Z^n} . \tag{2.45}$$

The main step in the computation is the calculation of the quantity $\mathsf{E}_0 S_v^{1\prime} C/Z^n$ by integration by parts. We advise the reader to study the elementary proof of Lemma 2.4.4 below as a preparation to this computation in a simpler setting. To apply the Gaussian integration by parts formula (A.17), we need to find a jointly Gaussian family $(g, z_1, \ldots, z_P)$ of r.v.s such that $g = S_v^{1\prime}$ and that C/Z^n is a function $F(z_1, \ldots, z_P)$ of $z_1, \ldots, z_P$. The first idea that comes to mind is to use for the r.v.s (z_p) the following family of variables, indexed by $\boldsymbol{\sigma}$ and ℓ,

$$\begin{aligned} z_{\boldsymbol{\sigma}}^\ell &= \sqrt{v} S_{M,t}(\boldsymbol{\sigma}, \xi_M^\ell) + \sqrt{1-v}\theta^\ell \\ &= \sqrt{v}\Bigl(\frac{1}{\sqrt{N}} \sum_{i<N} g_{i,M}\sigma_i + \sqrt{\frac{t}{N}} g_{N,M}\sigma_N + \sqrt{\frac{1-t}{N}}\xi_M^\ell\Bigr) \\ &\quad + \sqrt{1-v}(z\sqrt{q} + \xi^\ell\sqrt{1-q}) , \end{aligned}$$

where $\boldsymbol{\sigma} \in \Sigma_N$ takes all possible values and ℓ is an integer. Of course these variables depend on v but the dependence is kept implicit because we think now of v as fixed. We observe that

$$S_v^\ell = z_{\boldsymbol{\sigma}^\ell}^\ell , \tag{2.46}$$

so that we can think of C as a function of these quantities:

$$C = C_{\boldsymbol{\sigma}^1, \ldots, \boldsymbol{\sigma}^n} = F_{\boldsymbol{\sigma}^1, \ldots, \boldsymbol{\sigma}^n}((z_{\boldsymbol{\sigma}}^\ell)) , \tag{2.47}$$

where $F_{\boldsymbol{\sigma}^1, \ldots, \boldsymbol{\sigma}^n}$ is the function of the variables $x_{\boldsymbol{\sigma}}^\ell$ given by

$$F_{\boldsymbol{\sigma}^1, \ldots, \boldsymbol{\sigma}^n}((x_{\boldsymbol{\sigma}}^\ell)) = f(\boldsymbol{\sigma}^1, \ldots, \boldsymbol{\sigma}^n) u''(x_{\boldsymbol{\sigma}^1}^1) u'(x_{\boldsymbol{\sigma}^2}^2) \exp\Bigl(\sum_{\ell \le n} u(x_{\boldsymbol{\sigma}^\ell}^\ell)\Bigr) . \tag{2.48}$$

Condition (2.47) holds simply because to compute $F_{\boldsymbol{\sigma}^1, \ldots, \boldsymbol{\sigma}^n}((z_{\boldsymbol{\sigma}^\ell}^\ell))$, we substitute $z_{\boldsymbol{\sigma}^\ell}^\ell = S_v^\ell$ to $x_{\boldsymbol{\sigma}^\ell}^\ell$ in the previous formula. This construction however does not suffice, because Z cannot be considered as a function of the quantities $z_{\boldsymbol{\sigma}}^\ell$: the effect of the expectation E_ξ is that "the part depending on the r.v.s labeled ξ has been averaged out". The part of $z_{\boldsymbol{\sigma}}^\ell$ that does not depend on the r.v.s labeled ξ is simply

$$y_{\boldsymbol{\sigma}} = \sqrt{v}\Bigl(\frac{1}{\sqrt{N}} \sum_{i<N} g_{i,M}\sigma_i + \sqrt{\frac{t}{N}} g_{N,M}\sigma_N\Bigr) + \sqrt{1-v}\sqrt{q} z .$$

Defining

$$\xi_*^\ell = \sqrt{v}\sqrt{\frac{1-t}{N}}\xi_M^\ell + \sqrt{1-v}\sqrt{1-q}\xi^\ell ,$$

we then have

$$z_{\boldsymbol{\sigma}}^\ell = y_{\boldsymbol{\sigma}} + \xi_*^\ell .$$

It is now possible to express Z as a function of the r.v.s $y_{\boldsymbol{\sigma}}$. This is shown by the formula

$$Z = F_1((y_{\boldsymbol{\sigma}})) ,$$

where F_1 is the function of the variables $x_{\boldsymbol{\sigma}}$ given by

$$F_1((x_{\boldsymbol{\sigma}})) = \mathsf{E}_\xi \sum_{\boldsymbol{\sigma}} w_*(\boldsymbol{\sigma}, \boldsymbol{\xi}^1) \exp u(x_{\boldsymbol{\sigma}} + \xi_*^1) . \tag{2.49}$$

Let us now define

$$\begin{aligned} z_{\boldsymbol{\sigma}}^{\ell\prime} &= \frac{1}{2\sqrt{v}} S_{M,t}(\boldsymbol{\sigma}, \xi_M^\ell) - \frac{1}{2\sqrt{1-v}} \theta^\ell \\ &= \frac{1}{2\sqrt{v}} \left(\frac{1}{\sqrt{N}} \sum_{i<N} g_{i,M} \sigma_i + \sqrt{\frac{t}{N}} g_{N,M} \sigma_N + \sqrt{\frac{1-t}{N}} \xi_M^\ell \right) \\ &\quad - \frac{1}{2\sqrt{1-v}} (\sqrt{q} z + \sqrt{1-q} \xi^\ell) , \end{aligned}$$

so that $S_v^{\ell\prime} = z_{\boldsymbol{\sigma}^\ell}^{\ell\prime}$. The family of all the r.v.s $z_{\boldsymbol{\sigma}}^\ell, y_{\boldsymbol{\sigma}}, \xi_*^\ell$, and $z_{\boldsymbol{\sigma}}^{\ell\prime}$ is a Gaussian family, and this is the family we will use to apply the integration by parts formula. In the upcoming formulas, the reader should take great care to distinguish between the quantities $z_{\boldsymbol{\sigma}}^{\ell\prime}$ and $z_{\boldsymbol{\sigma}}^{\ell'}$ (The position of the $\prime$ is not the same).

We note the relations

$$\mathsf{E}(\theta^\ell)^2 = 1 = \mathsf{E}(S_{M,t}(\boldsymbol{\sigma}, \xi_M^\ell))^2 \ ; \ \ell \neq \ell' \Rightarrow \mathsf{E}\theta^\ell \theta^{\ell'} = q .$$

$$\ell \neq \ell' \Rightarrow \mathsf{E} S_{M,t}(\boldsymbol{\sigma}, \xi_M^\ell) S_{M,t}(\boldsymbol{\tau}, \xi_M^{\ell'}) = R^t(\boldsymbol{\sigma}, \boldsymbol{\tau}) := \frac{1}{N} \sum_{i<N} \sigma_i \tau_i + \frac{t}{N} \sigma_N \tau_N ,$$

so that

$$\mathsf{E} z_{\boldsymbol{\sigma}}^{\ell\prime} z_{\boldsymbol{\sigma}}^\ell = 0 \ ; \ \ell \neq \ell' \Rightarrow \mathsf{E} z_{\boldsymbol{\sigma}}^{\ell\prime} z_{\boldsymbol{\tau}}^{\ell'} = \frac{1}{2}(R^t(\boldsymbol{\sigma}, \boldsymbol{\tau}) - q) , \tag{2.50}$$

and

$$\mathsf{E} z_{\boldsymbol{\sigma}}^{\ell\prime} y_{\boldsymbol{\tau}} = \frac{1}{2}(R^t(\boldsymbol{\sigma}, \boldsymbol{\tau}) - q) . \tag{2.51}$$

We will simply use the integration by parts formula (A.17) and these relations to understand the form of the quantity

$$\mathsf{E}_0 S_v^{1\prime} \frac{C}{Z^n} = \mathsf{E}_0 z_{\boldsymbol{\sigma}^1}^{1\prime} \frac{F_{\boldsymbol{\sigma}^1, \ldots, \boldsymbol{\sigma}^n}((z_{\boldsymbol{\sigma}}^\ell))}{F_1((y_{\boldsymbol{\sigma}}))^n} . \tag{2.52}$$

Let us repeat that this integration by parts takes place given all the sources of randomness other than the r.v.s $g_{i,M}$, ξ_k^ℓ for $k < M$, z and ξ^ℓ (so that it is fine if f depends on some randomness independent of these). The exact result of the computation is not relevant now (it will be given in Chapter 9). For the present result we simply need the information that $d\nu_{t,v}(B_v f)/dv$ is a sum of terms of the type (using the notation $R^t_{\ell,\ell'} = R^t(\boldsymbol{\sigma}^\ell, \boldsymbol{\sigma}^{\ell'})$)

$$\nu_{t,v}(f(R^t_{\ell,\ell'} - q)A) , \tag{2.53}$$

where A is a monomial in the quantities $u'(S^m_v), u''(S^m_v), u^{(3)}(S^m_v)$ for $m \le n+2$. So, let us perform the integration by parts in (2.52):

$$\mathsf{E}_0 z^{1'}_{\boldsymbol{\sigma}^1} \frac{F_{\boldsymbol{\sigma}^1,\ldots,\boldsymbol{\sigma}^n}((z^\ell_{\boldsymbol{\sigma}}))}{F_1((y_{\boldsymbol{\sigma}}))^n} = \sum_{\boldsymbol{\tau},\ell} \mathsf{E}_0 z^{1'}_{\boldsymbol{\sigma}^1} z^\ell_{\boldsymbol{\tau}} \mathsf{E}_0 \frac{\partial F_{\boldsymbol{\sigma}^1,\ldots,\boldsymbol{\sigma}^n}}{\partial x^\ell_{\boldsymbol{\tau}}}((z^\ell_{\boldsymbol{\sigma}})) \frac{1}{F_1((y_{\boldsymbol{\sigma}}))^n}$$

$$- n \sum_{\boldsymbol{\tau}} \mathsf{E}_0 z^{1'}_{\boldsymbol{\sigma}^1} y_{\boldsymbol{\tau}} \mathsf{E}_0 \frac{\partial F_1}{\partial x_{\boldsymbol{\tau}}}((y_{\boldsymbol{\sigma}})) \frac{F_{\boldsymbol{\sigma}^1,\ldots,\boldsymbol{\sigma}^n}((z^\ell_{\boldsymbol{\sigma}}))}{F_1((y_{\boldsymbol{\sigma}}))^{n+1}} .$$

It is convenient to refer to the last term in the above (or similar) formula "as the term created by the denominator" when performing the integration by parts in (2.52). (It would be nice to remember this, since we will often use this expression in our future attempts at describing at a high level computations similar to the present one.) We first compute this term. We observe that

$$\frac{\partial F_1}{\partial x_{\boldsymbol{\tau}}} = \mathsf{E}_\xi w_*(\boldsymbol{\tau}, \boldsymbol{\xi}^1) u'(x_{\boldsymbol{\tau}} + \xi^1_*) \exp u(x_{\boldsymbol{\tau}} + \xi^1_*) .$$

Therefore using (2.51) we see that the term created by the denominator in (2.52) is

$$-\frac{n}{2} \mathsf{E}_0 \sum_{\boldsymbol{\tau}} (R^t(\boldsymbol{\sigma}^1, \boldsymbol{\tau}) - q) \frac{F_{\boldsymbol{\sigma}^1,\ldots,\boldsymbol{\sigma}^n}((z^\ell_{\boldsymbol{\sigma}})) \mathsf{E}_\xi w_*(\boldsymbol{\tau}, \boldsymbol{\xi}^1) u'(y_{\boldsymbol{\tau}} + \xi^1_*) \exp u(y_{\boldsymbol{\tau}} + \xi^1_*)}{F_1((y_{\boldsymbol{\sigma}}))^{n+1}} .$$

Since $y_{\boldsymbol{\tau}} + \xi^1_* = z^1_{\boldsymbol{\tau}}$, the contribution of this term to (2.44) is then

$$-\frac{n}{2} \mathsf{E} \sum_{\boldsymbol{\sigma}^1,\ldots,\boldsymbol{\sigma}^n,\boldsymbol{\tau}} w(R^t(\boldsymbol{\sigma}^1, \boldsymbol{\tau}) - q) \frac{F_{\boldsymbol{\sigma}^1,\ldots,\boldsymbol{\sigma}^n}((z^\ell_{\boldsymbol{\sigma}})) \mathsf{E}_\xi w_*(\boldsymbol{\tau}, \boldsymbol{\xi}^1) u'(z^1_{\boldsymbol{\tau}}) \exp u(z^1_{\boldsymbol{\tau}})}{F_1((y_{\boldsymbol{\sigma}}))^{n+1}} . \tag{2.54}$$

Now,

$$\mathsf{E}_\xi w_*(\boldsymbol{\tau}, \boldsymbol{\xi}^1) u'(z^1_{\boldsymbol{\tau}}) \exp u(z^1_{\boldsymbol{\tau}}) = \mathsf{E}_\xi w_*(\boldsymbol{\tau}, \boldsymbol{\xi}^{n+1}) u'(z^{n+1}_{\boldsymbol{\tau}}) \exp u(z^{n+1}_{\boldsymbol{\tau}}) ,$$

so that, changing the name of $\boldsymbol{\tau}$ into $\boldsymbol{\sigma}^{n+1}$, and since $w^{n+1}_* = w_*(\boldsymbol{\sigma}^{n+1}, \boldsymbol{\xi}^{n+1})$, the quantity (2.54) is equal to (using (2.46) in the second line)

$$= -\frac{n}{2} \mathsf{E} \sum_{\boldsymbol{\sigma}^1,\ldots,\boldsymbol{\sigma}^{n+1}} w(R^t_{1,n+1} - q) \frac{F_{\boldsymbol{\sigma}^1,\ldots,\boldsymbol{\sigma}^n}((z^\ell_{\boldsymbol{\sigma}})) \mathsf{E}_\xi w^{n+1}_* u'(z^{n+1}_{\boldsymbol{\sigma}^{n+1}}) \exp u(z^{n+1}_{\boldsymbol{\sigma}^{n+1}})}{F_1((y_{\boldsymbol{\sigma}}))^{n+1}}$$

$$= -\frac{n}{2} \mathsf{E} \sum_{\boldsymbol{\sigma}^1,\ldots,\boldsymbol{\sigma}^{n+1}} w(R^t_{1,n+1} - q) \frac{C \mathsf{E}_\xi w^{n+1}_* u'(S^{n+1}_v) \exp u(S^{n+1}_v)}{Z^{n+1}} .$$

In a last step we observe that in the above formula we can remove the expectation E_ξ. This is because the r.v.s labeled ξ that occur in this expectation (namely $\boldsymbol{\xi}^{n+1}$ and ξ^{n+1}) are independent of the other r.v.s labeled ξ that occur in C and w. In this manner we finally see that the contribution of this quantity to the computation of (2.42) is

$$-\frac{n}{2}\mathsf{E}\sum_{\boldsymbol{\sigma}^1,\ldots,\boldsymbol{\sigma}^{n+1}}\frac{C(R^t_{1,n+1}-q)ww_*^{n+1}u'(S_v^{n+1})\exp u(S_v^{n+1})}{Z^{n+1}}$$
$$=-\frac{n}{2}\nu_{t,v}\big(f(R^t_{1,n+1}-q)u''(S_v^1)u'(S_v^2)u'(S_v^{n+1})\big)\,.$$

In a similar manner we compute the contribution in (2.52) of the dependence of $F_{\boldsymbol{\sigma}^1,\ldots,\boldsymbol{\sigma}^n}$ on the variables $z^\ell_{\boldsymbol{\sigma}}$ at a given value of ℓ, i.e of the quantity

$$\sum_{\boldsymbol{\tau}}\mathsf{E}_0 z^{1\prime}_{\boldsymbol{\sigma}^1}z^\ell_{\boldsymbol{\tau}}\mathsf{E}_0\frac{\partial F_{\boldsymbol{\sigma}^1,\ldots,\boldsymbol{\sigma}^n}}{\partial x^\ell_{\boldsymbol{\tau}}}((z^\ell_{\boldsymbol{\sigma}}))\frac{1}{F_1((y_{\boldsymbol{\sigma}}))^n}\,. \tag{2.55}$$

We observe in particular from (2.48) that

$$\frac{\partial F_{\boldsymbol{\sigma}^1,\ldots,\boldsymbol{\sigma}^n}}{\partial x^\ell_{\boldsymbol{\tau}}}((z^\ell_{\boldsymbol{\sigma}}))=0$$

unless $\boldsymbol{\tau}=\boldsymbol{\sigma}^\ell$, so that the quantity (2.55) equals

$$\mathsf{E}_0 z^{1\prime}_{\boldsymbol{\sigma}^1}z^\ell_{\boldsymbol{\sigma}^\ell}\mathsf{E}_0\frac{\partial F_{\boldsymbol{\sigma}^1,\ldots,\boldsymbol{\sigma}^n}}{\partial x^\ell_{\boldsymbol{\sigma}^\ell}}((z^\ell_{\boldsymbol{\sigma}}))\frac{1}{F_1((y_{\boldsymbol{\sigma}}))^n}\,. \tag{2.56}$$

Since $\mathsf{E}z^{\ell\prime}_{\boldsymbol{\sigma}}z^\ell_{\boldsymbol{\sigma}}=0$ by (2.50) we see that for $\ell=1$ the contribution of this term is 0.

When $\ell\geq 3$, we have

$$\frac{\partial F_{\boldsymbol{\sigma}^1,\ldots,\boldsymbol{\sigma}^n}}{\partial x^\ell_{\boldsymbol{\tau}}}((x^\ell_{\boldsymbol{\sigma}}))=f(\boldsymbol{\sigma}^1,\ldots,\boldsymbol{\sigma}^n)u''(x^1_{\boldsymbol{\sigma}^1})u'(x^2_{\boldsymbol{\sigma}^2})u'(x^\ell_{\boldsymbol{\sigma}^\ell})\exp\Big(\sum_{\ell\leq n}u(x^\ell_{\boldsymbol{\sigma}^\ell})\Big)\,,$$

so that the term (2.55) is simply

$$\frac{1}{2}\nu_{t,v}\big(f(R^t_{1,\ell}-q)u''(S_v^1)u'(S_v^2)u'(S_v^\ell)\big)\,.$$

If $\ell=2$, there is another term because of the factor $u'(S_v^2)$, and this term is $\frac{1}{2}\nu_{t,v}\big(f(R^t_{1,2}-q)u''(S_v^1)u''(S_v^2)\big)$. So actually we have shown that

$$\begin{aligned}\nu_{t,v}(fS_v^{1\prime}u''(S_v^1)u'(S_v^2))&=\frac{1}{2}\nu_{t,v}\big(f(R^t_{1,2}-q)u''(S_v^1)u''(S_v^2)\big)\\&\quad+\frac{1}{2}\sum_{2\leq\ell\leq n}\nu_{t,v}\big(f(R^t_{1,\ell}-q)u''(S_v^1)u'(S_v^2)u'(S_v^\ell)\big)\\&\quad-\frac{n}{2}\nu_{t,v}\big(f(R^t_{1,n+1}-q)u''(S_v^1)u'(S_v^2)u'(S_v^{n+1})\big)\,.\end{aligned}$$

We strongly suggest to the enterprising reader to compute now all the other terms of (2.41). This is the best way to really understand the mechanism at work. There is no difficulty whatsoever, this just requires patience.

Calculations similar to the previous one will be needed again and again. We will not anymore explain them formally as above. Rather, we will give the result of the computation with possibly a few words of explanation. It is worth making now a simple observation that helps finding the result of such a computation. It is the fact that from (2.51) we have

$$\mathsf{E} z_{\boldsymbol{\sigma}}^{\ell'} y_{\boldsymbol{\tau}} = \mathsf{E} z_{\boldsymbol{\sigma}}^{\ell'} z_{\boldsymbol{\tau}}^{n+1} \,.$$

In a sense this means that when performing the integration by parts, we obtain the same result as if Z were actually a function of the variables $z_{\boldsymbol{\sigma}}^{n+1}$. It is useful to formulate this principle as a heuristic rule:

The result of the expectation E_ξ in the definition of Z is somehow "to shift the dependence of Z in S_v on a new replica" . (2.57)

When describing in the future the computation of a quantity such as $\nu_{t,v}(f S_v^{1\prime} u''(S_v^1) u'(S_v^2))$ by integration by parts, we will simply say: we integrate by parts using the relations

$$\mathsf{E} S_v^{\ell\prime} S_v^{\ell} = 0 \; ; \; \mathsf{E} S_v^{\ell\prime} S_v^{\ell'} = \frac{1}{2}(R_{\ell,\ell'}^t - q) \,, \tag{2.58}$$

and we will expect that the reader has understood enough of the algebraic mechanism at work to be able to check that the result of the computation is indeed the one we give, and the heuristic rule (2.57) should be precious for this purpose. There are two more such calculations in the present chapter, and the algebra in each is much simpler than in the present case. As a good start to develop the understanding of this mechanism, the reader should at the very least check the following two formulas involved in the computation of (2.41):

$$\begin{aligned}
&\nu_{t,v}\big(f S_v^{3\prime} u'(S_v^3) u'(S_v^1) u'(S_v^2)\big) \\
&= \frac{1}{2}\nu_{t,v}\big(f(R_{3,1}^t - q) u'(S_v^3) u''(S_v^1) u'(S_v^2)\big) \\
&+ \frac{1}{2}\nu_{t,v}\big(f(R_{3,2}^t - q) u'(S_v^3) u'(S_v^1) u''(S_v^2)\big) \\
&+ \frac{1}{2}\sum_{\ell \neq 3, \ell \le n} \nu_{t,v}\big(f(R_{3,\ell}^t - q) u'(S_v^3) u'(S_v^1) u'(S_v^2) u'(S_v^\ell)\big) \\
&- \frac{n}{2}\nu_{t,v}\big(f(R_{3,n+1}^t - q) u'(S_v^3) u'(S_v^1) u'(S_v^2) u'(S_v^{n+1})\big) \,,
\end{aligned}$$

and

$$\begin{aligned}
&\nu_{t,v}\big(fS_v^{n+1\prime}u'(S_v^{n+1})u'(S_v^1)u'(S_v^2)\big)\\
&= \frac{1}{2}\nu_{t,v}\big(f(R^t_{n+1,1}-q)u'(S_v^{n+1})u''(S_v^1)u'(S_v^2)\big)\\
&+ \frac{1}{2}\nu_{t,v}\big(f(R^t_{n+1,2}-q)u'(S_v^{n+1})u'(S_v^1)u''(S_v^2)\big)\\
&+ \frac{1}{2}\sum_{\ell\le n}\nu_{t,v}\big(f(R^t_{n+1,\ell}-q)u'(S_v^{n+1})u'(S_v^1)u'(S_v^2)u'(S_v^\ell)\big)\\
&- \frac{n+1}{2}\nu_{t,v}\big(f(R^t_{n+1,n+2}-q)u'(S_v^{n+1})u'(S_v^1)u'(S_v^2)u'(S_v^{n+2})\big)\,.
\end{aligned}$$

We bound a term (2.53) by

$$K(D)\nu_{t,v}(|f||R^t_{1,\ell'}-q|)\,,$$

and we write $|R^t_{\ell,\ell'}-q| \le |R_{\ell,\ell'}-q| + 1/N$ to obtain the inequality

$$\left|\frac{\mathrm{d}}{\mathrm{d}v}\nu_{t,v}(B_v f)\right| \le K(n,D)\Bigg(\sum_{1\le\ell<\ell'\le n+2}\nu_{t,v}(|f||R_{\ell,\ell'}-q|) + \frac{1}{N}\nu_{t,v}(|f|)\Bigg)\,. \tag{2.59}$$

To conclude we use Hölder's inequality. □

Exercise 2.3.3. Let us recall the notation $S^\ell_{k,t}$ of Proposition 2.2.3 and define

$$S^{\ell\prime}_{k,t} = \frac{1}{2\sqrt{N}}\left(\frac{g_k\varepsilon_\ell}{\sqrt{t}} - \frac{\xi_k^\ell}{\sqrt{1-t}}\right)\,,$$

so that (2.27) becomes

$$\frac{\mathrm{d}}{\mathrm{d}t}(-H^\ell_{N,M,t}) = \sum_{k\le M} S^{\ell\prime}_{k,t}u'(S^\ell_{k,t}) - \frac{\varepsilon_\ell Y}{2\sqrt{1-t}}\,.$$

Observe the relations

$$\mathsf{E}S^{\ell\prime}_{k,t}S^\ell_{k,t} = 0\,;\ \mathsf{E}S^{\ell\prime}_{k,t}S^{\ell'}_{k,t} = \frac{1}{2N}\varepsilon_\ell\varepsilon_{\ell'} \text{ if } \ell\ne\ell'\,;\ \mathsf{E}S^{\ell\prime}_{k,t}S^{\ell'}_{k',t} = 0 \text{ if } k\ne k'\,. \tag{2.60}$$

Get convinced that the previously described mechanism yields the formula (when $\ell \le n+1$)

$$\begin{aligned}
\nu_t(S^{\ell\prime}_{k,t}u'(S^\ell_{k,t})f) = \frac{1}{2N}\Bigg(&\sum_{\ell'\ne\ell,\ell'\le n+1}\nu_t(\varepsilon_\ell\varepsilon_{\ell'}u'(S^\ell_{k,t})u'(S^{\ell'}_{k,t})f)\\
&-(n+1)\nu_t(\varepsilon_\ell\varepsilon_{n+2}u'(S^\ell_{k,t})u'(S^{n+2}_{k,t})f)\Bigg)\,.
\end{aligned}$$

Then get convinced that the term I in (2.23) can be obtained "in one step" rather than by integrating by parts separately over the r.v.s $\xi_{k,\ell}$ and g_k as was done in the proof of Proposition 2.2.3.

To follow future computations it is really important to understand the difference between the situation (2.58) (where integration by parts "brings a factor $(R^t_{\ell,\ell'} - q)/2$ in each term") and the situation (2.60), where this integration by parts brings "a factor $\varepsilon_\ell \varepsilon_{\ell'}/2N$ in each term".

Let us point out that the constants $K(n, D)$ and $K(D)$ are simply avatars of our ubiquitous constant K, and they need not be the same at each occurrence. The only difference is that here we make explicit that these constants depend only on n and D (etc.) simply because this is easier to do when there are so few parameters. Of course, $K_1(D)$, etc. denote specific constants.

Lemma 2.3.4. *If $f \geq 0$ is a function on Σ_N^n we have*

$$\nu_{t,v}(f) \leq K(n, D)\nu_t(f) . \tag{2.61}$$

Proof. We use (2.40) with $B_v \equiv 1$, $\tau_1 = 1, \tau_2 = \infty$ to get

$$\left| \frac{\mathrm{d}}{\mathrm{d}v} \nu_{t,v}(f) \right| \leq K(n, D)\nu_{t,v}(f) .$$

We integrate and we use that $\nu_{t,1}(f) = \nu_t(f)$. □

Proposition 2.3.5. *Consider a function f on Σ_N^n. This function might be random, but it does not depend on the randomness of the variables $g_{i,M}, \xi^\ell_M, \xi^\ell$ or z. Then, whenever $1/\tau_1 + 1/\tau_2 = 1$, we have*

$$|\nu_t(f u'(S^1_{M,t}) u'(S^2_{M,t})) - \hat{r}\nu_t(f)| \leq K(n, D)\Big(\nu_t(|f|^{\tau_1})^{1/\tau_1} \nu_t(|R_{1,2} - q|^{\tau_2})^{1/\tau_2}$$
$$+ \frac{1}{N}\nu_t(|f|)\Big) . \tag{2.62}$$

This provides a good understanding of the term I of (2.23), provided we can find q such that the right-hand side is small.

Proof. We consider B_v as in Lemma 2.3.2, we write

$$|\nu_{t,1}(B_1 f) - \nu_{t,0}(B_0 f)| \leq \max_v \left| \frac{\mathrm{d}}{\mathrm{d}v} \nu_{t,v}(B_v f) \right| , \tag{2.63}$$

and we use (2.40) and (2.61) to get

$$|\nu_{t,1}(B_1 f) - \nu_{t,0}(B_0 f)| \leq \mathcal{B} , \tag{2.64}$$

where $\mathcal{B}$ is a term as in the right-hand side of (2.62). Thus in the case $B_v \equiv 1$, and since $\nu_{t,1} = \nu_t$, (2.37) and (2.64) imply that

$$|\nu_t(f) - \mathsf{E}\langle f \rangle_{t,\sim}| \leq \mathcal{B} . \tag{2.65}$$

In the case $B_v = u'(S^1_v)u'(S^2_v)$, (2.38) and (2.64) mean

$$\left|\nu_t\big(fu'(S^1_{M,t})u'(S^2_{M,t})\big) - \widehat{r}\,\mathsf{E}\langle f\rangle_{t,\sim}\right| \le \mathcal{B}$$

and combining with (2.65) finishes the proof. □

We now set $r = \alpha\widehat{r}$, and (2.62) implies

$$\begin{aligned}&\left|\alpha\nu_t\big(\varepsilon_\ell\varepsilon_{\ell'} fu'(S^\ell_{M,t})u'(S^{\ell'}_{M,t})\big) - r\nu_t(\varepsilon_\ell\varepsilon_{\ell'} f)\right|\\ &\le \alpha K(n,D)\left(\nu_t(|f|^{\tau_1})^{1/\tau_1}\nu_t(|R_{1,2}-q|^{\tau_2})^{1/\tau_2} + \frac{1}{N}\nu_t(|f|)\right).\end{aligned}$$

Looking again at the terms I and II of Proposition 2.2.3, we have proved the following.

Proposition 2.3.6. *Consider a function f on Σ_N^n (that does not depend on any of the r.v.s $g_{i,M}, \xi^\ell, \xi^\ell_M$ or z). Then, whenever $1/\tau_1 + 1/\tau_2 = 1$, we have*

$$|\nu_t'(f)| \le \alpha K(D,n)\left(\nu_t(|f|^{\tau_1})^{1/\tau_1}\nu_t(|R_{1,2}-q|^{\tau_2})^{1/\tau_2} + \frac{1}{N}\nu_t(|f|)\right). \tag{2.66}$$

The following is an obviously helpful way to relate ν and ν_t.

Lemma 2.3.7. *There exists a constant $K(D)$ with the following property. If $\alpha K(D) \le 1$, whenever $f \ge 0$ is a function on Σ_N^2 (that does not depend on any of the r.v.s $g_{i,M}, \xi^\ell, \xi^\ell_M$ or z), we have*

$$\nu_t(f) \le 2\nu(f)\,. \tag{2.67}$$

Proof. We use Proposition 2.3.6 with $\tau_1 = 1$ and $\tau_2 = \infty$ to see that

$$|\nu_t'(f)| \le \alpha K_1(D)\nu_t(f)\,,$$

from which (2.67) follows by integration if $\alpha K_1(D) \le \log 2$. □

2.4 The Replica Symmetric Solution

We recall the notation $\theta = z\sqrt{q} + \xi\sqrt{1-q}$ where z and ξ are independent standard Gaussian r.v.s, and that E_ξ denotes expectation in ξ only.

Theorem 2.4.1. *Given $D > 0$, there is a number $K(D)$ with the following property. Assume that the function u satisfies (2.7), i.e.*

$$\forall \ell \le 3\ , \qquad |u^{(\ell)}| \le D\,.$$

Then whenever $\alpha \le 1/K(D)$ the system of equations

$$q = \mathsf{E}\,\mathrm{th}^2(z\sqrt{r}) \quad ; \quad r = \alpha\mathsf{E}\left(\frac{\mathsf{E}_\xi u'(\theta)\exp u(\theta)}{\mathsf{E}_\xi \exp u(\theta)}\right)^2 \tag{2.68}$$

in the unknown q and r has a unique solution, and

$$\nu\big((R_{1,2}-q)^2\big) \le \frac{L}{N}\,. \tag{2.69}$$

Proof. Let us write the second equation of (2.68) as $r = \alpha\widehat{r} = \alpha\widehat{r}(q)$. Differentiation and integration by parts show that $|\widehat{r}'(q)| \le K(D)$ under (2.7). The function $r \mapsto \mathsf{E}\,\mathrm{th}^2(z\sqrt{r})$ has a bounded derivative; so the function $q \mapsto \psi(q) := \mathsf{E}\mathrm{th}^2(z\sqrt{\alpha\widehat{r}(q)})$ has a derivative $\le \alpha K_2(D)$. Therefore if $2\alpha K_2(D) \le 1$ there is a unique solution to the equation $q = \psi(q)$ because then the function $\psi(q)$ is valued in $[0,1]$ with a derivative $\le 1/2$.

Symmetry among sites yields

$$\nu\big((R_{1,2}-q)^2\big) = \nu(f) \tag{2.70}$$

where $f = (\varepsilon_1\varepsilon_2 - q)(R_{1,2} - q)$, and we write

$$\nu(f) \le \nu_0(f) + \sup_{0<t<1} |\nu_t'(f)| \,. \tag{2.71}$$

Since $q = \mathsf{E}\,\mathrm{th}^2(z\sqrt{r}) = \mathsf{E}\,\mathrm{th}^2 Y$, Lemma 2.2.2 implies

$$\nu_0((\varepsilon_1\varepsilon_2 - q)(R^-_{1,2} - q)) = (\mathsf{E}\,\mathrm{th}^2 Y - q)\nu_0(R^-_{1,2} - q) = 0\,,$$

and thus

$$\nu_0(f) = \frac{1}{N}\nu_0(1 - \varepsilon_1\varepsilon_2 q) = \frac{1}{N}(1 - q^2)\,. \tag{2.72}$$

To compute $\nu_t'(f)$, we use Proposition 2.3.6 with $n = 2$ and $\tau_1 = \tau_2 = 2$. Since $|f| < 2|R_{1,2} - q|$, we obtain

$$|\nu_t'(f)| \le \alpha K(D)\left(\nu_t\big((R_{1,2}-q)^2\big) + \frac{1}{N}\nu(|f|)\right)\,. \tag{2.73}$$

We substitute in (2.71) and use (2.67) to get the relation

$$\nu(f) = \nu\big((R_{1,2}-q)^2\big) \le \alpha K(D)\Big(\nu\big((R_{1,2}-q)^2\big) + \frac{1}{N}\nu(|f|)\Big) + \frac{1}{N}(1-q^2)\,,$$

so that since $|f| \le 4$ we obtain

$$\nu\big((R_{1,2}-q)^2\big) \le \alpha K(D)\nu\big((R_{1,2}-q)^2\big) + \frac{K(D)(\alpha+1)}{N}\,. \qquad \square$$

One should observe that in the above argument we never used the uniqueness of the solutions of the equations (2.68) to obtain (2.69), only their existence. In turn, uniqueness of these solutions follows from (2.69).

One may like to think of the present model as a kind of "square". There are two "spin systems", one that consists of the σ_i and one that consists of the S_k. These are coupled: the σ_i determine the S_k and these in turn determine the behavior of the σ_i. This philosophy undermines the first proof of Theorem 2.4.2 below.

From now on in this section, q and r always denote the solutions of (2.68). We recall the definition (2.11)

$$p_{N,M}(u) = \frac{1}{N}\mathsf{E}\log\sum_{\boldsymbol{\sigma}}\exp(-H_{N,M}(\boldsymbol{\sigma}))\ ,$$

and we define

$$p(u) = -\frac{1}{2}r(1-q) + \mathsf{E}\log(2\mathrm{ch}(z\sqrt{r})) + \alpha\mathsf{E}\log\mathsf{E}_\xi\exp u(z\sqrt{q}+\xi\sqrt{1-q})\ . \tag{2.74}$$

Theorem 2.4.2. *Under the conditions of Theorem 2.4.1 we have*

$$|p_{N,M}(u) - p(u)| \le \frac{K(D)}{N}\ . \tag{2.75}$$

We will present two proofs of this fact.

First proof of Theorem 2.4.2. We start with the most beautiful proof, which is somewhat challenging. It implements through interpolation the idea that "the quantities S_k behave like individual spins". We consider independent standard Gaussian r.v.s $z, (z_k)_{k\le M}, (z_i')_{i\le N}, (\xi_k)_{k\le M}$ and for $0 < s < 1$ the Hamiltonian

$$-H_{M,N,s} = \sum_{k\le M} u(\sqrt{s}S_k + \sqrt{1-s}\theta_k) + \sum_{i\le N}\sigma_i\sqrt{1-s}z_i'\sqrt{r} \tag{2.76}$$

where $\theta_k = z_k\sqrt{q} + \xi_k\sqrt{1-q}$. In this formula, we should think of z_i' and z_k as representing new randomness, and of ξ_k as representing "new spins", so that Gibbs averages are given by (2.19), and we define

$$p_{N,M,s} = \frac{1}{N}\mathsf{E}\log\mathsf{E}_\xi\sum_{\boldsymbol{\sigma}}\exp(-H_{M,N,s})\ .$$

The variables ξ_k are not the same as in Section 2.2; we could have denoted them by ξ_k' to insist on this fact, but we preferred simpler notation.

A key point of the present interpolation is that the equations giving the parameters q_s and r_s corresponding to the parameters q and r in the case $s = 1$ are now

$$q_s = \mathsf{E}\mathrm{th}^2\left(\sqrt{s}z\sqrt{r_s} + \sqrt{1-s}z'\sqrt{r}\right) \tag{2.77}$$

$$r_s = \alpha\mathsf{E}\left(\frac{\mathsf{E}_\xi u'(\theta_s)\exp u(\theta_s)}{\mathsf{E}_\xi\exp u(\theta_s)}\right)^2 \tag{2.78}$$

where

$$\theta_s = \sqrt{s}(z\sqrt{q_s} + \xi\sqrt{1-q_s}) + \sqrt{1-s}(z'\sqrt{q} + \xi'\sqrt{1-q})\ .$$

To understand the formula (2.77) one should first understand what happens if we include the action of a random external field in the Hamiltonian, i.e. we add a term $h\sum_{i\le N} g_i\sigma_i$ (where g_i are i.i.d. standard Gaussian) to the right-hand side of (2.6). Then there is nothing to change to the proof of Theorem 2.4.1; only the first formula of (2.68) becomes

$$q = \mathsf{E}\,\mathrm{th}^2(z\sqrt{r} + hg)\ , \tag{2.79}$$

where g, z are independent standard Gaussian r.v.s. We then observe that the last term in (2.76) is an external field, that creates the term $\sqrt{1-s}z'\sqrt{r}$ in (2.77). The second term in the definition of θ_s is created by the terms $\sqrt{1-s}\theta_k$ in the Hamiltonian (2.76), a source of randomness "inside u".

The values $q_s = q$, $r_s = r$ are solutions of the equations (2.77) and (2.78), because for these values $\sqrt{s}z\sqrt{q_s} + \sqrt{1-s}z'\sqrt{q}$ is distributed like $z\sqrt{q}$ (etc.). One could easily check that the solution of the system of equations (2.77) and (2.78) is unique when $\alpha K(D) \le 1$, but this is not needed.

We leave to the readers, as an excellent exercise for those who really want to master the present ideas, the task to prove (2.69) in the case of the Hamiltonian (2.76). Since we have already made the effort to understand the effect of the expectations E_ξ, there is really not much to change to the proof we gave.

So, with obvious notation, one has

$$\forall s \in [0,1]\ ,\ \nu_s\big((R_{1,2} - q)^2\big) \le \frac{L}{N}\ . \tag{2.80}$$

Let us define

$$S_{k,s} = \sqrt{s}S_k + \sqrt{1-s}\theta_k\ ;\ S'_{k,s} = \frac{1}{2\sqrt{s}}S_k - \frac{1}{2\sqrt{1-s}}\theta_k\ ,$$

so that

$$\begin{aligned}\frac{\mathrm{d}}{\mathrm{d}s}p_{N,M,s}(u) &= \frac{1}{N}\nu_s\left(\frac{\mathrm{d}}{\mathrm{d}s}(-H_{N,M,s})\right)\\ &= \frac{1}{N}\nu_s\Bigg(\sum_{k\le M} S'_{k,s}u'(S_{k,s}) - \frac{1}{2\sqrt{1-s}}\sum_{i\le N}\sigma_i z'_i\sqrt{r}\Bigg)\ .\end{aligned} \tag{2.81}$$

The next step is to integrate by parts. It should be obvious how to proceed for the integration by parts in z'_i; this gives

$$\frac{1}{N}\nu_s\Bigg(\frac{1}{2\sqrt{1-s}}\sum_{i\le N}\sigma_i z'_i\sqrt{r}\Bigg) = \frac{r}{2}(1 - \nu_s(R_{1,2}))\ .$$

Let us now explain how to compute $\nu_s(S'_{k,s}u'(S_{k,s}))$. Without loss of generality we assume $k = M$. We make explicit the dependence of the Hamiltonian on $S_{M,s}$ by introducing the Hamiltonian

$$-H_{M-1,N,s} = \sum_{k\le M-1} u(\sqrt{s}S_k + \sqrt{1-s}\theta_k) + \sum_{i\le N}\sigma_i\sqrt{1-s}z'_i\sqrt{r}\ .$$

Denoting by $\langle\cdot\rangle_\sim$ an average for this Hamiltonian, we then have

$$\nu_s(S'_{M,s}u'(S_{M,s})) = \mathsf{E}\frac{\langle S'_{M,s}u'(S_{M,s})\exp u(S_{M,s})\rangle_\sim}{\langle \exp u(S_{M,s})\rangle_\sim} . \tag{2.82}$$

Let us denote as usual by an upper index ℓ the fact "that the spins are in the ℓ-th replica". For example, (since we think of ξ_k as a spin) $\theta_k^\ell = z_k\sqrt{q} + \xi_k^\ell\sqrt{1-q}$ where ξ_k^ℓ are independent standard Gaussian r.v.s, and $S_{k,s}^\ell = \sqrt{s}S_k^\ell + \sqrt{1-s}\theta_k^\ell$, and let us observe the key relations (where the reader will not confuse $S_{M,s}^{\ell\prime}$ with $S_{M,s}^{\ell'}$)

$$\mathsf{E}S_{M,s}^{\ell\prime}S_{M,s}^{\ell} = 0 \; ; \; \ell \neq \ell' \Rightarrow \mathsf{E}S_{M,s}^{\ell\prime}S_{M,s}^{\ell'} = \frac{1}{2}(R_{\ell,\ell'} - q) .$$

Now we integrate by parts in (2.82). This integration by parts will take place given the randomness of $H_{M-1,N,s}$. We have explained in detail in the proof of Lemma 2.3.2 how to proceed. The present case is significantly simpler. There is only one term, "the term created by the denominator" (as defined page 168), and we obtain

$$\nu_s(S'_{M,s}u'(S_{M,s})) = -\frac{1}{2}\nu_s\big((R_{1,2}-q)u'(S_{M,s}^1)u'(S_{M,s}^2)\big) .$$

This illustrates again the principle (2.58) that the expectation E_ξ in the denominator "shifts the variables there to a new replica." Therefore we have found that

$$\frac{\mathrm{d}}{\mathrm{d}s}p_{N,M,s}(u) = -\frac{1}{2}\nu_s\Big((R_{1,2}-q)\frac{1}{N}\sum_{k\le M}u'(S_{k,s}^1)u'(S_{k,s}^2)\Big) - \frac{r}{2}(1-\nu_s(R_{1,2})) .$$

We will not use the fact that the contribution for each $k \le M$ is the same, but rather we regroup the terms as

$$\begin{aligned}\frac{\mathrm{d}}{\mathrm{d}s}p_{N,M,s}(u) &= -\frac{r}{2}(1-q)\\ &\quad - \frac{1}{2}\nu_s\bigg((R_{1,2}-q)\Big(\frac{1}{N}\sum_{k\le M}u'(S_{k,s}^1)u'(S_{k,s}^2) - r\Big)\bigg) .\end{aligned} \tag{2.83}$$

This formula should be compared to (1.65). There seems to be little hope to get any kind of positivity argument here. This is unfortunate because as of today, positivity arguments are almost our only tool to obtain low-temperature results.

We get, using the Cauchy-Schwarz inequality

$$\begin{aligned}\Big|\frac{\mathrm{d}}{\mathrm{d}s}p_{N,M,s}(u) + \frac{r}{2}(1-q)\Big| &\le \nu_s\big((R_{1,2}-q)^2\big)^{1/2}\\ &\quad\times \nu_s\bigg(\Big(\frac{1}{N}\sum_{k\le M}u'(S_{k,s}^1)u'(S_{k,s}^2) - r\Big)^2\bigg)^{1/2} .\end{aligned} \tag{2.84}$$

From (2.80) we see that the right-hand side is $\le K(D)/\sqrt{N}$; but to get the correct rate $K(D)/N$ (rather than $K(D)/\sqrt{N}$) in Theorem 2.4.2, we need to know the following, that is proved separately in Lemma 2.4.3 below:

$$\nu_s\biggl(\Bigl(\frac{1}{N}\sum_{k\le M} u'(S^1_{k,s})u'(S^2_{k,s}) - r\Bigr)^2\biggr) \le \frac{K(D)}{N} . \tag{2.85}$$

We combine with (2.80) to obtain from (2.84) that

$$\left|\frac{\mathrm{d}}{\mathrm{d}s} p_{N,M,s}(u) + \frac{r}{2}(1-q)\right| \le \frac{K(D)}{N}$$

so that, since $p_{N,M}(u) = p_{N,M,1}(u)$,

$$\left|p_{N,M}(u) + \frac{r}{2}(1-q) - p_{N,M,0}(u)\right| \le \frac{K(D)}{N} .$$

As the spins decouple in $p_{N,M,0}(u)$, the computation of this quantity is straightforward and this yields (2.75). □

Lemma 2.4.3. *Inequality (2.85) holds under the conditions of Theorem 2.4.1.*

Proof. Let us write

$$f = \frac{1}{N}\sum_{k\le M} u'(S^1_{k,s})u'(S^2_{k,s}) - r$$
$$f^- = \frac{1}{N}\sum_{k< M} u'(S^1_{k,s})u'(S^2_{k,s}) - r ,$$

so that, using symmetry between the values of $k \le M$,

$$\begin{aligned}\nu_s(f^2) &= \nu_s\bigl((\alpha u'(S^1_{M,s})u'(S^2_{M,s}) - r)f\bigr)\\ &\le \nu_s\bigl((\alpha u'(S^1_{M,s})u'(S^2_{M,s}) - r)f^-\bigr) + \frac{K(D)}{N} .\end{aligned} \tag{2.86}$$

We extend Proposition 2.3.5 to the present setting of the Hamiltonian (2.76) to get

$$\begin{aligned}&\bigl|\nu_s\bigl((\alpha u'(S^1_{M,s})u'(S^2_{M,s}) - r)f^-\bigr)\bigr|\\ &\le \alpha K(D)\Bigl(\nu_s\bigl((R_{1,2}-q)^2\bigr)^{1/2}\nu_s\bigl((f^-)^2\bigr)^{1/2} + \frac{1}{N}\Bigr) .\end{aligned}$$

Combining these, and since $2\sqrt{ab} \le a+b$, for $\alpha K(D) \le 1$ this yields

$$\nu_s(f^2) \le \frac{1}{2}\nu_s\bigl((R_{1,2}-q)^2\bigr) + \frac{1}{2}\nu_s\bigl((f^-)^2\bigr) + \frac{K(D)}{N}$$

and since $|f^2 - (f^-)^2| \le K(D)/N$ we get

$$\nu_s(f^2) \le \frac{1}{2}\nu_s\big((R_{1,2}-q)^2\big) + \frac{1}{2}\nu_s(f^2) + \frac{K(D)}{N} ,$$

which completes the proof using (2.80). □

To prepare for the second proof of Theorem 2.4.2, let us denote by $F(\alpha, r, q)$ the right-hand side of (2.74), i.e.

$$F(\alpha, r, q) = -\frac{1}{2}r(1-q) + \mathsf{E}\log(2\mathrm{ch}(z\sqrt{r})) + \alpha\mathsf{E}\log\mathsf{E}_\xi \exp u(\theta) ,$$

where $\theta = z\sqrt{q} + \xi\sqrt{1-q}$ and let us think of this quantity as a function of three unrelated variables. For convenience, we reproduce the equations (2.68):

$$q = \mathsf{E}\,\mathrm{th}^2(z\sqrt{r}) \quad ; \quad r = \alpha\mathsf{E}\left(\frac{\mathsf{E}_\xi u'(\theta)\exp u(\theta)}{\mathsf{E}_\xi \exp u(\theta)}\right)^2 . \tag{2.87}$$

Lemma 2.4.4. *The conditions (2.87) mean respectively that* $\partial F/\partial r = 0$, $\partial F/\partial q = 0$.

Proof. This is of course calculus, differentiation and integration by parts, but it would be nice to *really* understand why this is true. We give the proof in complete detail, but we suggest as a simple exercise that the reader tries first to figure out these details by herself.

Integration by parts yields

$$\frac{\partial F}{\partial r} = \frac{1}{2}\left(q - 1 + \frac{1}{\sqrt{r}}\mathsf{E}\,z\mathrm{th}z\sqrt{r}\right) = \frac{1}{2}\left(q - 1 + \mathsf{E}\frac{1}{\mathrm{ch}^2(z\sqrt{r})}\right)$$

so that $\partial F/\partial r = 0$ if

$$q = 1 - \mathsf{E}\frac{1}{\mathrm{ch}^2(z\sqrt{r})} = \mathsf{E}\,\mathrm{th}^2(z\sqrt{r}) .$$

Next, if

$$\theta = z\sqrt{q} + \xi\sqrt{1-q}, \; \theta' = \frac{z}{2\sqrt{q}} - \frac{\xi}{2\sqrt{1-q}} ,$$

we have

$$\frac{\partial F}{\partial q} = \frac{r}{2} + \frac{\alpha}{2}\mathsf{E}\left(\theta'\frac{u'(\theta)\exp u(\theta)}{\mathsf{E}_\xi \exp u(\theta)}\right) . \tag{2.88}$$

To integrate by parts, we observe that $F_1(z) = \mathsf{E}_\xi \exp u(\theta)$ does not depend on ξ and

$$\frac{\mathrm{d}F_1}{\mathrm{d}z} = \frac{\mathrm{d}}{\mathrm{d}z}\mathsf{E}_\xi \exp u(z\sqrt{q} + \xi\sqrt{1-q}) = \sqrt{q}\mathsf{E}_\xi\, u'(\theta)\exp u(\theta) .$$

We appeal to the integration by parts formula (A.17) to find, since $\mathsf{E}(\theta'\theta) = 0$, $\mathsf{E}(\theta' z) = 1/\sqrt{q}$ that

$$\begin{aligned}\mathsf{E}\left(\theta' \frac{u'(\theta)\exp u(\theta)}{F_1(z)}\right) &= -\mathsf{E}\left(\frac{1}{F_1(z)^2} u'(\theta)\exp u(\theta) \mathsf{E}_\xi(u'(\theta)\exp u(\theta))\right) \\ &= -\mathsf{E}\frac{(\mathsf{E}_\xi u'(\theta)\exp u(\theta))^2}{(\mathsf{E}_\xi \exp u(\theta))^2} ,\end{aligned}$$

so that by (2.88), $\partial F/\partial q = 0$ if and only if the second part of (2.87) holds. □

If q and r are now related by the conditions (2.87), for small α they are functions $q(\alpha)$ and $r(\alpha)$ of α (since, as shown by Theorem 1.4.1 the equations (2.87) have a unique solution). The quantity $F(\alpha, r(\alpha), q(\alpha))$ is function $F(\alpha)$ of α alone, and

$$\frac{\mathrm{d}F}{\mathrm{d}\alpha} = \frac{\partial F}{\partial \alpha} + \frac{\partial F}{\partial q}\frac{\mathrm{d}q}{\mathrm{d}\alpha} + \frac{\partial F}{\partial r}\frac{\mathrm{d}r}{\mathrm{d}\alpha} = \frac{\partial F}{\partial \alpha} ,$$

since $\partial F/\partial q = \partial F/\partial r = 0$ when $q = q(\alpha)$ and $r = r(\alpha)$. Therefore

$$F'(\alpha) = \mathsf{E}\log \mathsf{E}_\xi \exp u(\theta) . \tag{2.89}$$

Second proof of Theorem 2.4.2. We define $Z_{N,M} = \sum_{\boldsymbol{\sigma}} \exp(-H_{N,M}(\boldsymbol{\sigma}))$, and we note the identity

$$Z_{N,M+1} = Z_{N,M}\left\langle \exp u\left(\frac{1}{\sqrt{N}}\sum_{i\le N} g_{i,M+1}\sigma_i\right)\right\rangle$$

so that

$$p_{N,M+1}(u) - p_{N,M}(u) = \frac{1}{N}\mathsf{E}\log\left\langle \exp u\left(\frac{1}{\sqrt{N}}\sum_{i\le N} g_{i,M+1}\sigma_i\right)\right\rangle . \tag{2.90}$$

To compute the right-hand side of (2.90) we introduce

$$S_v = \sqrt{\frac{v}{N}}\sum_{i\le N} g_{i,M+1}\sigma_i + \sqrt{1-v}\,\theta ,$$

where $\theta = z\sqrt{q} + \xi\sqrt{1-q}$, where (I almost hesitate to say it again) z and ξ are independent standard Gaussian r.v.s, and where q is as in (2.68) for $\alpha = M/N$ (so that the value of q depends on M). We set

$$\varphi(v) = \mathsf{E}\log \mathsf{E}_\xi \langle \exp u(S_v)\rangle .$$

As usual E_ξ denotes expectation in all the r.v.s labeled ξ. Here this expectation is not built in the bracket $\langle\cdot\rangle$, in contrast with what we did e.g in (2.35), so that it must be written explicitly.

We note that

$$\varphi(1) = N(p_{N,M+1}(u) - p_{N,M}(u)) \ ; \ \varphi(0) = \mathsf{E} \log \mathsf{E}_\xi \exp u(\theta) \ .$$

With obvious notation we have

$$\varphi'(v) = \mathsf{E}\frac{\mathsf{E}_\xi \langle S'_v \exp u(S_v)\rangle}{\mathsf{E}_\xi \langle \exp u(S_v)\rangle} = \mathsf{E}\frac{\langle S'_v \exp u(S_v)\rangle}{\mathsf{E}_\xi \langle \exp u(S_v)\rangle} \ .$$

We then integrate by parts, exactly as in (2.82). This yields the formula

$$\varphi'(v) = -\frac{1}{2}\mathsf{E}\frac{\langle (R_{1,2}-q)u'(S_v^1)u'(S_v^2)\exp(u(S_v^1)+u(S_v^2))\rangle}{\mathsf{E}_\xi \langle \exp(u(S_v^1)+u(S_v^2))\rangle} \ , \tag{2.91}$$

where S_v^ℓ is defined as S_v, but replacing ξ by ξ^ℓ and $\boldsymbol{\sigma}$ by $\boldsymbol{\sigma}^\ell$. Now (2.69) implies

$$|\varphi'(v)| \le K(D)\nu(|R_{1,2}-q|) \le K(D)\nu\big((R_{1,2}-q)^2\big)^{1/2} \le \frac{K(D)}{\sqrt{N}} \ .$$

This bound unfortunately does not get the proper rate. To get the proper bound in $K(D)/N$ in (2.75) one must replace the bound

$$|\varphi(1) - \varphi(0)| \le \sup |\varphi'(v)|$$

by the bound

$$|\varphi(1) - \varphi(0) - \varphi'(0)| \le \sup |\varphi''(v)| \ . \tag{2.92}$$

A new differentiation and integration by parts in (2.91) bring out in each term a new factor $(R_{\ell,\ell'} - q)$, so that using (2.69) we now get

$$|\varphi''(v)| \le K(D)\nu\big((R_{1,2}-q)^2\big) \le \frac{K(D)}{N} \ .$$

As a special case of (2.91),

$$\varphi'(0) = -\frac{1}{2}\widehat{r}\nu(R_{1,2}-q) \ .$$

We shall prove later (when we learn how to prove central limit theorems in Chapter 9) the non-trivial fact that $|\nu(R_{1,2}-q)| \le K(D)/N$, and (2.92) then implies

$$\left| p_{N,M+1}(u) - p_{N,M}(u) - \frac{1}{N}\mathsf{E}\log \mathsf{E}_\xi \exp u(\theta)\right| \le \frac{K(D)}{N^2} \ . \tag{2.93}$$

One can then recover the value of $p_{N,M}(u)$ by summing these relations over M. This is a non-trivial task, since the value of q (and hence of θ) depends on M.

Let us recall the function $F(\alpha)$ of (2.89). It is tedious but straightforward to check that $F''(\alpha)$ remains bounded as $\alpha K(D) \leq 1$, so that (2.89) yields

$$\left| F\left(\frac{M+1}{N}\right) - F\left(\frac{M}{N}\right) - \frac{1}{N}\mathsf{E}\log \mathsf{E}_\xi \exp u(\theta)\right| \leq \frac{K(D)}{N^2} .$$

Comparing with (2.93) and summing over M then proves (2.75) (and even better, since the summation is over M, we get a bound $\alpha K(D)/N$). This completes the second proof of Theorem 2.4.2. □

It is worth noting that the first proof of Theorem 2.4.2 provides an easy way to discover the formula (2.74), but that this formula is much harder to guess if one uses the second proof. In some sense the first proof of Theorem 2.4.2 is more powerful and more elegant than the second proof. However we will meet situations (in Chapters 3 and 4) where it is not immediate to apply this method (and whether this is possible remains to be investigated). In these situations, we shall use instead the argument of the second proof of Theorem 2.4.2.

2.5 Exponential Inequalities

Our goal is to improve the control of $R_{1,2}-q$ from second to higher moments.

Theorem 2.5.1. *Given D, there is a number $K(D)$ such that if u satisfies (2.7), i.e. $|u^{(\ell)}| \leq D$ for all $0 \leq \ell \leq 3$ then for $\alpha K(D) \leq 1$, we have*

$$\forall\, k \geq 0\,, \quad \nu\big((R_{1,2}-q)^{2k}\big) \leq \left(\frac{64k}{N}\right)^k . \tag{2.94}$$

Proof. It goes by induction over k, and is nearly identical to that of Proposition 1.6.7.

For $1 \leq n \leq N$, we define $A_n = N^{-1}\sum_{n\leq i\leq N}(\sigma_i^1\sigma_i^2 - q)$, and the induction hypothesis is that for each $n \leq N$,

$$\nu(A_n^{2k}) \leq \left(\frac{64k}{N}\right)^k . \tag{2.95}$$

To perform the induction from k to $k+1$, we can assume $n < N$, for (2.95) holds if $n = N$. Using symmetry between sites yields

$$\nu(A_n^{2k+2}) = \frac{N-n+1}{N}\nu(f)\,,$$

where

$$f = (\varepsilon_1\varepsilon_2 - q)A_n^{2k+1} .$$

Thus

$$\nu(A_n^{2k+2}) \le |\nu_0(f)| + \sup_t |\nu_t'(f)| \,. \tag{2.96}$$

We first study the term $\nu_0(f)$. Consider

$$A' = \frac{1}{N} \sum_{n \le i \le N-1} (\sigma_i^1 \sigma_i^2 - q) \,.$$

Since by Lemma 2.2.2 we have $\nu_0((\varepsilon_1\varepsilon_2 - q)A'^{2k+1}) = 0$, using the inequality

$$|x^{2k+1} - y^{2k+1}| \le (2k+1)|x-y|(x^{2k} + y^{2k})$$

for $x = A_n$ and $y = A'$ we get, since $|x-y| \le 2/N$ and $|\varepsilon_1\varepsilon_2 - q| \le 2$,

$$|\nu_0(f)| \le \frac{4(2k+1)}{N} (\nu_0(A_n^{2k}) + \nu_0(A'^{2k})) \,.$$

We use (2.67), the induction hypothesis, and the observation that since $n < N$, we have

$$\nu(A'^{2k}) = \nu(A_{n+1}^{2k})$$

to obtain

$$|\nu_0(f)| \le \frac{16(2k+1)}{N} \left(\frac{64k}{N}\right)^k \le \frac{2k+1}{4(k+1)} \left(\frac{64(k+1)}{N}\right)^{k+1} \,. \tag{2.97}$$

To compute $\nu_t'(f)$ we use Proposition 2.3.6 with $n = 4, \tau_1 = (2k+2)/(2k+1)$, $\tau_2 = 2k+2$ and (2.67) to get

$$|\nu_t'(f)| \le \alpha K(D) \Big(\nu(A_n^{2k+2})^{1/\tau_1} \nu\big((R_{1,2} - q)^{2k+2}\big)^{1/\tau_2} + \frac{1}{N} \nu(|A_n|^{2k+1}) \Big) \,.$$

Using the inequality $x^{1/\tau_1} y^{1/\tau_2} \le x + y$ for $x = \nu(A_n^{2k+2})$ and $y = \nu((R_{1,2} - q)^{2k+2})$ this implies

$$|\nu_t'(f)| \le \alpha K(D) \Big(\nu(A_n^{2k+2}) + \nu\big((R_{1,2} - q)^{2k+2}\big) + \frac{1}{N} \nu(|A_n|^{2k+1}) \Big) \,.$$

Combining with (2.96) and (2.97) we get if $\alpha K(D) \le 1/4$,

$$\begin{aligned} \nu(A_n^{2k+2}) \le\; & \frac{1}{4} \left(\nu(A_n^{2k+2}) + \nu\big((R_{1,2} - q)^{2k+2}\big)\right) \\ & + \frac{2k+1}{4(k+1)} \left(\frac{64(k+1)}{N}\right)^{k+1} + \frac{1}{N} \nu(|A_n|^{2k+1}) \,. \end{aligned} \tag{2.98}$$

Since $|A_n| \le 2$ and hence $|A_n|^{2k+1} \le 2A_n^{2k}$, the induction hypothesis implies that the last term of (2.98) is at most

$$\frac{1}{32(k+1)} \left(\frac{64(k+1)}{N}\right)^{k+1} ,$$

so the sum of the last 2 terms is at most

$$\frac{1}{2}\left(\frac{64(k+1)}{N}\right)^{k+1}.$$

Since $A_1 = R_{1,2} - q$, considering first the case $n = 1$ provides the required inequality in that case. Using back this inequality in (2.98) provides the required inequality for all values of n. □

The following extends Lemma 2.4.3. Its proof is pretty similar to that of Theorem 2.5.1, and demonstrates the power of this approach. The reader who does not enjoy the argument should skip the forthcoming proof and make sure she does not miss the pretty Theorem 2.5.3. We denote by $K_0(D)$ the constant of Theorem 2.5.1.

Theorem 2.5.2. *Assume that u satisfies (2.7) for a certain number D. Then there is a number $K(D)$, depending on D only, with the following property. For $\alpha K_0(D) \le 1$ we have*

$$\forall k \ge 0\,, \quad \nu\left(\left(\frac{1}{N}\sum_{j\le M} u'(S_j^1)u'(S_j^2) - r\right)^{2k}\right) \le \left(\frac{\alpha k K(D)}{N}\right)^k . \tag{2.99}$$

Proof. We recall the definition of $\widehat{r}$ given by (2.36), i.e.

$$\widehat{r} = \mathsf{E}\left(\frac{\mathsf{E}_\xi u'(\theta)\exp u(\theta)}{\mathsf{E}_\xi \exp u(\theta)}\right)^2 ,$$

so that with the notation (2.87) we have $r = \alpha\widehat{r}$. For $1 \le n \le M$ we define

$$C_n = \frac{1}{M}\sum_{n\le j\le M}\left(u'(S_j^1)u'(S_j^2) - \widehat{r}\right).$$

Since $r = \alpha\widehat{r}$ and $1/N = \alpha/M$ the left-hand side of (2.99) is $\alpha^{2k}\nu(C_1^{2k})$.

We prove by induction over k that if $\alpha K_0(D) \le 1$ then for a suitable number $K_1(D)$ we have for $k \ge 1$ and any $n \le M$ that

$$\nu(C_n^{2k}) \le \left(\frac{kK_1(D)}{M}\right)^k . \tag{2.100}$$

Using this for $n = 1$ concludes the proof. For $k = 0$ (2.100) is true if one then understands the right-hand side of (2.99) as being 1. The reader disliking this can instead start the induction at $k = 1$. To prove the case $k = 1$ it suffices to repeat the proof of Lemma 2.4.3 (while keeping a tighter watch on the dependence on α). For the induction step from k to $k+1$ we can assume that $n < M$, and we use symmetry among the values of j to obtain

$$\nu(C_n^{2k+2}) = \nu(f^{\sim})\,, \tag{2.101}$$

where $f^{\sim} = (u'(S^1_M)u'(S^2_M) - \widehat{r})C_n^{2k+1}$. Let us define

$$C' = \frac{1}{M} \sum_{n \le j \le M-1} (u'(S^1_j)u'(S^2_j) - \widehat{r}) \,.$$

Using the inequality

$$|x^{2k+1} - y^{2k+1}| \le (2k+1)|x-y|(x^{2k} + y^{2k}) \tag{2.102}$$

for $x = C_n$ and $y = C'$, and since $|u'(S^1_M)u'(S^2_M) - \widehat{r}| \le 2D^2$, we obtain that for $f^* = (u'(S^1_M)u'(S^2_M) - \widehat{r})C'^{2k+1}$:

$$\nu(f^{\sim}) \le \nu(f^*) + \frac{2(2k+1)D^2}{M}(\nu(C_n^{2k}) + \nu(C'^{2k})) \,. \tag{2.103}$$

Since $n < M$, symmetry among the values of j implies $\nu(C'^{2k}) = \nu(C^{2k}_{n+1})$ and the induction hypothesis yields

$$\nu(f^{\sim}) \le \nu(f^*) + \frac{8(k+1)D^2}{M}\left(\frac{K_1(D)k}{M}\right)^k \,. \tag{2.104}$$

Next, we use (2.62) for $t = 1$, $f = C'^{2k+1}$ and $n = 2$. This is permitted because f does not depend on the randomness of ξ^ℓ_M, ξ^ℓ or $g_{i,M}$. We choose $\tau_1 = (2k+2)/(2k+1)$ and $\tau_2 = 2k+2$ to get

$$|\nu(f^*)| \le K_2(D)\left(\nu(C'^{2k+2})^{1/\tau_1}\nu\big((R_{1,2}-q)^{2k+2}\big)^{1/\tau_2} + \frac{1}{N}\nu(|C'|^{2k+1})\right) .$$

Since we work under the condition $\alpha K_0(D) \le 1$, we can as well assume that $\alpha \le 1$, so that $M \le N$ and

$$|\nu(f^*)| \le K_2(D)\left(\nu(C'^{2k+2})^{1/\tau_1}\nu\big((R_{1,2}-q)^{2k+2}\big)^{1/\tau_2} + \frac{1}{M}\nu(|C'|^{2k+1})\right) . \tag{2.105}$$

We recall the inequality $x^{1/\tau_1}y^{1/\tau_2} \le x + y$. Changing x to x/A and y to $A^{\tau_2/\tau_1}y$ in this inequality gives

$$x^{1/\tau_1}y^{1/\tau_2} \le \frac{x}{A} + A^{\tau_2/\tau_1}y \,.$$

Using this for $A = 2K_2(D)$, $x = \nu(C'^{2k+2})$ and $y = \nu((R_{1,2}-q)^{2k+2})$, we deduce from (2.105) that

$$|\nu(f^*)| \le \frac{1}{2}\nu(C'^{2k+2}) + K(D)^{2k+1}\nu\big((R_{1,2}-q)^{2k+2}\big) + \frac{K(D)}{M}\nu(|C'|^{2k+1}) \,. \tag{2.106}$$

We now use the inequality

$$|x^{2k+2} - y^{2k+2}| \le (2k+2)|x-y|(|x|^{2k+1} + |y|^{2k+1})$$

for $x = C'$ and $y = C_n$ to obtain

$$\nu(C'^{2k+2}) \le \nu(C_n^{2k+2}) + \frac{2(2k+2)D^2}{M}\left(\nu(|C'|^{2k+1}) + \nu(|C_n|^{2k+1})\right) .$$

We combine this with (2.106), we use that $|C_n|^{2k+1} \le 2D^2 C_n^{2k}$ and $|C'|^{2k+1} \le 2D^2 C'^{2k}$ and the induction hypothesis to get

$$\begin{aligned}|\nu(f^*)| \le \frac{1}{2}\nu(C_n^{2k+2}) &+ K(D)^{2k+2}\nu((R_{1,2}-q)^{2k+2}) \\ &+ \frac{(k+1)K(D)}{M}\left(\frac{K_1(D)k}{M}\right)^k ,\end{aligned}$$

and combining with (2.101) and (2.104) that

$$\begin{aligned}\nu(C_n^{2k+2}) \le \frac{1}{2}\nu(C_n^{2k+2}) &+ K(D)^{2k+2}\nu((R_{1,2}-q)^{2k+2}) \\ &+ \frac{(k+1)K(D)}{M}\left(\frac{K_1(D)k}{M}\right)^k .\end{aligned}$$

Finally we use (2.94) to conclude the proof that $\nu(C_n^{2k+2}) \le (K_1(D)(k+1)/M)^{k+1}$ if $K_1(D)$ has been chosen large enough. This completes the induction. □

The following central limit theorem describes the fluctuations of $p_{N,M}(u)$ (given by (2.11)). We recall that $a(k) = \mathsf{E}z^k$ where z is a standard Gaussian r.v. and that $O(k)$ denotes a quantity $A = A_N$ with $|A| \le KN^{-k/2}$ where K does not depend on N. We recall the notation $p(u)$ of (2.74),

$$p(u) = -\frac{1}{2}r(1-q) + \mathsf{E}\log(2\mathrm{ch}(z\sqrt{r})) + \alpha\mathsf{E}\log\mathsf{E}_\xi \exp u(z\sqrt{q} + \xi\sqrt{1-q}) .$$

Theorem 2.5.3. *Let*

$$b = \mathsf{E}(\log\mathrm{ch}(z\sqrt{r}))^2 - (\mathsf{E}\log\mathrm{ch}(z\sqrt{r}))^2 - qr .$$

Then for each $k \ge 1$ we have

$$\mathsf{E}(p_{N,M}(u) - p(u))^k = \left(\frac{b}{N}\right)^{k/2} a(k) + O(k+1) .$$

Proof. This argument resembles that in the proof of Theorem 1.4.11, and it would probably help the reader to review the proof of that theorem now. The present proof is organized a bit differently, avoiding the a priori estimate of Lemma 1.4.12. The interpolation method of the first proof of Theorem 2.4.2 is at the center of the argument, so the reader should feel comfortable with this proof in order to proceed. We recall the Hamiltonian (2.76) and we denote by $\langle\cdot\rangle_s$ an average for the corresponding Gibbs measure. In the proof $O(k)$ will denote a quantity $A = A_N$ such that $|A| \le KN^{-k/2}$ where K does not depend on N **or** s, and we will take for granted that Theorems 2.5.1 and 2.5.2 hold uniformly over s. (This fact is left as a good exercise for the reader.)

Consider the following quantities

$$A(s) = \frac{1}{N} \log \sum_{\boldsymbol{\sigma}} \mathsf{E}_\xi \exp(-H_{N,M,s}(\boldsymbol{\sigma}))$$

$$\mathrm{RS}(s) = \mathsf{E} \log 2\mathrm{ch}(z\sqrt{r}) + \alpha \mathsf{E} \log \mathsf{E}_\xi \exp u(z\sqrt{q} + \xi\sqrt{1-q}) - \frac{s}{2} r(1-q)$$

$$V(s) = A(s) - \mathrm{RS}(s)$$

$$b(s) = \mathsf{E}(\log \mathrm{ch}(z\sqrt{r}))^2 - (\mathsf{E} \log \mathrm{ch}(z\sqrt{r}))^2 - rqs \; .$$

The quantities $\mathsf{E}A(s)$, $\mathrm{RS}(s)$ and $b(s)$ are simply the quantities corresponding for the interpolating system respectively to the quantities $p_{N,M}(u)$, p_u, and b. Fixing k, we set

$$\psi(s) = \mathsf{E}V(s)^k \; .$$

We aim at proving by induction over k that $\psi(s) = (b(s)/N)^{k/2} a(k) + O(k+1)$, which, for $s = 1$, proves the theorem. Consider $\varphi(s,a) = \mathsf{E}(A(s)-a)^k$, so that $\psi(s) = \varphi(s, \mathrm{RS}(s))$ and by straightforward differentiation $\partial\varphi/\partial s$ is given by the quantity

$$\frac{k}{2N} \mathsf{E}\left(\left\langle \sum_{j \le M} \left(\frac{S_j}{\sqrt{s}} - \frac{\theta_j}{\sqrt{1-s}}\right) u'(S_{j,s}) - \sum_{i \le N} \frac{\sigma_i}{\sqrt{1-s}} z'_i \sqrt{r} \right\rangle_s (A(s)-a)^{k-1}\right),$$

where $S_{j,s} = \sqrt{s} S_j + \sqrt{1-s}\,\theta_j$. Next, defining $S^\ell_{j,s}$ as usual we claim that $\partial\varphi/\partial s = \mathrm{I} + \mathrm{II}$, where

$$\mathrm{I} = \frac{k}{2} \mathsf{E}\left(\left\langle -\frac{1}{N} \sum_{j \le M} (R_{1,2} - q) u'(S^1_{j,s}) u'(S^2_{j,s}) - r(1 - R_{1,2}) \right\rangle_s (A(s)-a)^{k-1}\right)$$

and II is the quantity

$$\frac{k(k-1)}{2N} \mathsf{E}\left(\left\langle \frac{1}{N} \sum_{j \le M} (R_{1,2} - q) u'(S^1_{j,s}) u'(S^2_{j,s}) - r R_{1,2} \right\rangle_s (A(s)-a)^{k-2}\right).$$

This follows by integrating by parts as in the proof of (2.83). The term I is created by the dependence of the bracket $\langle \cdot \rangle_s$ on the r.v.s S_j, θ_j and z'_i, and the term II by the dependence on these variables of $A(s)$. We note the obvious identity $\mathrm{I} = \mathrm{III} + \mathrm{IV}$ where

$$\mathrm{III} = -\frac{k}{2} \mathsf{E}\left(\left\langle (R_{1,2} - q)\left(\frac{1}{N} \sum_{j \le M} u'(S^1_{j,s}) u'(S^2_{j,s}) - r\right)\right\rangle_s (A(s)-a)^{k-1}\right)$$

and

$$\mathrm{IV} = -\frac{kr(1-q)}{2}\mathsf{E}((A(s)-a)^{k-1})\,.$$

Similarly we have also II = V + VI where V is the quantity

$$\frac{k(k-1)}{2N}\mathsf{E}\left(\left\langle (R_{1,2}-q)\left(\frac{1}{N}\sum_{j\le M} u'(S^1_{j,s})u'(S^2_{j,s})-r\right)\right\rangle_s (A(s)-a)^{k-2}\right)$$

and

$$\mathrm{VI} = -\frac{rq}{2N}k(k-1)\mathsf{E}((A(s)-a)^{k-2})\,.$$

Now,

$$\psi'(s) = \frac{\mathrm{d}}{\mathrm{d}s}\varphi(s,\mathrm{RS}(s)) = \frac{\partial\varphi}{\partial s}(s,\mathrm{RS}(s)) + \mathrm{RS}'(s)\frac{\partial\varphi}{\partial a}(s,\mathrm{RS}(s))\,. \tag{2.107}$$

Since $\mathrm{RS}'(s) = -r(1-q)/2$ and $\partial\varphi/\partial a(s,\mathrm{RS}(s)) = -k\mathsf{E}v(s)^{k-1}$, the second term of (2.107) cancels out with the term IV and we get

$$\psi'(s) = \mathrm{VII} + \mathrm{VIII} + \mathrm{IX} \tag{2.108}$$

where

$$\mathrm{VII} = -\frac{k}{2}\mathsf{E}\left(\left\langle (R_{1,2}-q)\left(\frac{1}{N}\sum_{j\le M} u'(S^1_{j,s})u'(S^2_{j,s})-r\right)\right\rangle_s V(s)^{k-1}\right)$$

$$\mathrm{VIII} = \frac{k(k-1)}{2N}\mathsf{E}\left(\left\langle (R_{1,2}-q)\left(\frac{1}{N}\sum_{j\le M} u'(S^1_{j,s})u'(S^2_{j,s})-r\right)\right\rangle_s V(s)^{k-2}\right)$$

$$\mathrm{IX} = -\frac{rq}{2N}k(k-1)\mathsf{E}V(s)^{k-2}\,.$$

The idea is that each of the factors $R_{1,2}-q$, $(N^{-1}\sum_{j\le M} u'(S^1_{j,s})u'(S^2_{j,s})-r)$ and $V(s)$ "counts as $N^{-1/2}$". This follows from Theorems 2.5.1 and 2.5.2 for the first two terms, but we have not proved it yet in the case of $V(s)$. (In the case of Theorem 1.4.11, the a priori estimate of Lemma 1.4.12 showed that $V(s)$ "counts as $N^{-1/2}$".) Should this be indeed the case, the terms VII and VIII will be of lower order $O(k+1)$. We turn to the proof that this is actually the case.

A first step is to show that

$$\mathrm{VII} \le \frac{K(k)}{N}(\mathsf{E}|V(s)|^k)^{\frac{k-1}{k}}\;;\quad \mathrm{VIII} \le \frac{K(k)}{N^2}(\mathsf{E}|V(s)|^k)^{\frac{k-2}{k}}\,. \tag{2.109}$$

In the case of VII, setting $A = R_{1,2}-q$ and

$$B = \frac{1}{N}\sum_{j\le M} u'(S^1_{j,s})u'(S^2_{j,s}) - r$$

we write, using Hölder's inequality and Theorems 2.5.1 and 2.5.2:

$$\begin{aligned}\mathsf{E}(\langle AB\rangle_s V(s)^{k-1}) &\le \mathsf{E}\langle A^{2k}\rangle_s^{1/2k}\mathsf{E}\langle B^{2k}\rangle_s^{1/2k}(\mathsf{E}|V(s)|^k)^{\frac{k-1}{k}}\\ &\le \frac{K(k)}{N}\mathsf{E}|V(s)|^k)^{\frac{k-1}{k}} .\end{aligned}$$

We proceed in a similar manner for VIII, i.e. we write that

$$\begin{aligned}\mathsf{E}(\langle AB\rangle_s V(s)^{k-1}) &\le \mathsf{E}\langle |A|^{k}\rangle_s^{1/k}\mathsf{E}\langle |B|^{k}\rangle_s^{1/k}(\mathsf{E}|V(s)|^k)^{\frac{k-2}{k}}\\ &\le \frac{K(k)}{N}(\mathsf{E}|V(s)|^k)^{\frac{k-2}{k}} ,\end{aligned}$$

and this proves (2.109).

Since $xy \le x^{\tau_1} + y^{\tau_2}$ for $\tau_2 = k/(k-2)$ and $\tau_1 = k/2$ we get

$$\frac{1}{N}(\mathsf{E}|V(s)|^k)^{\frac{k-2}{k}} \le \frac{1}{N^{k/2}} + \mathsf{E}|V(s)|^k .$$

This implies in particular

$$\mathrm{IX} \le \frac{K(k)}{N}(\mathsf{E}|V(s)|^k)^{\frac{k-2}{k}} \le K(k)\left(\frac{1}{N^{k/2}} + \mathsf{E}|V(s)|^k\right)$$

and

$$\mathrm{VIII} \le \frac{K(k)}{N}\left(\frac{1}{N^{k/2}} + \mathsf{E}|V(s)|^k\right) \le K(k)\left(\frac{1}{N^{k/2}} + \mathsf{E}|V(s)|^k\right) .$$

Next, we use that $xy \le x^{\tau_1} + y^{\tau_2}$ for $\tau_2 = k/(k-1)$ and $\tau_1 = k$ to get

$$\frac{1}{N}(\mathsf{E}|V(s)|^k)^{\frac{k-1}{k}} \le \frac{1}{N^k} + \mathsf{E}|V(s)|^k \le \frac{1}{N^{k/2}} + \mathsf{E}|V(s)|^k .$$

When k is even (so that $|V(s)|^k = V(s)^k$ and $\mathsf{E}|V(s)|^k = \psi(s)$) we have proved that

$$\psi'(s) \le K(k)\left(\frac{1}{N^{k/2}} + \psi(s)\right) . \tag{2.110}$$

Thus (2.110) and Lemma A.13.1 imply that

$$\psi(s) \le K(k)\left(\psi(0) + \frac{1}{N^{k/2}}\right) .$$

Since it is easy (as the spins decouple) to see that $\psi(0) \le K(k)N^{k/2}$, we have proved that for k even we have $\mathsf{E}V(s)^k = O(k)$. Since $\mathsf{E}|V(s)|^k \le (\mathsf{E}V(s)^{2k})^{1/2}$ this implies that $\mathsf{E}|V(s)|^k = O(k)$ for each k so that by (2.109) we have VII $= O(k+1)$ and VIII $= O(k+1)$. Thus (2.108) yields

$$\begin{aligned}\psi'(s) &= -\frac{rq}{2N}k(k-1)\mathsf{E}V(s)^{k-2} + O(k+1)\\ &= \frac{b'(s)}{N}\frac{k}{2}(k-1)\mathsf{E}V(s)^{k-2} + O(k+1) .\end{aligned}$$

As in Theorem 1.4.11, one then shows by induction over k that

$$\mathsf{E}V(s)^k = a(k)\left(\frac{b(s)}{N}\right)^{k/2} + O(k+1)\ ,$$

using that this is true for $s = 0$, which is again proved as in Theorem 1.4.11. □

Exercise 2.5.4. Rewrite the proof of Theorem 1.4.11 without using the a priori estimate of Lemma 1.4.12. This allows to cover the case where the r.v. h is not necessarily Gaussian.

Research Problem 2.5.5. (Level 1^+) Prove the result corresponding to Theorem 1.7.1 for the present model.

This problem has really two parts. The first (easier) part is to prove results for the present model. For this, the approach of "separating the numerator from the denominator" as explained in Section 9.1 seems likely to succeed. The second part (harder) is to find arguments that will carry over when we will have much less control over u as in Chapter 9. For this second part, the work is partially done in [100], but reaching only the rate $1/\sqrt{N}$ rather than the correct rate $1/N$.

Research Problem 2.5.6. (Level 2) For the present model prove the TAP equations.

These equations have two parts. One part expresses $\langle\sigma_i\rangle$ as a function of $(\langle u'(S_k)\rangle)_{k\leq M}$, and one part expresses $\langle u'(S_k)\rangle$ as a function of $(\langle\sigma_i\rangle)_{i\leq N}$. It is (perhaps) not too difficult to prove these equations when one has a good control over all derivatives of u, but it might be another matter to prove something as precise as Theorem 1.7.7 in the setting of Chapter 9.

2.6 Notes and Comments

The problems considered in this chapter are studied in [63] and [52].

It is predicted in [90] that the replica-symmetric solution holds up to α^*, so Problem 2.1.1 amounts to controlling the entire replica-symmetric (="high-temperature") region, typically a very difficult task.

It took a long time to discover the proof of Theorem 2.4.1. The weaker methods developed previously [148] for this model or for the SK and the Hopfield models just would not work. During this struggle, it became clear that the smart path method as used here was a better way to go for these three models.

3. The Shcherbina and Tirozzi Model

3.1 The Power of Convexity

In the present model the configuration space is $\mathbb{R}^N$, that is, the configuration $\boldsymbol{\sigma}$ can be any point in $\mathbb{R}^N$. Given another integer M, we will consider the Hamiltonian

$$-H_{N,M}(\boldsymbol{\sigma}) = \sum_{k\le M} u\left(\frac{1}{\sqrt{N}}\sum_{i\le N} g_{i,k}\sigma_i\right) + h\sum_{i\le N} g_i\sigma_i - \kappa\|\boldsymbol{\sigma}\|^2 \,. \tag{3.1}$$

Here $\|\boldsymbol{\sigma}\|^2 = \sum_{i\le N}\sigma_i^2$, $(g_{i,k})_{i\le N,k\le M}$ and $(g_i)_{i\le N}$ are independent standard Gaussian r.v.s and $\kappa > 0$, $h \ge 0$. We will always assume

$$u \le 0\ ,\ u \text{ is concave.} \tag{3.2}$$

To get a feeling for this Hamiltonian, let us think of u such that, for a certain number τ, $u(x) = -\infty$ if $x < \tau$ and $u(x) = 0$ if $x \ge \tau$. Then it is believable that the Hamiltonian (3.1) will teach us something about the region in $\mathbb{R}^N$ defined by

$$\forall k \le M\ ,\ \frac{1}{\sqrt{N}}\sum_{i\le N} g_{i,k}\sigma_i \ge \tau\,. \tag{3.3}$$

This region has a natural meaning: it is the intersection of M half-spaces of random directions, each of which is determined by an hyperplane at distance (about) τ from the origin. It is for the purpose of computing "the proportion" of the sphere $\mathbb{S}_N = \{\boldsymbol{\sigma}\ ;\ \|\boldsymbol{\sigma}\| = \sqrt{N}\}$ that belongs to the region (3.3) that the Hamiltonian (3.1) was introduced in [133]. This generalizes the problem considered in (0.2), where we had $\tau = 0$. The term $h\sum_{i\le N} g_i\sigma_i$ is not necessary for this computation, but it is not a real trouble either, and there is no reason to deprive the reader from the added charm it brings to the beautiful formulas the Hamiltonian (3.1) will create. We will always assume that $h \ge 0$. There are obvious connections between the present model and the model of Chapter 2. As in Chapter 2 the important case is when M is proportional to N.

This Hamiltonian is a convex function of $\boldsymbol{\sigma}$. In fact

M. Talagrand, *Mean Field Models for Spin Glasses*, Ergebnisse der Mathematik und ihrer Grenzgebiete. 3. Folge / A Series of Modern Surveys in Mathematics 54, DOI 10.1007/978-3-642-15202-3_3,

$$\forall \mathbf{x}, \mathbf{y} \in \mathbb{R}^N\ ,\ \frac{1}{2}(H_{N,M}(\mathbf{x}) + H_{N,M}(\mathbf{y})) - H_{N,M}\left(\frac{\mathbf{x}+\mathbf{y}}{2}\right) \geq \kappa \left\| \frac{\mathbf{x}-\mathbf{y}}{2} \right\|^2 . \tag{3.4}$$

The beauty of the present model is that it allows the use of powerful tools from convexity, from which a very strong control of the overlaps will follow. The overlaps are defined as usual, by

$$R_{\ell,\ell'} = \frac{\boldsymbol{\sigma}^\ell \cdot \boldsymbol{\sigma}^{\ell'}}{N} .$$

The case $\ell = \ell'$ is now of interest, $R_{\ell,\ell} = \|\boldsymbol{\sigma}^\ell\|^2/N$. Let us consider the Gibbs' measure G on $\mathbb{R}^N$ with Hamiltonian $H_{N,M}$, that is, for any subset B of $\mathbb{R}^N$,

$$G(B) = \frac{1}{Z_{N,M}} \int_B \exp(-H_{N,M}(\boldsymbol{\sigma}))\mathrm{d}\boldsymbol{\sigma}\ , \tag{3.5}$$

where $\mathrm{d}\boldsymbol{\sigma}$ denotes Lebesgue's measure and $Z_{N,M} = \int \exp(-H_{N,M}(\boldsymbol{\sigma}))\mathrm{d}\boldsymbol{\sigma}$ is the normalization factor. As usual, we denote by $\langle\cdot\rangle$ an average for this Gibbs measure, so that $G(B) = \langle \mathbf{1}_B \rangle$. We use the notation $\nu(f) = \mathsf{E}\langle f \rangle$.

The goal of this section is to prove the following.

Theorem 3.1.1. *Assume that for a certain number D we have*

$$\forall x\ ,\ u(x) \geq -D(1 + |x|)\ . \tag{3.6}$$

$$|u'| \leq D\ ; \qquad |u''| \leq D\ . \tag{3.7}$$

Then for $k \leq N/4$ we have

$$\nu\big((R_{1,1} - \nu(R_{1,1}))^{2k}\big) \leq \left(\frac{Kk}{N}\right)^k \tag{3.8}$$

$$\nu\big((R_{1,2} - \nu(R_{1,2}))^{2k}\big) \leq \left(\frac{Kk}{N}\right)^k , \tag{3.9}$$

where K does not depend on N or k.

There is of course nothing special in the value $N/4$ which is just a convenient choice. We could replace the condition $k \leq N/4$ by the condition $k \leq AN$ for any number A, with now a constant $K(A)$ depending on A.

The basic reason why in Theorem 3.1.1 one does not control all moments is that moments of high orders are very sensitive to what happens on very small sets or very rare events. For example moments of order about N are very sensitive to what happens on "events of size $\exp(-N/K)$". Controlling events that small is difficult, and is quite besides our main goal. Of course one can dream of an entire "large deviation theory" that would describe the extreme situations that can occur with such rarity. In the present model, and well as in the other models considered in the book, such a theory remains entirely to be built.

Theorem 3.1.1 asserts that the overlaps are nearly constant. For many of the systems studied in this book, it is a challenging task to prove that the overlaps are nearly constant, and this requires a "high-temperature" condition. In the present model, no such condition is necessary, so one might say that the system is always in a high-temperature state. One would expect that it is then a simple matter to completely understand this system, and in particular to compute

$$\lim_{N\to\infty, M/N\to\alpha} \frac{1}{N}\mathsf{E}\log\int\exp(-H_{N,M}(\boldsymbol{\sigma}))\mathrm{d}\boldsymbol{\sigma}\,. \tag{3.10}$$

This, however, does not seem to be the case. At the present time we know how to handle only very special situations, and the reasons for this will become apparent as the reader progresses through the present chapter.

Research Problem 3.1.2. (Level 2+). Under the conditions of Theorem 3.1.1, compute the limit in (3.10).

The fundamental fact about convexity theory is the following functional version of the Brunn-Minkowski theorem. A very clean proof can be found in ([93], Theorem 2.13). For the convenience of the reader, this proof is reproduced in Appendix A.15.

Theorem 3.1.3. *Consider non-negative functions U, V, W on $\mathbb{R}^N$ and a number $0 < s < 1$, and assume that for all $\mathbf{x}, \mathbf{y}$ in $\mathbb{R}^N$ we have*

$$W(s\mathbf{x} + (1-s)\mathbf{y}) \geq U(\mathbf{x})^s V(\mathbf{y})^{1-s}\,. \tag{3.11}$$

Then

$$\int W(\mathbf{x})\mathrm{d}\mathbf{x} \geq \left(\int U(\mathbf{x})\mathrm{d}\mathbf{x}\right)^s \left(\int V(\mathbf{x})\mathrm{d}\mathbf{x}\right)^{1-s}\,. \tag{3.12}$$

Consider sets A and B. The functions $U = \mathbf{1}_A$, $V = \mathbf{1}_B$ and $W = \mathbf{1}_{sA+(1-s)B}$ satisfy (3.11). Writing $\mathrm{Vol}A = \int_A \mathrm{d}\mathbf{x}$, we deduce from (3.12) that

$$\mathrm{Vol}(sA + (1-s)B) \geq (\mathrm{Vol}A)^s(\mathrm{Vol}B)^{1-s}\,, \tag{3.13}$$

the Brunn-Minkowski inequality.

B. Maurey discovered that Theorem 3.1.3 implies the following sweeping generalization of Theorem 1.3.4.

Theorem 3.1.4. *Consider a function H on $\mathbb{R}^N$, and assume that for some number $\kappa > 0$ we have (3.4) i.e.*

$$\forall \mathbf{x}, \mathbf{y} \in \mathbb{R}^N\,,\ \frac{1}{2}(H(\mathbf{x}) + H(\mathbf{y})) - H\left(\frac{\mathbf{x}+\mathbf{y}}{2}\right) \geq \kappa\left\|\frac{\mathbf{x}-\mathbf{y}}{2}\right\|^2\,.$$

Consider the probability measure μ on $\mathbb{R}^N$ with density proportional to $\exp(-H(\mathbf{x}))$ with respect to the Lebesgue measure. Then for any set $B \subset \mathbb{R}^N$ we have

$$\int \exp \frac{\kappa}{2} d^2(\mathbf{x}, B) \mathrm{d}\mu(\mathbf{x}) \le \frac{1}{\mu(B)} , \tag{3.14}$$

where $d(\mathbf{x}, B) = \inf\{d(\mathbf{x}, \mathbf{y}); \mathbf{y} \in B\}$ is the distance from $\mathbf{x}$ to B. Moreover, if f is a function on $\mathbb{R}^N$ with Lipschitz constant A, i.e. it satisfies

$$\forall \mathbf{x}, \mathbf{y} \in \mathbb{R}^N , \ |f(\mathbf{x}) - f(\mathbf{y})| \le A\|\mathbf{x} - \mathbf{y}\| , \tag{3.15}$$

then

$$\int \exp \frac{\kappa}{8A^2} \left(f(\mathbf{x}) - \int f \mathrm{d}\mu \right)^2 \mathrm{d}\mu(\mathbf{x}) \le 4 \tag{3.16}$$

and

$$\forall k \ge 1 , \ \int \left(f(\mathbf{x}) - \int f \mathrm{d}\mu \right)^{2k} \mathrm{d}\mu(\mathbf{x}) \le 4 \left(\frac{8kA^2}{\kappa} \right)^k . \tag{3.17}$$

The most striking feature of the inequalities (3.16) and (3.17) is that they do not depend on the dimension of the underlying space. When $H(\mathbf{x}) = \|\mathbf{x}\|^2/2$, μ is the canonical Gaussian measure and (3.17) recovers (1.47) (with worse constants).

Proof. Define the functions W, U, V as follows:

$$W(\mathbf{x}) = \exp(-H(\mathbf{x})) \ ; \ V(\mathbf{y}) = \exp\left(\frac{\kappa}{2} d(\mathbf{y}, B)^2 - H(\mathbf{y})\right)$$

and

$$\begin{array}{ll} U(\mathbf{x}) = 0 & \text{if } \mathbf{x} \notin B \\ U(\mathbf{x}) = \exp(-H(\mathbf{x})) & \text{if } \mathbf{x} \in B . \end{array}$$

These functions satisfy (3.11) with $s = 1/2$. Indeed, it suffices to consider the case where $\mathbf{x} \in B$, in which case (3.11) reduces to

$$-H\left(\frac{\mathbf{x} + \mathbf{y}}{2}\right) \ge \frac{1}{2}\left(-H(\mathbf{x}) - H(\mathbf{y}) + \frac{\kappa}{2} d(\mathbf{y}, B)^2\right) ,$$

which follows from (3.4) and the fact that $d(\mathbf{y}, B) \le \|\mathbf{x} - \mathbf{y}\|$. Then (3.12) holds, and for the previous choices it means exactly (3.14).

To prove (3.16) we consider a median m of f for μ, that is number m such that $\mu(\{f \le m\}) \ge 1/2$ and $\mu(\{f \ge m\}) \ge 1/2$. The set $B = \{f \le m\}$ then satisfies $\mu(B) \ge 1/2$ and since (3.15) implies

$$f(\mathbf{x}) \le m + Ad(\mathbf{x}, B)$$

it follows from (3.14) that

$$\int_{\{f\geq m\}} \exp \frac{\kappa}{2A^2}(f(\mathbf{x})-m)^2 \mathrm{d}\mu(\mathbf{x}) \leq 2\,. \tag{3.18}$$

Proceeding in a similar manner to control the integral over the set $\{f \leq m\}$ we get

$$\int \exp \frac{\kappa}{2A^2}(f(\mathbf{x})-m)^2 \mathrm{d}\mu(\mathbf{x}) \leq 4\,. \tag{3.19}$$

The convexity of the map $x \mapsto \exp x^2$ shows that

$$\exp \frac{1}{4}(x+y)^2 \leq \frac{1}{2}(\exp x^2 + \exp y^2)\,.$$

Since $|f(\mathbf{x})-f(\mathbf{y})| \leq |f(\mathbf{x})-m| + |f(\mathbf{y})-m|$ we deduce from (3.19) that

$$\int \exp \frac{\kappa}{8A^2}(f(\mathbf{x})-f(\mathbf{y}))^2 \mathrm{d}\mu(\mathbf{x})\mathrm{d}\mu(\mathbf{y}) \leq 4\,,$$

from which (3.16) follows using Jensen's inequality, averaging in $\mathbf{y}$ in the exponential rather than outside. To prove (3.17) we relate as usual exponential integrability and growth of moments. We write that if $x \geq 0$ we have $x^k/k! \leq \exp x$ so that

$$x^k \leq k^k \exp x \tag{3.20}$$

and hence

$$y^{2k} \leq \left(\frac{8kA^2}{\kappa}\right)^k \exp \frac{\kappa}{8A^2} y^2\,. \qquad \square$$

Let us point out that in Theorem 3.1.4 the function H can take the value $+\infty$. Equivalently, this theorem holds when μ is a probability on a convex set C with a density proportional to $\exp \psi(\boldsymbol{\sigma})$, where ψ satisfies

$$\frac{1}{2}(\psi(\mathbf{x})+\psi(\mathbf{y})) - \psi\left(\frac{\mathbf{x}+\mathbf{y}}{2}\right) \leq -\kappa \left\|\frac{\mathbf{x}-\mathbf{y}}{2}\right\|^2\,. \tag{3.21}$$

The argument that allows to deduce (3.16) from (3.19) is called a symmetrization argument. This argument proves also the following. For each number m, each function f and each probability μ we have

$$\int \left(f - \int f \mathrm{d}\mu\right)^{2k} \mathrm{d}\mu \leq 2^{2k} \int (f-m)^{2k} \mathrm{d}\mu\,. \tag{3.22}$$

To see this we simply write, using Jensen's inequality in the second line and that $(a+b)^{2k} \leq 2^{2k-1}(a^{2k}+b^{2k})$ in the third line,

$$\begin{aligned}\int\Big(f-\int f\mathrm{d}\mu\Big)^{2k}\mathrm{d}\mu &= \int\Big(f(x)-\int f(y)\mathrm{d}\mu(y)\Big)^{2k}\mathrm{d}\mu(x)\\ &\le \int (f(x)-f(y))^{2k}\,\mathrm{d}\mu(x)\mathrm{d}\mu(y)\\ &= \int ((f(x)-m)-(f(y)-m))^{2k}\,\mathrm{d}\mu(x)\mathrm{d}\mu(y)\\ &\le 2^{2k}\int (f-m)^{2k}\mathrm{d}\mu\,.\end{aligned}$$

The essential feature of the present model is that any realization of the Gibbs measure with Hamiltonian (3.1) satisfies (3.16) and (3.17). We will need to use (3.16) for functions such as $\|\mathbf{x}\|^2$ that are not Lipschitz on $\mathbb{R}^N$, but are Lipschitz when $\mathbf{x}$ is not too large. For this, it is useful to know that the Gibbs measure with Hamiltonian (3.1) essentially lives on a ball of radius about $\sqrt{N}$, and the next two lemmas prepare for this. In this chapter and the next, we will use many times the fact that

$$\int \exp(-t\|\boldsymbol{\sigma}\|^2)\mathrm{d}\boldsymbol{\sigma} = \Big(\int \exp(-tx^2)\mathrm{d}x\Big)^N = \Big(\frac{\pi}{t}\Big)^{N/2}\,. \tag{3.23}$$

Lemma 3.1.5. *Consider a probability μ on $\mathbb{R}^N$ such that for any subset B of $\mathbb{R}^N$ we have*

$$\mu(B) = \frac{1}{Z}\int_B \exp\Big(U(\boldsymbol{\sigma}) - \kappa\|\boldsymbol{\sigma}\|^2 + \sum_{i\le N} a_i\sigma_i\Big)\mathrm{d}\boldsymbol{\sigma}\,,$$

where $U \le 0$ and where Z is the normalizing factor. Then

$$\int \exp\frac{\kappa}{2}\|\boldsymbol{\sigma}\|^2\mathrm{d}\mu(\boldsymbol{\sigma}) \le \frac{1}{Z}\Big(\frac{2\pi}{\kappa}\Big)^{N/2}\exp\Big(\frac{1}{2\kappa}\sum_{i\le N}a_i^2\Big)\,.$$

Proof. Using the definition of μ in the first line, that $U \le 0$ in the second line, completing the squares in the third line and using (3.23) in the last line, we obtain

$$\begin{aligned}\int \exp\frac{\kappa}{2}\|\boldsymbol{\sigma}\|^2\mathrm{d}\mu(\boldsymbol{\sigma}) &= \frac{1}{Z}\int \exp\Big(U(\boldsymbol{\sigma}) - \frac{\kappa}{2}\|\boldsymbol{\sigma}\|^2 + \sum_{i\le N}a_i\sigma_i\Big)\mathrm{d}\boldsymbol{\sigma}\\ &\le \frac{1}{Z}\int \exp\Big(-\frac{\kappa}{2}\|\boldsymbol{\sigma}\|^2 + \sum_{i\le N}a_i\sigma_i\Big)\mathrm{d}\boldsymbol{\sigma}\\ &= \frac{1}{Z}\int \exp\Big(-\frac{\kappa}{2}\sum_{i\le N}\Big(\sigma_i - \frac{a_i}{\kappa}\Big)^2 + \frac{1}{2\kappa}\sum_{i\le N}a_i^2\Big)\mathrm{d}\boldsymbol{\sigma}\\ &= \frac{1}{Z}\exp\Big(\frac{1}{2\kappa}\sum_{i\le N}a_i^2\Big)\int \exp\Big(-\frac{\kappa}{2}\|\boldsymbol{\sigma}\|^2\Big)\mathrm{d}\boldsymbol{\sigma}\end{aligned}$$

$$= \frac{1}{Z}\left(\frac{2\pi}{\kappa}\right)^{N/2} \exp\left(\frac{1}{2\kappa}\sum_{i\le N} a_i^2\right).$$

This concludes the proof. □

In order to use Lemma 3.1.5 when μ is Gibbs' measure (3.5) we need an upper bound for $1/Z_{N,M}$.

Lemma 3.1.6. *Assume (3.6), that is $u(x) \ge -D(1+|x|)$ for a certain number D and all x. Then we have*

$$\frac{1}{Z_{N,M}} \le \left(\frac{\kappa}{\pi}\right)^{N/2} \exp D\left(M + \sqrt{\frac{M}{\kappa N}\sum_{i\le N,k\le M} g_{i,k}^2}\right).$$

Proof. The proof relies on the rotational invariance of the Gaussian measure γ on $\mathbb{R}^N$ of density $(\kappa/\pi)^{N/2}\exp(-\kappa\|\boldsymbol{\sigma}\|^2)$ with respect to Lebesgue's measure. For $\mathbf{x}\in\mathbb{R}^N$ we have

$$\int |\mathbf{x}\cdot\boldsymbol{\sigma}|\mathrm{d}\gamma(\boldsymbol{\sigma}) = \sqrt{\frac{1}{\pi\kappa}}\|\mathbf{x}\| \le \sqrt{\frac{1}{\kappa}}\|\mathbf{x}\|, \tag{3.24}$$

because the rotational invariance of γ reduces this to the case $N=1$.

Letting $\mathbf{g}_k = (g_{i,k})_{i\le N}$, we have

$$\begin{aligned}
Z_{N,M} &= \int \exp\left(\sum_{k\le M} u\left(\frac{\mathbf{g}_k\cdot\boldsymbol{\sigma}}{\sqrt{N}}\right) - \kappa\|\boldsymbol{\sigma}\|^2 + h\sum_{i\le N} g_i\sigma_i\right)\mathrm{d}\boldsymbol{\sigma} \\
&= \left(\frac{\pi}{\kappa}\right)^{N/2}\int \exp\left(\sum_{k\le M} u\left(\frac{\mathbf{g}_k\cdot\boldsymbol{\sigma}}{\sqrt{N}}\right) + h\sum_{i\le N} g_i\sigma_i\right)\mathrm{d}\gamma(\boldsymbol{\sigma}) \\
&\ge \left(\frac{\pi}{\kappa}\right)^{N/2}\exp\left(\int\left(\sum_{k\le M} u\left(\frac{\mathbf{g}_k\cdot\boldsymbol{\sigma}}{\sqrt{N}}\right) + h\sum_{i\le N} g_i\sigma_i\right)\right)\mathrm{d}\gamma(\boldsymbol{\sigma})\Bigg) \\
&= \left(\frac{\pi}{\kappa}\right)^{N/2}\exp\left(\sum_{k\le M}\int u\left(\frac{\mathbf{g}_k\cdot\boldsymbol{\sigma}}{\sqrt{N}}\right)\mathrm{d}\gamma(\boldsymbol{\sigma})\right),
\end{aligned}$$

using Jensen's inequality in the third line and since $\int \sigma_i \mathrm{d}\gamma(\boldsymbol{\sigma}) = 0$. Now, using (3.6) and (3.24) for $\mathbf{x} = \mathbf{g}_k$ yields

$$\begin{aligned}
\sum_{k\le M}\int u\left(\frac{\mathbf{g}_k\cdot\boldsymbol{\sigma}}{\sqrt{N}}\right)\mathrm{d}\gamma(\boldsymbol{\sigma}) &\ge -D\left(M + \frac{1}{\sqrt{\kappa}}\sum_{k\le M}\frac{\|\mathbf{g}_k\|}{\sqrt{N}}\right) \\
&\ge -D\left(M + \frac{1}{\sqrt{\kappa}}\sqrt{\frac{M}{N}}\sqrt{\sum_{k\le M}\|\mathbf{g}_k\|^2}\right),
\end{aligned}$$

using the Cauchy-Schwarz inequality. □

We will often assume that

$$\kappa \geq \kappa_0 , \; 0 \leq h \leq h_0 , \; M \leq 10N \tag{3.25}$$

where κ_0 and h_0 are given numbers. The condition $M \leq 10N$ is simply to avoid trivial complications, and it contains the case relevant to the computation of the part of the sphere $\mathbb{S}_N$ that belongs to the region (3.3).

In the entire chapter we make the convention that K denotes a number that might depend on κ_0, h_0, D but that does not depend on M and N or on any other parameter. This number does not need to be the same at each occurrence.

The following is an immediate consequence of Lemmas 3.1.5 and 3.1.6.

Corollary 3.1.7. *Under (3.6) and (3.25) we have*

$$\left\langle \exp \frac{\kappa}{2} \|\boldsymbol{\sigma}\|^2 \right\rangle \leq \exp K \left(N + \sqrt{\sum_{i \leq N, k \leq M} g_{i,k}^2} + \sum_{i \leq N} g_i^2 \right) . \tag{3.26}$$

We set

$$B^* = N + \sqrt{\sum_{i \leq N, k \leq M} g_{i,k}^2} + \sum_{i \leq N} g_i^2 . \tag{3.27}$$

It will help in all the forthcoming computations to think of B^* as being $\leq KN$ for all practical purposes. In other words, the event where this is not the case is so rare as being irrelevant for the questions we pursue. This will be made precise in Lemma 3.1.10 below.

With the notation (3.27) we rewrite (3.26) as

$$\left\langle \exp \frac{\kappa \|\boldsymbol{\sigma}\|^2}{2} \right\rangle \leq \exp K B^* . \tag{3.28}$$

This inequality is a sophisticated way to express that the Gibbs' measure "is basically supported by a ball of radius $K\sqrt{N}$". The following simple fact from Probability theory will help to exploit this inequality in terms of moments.

Lemma 3.1.8. *Consider a r.v. $X \geq 0$ and $C = \log \mathsf{E} \exp X$. Then for each k we have*

$$\mathsf{E} X^k \leq 2^k (k^k + C^k) . \tag{3.29}$$

Proof. By definition of C,

$$\mathsf{E} \exp(X - C) = 1$$

so that if $x^+ = \max(x, 0)$ we have $\exp(X - C)^+ \leq 1 + \exp(X - C)$ and hence

$$\mathsf{E} \exp(X - C)^+ \leq \mathsf{E}(1 + \exp(X - C)) = 2 .$$

Since by (3.20) we have $x^k \leq k^k e^x$ for $x \geq 0$, we get

$$\mathsf{E}(X - C)^{+k} \leq 2k^k .$$

Now $X \leq (X - C)^+ + C$ and $(a + C)^k \leq 2^{k-1}(a^k + C^k)$. $\square$

Corollary 3.1.9. *For $k \le 4N$ we have*

$$\langle \|\boldsymbol{\sigma}\|^{2k} \rangle \le (KB^*)^k \,. \tag{3.30}$$

As in the case of Theorem 3.1.1, there is nothing specific here in the choice of the number 4 in the inequality $k \le 4N$. We can replace the condition $k \le 4N$ by the condition $k \le AN$ for any number A (with a constant $K(A)$ depending on A). The same comment applies to many results of this section. Let us also note that the fact that (3.30) holds for each $k \le 4N$ is equivalent to saying that it holds for $k = 4N$, by Hölder's inequality.

Proof. We use (3.29) in the probability space given by Gibbs' measure. If $X = \kappa\|\boldsymbol{\sigma}\|^2/2$, then (3.28) implies $\log\langle \exp X\rangle \le KB^*$ and (3.29) then implies $\langle X^k\rangle \le 2^k(k^k + (KB^*)^k)$. Since $k \le 4N \le 4B^*$, we finally get

$$\langle X^k\rangle \le 2^k((4B^*)^k + (KB^*)^k) \le ((8+2K)B^*)^k \,, \tag{3.31}$$

and this finishes the proof. □

As the reader is getting used to the technique of denoting by the letter K an unspecified constant, we will soon no longer fully detail trivial bounds such as (3.31). Rather we will simply write "since $k \le 4N \le 4B^*$ we have $\langle X^k\rangle \le 2^k(k^k + (KB^*)^k) \le (KB^*)^k$".

Lemma 3.1.10. *For $k \le N$ we have*

$$\mathsf{E}B^{*k} \le (KN)^k \,. \tag{3.32}$$

Proof. Using that $2\sqrt{x} \le x/a + a$ for $x = \sum_{i\le N, k\le M} g_{i,k}^2$ and $a = N$, and then using (A.11) and independence, we get

$$\begin{aligned} \mathsf{E}\exp\frac{B^*}{4} &\le \exp\Big(\frac{N}{4} + \frac{N}{2} + \frac{1}{8}\sum_{i\le N, k\le M} g_{i,k}^2 + \frac{1}{4}\sum_{i\le N} g_i^2\Big) \\ &\le \exp\Big(\frac{3N}{4}\Big)\Big(\frac{1}{\sqrt{1-1/4N}}\Big)^{MN}\Big(\frac{1}{\sqrt{1-1/2}}\Big)^{N} \le \exp LN \,, \end{aligned} \tag{3.33}$$

and we use (3.29) for $X = B^*/4$. □

After these preliminaries, we turn to the central argument, the use of Theorem 3.1.4 to control the overlaps. The idea is simply that since Gibbs' measure is essentially supported by a ball of radius $\sqrt{B^*}$ centered at the origin, we can basically pretend that the functions $R_{1,2}$ and $R_{1,1}$ have a Lipschitz constant $\le \sqrt{B^*/N}$ and use (3.17).

Theorem 3.1.11. *For $k \le N$ we have*

$$\langle (R_{1,2} - \langle R_{1,2}\rangle)^{2k}\rangle \le \Big(\frac{KkB^*}{N^2}\Big)^k \,, \tag{3.34}$$

$$\langle (R_{1,1} - \langle R_{1,1}\rangle)^{2k}\rangle \le \Big(\frac{KkB^*}{N^2}\Big)^k \,. \tag{3.35}$$

Proof. We write $\mathbf{b} = \langle \boldsymbol{\sigma} \rangle$, so that $\langle R_{1,2} \rangle = \|\mathbf{b}\|^2/N$, and

$$|R_{1,2} - \langle R_{1,2} \rangle| \le \left| \frac{\boldsymbol{\sigma}^1 \cdot \boldsymbol{\sigma}^2}{N} - \frac{\boldsymbol{\sigma}^1 \cdot \mathbf{b}}{N} \right| + \left| \frac{\boldsymbol{\sigma}^1 \cdot \mathbf{b}}{N} - \frac{\mathbf{b} \cdot \mathbf{b}}{N} \right| . \tag{3.36}$$

If we fix $\boldsymbol{\sigma}^1$, the map $f : \mathbf{x} \mapsto \boldsymbol{\sigma}^1 \cdot \mathbf{x}/N$ satisfies (3.15) with $A = \|\boldsymbol{\sigma}^1\|/N$, so that by (3.17) we get

$$\int \left(\frac{\boldsymbol{\sigma}^1 \cdot \boldsymbol{\sigma}^2}{N} - \frac{\boldsymbol{\sigma}^1 \cdot \mathbf{b}}{N} \right)^{2k} \mathrm{d}G(\boldsymbol{\sigma}^2) \le \left(\frac{Kk\|\boldsymbol{\sigma}^1\|^2}{N^2} \right)^k ,$$

and therefore, integrating the previous inequality for $\boldsymbol{\sigma}^1$ with respect to G,

$$\int \left(\frac{\boldsymbol{\sigma}^1 \cdot \boldsymbol{\sigma}^2}{N} - \frac{\boldsymbol{\sigma}^1 \cdot \mathbf{b}}{N} \right)^{2k} \mathrm{d}G(\boldsymbol{\sigma}^1)\mathrm{d}G(\boldsymbol{\sigma}^2) \le \left(\frac{KkB^*}{N^2} \right)^k$$

using (3.30). The second term on the right-hand side of (3.36) is handled similarly, using now that $\|\mathbf{b}\|^{2k} \le (KB^*)^k$ by (3.30) and Jensen's inequality.

To prove (3.35), let us consider a parameter a to be chosen later and let

$$f(\boldsymbol{\sigma}) = \min(\|\boldsymbol{\sigma}\|^2/N, a^2/N) = (\min(\|\boldsymbol{\sigma}\|, a))^2/N .$$

This function satisfies (3.15) for $A = 2a/N$, so that by (3.17) we get

$$\langle (f - \langle f \rangle)^{2k} \rangle \le \left(\frac{Ka^2k}{N^2} \right)^k . \tag{3.37}$$

Let $\varphi(\boldsymbol{\sigma}) = \|\boldsymbol{\sigma}\|^2/N - f(\boldsymbol{\sigma})$, so that

$$|\varphi(\boldsymbol{\sigma})| \le \frac{\|\boldsymbol{\sigma}\|^2}{N} \mathbf{1}_{\{\|\boldsymbol{\sigma}\| \ge a\}} ,$$

and, using (3.22) for $m = 0$ in the first inequality and the Cauchy-Schwarz inequality in the second line,

$$\begin{aligned} \langle (\varphi - \langle \varphi \rangle)^{2k} \rangle \le 2^{2k} \langle \varphi^{2k} \rangle &\le 2^{2k} \left\langle \left(\frac{\|\boldsymbol{\sigma}\|^2}{N} \right)^{2k} \mathbf{1}_{\{\|\boldsymbol{\sigma}\| \ge a\}} \right\rangle \\ &\le 2^{2k} \left\langle \left(\frac{\|\boldsymbol{\sigma}\|^2}{N} \right)^{4k} \right\rangle^{1/2} \langle \mathbf{1}_{\{\|\boldsymbol{\sigma}\| \ge a\}} \rangle^{1/2} . \end{aligned} \tag{3.38}$$

Using (3.30) (for $k' = 4k \le 4N$ rather than for k) we obtain

$$\left\langle \left(\frac{\|\boldsymbol{\sigma}\|^2}{N} \right)^{4k} \right\rangle^{1/2} \le \left(\frac{KB^*}{N} \right)^{2k} \tag{3.39}$$

and using (3.28) we see that if "we choose $a = K\sqrt{B^*}$" then

$$\langle \mathbf{1}_{\{\|\boldsymbol{\sigma}\| \geq a\}} \rangle \leq \exp(-2B^*) . \tag{3.40}$$

Again, here, to understand what this means the reader must keep in mind that the letter K might denote different constants at different occurrences. The complete argument is that if

$$\left\langle \exp \frac{\kappa \|\boldsymbol{\sigma}\|^2}{2} \right\rangle \leq \exp K_1 B^* ,$$

then

$$\langle \mathbf{1}_{\{\|\boldsymbol{\sigma}\| \geq a\}} \rangle \leq \exp\Big(K_1 B^* - \frac{\kappa a^2}{2} \Big) ,$$

so that (3.40) holds for $a = K_2 \sqrt{B^*}$ whenever $K_2 \geq \sqrt{2(K_1 + 2)/\kappa}$.

Therefore with this choice of a we have, plugging (3.40) and (3.39) into (3.38),

$$\langle (\varphi - \langle \varphi \rangle)^{2k} \rangle \leq \exp(-B^*) \left(\frac{KB^*}{N} \right)^{2k} .$$

Since $R_{1,1} = \|\boldsymbol{\sigma}\|^2/N = f + \varphi$, using that $(x+y)^{2k} \leq 2^{2k}(x^{2k} + y^{2k})$ and (3.37) we get the estimate

$$\begin{aligned} \langle (R_{1,1} - \langle R_{1,1} \rangle)^{2k} \rangle &\leq 2^{2k} \big(\langle (f - \langle f \rangle)^{2k} \rangle + \langle (\varphi - \langle \varphi \rangle)^{2k} \rangle \big) \\ &\leq \left(\frac{KB^* k}{N^2} \right)^k + \exp(-B^*) \left(\frac{KB^*}{N} \right)^{2k} . \end{aligned}$$

We deduce from (3.20) that

$$\exp(-y) \leq \left(\frac{k}{y} \right)^k \tag{3.41}$$

so that

$$\exp(-B^*) \left(\frac{KB^*}{N} \right)^{2k} \leq \left(\frac{k}{B^*} \right)^k \left(\frac{KB^*}{N} \right)^{2k} = \left(\frac{K^2 B^* k}{N^2} \right)^k ,$$

and the result follows. □

Combining with Lemma 3.1.10 we get the following.

Proposition 3.1.12. *For $k \leq N$ we have*

$$\mathsf{E} \langle (R_{1,2} - \langle R_{1,2} \rangle)^{2k} \rangle \leq \left(\frac{Kk}{N} \right)^k \tag{3.42}$$

$$\mathsf{E} \langle (R_{1,1} - \langle R_{1,1} \rangle)^{2k} \rangle \leq \left(\frac{Kk}{N} \right)^k . \tag{3.43}$$

A further remarkable property to which we turn now is that the random quantities $\langle R_{1,2}\rangle$ and $\langle R_{1,1}\rangle$ are nearly constant. A general principle that we will study later (the Ghirlanda-Guerra identities) implies that (in some sense) this "near constancy" is an automatic consequence of Proposition 3.1.12. On the other hand, in the specific situation considered here, Shcherbina and Tirozzi discovered ([134]) a special argument that gives a much better rate of convergence than general principles. The idea is that if we think of $\langle R_{1,2}\rangle$ and $\langle R_{1,1}\rangle$ as functions of the Gaussian r.v.s $(g_{i,k})_{i\le N,k\le M}$ and $(g_i)_{i\le N}$, they are essentially Lipschitz functions with Lipschitz constant of order $1/\sqrt{N}$; so that we can use (3.17). We need however to work a bit more before we can show this. We recall the notation $\mathbf{b} = \langle \boldsymbol{\sigma}\rangle$ and we will use not only (3.6) but also (3.7).

Lemma 3.1.13. *For any random function f on $\mathbb{R}^N$ we have*

$$\|\langle(\boldsymbol{\sigma}-\mathbf{b})f(\boldsymbol{\sigma})\rangle\| \le K\langle f^2\rangle^{1/2}\,. \tag{3.44}$$

Consequently

$$\langle(\boldsymbol{\sigma}^1-\mathbf{b})\cdot(\boldsymbol{\sigma}^2-\mathbf{b})f(\boldsymbol{\sigma}^1)f(\boldsymbol{\sigma}^2)\rangle = \|\langle(\boldsymbol{\sigma}-\mathbf{b})f(\boldsymbol{\sigma})\rangle\|^2 \le K\langle f^2\rangle\,. \tag{3.45}$$

Here the function f is permitted to depend on the randomness $g_{i,k}, g_i$.

Proof. For any $\mathbf{y}\in\mathbb{R}^N$, using the Cauchy-Schwarz inequality, we see that

$$\begin{aligned}\langle(\boldsymbol{\sigma}-\mathbf{b})f(\boldsymbol{\sigma})\rangle\cdot\mathbf{y} &= \langle(\boldsymbol{\sigma}-\mathbf{b})\cdot\mathbf{y}f(\boldsymbol{\sigma})\rangle\\ &\le \langle((\boldsymbol{\sigma}-\mathbf{b})\cdot\mathbf{y})^2\rangle^{1/2}\langle f^2\rangle^{1/2}\,.\end{aligned}$$

We then use (3.17) for $\overline{f}(\boldsymbol{\sigma}) = \boldsymbol{\sigma}\cdot\mathbf{y}$ and $k=1$ to see that $\langle((\boldsymbol{\sigma}-\mathbf{b})\cdot\mathbf{y})^2\rangle \le K\|\mathbf{y}\|^2$, so that combining with the above we get

$$\langle(\boldsymbol{\sigma}-\mathbf{b})f(\boldsymbol{\sigma})\rangle\cdot\mathbf{y} \le K\|\mathbf{y}\|\langle f^2\rangle^{1/2}\,.$$

Since this holds for any value of $\mathbf{y}$, (3.44) follows. □

We denote by

$$B' \text{ the operator norm of the matrix } (g_{i,k})_{i\le N,k\le M}\,, \tag{3.46}$$

so that for any sequences $(x_i)_{i\le N}$ and $(y_k)_{k\le M}$ we have

$$\sum_{i\le N,k\le M} g_{i,k}x_iy_k \le B'\Big(\sum_{i\le N} x_i^2\Big)^{1/2}\Big(\sum_{k\le M} y_k^2\Big)^{1/2}\,,$$

and, equivalently,

$$\Big(\sum_{i\le N}\Big(\sum_{k\le M} g_{i,k}y_k\Big)^2\Big)^{1/2} \le B'\Big(\sum_{k\le M} y_k^2\Big)^{1/2}\,. \tag{3.47}$$

It is useful to think that for all practical purposes we have $B'^2 \le KN$, as is shown in Lemma A.9.1.

We recall the standard notation $S_k = N^{-1/2}\sum_{i\le N} g_{i,k}\sigma_i$.

Lemma 3.1.14. *Given any* $\mathbf{y} = (y_k)_{k\le M} \in \mathbb{R}^M$, *the function*

$$\boldsymbol{\sigma} \mapsto \overline{f}(\boldsymbol{\sigma}) = \sum_{k\le M} u'(S_k) y_k$$

has a Lipschitz constant $A \le KB'\big(\sum_{k\le M} y_k^2\big)^{1/2}/\sqrt{N} = KB'\|\mathbf{y}\|/\sqrt{N}$.

Proof. Since

$$\frac{\partial}{\partial\sigma_i}\overline{f}(\boldsymbol{\sigma}) = \frac{1}{\sqrt{N}}\sum_{k\le M} g_{i,k} u''(S_k) y_k \ ,$$

the length of the gradient of $\overline{f}$ is

$$\left(\frac{1}{N}\sum_{i\le N}\left(\sum_{k\le M} g_{i,k}u''(S_k)y_k\right)^2\right)^{1/2} \le \frac{KB'}{\sqrt{N}}\left(\sum_{k\le M} y_k^2\right)^{1/2}$$

by (3.47) and since $|u''(S_k)| \le D$. □

Lemma 3.1.15. *Let us denote by* $\mathbf{U} = \mathbf{U}(\boldsymbol{\sigma})$ *the* M*-dimensional vector* $(u'(S_k))_{k\le M}$. *Then for any random function* f *on* $\mathbb{R}^N$ *we have*

$$\|\langle(\mathbf{U} - \langle\mathbf{U}\rangle)f\rangle\| \le \frac{KB'}{\sqrt{N}}\langle f^2\rangle^{1/2} \tag{3.48}$$

and, consequently

$$\langle(\mathbf{U}(\boldsymbol{\sigma}^1) - \langle\mathbf{U}\rangle)(\mathbf{U}(\boldsymbol{\sigma}^2) - \langle\mathbf{U}\rangle)f(\boldsymbol{\sigma}^1)f(\boldsymbol{\sigma}^2)\rangle \le \frac{KB'^2}{N}\langle f^2\rangle \ . \tag{3.49}$$

Proof. It is identical to the proof of Lemma 3.1.13. If $\mathbf{y} \in \mathbb{R}^M$, then

$$\begin{aligned}\langle(\mathbf{U} - \langle\mathbf{U}\rangle)f\rangle\cdot\mathbf{y} &= \langle(\mathbf{U} - \langle\mathbf{U}\rangle)\cdot\mathbf{y}f\rangle\\ &\le \langle((\mathbf{U} - \langle\mathbf{U}\rangle)\cdot\mathbf{y})^2\rangle^{1/2}\langle f^2\rangle^{1/2} \ .\end{aligned}$$

Using Lemma 3.1.14 and applying (3.17) to $\overline{f}(\boldsymbol{\sigma}) = \mathbf{U}(\boldsymbol{\sigma})\cdot\mathbf{y}$, we obtain that $\langle((\mathbf{U} - \langle\mathbf{U}\rangle)\cdot\mathbf{y})^2\rangle^{1/2} \le KB'\|\mathbf{y}\|/\sqrt{N}$. Therefore

$$\langle(\mathbf{U} - \langle\mathbf{U}\rangle)f\rangle\cdot\mathbf{y} \le \frac{KB'\|\mathbf{y}\|}{\sqrt{N}}\langle f^2\rangle^{1/2} \ ,$$

and this yields (3.48). Furthermore, the left-hand side of (3.49) is the square of the left-hand side of (3.48). □

Proposition 3.1.16. *Let us denote by* $\boldsymbol{\nabla}$ *the gradient of* $\langle R_{1,1}\rangle$ *(resp.* $\langle R_{1,2}\rangle$*) when this quantity is seen as a function of the numbers* $(g_{i,k})_{i\le N,k\le M}$ *and* $(g_i)_{i\le N}$. *Then, recalling the quantities* B^* *of (3.27) and* B' *of (3.46) we have*

$$\|\boldsymbol{\nabla}\|^2 \le K\left(\frac{B^*}{N^2} + \frac{B^{*2}B'^2}{N^4}\right) \ .$$

If we think that B and B'^2 are basically of order N, this shows that $\|\boldsymbol{\nabla}\|^2$ is about $1/N$, i.e. that the functions $R_{1,2}$ and $R_{1,1}$ have Lipschitz constants about $1/\sqrt{N}$.

Proof. With the customary abuse of notation we have

$$\begin{aligned}\frac{\partial}{\partial g_{i,k}}\langle R_{1,1}\rangle &= \frac{1}{\sqrt{N}}(\langle R_{1,1}\sigma_i^1 u'(S_k^1)\rangle - \langle R_{1,1}\rangle\langle\sigma_i^1 u'(S_k^1)\rangle)\\ &= \frac{1}{\sqrt{N}}(\langle f(\boldsymbol{\sigma}^1)\sigma_i^1 u'(S_k^1)\rangle)\,,\end{aligned}$$

where $f(\boldsymbol{\sigma}^1) = R_{1,1} - \langle R_{1,1}\rangle$. We define $\dot{\sigma}_i^1 = \sigma_i^1 - \langle\sigma_i\rangle$ and $\dot{u}'(S_k^1) = u'(S_k^1) - \langle u'(S_k^1)\rangle$. Since $\langle f\rangle = 0$ the identity

$$\langle f(\boldsymbol{\sigma}^1)\sigma_i^1 u'(S_k^1)\rangle = \langle f(\boldsymbol{\sigma}^1)\dot{\sigma}_i^1 u'(S_k^1)\rangle + \langle\sigma_i\rangle\langle f(\boldsymbol{\sigma}^1)\dot{u}'(S_k^1)\rangle$$

holds. Thus

$$\sum_{i,k}\left(\frac{\partial}{\partial g_{ik}}\langle R_{1,1}\rangle\right)^2 = \frac{1}{N}\sum_{i,k}\langle f(\boldsymbol{\sigma}^1)\sigma_i^1 u'(S_k^1)\rangle^2 \le 2(\mathrm{I}+\mathrm{II})\,,$$

where, using replicas

$$\begin{aligned}\mathrm{I} &= \frac{1}{N}\sum_{i\le N,k\le M}\langle f(\boldsymbol{\sigma}^1)\dot{\sigma}_i^1 u'(S_k^1)\rangle^2\\ &= \frac{1}{N}\sum_{i\le N,k\le M}\langle f(\boldsymbol{\sigma}^1)f(\boldsymbol{\sigma}^2)\dot{\sigma}_i^1\dot{\sigma}_i^2 u'(S_k^1)u'(S_k^2)\rangle\\ &= \frac{1}{N}\sum_{k\le M}\langle f(\boldsymbol{\sigma}^1)f(\boldsymbol{\sigma}^2)(\boldsymbol{\sigma}^1-\mathbf{b})\cdot(\boldsymbol{\sigma}^2-\mathbf{b})u'(S_k^1)u'(S_k^2)\rangle\,,\end{aligned}$$

and

$$\begin{aligned}\mathrm{II} &= \left(\frac{1}{N}\sum_{i\le N}\langle\sigma_i\rangle^2\right)\sum_{k\le M}\langle f(\boldsymbol{\sigma}^1)\dot{u}'(S_k^1)\rangle^2\\ &= \left(\frac{1}{N}\sum_{i\le N}\langle\sigma_i\rangle^2\right)\sum_{k\le M}\langle f(\boldsymbol{\sigma}^1)f(\boldsymbol{\sigma}^2)\dot{u}'(S_k^1)\dot{u}'(S_k^2)\rangle\\ &= \left(\frac{1}{N}\sum_{i\le N}\langle\sigma_i\rangle^2\right)\langle f(\boldsymbol{\sigma}^1)f(\boldsymbol{\sigma}^2)(\mathbf{U}(\boldsymbol{\sigma}^1)-\langle\mathbf{U}\rangle)\cdot(\mathbf{U}(\boldsymbol{\sigma}^2)-\langle\mathbf{U}\rangle)\rangle\,.\end{aligned}$$

Using (3.35) for $k=1$ we get

$$\langle f^2\rangle \le \frac{KB}{N^2}\,. \tag{3.50}$$

We use (3.45) with $f(\boldsymbol{\sigma}^1)u'(S_k^1)$ instead of f to get, since $|u'|\le D$ and $M\le N$,

$$\mathrm{I} \le \frac{KM}{N}\langle f^2\rangle \le \frac{KB^*}{N^2} \,.$$

We note that (3.30) used for $k = 1$ implies

$$\langle R_{1,2}\rangle = \frac{1}{N}\sum_{i\le N}\langle\sigma_i\rangle^2 \le \frac{1}{N}\sum_{i\le N}\langle\sigma_i^2\rangle = \frac{1}{N}\langle\|\boldsymbol{\sigma}\|^2\rangle \le \frac{KB^*}{N} \,. \tag{3.51}$$

We use (3.49) and (3.50) to get

$$\mathrm{II} \le \left(\frac{1}{N}\sum_{i\le N}\langle\sigma_i\rangle^2\right)\frac{KB'^2}{N}\langle f^2\rangle \le \frac{KB'^2B^{*2}}{N^4} \,.$$

We take care in a similar manner of the term $\partial\langle R_{1,1}\rangle/\partial g_i$, and the case of $R_{1,2}$ is similar. □

Proposition 3.1.17. *For $k \le N/4$ we have*

$$\mathsf{E}(\langle R_{1,1}\rangle - \mathsf{E}\langle R_{1,1}\rangle)^{2k} \le \left(\frac{Kk}{N}\right)^k \tag{3.52}$$

$$\mathsf{E}(\langle R_{1,2}\rangle - \mathsf{E}\langle R_{1,2}\rangle)^{2k} \le \left(\frac{Kk}{N}\right)^k \,. \tag{3.53}$$

Proof. We consider the space $\mathbb{R}^{N\times M} \times \mathbb{R}^N$, in which we denote the generic point by $\mathbf{g} = ((g_{i,k})_{i\le N,k\le M}, (g_i)_{i\le N})$. We provide this space with the canonical Gaussian measure γ. Integration with respect to this measure means that we take expectation in the $(g_{i,k}), (g_i)$ seen as independent standard Gaussian r.v.s. Let us consider the convex set

$$C = \{\mathbf{g};\ B^* \le LN;\ B'^2 \le LN\} \,,$$

where we have chosen the number L large enough that

$$\mathsf{P}(C^c) \le L\exp(-N) \,. \tag{3.54}$$

(To see that this is possible we recall Lemma A.9.1 and that $\mathsf{E}\exp B^*/4 \le \exp LN$ by (3.33).) Let us think of $\langle R_{1,1}\rangle$ (resp. $\langle R_{1,2}\rangle$) as a function $f(\mathbf{g})$, so that by Proposition 3.1.16, on C the gradient $\boldsymbol{\nabla} f$ of f satisfies $\|\boldsymbol{\nabla} f\|^2 \le K/N$, and since C is convex f satisfies (3.15) on C with $A = K/\sqrt{N}$.

Consider the probability measure γ' on C with density proportional to

$$W = \exp\left(-\frac{1}{2}\sum_{i\le N,k\le M} g_{i,k}^2 - \frac{1}{2}\sum_{i\le N} g_i^2\right) \,. \tag{3.55}$$

By (3.17) we have

$$\forall k \geq 1 \, , \quad \int (f-m)^{2k} \mathrm{d}\gamma' \leq \left(\frac{Kk}{N}\right)^k \, , \tag{3.56}$$

where $m = \int f \mathrm{d}\gamma'$. The rest of the proof consists simply in checking as expected that the set C^c is so small that (3.52) and (3.53) follow from (3.56). This is tedious and occupies the next half page. By definition of γ', for any function h we have

$$\int h \mathrm{d}\gamma' = \frac{\int_C hW \mathrm{d}\boldsymbol{\sigma}}{\int_C W \mathrm{d}\boldsymbol{\sigma}} = \frac{\int_C h \mathrm{d}\gamma}{\gamma(C)} \, .$$

Thus

$$\mathsf{E}\big(\mathbf{1}_C (f-m)^{2k}\big) = \int_C (f-m)^{2k} \mathrm{d}\gamma = \gamma(C) \int_C (f-m)^{2k} \mathrm{d}\gamma' \leq \left(\frac{Kk}{N}\right)^k \, ,$$

and

$$\begin{aligned} \mathsf{E}(f-m)^{2k} &= \mathsf{E}\big(\mathbf{1}_C (f-m)^{2k}\big) + \mathsf{E}\big(\mathbf{1}_{C^c} (f-m)^{2k}\big) \\ &\leq \left(\frac{Kk}{N}\right)^k + \mathsf{E}\big(\mathbf{1}_{C^c} (f-m)^{2k}\big) \\ &\leq \left(\frac{Kk}{N}\right)^k + \mathsf{P}(C^c)^{1/2} \big(\mathsf{E}(f-m)^{4k}\big)^{1/2} \, . \end{aligned}$$

Using (3.51), we see that $|f| \leq KB^*/N$, and since γ' is supported by C and $B^* \leq LN$ on C we have $|m| = |\int f \mathrm{d}\gamma'| \leq K$. Also $(\mathsf{E}f^{4k})^{1/2} \leq K^k$ by (3.30) and (3.32). Therefore $(\mathsf{E}(f-m)^{4k})^{1/2} \leq K^k$. Hence, recalling that by (3.54) we have $\mathsf{P}(C^c) \leq \exp(-N)$ and using that $\exp(-N/2) \leq (2k/N)^k$ by (3.41) we obtain

$$\mathsf{E}(f-m)^{2k} \leq \left(\frac{Kk}{N}\right)^k + L \exp\Big(-\frac{N}{2}\Big) K^k \leq \left(\frac{Kk}{N}\right)^k$$

for $k \leq N$. The conclusion follows by the symmetrization argument (3.22).

□

Combining Propositions 3.1.12 and 3.1.17, we have proved the following.

Theorem 3.1.18. *For $k \leq N/4$, and assuming (3.6) and (3.7) we have*

$$\nu\big((R_{1,1} - \nu(R_{1,1}))^{2k}\big) \leq \left(\frac{Kk}{N}\right)^k \tag{3.57}$$

$$\nu\big((R_{1,2} - \nu(R_{1,2}))^{2k}\big) \leq \left(\frac{Kk}{N}\right)^k \, , \tag{3.58}$$

where K depends only on κ_0, h_0 and D.

3.2 The Replica-Symmetric Equations

Theorem 3.1.18 brings forward the importance of the numbers

$$q = q_{N,M} = \nu(R_{1,2}) \, ; \; \rho = \rho_{N,M} = \nu(R_{1,1}) \, . \tag{3.59}$$

The notation, similar to that of the previous chapter, should not hide that the procedure is different. The numbers q and ρ are not defined through a system of equations, but by "the physical system". They depend on N and M. It would help to remember the definition (3.59) now. The purpose of the present section is to show that q and ρ nearly satisfy the system of "replica-symmetric" equations (3.69), (3.76) and (3.104) below. These equations should in principle allow the computation of q and ρ.

Since the cavity method, i.e. the idea of "bringing forward the influence of the last spin" was successful in previous chapters, let us try it here. The following approach is quite close to that of Section 2.2 so some familiarity with that section would certainly help the reader who wishes to follow all the details. Consider two numbers r and $\overline{r}$, with $\overline{r} \le r$. Consider a centered Gaussian r.v. Y, independent of all the other r.v.s already considered, with $\mathsf{E}Y^2 = r$, and consider $0 < t < 1$. We write

$$S_{k,t} = \frac{1}{\sqrt{N}} \sum_{i \le N-1} g_{i,k}\sigma_i + \sqrt{\frac{t}{N}} g_{N,k}\sigma_N \, , \tag{3.60}$$

and we consider the Hamiltonian

$$\begin{aligned} -H_{N,M,t}(\boldsymbol{\sigma}) = \sum_{k \le M} u(S_{k,t}(\boldsymbol{\sigma})) - \kappa \|\boldsymbol{\sigma}\|^2 + h \sum_{i \le N} g_i \sigma_i \\ + \sigma_N \sqrt{1-t} Y - (1-t)(r - \overline{r}) \frac{\sigma_N^2}{2} \, . \end{aligned} \tag{3.61}$$

Comparing (3.60) with (2.15) we observe that now we do not have the last term $\sqrt{(1-t)/N}\xi_k$ of (2.15). The purpose of this term was to ensure that the variance of the quantity (2.15) does not depend on t. Since it is no longer true that $\sigma_N^2 = 1$ we can no longer use the same device here. Fortunately, as already pointed out, this device was not essential. The last term in the Hamiltonian (3.61) also accounts in a more subtle way for the fact that it is not true that $\sigma_N^2 = 1$.

We denote an average for the Gibbs measure with Hamiltonian (3.61) by $\langle \cdot \rangle_t$, and we write $\nu_t(f) = \mathsf{E}\langle f \rangle_t$, $\nu_t'(f) = \mathrm{d}\nu_t(f)/\mathrm{d}t$. We recall the notation $\varepsilon_\ell = \sigma_N^\ell$.

Please do not be discouraged by the upcoming formula. Very soon it will be clear to you that Proposition 3.2.1 is no more complicated to use than Proposition 2.2.3.

Proposition 3.2.1. *If f is a function on $(\mathbb{R}^N)^n$, then for $\alpha = M/N$ we have*

$$\nu_t'(f) = \mathrm{I} + \mathrm{II} + \mathrm{III} + \mathrm{IV} + \mathrm{V}\,,$$

where

$$\mathrm{I} = \frac{\alpha}{2}\Bigg(\sum_{\ell\le n} \nu_t\Big(\varepsilon_\ell^2\big(u'^2(S_{M,t}^\ell) + u''(S_{M,t}^\ell)\big)f\Big) - n\nu_t\Big(\varepsilon_{n+1}^2\big(u'^2(S_{M,t}^{n+1}) + u''(S_{M,t}^{n+1})\big)f\Big)\Bigg) \tag{3.62}$$

$$\begin{aligned}\mathrm{II} = \alpha\Bigg(&\sum_{1\le\ell<\ell'\le n} \nu_t\Big(\varepsilon_\ell\varepsilon_{\ell'}u'(S_{M,t}^\ell)u'(S_{M,t}^{\ell'})f\Big)\\ &- n\sum_{\ell\le n}\nu_t\Big(\varepsilon_\ell\varepsilon_{n+1}u'(S_{M,t}^\ell)u'(S_{M,t}^{n+1})f\Big)\\ &+ \frac{n(n+1)}{2}\nu_t\Big(\varepsilon_{n+1}\varepsilon_{n+2}u'(S_{M,t}^{n+1})u'(S_{M,t}^{n+2})f\Big)\Bigg)\end{aligned} \tag{3.63}$$

$$\begin{aligned}\mathrm{III} = -r\Bigg(&\sum_{1\le\ell<\ell'\le n}\nu_t(\varepsilon_\ell\varepsilon_{\ell'}f) - n\sum_{\ell\le n}\nu_t(\varepsilon_\ell\varepsilon_{n+1}f)\\ &+ \frac{n(n+1)}{2}\nu_t(\varepsilon_{n+1}\varepsilon_{n+2}f)\Bigg)\end{aligned} \tag{3.64}$$

$$\mathrm{IV} = -\frac{r}{2}\Bigg(\sum_{\ell\le n}\nu_t(\varepsilon_\ell^2 f) - n\nu_t(\varepsilon_{n+1}^2 f)\Bigg) \tag{3.65}$$

$$\mathrm{V} = \frac{1}{2}(r-\overline{r})\Bigg(\sum_{\ell\le n}\nu_t(\varepsilon_\ell^2 f) - n\nu_t(\varepsilon_{n+1}^2 f)\Bigg)\,. \tag{3.66}$$

We do not merge in this statement the similar terms IV and V since it is then easier to explain why the formula is true.

Proof. Of course this is obtained by differentiation and integration by parts. Probably the best way to understand this formula is to compare it with Proposition 2.2.3. The term V is simply created by the last term of (3.61); the term IV, created when integrating by parts in Y, was invisible in (2.23) because there $\varepsilon_\ell^2 = 1$. The really new feature is the term I, which is created by the fact that the variances of the quantities $S_{k,t}$ are not constant. It is exactly to avoid this term in Proposition 2.2.3 that we introduced the last term $\sqrt{(1-t)/N}\xi_k$ in the quantities of (2.15), see Exercise 2.2.4. □

The reason why the formula of Proposition 3.2.1 is manageable is exactly the same why the formula of Proposition 2.2.3 is manageable. Quantities such as

$$u'(S_{M,t}^\ell)u'(S_{M,t}^{\ell'})$$

can be replaced by their averages

$$\frac{1}{M}\sum_{k\le M} u'(S^{\ell}_{k,t})u'(S^{\ell'}_{k,t})\,,$$

and we can expect these to behave like constants. If we make the proper choice for r and $\overline{r}$, the terms II and III will nearly cancel each other, while the term I will nearly cancel out with IV + V. For these choices (that will not be very hard to guess) we will have that $\nu_t'(f) \simeq 0$, i.e. $\nu(f) \simeq \nu_0(f)$. The strategy to prove the replica-symmetric equations will then (predictably) be as follows. Using symmetry between sites, we have $\rho = \nu(R_{1,1}) = \nu(\varepsilon_1^2)$, and $\nu(\varepsilon_1^2) \simeq \nu_0(\varepsilon_1^2)$ is easy to compute because the last spin decouples for ν_0.

Before we start the derivation of the replica-symmetric equations, let us try to describe the overall strategy. This is best done by comparison with the situation of Chapter 2. There, to compute the quantity q, that contained information about the spins σ_i, we needed an auxiliary quantity r, that contained information about the "spins" S_k. We could express r as a function of q and then q as a function of r. Now we have two quantities q and ρ that contain information about the spins σ_i. To determine them we will need the two auxiliary quantities r and $\overline{r}$, which "contain information about the spins S_k". We will express r and $\overline{r}$ as functions of q and r, and in a second stage we will express r and $\overline{r}$ as functions of q and ρ, and reach a system of four equations with four unknown.

We now define r and $\overline{r}$ as functions of q and ρ. Of course the forthcoming formulas have been guessed by analyzing the "cavity in M" arguments of Chapter 2. Consider independent standard Gaussian r.v.s ξ, z. Consider numbers $0 \le x < y$, the r.v. $\theta = z\sqrt{x} + \xi\sqrt{y-x}$, and define

$$\begin{aligned}\Psi(x,y) &= \alpha\mathsf{E}\left(\frac{\mathsf{E}_\xi(u'(\theta)\exp u(\theta))}{\mathsf{E}_\xi \exp u(\theta)}\right)^2 \\ &= \frac{\alpha}{y-x}\mathsf{E}\left(\frac{\mathsf{E}_\xi\left(\xi \exp u(\theta)\right)}{\mathsf{E}_\xi \exp u(\theta)}\right)^2 ,\end{aligned} \tag{3.67}$$

using integration by parts (of course as usual E_ξ denotes averaging in ξ only). We also define

$$\begin{aligned}\overline{\Psi}(x,y) &= \alpha\mathsf{E}\,\frac{\mathsf{E}_\xi\left((u''(\theta)+u'^2(\theta))\exp u(\theta)\right)}{\mathsf{E}_\xi \exp u(\theta)} \\ &= \frac{\alpha}{y-x}\mathsf{E}\,\frac{\mathsf{E}_\xi\left((\xi^2-1)\exp u(\theta)\right)}{\mathsf{E}_\xi \exp u(\theta)} ,\end{aligned} \tag{3.68}$$

integrating by parts twice. We set

$$r = \Psi(q,\rho);\ \overline{r} = \overline{\Psi}(q,\rho)\,. \tag{3.69}$$

This makes sense because by the Cauchy-Schwarz inequality $R_{1,2} \le R_{1,1}^{1/2}R_{2,2}^{1/2}$ and thus $q = \nu(R_{1,2}) \le \rho = \nu(R_{1,1})$. We also observe that from the first line of (3.67) and (3.68) we have $r, \overline{r} \le K(D)$. We first address a technical point by proving that $\overline{r} \le r$.

Lemma 3.2.2. *Consider a number $c > 0$ and a concave function w. Assume that $w'' \leq -c < 0$. Consider the unique point x^* where $w'(x^*) = 0$. Then*

$$c \int (x - x^*)^2 \exp w(x) \mathrm{d}x \leq \int \exp w(x) \mathrm{d}x \,. \tag{3.70}$$

Proof. We have $w'(x) \geq -c(x - x^*)$ for $x \leq x^*$ and $w'(x) \leq -c(x - x^*)$ for $x \geq x^*$, so that $c(x - x^*)^2 \leq -w'(x)(x - x^*)$. Hence

$$\begin{aligned} c \int (x - x^*)^2 \exp w(x) \mathrm{d}x &\leq \int -w'(x)(x - x^*) \exp w(x) \mathrm{d}x \\ &= \int \exp w(x) \mathrm{d}x \,, \end{aligned}$$

by integration by parts. □

Lemma 3.2.3. *We have $\overline{r} \leq r$.*

Proof. If v is a concave function, using (3.70) for $w(x) = v(x) - x^2/2$ and $c = 1$ implies that if ξ is a standard Gaussian r.v., then

$$\mathsf{E}((\xi - x^*)^2 \exp v(\xi)) \leq \mathsf{E} \exp v(\xi) \,. \tag{3.71}$$

Minimization of the left-hand side over x^* yields

$$\mathsf{E}(\xi^2 \exp v(\xi)) - \frac{\big(\mathsf{E}(\xi \exp v(\xi))\big)^2}{\mathsf{E} \exp v(\xi)} \leq \mathsf{E} \exp v(\xi) \tag{3.72}$$

i.e.

$$\frac{\mathsf{E}((\xi^2 - 1) \exp v(\xi))}{\mathsf{E} \exp v(\xi)} \leq \left(\frac{\mathsf{E}(\xi \exp v(\xi))}{\mathsf{E} \exp v(\xi)} \right)^2 . \tag{3.73}$$

Now we fix z and we use this inequality for the function $v(x) = u(z\sqrt{q} + x\sqrt{\rho - q})$. Combining with (3.67) and (3.68) yields the result. □

We are now in a position to guess how to express q and ρ as functions of r and $\overline{r}$.

Proposition 3.2.4. *We have*

$$\rho = \frac{1}{2\kappa + r - \overline{r}} + \frac{r + h^2}{(2\kappa + r - \overline{r})^2} + \delta_1 \tag{3.74}$$

$$q = \frac{r + h^2}{(2\kappa + r - \overline{r})^2} + \delta_2 \,, \tag{3.75}$$

with $\delta_1 \leq |\nu(\varepsilon_1^2) - \nu_0(\varepsilon_1^2)|$ and $\delta_2 \leq |\nu(\varepsilon_1 \varepsilon_2) - \nu_0(\varepsilon_1 \varepsilon_2)|$.

Presumably δ_1 and δ_2 will be small when N is large, so that $(q, \rho, r, \overline{r})$ is a near solution of the system of equations (3.69) together with

$$\rho = \frac{1}{2\kappa + r - \overline{r}} + \frac{r + h^2}{(2\kappa + r - \overline{r})^2} \tag{3.76}$$

$$q = \frac{r + h^2}{(2\kappa + r - \overline{r})^2} \,. \tag{3.77}$$

These four equations are the "replica-symmetric" equations of the present model. Please note that (3.76) and (3.77) are exact equations, in contrast with (3.74) and (3.75). When we write the equations (3.69), (3.76) and (3.77), we think of $q, \rho, r, \overline{r}$ as variables, while in (3.74) and (3.75) they are given by (3.59). This follows our policy that a bit of informality is better than bloated notation. This will not be confusing. Until the end of this section, q and ρ keep the meaning (3.59), and afterwards we will revert to the notation $q_{N,M}$ and $\rho_{N,M}$.

Proof. Symmetry between sites entails $\rho = \nu(R_{1,1}) = \nu(\varepsilon_1^2)$, $q = \nu(R_{1,2}) = \nu(\varepsilon_1\varepsilon_2)$, so it suffices to show that $\nu_0(\varepsilon_1^2)$ is given by the right-hand side of (3.76) and $\nu_0(\varepsilon_1\varepsilon_2)$ is given by the right-hand side of (3.77).

We observe that, for ν_0, the last spin decouples from the others (which is a major reason behind the definition of ν_0) so that

$$\nu_0(\varepsilon_1^2) = \mathsf{E}\frac{1}{Z}\int \varepsilon^2 \exp\Big(\varepsilon(Y + hg_N) - \frac{\varepsilon^2}{2}(2\kappa + r - \overline{r})\Big)\mathrm{d}\varepsilon \tag{3.78}$$

$$\nu_0(\varepsilon_1\varepsilon_2) = \mathsf{E}\left(\frac{1}{Z}\int \varepsilon \exp\Big(\varepsilon(Y + hg_N) - \frac{\varepsilon^2}{2}(2\kappa + r - \overline{r})\Big)\mathrm{d}\varepsilon\right)^2 , \tag{3.79}$$

where

$$Z = \int \exp\Big(\varepsilon(Y + hg_N) - \frac{\varepsilon^2}{2}(2\kappa + r - \overline{r})\Big)\mathrm{d}\varepsilon \,.$$

We compute these Gaussian integrals as follows. If z is a centered Gaussian r.v., and d is a number, writing $z^2 e^{dz} = z(ze^{dz})$, integration by parts yields

$$\mathsf{E}\, z^2 e^{dz} = \mathsf{E}\, z^2(\mathsf{E}\, e^{dz} + d\mathsf{E}\, ze^{dz})$$

$$\mathsf{E}\, ze^{dz} = d\mathsf{E}\, z^2 \mathsf{E}\, e^{dz} \,.$$

Thus

$$\frac{\mathsf{E}\, z^2 e^{dz}}{\mathsf{E}\, e^{dz}} = \mathsf{E}\, z^2 + d^2(\mathsf{E}\, z^2)^2 \,.$$

Using this for $d = Y + hg_N$, $\mathsf{E}\, z^2 = 1/(2\kappa + r - \overline{r})$ we get

$$\langle \varepsilon_1^2 \rangle_0 = \frac{1}{2\kappa + r - \overline{r}} + \frac{(Y + hg_N)^2}{(2\kappa + r - \overline{r})^2} \,,$$

and, taking expectation,

$$\nu_0(\varepsilon_1^2) = \frac{1}{2\kappa + r - \overline{r}} + \frac{r + h^2}{(2\kappa + r - \overline{r})^2}, \tag{3.80}$$

and we compute $\nu_0(\varepsilon_1\varepsilon_2)$ similarly. □

We now start the real work, the proof that when N is large, δ_1 and δ_2 in Proposition 3.2.4 are small. In order to have a chance to make estimates using Proposition 3.2.1, we need some integrability properties of $\varepsilon = \sigma_N$, and we address this technical point first. We will prove an exponential inequality, which is quite stronger than what we really need, but the proof is not any harder than that of weaker statements. We start by a general principle.

Lemma 3.2.5. *Consider a concave function $T(\boldsymbol{\sigma}) \le 0$ on $\mathbb{R}^N$, numbers $(a_i)_{i \le N}$, numbers $\kappa, \kappa' > 0$ and a convex subset C of $\mathbb{R}^N$. Consider the probability measure G on $\mathbb{R}^N$ given by*

$$\forall B\,,\quad G(B) = \frac{1}{Z}\int_{B\cap C} \exp\Big(T(\boldsymbol{\sigma}) - \kappa\|\boldsymbol{\sigma}\|^2 - \kappa'\sigma_N^2 + \sum_{i\le N} a_i\sigma_i\Big)\mathrm{d}\boldsymbol{\sigma}\,, \tag{3.81}$$

where Z is the normalizing factor. Let us denote by $\boldsymbol{\rho}$ the generic point of $\mathbb{R}^{N-1}$, so that, keeping the notation $\sigma_N = \varepsilon$, we write $\boldsymbol{\sigma} = (\boldsymbol{\rho}, \varepsilon)$. Consider the projection C' of C on the last component of $\mathbb{R}^N$, that is

$$C' = \{\varepsilon \in \mathbb{R}\,;\, \exists \boldsymbol{\rho} \in \mathbb{R}^{N-1}, (\boldsymbol{\rho}, \varepsilon) \in C\}\,.$$

Consider the function f on C' defined by

$$f(\varepsilon) = \log \int_{(\boldsymbol{\rho},\varepsilon)\in C} \exp\Big(T(\boldsymbol{\sigma}) - \kappa\sum_{i\le N-1}\sigma_i^2 + \sum_{i\le N-1} a_i\sigma_i\Big)\mathrm{d}\boldsymbol{\rho}\,. \tag{3.82}$$

Then this function is concave and the law μ of σ_N under G is the probability measure on C' with density proportional to $\exp w(x)$, where

$$w(x) = f(x) - (\kappa + \kappa')x^2 + a_N x\,.$$

Proof. Let us define

$$F(\boldsymbol{\sigma}) = T(\boldsymbol{\sigma}) - \kappa\sum_{i\le N-1}\sigma_i^2 + \sum_{i\le N-1} a_i\sigma_i\,,$$

so that (3.82) simply means that

$$f(\varepsilon) = \log\int_{(\boldsymbol{\rho},\varepsilon)\in C} \exp F(\boldsymbol{\sigma})\mathrm{d}\boldsymbol{\rho}\,. \tag{3.83}$$

The definition of μ as the law of σ_N under G implies that for any function v,

$$\int v(x)\mathrm{d}\mu(x) = \frac{1}{Z}\int_C v(\varepsilon)\exp\Big(T(\boldsymbol{\sigma}) - \kappa\|\boldsymbol{\sigma}\|^2 - \kappa'\sigma_N^2 + \sum_{i\le N} a_i\sigma_i\Big)\mathrm{d}\boldsymbol{\sigma}\,. \tag{3.84}$$

where Z is the normalizing factor. Now, since $\sigma_N = \varepsilon$, we have

$$T(\boldsymbol{\sigma}) - \kappa\|\boldsymbol{\sigma}\|^2 - \kappa'\sigma_N^2 + \sum_{i\le N} a_i\sigma_i = F(\boldsymbol{\sigma}) - (\kappa+\kappa')\varepsilon^2 + a_N\varepsilon \,,$$

Integration in $\boldsymbol{\rho}$ first in the right-hand side of (3.84) gives

$$\begin{aligned}\int v(x)\mathrm{d}\mu(x) &= \frac{1}{Z}\int_{C'} v(\varepsilon)\left(\int_{(\boldsymbol{\rho},\varepsilon)\in C} \exp F(\boldsymbol{\sigma})\mathrm{d}\boldsymbol{\rho}\right)\exp(-(\kappa+\kappa')\varepsilon^2 + a_N\varepsilon)\mathrm{d}\varepsilon \\ &= \frac{1}{Z}\int_{C'} v(\varepsilon)\exp(f(\varepsilon) - (\kappa+\kappa')\varepsilon^2 + a_N\varepsilon)\mathrm{d}\varepsilon \\ &= \frac{1}{Z}\int_{C'} v(\varepsilon)\exp w(\varepsilon)\mathrm{d}\varepsilon \,.\end{aligned}$$

This proves that μ has a density $\exp w(\varepsilon)$. To finish the proof it suffices to show that f is concave. Let us write

$$C(\varepsilon) = \{\boldsymbol{\rho} \in \mathbb{R}^{N-1}; (\boldsymbol{\rho},\varepsilon) \in C\} \,,$$

so that recalling (3.83) we get

$$\exp f(\varepsilon) = \int_{C(\varepsilon)} \exp F(\boldsymbol{\sigma})\mathrm{d}\boldsymbol{\rho} = \int \mathbf{1}_{C(\varepsilon)}(\boldsymbol{\rho})\exp F(\boldsymbol{\sigma})\mathrm{d}\boldsymbol{\rho} \,.$$

Fixing $\varepsilon_1, \varepsilon_2 \in \mathbb{R}$ and $0 < s < 1$, we define the functions

$$\begin{aligned}W(\boldsymbol{\rho}) &= \mathbf{1}_{C(s\varepsilon_1+(1-s)\varepsilon_2)}(\boldsymbol{\rho})\exp F(\boldsymbol{\rho}, s\varepsilon_1 + (1-s)\varepsilon_2) \\ U(\boldsymbol{\rho}) &= \mathbf{1}_{C(\varepsilon_1)}(\boldsymbol{\rho})\exp F(\boldsymbol{\rho},\varepsilon_1) \\ V(\boldsymbol{\rho}) &= \mathbf{1}_{C(\varepsilon_2)}(\boldsymbol{\rho})\exp F(\boldsymbol{\rho},\varepsilon_2) \,.\end{aligned}$$

We observe that (3.11) holds by concavity of F and we simply use (3.12) to obtain that $f(s\varepsilon_1+(1-s)\varepsilon_2) \ge sf(\varepsilon_1)+(1-s)f(\varepsilon_2)$. (The argument actually proves the general fact that the marginal of a log-concave density function is log-concave.) □

We return to the problem of controlling $\langle\sigma_N\rangle$.

Lemma 3.2.6. *Under (3.25) we have*

$$\nu_t\left(\exp\frac{\sigma_N^2}{K}\right) \le K \,.$$

Let us remind the reader that K depends only on κ_0, h_0 and D, so in particular it is *does not depend on* t.

The proof will use several times the following simple observation. If two quantities f_1, f_2 satisfy $\nu_t(\exp(f_1^2/K)) \le K$ and $\nu_t(\exp(f_2^2/K)) \le K$ then

$\nu_t(\exp((f_1+f_2)^2/K)) \le K$ (of course for a different K). This follows from the convexity of the function $x \mapsto \exp x^2$.

Proof. The Gibbs measure corresponding to the Hamiltonian (3.61) is given by the formula (3.81) for $C = \mathbb{R}^N$, $T(\boldsymbol{\sigma}) = \sum_{k\le M} u(S_{k,t}(\boldsymbol{\sigma}))$, $a_i = hg_i$ if $i < N$, $a_N = hg_N + \sqrt{1-t}Y$ and $\kappa' = (1-t)(r-\overline{r})/2$. Lemma 3.2.5 implies that the function

$$f(\sigma_N) = \log \int \exp\Bigl(\sum_{k\le M} u(S_{k,t}) - \kappa \sum_{i\le N-1} \sigma_i^2 + h \sum_{i\le N-1} g_i\sigma_i\Bigr)\mathrm{d}\boldsymbol{\rho} \tag{3.85}$$

is concave, and that the law of σ_N under $\langle\cdot\rangle_t$ has a density proportional to $\exp w(x)$, where

$$w(x) = f(x) - \kappa(t)x^2 + Y_t x$$

for $\kappa(t) = \kappa + (1-t)(r-\overline{r})/2$ and $Y_t = \sqrt{1-t}Y + hg_N$. We note that since $r \ge \overline{r}$ we have, recalling (3.25), that $\kappa(t) \ge \kappa \ge \kappa_0$.

Consider the point x^* where the concave function $w(x)$ is maximum (the dependence of this point on t is kept implicit). It follows from (3.70) that $\langle(\sigma_N - x^*)^2\rangle_t \le 1/2\kappa_0$, so that $\langle|\sigma_N - x^*|\rangle_t \le 1/\sqrt{2\kappa_0}$ and $|\langle\sigma_N\rangle_t - x^*| \le 1/\sqrt{2\kappa_0}$, and therefore

$$\langle\sigma_N\rangle_t \le \frac{1}{\sqrt{2\kappa_0}} + |x^*|\,.$$

Now, since $w''(x) \le -2\kappa$, we have $|w'(x^*) - w'(0)| \ge 2\kappa|x^*|$, and since $w'(x^*) = 0$ this shows that $|x^*| \le |w'(0)|/2\kappa$. Since $|w'(0)| = |f'(0) + Y_t| \le |f'(0)| + |Y_t|$, we have shown that

$$|\langle\sigma_N\rangle_t| \le \frac{1}{\sqrt{2\kappa_0}} + \frac{1}{2\kappa_0}\bigl(|Y_t| + |f'(0)|\bigr)\,. \tag{3.86}$$

Also, it follows from (3.16) that

$$\Bigl\langle \exp\frac{(\sigma_N - \langle\sigma_N\rangle_t)^2}{K}\Bigr\rangle_t \le 4\,, \tag{3.87}$$

so that it suffices to prove that $\mathsf{E}\exp\langle\sigma_N\rangle_t^2/K \le K$, and, by (3.86) it suffices to prove that $\mathsf{E}\exp(f'(0)^2/K) \le K$. We compute $f'(0)$ by differentiating (3.85). We observe that the only dependence of the right-hand side on σ_N is through the terms $u(S_{k,t})$ and that

$$\frac{\partial S_{k,t}}{\partial\sigma_N}\Bigr|_{\sigma_N=0} = \frac{\sqrt{t}}{\sqrt{N}} S_{k,0}\,,$$

where

$$S_{k,0} = \frac{1}{\sqrt{N}} \sum_{i\le N-1} g_{i,k}\sigma_i \quad (= S_{k,t}|_{t=0})\,.$$

Therefore we get

$$f'(0) = \frac{\sqrt{t}}{\sqrt{N}} \sum_{k \le M} g_{N,k} \langle u'(S_{k,0}) \rangle \ ,$$

where $\langle \cdot \rangle$ is a certain Gibbs average that does not depend on the r.v.s $g_{N,k}$. Let us denote by E_0 expectation in the r.v.s $g_{N,k}$ only. Then, since $|u'| \le D$,

$$\mathsf{E}_0 f'(0)^2 = \frac{t}{N} \sum_{k \le M} \langle u'(S_{k,0}) \rangle^2 \le \alpha D^2 \ ,$$

and thus by (A.11) we have

$$\mathsf{E}_0 \exp \frac{f'(0)^2}{4\alpha D^2} \le \frac{1}{\sqrt{1 - 1/2}} \le 2 \ .$$

Therefore

$$\mathsf{E} \exp \frac{f'(0)^2}{4\alpha D^2} = \mathsf{E}\mathsf{E}_0 \exp \frac{f'(0)^2}{4\alpha D^2} \le 2 \ . \qquad \square$$

Despite this excellent control, the fact that σ_N is not bounded does create hardship. For example, it does not seem possible to use the argument of Lemma 2.3.7 to compare $\nu(f)$ and $\nu_t(f)$ when $f \ge 0$.

We turn to the study of terms I and II of Proposition 3.2.1. Let us consider the Hamiltonian

$$\begin{aligned} - H_{N,M-1,t}(\boldsymbol{\sigma}) = & \sum_{k \le M-1} u(S_{k,t}(\boldsymbol{\sigma})) - \kappa \|\boldsymbol{\sigma}\|^2 + h \sum_{i \le N} g_i \sigma_i \\ & + \sigma_N \sqrt{1-t} Y - (1-t)(r - \overline{r}) \frac{\sigma_N^2}{2} \ . \end{aligned} \tag{3.88}$$

The difference with (3.61) is that the summation is over $k \le M - 1$ rather than over $k \le M$. We denote by $\langle \cdot \rangle_{t,\sim}$ an average for the corresponding Gibbs measure.

We consider standard Gaussian r.v.s $z, (\xi^\ell)$ that are independent of all the other r.v.s already considered, and we set

$$\theta^\ell = z\sqrt{q} + \xi^\ell \sqrt{\rho - q} \ . \tag{3.89}$$

For $0 \le v \le 1$ we define

$$S_v^\ell = \sqrt{v} S_{M,t}^\ell + \sqrt{1-v}\theta^\ell \ . \tag{3.90}$$

The dependence on t is kept implicit; when using S_v^ℓ we think of t (and M) as being fixed. We then define

$$\nu_{t,v}(f) = \mathsf{E}\frac{\langle f \exp(\sum_{\ell\le n} u(S_v^\ell))\rangle_{t,\sim}}{\langle \mathsf{E}_\xi \exp u(S_v^1)\rangle_{t,\sim}^n} . \tag{3.91}$$

Here as usual E_ξ means expectation in all the r.v.s labeled ξ, or, equivalently here, in ξ^1. This really is the same as definition (2.35). The notation is a bit different (there is an expectation E_ξ in the denominator) simply because in (2.35) we made the convention that this expectation E_ξ was "built-in" the average $\langle\cdot\rangle_{t,\sim}$ and we do not do it here (for the simple reason that we do not want to have to remind the reader of this each time we write a similar formula). Obviously we have

$$\nu_{t,1}(f) = \nu_t(f) .$$

The magic of the definition of $\nu_{t,v}$ is revealed by the following, whose proof is nearly identical to that of Lemma 2.3.1.

Lemma 3.2.7. *Consider a function f on Σ_N^n. Then we have*

$$\nu_{t,0}(f) = \mathsf{E}\langle f\rangle_{t,\sim} , \tag{3.92}$$

$$\alpha\nu_{t,0}(u'(S_0^1)u'(S_0^2)f) = r\mathsf{E}\langle f\rangle_{t,\sim} = r\nu_{t,0}(f) \tag{3.93}$$

and

$$\alpha\nu_{t,0}((u''(S_0^1) + u'(S_0^1)^2)f) = \overline{r}\mathsf{E}\langle f\rangle_{t,\sim} = \overline{r}\nu_{t,0}(f) . \tag{3.94}$$

Throughout the rest of the chapter we reinforce (3.7) into

$$\forall\ell\ ,\ 1\le\ell\le 4\ ,\ |u^{(\ell)}|\le D . \tag{3.95}$$

Lemma 3.2.8. *If B_v is one of the following: 1, $u'(S_v^1)u'(S_v^2)$, $u'^2(S_v^1) + u''(S_v^1)$, then for a function f on Σ_N^n, we have*

$$\left|\frac{\mathrm{d}}{\mathrm{d}v}\nu_{t,v}(fB_v)\right| \le K(n,\kappa_0,h_0,D)\Bigg(\sum_{\ell\le n+1}\nu_{t,v}(|f||R_{\ell,\ell}-\rho|)$$

$$+\sum_{1\le\ell<\ell'\le n+2}\nu_{t,v}(|f||R_{\ell,\ell'}-q|) + \frac{1}{N}\nu_{t,v}(|f|)\Bigg) . \tag{3.96}$$

The proof is nearly identical to that of (2.59) (except that one does not use Hölder's inequality in the last step). The new feature is that there are more terms when we integrate by parts. Defining

$$S_v^{\ell\prime} = \frac{1}{2\sqrt{v}}S_{M,t}^\ell - \frac{1}{2\sqrt{1-v}}\theta^\ell ,$$

we then have

$$\mathsf{E}S_v^{\ell'}S_v^{\ell} = \frac{1}{2}(R_{\ell,\ell} - \rho)\ ,$$

while in (2.59) we had $\mathsf{E}S_v^{\ell'}S_v^{\ell} = 0$. This is what creates the new terms $\nu_{t,v}(|f||R_{\ell,\ell} - \rho|)$ in (3.96) compared to (2.59). In (2.59) these terms do not occur because there $R_{\ell,\ell} = 1 = \rho$.

We have proved in Theorem 3.1.18 that, with respect to ν, we have $R_{1,2} \simeq q$ and $R_{1,2} \simeq \rho$. If the same is true with respect to $\nu_{t,v}$ then (3.96) will go a long way to fulfill our program that the terms of Proposition 3.2.1 nearly cancel out.

The first step will be to prove that in the bound (3.96) we can replace $\nu_{t,v}$ by ν_t in the right-hand side. (So that it to use this bound it will suffice to know that $R_{1,2} \simeq q$ and $R_{1,2} \simeq \rho$ for ν_t). Unfortunately we cannot immediately write a differential inequality such as $|\mathrm{d}\nu_{t,v}(f)/\mathrm{d}v| \le K(n)\nu_{t,v}(f)$ when $f \ge 0$ because it is not true that the quantities $|R_{\ell,\ell} - \rho|$ and $|R_{\ell,\ell'} - q|$ are bounded. But it is true that they are "almost bounded" in the sense that they are bounded outside an exponentially small set, namely that we can find K for which

$$\nu_{t,v}(\mathbf{1}_{\{|R_{1,2}-q|\ge K\}}) \le \exp(-4N) \tag{3.97}$$

$$\nu_{t,v}(\mathbf{1}_{\{|R_{1,1}-\rho|\ge K\}}) \le \exp(-4N)\ . \tag{3.98}$$

The reader wishing to skip the proof of this purely technical point can jump ahead to (3.105) below. To prove these inequalities, we observe from (3.91) that when f is a function on Σ_N (that does not depend on the r.v.s ξ^ℓ) then

$$\nu_{t,v}(f) = \mathsf{E}\langle f\rangle_{t,v}\ ,$$

where $\langle\cdot\rangle_{t,v}$ is a Gibbs average for the Hamiltonian

$$-H(\boldsymbol{\sigma}) = \sum_{k<M} u(S_{k,t}(\boldsymbol{\sigma})) + u_v(\sqrt{v}S_{M,t}(\boldsymbol{\sigma}) + \sqrt{1-v}\sqrt{q}z) - \kappa\|\boldsymbol{\sigma}\|^2$$
$$+ h\sum_{i\le N} g_i\sigma_i + \sigma_N\sqrt{1-t}Y - (1-t)(r-\overline{r})\frac{\sigma_N^2}{2}\ , \tag{3.99}$$

and where the function u_v is given by

$$u_v(x) = \log \mathsf{E}\exp u(x + \xi\sqrt{1-v}\sqrt{\rho - q})\ .$$

This function is concave because a marginal of the log-concave function is log-concave, as was shown in the proof of Lemma 3.2.5, and since $\exp u_v(x)$ is the marginal of the log concave function

$$(x,y) \mapsto \exp(u(x) + y\sqrt{1-v}\sqrt{\rho-q} - y^2/2)\ .$$

Another proof of the concavity of u_v is as follows. Writing $X = x + \xi\sqrt{1-v}\sqrt{\rho-q}$, the concavity of u_v i.e. the fact $u_v'' \le 0$ means that

$$\frac{\mathsf{E}((u'(X)^2 + u''(X))e^{u(X)})}{\mathsf{E}e^{u(X)}} \leq \left(\frac{\mathsf{E}u'(X)e^{u(X)}}{\mathsf{E}e^{u(X)}}\right)^2 ,$$

an inequality that we can prove by applying (3.73) to the function $v(\xi) = u(x + \xi\sqrt{1-v}\sqrt{\rho - q})$ and integration by parts as in (3.67) and (3.68).

There is nothing to change to the proof of Lemma 3.2.6 to obtain

$$\nu_{t,v}\left(\exp\frac{\sigma_N^2}{K}\right) \leq K . \tag{3.100}$$

Again K is as usual in this chapter, depending only on D and the quantities κ_0, h_0 of (3.25) and in particular it *does not depend* on t or v. There is very little to change to the proof of (3.28) to get

$$\left\langle \exp\frac{\kappa}{2}\|\boldsymbol{\sigma}\|^2\right\rangle_{t,v} \leq \exp KB^* , \tag{3.101}$$

where $\langle\cdot\rangle_{t,v}$ denotes an average for the Gibbs measure with Hamiltonian (3.99). We now prove that

$$\nu_{t,v}\left(\exp\frac{\|\boldsymbol{\sigma}\|^2}{K}\right) \leq \exp LN . \tag{3.102}$$

For this, we recall that $\mathsf{E}\exp(B^*/4) \leq \exp LN$ by (3.33), and we denote by K_0 the constant K in (3.101). We define $K_1 = 8K_0/\kappa$. Using Hölder's inequality in the first inequality and (3.101) in the second inequality we get

$$\begin{aligned}\nu_{t,v}\left(\exp\frac{\|\boldsymbol{\sigma}\|^2}{K_1}\right) = \mathsf{E}\left\langle \exp\frac{\|\boldsymbol{\sigma}\|^2}{K_1}\right\rangle_{t,v} &\leq \mathsf{E}\left\langle \exp\frac{\kappa}{2}\|\boldsymbol{\sigma}\|^2\right\rangle_{t,v}^{2/\kappa K_1} \\ &\leq \mathsf{E}\exp\left(\frac{2K_0}{\kappa K_1}B^*\right) = \mathsf{E}\exp\left(\frac{B^*}{4}\right) \leq \exp LN .\end{aligned}$$

It follows (for yet another constant K) that

$$\nu_{t,v}(\mathbf{1}_{\{\|\boldsymbol{\sigma}\|^2 \geq KN\}}) \leq \exp(-4N) .$$

Since

$$N|R_{1,2}| = |\boldsymbol{\sigma}^1 \cdot \boldsymbol{\sigma}^2| \leq \|\boldsymbol{\sigma}^1\|\|\boldsymbol{\sigma}^2\| ,$$

we have

$$\begin{aligned}|R_{1,2}| \geq t &\Rightarrow (Nt)^2 \leq \|\boldsymbol{\sigma}^1\|^2\|\boldsymbol{\sigma}^2\|^2 \\ &\Rightarrow \|\boldsymbol{\sigma}^1\|^2 \geq tN \text{ or } \|\boldsymbol{\sigma}^2\|^2 \geq tN ,\end{aligned}$$

and it follows that

$$\nu_{t,v}(\mathbf{1}_{\{R_{1,1} \geq K\}}) \leq 2\exp(-4N) \; ; \; \nu_{t,v}(\{|R_{1,2}| \geq K\}) \leq 2\exp(-4N) . \tag{3.103}$$

Since by (3.28) and (3.32) we have $|\rho| \leq K$ and $|q| \leq K$, (3.97) and (3.98) follow. Let us also note from (3.102) by a similar argument that

$$\nu_{t,v}\big((R_{1,2}-q)^8\big) \leq K \; ; \; \nu_{t,v}\big((R_{1,1}-\rho)^8\big) \leq K \; , \tag{3.104}$$

where of course there is nothing magic in the choice of the number 8.

It seems unlikely that what happens in the exponentially small set where $|R_{\ell,\ell'}-q|$ and $|R_{\ell,\ell}-\rho|$ might be large could be troublesome; nonetheless we must spend a few lines to check it. We recall that for a function f^* we write $\nu(f^*)^{1/4}$ rather than $(\nu(f^*))^{1/4}$ (etc.). We have

$$\begin{aligned}\nu_{t,v}(|f||R_{\ell,\ell}-\rho|) &\leq \nu_{t,v}(|f||R_{\ell,\ell}-\rho|\mathbf{1}_{\{|R_{\ell,\ell}-\rho|\leq K\}}) \\ &\quad + \nu_{t,v}(|f||R_{\ell,\ell}-\rho|\mathbf{1}_{\{|R_{\ell,\ell}-\rho|>K\}}) \\ &\leq K\nu_{t,v}(|f|) \\ &\quad + \nu_{t,v}(f^2)^{1/2}\nu_{t,v}\big((R_{\ell,\ell}-\rho)^4\big)^{1/4}\nu_{t,v}(\mathbf{1}_{\{|R_{\ell,\ell}-\rho|>K\}})^{1/4} \\ &\leq K\nu_{t,v}(|f|)+K\exp(-N)\nu_{t,v}(f^2)^{1/2} \; ,\end{aligned}$$

using (3.104) and (3.98). We then proceed in a similar manner for $|R_{\ell,\ell'}-q|$.

In this fashion, we deduce from (3.96) that, if f is any function on Σ_N^4 then

$$\left|\frac{\mathrm{d}}{\mathrm{d}v}\nu_{t,v}(f)\right| \leq K\nu_{t,v}(|f|)+K\exp(-N)\sup_v \nu_{t,v}(f^2)^{1/2} \; . \tag{3.105}$$

Lemma 3.2.9. *Consider a function $f^* \geq 0$ on Σ_N^4. Then*

$$\nu_{t,v}(f^*) \leq K\big(\nu_t(f^*)+\exp(-N)\sup_v \nu_{t,v}(f^{*2})^{1/2}\big) \; . \tag{3.106}$$

Proof. This follows from (3.105) and Lemma A.13.1. □

Proposition 3.2.10. *If $f=\varepsilon_1^2$ or $f=\varepsilon_1\varepsilon_2$, we have*

$$|\nu_t'(f)| \leq K\left(\nu_t\big((R_{1,2}-q)^2\big)^{1/2}+\nu_t\big((R_{1,1}-\rho)^2\big)^{1/2}+\frac{1}{N}\right) \; . \tag{3.107}$$

Proof. The idea is that we reproduce the proof of (2.66), using Proposition 3.2.1 instead of Proposition 2.2.3 and using Lemma 3.2.7 instead of Lemma 2.3.1, Lemma 3.2.9 being an appropriate substitute for Lemma 2.3.4. More specifically, computing $\nu_t'(f)$ through Proposition 3.2.1, and denoting by $\mathcal{R}$ a quantity such that $|\mathcal{R}|$ is bounded by the right-hand of (3.107) (with possibly a different value of K), we will prove that

$$\begin{aligned}\alpha\nu_t(\varepsilon_\ell^2(u'^2(S_{M,t}^\ell)+u''(S_{M,t}^\ell))f) &= \overline{r}\nu_0(\varepsilon_\ell^2 f)+\mathcal{R} \; ; \qquad (3.108)\\ \nu_t(\varepsilon_\ell^2 f) &= \nu_0(\varepsilon_\ell^2 f)+\mathcal{R} \; ; \\ \alpha\nu_t(\varepsilon_\ell\varepsilon_{\ell'}u'(S_{M,t}^\ell)u'(S_{M,t}^{\ell'})f) &= r\nu_0(\varepsilon_\ell\varepsilon_{\ell'}f)+\mathcal{R} \; ; \\ \nu_t(\varepsilon_\ell\varepsilon_{\ell'}f) &= \nu_0(\varepsilon_\ell\varepsilon_{\ell'}f)+\mathcal{R} \; .\end{aligned}$$

We will prove only (3.108), since the proof of the other relations is entirely similar. Let $\varphi(v) = \alpha\nu_{t,v}(\varepsilon_\ell\varepsilon_{\ell'}u'(S_v^\ell)u'(S_v^{\ell'})f)$. Lemma 3.2.7 implies

$$|\varphi(1) - \varphi(0)| = |\alpha\nu_t(\varepsilon_\ell\varepsilon_{\ell'}u'(S_{M,t}^\ell)u'(S_{M,t}^{\ell'})f) - r\nu_0(\varepsilon_\ell\varepsilon_{\ell'}f)| . \tag{3.109}$$

On the other hand, $|\varphi(1)-\varphi(0)| \le \sup_v |\varphi'(v)|$, and by (3.96) (used for $\varepsilon_\ell\varepsilon_{\ell'}f$ rather than f) we obtain

$$\begin{aligned}|\varphi'(v)| \le K\Bigg(&\sum_{\ell_1\le 3}\nu_{t,v}(|\varepsilon_\ell\varepsilon_{\ell'}f||R_{\ell_1,\ell_1}-\rho|)\\ &+\sum_{1\le\ell_1<\ell_2\le 4}\nu_{t,v}(|\varepsilon_\ell\varepsilon_{\ell'}f||R_{\ell_1,\ell_2}-q|)+\frac{1}{N}\nu_{t,v}(|\varepsilon_\ell\varepsilon_{\ell'}f|)\Bigg) . \end{aligned}\tag{3.110}$$

Now since $f = \varepsilon_1\varepsilon_2$ or $f = \varepsilon_1^2$, using Hölder's inequality and then (3.100) we get $\nu_{t,v}((\varepsilon_\ell\varepsilon_{\ell'}f)^2) \le \nu_{t,v}(\varepsilon_1^4) \le K$. Using the Cauchy-Schwarz inequality we then deduce from (3.109) and (3.110) that

$$\begin{aligned}&|\alpha\nu_t(\varepsilon_\ell\varepsilon_{\ell'}u'(S_{M,t}^\ell)u'(S_{M,t}^{\ell'})f) - r\nu_t(\varepsilon_\ell\varepsilon_{\ell'}f)|\\ &\le K\sup_v\left(\nu_{t,v}\big((R_{1,1}-\rho)^2\big)^{1/2}+\nu_{t,v}\big((R_{1,2}-q)^2\big)^{1/2}+\frac{1}{N}\right) .\end{aligned}$$

We finally conclude with (3.106) and (3.104), used for $f = R_{1,2} - q$ or $f = R_{1,1}$. □

We know from Theorem 3.1.18 that $\nu((R_{1,2} - q)^2)^{1/2} \le K/\sqrt{N}$ and $\nu((R_{1,1}-\rho)^2)^{1/2} \le K/\sqrt{N}$, so in the right-hand side of (3.107), we would like to replace ν_t by ν. Unfortunately, since σ_N is not bounded, it is unclear how one could prove a differential inequality such as $|\nu_t'(f)| \le K\nu_t(f)$ to relate ν_t and ν. The crucial observation to bypass this difficulty is that Theorem 3.1.18 holds uniformly over the functionals ν_t (with the same proof), so that, if we set

$$q_t = \nu_t(R_{1,2}) \; ; \; \rho_t = \nu_t(R_{1,1}) ,$$

we have in particular

$$\nu_t\big((R_{1,1}-\rho_t)^4\big) \le \frac{K}{N^2} \; ; \; \nu_t\big((R_{1,2}-q_t)^4\big) \le \frac{K}{N^2} . \tag{3.111}$$

Therefore it is of interest to bound $q - q_t$ and $\rho - \rho_t$.

Lemma 3.2.11. *We have*

$$|q_t - q| \le \frac{K}{N} \; ; \; |\rho_t - \rho| \le \frac{K}{N} . \tag{3.112}$$

Proof. Since $q = q_1$ it suffices to prove that $q_t' = \mathrm{d}q_t/\mathrm{d}t$ satisfies $|q_t'| \le K/N$ (and similarly for ρ). Since $q_t = \nu_t(R_{1,2})$, $q_t' = \nu_t'(R_{1,2})$ is given by Proposition 3.2.1 for $f = R_{1,2}$.

A key observation is that the five terms of this proposition cancel out (as they should!) if f is a constant, i.e. is not random and does not depend on $\boldsymbol{\sigma}^1, \boldsymbol{\sigma}^2, \ldots$. Therefore to evaluate $\nu_t'(R_{1,2})$ we can in each of these terms replace $f = R_{1,2}$ by $R_{1,2} - q_t$, because the contributions of q_t to the various terms cancel out.

The point of doing this is that the quantity $R_{1,2} - q_t$ is small (for ν_t) as is shown by (3.112), and therefore each of the terms of Proposition 3.2.1 is at most $K/\sqrt{N}$. This is seen by using that $|u'| \le D, |u''| \le D$, Hölder's inequality, (3.112) (and (3.100) to take care of the terms $\varepsilon_\ell \varepsilon_{\ell'}$).

This argument is enough to prove (3.112) with a bound $K/\sqrt{N}$ rather than K/N. This is all what is required to prove Proposition 3.2.12 below.

The rest of this proof describes the extra work required to reach the correct rate K/N (just for the beauty of it).

We proceed as in the proof of Proposition 3.2.10 (with now $f = R_{1,2} - q_t$) but in the right-hand side of (3.110) we use Hölder's inequality as in

$$\nu_{t,v}(|\varepsilon_\ell \varepsilon_{\ell'} f(R_{\ell_1,\ell_1} - \rho)|) \le \nu_{t,v}((\varepsilon_\ell \varepsilon_{\ell'})^2)^{1/2} \nu_{t,v}(f^4)^{1/4} \nu_{t,v}\big((R_{\ell_1,\ell_1} - \rho)^4\big)^{1/4} ,$$

and, since $f = R_{1,2} - q_t$, we get

$$\begin{aligned} |q_t'| &\le \sup_v \Bigg(\nu_{t,v}\big((R_{1,2} - q_t)^4\big)^{1/4} \\ &\quad \times \Big(\nu_{t,v}\big((R_{1,2} - q)^4\big)^{1/4} + \nu_{t,v}\big((R_{1,1} - \rho)^4\big)^{1/4} + \frac{1}{N} \Big) \Bigg) \\ &\le \frac{K}{\sqrt{N}} \sup_v \Big(\nu_{t,v}\big((R_{1,2} - q)^4\big)^{1/4} + \nu_{t,v}\big((R_{1,1} - \rho)^4\big)^{1/4} + \frac{1}{N} \Big) , \end{aligned}$$

using (3.111) in the second line. Using (3.106) and (3.104) we get

$$|q_t'| \le \frac{K}{\sqrt{N}} \Big(\nu_t\big((R_{1,2} - q)^4\big)^{1/4} + \nu_t\big((R_{1,1} - \rho)^4\big)^{1/4} + \frac{1}{N} \Big) .$$

Using (3.111) and the triangle inequality we obtain

$$\nu_t\big((R_{1,2} - q)^4\big)^{1/4} \le |q - q_t| + \nu_t\big((R_{1,2} - q_t)^4\big)^{1/4} \le |q - q_t| + \frac{K}{\sqrt{N}} \tag{3.113}$$

$$\nu_t\big((R_{1,1} - \rho)^4\big)^{1/4} \le |\rho - \rho_t| + \nu_t\big((R_{1,1} - \rho_t)^4\big)^{1/4} \le |\rho - \rho_t| + \frac{K}{\sqrt{N}} , \tag{3.114}$$

and we reach that

$$|q_t'| \le \frac{K}{\sqrt{N}} (|q - q_t| + |\rho - \rho_t|) + \frac{K}{N} .$$

Similarly we get

$$|\rho'_t| \le \frac{K}{\sqrt{N}}(|q - q_t| + |\rho - \rho_t|) + \frac{K}{N} ,$$

so that if $\psi(t) = |q - q_t| + |\rho - \rho_t|$ the right derivative $\psi'(t)$ satisfies

$$|\psi'(t)| \le K\left(\frac{\psi(t)}{\sqrt{N}} + \frac{1}{N}\right) \le K\psi(t) + \frac{K}{N} .$$

Since $\psi(1) = 0$, Lemma A.13.1 shows that $\psi(t) \le K/N$. □

Using (3.113), (3.114) and (3.112) we get $\nu_t((R_{1,2} - q)^4)^{1/4} \le K/N$ and $\nu_t((R_{1,1} - \rho)^4)^{1/4} \le K/N$, so that combining with (3.107) we have proved that $|\nu'_t(f)| \le K/N$ for $f = \varepsilon_1^2$ or $f = \varepsilon_1\varepsilon_2$, and therefore the following.

Proposition 3.2.12. *We have*

$$|\nu(\varepsilon_1^2) - \nu_0(\varepsilon_1^2)| \le \frac{K}{\sqrt{N}} \; ; \; |\nu(\varepsilon_1\varepsilon_2) - \nu_0(\varepsilon_1\varepsilon_2)| \le \frac{K}{\sqrt{N}} . \tag{3.115}$$

Combining with Proposition 3.2.4, this shows that (q, ρ) is a solution of the system of replica-symmetric equations (3.69), (3.76) and (3.104) "with accuracy $K/\sqrt{N}$". Letting $N \to \infty$ this proves in particular that this system does have a solution, which did not seem obvious beforehand.

Let us consider the function

$$F(q,\rho) = \alpha \mathsf{E} \log \mathsf{E}_\xi \exp u(z\sqrt{q} + \xi\sqrt{\rho - q}) + \frac{1}{2}\frac{q}{\rho - q} + \frac{1}{2}\log(\rho - q) - \kappa\rho + \frac{h^2}{2}(\rho - q) , \tag{3.116}$$

which is defined for $0 \le q < \rho$. It is elementary (calculus and integration by parts as in Lemma 2.4.4) to show that the conditions $\partial F/\partial\rho = 0 = \partial F/\partial q$ mean that (3.69), (3.76) and (3.104) are satisfied.

We would like to prove that for large N the quantity

$$\frac{1}{N}\mathsf{E}\log\int \exp(-H_{N,M,t}(\boldsymbol{\sigma}))\mathrm{d}(\boldsymbol{\sigma}) \tag{3.117}$$

is nearly $F(q,\rho) + \log(2e\pi)/2$. Unfortunately we see no way to do this unless we know something about the uniqueness of the solutions of (3.69), (3.76), (3.104).

Research Problem 3.2.13. (Level 2) Find general conditions under which the equations (3.69), (3.76), (3.104) have a unique solution.

As we will show in the next section, Shcherbina and Tirozzi managed to solve this problem in a very important case. Before we turn to this, we must however address the taste of unfinished work left by Proposition 3.2.12. We turn to the proof of the correct result.

Theorem 3.2.14. *We have*

$$|\nu(\varepsilon_1^2) - \nu_0(\varepsilon_1^2)| \le \frac{K}{N} \; ; \; |\nu(\varepsilon_1\varepsilon_2) - \nu_0(\varepsilon_1\varepsilon_2)| \le \frac{K}{N} \, . \tag{3.118}$$

Consequently, (q, ρ) is a solution of the equations (3.69), (3.74) and (3.75) "with accuracy K/N". Of course improving (3.113) into (3.118) is really a side story; but it is not very difficult, so we cannot resist the pleasure of doing it.

Proof. The proof we give is complete but sketchy, and filling in all details should be a nice exercise for the motivated reader. We will obtain the estimates

$$|\nu_t'(\varepsilon_1^2)| \le \frac{K}{N} \; ; \; |\nu_t'(\varepsilon_1\varepsilon_2)| \le \frac{K}{N} \, .$$

For this we will prove that when using Proposition 3.2.1 the cancellation of the various terms occurs with accuracy K/N. Consider $f = \varepsilon_1^2$ or $f = \varepsilon_1\varepsilon_2$, and B_v as in Lemma 3.2.8. To prove that in Proposition 3.2.1 this cancellation of the various terms occurs with accuracy K/N we have to show that

$$|\nu_{t,1}(B_1 f) - \nu_{t,0}(B_0 f)| \le \frac{K}{N} \, . \tag{3.119}$$

To prove this we replace the first order estimate

$$|\nu_{t,1}(B_1 f) - \nu_{t,0}(B_0 f)| \le \sup_{0<v<1} \left| \frac{\mathrm{d}}{\mathrm{d}v} \nu_{t,v}(B_v f) \right|$$

that we used in the proof of Proposition 3.2.10 by a second order estimate

$$\left| \nu_{t,1}(B_1 f) - \nu_{t,0}(B_0 f) - \frac{\mathrm{d}}{\mathrm{d}v} \nu_{t,v}(B_v f) \Big|_{v=0} \right| \le \sup_{0<v<1} \left| \frac{\mathrm{d}^2}{\mathrm{d}v^2} \nu_{t,v}(B_v f) \right| . \tag{3.120}$$

Differentiating in v once creates terms that each contains a factor $R_{\ell,\ell} - \rho$ or $R_{\ell,\ell'} - q$. Differentiating twice brings a second such factor in each term. We know (3.111) (and a similar result for higher powers) and (3.112). Using (3.106) shows that the right-hand side of (3.120) is $\le K/N$. Thus to prove (3.119) the issue is to prove that

$$\left| \frac{\mathrm{d}}{\mathrm{d}v} \nu_{t,v}(B_v f) \Big|_{v=0} \right| \le \frac{K}{N} \, .$$

Computation of this derivative shows that it is a sum of terms $A\mathsf{E}\langle f(R_{\ell,\ell'} - q)\rangle_{t,\sim}$ or $A\mathsf{E}\langle f(R_{\ell,\ell} - \rho)\rangle_{t,\sim}$ (and of a lower order term due to the difference between $R^t_{\ell,\ell'}$ and $R_{\ell,\ell}$), where A is a quantity which does not depend on N. Therefore it suffices to show that

$$|\mathsf{E}\langle f(R_{\ell,\ell} - \rho)\rangle_{t,\sim}| \le \frac{K}{N} \; ; \; |\mathsf{E}\langle f(R_{\ell,\ell'} - q)\rangle_{t,\sim}| \le \frac{K}{N} \, .$$

We have

$$|\nu_t(f(R_{\ell,\ell'}-q)) - \mathsf{E}\langle f(R_{\ell,\ell'}-q)\rangle_{t,\sim}| \le \frac{K}{N}\,,$$

by proceeding as in (2.65) because, as expected, the extra factor $R_{\ell,\ell'}-q$ allows one to gain a factor $1/\sqrt{N}$ (and similarly for $R_{\ell,\ell}-\rho$). Therefore to prove (3.119) it suffices to prove that

$$|\nu_t(f(R_{\ell,\ell'}-q))| \le \frac{K}{N}\ ;\ |\nu_t(f(R_{\ell,\ell}-\rho))| \le \frac{K}{N}\,.$$

In the same manner that we have proved the inequality $|\nu_t'(f)| \le K/\sqrt{N}$, we show now that $|\nu_t'(f(R_{\ell,\ell'}-q))| \le K/N$ and $|\nu_t'(f(R_{\ell,\ell}-\rho))| \le K/N$ (gaining a factor $1/\sqrt{N}$ because of the extra term $R_{\ell,\ell'}-q$ or $R_{\ell,\ell}-\rho$) so the issue is to prove that

$$|\nu(f(R_{\ell,\ell'}-q))| \le \frac{K}{N}\ ;\ |\nu(f(R_{\ell,\ell}-\rho))| \le \frac{K}{N}\,.$$

By symmetry among sites, when $f=\varepsilon_1^2$,

$$\nu(\varepsilon_1^2(R_{\ell,\ell'}-q)) = \nu(R_{1,1}(R_{\ell,\ell'}-q)) = \nu((R_{1,1}-\rho)(R_{\ell,\ell'}-q))$$

since $\nu(R_{\ell,\ell'}-q)=0$. Using Theorem 3.1.18 for $k=1$ and the Cauchy-Schwarz inequality we then obtain that $\nu(\varepsilon_1^2(R_{\ell,\ell'}-q)) \le K/N$. The case of $f=\varepsilon_1\varepsilon_2$ is similar. □

3.3 Controlling the Solutions of the RS Equations

We recall the notation (3.59) (where the function u is implicit)

$$q_{N,M} = \nu(R_{1,2});\ \rho_{N,M} = \nu(R_{1,1})\,.$$

In Section 3.2 these were denoted simply q and ρ, but we now find **it more convenient to denote in this section by q and ρ two "variables"** with $0 \le q < \rho$.

As pointed out in Section 3.1, the case where $\exp u(x) = \mathbf{1}_{\{x\ge\tau\}}$ is of special interest. In this case, we will prove that the system of equations (3.69), (3.76) and (3.104) has a unique solution. The function (3.116) takes the form

$$\begin{aligned} F(q,\rho) = {} & \alpha\mathsf{E}\log\mathsf{P}_\xi(z\sqrt{q}+\xi\sqrt{\rho-q}\ge\tau) + \frac{1}{2}\frac{q}{\rho-q} \\ & + \frac{1}{2}\log(\rho-q) - \kappa\rho + \frac{h^2}{2}(\rho-q)\,. \end{aligned} \tag{3.121}$$

We observe that

$$\mathsf{P}_\xi(z\sqrt{q}+\xi\sqrt{\rho-q}\geq\tau)=\mathcal{N}\left(\frac{\tau-z\sqrt{q}}{\sqrt{\rho-q}}\right),$$

where

$$\mathcal{N}(x)=\mathsf{P}(\xi\geq x)\,.$$

We now fix once and for all $\tau\geq 0$.

Theorem 3.3.1. *If $\alpha<2$ and F is given by (3.121) there is a unique solution $q_0=q_0(\alpha,\kappa,h)$, $\rho_0=\rho_0(\alpha,\kappa,h)$ to the equations*

$$\frac{\partial F}{\partial q}(q_0,\rho_0)=0=\frac{\partial F}{\partial \rho}(q_0,\rho_0)\,. \tag{3.122}$$

We define

$$\begin{aligned}\mathrm{RS}_0(\alpha)&=\alpha\mathsf{E}\log\mathcal{N}\left(\frac{\tau-z\sqrt{q_0}}{\sqrt{\rho_0-q_0}}\right)\\&+\frac{1}{2}\frac{q_0}{\rho_0-q_0}+\frac{1}{2}\log(\rho_0-q_0)-\kappa\rho_0+\frac{h^2}{2}(\rho_0-q_0)\,.\end{aligned} \tag{3.123}$$

The reader will recognize that this is $F(q_0,\rho_0)$, where F is defined in (3.121). The value of κ and h will be kept implicit. (We recall that τ has been fixed once and for all.) The main result of this chapter is as follows.

Theorem 3.3.2. *Consider $\alpha_0<2$, $0<\kappa_0<\kappa_1$, $h_0>0$, $\varepsilon>0$. Then we can find $\varepsilon'>0$ with the following property. Consider any concave function $u\leq 0$, with the following properties:*

$$x\geq\tau\ \Rightarrow\ u(x)=0 \tag{3.124}$$

$$\exp u(\tau-\varepsilon')\leq\varepsilon' \tag{3.125}$$

u is four times differentiable and $|u^{(\ell)}|$ is bounded for $1\leq\ell\leq 4$. (3.126)

Then for N large enough, and if $H_{N,M}$ denotes the Hamiltonian (3.1) we have

$$\left|\mathsf{E}\frac{1}{N}\log\int\exp(-H_{N,M}(\boldsymbol{\sigma}))\mathrm{d}\boldsymbol{\sigma}-\left(\mathrm{RS}_0\left(\frac{M}{N}\right)+\frac{1}{2}\log(2e\pi)\right)\right|\leq\varepsilon \tag{3.127}$$

whenever $\kappa_0\leq\kappa\leq\kappa_1$, $h\leq h_0$, $M/N\leq\alpha_0$.

In particular, we succeed in computing

$$\lim_{u\to\mathbf{1}_{\{x\geq\tau\}}}\ \lim_{N\to\infty,M/N\to\alpha}\mathsf{E}\frac{1}{N}\log\int\exp(-H_{N,M}(\boldsymbol{\sigma}))\mathrm{d}\boldsymbol{\sigma}\,. \tag{3.128}$$

In Volume II we will prove the very interesting fact that the limits can be interchanged, solving the problem of computing the "part of the sphere $\mathbb{S}_N$ that belongs to the intersection of M random half-spaces".

Besides Theorem 3.3.1, the proof of Theorem 3.3.2 requires the following.

Proposition 3.3.3. *Consider $\kappa_0 > 0$, $\alpha_0 < 2$ and $h_0 > 0$. Then we can find a number C, depending only on κ_0, α_0 and h_0, such that if $\kappa \geq \kappa_0$, $\alpha \leq \alpha_0$ and $h \leq h_0$, then for any concave function $u \leq 0$ that satisfies (3.124), and whenever q and ρ satisfy the system of equations (3.69), (3.76) and (3.104), we have*

$$q, \rho \leq C \; ; \; \frac{1}{\rho - q} \leq C \, . \tag{3.129}$$

We recall that the numbers q_0 and ρ_0 are given by (3.122).

Corollary 3.3.4. *Given $0 < \kappa_0 < \kappa_1$, $\alpha_0 < 2$, $h_0 > 0$ and $\varepsilon > 0$, we can find a number $\varepsilon' > 0$ such that whenever the concave function $u \leq 0$ satisfy (3.124) and (3.125), whenever $\kappa_0 \leq \kappa \leq \kappa_1$, $h \leq h_0$ and $\alpha \leq \alpha_0$, given any numbers $0 \leq q \leq \rho$ that satisfy the equations (3.69), (3.76) and (3.104) we have*

$$|q - q_0| \leq \varepsilon \; ; \; |\rho - \rho_0| \leq \varepsilon \, .$$

It is here that Theorem 3.3.1 is really needed. Without it, it seems very difficult to control q and ρ.

Proof. This is a simple compactness argument now that we know (3.129). We simply sketch the proof of this "soft" argument. Assume for contradiction that we can find a sequence $\varepsilon'_n \to 0$, a sequence u_n of functions that satisfies (3.124) and (3.125) for ε'_n rather than ε', numbers $\kappa_0 \leq \kappa_n \leq \kappa_1$, $h_n \leq h_0$, $\alpha_n \leq \alpha_0$, numbers q_n and ρ_n that satisfy the corresponding equations (3.69), (3.76) and (3.104), and are such that $|q_n - q_0| \geq \varepsilon$ and $|\rho_n - \rho_0| \geq \varepsilon$. By Proposition 3.3.3 we have $q_n, \rho_n, 1/(q_n - \rho_n) \leq C$. This boudedness permits us to take converging subsequences. So, without loss of generality we can assume that the sequences κ_n, h_n, α_n, q_n and ρ_n have limits called κ, h, α, q and ρ respectively. Moreover $1/(q - \rho) < C$, so in particular $\rho < q$. Finally we have $|q - q_0| \geq \varepsilon$ and $|\rho - \rho_0| \geq \varepsilon$. If one writes explicitly the equations (3.122), it is obvious from the fact that (q_n, ρ_n) is a solution to the equations (3.69), (3.76) and (3.104) (for κ_n and h_n rather than for κ and h) that (q, ρ) is a solution to these equations. But this is absurd, since by Theorem 3.3.1 one must then have $q = q_0$ and $\rho = \rho_0$. □

Once this has been obtained the proof of Theorem 3.3.2 is easy following the approach of the second proof of Theorem 2.4.2, so we complete it first.

We recall the bracket $\langle \cdot \rangle_{t,\sim}$ associated with the Hamiltonian (3.88). To lighten notation we write $\langle \cdot \rangle_\sim$ rather than $\langle \cdot \rangle_{1,\sim}$.

Lemma 3.3.5. *Assume that the function u satisfies (3.7). Writing $g_i = g_{i,M}$, we have*

$$\begin{aligned} &\mathsf{E} \log \left\langle \exp u \left(\frac{1}{\sqrt{N}} \sum_{i \leq N} g_i \sigma_i \right) \right\rangle_\sim \\ &= \mathsf{E} \log \mathsf{E}_\xi \exp u(z\sqrt{q_{N,M}} + \xi \sqrt{\rho_{N,M} - q_{N,M}}) + \mathcal{R} \, , \end{aligned}$$

where

$$|\mathcal{R}| \leq \frac{K(\kappa_0, h_0, D)}{\sqrt{N}} .$$

Now that we have proved Theorem 3.1.18 this is simply an occurrence of the general principle explained in Section 1.5. We compare a quantity of the type (1.140) with the corresponding quantity (1.141) when $f(x) = \exp u(x)$ and $w(x) = \log x$, when μ is Gibbs' measure.

Proof. We consider

$$S_v = \sqrt{\frac{v}{N}} \sum_{i \leq N} g_i \sigma_i + \sqrt{1-v}(z\sqrt{q_{N,M}} + \xi\sqrt{\rho_{N,M} - q_{N,M}})$$

and

$$\varphi(v) = \mathsf{E} \log \mathsf{E}_\xi \langle \exp u(S_v) \rangle_\sim .$$

We differentiate and integrate by parts to obtain:

$$\begin{aligned} |\varphi'(v)| &\leq K(D)\mathsf{E} \frac{\langle (|R_{1,1} - \rho_{N,M}| + |R_{1,2} - q_{N,M}|) \exp(u(S_v^1) + u(S_v^2)) \rangle_\sim}{(\mathsf{E}_\xi \langle \exp u(S_v) \rangle_\sim)^2} \\ &= K(D)\nu_{1,v}(|R_{1,1} - \rho_{N,M}| + |R_{1,2} - q_{N,M}|) . \end{aligned}$$

We use (3.106) with $t = 1$ to get

$$|\varphi'(v)| \leq K\nu(|R_{1,1} - \rho_{N,M}| + |R_{1,2} - q_{N,M}|) + K \exp(-N)$$

and we conclude with Theorem 3.1.18. □

Lemma 3.3.6. *We have*

$$\frac{\mathrm{dRS}_0}{\mathrm{d}\alpha}(\alpha) = \mathsf{E} \log \mathcal{N}\left(\frac{\tau - z\sqrt{q_0}}{\sqrt{\rho_0 - q_0}}\right) = \mathsf{E} \log \mathsf{P}_\xi(z\sqrt{q_0} + \xi\sqrt{\rho_0 - q_0} \geq \tau) . \tag{3.130}$$

Proof. Obvious by (3.122). □

Proof of Theorem 3.3.2. Let us write

$$p_{N,M} = \mathsf{E} \frac{1}{N} \log \int \exp(-H_{N,M}(\boldsymbol{\sigma}))\mathrm{d}\boldsymbol{\sigma} \tag{3.131}$$

and let us first consider the case $M = 0$. In that case

$$\int \exp\left(-\kappa\|\boldsymbol{\sigma}\|^2 + h \sum_{i \leq N} g_i \sigma_i\right) \mathrm{d}\boldsymbol{\sigma} = \left(\frac{\pi}{\kappa}\right)^{N/2} \exp \frac{h^2}{4\kappa} \sum_{i \leq N} g_i^2$$

and

$$p_{N,M} = \frac{1}{2} \log\left(\frac{\pi}{\kappa}\right) + \frac{h^2}{4\kappa} .$$

When $\alpha = 0$, we have $r = \overline{r} = 0$, so (3.76) and (3.77) yield $\rho_0 - q_0 = 1/(2\kappa)$, $q_0 = h^2/(4\kappa^2)$, and thus by straight forward algebra,

$$\mathrm{RS}_0(0) = \frac{1}{2}\log\left(\frac{1}{2\kappa}\right) - \frac{1}{2} + \frac{h^2}{4\kappa} ,$$

so that (3.127) holds in that case. Next, we observe that

$$p_{N,M} - p_{N,M-1} = \frac{1}{N}\mathsf{E}\log\left\langle \exp u\left(\frac{1}{\sqrt{N}}\sum g_{i,M}\sigma_i\right)\right\rangle_\sim .$$

Informally, the rest of the proof goes as follows. By Lemma 3.3.5 we have

$$\begin{aligned}&\mathsf{E}\log\left\langle \exp u\left(\frac{1}{\sqrt{N}}\sum g_{i,M}\sigma_i\right)\right\rangle_\sim \\ &\simeq \mathsf{E}\log \mathsf{E}_\xi \exp u(z\sqrt{q_{N,M}} + \xi\sqrt{\rho_{N,M} - q_{N,M}}) .\end{aligned}$$

Now by Propositions 3.2.4 and 3.2.12 the numbers $q_{N,M}$ and $\rho_{N,M}$ are near solutions of the system of equations (3.69), (3.76) and (3.104).

As $N \to \infty$, (u being fixed) these quantities become close (uniformly on α) to a true solution of these equations. Thus, by Corollary 3.3.4, and provided u satisfies (3.124) and (3.125) and ε' is small enough, we have $q_{N,M} \simeq q_0$ $(= q_0(M/N,\kappa,h))$ and $\rho_{N,M} \simeq \rho_0$ $(= \rho_0(M/N,\kappa,h))$ and thus

$$\begin{aligned}&\mathsf{E}\log \mathsf{E}_\xi \exp u(z\sqrt{q_{N,M}} + \xi\sqrt{\rho_{N,M} - q_{N,M}}) \\ &\simeq \mathsf{E}\log \mathsf{E}_\xi u(z\sqrt{q_0} + \xi\sqrt{\rho_0 - q_0}) \simeq \mathsf{E}\log \mathsf{P}_\xi(z\sqrt{q_0} + \xi\sqrt{\rho_0 - q_0} \ge \tau) ,\end{aligned}$$

using again (3.124) and (3.125). Now (3.130) implies

$$\begin{aligned}\frac{1}{N}\mathsf{E}\log \mathsf{P}_\xi(z\sqrt{q_0} + \xi\sqrt{\rho_0 - q_0} \ge \tau) &\simeq \int_{(M-1)/N}^{M/N} \frac{\mathrm{d}}{\mathrm{d}\alpha}\mathrm{RS}_0(\alpha)\mathrm{d}\alpha \\ &= \mathrm{RS}_0\left(\frac{M}{N},\kappa,h\right) - \mathrm{RS}_0\left(\frac{M-1}{N},\kappa,h\right) .\end{aligned}$$

This chain of approximations yields

$$p_{N,M} - p_{N,M-1} \simeq \mathrm{RS}_0\left(\frac{M}{N},\kappa,h\right) - \mathrm{RS}_0\left(\frac{M-1}{N},\kappa,h\right) ,$$

where $\simeq$ means with error $\le \varepsilon/N$. Summation over M of these relations together with the case $M = 0$ yields the desired result.

It is straightforward to write an "ε-δ proof" following the previous scheme, so there seems to be no point in doing it here. □

Our next goal is the proof of Proposition 3.3.3, that will reveal how the initial condition $\alpha_0 < 2$ comes into play. Preparing for this proof, we consider the function

$$\mathcal{A}(x) = -\frac{\mathrm{d}}{\mathrm{d}x}\log \mathcal{N}(x) = \frac{1}{\sqrt{2\pi}}\frac{e^{-x^2/2}}{\mathcal{N}(x)} , \tag{3.132}$$

about which we collect simple facts.

Lemma 3.3.7. *We have*

$$\mathcal{A}(v) \geq v \tag{3.133}$$

$$\mathcal{A}'(v) = \mathcal{A}(v)^2 - v\mathcal{A}(v) \geq 0 \tag{3.134}$$

$$v\mathcal{A}(v)\mathcal{A}'(v) \leq \mathcal{A}(v)^2 \tag{3.135}$$

$$v\mathcal{A}(v) \leq 1 + v^2\,. \tag{3.136}$$

Proof. To prove (3.133) we can assume $v \geq 0$. Then

$$v \int_v^\infty \exp\left(-\frac{t^2}{2}\right) \mathrm{d}t \leq \int_v^\infty t \exp\left(-\frac{t^2}{2}\right) \mathrm{d}t = \exp\left(-\frac{v^2}{2}\right)$$

which is (3.133). The equality in (3.134) is straightforward, and the inequality follows from (3.133) since $\mathcal{A}(v) \geq 0$.

Now (3.134) implies

$$v\mathcal{A}(v)\mathcal{A}'(v) = \mathcal{A}(v)^2(v\mathcal{A}(v) - v^2)\,,$$

so that (3.135) is equivalent to (3.136). Integrating by parts,

$$\int_v^\infty t^2 \exp\left(-\frac{t^2}{2}\right) \mathrm{d}t = v \exp\left(-\frac{v^2}{2}\right) + \sqrt{2\pi}\mathcal{N}(v)$$

$$\int_v^\infty t \exp\left(-\frac{t^2}{2}\right) \mathrm{d}t = \exp\left(-\frac{v^2}{2}\right)$$

so that expanding the square and using the previous equalities we get

$$0 \leq \int_v^\infty (t-v)^2 \exp\left(-\frac{t^2}{2}\right) \mathrm{d}t = (1+v^2)\sqrt{2\pi}\mathcal{N}(v) - v \exp\left(-\frac{v^2}{2}\right)\,.$$

This proves (3.136). □

let us observe that (3.133) and (3.136) mean that when $x \geq 0$

$$\frac{x}{1+x^2}\frac{1}{\sqrt{2\pi}} \exp\left(-\frac{x^2}{2}\right) \leq \mathsf{P}(\xi \geq x) \leq \frac{1}{x\sqrt{2\pi}} \exp\left(-\frac{x^2}{2}\right)\,, \tag{3.137}$$

which becomes quite accurate as $x \to \infty$.

Lemma 3.3.8. *Consider numbers $0 \leq q < \rho$ and a concave function $u \leq 0$ with $u(x) = 0$ for $x \geq \tau$. Consider independent standard Gaussian r.v.s z and ξ and set $\theta = z\sqrt{q} + \xi\sqrt{\rho - q}$ and $Y = (\tau - z\sqrt{q})/\sqrt{\rho - q}$. Then*

$$\left(\frac{\mathsf{E}_\xi\, \xi \exp u(\theta)}{\mathsf{E}_\xi \exp u(\theta)}\right)^2 \leq \begin{cases} Y^2 + L & \text{if } \; Y \geq 0 \\ L & \text{if } \; Y \leq 0\,, \end{cases} \tag{3.138}$$

where L is a universal constant.

Proof. We observe that, integrating by parts and since $u' \geq 0$,

$$\mathsf{E}_\xi\, \xi \exp u(\theta) = \mathsf{E}_\xi \sqrt{\rho - q}\, u'(\theta) \exp u(\theta) \geq 0\,. \tag{3.139}$$

Consider first the case where $Y \geq 0$. Let $U = \mathsf{E}_\xi(\mathbf{1}_{\{\xi<Y\}} \exp u(\theta))$. Then, since $u \leq 0$,

$$\begin{aligned}\mathsf{E}_\xi\, \xi \exp u(\theta) &= \mathsf{E}_\xi(\xi \mathbf{1}_{\{\xi<Y\}} \exp u(\theta)) + \mathsf{E}_\xi(\xi \mathbf{1}_{\{\xi\geq Y\}} \exp u(\theta)) \\ &\leq YU + \mathsf{E}_\xi(\xi \mathbf{1}_{\{\xi\geq Y\}}) = YU + \frac{1}{\sqrt{2\pi}} \exp\left(-\frac{Y^2}{2}\right). \end{aligned} \tag{3.140}$$

Since $u(\theta) = 0$ for $\xi \geq Y$, (3.137) implies

$$\mathsf{E}_\xi(\mathbf{1}_{\{\xi\geq Y\}} \exp u(\theta)) = \mathsf{P}_\xi(\xi \geq Y) \geq \frac{Y}{1+Y^2} \frac{1}{\sqrt{2\pi}} \exp\left(-\frac{Y^2}{2}\right),$$

and thus

$$\begin{aligned}\mathsf{E}_\xi \exp u(\theta) &= U + \mathsf{E}_\xi(\mathbf{1}_{\{\xi\geq Y\}} \exp u(\theta)) \\ &\geq U + \frac{Y}{1+Y^2} \frac{1}{\sqrt{2\pi}} \exp\left(-\frac{Y^2}{2}\right). \end{aligned} \tag{3.141}$$

Combining (3.140) and (3.141) we get

$$0 \leq \frac{\mathsf{E}_\xi\, \xi \exp u(\theta)}{\mathsf{E}_\xi \exp u(\theta)} \leq \frac{\sqrt{2\pi} YU + \exp\left(-\frac{Y^2}{2}\right)}{\sqrt{2\pi} U + \frac{Y}{1+Y^2} \exp\left(-\frac{Y^2}{2}\right)}\,. \tag{3.142}$$

It is elementary that for numbers $a, b > 0$ we have

$$\frac{aY + b}{a + \frac{Y}{1+Y^2} b} \leq Y + \frac{1}{Y}\,.$$

Combining with (3.142) yields

$$0 \leq \frac{\mathsf{E}_\xi\, \xi \exp u(\theta)}{\mathsf{E}_\xi \exp u(\theta)} \leq Y + \frac{1}{Y}\,.$$

Taking squares proves (3.138) when $Y \geq 1$. When $Y \leq 1$ (and Y is not necessarily ≥ 0) since $u \leq 0$ we have

$$0 \leq \mathsf{E}_\xi\, \xi \exp u(\theta) \leq \mathsf{E}|\xi| = \sqrt{\frac{2}{\pi}}$$

and, since $u(x) = 0$ for $x \geq \tau$,

$$\mathsf{E}_\xi \exp u(\theta) \geq \mathsf{P}_\xi(\xi \geq Y) \geq \mathsf{P}_\xi(\xi \geq 1)$$

and this finishes the proof. □

We bring forward the following trivial fact which seems to be at the root of the condition "$\alpha \leq 2$".

Lemma 3.3.9. *If z is a standard Gaussian r.v.,*

$$\lim_{b\to 0} \mathsf{E}((z-b)^2 \mathbf{1}_{\{z\le b\}}) = \frac{1}{2} \, . \tag{3.143}$$

The following is also straightforward, where we recall (3.67) and (3.68).

Lemma 3.3.10. *Given $\rho \ge q$, consider the function*

$$v(x) = \log \mathsf{E}_\xi \exp u(x + \xi\sqrt{\rho - q}) \, .$$

Then, recalling the definitions (3.67) and (3.68) of Ψ and $\overline{\Psi}$, we have $v' \ge 0$ and

$$\Psi(\rho, q) = \alpha \mathsf{E} v'(z\sqrt{q})^2 \; ; \; \overline{\Psi}(\rho, q) - \Psi(\rho, q) = \alpha \mathsf{E} v''(z\sqrt{q}) \, . \tag{3.144}$$

Proof of Proposition 3.3.3. From (3.76) and (3.77) we have

$$\frac{1}{\rho - q} = 2\kappa + r - \overline{r} = 2\kappa - \alpha \mathsf{E} v''(z\sqrt{q}) \, , \tag{3.145}$$

where the last equality uses that $\overline{r} - r = \overline{\Psi}(\rho, q) - \Psi(\rho, q)$ and (3.144). Integration by parts yields

$$-\mathsf{E} v''(z\sqrt{q}) = -\frac{1}{\sqrt{q}} \mathsf{E} z v'(z\sqrt{q}) \, .$$

A direct computation proves that $v' \ge 0$ since $u' \ge 0$. Hence, using the Cauchy-Schwarz inequality in the second line,

$$\begin{aligned} -\mathsf{E} z v'(z\sqrt{q}) &\le -\mathsf{E} z \mathbf{1}_{\{z\le 0\}} v'(z\sqrt{q}) \\ &\le (\mathsf{E} z^2 \mathbf{1}_{\{z\le 0\}})^{1/2} (\mathsf{E} v'(z\sqrt{q})^2)^{1/2} \\ &= \frac{1}{\sqrt{2}} (\mathsf{E} v'(z\sqrt{q})^2)^{1/2} \, . \end{aligned}$$

Thus, combining the previous relations we obtain

$$\frac{1}{\rho - q} \le 2\kappa + \frac{\alpha}{\sqrt{2}\sqrt{q}} (\mathsf{E} v'(z\sqrt{q})^2)^{1/2} \, . \tag{3.146}$$

On the other hand, (3.144) implies

$$r = \Psi(\rho, q) = \alpha \mathsf{E} v'(z\sqrt{q})^2$$

and from (3.75) and the first part of (3.145) we deduce

$$q = \frac{r + h^2}{(2\kappa + r - \overline{r})^2} \ge \frac{r}{(2\kappa + r - \overline{r})^2} = (\rho - q)^2 r = \alpha (\rho - q)^2 \mathsf{E} v'(z\sqrt{q})^2 \, .$$

This inequality can be rewritten as

$$\frac{\alpha}{\sqrt{2}\sqrt{q}}(\mathsf{E}v'(z\sqrt{q})^2)^{1/2} \leq \sqrt{\frac{\alpha}{2}}\frac{1}{\rho-q}\,,$$

and combining with (3.146) yields

$$\frac{1}{\rho-q} \leq 2\kappa + \sqrt{\frac{\alpha}{2}}\frac{1}{\rho-q}\,,$$

so that $(\rho-q)^{-1} \leq 2\kappa_0(1-\sqrt{\alpha_0/2})^{-1}$. The exact form of the right-hand side is not relevant, but this shows that $1/(\rho-q)$ is bounded by a number depending only on α_0 and κ_0.

We now try to bound similarly ρ and q. Since $\overline{r} \leq r$, we have $\rho - q \leq (2\kappa)^{-1} \leq (2\kappa_0)^{-1}$, so the issue is to bound q. Using (3.104) and (3.145) again, and since $r \geq \overline{r}$,

$$q = \frac{r+h^2}{(2\kappa + r - \overline{r})^2} \leq \frac{h_0^2}{(2\kappa_0)^2} + r(\rho-q)^2\,. \tag{3.147}$$

Recalling (3.67) and using Lemma 3.3.8 we get

$$\begin{aligned} r = \Psi(\rho,q) &= \frac{\alpha}{\rho-q}\mathsf{E}\left(\frac{\mathsf{E}_\xi\,\xi\exp u(\theta)}{\mathsf{E}_\xi\exp u(\theta)}\right)^2 \\ &< \frac{\alpha}{\rho-q}(L+\mathsf{E}(Y^2\mathbf{1}_{\{Y\geq 0\}})) \end{aligned} \tag{3.148}$$

where

$$Y = \frac{\tau - z\sqrt{q}}{\sqrt{\rho-q}} = -\sqrt{\frac{q}{\rho-q}}\Big(z-\frac{\tau}{\sqrt{q}}\Big)$$

so that Y satisfies

$$\mathsf{E}(Y^2\mathbf{1}_{\{Y\geq 0\}}) = \frac{q}{\rho-q}\mathsf{E}\left(\Big(z-\frac{\tau}{\sqrt{q}}\Big)^2\mathbf{1}_{\{z\leq\tau/\sqrt{q}\}}\right)\,.$$

Since $\alpha_0 < 2$ we can find $a > 1/2$ with $a\alpha_0 < 1$. Then by (3.143) there is a number $q(\tau,a)$ satisfying

$$q \geq q(\tau,a) \Rightarrow \mathsf{E}\left(\Big(z-\frac{\tau}{\sqrt{q}}\Big)^2\mathbf{1}_{\{z\leq\tau/\sqrt{q}\}}\right) < a \Rightarrow \mathsf{E}(Y^2\mathbf{1}_{\{Y\geq 0\}}) \leq \frac{aq}{\rho-q}\,.$$

Thus, using (3.148) we get, using also that $\rho - q \leq 1/(2\kappa_0)$ in the second inequality,

$$q \geq q(\tau,a) \Rightarrow r(\rho-q)^2 \leq \alpha L(\rho-q) + a\alpha q \leq \frac{\alpha L}{2\kappa_0} + a\alpha q$$

and combining with (3.147) yields

$$q \geq q(\tau, a) \Rightarrow q \leq \frac{h_0^2}{(2\kappa_0)^2} + \frac{\alpha L}{2\kappa_0} + a\alpha q \ ,$$

so that

$$q \geq q(\tau, a) \Rightarrow (1 - a\alpha)q \leq \frac{h_0^2}{(2\kappa_0)^2} + \frac{\alpha L}{2\kappa_0} \ .$$

Since $a\alpha \leq a\alpha_0 < 1$, this proves that q (and hence ρ) is bounded by a number depending only on h_0, κ_0 and α_0. □

It remains to prove Theorem 3.3.1. The proof is unrelated to the methods of this work. While it is not difficult to follow line by line, the author cannot really explain why it works (or how Shcherbina and Tirozzi could ever find it). The need for a more general and enlightening approach is rather keen here.

We make the change of variable $x = q/(\rho - q)$, so that $q = x\rho/(1+x)$, $\rho - q = \rho/(1+x)$, and

$$F(q,\rho) = G(x,\rho) := \alpha \, \mathsf{E} \log \mathcal{N}\left(\frac{\tau\sqrt{1+x}}{\sqrt{\rho}} - z\sqrt{x}\right) + \frac{x}{2} + \frac{1}{2}\log\rho - \frac{1}{2}\log(1+x) - \kappa\rho + \frac{h^2\rho}{2(1+x)} \ . \tag{3.149}$$

Proposition 3.3.11. *For $x > 0$ and $\rho > 0$ we have*

$$\frac{\partial^2 G}{\partial \rho^2} < 0 \ ; \quad \frac{\partial}{\partial x}\left(\frac{x+1}{x}\frac{\partial G}{\partial x}\right) > 0 \ . \tag{3.150}$$

Corollary 3.3.12. *a) Given $\rho > 0$ there exists at most one value x_1 such that $(\partial G/\partial x)(x_1, \rho) = 0$. If such a value exists, the function $x \mapsto G(q,\rho)$ attains its minimum at x_1.*

b) Given $\rho > 0$ there exists at most one value q_1 such that $(\partial F/\partial q)(q_1,\rho) = 0$. If such a value exists, the function $q \mapsto F(q,\rho)$ attains its minimum at q_1.

Proof. a) By the second part of (3.150) we have $\partial G(x,\rho)/\partial x < 0$ for $x < x_1$ while $\partial G(x,\rho)/\partial x > 0$ for $x > x_1$.

b) Follows from a) since at given ρ the change of variable $x = q/(\rho - q)$ is monotonic. □

Proof of Theorem 3.3.1. Suppose that we have $\partial G/\partial x = 0$ and $\partial G/\partial \rho = 0$ at the points (x_1, ρ_1) and (x_2, ρ_2). Then, since $\partial^2 G/\partial \rho^2 < 0$, we have $G(x_2, \rho_1) < G(x_2, \rho_2)$ unless $\rho_2 = \rho_1$. By the first part of Corollary 3.3.12 used for $\rho = \rho_1$ we have $G(x_1, \rho_1) < G(x_2, \rho_1)$ unless $x_1 = x_2$. So $G(x_1, \rho_1) < G(x_2, \rho_2)$ unless $(x_1, \rho_1) = (x_2, \rho_2)$. Reversing the argument shows that $(x_1, \rho_1) = (x_2, \rho_2)$. □

We write $W = \tau\sqrt{1+x}/\sqrt{\rho} - z\sqrt{x}$.

Lemma 3.3.13. *Recalling the definition (3.132) of the function $\mathcal{A}(x)$, we have*

$$2\frac{\partial G}{\partial \rho} = \alpha\tau\sqrt{1+x}\rho^{-3/2}\mathsf{E}\,\mathcal{A}(W) + \frac{1}{\rho} - 2\kappa + \frac{h^2}{1+x} \tag{3.151}$$

and

$$2\left(\frac{x+1}{x}\right)\frac{\partial G}{\partial x} = -\frac{\alpha}{x}\mathsf{E}\mathcal{A}^2(W) + 1 - \frac{h^2\rho}{x(1+x)}\,. \tag{3.152}$$

Proof. We differentiate (3.149) in ρ to obtain (3.151). To prove (3.152) we differentiate (3.149) in x to obtain

$$\begin{aligned} 2\frac{\partial G}{\partial x} &= -\alpha\mathsf{E}\Big(\Big(-\frac{z}{\sqrt{x}} + \frac{\tau}{\sqrt{1+x}\sqrt{\rho}}\Big)\mathcal{A}(W)\Big) \\ &\quad + 1 - \frac{1}{1+x} - \frac{h^2\rho}{(1+x)^2}\,. \end{aligned} \tag{3.153}$$

Now, by integration by parts and (3.134)

$$\begin{aligned} \mathsf{E}\left(\frac{z}{\sqrt{x}}\mathcal{A}(W)\right) &= -\mathsf{E}\left(\mathcal{A}'(W)\right) = \mathsf{E}\left(W\mathcal{A}(W)\right) - \mathsf{E}\,\mathcal{A}(W)^2 \\ &= \mathsf{E}\left(\left(\frac{\tau\sqrt{1+x}}{\sqrt{\rho}} - z\sqrt{x}\right)\mathcal{A}(W)\right) - \mathsf{E}\,\mathcal{A}(W)^2 \end{aligned}$$

and thus

$$(1+x)\,\mathsf{E}\left(\frac{z}{\sqrt{x}}\mathcal{A}(W)\right) = \frac{\tau\sqrt{1+x}}{\sqrt{\rho}}\mathsf{E}\,\mathcal{A}(W) - \mathsf{E}\,\mathcal{A}(W)^2\,. \tag{3.154}$$

Plugging back this value in (3.153) yields (3.152). □

Proof of Proposition 3.3.11. We differentiate (3.151) in ρ to obtain

$$2\frac{\partial^2 G}{\partial \rho^2} = -\frac{3}{2}\frac{\alpha\tau\sqrt{1+x}}{\rho^{5/2}}\mathsf{E}\,\mathcal{A}(W) - \alpha\left(\frac{\tau\sqrt{1+x}}{\rho^{3/2}}\right)^2\mathsf{E}\mathcal{A}'(W) - \frac{1}{\rho^2}$$

and this is ≤ 0 since $\mathcal{A}' \geq 0$. We differentiate (3.152) in x to get

$$\begin{aligned} 2\frac{\partial}{\partial x}\left(\frac{1+x}{x}\frac{\partial G}{\partial x}\right) &= -\frac{\alpha}{x}\mathsf{E}\Big(\Big(\frac{\tau}{\sqrt{1+x}\sqrt{\rho}} - \frac{z}{\sqrt{x}}\Big)\mathcal{A}(W)\mathcal{A}'(W)\Big) \\ &\quad + \frac{\alpha}{x^2}\mathsf{E}\,\mathcal{A}(W)^2 + \frac{h^2\rho}{x^2} - \frac{h^2\rho}{(1+x)^2}\,. \end{aligned} \tag{3.155}$$

Now, we observe the identity

$$\frac{\tau}{\sqrt{1+x}\sqrt{\rho}} - \frac{z}{\sqrt{x}} = \frac{W}{x} - \frac{\tau}{x\sqrt{1+x}\sqrt{\rho}}\,,$$

so that (3.155) yields

$$2\frac{\partial}{\partial x}\left(\frac{1+x}{x}G\right) = \frac{\alpha\tau}{x^2\sqrt{1+x\sqrt{\rho}}}\mathsf{E}\left(\mathcal{A}(W)\mathcal{A}'(W)\right) \tag{3.156}$$
$$+\frac{\alpha}{x^2}\mathsf{E}\left(\mathcal{A}(W)^2 - W\mathcal{A}(W)\mathcal{A}'(W)\right) + h^2\rho\left(\frac{1}{x^2} - \frac{1}{(1+x)^2}\right)$$

and all the terms are ≥ 0 by (3.135) and since $x \geq 0$. □

3.4 Notes and Comments

The results of this chapter are essentially proved in [133] and [134]. The way Shcherbina and Tirozzi [133] obtain the replica-symmetric equations seems genuinely different from what I do. It would be nice to make explicit the rationale behind their approach, but as I am allergic to the style in which their papers are written I find it much easier to discover my own proofs than to decipher their work.

Instead of basing the approach on Theorem 3.1.4 one can also use the Brascamp-Lieb inequalities [37]. In dimension 1, the inequality is stated in (3.158) below. In more dimensions, it is convenient to use a simplified form of these inequalities as follows. Consider a measure μ on $\mathbb{R}^N$ as in Theorem 3.1.4. Consider a function f on $\mathbb{R}^N$ (that need not be Lipschitz). Then, if $\boldsymbol{\nabla} f$ denotes the gradient of f,

$$\int f^2\,\mathrm{d}\mu - \left(\int f \mathrm{d}\mu\right)^2 \leq \frac{1}{\kappa}\int \|\boldsymbol{\nabla} f\|^2 \mathrm{d}\mu\,. \tag{3.157}$$

This inequality can be iterated. If f has a Lipschitz constant A as in (3.15) then $\|\boldsymbol{\nabla} f\| \leq A$. Using (3.157) for $\exp(\lambda f/2)$ rather than f yields

$$\int \exp\lambda f\,\mathrm{d}\mu - \left(\int \exp\left(\frac{\lambda f}{2}\right)\mathrm{d}\mu\right)^2 \leq \frac{\lambda^2 A^2}{\kappa}\int \exp\lambda f\,\mathrm{d}\mu$$

so that if $\lambda^2 A^2 \leq \kappa$,

$$\int \exp\lambda f\,\mathrm{d}\mu \leq \frac{1}{1-\frac{\lambda^2 A^2}{\kappa}}\left(\int \exp\frac{\lambda f}{2}\,\mathrm{d}\mu\right)^2 .$$

By iteration we get

$$\int \exp\lambda f\,\mathrm{d}\mu \leq \prod_{0\leq \ell<k}\left(\frac{1}{1-\frac{\lambda^2 A^2}{\kappa 2^{2\ell}}}\right)^{2^\ell}\left(\int \exp\left(\frac{\lambda f}{2^k}\right)\mathrm{d}\mu\right)^{2^k}$$

so that when $\lambda^2 A^2 \leq \kappa/2$ this implies

$$\int \exp \lambda f \, d\mu \le L \left(\int \exp \left(\frac{\lambda f}{2^k} \right) d\mu \right)^{2^k} .$$

Now the inequality $|e^x - x - 1| \le x^2 e^x$ shows that if $\int f \, d\mu = 0$, the right-hand side goes to L as $k \to \infty$, so that if f has a Lipschitz constant A, we have

$$\int \exp \lambda \left(f - \int f \, d\mu \right) d\mu \le L$$

whenever $|\lambda| = \sqrt{\kappa}/2A$, a result that is not far from (3.16), at least for our purposes here. The reader should consult [11] for more on the relations between these different inequalities.

Inequality (3.157) follows from a nice general result, namely that in dimension 1,

$$\int f^2 \, d\mu - \left(\int f \, d\mu \right)^2 \le \int \frac{f'^2}{H''} \, d\mu . \tag{3.158}$$

(The Brascamp-Lieb inequality in dimension 1.) For $f(x) = x$, $-H(x) = u(x) - x^2/2$ this implies (3.72); the proof we give is just a simplified proof of (3.158) in the special case we need.

4. The Hopfield Model

4.1 Introduction: The Curie-Weiss Model

We go back to the case where the spins take values in $\{-1,1\}$. The Curie-Weiss model is the "canonical" model for mean-field (deterministic) ferromagnetic interaction, i.e. interaction where the spins tend to align with each other. The simplest Hamiltonian that will achieve this will contain a term $\sigma_i\sigma_j$ for each pair of spins, so it will be (proportional to) $\sum_{i<j}\sigma_i\sigma_j$. Equivalently, we consider the Hamiltonian

$$-H_N(\boldsymbol{\sigma}) = \frac{\beta N}{2}\left(\frac{1}{N}\sum_{i\le N}\sigma_i\right)^2 = \frac{\beta}{2N}\left(\sum_{i\le N}\sigma_i\right)^2 . \tag{4.1}$$

This is a simple, almost trivial model, that can be studied in considerable detail (see [54]). It is not our purpose to do this, but we will explain the basic facts that are relevant to this chapter. The partition function is given by

$$Z_N(\beta) = \sum_{\boldsymbol{\sigma}} \exp(-H_N(\boldsymbol{\sigma})) = \sum_{|k|\le N} A_k \exp\left(\frac{\beta}{2N}k^2\right) , \tag{4.2}$$

where

$$A_k = \operatorname{card}\left\{\boldsymbol{\sigma}\in\Sigma_N \; ; \; \sum_{i\le N}\sigma_i = k\right\} .$$

Consider the function

$$\mathcal{I}(t) = \frac{1}{2}\big((1+t)\log(1+t) + (1-t)\log(1-t)\big) , \tag{4.3}$$

which is defined for $-1 < t < 1$, and can be defined for $-1 \le t \le 1$ by setting $\mathcal{I}(-1) = \mathcal{I}(1) = \log 2$. We recall (see (A.29)) that

$$A_k \le 2^N \exp\left(-N\mathcal{I}\left(\frac{k}{N}\right)\right) , \tag{4.4}$$

so by (4.2) we have, bounding the sum in the right-hand side by the number of terms (i.e. $2N+1$) times the largest term,

M. Talagrand, *Mean Field Models for Spin Glasses*, Ergebnisse der Mathematik und ihrer Grenzgebiete. 3. Folge / A Series of Modern Surveys in Mathematics 54, DOI 10.1007/978-3-642-15202-3_4, © Springer-Verlag Berlin Heidelberg 2011

$$Z_N(\beta) \le (2N+1)2^N \exp\left(N \max_t \left(\frac{\beta t^2}{2} - \mathcal{I}(t)\right)\right) . \tag{4.5}$$

Also, by (A.30), when $N+k$ is even we have

$$A_k \ge \frac{2^N}{L\sqrt{N}} \exp\left(-N\mathcal{I}\left(\frac{k}{N}\right)\right) , \tag{4.6}$$

and thus

$$Z_N(\beta) \ge \max_{k+N \text{ even}} \frac{2^N}{L\sqrt{N}} \exp\left(-N\mathcal{I}\left(\frac{k}{N}\right)\right) \exp\left(\frac{\beta}{2N}k^2\right) .$$

Finally we get

$$\frac{1}{N} \log Z_N(\beta) = \log 2 + \max_{t \in [-1,1]} \left(\frac{\beta t^2}{2} - \mathcal{I}(t)\right) + o(1) , \tag{4.7}$$

where $o(1)$ is a quantity such that $o(1) \to 0$ as $N \to \infty$. The function $\beta t^2/2 - \mathcal{I}(t)$ attains its maximum at a point t such that

$$\beta t = \mathcal{I}'(t) = \frac{1}{2} \log\left(\frac{1+t}{1-t}\right) \tag{4.8}$$

or, equivalently,

$$\frac{1+t}{1-t} = \exp(2\beta t) ,$$

i.e.

$$t = \frac{\exp(2\beta t) - 1}{\exp(2\beta t) + 1} = \text{th}\beta t . \tag{4.9}$$

If $\beta \le 1$, the only root of (4.9) is $t = 0$. For $\beta > 1$, there is a unique root $m^* > 0$. That is, $m^* = m^*(\beta)$ satisfies

$$\text{th}\beta m^* = m^* . \tag{4.10}$$

Since $\text{th}x = x - x^3/3 + x^3 o(x)$, where $o(x) \to 0$ as $x \to 0$, (4.10) implies

$$\beta m^* - \frac{\beta^3 m^{*3}}{3} + \beta^3 m^{*3} o(\beta m^*) = m^* ,$$

so that

$$m^*(\beta) \sim \sqrt{3(\beta - 1)} \quad \text{as} \quad \beta \to 1_+ . \tag{4.11}$$

We define

$$b^* = \frac{\beta m^{*2}}{2} - \mathcal{I}(m^*) \tag{4.12}$$

so (4.7) reads

$$\frac{1}{N}\log Z_N(\beta) = \log 2 + b^* + o(1) . \tag{4.13}$$

When $\beta > 1$, as $N \to \infty$, Gibbs' measure is essentially supported by the set of configurations $\boldsymbol{\sigma}$ for which $N^{-1}\sum_{i\le N}\sigma_i \simeq \pm m^*$. This is because for a subset U of $\mathbb{R}$,

$$G_N\left(\left\{\boldsymbol{\sigma}\ ;\ \frac{1}{N}\sum_{i\le N}\sigma_i \in U\right\}\right) = Z_N^{-1}(\beta)\sum \exp(-H_N(\boldsymbol{\sigma})) ,$$

where the summation is over all sequences for which $N^{-1}\sum_{i\le N}\sigma_i \in U$. Thus, using (4.4), and bounding the sum in the second line by $(2N+1)$ times a bound for the largest term,

$$\begin{aligned} G_N\left(\left\{\boldsymbol{\sigma}; \frac{1}{N}\sum_{i\le N}\sigma_i \in U\right\}\right) &= \frac{1}{Z_N(\beta)}\sum_{k/N\in U} A_k \exp\frac{\beta N}{2}\left(\frac{k}{N}\right)^2 \\ &\le \frac{2^N}{Z_N(\beta)}(2N+1)\exp\left(N\sup_{t\in U}\left(\frac{\beta t^2}{2} - \mathcal{I}(t)\right)\right) . \end{aligned} \tag{4.14}$$

If we take

$$U = \{t\ ;\ |t \pm m^*| \ge \varepsilon\}$$

where ε is given (does not depend on N) then

$$\sup_{t\in U}\left(\frac{\beta t^2}{2} - \mathcal{I}(t)\right) < \max_{t\in[0,1]}\left(\frac{\beta t^2}{2} - \mathcal{I}(t)\right) ,$$

so (4.7) shows that the right-hand side of (4.14) goes to zero as $N \to \infty$.

Thus, (when weighted with Gibbs' measure), the set of configurations is made of two different pieces: the configurations for which $N^{-1}\sum_{i\le N}\sigma_i \simeq m^*$ and those for which $N^{-1}\sum_{i\le N}\sigma_i \simeq -m^*$. The global invariance of Gibbs' measure by the transformation $\boldsymbol{\sigma} \mapsto -\boldsymbol{\sigma}$ shows that these two pieces have the same weight. The system "spontaneously breaks down in two states".

This situation changes drastically if one adds an external field, i.e. one considers the Hamiltonian

$$-H_N(\boldsymbol{\sigma}) = \frac{\beta N}{2}\left(\frac{1}{N}\sum_{i\le N}\sigma_i\right)^2 + h\sum_{i\le N}\sigma_i , \tag{4.15}$$

where $h > 0$. To see where Gibbs' measure lies, one should now maximize

$$f(t) = \frac{\beta}{2}t^2 - \mathcal{I}(t) + th .$$

This maximum is attained at a point $0 < t < 1$ because $f(t) > f(-t)$ for $t > 0$; this point t must satisfy $\beta t + h = \mathcal{I}'(t)$, i.e.

$$t = \operatorname{th}(\beta t + h) , \tag{4.16}$$

and we will see that there is a unique positive root to this equation. The external field "breaks the symmetry between the two states".

Consider now a random sequence $(\eta_i)_{i\le N}$, $\eta_i = \pm 1$, and the random Hamiltonian

$$-H_N(\boldsymbol{\sigma}) = \frac{\beta}{2N}\Big(\sum_{i\le N} \eta_i \sigma_i\Big)^2 . \tag{4.17}$$

The randomness is not intrinsic, and can be removed by setting $\sigma'_i = \eta_i \sigma_i$. We can describe the Hamiltonian (4.17) by saying that it tends to align the spins $(\sigma_i)_{i\le N}$ with the sequence $(\eta_i)_{i\le N}$ or the sequence $(-\eta_i)_{i\le N}$ rather than with the sequences $(1,\dots,1)$ or $(-1,\dots,-1)$.

The situation is much more interesting if we put in the Hamiltonian several terms that "pull in different directions". Consider numbers $(\eta_{i,k})_{i\le N, k\le M}$, $\eta_{i,k} = \pm 1$, and the Hamiltonian

$$-H_{N,M}(\boldsymbol{\sigma}) = \frac{\beta N}{2}\sum_{k\le M}\Big(\frac{1}{N}\sum_{i\le N}\eta_{i,k}\sigma_i\Big)^2 . \tag{4.18}$$

We will always write

$$m_k = m_k(\boldsymbol{\sigma}) = \frac{1}{N}\sum_{i\le N}\eta_{i,k}\sigma_i . \tag{4.19}$$

When $\beta > 1$, the effect of the term $\beta N m_k^2/2$ of (4.18) is to tend to align the sequence $\boldsymbol{\sigma}$ with the sequence $(\eta_{i,k})_{i\le N}$ or the sequence $(-\eta_{i,k})_{i\le N}$. If the sequences $(\eta_{i,k})_{i\le N}$ are really different as k varies, this creates conflict. For this reason the case $\beta > 1$ seems the most interesting.

The Hopfield model is the system with random Hamiltonian (4.18), when the numbers $\eta_{i,k}$ are independent Bernoulli r.v.s, that is are such that $\mathsf{P}(\eta_{i,k} = \pm 1) = 1/2$. It simplifies notation to observe that, equivalently, one can assume

$$\begin{cases} \eta_{i,1} = 1 \quad \forall i \le N \,;\ \text{the numbers } (\eta_{i,k})_{i\le N, 2\le k\le M} \\ \text{are independent r.v.s with } \mathsf{P}(\eta_{i,k} = \pm 1) = 1/2 \,. \end{cases} \tag{4.20}$$

This assumption *is made throughout this chapter* and Chapter 10. The Hopfield model is already of interest if we fix M and let $N \to \infty$. We shall however focus on the more challenging case where $N \to \infty$, $M \to \infty$, $M/N \to \alpha, \alpha > 0$.

The Hopfield model (with Hamiltonian (4.18), that is, without external field) has a "high -temperature phase" somewhat similar to the phase $\beta < 1$, $h = 0$ of the SK model. This phase occurs in the region

$$\beta(1+\sqrt{\alpha}) < 1 \tag{4.21}$$

and it is quite interesting to see how this condition occurs. We will refer the reader to Section 2 of [142] for this, because this study does not use the cavity method and is somewhat distinct from the main theme we pursue here.

Another topic of interest is the "zero-temperature" problem, i.e. the study of the (random) function

$$\boldsymbol{\sigma} \mapsto -H_N(\boldsymbol{\sigma})$$

on Σ_N. We will not study this topic either because we feel that the current results in this direction are too far from their optimal form. We refer the reader to [112], [97], [56] for increasingly more sophisticated results.

Compared with the SK model, the Hopfield model brings two kinds of new features. One is the ferromagnetic interaction (4.1). For $\beta > 1$ and β close to one, this interaction creates difficulties that arise from the fact that the root of the equation $m^* = \mathrm{th}\beta m^*$ is "not so stable", in the sense that the slope of the tangent to the function $t \mapsto \mathrm{th}\beta t$ at $t = m^*$ gets close to 1 as $\beta \to 1$. This simple fact creates many of the technical difficulties inherent to the Hopfield model. Another difference between the Hopfield model and the SK model is that the nature of the disorder is not exactly the same.

It would be more pedagogical, before attacking the Hopfield model, to study a disordered system that presents the difficulties due to the ferromagnetic interaction, but with a familiar disorder. Such a model exists. It is the SK model with ferromagnetic interaction. The Hamiltonian is

$$-H_N(\boldsymbol{\sigma}) = \frac{\beta_1 N}{2}\Big(\frac{1}{N}\sum_{i\le N}\sigma_i\Big)^2 + \frac{\beta_2}{\sqrt{N}}\sum_{i<j} g_{ij}\sigma_i\sigma_j + h\sum_{i\le N}\sigma_i \,. \tag{4.22}$$

Space (and energy!) limitations do not allow the study of this model here.

Research Problem 4.1.1. (Level 1) Extend the results of Chapter 1 to the Hamiltonian (4.22).

What is really interesting is not to study this model for β_1, β_2 small, but, given β_1 (possibly large) to study the system for β_2 as large as possible. The "replica-symmetric" equations for this model are

$$\mu = \mathsf{E}\,\mathrm{th}(\beta_2 z\sqrt{q} + \beta_1\mu + h) \tag{4.23}$$

$$q = \mathsf{E}\,\mathrm{th}^2(\beta_2 z\sqrt{q} + \beta_1\mu + h) \,. \tag{4.24}$$

Throughout the chapter we will consider the Hopfield model with external field, so that the Hamiltonian is

$$\begin{aligned} -H_{N,M}(\boldsymbol{\sigma}) &= \frac{N\beta}{2}\sum_{k\le M}\Big(\frac{1}{N}\sum_{i\le N}\eta_{i,k}\sigma_i\Big)^2 + h\sum_{i\le N}\sigma_i \\ &= \frac{N\beta}{2}\sum_{k\le M} m_k^2 + Nhm_1 \,. \end{aligned} \tag{4.25}$$

It is of course important that we have chosen $\eta_{i,1} = 1$, so that the external field "pulls in the same direction as m_1". Among the values of k, when $h \neq 0$

we can expect the value $k = 1$ to play a special role. Without loss of generality we can and do assume $h \geq 0$.

We observe that the function $f(x) = \mathrm{th}(\beta x + h)$ is concave for $x \geq 0$. If $h > 0$ we have $f(0) > 0$. If $h = 0$ and $\beta > 1$ we have $f(0) = 0$ and $f'(0) > 1$. Thus if $\beta > 1$ there is a unique positive solution to (4.16). Throughout the chapter, we denote by $m^* = m^*(\beta, h)$ this solution, i.e.

$$m^* = \mathrm{th}(\beta m^* + h) . \tag{4.26}$$

We set

$$b^* = \log \mathrm{ch}(\beta m^* + h) - \frac{\beta}{2} m^{*2} . \tag{4.27}$$

The expression of b^* given here is appropriate for the proof of Lemma 4.1.2 below. It is not obvious that this is the same as the value (4.12), which, in the presence of the external field, is

$$\frac{\beta m^{*2}}{2} + m^* h - \mathcal{I}(m^*) . \tag{4.28}$$

To prove the equality of (4.27) and (4.28), we observe that (A.26) implies

$$\mathcal{I}(x) = \max_{\lambda}(\lambda x - \log \mathrm{ch}\lambda)$$

and that the maximum is obtained for $\mathrm{th}\lambda = x$, so that, if $x = m^* = \mathrm{th}(m^*\beta + h)$, λ is $m^*\beta + h$ and hence $\mathcal{I}(m^*) = m^{*2}\beta + m^* h - \log \mathrm{ch}(m^*\beta + h)$, so that the quantities (4.27) and (4.28) coincide.

Lemma 4.1.2. *If* $\beta > 1$ *we have*

$$|\log Z_{N,1} - N(b^* + \log 2)| \leq K(\beta, h) .$$

Of course $Z_{N,M} = Z_{N,M}(\beta, h)$ denotes the partition function of the Hamiltonian (4.25). Thus $Z_{N,1}$ is the partition function of the Curie-Weiss model with external field. The proof serves as an introduction to the method of Section 4.3. It is much more effective and accurate than the more natural method leading to (4.13). The result is also true for $\beta < 1$ if we define m^* by (4.26) for $h > 0$ and $m^* = 0$ for $h = 0$. This is left as an exercise.

Proof. We start with the identity (see (A.6))

$$\mathsf{E} \exp ag = \exp \frac{a^2}{2}$$

whenever g is standard Gaussian r.v., so that

$$Z_{N,1} = \sum_{\boldsymbol{\sigma}} \mathsf{E} \exp(\sqrt{N\beta} g m_1 + N h m_1) .$$

Now, since $m_1 = N^{-1} \sum_{i \leq N} \sigma_i$ we have, using (1.30) in the second equality,

$$\sum_{\boldsymbol{\sigma}} \exp(\sqrt{N\beta} g m_1 + N h m_1) = \sum_{\boldsymbol{\sigma}} \exp\left(\sum_{i \le N} \sigma_i \left(\sqrt{\frac{\beta}{N}} g + h\right)\right)$$
$$= 2^N \operatorname{ch}\left(\sqrt{\frac{\beta}{N}} g + h\right)^N ,$$

and therefore

$$Z_{N,1} = 2^N \mathsf{E} \operatorname{ch}\left(\sqrt{\frac{\beta}{N}} g + h\right)^N .$$

Thus

$$Z_{N,1} = 2^N \frac{1}{\sqrt{2\pi}} \int_{\mathbb{R}} \exp\left(N \log \operatorname{ch}\left(\sqrt{\frac{\beta}{N}} t + h\right) - \frac{t^2}{2}\right) \mathrm{d}t$$
$$= 2^N \sqrt{\frac{N\beta}{2\pi}} \int_{\mathbb{R}} \exp N\left(\log \operatorname{ch}(\beta z + h) - \frac{\beta z^2}{2}\right) \mathrm{d}z$$

with the change of variable $t = \sqrt{N\beta} z$. The function $z \mapsto \log \operatorname{ch}(\beta z + h) - \beta z^2/2$ attains its maximum at the point z such that $\operatorname{th}(\beta z + h) = z$, i.e. $z = m^*$, and this maximum is b^*. Thus

$$Z_{N,1} = 2^N \exp(N b^*) A_N \tag{4.29}$$

where

$$A_N = \sqrt{\frac{N\beta}{2\pi}} \int_{\mathbb{R}} \exp N\psi(z) \mathrm{d}z ,$$

for

$$\psi(z) = \log \operatorname{ch}(\beta z + h) - \frac{\beta z^2}{2} - b^* .$$

To finish the proof we will show that there is a number K such that

$$\psi(z) \le -\frac{1}{K}(z - m^*)^2 . \tag{4.30}$$

Making the change of variable $z = m^* + x/\sqrt{N}$ then implies easily that $\log A_N$ stays bounded as $N \to \infty$, and (4.29) concludes the proof.

The proof of (4.30) is elementary and tedious. We observe that the function ψ satisfies $\psi(m^*) = \psi'(m^*) = 0$. Also, the function

$$\psi'(z) = \beta(\operatorname{th}(\beta z + h) - z)$$

is strictly concave for $z \ge 0$ and is ≥ 0 for $z < m^*$, so that its derivative at $z = m^*$ must be < 0, i.e. $\psi''(m^*) < 0$. This proves that (4.30) holds for z close to m^*. Next, we observe that $\psi(z) < 0$ if $z \ne m^*$. For $z \ge 0$ this follows from the fact that $\psi'(z) > 0$ for $0 < z < m^*$ while $\psi'(z) < 0$ for $z > m^*$, and for $z < 0$ this follows from the fact that $\psi(z) < \psi(-z) \le 0$.

Since $\psi(z) < -\beta(z - m^*)^2/4$ for large z, it follows that (4.30) holds for all values of z. □

The Hopfield model has a kind of singularity for $\beta = 1$. In that case, some understanding has been gained only when $M/N \to 0$, see [154] and the references therein to earlier work. These results again do not rely on the cavity method and are not reproduced here. Because of that singularity, we study the Hopfield model only for $\beta \neq 1$. Our efforts in the next sections concentrate on the most interesting case, i.e. $\beta > 1$. We will explain why the case $\beta < 1$ is several orders of magnitude easier than the case $\beta > 1$. It is still however not trivial. This is because the special methods that allow the control of the Hopfield model *without* external field under the condition (4.21) break down in the presence of an external field.

When studying the Hopfield model, we will think of N and M as large but fixed. Throughout the chapter we write

$$\alpha = \frac{M}{N} \, .$$

The model then depends on the parameters (N, α, β, h).

Exercise 4.1.3. Prove that there exists a large enough universal constant L such that one can control the Hopfield model with external field in a region of the type $\beta < 1, \alpha \leq (1 - \beta)^2/L$.

Of course this exercise should be completed only after reading some of the present chapter, and in particular Theorem 4.2.4 below. On the other hand, even if $\beta < 1$, when $h \neq 0$, reaching the largest possible value of α for which there is "high-temperature" behavior is likely to be a level 3 problem.

4.2 Local Convexity and the Hubbard-Stratonovitch Transform

We recall the Hamiltonian (4.25):

$$-H_{N,M}(\boldsymbol{\sigma}) = \frac{N\beta}{2} \sum_{k \leq M} m_k^2(\boldsymbol{\sigma}) + Nhm_1(\boldsymbol{\sigma}) \, . \tag{4.31}$$

Since it is defined entirely as a function of the quantities $(m_k(\boldsymbol{\sigma}))_{k \leq M}$ (defined in (4.19)), these should be important objects.

Consider the image G' of the Gibbs measure G under the random map

$$\boldsymbol{\sigma} \mapsto \mathbf{m}(\boldsymbol{\sigma}) := (m_k(\boldsymbol{\sigma}))_{k \leq M} \, . \tag{4.32}$$

This is a random probability measure on $\mathbb{R}^M$.

As an auxiliary tool, we consider the probability measure γ on $\mathbb{R}^M$, of density $W\exp(-\beta N\|\mathbf{z}\|^2/2)$ with respect to Lebesgue measure on $\mathbb{R}^M$, where $\|\mathbf{z}\|$ is the Euclidean norm of $\mathbf{z}$ and W is the normalizing factor

$$W = (N\beta/2\pi)^{M/2} . \tag{4.33}$$

(This notation W will be used throughout the present chapter.) It will be useful to replace G' by its convolution $\overline{G} = G' * \gamma$ with γ. This method is called the Hubbard-Stratonovitch transform. It is an elaboration of the trick used in Lemma 4.1.2.

It is useful to think of $\overline{G}$ as a small perturbation of G', an idea that we will make precise later. The reason why $\overline{G}$ is more convenient than G' is that it has a simple density with respect to Lebesgue measure. To see this, we consider the vector

$$\boldsymbol{\eta}_i = (\eta_{i,k})_{k\le M}$$

of $\mathbb{R}^M$, and the (random) function ψ on $\mathbb{R}^M$ given by

$$\psi(\mathbf{z}) = -\frac{N\beta}{2}\|\mathbf{z}\|^2 + \sum_{i\le N}\log\operatorname{ch}(\beta\boldsymbol{\eta}_i\cdot\mathbf{z}+h) , \tag{4.34}$$

where of course $\|\mathbf{z}\|^2 = \sum_{k\le M} z_k^2$ and $\boldsymbol{\eta}_i\cdot\mathbf{z} = \sum_{k\le M}\eta_{i,k}z_k$. This function ψ is a multidimensional generalization of the function $\log\operatorname{ch}(\beta z+h) - \beta^2z^2/2$ used in the proof of Lemma 4.1.2.

Lemma 4.2.1. *The probability $\overline{G}$ has a density*

$$W2^N Z_{N,M}^{-1}\exp\psi(\mathbf{z})$$

with respect to Lebesgue's measure, where $Z_{N,M}$ is the partition function,

$$Z_{N,M} = \sum_{\boldsymbol{\sigma}}\exp(-H_{N,M}(\boldsymbol{\sigma})) .$$

Proof. If we consider the positive measure δ consisting of a mass a at a point $\mathbf{x}$, the density of $\delta*\gamma$ at a point $\mathbf{z}$ is given by

$$aW\exp\left(-\frac{\beta N}{2}\|\mathbf{z}-\mathbf{x}\|^2\right) .$$

Since for each $\boldsymbol{\sigma}$ the probability measure G' gives mass

$$\frac{1}{Z_{N,M}}\exp(-H_{N,M}(\boldsymbol{\sigma})) = \frac{1}{Z_{N,M}}\exp\left(\frac{N\beta}{2}\|\mathbf{m}(\boldsymbol{\sigma})\|^2 + Nhm_1(\boldsymbol{\sigma})\right)$$

to the point $\mathbf{m}(\boldsymbol{\sigma}) = (m_k(\boldsymbol{\sigma}))_{k\le M}$, the density at $\mathbf{z}$ of $G'*\gamma$ is

$$\frac{1}{Z_{N,M}}\sum_{\boldsymbol{\sigma}} W\exp\left(\frac{N\beta}{2}\|\mathbf{m}(\boldsymbol{\sigma})\|^2 + Nhm_1(\boldsymbol{\sigma}) - \frac{N\beta}{2}\|\mathbf{z}-\mathbf{m}(\boldsymbol{\sigma})\|^2\right) .$$

This is

$$\frac{W}{Z_{N,M}} \exp\left(-\frac{N\beta}{2}\|\mathbf{z}\|^2\right) \sum_{\boldsymbol{\sigma}} \exp(N\beta\mathbf{z}\cdot\mathbf{m}(\boldsymbol{\sigma}) + Nhm_1(\boldsymbol{\sigma})) \ .$$

Now

$$\begin{aligned} N\beta\mathbf{z}\cdot\mathbf{m}(\boldsymbol{\sigma}) + Nhm_1(\boldsymbol{\sigma}) &= \beta\sum_{k\le M} z_k\left(\sum_{i\le N}\eta_{i,k}\sigma_i\right) + h\sum_{i\le N}\sigma_i \\ &= \sum_{i\le N}\sigma_i\beta\sum_{k\le M} z_k\eta_{i,k} + h\sum_{i\le N}\sigma_i \\ &= \sum_{i\le N}\sigma_i(\beta\boldsymbol{\eta}_i\cdot\mathbf{z} + h) \ , \end{aligned}$$

and therefore

$$\begin{aligned} \sum_{\boldsymbol{\sigma}} \exp(N\beta\mathbf{z}\cdot\mathbf{m}(\boldsymbol{\sigma}) + Nhm_1(\boldsymbol{\sigma})) &= 2^N\prod_{i\le N}\mathrm{ch}(\beta\boldsymbol{\eta}_i\cdot\mathbf{z} + h) \\ &= 2^N\exp\sum_{i\le N}\log\mathrm{ch}(\beta\boldsymbol{\eta}_i\cdot\mathbf{z} + h) \ , \end{aligned}$$

which finishes the proof. □

In the present chapter we largely follow an approach invented by Bovier and Gayrard. The basic idea is to use the tools of Section 3.1 to control the overlaps. This approach is made possible by the following convexity property, that was also discovered by Bovier and Gayrard. We denote by $(\mathbf{e}_k)_{k\le M}$ the canonical basis of $\mathbb{R}^M$.

Let us recall that everywhere in this chapter we write $\alpha = M/N$.

Theorem 4.2.2. *There exists a number L with the following property. Given $\beta > 1$, there exists a number $\kappa > 0$ with the following property. Assume that $\alpha \le m^{*4}/L\beta$. Then there exists a number K such that with probability $\ge 1 - K\exp(-N/K)$, the function $\mathbf{z} \mapsto \psi(\mathbf{z}) + \kappa N\|\mathbf{z}\|^2$ is concave in the region*

$$\left\{\mathbf{z}\ ;\ \|\mathbf{z} - m^*\mathbf{e}_1\| \le \frac{m^*}{L(1+\log\beta)}\right\} . \tag{4.35}$$

Here, and everywhere in this chapter, K denotes a number that does not depend on N or M (so that K never depends on $\alpha = M/N$). In the present case, K depends only on β and h. As usual the letter L denotes a universal constant, that certainly need not be the same at each occurrence. We will very often omit the sentence "There exists a number L with the following property" and the sentence "There exists a number K" in subsequent statements.

The point of Theorem 4.2.2 is that the function $\|\mathbf{z}\|^2$ is convex, so that the meaning of this theorem is that in the region (4.35) the function ψ is sufficiently concave that it will satisfy (3.21), opening the way to the use of Theorem 3.1.4. The conditions $\alpha \leq m^{*4}/L\beta$ and (4.35) are by no means intuitive, but are the result of a careful analysis.

Even though Theorem 4.2.2 will not be used before Section 4.5 we will present the proof now, since it is such a crucial result for the present approach (or the other hand, when we return to the study of the Hopfield model in Chapter 10 this result will no longer be needed). We must not hide the fact that this proof uses ideas from probability theory, which, while elementary, have been pushed quite far. This is also the case of the results of Section 4.3. These proofs contain no "spin glasses ideas". Therefore the reader who finds these proofs difficult should simply skip them all. In Section 4.4 page 272, matters become quite easier.

Throughout the book we will use the letter Ω to denote an event (so we do not follow the standard probability notation, which is to denote by Ω the entire underlying probability space).

Definition 4.2.3. *We say that an event Ω occurs* with overwhelming probability *if $P(\Omega) \geq 1 - K\exp(-N/K)$ where K does not depend on N or M.*

Of course the event $\Omega = \Omega_{N,M}$ depends on N and M, so it would be more formal to say that "a family of events $\Omega_{N,M}$ occurs with overwhelming probability", but it seems better to be a bit informal than pedantic.

Using Definition 4.2.3, the second sentence of Theorem 4.2.2 reformulates as "With overwhelming probability the function $\mathbf{z} \mapsto \psi(\mathbf{z}) + \kappa N\|\mathbf{z}\|^2$ is concave in the region (4.35)."

Maybe it will help to mention that one of our goals is, given β and h, to control the Hopfield model *uniformly* over all the values of M and N with $\alpha = M/N \leq \alpha_0(\beta, h)$ for a certain number $\alpha_0(\beta, h)$ (as large as we can achieve). This will be technically important in Section 10.8, and is one of the reasons why we insist that K does not depend on N or M.

As a warm-up, and in order to make the point that things are so much simpler when $\beta < 1$ we shall prove the following.

Theorem 4.2.4. *Given $\beta < 1$, there exists a number $\kappa > 0$ with the following property. Assume that $\alpha \leq (\beta - 1)^2/L$. Then with overwhelming probability the function $\psi(\mathbf{z}) + \kappa N\|\mathbf{z}\|^2$ is concave.*

Again, we have omitted the sentence "There exists a number L such that the following holds". In Theorem 4.2.4 the constant K implicit in the words "with overwhelming probability" depends only on β.

To prove that a function φ is concave in a convex domain, we prove that at each point $\mathbf{w}$ of this domain the second differential $D^2_{\mathbf{w}}$ of φ satisfies

$D^2_{\mathbf{w}}(\mathbf{v},\mathbf{v}) \le 0$ for each vector $\mathbf{v}$. If differentials are not familiar to you, the quantity $D^2_{\mathbf{w}}(\mathbf{v},\mathbf{v})$ is simply the second derivative at $t = 0$ of the function $t \mapsto \varphi(\mathbf{w} + t\mathbf{v})$.

Proof of Theorem 4.2.4. Let us set $\mathbf{z} = m^*\mathbf{e}_1 + \mathbf{w}$, and denote by $D^2_{\mathbf{w}}$ the second differential of ψ at the point $\mathbf{z}$, so that, for $\mathbf{v} \in \mathbb{R}^M$ (and since $\eta_{i,1} = 1$ for each i),

$$\begin{aligned} D^2_{\mathbf{w}}(\mathbf{v},\mathbf{v}) &= -\beta N\|\mathbf{v}\|^2 + \beta^2 \sum_{i\le N} \frac{1}{\mathrm{ch}^2(\beta m^* + h + \beta\boldsymbol{\eta}_i\cdot\mathbf{w})}(\boldsymbol{\eta}_i\cdot\mathbf{v})^2 \\ &\le \beta\Big(-N\|\mathbf{v}\|^2 + \beta\sum_{i\le N}(\boldsymbol{\eta}_i\cdot\mathbf{v})^2\Big) . \end{aligned} \tag{4.36}$$

It follows from Corollary A.9.4 that (there exists a number L such that) if $M \le (1-\beta)^2 N/L$, with overwhelming probability one has

$$\forall \mathbf{v} \,,\ \sum_{i\le N}(\boldsymbol{\eta}_i\cdot\mathbf{v})^2 \le N(1 + (1-\beta))\|\mathbf{v}\|^2 \,,$$

and therefore $D^2_{\mathbf{w}}(\mathbf{v},\mathbf{v}) \le -\beta(1-\beta)^2 N\|\mathbf{v}\|^2$, so that if $\kappa = \beta(1-\beta)^2$, the function $\psi(\mathbf{z}) + \kappa N\|\mathbf{z}\|^2$ is concave. □

To continue the study of the case $\beta < 1$ and complete Exercise 4.1.3, the reader can go directly to Section 4.5. Through the rest of this chapter, we assume that $\beta > 1$.

Before the real work starts, we need some simple facts about the behavior of m^*. These facts are needed to get a qualitatively correct dependence of α on β, but are otherwise not fundamental.

Lemma 4.2.5. *The quantity $m^*(\beta,h)$ increases as β or h increases. Moreover we have*

$$\beta \ge 2 \Rightarrow \beta(1 - m^{*2}) \le L\exp(-\beta/L) \tag{4.37}$$

$$\frac{m^{*2}}{L} \le a^* := 1 - \beta(1 - m^{*2}) \le m^{*2} \,. \tag{4.38}$$

Proof. We observe that if $z \ge 0$ then

$$z \le \mathrm{th}(\beta z + h) \Longleftrightarrow z \le m^*(\beta,h) \,. \tag{4.39}$$

Now if $\beta' \ge \beta$ and $h' \ge h$ we have

$$m^*(\beta,h) = \mathrm{th}(\beta m^*(\beta,h) + h) \le \mathrm{th}(\beta' m^*(\beta,h) + h') \,,$$

and therefore $m^*(\beta,h) \le m^*(\beta',h')$ by (4.39), so that the quantity $m^*(\beta,h)$ increases as β or h increases. To prove (4.37) we observe that $m^* = m^*(\beta,h) \ge m^*(2,0)$ and hence

$$m^* = \mathrm{th}(\beta m^* + h) \ge \mathrm{th}(\beta m^*(2,0))$$

and consequently,

$$\beta(1-m^{*2}) \le \frac{\beta}{\mathrm{ch}^2(\beta m^*(2,0))} \le L\exp\left(-\frac{\beta}{L}\right) .$$

The right-hand side inequality of (4.38) holds since $1-\beta(1-m^{*2}) \le 1-(1-m^{*2}) = m^{*2}$. To prove the left-hand side inequality of (4.38), we observe that, since this inequality is equivalent to $\beta - 1 \le (\beta - 1/L)m^{*2}$, and since $m^*(\beta,h)$ increases with h, we can assume $h=0$. We then observe that for $x>0$ we have

$$1-\frac{x}{\mathrm{th}x\mathrm{ch}^2x} = 1-\frac{2x}{\mathrm{sh}(2x)} = \frac{\mathrm{sh}(2x)-2x}{\mathrm{sh}(2x)} \ge \frac{1}{L}\frac{x^2}{1+x^2} ,$$

as is seen by studying the behavior at $x\to 0$ and $x\to\infty$. Taking $x=\beta m^*$ where $m^* = m^*(\beta,0)$, we get, since $m^* = \mathrm{th}\beta m^*$,

$$\begin{aligned} 1-\beta(1-m^{*2}) &= 1-\frac{\beta}{\mathrm{ch}^2\beta m^*} = 1-\frac{\beta m^*}{m^*\mathrm{ch}^2\beta m^*} = 1-\frac{\beta m^*}{\mathrm{th}\beta m^*\mathrm{ch}^2\beta m^*} \\ &= 1-\frac{x}{\mathrm{th}x\mathrm{ch}^2x} \ge \frac{1}{L}\frac{x^2}{1+x^2} \ge \frac{1}{L}\frac{\beta^2 m^{*2}}{(1+\beta^2)} \ge \frac{1}{L}m^{*2} , \end{aligned}$$

using that $1+x^2 \le 1+\beta^2$. □

Theorem 4.2.2 asserts that with overwhelming probability we control the Hessian (= the second differential) of ψ over the entire region (4.35). Controlling the Hessian at a given point with overwhelming probability is easy, but controlling at the same time every point of a region is distinctly more difficult, and it is not surprising that this should require significant work. The key to our approach is the following.

Proposition 4.2.6. *We can find numbers L and L_1 with the following property. Consider $0<a<1$ and $b \ge L_1\sqrt{\log(2/a)}$. Assume that $\alpha \le a^2$. Then the following event occurs with probability $\ge 1-L\exp(-Na^2)$: for each $\mathbf{w},\mathbf{w}\in\mathbb{R}^M$, we have*

$$\sum_{i\in J(\mathbf{w})} (\boldsymbol{\eta}_i\cdot\mathbf{v})^2 \le LNa^2\|\mathbf{v}\|^2 , \tag{4.40}$$

where

$$J(\mathbf{w}) = \{i\le N\ ;\ |\boldsymbol{\eta}_i\cdot\mathbf{w}| \ge b\|\mathbf{w}\|\} . \tag{4.41}$$

To understand this statement, we note that $\mathsf{E}(\boldsymbol{\eta}_i\cdot\mathbf{z})^2 = \|\mathbf{z}\|^2$, so that $\mathsf{E}\sum_{i\le N}(\boldsymbol{\eta}_i\cdot\mathbf{v})^2 = N\|\mathbf{v}\|^2$. Also when $b\gg 1$, and since $\mathsf{E}(\boldsymbol{\eta}_i\cdot\mathbf{w})^2 = \|\mathbf{w}\|^2$, it is rare that $|\boldsymbol{\eta}_i\cdot\mathbf{w}| \ge b\|\mathbf{w}\|$, so the set $J(\mathbf{w})$ has a tendency to be a rather small subset of $\{1,\ldots,N\}$, and it is much easier to control in (4.40) the sum over $J(\mathbf{w})$ rather than the sum over $\{1,\ldots,N\}$. The difficulty of course is to find a statement that holds for all $\mathbf{v}$ and $\mathbf{w}$.

Proof of Theorem 4.2.2. As in the proof of Theorem 4.2.4 we set $\mathbf{z} = m^*\mathbf{e}_1 + \mathbf{w}$, we denote by $D^2_{\mathbf{w}}$ the second differential of ψ at the point $\mathbf{z}$, and we recall (4.36):

$$D^2_{\mathbf{w}}(\mathbf{v},\mathbf{v}) = -\beta N\|\mathbf{v}\|^2 + \beta^2 \sum_{i\le N} \frac{1}{\mathrm{ch}^2(\beta m^* + h + \beta\boldsymbol{\eta}_i\cdot\mathbf{w})}(\boldsymbol{\eta}_i\cdot\mathbf{v})^2 . \tag{4.42}$$

In contrast with the case $\beta < 1$ we must now take advantage of the fact that the denominators have a tendency to be > 1, or even $\gg 1$ for large β. The difficulty is that some of the terms $\beta m^* + h + \beta\boldsymbol{\eta}_i\cdot\mathbf{w}$ might be close to 0, in which case $\mathrm{ch}^2(\beta m^* + h + \beta\boldsymbol{\eta}_i\cdot\mathbf{w})$ is not large. We have to show somehow that these terms do not contribute too much. The strategy is easier to understand when β is not close to 1. In that case, the only terms that can be troublesome are those for which $\beta m^* + h + \beta\boldsymbol{\eta}_i\cdot\mathbf{w}$ might be close to 0 (for otherwise $\mathrm{ch}^2(\beta m^* + h + \beta\boldsymbol{\eta}_i\cdot\mathbf{w}) \gg 1$) and these are such that $\boldsymbol{\eta}_i\cdot\mathbf{w} \le -m^*/2$ and in particular $|\boldsymbol{\eta}_i\cdot\mathbf{w}| \ge m^*/2$. Proposition 4.2.6 is perfectly appropriate to control these terms (as it should, since this is why it was designed).

We first consider the case $\beta \ge 2$. In that case (following the argument of (4.37)), since $m^* \ge m^*(2,0)$, we have

$$\beta^2 \frac{1}{\mathrm{ch}^2(\beta m^*/2 + h)} \le L\exp\left(-\frac{\beta}{L}\right) ,$$

and thus

$$D^2_{\mathbf{w}}(\mathbf{v},\mathbf{v}) \le -\beta N\|\mathbf{v}\|^2 + L\exp\left(-\frac{\beta}{L}\right)\sum_{i\le N}(\boldsymbol{\eta}_i\cdot\mathbf{v})^2$$
$$+ \beta^2 \sum_{i\le N} \mathbf{1}_{\{|\boldsymbol{\eta}_i\cdot\mathbf{w}|\ge m^*/2\}}(\boldsymbol{\eta}_i\cdot\mathbf{v})^2 . \tag{4.43}$$

To control the second term in the right-hand side, we note that by Corollary A.9.4, with overwhelming probability we have (whenever $M \le N$)

$$\forall \mathbf{v},\ \sum_{i\le N}(\boldsymbol{\eta}_i\cdot\mathbf{v})^2 \le L\|\mathbf{v}\|^2 . \tag{4.44}$$

Next, denoting by L_0 the constant of (4.40), we set $L_2 = 2L_0$, so that if we define a by $a^2 = 1/L_2\beta$, the right-hand side of (4.40) is

$$L_0 N a^2\|\mathbf{v}\|^2 = \frac{L_0}{L_2\beta}N\|\mathbf{v}\|^2 = \frac{N\|\mathbf{v}\|^2}{2\beta} .$$

Moreover since $\beta > 2$ there exists a universal constant L_3 such that

$$L_3\sqrt{\log\beta} \ge L_1\sqrt{\log(2/a)} .$$

Thus we can use Proposition 4.2.6 with a as above and $b = L_3\sqrt{\log\beta}$. We observe that

$$\|\mathbf{w}\| \le \frac{m^*}{2b} \quad \text{and} \quad |\boldsymbol{\eta}_i \cdot \mathbf{w}| \ge \frac{m^*}{2} \Rightarrow |\boldsymbol{\eta}_i \cdot \mathbf{w}| \ge b\|\mathbf{w}\| \,.$$

It then follows from Proposition 4.2.6 that if $M \le Na^2 = N/L\beta$ (i.e. $\alpha \le 1/L\beta$), then with overwhelming probability, the following occurs:

$$\forall \mathbf{w}, \quad \|\mathbf{w}\| \le m^*/2b = m^*/L\sqrt{\log\beta} \Rightarrow$$

$$\sum_{i \le N} \mathbf{1}_{\{|\boldsymbol{\eta}_i \cdot \mathbf{w}| \ge m^*/2\}} (\boldsymbol{\eta}_i \cdot \mathbf{v})^2 \le \frac{N}{2\beta} \|\mathbf{v}\|^2 \,,$$

and (4.36) yields

$$D^2_{\mathbf{w}}(\mathbf{v},\mathbf{v}) \le -N\left(\frac{\beta}{2} - L\exp\left(-\frac{\beta}{L}\right)\right)\|\mathbf{v}\|^2 \,.$$

Therefore, when β is large enough, say $\beta \ge \beta_0$, we have shown that if $\alpha \le 1/L\beta$, with overwhelming probability we have

$$\|\mathbf{w}\| \le \frac{m^*}{L\sqrt{\log\beta}} \Rightarrow D^2_{\mathbf{w}}(\mathbf{v},\mathbf{v}) \le -\frac{N\beta_0}{4}\|\mathbf{v}\|^2 \,. \tag{4.45}$$

We now turn to the case $1 < \beta \le \beta_0$. We will as before consider separately the terms for which $|\boldsymbol{\eta}_i \cdot \mathbf{w}| \ge cm^*$ where $0 < c < 1$ is a parameter $0 < c < 1/2\beta_0$ (< 1) to be determined later. We first prove the inequality

$$\begin{aligned}\frac{1}{\mathrm{ch}^2(\beta(m^*+x)+h)} &\le \frac{1}{\mathrm{ch}^2(\beta m^*+h)} + 2m^{*2}\beta c + m^{*2}\mathbf{1}_{\{|x|\ge cm^*\}} \\ &= 1 - m^{*2} + 2m^{*2}\beta c + m^{*2}\mathbf{1}_{\{|x|\ge cm^*\}} \,. \end{aligned} \tag{4.46}$$

This is obvious if $|x| \ge cm^*$ because then the right-hand side is ≥ 1. This is also obvious if $x \ge 0$ because this is true for $x = 0$ and the function $f(x) = \mathrm{ch}^{-2}(\beta(m^*+x)+h)$ decreases. Now,

$$f'(x) = -\frac{2\beta\mathrm{th}(\beta(m^*+x)+h)}{\mathrm{ch}^2(\beta(m^*+x)+h)} \,,$$

so that for $-m^* \le x \le 0$ we have

$$|f'(x)| \le 2\beta\mathrm{th}(\beta(m^*+x)+h) \le 2\beta\mathrm{th}(\beta m^*+h) = 2\beta m^* \,,$$

and thus for $-cm^* \le x \le 0$ we get

$$f(x) \le f(0) + 2\beta m^{*2} c = 1 - m^{*2} + 2\beta m^{*2} c \,.$$

Therefore (4.46) also holds for $|x| \le cm^*$, and is proved in every case.

We define

$$d = 1 - m^{*2} + 2\beta m^{*2} c \,,$$

and we note that since $c < 1/2\beta_0$ and $\beta < \beta_0$ we have

$$d < 1\,.$$

Now (4.46) implies

$$\frac{1}{\operatorname{ch}^2(\beta(m^* + \boldsymbol{\eta}_i \cdot \mathbf{w}) + h)} \le d + m^{*2}\mathbf{1}_{\{|\boldsymbol{\eta}_i \cdot \mathbf{w}| \ge cm^*\}}\,, \tag{4.47}$$

and we deduce from (4.42) that

$$D^2_{\mathbf{w}}(\mathbf{v},\mathbf{v}) \le \mathrm{I} + \mathrm{II} \tag{4.48}$$

where

$$\mathrm{I} = -\beta N\|\mathbf{v}\|^2 + \beta^2 d \sum_{i\le N} (\boldsymbol{\eta}_i \cdot \mathbf{v})^2$$

$$\mathrm{II} = \beta^2 m^{*2} \sum_{i\le N} \mathbf{1}_{\{|\boldsymbol{\eta}_i \cdot \mathbf{w}| \ge cm^*\}} (\boldsymbol{\eta}_i \cdot \mathbf{v})^2\,.$$

Consider a parameter $\rho > 0$, to be fixed later. It follows from Corollary A.9.4 that if $\alpha \le \rho^2/L$, with overwhelming probability we have

$$\forall \mathbf{v}, \ \sum_{i\le N} (\boldsymbol{\eta}_i \cdot \mathbf{v})^2 \le N(1+\rho)\|\mathbf{v}\|^2\,,$$

and consequently

$$\mathrm{I} \le -\beta N\|\mathbf{v}\|^2 (1 - \beta d(1+\rho))\,. \tag{4.49}$$

By (4.38), we have $1 - \beta(1 - m^{*2}) \ge m^{*2}/L$, so that, recalling the definition of d, that $d \le 1$, and that $\beta \le \beta_0$,

$$\begin{aligned} 1 - \beta d(1+\rho) &\ge 1 - \beta d - \beta_0 \rho \\ &= 1 - \beta(1 - m^{*2}) - 2\beta^2 m^{*2} c - \beta_0 \rho \\ &\ge 1 - \beta(1 - m^{*2}) - 2\beta_0^2 m^{*2} c - \beta_0 \rho \\ &\ge \frac{m^{*2}}{L_0} - 2\beta_0^2 m^{*2} c - \beta_0 \rho\,. \end{aligned}$$

We make the choices

$$\rho = \frac{m^{*2}}{4\beta_0 L_0}\,;\quad c = \frac{1}{8\beta_0^2 L_0}\,,$$

so that $1 - \beta d(1+\rho) \ge m^{*2}/2L_0$ and we see that provided that

$$\alpha \le \frac{\rho^2}{L} = \frac{m^{*4}}{L'}\,,$$

with overwhelming probability we have

$$\mathrm{I} \leq -\beta N\|\mathbf{v}\|^2 \frac{m^{*2}}{2L_0} . \tag{4.50}$$

To take care of the term II we use Proposition 4.2.6 again. We choose $a = 1/L_4$, where $L_4 = 2\beta_0 L_0$. We can then apply Proposition 4.2.6 for $b = L_1\sqrt{\log(2/a)}$ $(= L_5)$. Then, since L_0 is the constant in (4.40), the right-hand side of this inequality is $N\|\mathbf{v}\|^2/4\beta_0^2 L_0$. Since when $|\boldsymbol{\eta} \cdot \mathbf{w}| \geq cm^*$ and $\|\mathbf{w}\| \leq m^* c/b = m^*/L$ then $|\boldsymbol{\eta} \cdot \mathbf{w}| \geq b\|\mathbf{w}\|$, this proves that if $M \leq a^2 N = N/L_4^2$ then with overwhelming probability $\mathrm{II} \leq N\|\mathbf{v}\|^2 m^{*2}/4L_0$ whenever $\|\mathbf{w}\| \leq m^* c/b = m^*/L$. Consequently, combining with (4.50) we have shown that if $\beta \leq \beta_0$ and $\alpha \leq m^{*4}/L'$, then with overwhelming probability

$$\|\mathbf{w}\| \leq \frac{m^*}{L} \Rightarrow D_{\mathbf{w}}(\mathbf{v},\mathbf{v}) \leq -\beta N\|\mathbf{v}\|^2 \frac{m^{*2}}{4L_0} .$$

Combining with (4.45) we have completed the proof, because if the constant L in (4.35) is large enough, the region this condition defines is included into the region we have controlled. This is obvious by distinguishing the cases $\beta \geq \beta_0$ and $\beta \leq \beta_0$. □

Proof of Proposition 4.2.6. Consider the largest integer $N_0 \leq N$ with $N_0 \log(eN/N_0) \leq Na^2$. In Proposition A.9.5 it is shown that the following event occurs with probability $\geq 1 - L\exp(-Na^2)$:

$$\forall J \subset \{1, \cdots, N\}, \quad \mathrm{card} J \leq N_0, \quad \forall \mathbf{w} \in \mathbb{R}^M,$$

$$\sum_{i \in J} (\boldsymbol{\eta}_i \cdot \mathbf{w})^2 \leq \|\mathbf{w}\|^2 (N_0 + L\max(Na^2, \sqrt{NN_0}a)) . \tag{4.51}$$

This statement is of similar nature as (4.51), except that we have a cardinality restriction on the index set J instead of specifying that it is of the type $J(\mathbf{w})$ as defined by (4.41). The core of the proof is to show that when (4.51) holds, then for each w we have $\mathrm{card} J(\mathbf{w}) < N_0$, after which a straightforward use of (4.51) will imply (4.40).

To control the cardinality of $J(\mathbf{w})$ suppose, if possible, that there exists $J \subset J(\mathbf{w})$ with $\mathrm{card} J = N_0$. Then since $|\boldsymbol{\eta}_i \cdot \mathbf{w}| \geq b\|\mathbf{w}\|$ for $i \in J$ we have

$$\sum_{i \in J} (\boldsymbol{\eta}_i \cdot \mathbf{w})^2 \geq b^2 N_0 \|\mathbf{w}\|^2 ,$$

and, comparing with (4.51), we see that

$$b^2 N_0 \leq N_0 + L\max(Na^2, \sqrt{NN_0}a) ,$$

and therefore

$$(b^2 - 1)N_0 \leq L\max(Na^2, \sqrt{NN_0}a) . \tag{4.52}$$

The idea of the proof is to show that this bound on N_0 contradicts the definition of N_0, by forcing N_0 to be too small. It is clear that, given a, this is the case if b is large enough, but to get values of b of the right order one has to work a bit. Assuming without loss of generality $b \geq 2$, we have $b^2 - 1 \geq b^2/2$, so that (4.52) implies

$$b^2 N_0 \leq L \max(Na^2, \sqrt{NN_0}a) .$$

Thus we have either $N_0 \leq LNa^2/b^2$ or else $b^2 N_0 \leq L\sqrt{NN_0}a$, i.e.

$$N_0 \leq LN\frac{a^2}{b^4} \leq LN\frac{a^2}{b^2} .$$

Therefore we always have $N_0 \leq L_6 Na^2/b^2$. We show now that we can choose the constant L_1 large enough so that

$$b \geq L_1\sqrt{\log(2/a)} \Longrightarrow \frac{L_6}{b^2}\log\left(\frac{eb^2}{L_6 a^2}\right) \leq \frac{1}{2} . \tag{4.53}$$

To see that such a number L_1 exists we can assume $L_6 \geq e$ and we observe that $\log(eb^2/L_6a^2) \leq 2\log b + 2\log(2/a)$. We moreover take L_1 large enough such that we also have $L_6a^2/b^2 \leq L_6/b^2 \leq 1$.

Since the function $x \mapsto x\log(eN/x)$ increases for $x \leq N$, and since $N_0 \leq L_6Na^2/b^2 \leq N$, when $b \geq L_1\sqrt{\log(2/a)}$ we deduce from (4.53) that

$$N_0 \log\left(\frac{eN}{N_0}\right) \leq L_6 N\frac{a^2}{b^2}\log\left(\frac{eb^2}{L_6a^2}\right) \leq \frac{Na^2}{2} ,$$

and therefore since $N_0 + 1 \leq 2N_0$ we have

$$(N_0 + 1)\log\frac{eN}{N_0 + 1} \leq Na^2 .$$

But this contradicts the definition of N_0.

Thus we have shown that $\mathrm{card}J(\mathbf{w}) < N_0$. Then, by (4.51), and since $N_0 \leq Na^2$ we get

$$\sum_{i\in J(\mathbf{w})} (\boldsymbol{\eta}_i \cdot \mathbf{v})^2 \leq LNa^2\|\mathbf{v}\|^2. \qquad \square$$

4.3 The Bovier-Gayrard Localization Theorem

Theorem 4.2.2 can be really useful only if the region (4.35) is actually relevant for the computation of G'. This is what we shall prove in this section.

Before we state the main result, we introduce some terminology, that matches the spirit of Definition 4.2.3.

Definition 4.3.1. *We say that a set A of $\mathbb{R}^M$ is* negligible *if*

$$\mathsf{E}\, G'(A) \leq K\exp\left(-\frac{N}{K}\right) \tag{4.54}$$

where K does not depend on N, M. We say that G' is essentially supported *by A if $A^c = \mathbb{R}^M \backslash A$ is negligible.*

Of course the set A also depends on M and N, so it would be more formal to say that "a family $A_{N,M}$ of sets is negligible if $\mathsf{E}\,G(A_{N,M}) \leq K\exp(-N/K)$, where K does not depend on M or N".

In a similar manner, we will say that a subset A of Σ_N is negligible if $\mathsf{E}\,G(A) \leq K\exp(-N/K)$, where K does not depend on N or M.

Theorem 4.3.2. *(The Bovier-Gayrard localization theorem.) Consider $\beta > 1$, $h \geq 0$ and $\rho_0 \leq m^*/2$. If $\alpha \leq m^{*2}\rho_0^2/L$, then G' is essentially supported by the union of the $2M$ balls in $\mathbb{R}^M$ of radius ρ_0 centered at the points $\pm m^*\mathbf{e}_k$, $k \leq M$.*

It is useful to note that the balls of this theorem are disjoint since $\rho_0 \leq m^*/2$.

To reformulate Theorem 4.3.2, if we consider the set

$$A = \{\mathbf{z} \in \mathbb{R}^M \; ; \; \forall k \leq M \; , \; \forall \tau = \pm 1 \; , \; \|\mathbf{z} - \tau m^* \varepsilon_k\| > \rho_0\} \; ,$$

then when $\alpha \leq m^{*2}\rho_0^2/L$ we have $\mathsf{E}G'(A) \leq K\exp(-K/N)$ where K depends only on β, h and ρ_0 but certainly not on α or N.

It is intuitive that something of the type of Theorem 4.3.2 should happen when $h = 0$. (The case $h > 0$ is discussed in Section 4.4.) Each of the terms $m_k(\boldsymbol{\sigma})$ in the Hamiltonian attempts to create a Curie-Weiss model "in the direction of $(\eta_{i,k})_{i\leq N}$"; and in such a model $m_k(\boldsymbol{\sigma}) \simeq \pm m^*$. What Theorem 4.3.2 says is that if there are not too many such terms, for (nearly) each configuration, one of these terms wins over the others. For one k (depending on the configuration) we have $|m_k(\boldsymbol{\sigma}) \pm m^*| \leq \rho_0$, and for $k' \neq k$, $|m_{k'}(\boldsymbol{\sigma})| \leq \rho_0$ is smaller. What is not intuitive is how large α can be taken. It is a separate problem to know whether the same k "wins" independently of the configuration.

The Bovier-Gayrard localization theorem is a deep fact, that will require significant work. The methods are of interest, but they are not related to the main theme of the book (the cavity method) and will be used only in this section. Therefore the reader mostly interested in following the main story should skip this material.

We recall the probability γ introduced page 245. That is, γ has density $W\exp(-\beta N\|\mathbf{z}\|^2/2)$ with respect to Lebesgue measure on $\mathbb{R}^M$, where W is given by (4.33).

We first elaborate on the idea that $\overline{G} = G' * \gamma$ is a small perturbation of G'. One reason is that if $\alpha \leq \beta\rho^2/4$, then γ is essentially supported by the ball centered at the origin, of radius ρ. To see this, we observe that, by change of variable,

$$\begin{aligned}\int_{\mathbb{R}^M} \exp\left(\frac{\beta N}{4}\|\mathbf{z}\|^2\right) \mathrm{d}\gamma(\mathbf{z}) &= W\int_{\mathbb{R}^M} \exp\left(-\frac{\beta N}{4}\|\mathbf{z}\|^2\right) \mathrm{d}\mathbf{z} \\ &= 2^{M/2}W\int_{\mathbb{R}^M} \exp\left(-\frac{\beta N}{2}\|\mathbf{z}\|^2\right) \mathrm{d}\mathbf{z} = 2^{M/2} \; .\end{aligned}$$

Thus,

$$\gamma(\{\|\mathbf{z}\|^2 \geq \rho^2\}) \exp\left(\frac{N\beta}{4}\rho^2\right) \leq 2^{M/2} \leq \exp\left(\frac{\alpha N}{2}\right)$$

and, since $\alpha \leq \beta\rho^2/4$, we get

$$\gamma(\{\|\mathbf{z}\|^2 \geq \rho^2\}) \leq \exp\left(-\frac{N}{4}(\beta\rho^2 - 2\alpha)\right) \leq \exp\left(-\frac{N\beta\rho^2}{8}\right) . \qquad (4.55)$$

This inequality shows in particular by taking $\rho = 2\sqrt{\alpha/\beta}$ that if

$$B = \left\{\mathbf{z}\ ;\ \|\mathbf{z}\| \leq 2\sqrt{\frac{\alpha}{\beta}}\right\} \qquad (4.56)$$

then $\gamma(B) \geq 1/L$ (observe that $\alpha N = M \geq 1$, so that $N\beta\rho^2/8 = M/2 \geq 1/2$). Thus, given any subset A of $\mathbb{R}^M$, we have

$$\overline{G}(A+B) = G' \otimes \gamma(\{(\mathbf{x},\mathbf{y})\ ;\ \mathbf{x}+\mathbf{y} \in A+B\}) \geq G'(A)\gamma(B) ,$$

and hence

$$G'(A) \leq L\overline{G}(A+B) . \qquad (4.57)$$

To prove that a set A is negligible for G' it therefore suffices to prove that $A+B$ is negligible for $\overline{G}$. Consequently if $\overline{G}$ is essentially supported by a set C, then G' is essentially supported by $C+B$. This is because the complement A of $C+B$ is such that $A+B$ is contained in the complement of C so that it is negligible for $\overline{G}$ and hence A is negligible for G'. In particular when $\overline{G}$ is essentially supported by the union C of the balls of radius $\rho_0/2$ centered at the points $\pm m^*\mathbf{e}_k$, then G' is essentially supported by $C+B$. When $\alpha \leq m^{*2}\rho_0^2/16$, we have $2\sqrt{\alpha/\beta} \leq \rho_0/2$ and hence $C+B$ is contained in the union of the balls of radius ρ_0 centered at the points $\pm m^*\mathbf{e}_k$. Thus it suffices to prove Theorem 4.3.2 for $\overline{G}$ rather than for G'.

As a consequence of Lemma 4.2.1, for a subset A of $\mathbb{R}^M$, the identity

$$\overline{G}(A) = \frac{W\int_A \exp\psi(\mathbf{z})d\mathbf{z}}{2^{-N}Z_{N,M}} \qquad (4.58)$$

holds, and the strategy to prove that A is negligible is simply to prove that typically the numerator in (4.58) is much smaller than the denominator. For this it certainly helps to bound the denominator from below. As is often the case in this chapter, we need different arguments when β is close to 1 and when β is away from 1. Of course the choice of the number 2 below is very much arbitrary.

Proposition 4.3.3. *If* $1 < \beta \leq 2$ *and* $\alpha \leq m^{*4}$, *we have*

$$2^{-N}Z_{N,M} \geq \left(\frac{1}{La^*}\right)^{M/2} \exp Nb^* \qquad (4.59)$$

where

$$b^* = \log \mathrm{ch}(\beta m^* + h) - \beta m^{*2}/2 \ , \ \ a^* = 1 - \beta(1 - m^{*2}) \ .$$

This bound is true for any realization of the randomness. It somewhat resembles Lemma 4.1.2.

Before we start the proof, we mention the following elementary fact. The function $\xi(x) = \log \mathrm{ch} x$ satisfies

$$\xi'(x) = \mathrm{th} x; \ \ \xi''(x) = \frac{1}{\mathrm{ch}^2 x}; \ \ \xi'''(x) = -2\frac{\mathrm{th} x}{\mathrm{ch}^2 x}; \ \ |\xi^{(4)}(x)| \le 4 \ . \tag{4.60}$$

Proof of Proposition 4.3.3. Since $\overline{G}$ is a probability, Lemma 4.2.1 shows that

$$2^{-N} Z_{N,M} = W \int \exp \psi(\mathbf{z}) \mathrm{d}\mathbf{z} \ ,$$

so that if we set $\psi^{\sim}(\mathbf{v}) = \psi(m^* \mathbf{e}_1 + \mathbf{v})$ then

$$2^{-N} Z_{N,M} = W \int \exp \psi^{\sim}(\mathbf{v}) \mathrm{d}\mathbf{v} \ . \tag{4.61}$$

Now, since we assume $\eta_{i,1} = 1$, we have

$$\boldsymbol{\eta}_i \cdot (m^* \mathbf{e}_1 + \mathbf{v}) = m^* + \boldsymbol{\eta}_i \cdot \mathbf{v} \ ,$$

so that, setting $b = \beta m^* + h$, we get,

$$\begin{aligned}
\psi^{\sim}(\mathbf{v}) &= \psi(m^* \mathbf{e}_1 + \mathbf{v}) \\
&= -\frac{N\beta}{2} \|m^* \mathbf{e}_1 + \mathbf{v}\|^2 + \sum_{i \le N} \log \mathrm{ch}(\beta \boldsymbol{\eta}_i \cdot (m^* \mathbf{e}_1 + \mathbf{v}) + h) \\
&= -\frac{N\beta}{2} \|m^* \mathbf{e}_1 + \mathbf{v}\|^2 + \sum_{i \le N} \log \mathrm{ch}(b + \beta \boldsymbol{\eta}_i \cdot \mathbf{v}) \ .
\end{aligned} \tag{4.62}$$

We make an order 4 Taylor expansion of log ch around $b\,(= \beta m^* + h)$. This yields

$$\begin{aligned}
\psi^{\sim}(\mathbf{v}) = & -\frac{N\beta}{2} m^{*2} - \frac{N\beta}{2} \|\mathbf{v}\|^2 - N\beta m^* \mathbf{v} \cdot \mathbf{e}_1 \\
& + N \log \mathrm{ch} b + \beta \mathrm{th} b \sum_{i \le N} \boldsymbol{\eta}_i \cdot \mathbf{v} + \frac{\beta^2}{2 \mathrm{ch}^2 b} \sum_{i \le N} (\boldsymbol{\eta}_i \cdot \mathbf{v})^2 \\
& - \frac{\beta^3}{3} \frac{\mathrm{th} b}{\mathrm{ch}^2 b} \sum_{i \le N} (\boldsymbol{\eta}_i \cdot \mathbf{v})^3 + \frac{\beta^4}{6} \sum_{i \le N} R_i(\mathbf{v}) (\boldsymbol{\eta}_i \cdot \mathbf{v})^4
\end{aligned} \tag{4.63}$$

where $|R_i(\mathbf{v})| \le 1$.

The idea of the proof is to simplify (4.63) by averaging over rotations. If U denotes a rotation of $\mathbb{R}^M$, the invariance of Lebesgue's measure by U shows from (4.61) that

$$2^{-N} Z_{N,M} = W \int \exp \psi^{\sim}(U(\mathbf{v}))\mathrm{d}\mathbf{v} \tag{4.64}$$

so that, if $\mathrm{d}U$ denotes Haar measure on the group of rotations, we have

$$\begin{aligned} 2^{-N} Z_{N,M} &= W \int \exp \psi^{\sim}(U(\mathbf{v}))\mathrm{d}U\mathrm{d}\mathbf{v} \\ &\geq W \int \exp \left(\int \psi^{\sim}(U(\mathbf{v}))\mathrm{d}U \right) \mathrm{d}\mathbf{v} \end{aligned} \tag{4.65}$$

by Jensen's inequality.

Given a vector $\mathbf{x}$ of $\mathbb{R}^M$, we have for a certain constant c_p that

$$\int (\mathbf{x} \cdot U(\mathbf{v}))^p \mathrm{d}U = c_p \|\mathbf{x}\|^p \|\mathbf{v}\|^p \tag{4.66}$$

because the left-hand side depends only on $\|\mathbf{x}\|$ and $\|\mathbf{v}\|$.

To compute the quantity c_p, we consider a vector $\mathbf{g} = (g_k)_{k\leq M}$ where (g_k) are independent standard Gaussian r.v.s. We apply (4.66) to $\mathbf{x} = \mathbf{g}$ and we take expectation. We observe that $\mathbf{g} \cdot U(\mathbf{v})$ is a Gaussian r.v. of variance $\|U(\mathbf{v})\|^2$ so that

$$\mathsf{E}((\mathbf{g} \cdot U(\mathbf{v}))^p) = \|U(\mathbf{v})\|^p \mathsf{E}\, g^p = \|\mathbf{v}\|^p \mathsf{E}\, g^p ,$$

where g is a standard Gaussian r.v. Thus we obtain

$$c_p = \frac{\mathsf{E}\, g^p}{\mathsf{E}\, \|\mathbf{g}\|^p} .$$

In particular, $c_p = 0$ when p is odd, $c_2 = 1/M$, and, since

$$\mathsf{E}\, \|\mathbf{g}\|^4 \geq (\mathsf{E}\, \|\mathbf{g}\|^2)^2 = M^2$$

and $\mathsf{E} g^4 = 3$ we get $c_4 \leq 3/M^2$.

We observe that $\|\boldsymbol{\eta}_i\|^2 = M$, so that using (4.66) for $\mathbf{x} = \boldsymbol{\eta}_i$, (4.63) implies

$$\begin{aligned} \int \psi^{\sim}(U(\mathbf{v}))\mathrm{d}U &\geq N \left(\log \mathrm{ch} b - \frac{\beta m^{*2}}{2} \right) \\ &\quad - \frac{N\beta}{2} \left(1 - \frac{\beta}{\mathrm{ch}^2 b} \right) \|\mathbf{v}\|^2 - \frac{N\beta^4}{2} \|\mathbf{v}\|^4 . \end{aligned} \tag{4.67}$$

Since $b = \beta m^* + h$, we have $\mathrm{th} b = m^*$ and thus

$$1 - \frac{\beta}{\mathrm{ch}^2 b} = 1 - \beta(1 - \mathrm{th}^2 b) = 1 - \beta(1 - m^{*2}) = a^* ,$$

so that from (4.65), and since $b^* = \log \mathrm{ch} b - \beta m^{*2}/2$,

$$2^{-N} Z_{N,M} \geq (\exp N b^*) W \int \exp\left(-\frac{N\beta}{2} a^* \|\mathbf{v}\|^2 - \frac{N\beta^4}{2}\|\mathbf{v}\|^4\right) \mathrm{d}\mathbf{v}$$

$$= \left(\frac{1}{a^*}\right)^{M/2} (\exp N b^*) W \int \exp\left(-\frac{N\beta}{2}\|\mathbf{v}\|^2 - \frac{N\beta^4 \|\mathbf{v}\|^4}{2a^{*2}}\right) \mathrm{d}\mathbf{v}$$

by change of variable. Therefore, the definition of γ implies

$$2^{-N} Z_{N,M} \geq \left(\frac{1}{a^*}\right)^{M/2} (\exp N b^*) \int \exp\left(-\frac{N\beta^4}{2}\frac{\|\mathbf{v}\|^4}{a^{*2}}\right) \mathrm{d}\gamma(\mathbf{v}) \, .$$

Recalling also that B of (4.56), we get

$$\int \exp\left(-\frac{N\beta^4}{2}\frac{\|\mathbf{v}\|^4}{a^{*2}}\right) \mathrm{d}\gamma(\mathbf{v}) \geq \gamma(B) \exp\left(-\frac{N\beta^4}{2a^{*2}}\left(2\sqrt{\frac{\alpha}{\beta}}\right)^4\right)$$

$$\geq \frac{1}{L}\exp\left(-\frac{LN\beta^2\alpha^2}{a^{*2}}\right)$$

$$= \frac{1}{L}\exp\left(-\frac{LM\beta^2\alpha}{a^{*2}}\right) .$$

Recalling that $m^{*4} \leq L a^{*2}$ by (4.38), and since we assume that $\beta \leq 2$ we have

$$\alpha \leq m^{*4} \leq L a^{*2} \Rightarrow 2^{-N} Z_{N,M} \geq \left(\frac{1}{L a^*}\right)^{M/2} \exp N b^* \, . \tag{4.68}$$

□

When $\beta \geq 2$, we will use a different bound. We will use the vector $\boldsymbol{\theta} = (\theta_k)_{k \leq M}$ given by

$$\boldsymbol{\theta} = \frac{m^*}{N} \sum_{1 \leq i \leq N} (\boldsymbol{\eta}_i - \mathbf{e}_1) \, , \tag{4.69}$$

so that $\theta_1 = 0$, whereas for $2 \leq k \leq M$ we have

$$\theta_k = \frac{m^*}{N} \sum_{1 \leq i \leq N} \eta_{i,k} \, .$$

Proposition 4.3.4. *We have*

$$2^{-N} Z_{N,M} \geq \exp\left(N b^* + \frac{N\beta}{2}\|\boldsymbol{\theta}\|^2\right) . \tag{4.70}$$

This bound is true for any realization of the disorder and every value of β, M and N. Since $\|\boldsymbol{\theta}\|^2$ is about $\alpha m^*/N$ this is much worse when $\beta \le 2$ than the bound (4.59) when $La^* < 1$.

Proof. The convexity of the function $\log \mathrm{ch}$, and the fact that since $b = \beta m^* + h$, we have $\mathrm{th} b = m^*$ imply that $\log \mathrm{ch}(b + x) \ge \log \mathrm{ch} b + m^* x$. Therefore (4.62) implies

$$\begin{aligned}\psi^{\sim}(\mathbf{v}) &\ge -\frac{N\beta}{2}\|m^*\mathbf{e}_1 + \mathbf{v}\|^2 + N \log \mathrm{ch} b + m^* \sum_{i\le N} \beta \boldsymbol{\eta}_i \cdot \mathbf{v} \\ &= Nb^* - \frac{N\beta}{2}\|\mathbf{v}\|^2 + \beta m^* \Big(\sum_{i\le N} (\boldsymbol{\eta}_i - \mathbf{e}_1) \cdot \mathbf{v}\Big) \\ &= Nb^* - \frac{N\beta}{2}\|\mathbf{v}\|^2 + N\beta \boldsymbol{\theta} \cdot \mathbf{v} \\ &= Nb^* + \frac{N\beta}{2}\|\boldsymbol{\theta}\|^2 - \frac{N\beta}{2}\|\mathbf{v} - \boldsymbol{\theta}\|^2 .\end{aligned}$$

Thus

$$\begin{aligned}W \int \exp \psi^{\sim}(\mathbf{v}) d\mathbf{v} &\ge \exp\Big(Nb^* + \frac{N\beta}{2}\|\boldsymbol{\theta}\|^2\Big) W \int \exp\Big(-\frac{N\beta}{2}\|\mathbf{v} - \boldsymbol{\theta}\|^2\Big) d\mathbf{v} \\ &= \exp\Big(Nb^* + \frac{N\beta}{2}\|\boldsymbol{\theta}\|^2\Big) ,\end{aligned}$$

and the result follows from (4.61). □

We have the following convenient consequence of Propositions 4.3.3 and 4.3.4: whenever $\alpha \le m^{*4}$,

$$2^{-N} Z_{N,M} \ge \Big(\frac{1}{La^*}\Big)^{M/2} \exp b^* N . \tag{4.71}$$

Indeed, if $\beta \le 2$ this follows from Proposition 4.3.3, while if $\beta \ge 2$, by Proposition 4.3.4 we have $2^{-N} Z_{M,N} \ge \exp b^* N$, and, since a^* remains bounded away from 0 as $\beta \ge 2$, we simply take L large enough that then $La^* \ge 1$. The bound (4.71) does not however capture (4.70).

We turn to the task of finding upper bounds for the numerator of (4.58). For this we will have to find an upper bound for ψ. We will use two rather distinct bounds, the first of which will rely on the following elementary fact.

Lemma 4.3.5. *The function*

$$\varphi(x) = \log \mathrm{ch}(\beta\sqrt{x} + h) \tag{4.72}$$

is concave. Moreover, if $x \le 2$ then

$$\varphi''(x) \le -\frac{\beta}{L} . \tag{4.73}$$

Proof. Setting $y = \beta\sqrt{x} + h$, computation shows that

$$\begin{aligned}\varphi''(x) &= \frac{\beta}{4x^{3/2}}\left(\frac{\beta\sqrt{x}}{\mathrm{ch}^2 y} - \mathrm{th} y\right) \le \frac{\beta}{4x^{3/2}}\left(\frac{y}{\mathrm{ch}^2 y} - \mathrm{th} y\right)\\ &= -\frac{\beta}{8x^{3/2}}\left(\frac{\mathrm{sh}2y - 2y}{\mathrm{ch}^2 y}\right) \le 0\,.\end{aligned}$$

Moreover, distinguishing the cases $y \le 1$ and $y \ge 1$, we obtain

$$\frac{\mathrm{sh}2y - 2y}{\mathrm{ch}^2 y} \ge \frac{1}{L}\min(1, y^3) \ge \frac{1}{L}\min(1, x^{3/2})\,.$$

The result follows. □

Before we prove our first localization result, let us make a simple observation that we will use many times: to prove that a set A is negligible, it suffices to show that with overwhelming probability we have $\overline{G}(A) \le K\exp(-N/K)$. This is because for any set A and any $\varepsilon > 0$, we have

$$\mathsf{E}\,\overline{G}(A) \le \mathsf{P}(\overline{G}(A) \ge \varepsilon) + \varepsilon$$

since $\overline{G}(A) \le 1$.

Proposition 4.3.6. *If $\alpha \le m^{*4}/L$, the set*

$$A = \{\mathbf{z} \in \mathbb{R}^M\ ;\ \|\mathbf{z}\| \ge 2m^*\}$$

is negligible for $\overline{G}$, that is

$$\mathsf{E}\,\overline{G}(A) \le K\exp\left(-\frac{N}{K}\right)\,,$$

where K depends only on β and h.

Proof. We write

$$\sum_{i\le N}\log\mathrm{ch}(\beta\boldsymbol{\eta}_i\cdot\mathbf{z} + h) \le \sum_{i\le N}\varphi((\boldsymbol{\eta}_i\cdot\mathbf{z})^2) \le N\varphi\Big(\frac{1}{N}\sum_{i\le N}(\boldsymbol{\eta}_i\cdot\mathbf{z})^2\Big) \tag{4.74}$$

by concavity of φ.

Using Corollary A.9.4 we see that provided

$$\alpha \le \frac{a^{*2}}{L} \tag{4.75}$$

then the event

$$\forall\mathbf{z}\in\mathbb{R}^M,\qquad \frac{1}{N}\sum_{i\le N}(\boldsymbol{\eta}_i\cdot\mathbf{z})^2 \le \Big(1 + \frac{a^*}{8}\Big)\|\mathbf{z}\|^2 \tag{4.76}$$

occurs with overwhelming probability (that is, the probability of failure is at most $K \exp(-N/K)$, where K depends on β, h only). When the event (4.76) occurs, (4.74) implies

$$\begin{aligned}\psi(\mathbf{z}) &\le -\frac{N\beta}{2}\|\mathbf{z}\|^2 + N\varphi\Big(\frac{1}{N}\sum_{i\le N}(\boldsymbol{\eta}_i\cdot\mathbf{z})^2\Big) \\ &\le -\frac{N\beta}{2}\|\mathbf{z}\|^2 + N\varphi\left(\left(1+\frac{a^*}{8}\right)\|\mathbf{z}\|^2\right) . \qquad (4.77)\end{aligned}$$

Let us consider the function $f(t) = \log \operatorname{ch}(\beta t + h) - \beta t^2/2$, so that $\varphi(x) = f(\sqrt{x}) + \beta x/2$ and (4.77) means

$$\psi(\mathbf{z}) \le \frac{N\beta a^*}{16}\|\mathbf{z}\|^2 + Nf\Big(\sqrt{1+\frac{a^*}{8}}\|\mathbf{z}\|\Big) . \qquad (4.78)$$

The second derivative of f is $f''(t) = \beta^2 \operatorname{ch}^{-2}(\beta t + h) - \beta$, which decreases as t increases from 0. Moreover

$$f''(m^*) = -\beta\left(1 - \frac{\beta}{\operatorname{ch}^2(\beta m^* + h)}\right) = -\beta a^* ,$$

where $a^* = 1 - \beta(1 - m^{*2})$. For $t \ge m^*$, we have $f''(t) \le f''(m^*) = -\beta a^*$. Since $f(m^*) = b^*$, $f'(m^*) = 0$, for $t \ge m^*$ we get

$$f(t) \le b^* - \frac{\beta}{2}a^*(t - m^*)^2 .$$

Thus for $t \ge 2m^*$, and since then $t - m^* \ge t/2$ and thus $(t - m^*)^2 \ge t^2/4$, we have

$$f(t) \le b^* - \frac{\beta a^*}{8}t^2 ,$$

and therefore

$$f\Big(\sqrt{1+\frac{a^*}{8}}\|\mathbf{z}\|\Big) \le b^* - \frac{\beta a^*}{8}\left(1+\frac{a^*}{8}\right)\|\mathbf{z}\|^2 \le b^* - \frac{\beta a^*}{8}\|\mathbf{z}\|^2 .$$

It then follows from (4.78) that $\psi(\mathbf{z}) \le Nb^* - N\beta a^*\|\mathbf{z}\|^2/16$ for $\|\mathbf{z}\| \ge 2m^*$. Thus, under (4.75), with overwhelming probability we have

$$\begin{aligned}\int_A \psi(\mathbf{z})d\mathbf{z} &\le \exp Nb^* \int_A \exp\left(-\frac{N\beta}{8}a^*\|\mathbf{z}\|^2\right) d\mathbf{z} \qquad (4.79) \\ &\le \exp\left(N\left(b^* - \frac{\beta a^* m^{*2}}{4}\right)\right)\int_A \exp\left(-\frac{N\beta}{16}a^*\|\mathbf{z}\|^2\right) d\mathbf{z}\end{aligned}$$

because $\|\mathbf{z}\|^2 \ge 4m^{*2}$ on A. Now, by change of variable,

$$\int \exp\left(-\frac{N\beta}{16}a^*\|\mathbf{z}\|^2\right)\mathrm{d}\mathbf{z} = \left(\frac{8}{a^*}\right)^{M/2}\int \exp\left(-\frac{N\beta}{2}\|\mathbf{z}\|^2\right)\mathrm{d}\mathbf{z}$$
$$= \left(\frac{8}{a^*}\right)^{M/2} W^{-1}\,,$$

so that

$$\int_A \psi(\mathbf{z})\mathrm{d}\mathbf{z} \le \exp\left(N\left(b^* - \frac{\beta a^* m^{*2}}{8}\right)\right)\left(\frac{8}{a^*}\right)^{M/2} W^{-1}$$
$$\le \exp\left(N\left(b^* - \frac{\beta m^{*4}}{L}\right)\right)\left(\frac{L}{a^*}\right)^{M/2} W^{-1}$$

since $a^* \ge m^{*2}/L$ by (4.38). Combining with (4.58) and (4.59), we deduce that with overwhelming probability it holds

$$\overline{G}(A) \le L^M \exp\left(-N\frac{\beta m^{*4}}{L}\right) \le \exp N\left(\alpha L_7 - \frac{\beta m^{*4}}{L_7}\right) \le \exp\left(-N\frac{\beta m^{*4}}{2L_7}\right),$$

provided $\alpha \le m^{*4}/2L_7^2$. This completes the proof. □

This preliminary result is interesting in itself, and will be very helpful since from now on we need to be concerned only with the values of $\mathbf{z}$ such that $|\mathbf{z}| \le 2m^*$.

Our further results will be based on the following upper bound for ψ; it is this bound that is the crucial fact.

Lemma 4.3.7. *We have*

$$\psi(\mathbf{z}) \le Nb^* + \frac{\beta}{2}\left(\sum_{i\le N}(\boldsymbol{\eta}_i\cdot\mathbf{z})^2 - N\|\mathbf{z}\|^2\right)$$
$$-\frac{\beta}{L}\sum_{i\le N}\min\bigl(1, ((\boldsymbol{\eta}_i\cdot\mathbf{z})^2 - m^{*2})^2\bigr)\,. \tag{4.80}$$

The last term in (4.80) has a crucial influence. There are two main steps to use this term. First, we will learn to control it from below uniformly on large balls. This control will be achieved by proving that with overwhelming probability at every point of the ball this term is not too much smaller than its expectation. In a second but separate step, we will show that this expectation cannot be small unless $\mathbf{z}$ is close to one of the points $\pm m^*\mathbf{e}_i$. Therefore with overwhelming probability this last term can be small only if $\mathbf{z}$ is close to one of the points $\pm m^*\mathbf{e}_i$, and this explains why Gibbs' measure concentrates near these points.

There is some simple geometry behind the behavior of the expectation of the last term of (4.80). If we forget the minimum and consider simply the average

$$\frac{1}{N}\sum_{i\le N}\left((\boldsymbol{\eta}_i\cdot\mathbf{z})^2-m^{*2}\right)^2 ,$$

its expectation is precisely

$$(\|\mathbf{z}\|^2-m^{*2})^2+\sum_{k\ne\ell} z_k^2 z_\ell^2 .$$

As a warm-up before the real proof, the reader should convince herself that this quantity can be small only if one of the z_k's is approximately $\pm m^*$ and the rest are nearly zero.

Proof of Lemma 4.3.7. We recall the function φ of Lemma 4.3.5. Assuming for definiteness that $x\ge m^{*2}$, this lemma implies

$$\begin{aligned}\varphi(x)&=\varphi(m^{*2})+\varphi'(m^{*2})(x-m^{*2})+\int_{m^{*2}}^{x}(x-t)\varphi''(t)\mathrm{d}t\\&\le\varphi(m^{*2})+\varphi'(m^{*2})(x-m^{*2})-\frac{\beta}{L}\int_{m^{*2}}^{\min(x,2)}(x-t)\mathrm{d}t\\&\le\varphi(m^{*2})+\varphi'(m^{*2})(x-m^{*2})-\frac{\beta}{L}\min\bigl(1,(x-m^{*2})^2\bigr) .\end{aligned}\tag{4.81}$$

Now,

$$\varphi'(m^{*2})=\frac{\beta}{2m^*}\mathrm{th}(\beta m^*+h)=\frac{\beta}{2}$$

and

$$\varphi(m^{*2})-\frac{\beta}{2}m^{*2}=\log\mathrm{ch}(\beta m^*+h)-\frac{\beta}{2}m^{*2}=b^* ,$$

so that (4.81) implies

$$\varphi(x)\le b^*+\frac{\beta}{2}x-\frac{\beta}{L}\min\bigl(1,(x-m^{*2})^2\bigr) .$$

Using this for $x=(\boldsymbol{\eta}_i\cdot\mathbf{z})^2$, summing over $i\le N$ and using the first inequality in (4.74) yields the result. □

To perform the program outlined after the statement of Lemma 4.3.7, it helps to introduce a kind of truncation. Given a parameter $d\le 1$, we write

$$R_d(\mathbf{z})=\mathsf{E}\min\bigl(d,((\boldsymbol{\eta}_i\cdot\mathbf{z})^2-m^{*2})^2\bigr) ,\tag{4.82}$$

a quantity which is of course does not depend on i.

Proposition 4.3.8. *Consider a ball B of $\mathbb{R}^M$, of radius ρ, and assume that*

$$B\subset\{\mathbf{z}\ ;\ \|\mathbf{z}\|\le 2m^*\} .\tag{4.83}$$

Then, for each $\varepsilon>0$, with overwhelming probability, we have

$$\forall \mathbf{z} \in B, \quad \sum_{i \le N} \min\big(d, ((\boldsymbol{\eta}_i \cdot \mathbf{z})^2 - m^{*2})^2\big)$$
$$\ge \frac{N}{4}\left(R_d(\mathbf{z}) - 4\alpha \log\left(1 + \frac{\rho}{\varepsilon}\right) - L\varepsilon m^* \sqrt{d}\right) . \qquad (4.84)$$

This is the first part of our program, showing that the last term of (4.80) is not too much below its expectation. The strange logarithm in the right-hand side will turn to be harmless, because we will always choose ρ and ε of the same order.

The proof of Proposition 4.3.8 itself has two steps. In the first we will show that the left-hand side of (4.84) can be controlled uniformly over many points. In the second we will show that this implies uniform control over B.

Lemma 4.3.9. *Consider a finite subset A of $\mathbb{R}^M$ and $C > 1$. Then, with probability at least $1 - 1/C$ we have*

$$\forall \mathbf{z} \in A, \sum_{i \le N} \min\big(d, ((\boldsymbol{\eta}_i \cdot \mathbf{z})^2 - m^{*2})^2\big) \ge \frac{N}{4} R_d(\mathbf{z}) - \log \operatorname{card} A - \log C \ . \quad (4.85)$$

Proof. Let

$$A_1 = \{\mathbf{z} \in A \ ; \ \frac{N}{4} R_d(\mathbf{z}) \ge \log \operatorname{card} A + \log C\} \ .$$

To prove (4.85), it suffices to achieve control over $\mathbf{z} \in A_1$. Let us fix $\mathbf{z}$ in A_1. The r.v.s

$$X_i = \min\big(d, ((\boldsymbol{\eta}_i \cdot \mathbf{z})^2 - m^{*2})^2\big)$$

are i.i.d., $0 \le X_i \le 1$, $\mathsf{E} X_i = R_d(\mathbf{z})$. We prove an elementary exponential inequality about these variables. Since $\exp(-x) \le 1 - x/2 \le \exp(-x/2)$ for $x \le 1$, we have

$$\mathsf{E} \exp(-X_i) \le 1 - \frac{\mathsf{E} X_i}{2} \le \exp\left(-\frac{\mathsf{E} X_i}{2}\right) = \exp\left(-\frac{R_d(\mathbf{z})}{2}\right)$$

and thus

$$\mathsf{E} \exp\Big(-\sum_{i \le N} X_i\Big) \le \exp\left(-\frac{N R_d(\mathbf{z})}{2}\right) ,$$

so that

$$\mathsf{P}\Big(\sum_{i \le N} X_i \le \frac{N R_d(\mathbf{z})}{4}\Big) \exp\left(-\frac{N R_d(\mathbf{z})}{4}\right) \le \exp\left(-\frac{N R_d(\mathbf{z})}{2}\right) ,$$

and

$$\mathsf{P}\Big(\sum_{i \le N} X_i \le \frac{N R_d(\mathbf{z})}{4}\Big) \le \exp\left(-\frac{N R_d(\mathbf{z})}{4}\right) \le \frac{1}{C \operatorname{card} A}$$

since $\mathbf{z} \in A_1$. Thus, with probability at least $1 - 1/C$, we have

$$\forall \mathbf{z} \in A_1 \,, \sum_{i \le N} \min\big(d, ((\boldsymbol{\eta}_i \cdot \mathbf{z})^2 - m^{*2})^2\big) \ge \frac{N R_d(\mathbf{z})}{4} \,. \qquad \square$$

Next, we relate what happens for two points close to each other.

Lemma 4.3.10. *We have*

$$|R_d(\mathbf{z}_1) - R_d(\mathbf{z}_2)| \le L\sqrt{d}\|\mathbf{z}_1 - \mathbf{z}_2\|(\|\mathbf{z}_1\| + \|\mathbf{z}_2\|) \,. \tag{4.86}$$

Moreover, with overwhelming probability it is true that for any $\mathbf{z}_1$ *and* $\mathbf{z}_2$ *in* $\mathbb{R}^M$ *we have*

$$\Big|\sum_{i \le N} \min\big(d, ((\boldsymbol{\eta}_i \cdot \mathbf{z}_1)^2 - m^{*2})^2\big) - \sum_{i \le N} \min\big(d, ((\boldsymbol{\eta}_i \cdot \mathbf{z}_2)^2 - m^{*2})^2\big)\Big|$$
$$\le NL\sqrt{d}\|\mathbf{z}_1 - \mathbf{z}_2\|(\|\mathbf{z}_1\| + \|\mathbf{z}_2\|) \,. \tag{4.87}$$

Proof. We start with the observation that, since $d \le 1$,

$$\begin{aligned} &|\min(d, x^2) - \min(d, y^2)| \\ &= |\min(\sqrt{d}, |x|)^2 - \min(\sqrt{d}, |y|)^2| \\ &= |\min(\sqrt{d}, |x|) - \min(\sqrt{d}, |y|)|(\min(\sqrt{d}, |x|) + \min(\sqrt{d}, |y|)) \\ &\le 2\sqrt{d}|\min(\sqrt{d}, |x|) - \min(\sqrt{d}, |x|)| \le 2\sqrt{d}|x - y| \,, \end{aligned} \tag{4.88}$$

and thus the left-hand side of (4.87) is bounded by

$$\begin{aligned} &2\sqrt{d}\sum_{i \le N} |(\boldsymbol{\eta}_i \cdot \mathbf{z}_1)^2 - (\boldsymbol{\eta}_i \cdot \mathbf{z}_2)^2| \\ &\le 2\sqrt{d}\sum_{i \le N} |\boldsymbol{\eta}_i \cdot (\mathbf{z}_1 - \mathbf{z}_2)|(|\boldsymbol{\eta}_i \cdot \mathbf{z}_1| + |\boldsymbol{\eta}_i \cdot \mathbf{z}_2|) \\ &\le 2\sqrt{d}\Big(\sum_{i \le N} (\boldsymbol{\eta}_i \cdot (\mathbf{z}_1 - \mathbf{z}_2))^2\Big)^{1/2} \\ &\quad \times \Big[\Big(\sum_{i \le N} (\boldsymbol{\eta}_i \cdot \mathbf{z}_1)^2\Big)^{1/2} + \Big(\sum_{i \le N} (\boldsymbol{\eta}_i \cdot \mathbf{z}_2)^2\Big)^{1/2}\Big] \,. \end{aligned} \tag{4.89}$$

Taking expectation and using the Cauchy-Schwarz inequality proves (4.86) since the expectation of the left-hand side of (4.87) is $N|R_d(\mathbf{z}_1) - R_d(\mathbf{z}_2)|$. Moreover, with overwhelming probability (4.44) holds, and then (4.87) is a simple consequence of (4.89). $\square$

Proof of Proposition 4.3.8. It is shown in Proposition A.8.1 that we can find a subset A of B such that

$$\operatorname{card} A \le \Big(1 + \frac{\rho}{\varepsilon}\Big)^M$$

and such that each point of B is within distance 2ε of A. We apply Lemma 4.3.9 with $C = \exp(N\varepsilon m^* \sqrt{d})$. We observe that given $\mathbf{z}_2$ in B, there exists $\mathbf{z}_1$ in A with $\|\mathbf{z}_2 - \mathbf{z}_1\| \le 2\varepsilon$, and $\|\mathbf{z}_1\|, \|\mathbf{z}_2\| \le 2m^*$. We then apply (4.87) and (4.86) to obtain the result. The choice of C is no magic, the exponent is simply small enough that $\log C$ is about the error term $LN\varepsilon m^* \sqrt{d}$ produced by Lemma 4.3.10. □

To use efficiently (4.84), we need to understand the geometric nature of $R_d(\mathbf{z})$. We will show that this quantity is small only when $\mathbf{z}$ is close to a point $\pm m^* \mathbf{e}_k$.

Lemma 4.3.11. *Consider a number* $0 \le \xi \le 1$. *Assume that*

$$\forall k \le M, \quad \|\mathbf{z} \pm m^* \mathbf{e}_k\| \ge \xi m^* . \tag{4.90}$$

Then if

$$d = \xi^2 m^{*4} , \tag{4.91}$$

we have

$$R_d(\mathbf{z}) \ge \frac{\xi^2}{L} m^{*4} . \tag{4.92}$$

The proof relies on the following probabilistic estimate.

Lemma 4.3.12. *We have*

$$R_d(\mathbf{z}) \ge \frac{1}{L} \min\Big(d, (\|\mathbf{z}\|^2 - m^{*2})^2 + \sum_{k \ne \ell} z_k^2 z_\ell^2 \Big) . \tag{4.93}$$

Proof of Lemma 4.3.11. Using (4.93), it is enough to prove that if

$$(\|\mathbf{z}\|^2 - m^{*2})^2 + \sum_{k \ne \ell} z_k^2 z_\ell^2 \le \frac{\xi^2 m^{*4}}{16} \tag{4.94}$$

then we can find $k \le M$ and $\tau = \pm 1$ such that

$$\|\mathbf{z} - \tau m^* \mathbf{e}_k\| < \xi m^* . \tag{4.95}$$

First, we observe from (4.94) that

$$\big|\|\mathbf{z}\|^2 - m^{*2}\big| \le \frac{\xi m^{*2}}{4} \tag{4.96}$$

so that

$$\big|\|\mathbf{z}\| - m^*\big| = \frac{\big|\|\mathbf{z}\|^2 - m^{*2}\big|}{\|\mathbf{z}\| + m^*} \le \frac{\xi m^*}{4} . \tag{4.97}$$

Next, (4.94) implies

$$\frac{\xi^2 m^{*4}}{16} \geq \sum_{k\neq\ell} z_k^2 z_\ell^2 = \|\mathbf{z}\|^4 - \sum_{\ell\leq M} z_\ell^4$$
$$\geq \|\mathbf{z}\|^4 - (\max_{\ell\leq M} z_\ell^2) \sum_{\ell\leq M} z_\ell^2$$
$$= \|\mathbf{z}\|^2(\|\mathbf{z}\|^2 - \max_{\ell\leq M} z_\ell^2) . \tag{4.98}$$

Consider k such that $z_k^2 = \max_{\ell\leq M} z_\ell^2$. Then, since $\|\mathbf{z}\|^2 \geq 3m^{*2}/4$ by (4.96), we have from (4.98) that

$$\|\mathbf{z}\|^2 - z_k^2 \leq \frac{\xi^2 m^{*4}}{16\|\mathbf{z}\|^2} \leq \frac{\xi^2 m^{*2}}{12} .$$

Now $\|\mathbf{z}\|^2 - z_k^2 = \sum_{\ell\neq k} z_\ell^2 = \|\mathbf{z} - z_k\mathbf{e}_k\|^2$, so that

$$\|\mathbf{z} - z_k\mathbf{e}_k\| \leq \frac{\xi m^*}{3} \tag{4.99}$$

and consequently

$$|\|\mathbf{z}\| - |z_k|| \leq \frac{\xi m^*}{3} . \tag{4.100}$$

Moreover, if $\tau = \mathrm{sign} z_k$, we have

$$\|z_k\mathbf{e}_k - \tau m^*\mathbf{e}_k\| = |z_k - \tau m^*| = ||z_k| - m^*|$$
$$\leq |\|\mathbf{z}\| - m^*| + |\|\mathbf{z}\| - |z_k|| \leq \left(\frac{1}{4} + \frac{1}{3}\right) \xi m^* , \tag{4.101}$$

using (4.97) and (4.100). Combining with (4.99) proves (4.95). □

Proof of Lemma 4.3.12. We consider the r.v.s

$$X = ((\boldsymbol{\eta}_i \cdot \mathbf{z})^2 - m^{*2})^2 , \tag{4.102}$$

so the Paley-Zygmund inequality (A.80) implies

$$\mathsf{P}\left(X \geq \frac{1}{2}\mathsf{E}\, X\right) \geq \frac{1}{4}\frac{(\mathsf{E}\, X)^2}{\mathsf{E}\,(X^2)} , \tag{4.103}$$

and thus

$$\mathsf{E} \min(d, X) \geq \min\left(d, \frac{\mathsf{E}\, X}{2}\right) \mathsf{P}\left(X \geq \frac{\mathsf{E}\, X}{2}\right)$$
$$\geq \min\left(d, \frac{\mathsf{E}\, X}{2}\right) \frac{(\mathsf{E}\, X)^2}{4\mathsf{E}\,(X^2)} . \tag{4.104}$$

We have

$$X = (U + a)^2 , \tag{4.105}$$

where $a = \|\mathbf{z}\|^2 - m^{*2}$,

$$U = \sum_{k \neq \ell} \eta_{i,k}\eta_{i,\ell} z_k z_\ell \ ,$$

so that $\mathsf{E}\, U = 0$, $\mathsf{E}\, U^2 = \sum_{k \neq \ell} z_k^2 z_\ell^2$ and thus

$$\mathsf{E}\, X = \sum_{k \neq \ell} z_k^2 z_\ell^2 + a^2 \ . \tag{4.106}$$

It can be checked simply by force (expansion) that

$$\mathsf{E}\, U^4 \leq L(\mathsf{E}\, U^2)^2 \ , \tag{4.107}$$

but there are much nicer arguments to do this [14]. From (4.105) it follows that

$$\begin{aligned} \mathsf{E}\, X^2 &= \mathsf{E}\, U^4 + 4a\mathsf{E}\, U^3 + 6a^2 \mathsf{E}\, U^2 + 4a^3 \mathsf{E}\, U + a^4 \\ &\leq \mathsf{E}\, U^4 + 4a(\mathsf{E}\, U^4)^{3/4} + 6a^2 \mathsf{E}\, U^2 + a^4 \\ &\leq L((\mathsf{E}\, U^2)^2 + a(\mathsf{E}\, U^2)^{3/2} + a^2 \mathsf{E}\, U^2 + a^4) \end{aligned}$$

using (4.107). Using that $ab \leq a^4 + b^{4/3}$ for $b = (\mathsf{E}\, U^2)^{3/2}$, we get $\mathsf{E}\, X^2 \leq L(\mathsf{E}\, U^2 + a^2)^2 = L(\mathsf{E}\, X)^2$ and (4.104) implies

$$\mathsf{E}\, \min(d, X) \geq \frac{1}{L} \min\left(d, \frac{\mathsf{E}\, X}{2}\right) \ . \qquad \square$$

We now put together the different pieces, and we state our main tool for the proof of Theorem 4.3.2.

Proposition 4.3.13. *Assume that $\alpha \leq m^{*4}/L$. Consider a ball B of $\mathbb{R}^M$ of radius ρ, and assume that*

$$B \subset \{\|\mathbf{z}\| \ ; \ \|\mathbf{z}\| \leq 2m^*\} \ .$$

Consider a subset A of B and assume that for some number $\xi < 1$,

$$\mathbf{z} \in A \Rightarrow \forall k \leq M \ , \ \|\mathbf{z} \pm m^* \mathbf{e}_k\| \geq \xi m^* \ .$$

Then, with overwhelming probability, we have

$$\begin{aligned} \overline{G}(A) \leq (La^*)^{M/2} W \int_A \exp\Bigg(\frac{\beta}{2}\Bigg(\sum_{i \leq N} (\boldsymbol{\eta}_i \cdot \mathbf{z})^2 - N\|\mathbf{z}\|^2\Bigg) \\ - \frac{N\beta}{L}\left(\xi^2 m^{*4} - L\alpha \log\left(1 + \frac{L\rho}{\xi m^*}\right)\right)\Bigg) \mathrm{d}\mathbf{z} \ . \end{aligned} \tag{4.108}$$

Proof. We take $d = \xi^2 m^{*4}$. We recall (4.80) and we consider Proposition 4.3.8 and Lemma 4.3.11 to see that, given $\varepsilon > 0$, with overwhelming probability we have

$$\forall \mathbf{z} \in A\,,\ \psi(\mathbf{z}) \le Nb^* + \frac{\beta}{2}\Big(\sum_{i\le N}(\boldsymbol{\eta}_i\cdot\mathbf{z})^2 - N\|\mathbf{z}\|^2\Big)$$
$$-\frac{N\beta}{L}\Big(\xi^2 m^{*4} - L\alpha\log\Big(1+\frac{\rho}{\varepsilon}\Big) - L_7\varepsilon\xi m^{*3}\Big)\,.$$

We choose $\varepsilon = \xi m^*/2L_7$ and the result then follows from (4.58) and (4.71). □

Proof of Theorem 4.3.2. First we consider the case where

$$B = \{\mathbf{z}\,;\ \|\mathbf{z}\| \le 2m^*\}$$

$$A = \Big\{\mathbf{z}\,;\ \|\mathbf{z}\| \le 2m^*\,,\ \forall k \le M\,,\ \|\mathbf{z} \pm m^*\mathbf{e}_k\| \ge \frac{1}{2}m^*\Big\}\,. \tag{4.109}$$

Thus, we can use (4.108) with $\rho = 2m^*$ and $\xi = 1/2$. Consider a number $0 < c < 1$, to be determined later. Using Corollary A.9.4 we see that if $\alpha \le c^2/L$, with overwhelming probability we have

$$\overline{G}(A) \le (La^*)^{M/2} W \int_A \exp N\beta\Big(cL_8\|\mathbf{z}\|^2 - \frac{m^{*4}}{L_8} + L_8\alpha\Big)\,d\mathbf{z}\,. \tag{4.110}$$

It appears that a good choice for c is $c = m^{*2}/16L_8^2$, so that

$$\|\mathbf{z}\| \le 2m^* \Rightarrow cL_8\|\mathbf{z}\|^2 - \frac{m^{*4}}{L_8} \le -\frac{m^{*4}}{2L_8} - cL_8\|\mathbf{z}\|^2$$

and thus (4.110) yields that if $\alpha \le m^{*4}/4L_8^2$, with overwhelming probability we have

$$\overline{G}(A) \le (La^*)^{M/2} W \exp\Big(-\frac{N\beta m^{*4}}{L}\Big)\int \exp\Big(-\frac{N\beta m^{*2}}{L}\|\mathbf{z}\|^2\Big)\,d\mathbf{z}$$
$$\le \Big(\frac{La^*}{m^{*2}}\Big)^{M/2}\exp\Big(-\frac{N\beta m^{*4}}{L}\Big)\,. \tag{4.111}$$

Since $a^* \le Lm^{*2}$ by (4.38) we get

$$\Big(\frac{La^*}{m^{*2}}\Big)^{M/2} \le L^M = L^{\alpha N} \le \exp L\alpha N\,,$$

so that for $\alpha \le m^{*4}/L$ with overwhelming probability we have $\overline{G}(A) \le \exp(-N\beta m^{*4}/L)$. This implies that A is negligible.

At this stage, we know that $\overline{G}$ is essentially supported by the sets $\|\mathbf{z} \pm m^*\mathbf{e}_k\| \le m^*/2$. To go beyond this, the difficulty in using (4.108) is to control the term $\sum_{i\leq N}(\boldsymbol{\eta}_i\cdot\mathbf{z})^2 - N\|\mathbf{z}\|^2$. A simple idea is that it is easier to control this term when we know that $\mathbf{z}$ is not too far from a point $\pm m^*\mathbf{e}_k$.

Given an integer $\ell\geq 1$, given $\tau=\pm 1$, given $k\leq M$, we want to apply (4.108) to the sets

$$B=\{\mathbf{z}\,;\ \ \|\mathbf{z}-\tau m^*\mathbf{e}_k\|\leq 2^{-\ell}m^*\}\,,$$

$$A=A_{k,\tau}=\{\mathbf{z}\,;\ \ 2^{-\ell-1}m^*\leq\|\mathbf{z}-\tau m^*\mathbf{e}_k\|\leq 2^{-\ell}m^*\}\,. \tag{4.112}$$

Thus, we can use (4.108) with $\xi=2^{-\ell-1}$, $\rho=2^{-\ell}m^*$. We set $\mathbf{v}=\mathbf{z}-\tau m^*\mathbf{e}_k$, so that

$$\sum_{i\leq N}(\boldsymbol{\eta}_i\cdot\mathbf{z})^2-N\|\mathbf{z}\|^2=\sum_{i\leq N}(\boldsymbol{\eta}_i\cdot\mathbf{v})^2-N\|\mathbf{v}\|^2 \tag{4.113}$$
$$+\,2\tau m^*\left(\sum_{i\leq N}(\boldsymbol{\eta}_i\cdot\mathbf{v})(\boldsymbol{\eta}_i\cdot\mathbf{e}_k)-N\mathbf{v}\cdot\mathbf{e}_k\right),$$

where we have used that $|\boldsymbol{\eta}_i\cdot\mathbf{e}_k|=1$. Consider now a parameter c to be chosen later (depending only on ℓ,β,h). Corollary A.9.4 implies that if $\alpha\leq c^2/L$, with overwhelming probability the quantity (4.113) is at most

$$Nc(\|\mathbf{v}\|^2+m^*\|\mathbf{v}\|)\,,$$

and (4.108) implies

$$\overline{G}(A)\leq(La^*)^{M/2}W\int_{\{\|\mathbf{v}\|\leq 2^{-\ell}m^*\}}\exp N\frac{\beta}{2}C(\mathbf{v})d\mathbf{v}\,, \tag{4.114}$$

where

$$C(\mathbf{v})=c(\|\mathbf{v}\|^2+m^*\|\mathbf{v}\|)-\frac{m^{*4}2^{-2\ell}}{L}+L\alpha\,.$$

Since $\alpha\leq c^2$ and $\|\mathbf{v}\|\leq 2^{-\ell}m^*$, we get

$$C(\mathbf{v})\leq L_9c(2^{-\ell}m^{*2})-\frac{m^{*4}2^{-2\ell}}{L_9}+L_9c^2\,.$$

It is a good idea to take $c=2^{-\ell}m^{*2}/4L_9^2$ because then, for $\|\mathbf{v}\|\leq 2^{-\ell}m^*$,

$$C(\mathbf{v})\leq-\frac{m^{*4}2^{-2\ell}}{2L_9}\leq\frac{m^{*4}2^{-2\ell}}{4L_9}-\frac{m^{*2}}{4L_9}\|\mathbf{v}\|^2\,,$$

and, since $a^*\leq Lm^{*2}$, (4.114) gives

$$\begin{aligned}\overline{G}(A)&\leq(La^*)^{M/2}W\exp\left(-\frac{N\beta m^{*4}2^{-2\ell}}{L}\right)\int\exp\left(-\frac{N\beta m^{*2}}{L}\|\mathbf{v}\|^2\right)d\mathbf{v}\\&=\left(\frac{La^*}{m^{*2}}\right)^{M/2}\exp\left(-\frac{N\beta m^{*4}2^{-2\ell}}{L}\right)\\&\leq L^{M/2}\exp\left(-\frac{N\beta m^{*4}2^{-2\ell}}{L}\right)\\&=\exp\left(ML_{10}-\frac{N\beta m^{*4}2^{-2\ell}}{L_{10}}\right)\leq K\exp\left(-\frac{N}{K}\right)\end{aligned} \tag{4.115}$$

provided $\alpha \le 2^{-2\ell} m^{*4}/2L_{10}^2$. Thus, if

$$A_\ell^* = \bigcup_{k,\tau} A_{k,\tau} = \{\mathbf{z} \; ; \; \exists k \le M, \tau = \pm 1, \; 2^{-\ell-1} m^* \le \|\mathbf{z} - \tau m^* \mathbf{e}_k\| \le 2^{-\ell} m^*\} \, ,$$

we have $G(A_\ell^*) \le KM \exp(-N/K) \le K' \exp(-N/2K)$ since $M \le N$. Summarizing, if $\alpha \le m^{*2}\rho_0^2/L$, the set A_ℓ^* is negligible whenever $\ell \ge 0$ and $m^* 2^{-\ell} \ge \rho_0$. Combining with Proposition 4.3.6, we have proved Theorem 4.3.2. □

4.4 Selecting a State with an External Field

We recall the notation $\mathbf{m}(\boldsymbol{\sigma}) = (m_k(\boldsymbol{\sigma}))_{k \le M}$ of (4.32). A consequence of Theorem 4.3.2 is as follows. Consider, for $k \le M$, the set

$$C_k = \left\{ \boldsymbol{\sigma} \; ; \; \|\mathbf{m}(\boldsymbol{\sigma}) \pm m^* \mathbf{e}_k\| \le \frac{m^*}{4} \right\} . \tag{4.116}$$

Then, if $\alpha \le m^{*4}/L$, Gibbs' measure is essentially supported by the union of the sets C_k, as $k \le M$.

Conjecture 4.4.1. (Level 2) Assume that $h = 0$. If M is large, prove that, for the typical disorder, there is a $k \le M$ such that $G(C_k)$ is nearly 1.

The paper [30] contains some results relevant to this conjecture.

Research Problem 4.4.2. (Level 2) More generally, when $h = 0$, understand precisely the properties of the random sequence $(G(C_k))_{k \le M}$.

Our goal in the present section is to prove the following.

Theorem 4.4.3. *If $\beta > 1$, $h > 0$, then for $\alpha \le m^{*4}/L$, the set*

$$A = \left\{ \boldsymbol{\sigma} \; ; \; \|\mathbf{m}(\boldsymbol{\sigma}) - m^* \mathbf{e}_1\| \ge \frac{m^*}{4} \right\} \tag{4.117}$$

is negligible.

This means of course that $\mathsf{E} G(A) \le K \exp(-N/K)$ where K depends only on β, h.

The content of Theorem 4.4.3 is as follows. We know from Theorem 4.3.2 that Gibbs' measure is essentially supported by the union of $2M$ balls centered at $\pm m^* \mathbf{e}_k$ with a certain radius. When $h \ne 0$, *however small*, only the ball centered at $m^* \mathbf{e}_1$ matters. A precise consequence is as follows.

Theorem 4.4.4. *Consider $\beta > 1$, $h > 0$ and $\rho_0 \le m^*/2$. If $\alpha \le m^{*2}\rho_0^2/L$, then G' is essentially supported by the ball in $\mathbb{R}^M$ of radius ρ_0 centered at the point $m^*\mathbf{e}_1$. Equivalently, the set*

$$\{\boldsymbol{\sigma}\ ;\ \|\mathbf{m}(\boldsymbol{\sigma}) - m^*\mathbf{e}_1\| \ge \rho_0\} \tag{4.118}$$

is negligible.

Proof. By Theorem 4.3.2 the set

$$\{\boldsymbol{\sigma}\ ;\ \forall k \le M, \|\mathbf{m}(\boldsymbol{\sigma}) \pm m^*\mathbf{e}_k\| \ge \rho_0\}$$

is negligible. The union of this set and the set (4.117) is the set (4.118), which is therefore negligible. □

For $k \le M$ and $\tau = \pm 1$ we consider the sets

$$B_{k,\tau} = \left\{\boldsymbol{\sigma}\ ;\ \|\mathbf{m}(\boldsymbol{\sigma}) - \tau m^*\mathbf{e}_k\| \le \frac{m^*}{4}\right\} \tag{4.119}$$

$$C_{k,\tau} = \left\{\boldsymbol{\sigma}\ ;\ \tau m_k(\boldsymbol{\sigma}) \ge \frac{3m^*}{4}\right\} . \tag{4.120}$$

Let us denote by $H^0_{N,M}(\boldsymbol{\sigma})$ the Hamiltonian (4.18), which corresponds to the case $h = 0$. We define

$$S(k,\tau) = \sum_{\boldsymbol{\sigma} \in C_{k,\tau}} \exp(-H^0_{N,M}(\boldsymbol{\sigma})) . \tag{4.121}$$

The crucial property is as follows.

Lemma 4.4.5. *There exists a number a such that for each $k \le M$, each $\tau = \pm 1$, we have*

$$0 \le u \le 1 \Rightarrow \mathsf{P}\left(\left|\frac{1}{N}\log S(k,\tau) - a\right| \ge u\right) \le K \exp\left(-\frac{Nu^2}{K}\right) . \tag{4.122}$$

It suffices to prove this for $k = \tau = 1$, because the r.v.s $S(k,\tau)$ all have the same distribution. This inequality relies on a "concentration of measure" principle that is somewhat similar to Theorem 1.3.4. This principle, which has received numerous applications, is explained in great detail in Section 6 of [140], and Lemma 4.4.5 is proved exactly as Theorem 6.8 there. The author believes that learning properly this principle is well worth the effort, and that this is better done by reading [140] than by reading only the proof of Lemma 4.4.5. Thus, Lemma 4.4.5 will be one of the few exceptions to our policy of giving complete self-contained proofs.

Proof of Theorem 4.4.3. We have

$$G(C_{1,1}) = \frac{1}{Z_{N,M}} \sum_{\boldsymbol{\sigma} \in C_{1,1}} \exp(-H_{N,M}(\boldsymbol{\sigma})) .$$

For $\boldsymbol{\sigma}$ in $C_{1,1}$, we have $hm_1(\boldsymbol{\sigma}) \geq 3hm^*/4$, so that $-H_{N,M}(\boldsymbol{\sigma}) \geq -H^0_{N,M}(\boldsymbol{\sigma}) + 3Nhm^*/4$, and

$$1 \geq G(C_{1,1}) \geq \exp\left(\frac{3}{4}Nhm^*\right) \frac{S(1,1)}{Z_{N,M}} . \tag{4.123}$$

If $(k,\tau) \neq (1,1)$, we have $hm_1(\boldsymbol{\sigma}) \leq hm^*/4$ for $\boldsymbol{\sigma}$ in $B_{k,\tau}$ so that

$$\begin{aligned} \sum_{\boldsymbol{\sigma} \in B_{k,\tau}} \exp(-H_{N,M}(\boldsymbol{\sigma})) &\leq \exp\left(\frac{1}{4}Nhm^*\right) \sum_{\boldsymbol{\sigma} \in B_{k,\tau}} \exp(-H^0_{N,M}(\boldsymbol{\sigma})) \\ &\leq \exp\left(\frac{1}{4}Nhm^*\right) S(k,\tau) \end{aligned}$$

and thus

$$G(B_{k,\tau}) \leq \exp\left(\frac{1}{4}Nhm^*\right) \frac{S(k,\tau)}{Z_{N,M}} . \tag{4.124}$$

Taking $u = \min(1, hm^*/8)$ in Lemma 4.4.5 shows that with overwhelming probability we have

$$S(1,1) \geq \exp\left(Na - \frac{1}{8}Nhm^*\right)$$

$$\forall\, k,\tau, \quad S(k,\tau) \leq \exp\left(Na + \frac{1}{8}Nhm^*\right)$$

and thus

$$\forall\, k,\tau, \quad S(k,\tau) \leq \exp\left(\frac{1}{4}Nhm^*\right) S(1,1) .$$

Combining with (4.123) and (4.124) yields that, with overwhelming probability,

$$(k,\tau) \neq (1,1) \Rightarrow G(B_{k,\tau}) \leq \exp\left(-\frac{1}{4}Nhm^*\right)$$

so that $B_{k,\tau}$ is negligible. Combining with Theorem 4.3.2 finishes the proof. □

4.5 Controlling the Overlaps

From now on we assume $h > 0$, and we recall Theorem 4.4.4: given $\rho_0 \leq m^*/2$, if $\alpha \leq m^{*2}\rho_0^2/L_8$, then G' is essentially supported by the set

$$B_1 = \{\mathbf{z}\ ;\ \|\mathbf{z} - m^*\mathbf{e}_1\| \leq \rho_0\} .$$

Moreover (assuming without loss of generality that $L_8 \geq 4$) if follows from (4.55) that γ is essentially supported by the set

$$B_2 = \{\mathbf{z} \; ; \; \|\mathbf{z}\| \leq \rho_0\} \; .$$

Therefore $\overline{G} = G' * \gamma$ is essentially supported by

$$B = B_1 + B_2 = \{\mathbf{z} \; ; \; \|\mathbf{z} - m^* \mathbf{e}_1\| \leq 2\rho_0\} \; .$$

Using this for $\rho_0 = m^*/L\sqrt{\beta}$ proves that if $\alpha \leq m^{*4}/L\beta$, then $\overline{G}$ is essentially supported by the set

$$B = \left\{\mathbf{z} \; ; \; \|\mathbf{z} - m^* \mathbf{e}_1\| \leq \frac{m^*}{L\sqrt{\beta}}\right\} . \tag{4.125}$$

Combining with Theorem 4.2.2 yields that moreover there exists $\kappa > 0$ such that, with overwhelming probability, the function $\mathbf{z} \mapsto \psi(\mathbf{z}) + \kappa N \|\mathbf{z}\|^2$ is concave on B (so that ψ satisfies (3.21)). This will permit us to control the model in the region where $\alpha \leq m^{*4}/L\beta$. (In Chapter 10 we will be able to control a larger region using a different approach.)

Consider the random probability measure G^* on B which has a density proportional to $\exp \psi(\mathbf{z})$ with respect to Lebesgue's measure on B. Then we should be able to study G^* using the tools of Section 3.1. Moreover, we can expect that $\overline{G}$ and G^* are "exponentially close", so we should be able to transfer our understanding of G^* to $\overline{G}$, and then to G'. We will address this technical point later, and we start the study of G^*. As usual, one can expect the overlaps to play a decisive part. The coordinate z_1 plays a special rôle, so we exclude it from the following definition of the overlap of two configurations

$$U_{\ell,\ell'} = U_{\ell,\ell'}(\mathbf{z}^\ell, \mathbf{z}^{\ell'}) = \sum_{2 \leq k \leq M} z_k^\ell z_k^{\ell'} \; . \tag{4.126}$$

There is no factor $1/N$ because we are here in a different normalization than in the previous chapter. We will also write

$$U_{1,1} = U_{1,1}(\mathbf{z}) = \sum_{2 \leq k \leq M} (z_k)^2 \; .$$

As a first goal, we would like to show that for $k \geq 1$,

$$\mathsf{E}\big\langle (U_{1,1} - \mathsf{E}\langle U_{1,1}\rangle^*)^{2k} \big\rangle^* \tag{4.127}$$

is small, and we explain the beautiful argument of Bovier and Gayrard which proves this. For $\lambda \geq 0$, consider the probability G_λ on B that has a density proportional to $\exp(\psi(\mathbf{z}) + \lambda N U_{1,1}(\mathbf{z}))$ with respect to Lebesgue's measure on B; and denote by $\langle \cdot \rangle_\lambda^*$ an average for this probability, so that $\langle \cdot \rangle^* = \langle \cdot \rangle_0^*$. The function

$$\mathbf{z} \mapsto \lambda U_{1,1}(\mathbf{z}) - \frac{\kappa}{2}\|\mathbf{z}\|^2 = -\frac{\kappa}{2}z_1^2 - \sum_{2 \le k \le M} \left(\frac{\kappa}{2} - \lambda\right) z_k^2$$

is concave for $\lambda \le \kappa/2$. Therefore, for every $\lambda \le \kappa/2$, the function

$$\mathbf{z} \mapsto \psi(\mathbf{z}) + \lambda N U_{1,1}(\mathbf{z}) + \frac{\kappa}{2} N \|\mathbf{z}\|^2 \tag{4.128}$$

is concave whenever the function $\mathbf{z} \mapsto \psi(\mathbf{z}) + \kappa N \|\mathbf{z}\|^2$ is concave, and this occurs with overwhelming probability. Also, since $\|\mathbf{z}\| \le 2$ for $\mathbf{z} \in B$,

$$\forall \mathbf{x}, \mathbf{y} \in B, \quad |U_{1,1}(\mathbf{x}) - U_{1,1}(\mathbf{y})| \le 4\|\mathbf{x} - \mathbf{y}\| .$$

When the function (4.128) is concave, we can use (3.17) (with $N\kappa/2$ instead of κ) to get

$$\forall k \ge 1 \ , \ \langle (U_{1,1} - \langle U_{1,1} \rangle_\lambda^*)^{2k} \rangle_\lambda^* \le \left(\frac{Kk}{N}\right)^k , \tag{4.129}$$

where K depends on κ only, and hence on β only.

The next step is to control the fluctuations of $\langle U_{1,1} \rangle^*$. For this we consider the random function

$$\varphi(\lambda) = \frac{1}{N} \log \int_B \exp(\psi(\mathbf{z}) + \lambda N U_{1,1}(\mathbf{z})) d\mathbf{z} \tag{4.130}$$

so that it is straightforward to obtain that

$$\varphi'(\lambda) = \langle U_{1,1} \rangle_\lambda^* \ ; \quad \varphi''(\lambda) = N \langle (U_{1,1} - \langle U_{1,1} \rangle_\lambda^*)^2 \rangle_\lambda^* . \tag{4.131}$$

Thus φ is convex since $\varphi'' \ge 0$. Taking $k = 1$ in (4.129) yields

$$\lambda < \frac{\kappa}{2} \Rightarrow \varphi''(\lambda) \le K \text{ with overwhelming probability.} \tag{4.132}$$

Also, since on B we have $|U_{1,1}| \le L$, relation (4.131) implies that $\varphi''(\lambda) \le LN$ and (4.132) that

$$\lambda < \frac{\kappa}{2} \Rightarrow \mathsf{E}\varphi''(\lambda) \le K . \tag{4.133}$$

We will deduce the fact that $\varphi'(0)$ has small fluctuations from (4.132) and the fact that $\varphi(\lambda)$ has small fluctuations. We write

$$\overline{\varphi}(\lambda) = \mathsf{E}\varphi(\lambda) ,$$

and we show first that φ has small fluctuations.

Lemma 4.5.1. *We have*

$$\forall k > 1, \ \mathsf{E}(\varphi(\lambda) - \overline{\varphi}(\lambda))^{2k} \le \left(\frac{Kk}{N}\right)^k . \tag{4.134}$$

This is really another occurrence of the principle behind Lemma 4.4.5. We give enough details so that the reader interested in the abstract principle of [120], Section 6, should find its application to the present situation immediate. Consider $(\mathbf{x}_i)_{i\le N}$, $\mathbf{x}_i \in \mathbb{R}^M$, and define

$$F(\mathbf{x}_1,\ldots,\mathbf{x}_N) = \frac{1}{N}\log\int_B \exp\Bigg(-\frac{N\beta}{2}\|\mathbf{z}\|^2 + \sum_{i\le N}\log\mathrm{ch}(\beta\mathbf{x}_i\cdot\mathbf{z}+h) + \lambda N U_{1,1}(\mathbf{z})\Bigg)\mathrm{d}\mathbf{z}\,.$$

This function has the following two remarkable properties: first, given any number b, the set $\{F \le b\}$ is a convex set in $\mathbb{R}^{N\times M}$. This follows from the convexity of the function log ch and Hölder's inequality. Second, its Lipschitz constant is $\le K/\sqrt{N}$. Indeed,

$$\frac{\partial F}{\partial x_{i,k}} = \frac{\beta}{N}\langle z_k \mathrm{th}(\beta\mathbf{x}_i\cdot\mathbf{z}+h)\rangle\,,$$

where the meaning of the average $\langle\cdot\rangle$ should be obvious. Thus

$$\left(\frac{\partial F}{\partial x_{i,k}}\right)^2 \le \frac{\beta^2}{N^2}\langle z_k^2\rangle\,,$$

and since for $\mathbf{z}\in B$ we have $\sum_{k\le M} z_k^2 \le L$ this yields

$$\sum_{i\le N,k\le M}\left(\frac{\partial F}{\partial x_{i,k}}\right)^2 \le \frac{L\beta^2}{N}\,,$$

i.e. the gradient of F has a norm $\le K/\sqrt{N}$. The abstract principle of [120] implies then that

$$\forall u>0,\quad P(|\varphi(\lambda)-a|\ge u) \le \exp\left(-\frac{Nu^2}{K}\right)\,,$$

where a is the median of $\varphi(\lambda)$. Therefore by (A.35) we have $\mathsf{E}(\varphi(\lambda)-a)^{2k} \le (Kk/N)^k$ for $k\ge 1$, and (4.134) follows by the symmetrization argument (3.22). □

The following gives an elementary method to control the fluctuations of the derivative of a random convex function when we control the fluctuations of the function and the size of its second derivative.

Lemma 4.5.2. *Consider $\lambda > 0$. Consider a random convex function $\varphi : [0,\lambda_0] \to \mathbb{R}$ that satisfies the following conditions, where δ, C_0, C_1, C_2 are numbers, where $k\ge 1$ is a given integer and where $\overline{\varphi} = \mathsf{E}\varphi$:*

$$|\varphi'| \le C_0 \tag{4.135}$$

$$\overline{\varphi}'' \le C_1 \tag{4.136}$$

$$\varphi'' \le C_1 \textit{ with probability } \ge 1-\delta \tag{4.137}$$

$$\mathsf{E}(\varphi(\lambda) - \overline{\varphi}(\lambda))^{2k} \le C_2^k \,. \tag{4.138}$$

Then when $C_2 \le \lambda_0^4 C_1^2$ *we have*

$$\mathsf{E}(\varphi'(0) - \overline{\varphi}'(0))^{2k} \le L^k(\delta C_0^{2k} + C_1^k C_2^{k/2}) \,. \tag{4.139}$$

Proof. Since φ is convex we have $\varphi'' \ge 0$ so when $\varphi'' \le C_1$ we have $|\varphi''| \le C_1$, and for $x \ge 0$ we have $|\varphi'(x) - \varphi'(0)| \le C_1 x$. Integration of this inequality for $0 \le x \le \lambda$ (where $\lambda < \lambda_0$) yields

$$\left| \varphi'(0) - \frac{\varphi(\lambda) - \varphi(0)}{\lambda} \right| \le \frac{C_1 \lambda}{2} \,.$$

For the same reason, using now (4.136), we get

$$\left| \overline{\varphi}'(0) - \frac{\overline{\varphi}(\lambda) - \overline{\varphi}(0)}{\lambda} \right| \le \frac{C_1 \lambda}{2} \,,$$

and therefore

$$|\varphi'(0) - \overline{\varphi}'(0)| \le C_1\lambda + \frac{|\varphi(\lambda) - \overline{\varphi}(\lambda)|}{\lambda} + \frac{|\varphi(0) - \overline{\varphi}(0)|}{\lambda} \,,$$

so that

$$(\varphi'(0) - \overline{\varphi}'(0))^{2k} \le 3^k \Big((C_1\lambda)^{2k} + \frac{1}{\lambda^{2k}} (\varphi(\lambda) - \overline{\varphi}(\lambda))^{2k} + \frac{1}{\lambda^{2k}} (\varphi(0) - \overline{\varphi}(0))^{2k} \Big) \,. \tag{4.140}$$

Recalling that (4.140) might fail with probability δ, taking expectation and using (4.135) and (4.138) we obtain that for $\lambda \le \lambda_0$

$$\mathsf{E}(\varphi'(0) - \overline{\varphi}'(0))^{2k} \le L^k \left(\delta C_0^{2k} + (C_1\lambda)^{2k} + \frac{1}{\lambda^{2k}} C_2^k \right) \,.$$

Choosing now $\lambda = C_2^{1/4} C_1^{-1/2}$, this yields that whenever $\lambda \le \lambda_0$, we have

$$\mathsf{E}(\varphi'(0) - \overline{\varphi}'(0))^{2k} \le L^k(\delta C_0^{2k} + C_1^k C_2^{k/2}) \,. \qquad \square$$

Corollary 4.5.3. *For* $k \ge 1$ *we have*

$$\mathsf{E}(\langle U_{1,1}\rangle^* - \mathsf{E}\langle U_{1,1}\rangle^*)^{2k} \le \left(\frac{Kk}{N}\right)^{k/2} \,. \tag{4.141}$$

Comment. We expect that (4.141) holds with the better bound $(Kk/N)^k$. It does not seem to be possible to prove this by the arguments of this section, that create an irretrievable loss of information.

Proof. We are going to apply Lemma 4.5.2 to the function (4.130) with $\lambda_0 = \kappa/2$. Since $|U_{1,1}| \le L$ on B the first part of (4.131) implies that $|\varphi'| \le L$ so that (4.135) holds for $C_0 = L$. We see from (4.132) and (4.133) that (4.136) and (4.137) hold for $C_1 = K$ and $\delta = K\exp(-N/K)$, and from Lemma 4.5.1 that (4.138) holds with $C_2 = Kk/N$. We conclude from (4.139) that, provided $C_2 \le \lambda_0^4 C_1^2$, and in particular whenever $Kk/N \le \kappa^4 K_1$, we have

$$\mathsf{E}(\langle U_{1,1}\rangle^* - \mathsf{E}\langle U_{1,1}\rangle^*)^{2k} \le K^k \left(\exp\left(-\frac{N}{K_2}\right) + \left(\frac{Kk}{N}\right)^{k/2}\right) .$$

This implies that (4.141) holds whenever $\exp(-N/K_2) \le (Kk/N)^{k/2}$. This occurs provided $k \le N/K_3$. To handle the case $k \ge N/K_3$, we simply write that

$$\mathsf{E}(\langle U_{1,1}\rangle^* - \mathsf{E}\langle U_{1,1}\rangle^*)^{2k} \le L^{2k} \le \left(\frac{L^4 K_3 k}{N}\right)^{k/2} .$$

□

Proposition 4.5.4. *For $k \ge 1$ we have*

$$\mathsf{E}\langle (U_{1,1} - \mathsf{E}\langle U_{1,1}\rangle^*)^{2k}\rangle \le \left(\frac{Kk}{N}\right)^{k/2} . \tag{4.142}$$

Proof. Inequality (4.129) might fail with probability $\le K\exp(-K/N)$, but since $|U_{1,1}| \le L$, taking expectation in this inequality we get

$$\mathsf{E}\big\langle (U_{1,1} - \langle U_{1,1}\rangle^*)^{2k}\big\rangle \le \left(\frac{Kk}{N}\right)^k + L^{2k} K\exp(-N/K) ,$$

and we show as in Corollary 4.5.3 that this implies in fact that

$$\mathsf{E}\big\langle (U_{1,1} - \langle U_{1,1}\rangle^*)^{2k}\big\rangle \le \left(\frac{Kk}{N}\right)^k .$$

Combining with (4.141) completes the proof. □

Proposition 4.5.5. *For $k \ge 1$, we have*

$$\forall j \le M, \;\; \mathsf{E}\big\langle (z_j - \mathsf{E}\langle z_j\rangle^*)^{2k}\big\rangle^* \le \left(\frac{Kk}{N}\right)^{k/2} . \tag{4.143}$$

Proof. Identical to that of Proposition 4.5.4, using now the Gibbs measure with density proportional to $\exp(\psi(\mathbf{z}) + \lambda N z_j)$. The proof can be copied verbatim replacing $U_{1,1}$ by z_j. □

Proposition 4.5.6. *For $k \geq 1$, we have*

$$\mathsf{E}\langle (U_{1,2} - \mathsf{E}\langle U_{1,2}\rangle^*)^{2k}\rangle^* \leq \left(\frac{Kk}{N}\right)^{k/2} . \tag{4.144}$$

Proof. We consider the Gibbs measure on $B \times B$ with density proportional to

$$\exp(\psi(\mathbf{z}_1) + \psi(\mathbf{z}_2) + \lambda N U_{1,2}) . \tag{4.145}$$

We observe that for $\lambda \leq \kappa$ the function

$$(\mathbf{z}^1, \mathbf{z}^2) \mapsto \lambda U_{1,2} - \frac{\kappa}{2}(\|\mathbf{z}^1\|^2 + \|\mathbf{z}^2\|^2)$$

is concave. This is because at every point its second differential D^2 satisfies

$$D^2((\mathbf{v}^1, \mathbf{v}^2), (\mathbf{v}^1, \mathbf{v}^2)) = 2\lambda \sum_{2 \leq k \leq M} v_k^1 v_k^2 - \kappa(\|\mathbf{v}^1\|^2 + \|\mathbf{v}^2\|^2) \leq 0 ,$$

using that $2v_k^1 v_k^2 \leq (v_k^1)^2 + (v_k^2)^2$. The proof is then identical to that of Proposition 4.5.4. □

We now turn to the task of transferring our results from G^* to $\overline{G}$ and then to G'. One expects that these measures are very close to each other; still we must check that the exponentially small set B^c does not create trouble. Such lackluster technicalities occupy the rest of this section. Let us denote by $\langle\cdot\rangle^-$ an average for $\overline{G}$. By definition of $\langle\cdot\rangle^*$, for a function f on $\mathbb{R}^M$ we have $\langle \mathbf{1}_B f\rangle^- = \overline{G}(B)\langle f\rangle^*$, so that

$$\langle f\rangle^- = \langle \mathbf{1}_B f\rangle^- + \langle \mathbf{1}_{B^c} f\rangle^- = \overline{G}(B)\langle f\rangle^* + \langle \mathbf{1}_{B^c} f\rangle^- .$$

Taking expectation and using the Cauchy-Schwarz inequality in the last term shows that when $f \geq 0$ it holds

$$\mathsf{E}\langle f\rangle^- \leq \mathsf{E}\langle f\rangle^* + (\mathsf{E}G(B^c))^{1/2}(\mathsf{E}\langle f^2\rangle^-)^{1/2} ,$$

and, in particular, since $\overline{G}$ is essentially supported by B,

$$\mathsf{E}\langle f\rangle^- \leq \mathsf{E}\langle f\rangle^* + K\exp\left(-\frac{N}{K}\right)(\mathsf{E}\langle f^2\rangle^-)^{1/2} . \tag{4.146}$$

To use (4.146) constructively it suffices to show that $\mathsf{E}\langle f^2\rangle^-$ is not extremely large. Of course one never doubts that this is the case for the functions we are interested in, but this has to be checked nonetheless. Inequality (4.151) below will take care of this.

Lemma 4.5.7. *The quantity*

$$T = \max_{\boldsymbol{\sigma}} \sum_{j \leq M} m_j^2(\boldsymbol{\sigma}) \tag{4.147}$$

satisfies

$$\forall k\,, \quad \mathsf{E}T^k \le L^k\left(1+\left(\frac{k}{N}\right)^k\right)\,, \tag{4.148}$$

and therefore

$$\forall k \le N\,, \quad \mathsf{E}T^k \le L^k\,. \tag{4.149}$$

This means that for all practical purposes, one can think of T as being bounded. This quantity occurs in many situations.

Proof. We have

$$\exp\frac{NT}{4} \le \sum_{\boldsymbol{\sigma}} \exp\frac{N}{4}\sum_{j\le M} m_j^2(\boldsymbol{\sigma})$$

and therefore

$$\begin{aligned}\mathsf{E}\exp\frac{NT}{4} &\le \sum_{\boldsymbol{\sigma}} \mathsf{E}\exp\frac{N}{4}\sum_{j\le M} m_j^2(\boldsymbol{\sigma})\\ &= \sum_{\boldsymbol{\sigma}}\prod_{j\le M}\mathsf{E}\exp\frac{N}{4}m_j^2(\boldsymbol{\sigma}) \le 2^{N+M}\,,\end{aligned} \tag{4.150}$$

using independence and (A.24) to see that $\mathsf{E}\exp(Nm_j^2(\boldsymbol{\sigma})/4) \le 2$. We use Lemma 3.1.8 with $X = NT/4$ to get, since $M \le N$, that

$$\mathsf{E}X^k \le 2^k(k^k + (LN)^k) \le (LN)^k + (Lk)^k\,. \qquad \square$$

Corollary 4.5.8. *For any number C that does not depend on N we have $\mathsf{E}\exp CT \le K(C)$.*

Proof. We can either use (4.150) and Hölder's inequality or expand the exponential as a power series and use (4.148). $\square$

Lemma 4.5.9. *We have*

$$\forall k \le N\,, \quad \mathsf{E}\big\langle \|\mathbf{z}\|^{2k}\big\rangle^- \le L^k\,. \tag{4.151}$$

Proof. Since $\overline{G}$ is the convolution of G' and γ, we have

$$\begin{aligned}\big\langle \|\mathbf{z}\|^{2k}\big\rangle^- &= \int \|\mathbf{x}+\mathbf{y}\|^{2k}\mathrm{d}G'(\mathbf{x})\mathrm{d}\gamma(\mathbf{y})\\ &\le 2^{2k}\left(\int \|\mathbf{x}\|^{2k}\mathrm{d}G'(\mathbf{x}) + \int \|\mathbf{y}\|^{2k}\mathrm{d}\gamma(\mathbf{y})\right)\,.\end{aligned} \tag{4.152}$$

Since

$$\int \exp(\beta N\|\mathbf{y}\|^2/4)\mathrm{d}\gamma(\mathbf{y}) = W\int \exp(-\beta N\|\mathbf{y}\|^2/4)\mathrm{d}\mathbf{y} = 2^{M/2}\,,$$

and since $\exp x^2 \geq x^{2k}/k!$, taking $k = N$ implies

$$\frac{(N\beta)^N}{4^N N!} \int \|\mathbf{y}\|^{2N} \mathrm{d}\gamma(\mathbf{y}) \leq 2^{M/2} ,$$

so that $\int \|\mathbf{y}\|^{2N} \mathrm{d}\gamma(\mathbf{y}) \leq L^N$, and in particular $\int \|\mathbf{y}\|^{2k} \mathrm{d}\gamma(\mathbf{y}) \leq L^k$ for $k \leq N$ by Hölder's inequality.

By definition G' is the image of G under that map $\boldsymbol{\sigma} \mapsto \mathbf{m}(\boldsymbol{\sigma}) = (m_k(\boldsymbol{\sigma}))_{k \leq M}$, so that

$$G'(\{\mathbf{x} \in \mathbb{R}^M \; ; \; \|\mathbf{x}\|^2 > T\}) = 0 , \tag{4.153}$$

and hence $\int \|\mathbf{x}\|^{2k} \mathrm{d}G'(\mathbf{x}) \leq T^k$. The result then follows from (4.149). □

Proposition 4.5.10. *For $k \leq N$ we have*

$$\mathsf{E}\big\langle (U_{1,1} - \mathsf{E}\langle U_{1,1}\rangle^-)^{2k} \big\rangle^- \leq \left(\frac{Kk}{N}\right)^{k/2} \tag{4.154}$$

$$\forall j \leq M, \;\; \mathsf{E}\big\langle (z_j - \mathsf{E}\langle z_j\rangle^-)^{2k} \big\rangle^- \leq \left(\frac{Kk}{N}\right)^{k/2} \tag{4.155}$$

$$\mathsf{E}\big\langle (U_{1,2} - \mathsf{E}\langle U_{1,2}\rangle^-)^{2k} \big\rangle^- \leq \left(\frac{Kk}{N}\right)^{k/2} . \tag{4.156}$$

Proof. Condition (4.151) implies that for $k \leq N$ we have

$$\mathsf{E}\big\langle (U_{1,1} - \mathsf{E}\langle U_{1,1}\rangle^*)^{2k} \big\rangle^- \leq L^k ,$$

so that (4.142) and (4.146) imply

$$\mathsf{E}\big\langle (U_{1,1} - \mathsf{E}\langle U_{1,1}\rangle^*)^{2k} \big\rangle^- \leq \left(\frac{Kk}{N}\right)^{k/2} + K \exp\Big(-\frac{K}{N}\Big) L^k \leq \left(\frac{K'k}{N}\right)^{k/2}$$

for $k \leq N$. This yields in particular

$$|\mathsf{E}\langle U_{1,1}\rangle^- - \mathsf{E}\langle U_{1,1}\rangle^*| \leq \frac{K}{\sqrt{N}}$$

and (4.154). The proof of (4.155) is similar, and only a small adaptation of (4.146) to the case of 2 replicas is required to prove (4.156) using the same scheme. □

The measure $\overline{G}$ itself is a technical tool. What we are really looking for is information about G', and we are ready to prove it. We denote by $\langle \cdot \rangle'$ an average for G'.

Proposition 4.5.11. *For $k \le N$ we have*

$$\mathsf{E}\big\langle (U_{1,1} - \mathsf{E}\langle U_{1,1}\rangle')^{2k}\big\rangle' \le \left(\frac{Kk}{N}\right)^{k/2} \tag{4.157}$$

$$\forall j \le M,\quad \mathsf{E}\big\langle (z_j - \mathsf{E}\langle z_j\rangle')^{2k}\big\rangle' \le \left(\frac{Kk}{N}\right)^{k/2} \tag{4.158}$$

$$\mathsf{E}\big\langle (U_{1,2} - \mathsf{E}\langle U_{1,2}\rangle')^{2k}\big\rangle' \le \left(\frac{Kk}{N}\right)^{k/2} . \tag{4.159}$$

Proof. The basic reason this follows from Proposition 4.5.10 is that "convolution spreads out the measure", so that statements of Proposition 4.5.10 are stronger than corresponding statements of Proposition 4.5.11. For $\mathbf{x}, \mathbf{y}$ in $\mathbb{R}^M$, let us write

$$(\mathbf{x}, \mathbf{y}) = \sum_{2 \le j \le M} x_j y_j ,$$

so that $U_{1,1}(\mathbf{z}) = (\mathbf{z}, \mathbf{z})$. Then, since for all $\mathbf{x}$ we have $\int (\mathbf{x}, \mathbf{y}) \mathrm{d}\gamma(\mathbf{y}) = 0$, and since $\overline{G}$ is the convolution of G' and γ,

$$\begin{aligned}\langle U_{1,1}\rangle^- &= \int (\mathbf{x}+\mathbf{y}, \mathbf{x}+\mathbf{y}) \mathrm{d}G'(\mathbf{x}) \mathrm{d}\gamma(\mathbf{y}) = \int (\mathbf{x}, \mathbf{x}) \mathrm{d}G'(\mathbf{x}) + C \\ &= \langle U_{1,1}\rangle' + C ,\end{aligned}$$

where $C = \int (\mathbf{y}, \mathbf{y}) \mathrm{d}\gamma(\mathbf{y})$ is non-random. Thus, using (4.154),

$$\mathsf{E}(\langle U_{1,1}\rangle^- - \mathsf{E}\langle U_{1,1}\rangle^-)^{2k} = \mathsf{E}(\langle U_{1,1}\rangle' - \mathsf{E}\langle U_{1,1}\rangle')^{2k} \le \left(\frac{Kk}{N}\right)^{k/2} . \tag{4.160}$$

Next,

$$\begin{aligned}&\big\langle (U_{1,1} - U_{2,2})^{2k}\big\rangle^- = \\ &\int \big((\mathbf{x}^1+\mathbf{y}^1, \mathbf{x}^1+\mathbf{y}^1) - (\mathbf{x}^2+\mathbf{y}^2, \mathbf{x}^2+\mathbf{y}^2)\big)^{2k} \mathrm{d}G'(\mathbf{x}^1) \mathrm{d}G'(\mathbf{x}^2) \mathrm{d}\gamma(\mathbf{y}^1) \mathrm{d}\gamma(\mathbf{y}^2) \\ &\ge \int ((\mathbf{x}^1, \mathbf{x}^1) - (\mathbf{x}^2, \mathbf{x}^2))^{2k} \mathrm{d}G'(\mathbf{x}^1) \mathrm{d}G'(\mathbf{x}^2) ,\end{aligned} \tag{4.161}$$

by using Jensen's inequality to integrate in γ inside the power $(\cdot)^{2k}$ rather than outside, and using again the fact that $\int (\mathbf{x}, \mathbf{y}) \mathrm{d}\gamma(\mathbf{y}) = 0$. Thus, applying Jensen's inequality in the second inequality below, we get

$$\big\langle (U_{1,1} - U_{2,2})^{2k}\big\rangle^- \ge \big\langle (U_{1,1} - U_{2,2})^{2k}\big\rangle' \ge \big\langle (U_{1,1} - \langle U_{1,1}\rangle')^{2k}\big\rangle' .$$

Since $(U_{1,1} - U_{2,2})^{2k} \le 2^{2k}((U_{1,1} - \mathsf{E}\langle U_{1,1}\rangle^-)^{2k} + (U_{2,2} - \mathsf{E}\langle U_{1,1}\rangle^-)^{2k})$, using (4.154) yields

$$\mathsf{E}\big\langle (U_{1,1} - \langle U_{1,1}\rangle')^{2k}\big\rangle' \le \left(\frac{Kk}{N}\right)^{k/2} .$$

Combining with (4.160) proves (4.157); the rest is similar. □

The following improves on (4.158) when $j \geq 2$.

Proposition 4.5.12. *For $2 \leq j \leq M$ and $k \leq N$ we have*

$$\mathsf{E}\langle z_j^{2k} \rangle' \leq \left(\frac{Kk}{N} \right)^{k/2} . \tag{4.162}$$

Proof. Using (4.158) it suffices to see that $|\mathsf{E}\langle z_j \rangle'| \leq K/\sqrt{N}$. Using symmetry between sites,

$$(\mathsf{E}\langle z_j \rangle')^2 \leq \mathsf{E}\langle z_j^2 \rangle' = \frac{1}{M-1} \mathsf{E} \left\langle \sum_{2 \leq j \leq M} z_j^2 \right\rangle' .$$

It follows from (A.56) (used for $a = 1$) that with overwhelming probability $T = \max_{\boldsymbol{\sigma}} \sum_{j \leq M} m_j^2(\boldsymbol{\sigma}) \leq LM/N$. Using (4.149) and the Cauchy-Schwarz inequality to control the expectation of T on the rare event where this fails, we obtain that $\mathsf{E}T \leq LM/N$. Since $\langle \sum_{2 \leq j \leq M} z_j^2 \rangle' \leq T$ by (4.153), this concludes the proof. □

4.6 Approximate Integration by Parts and the Replica-Symmetric Equations

We denote by $\langle \cdot \rangle$ an average for the Gibbs measure with Hamiltonian (4.25), and we write $\nu(f) = \mathsf{E}\langle f \rangle$.

Since G' is the image of the Gibbs measure under the map $\boldsymbol{\sigma} \mapsto \mathbf{m}(\boldsymbol{\sigma}) = (m_k(\boldsymbol{\sigma}))_{k \leq M}$, for a function f on $\mathbb{R}^M$ we have $\langle f \rangle' = \langle f(m_1(\boldsymbol{\sigma}), \ldots, m_M(\boldsymbol{\sigma})) \rangle$, and similar formulas hold for replicas.

We define the following quantities

$$\mu = \nu(m_1(\boldsymbol{\sigma})) = \mathsf{E}\langle z_1 \rangle' \tag{4.163}$$

$$\rho = \nu\left(\sum_{2 \leq k \leq M} m_k^2(\boldsymbol{\sigma}) \right) = \mathsf{E}\langle U_{1,1} \rangle' \tag{4.164}$$

$$r = \nu\left(\sum_{2 \leq k \leq M} m_k(\boldsymbol{\sigma}^1) m_k(\boldsymbol{\sigma}^2) \right) = \mathsf{E}\langle U_{1,2} \rangle' \tag{4.165}$$

$$q = \nu(R_{1,2}) . \tag{4.166}$$

As in (3.59) these quantities depend on (β, h) and M, N, although this is not indicated by the notation. The purpose of this section is to show that these fundamental quantities attached to the system nearly satisfy the following system of (exact) equations:

$$\mu = \mathsf{E}\mathrm{th}(\beta z\sqrt{r} + \beta\mu + h)\ ; q = \mathsf{E}\mathrm{th}^2(\beta z\sqrt{r} + \beta\mu + h)\ ; r(1-\beta(1-q))^2 = \alpha q \tag{4.167}$$

and

$$(\rho - r)(1 - \beta(1-q)) = \alpha(1-q)\ , \tag{4.168}$$

where as usual $\alpha = M/N$ and z is a standard Gaussian r.v. The equations (4.167) are called the replica-symmetric equations. To pursue the study of the Hopfield model it seems then required to show that the system of replica-symmetric equations determine the values of μ, r and q. This task is in principle elementary, but it is quite tedious and is deferred to Volume II. For the time being, our study of the Hopfield model will culminate with the proof that the quantities (4.163) to (4.165) nearly satisfy the replica-symmetric equations (4.167). The correct result is that the replica-symmetric equations are satisfied "with accuracy K/N". The methods of this chapter do not seem to be able to reach better than a rate $K/\sqrt{N}$, for the reasons stated after Proposition 4.5.4. Even reaching that rate requires significant work. We have made the choice to prove in this section that the replica-symmetric equations hold with rate $K/N^{1/4}$, even though the proof that the equations "just hold in the limit" (without a rate) is simpler. Besides the fact that this choice is coherent with the use of the quantitative methods that form the core of this work, it is really a pretense to learn the fundamental technique of approximate integration by parts that we will use a great many times later.

Before we start the proof we observe that we can reformulate Propositions 4.5.11 and 4.5.12 as follows.

Proposition 4.6.1. *For $k \le N$, we have*

$$\nu\left(\left(\sum_{2\le j\le M} m_j(\boldsymbol{\sigma})^2 - \rho\right)^{2k}\right) \le \left(\frac{Kk}{N}\right)^{k/2}\ ; \tag{4.169}$$

$$\nu\left(\left(\sum_{2\le j\le M} m_j(\boldsymbol{\sigma}^1)m_j(\boldsymbol{\sigma}^2) - r\right)^{2k}\right) \le \left(\frac{Kk}{N}\right)^{k/2}\ ; \tag{4.170}$$

$$\text{for all } 2 \le j \le M\ ,\ \nu(m_j(\boldsymbol{\sigma})^{2k}) \le \left(\frac{Kk}{N}\right)^{k/2}\ . \tag{4.171}$$

Given $\boldsymbol{\sigma} = (\sigma_1, \ldots, \sigma_N)$, we write $\boldsymbol{\rho} = (\sigma_1, \ldots, \sigma_{N-1}) \in \Sigma_{N-1}$, and

$$n_k = n_k(\boldsymbol{\sigma}) = n_k(\boldsymbol{\rho}) = \frac{1}{N}\sum_{i\le N-1} \eta_{i,k}\sigma_i\ . \tag{4.172}$$

We note that

$$m_k(\boldsymbol{\sigma}) = n_k(\boldsymbol{\sigma}) + \frac{\eta_k\sigma_N}{N}\ , \tag{4.173}$$

where for simplicity we write η_k rather than $\eta_{N,k}$.

Lemma 4.6.2. *We have*

$$\mu = \nu(\sigma_N) ; \tag{4.174}$$

$$q = \nu(\sigma_N^1 \sigma_N^2) ; \tag{4.175}$$

$$\rho = \frac{M-1}{N} + \nu\Big(\sigma_N \sum_{2 \le k \le M} \eta_k n_k(\boldsymbol{\sigma})\Big) ; \tag{4.176}$$

$$r = \frac{M-1}{N} q + \nu\Big(\sigma_N^1 \sum_{2 \le k \le M} \eta_k n_k(\boldsymbol{\sigma}^2)\Big) . \tag{4.177}$$

Proof. Using (4.173) and symmetry among sites yields

$$\nu(m_k^2(\boldsymbol{\sigma})) = \nu(\eta_k \sigma_N m_k(\boldsymbol{\sigma})) = \frac{1}{N} + \nu(\sigma_N \eta_k n_k(\boldsymbol{\sigma})) ,$$

from which (4.176) follows by summation over $2 \le k \le M$. Relation (4.177) is similar and the rest is obvious. □

To use these formulas, we make the dependence of the Hamiltonian on σ_N explicit. We define

$$-H_{N-1,M}(\boldsymbol{\rho}) = \frac{N\beta}{2} \sum_{1 \le k \le M} n_k^2(\boldsymbol{\rho}) + Nhn_1(\boldsymbol{\rho}) .$$

(Despite the notation, this is not exactly the Hamiltonian of an $(N-1)$-spin system; more specifically, this is the Hamiltonian of an $(N-1)$-spin system where β has been replaced by $\beta N/(N-1)$.) Using (4.173) in the definition of $H_{N,M}$ and expending the squares shows that

$$-H_{N,M}(\boldsymbol{\sigma}) = -H_{N-1,M}(\boldsymbol{\rho}) + \beta\sigma_N \sum_{1 \le k \le M} \eta_k n_k(\boldsymbol{\rho}) + \sigma_N h , \tag{4.178}$$

ignoring the constant $\beta M/(2N)$ that plays no role. The strategy we will follow should come as no surprise. We will express the averages $\langle\cdot\rangle$ in Lemma 4.6.2 using the Hamiltonian (4.178). We will bet that the quantities

$$\sum_{2 \le k \le M} \eta_k n_k(\boldsymbol{\rho}) \tag{4.179}$$

have a Gaussian behavior, and to bring this out we will interpolate them with suitable Gaussian r.v.s. The reader may observe that in (4.179) the sum is over $2 \le k \le M$. The quantity $n_1(\boldsymbol{\rho})$ requires a special treatment. The idea is that for $k \ge 2$ the quantity $n_k(\boldsymbol{\rho})$ should be very small, allowing the quantity (4.179) to have a Gaussian behavior. On the other hand, one should think of the quantity $n_1(\boldsymbol{\rho})$ as $n_1(\boldsymbol{\rho}) \simeq \mu \ne 0$.

Given replicas, $\boldsymbol{\rho}^1, \dots, \boldsymbol{\rho}^n$, we write $n_k^\ell = n_k(\boldsymbol{\rho}^\ell)$, and given a parameter t we define

$$g_t^\ell = \sqrt{t} \sum_{2 \le k \le M} \eta_k n_k^\ell + \sqrt{1-t}(z\sqrt{r} + \xi^\ell \sqrt{\rho - r}) , \tag{4.180}$$

where z, ξ^ℓ, $\ell \ge 1$ are independent standard Gaussian r.v.s. Denoting by $\langle \cdot \rangle_-$ an average for the Gibbs measure with Hamiltonian $H_{N-1,M}$, and given a function $f = f(\boldsymbol{\sigma}^1, \ldots, \boldsymbol{\sigma}^n)$, that might be random, we define

$$\nu_t(f) = \mathsf{E} \frac{\langle \mathrm{Av}_{\varepsilon_1,\ldots,\varepsilon_n=\pm 1} f \mathcal{E}_t \rangle_-}{\mathsf{E}_\xi \langle \mathrm{Av}_{\varepsilon_1,\ldots,\varepsilon_n=\pm 1} \mathcal{E}_t \rangle_-} , \tag{4.181}$$

where $\varepsilon_\ell = \sigma_N^\ell$, E_ξ denotes as usual expectation in $\xi^1, \ldots, \xi^n$, and where

$$\mathcal{E}_t = \exp \sum_{\ell \le n} \varepsilon_\ell \big(\beta(g_t^\ell + t n_1^\ell + (1-t)\mu) + h\big) . \tag{4.182}$$

As already pointed out, the quantity n_1 receives special treatment compared to the quantities n_k, $2 \le k \le M$, and the previous interpolation implements the idea that $n_1^\ell \simeq \mu$.

Given a function $f = f(\boldsymbol{\sigma}^1, \ldots, \boldsymbol{\sigma}^n, x_1, \ldots, x_n)$ we write

$$f_t = f(\boldsymbol{\sigma}^1, \ldots, \boldsymbol{\sigma}^n, g_t^1, \ldots, g_t^n) . \tag{4.183}$$

We shall show that for the four choices of f occurring in Lemma 4.6.2 we have $\nu_0(f_0) \simeq \nu_1(f_1)$, where $\simeq$ means that the error is $\le KN^{-1/4}$. This will provide the desired equations for μ, ρ, r, q. The computation of $\nu_0(f_0)$ is fun, so we do it first. We write

$$Y = \beta z \sqrt{r} + \beta\mu + h .$$

Lemma 4.6.3. *a) If* $f(\boldsymbol{\sigma}^1) = \sigma_N^1$ *then*

$$\nu_0(f_0) = \mathsf{E}\, \mathrm{th} Y . \tag{4.184}$$

b) If $f(\boldsymbol{\sigma}^1, \boldsymbol{\sigma}^2) = \sigma_N^1 \sigma_N^2$ *then*

$$\nu_0(f_0) = \mathsf{E}\, \mathrm{th}^2 Y . \tag{4.185}$$

c) If $f(\boldsymbol{\sigma}^1, x_1) = \sigma_N^1 x_1$ *then*

$$\nu_0(f_0) = \beta r(1 - \mathsf{E}\, \mathrm{th}^2 Y) + \beta(\rho - r) . \tag{4.186}$$

d) If $f(\boldsymbol{\sigma}^1, x_1, x_2) = \sigma_N^1 x_2$ *then*

$$\nu_0(f_0) = \beta(\rho - r) q + \beta r(1 - \mathsf{E}\, \mathrm{th}^2 Y) . \tag{4.187}$$

Proof. Let $Y' = Y + \beta\sqrt{\rho - r}\,\xi$. Then

$$\mathsf{E}_\xi \mathrm{sh} Y' = \exp \frac{\beta(\rho - r)}{2} \mathrm{sh} Y$$

and similarly for $\mathsf{E}_\xi \mathrm{ch} Y'$. Since

$$\nu_0(\sigma_N^1) = \mathsf{E}\,\frac{\mathrm{sh}Y'}{\mathsf{E}_\xi \mathrm{ch}Y'} = \mathsf{E}\,\frac{\mathsf{E}_\xi \mathrm{sh}Y'}{\mathsf{E}_\xi \mathrm{ch}Y'}\,,$$

this makes (4.184) obvious, and (4.185) is similar. So we prove (4.186). Now

$$\nu_0(f_0) = \mathsf{E}\,\frac{\mathsf{E}_\xi(z\sqrt{r} + \xi\sqrt{\rho - r})\mathrm{sh}Y'}{\mathsf{E}_\xi \mathrm{ch}Y'}\,,$$

and by integration by parts

$$\mathsf{E}_\xi \xi\sqrt{\rho - r}\,\mathrm{sh}Y' = \beta(\rho - r)\mathsf{E}_\xi \mathrm{ch}Y'\,.$$

Thus, integrating by parts in the second equality,

$$\nu_0(f_0) = \beta(\rho - r) + \mathsf{E}\,z\sqrt{r}\,\mathrm{th}Y = \beta(\rho - r) + \beta r \mathsf{E}\,\frac{1}{\mathrm{ch}^2 Y}\,,$$

and the conclusion follows since $1 - \mathrm{th}^2 Y = 1/\mathrm{ch}^2 Y$. The proof of (4.187) is similar. □

In the reminder of the chapter we shall prove that $\nu(f_1) = \nu_1(f_1) \simeq \nu_0(f_0)$ for the functions of Lemma 4.6.3. Before doing this, we explain why this implies that (μ, q, r) is nearly a solution of the system of equations (4.167). Combining the relation $\nu(f_1) = \nu_1(f_1) \simeq \nu_0(f_0)$ with Lemma 4.6.2 proves that the relations

$$\mu \simeq \mathsf{E}\,\mathrm{th}Y\,;\, q \simeq \mathsf{E}\,\mathrm{th}^2 Y \tag{4.188}$$

$$\rho \simeq \alpha + \beta r(1 - q) + \beta(\rho - r)\,; \tag{4.189}$$

$$r \simeq \alpha q + \beta(\rho - r)q + \beta r(1 - q) \tag{4.190}$$

hold. Subtraction of the last two relations gives

$$(\rho - r)(1 - \beta(1 - q)) \simeq \alpha(1 - q)\,. \tag{4.191}$$

We rewrite (4.190) as

$$r(1 - \beta(1 - q)) \simeq \alpha q + \beta(\rho - r)q\,,$$

and we multiply by $1 - \beta(1 - q)$ to get

$$r(1 - \beta(1 - q))^2 \simeq \alpha q(1 - \beta(1 - q)) + \beta q(\rho - r)(1 - \beta(1 - q))\,.$$

Using (4.191) in the second term in the right-hand side then yields

$$r(1 - \beta(1 - q))^2 \simeq \alpha q\,. \tag{4.192}$$

This shows as promised that (μ, q, r) is nearly a solution of the system of equations (4.167).

We turn to the comparison of $\nu(f_1)$ and $\nu_0(f_0)$, with the goal of proving that for the functions of Lemma 4.6.3 these two quantiles are nearly equal. As expected this will be done by controlling the derivative of the function $t \mapsto \nu_t(f_t)$. We define

$$g_t^{\ell\prime} = \frac{1}{2\sqrt{t}}\Big(\sum_{2\le k\le M} \eta_k n_k^\ell\Big) - \frac{1}{2\sqrt{1-t}}(z\sqrt{r} + \xi^\ell\sqrt{\rho - r}) \,. \tag{4.193}$$

Lemma 4.6.4. *We have*

$$\frac{\mathrm{d}}{\mathrm{d}t}\nu_t(f_t) = \mathrm{I} + \mathrm{II} + \mathrm{III} \tag{4.194}$$

where

$$\mathrm{I} = \sum_{\ell\le n} \nu_t\left(g_t^{\ell\prime}\frac{\partial f_t}{\partial x_\ell}\right) \tag{4.195}$$

$$\mathrm{II} = \beta\Big(\sum_{\ell\le n} \nu_t(\varepsilon_\ell g_t^{\ell\prime} f_t) - n\nu_t(\varepsilon_{n+1} g_t^{n+1\prime} f_t)\Big) \tag{4.196}$$

$$\mathrm{III} = \beta\sum_{\ell\le n} \nu_t(\varepsilon_\ell(n_1^\ell - \mu)f_t) - n\nu_t(\varepsilon_{n+1}(n_1^{n+1} - \mu)f_t) \,. \tag{4.197}$$

Here of course $\varepsilon_\ell = \sigma_N^\ell$ and

$$\frac{\partial f_t}{\partial x_\ell} = \frac{\partial f}{\partial x_\ell}(\boldsymbol{\sigma}^1, \dots, \boldsymbol{\sigma}^n, g_t^1, \dots, g_t^n) \,.$$

Proof. This looks complicated, but this is straightforward differentiation. There are 3 separate reasons why $\nu_t(f_t)$ depends on t. First, f_t depends on t through g_t^ℓ, and this creates the term I. Second, $\nu_t(f_t)$ depends on t because in (4.181) the term $\mathcal{E}_t$ depends on t through g_t^ℓ, and this creates the term II. Finally, $\nu_t(f_t)$ depends on t because in (4.181) the term $\mathcal{E}_t$ depends on t through the quantity $(1-t)\mu$, and this creates the term III. Let us also mention that for clarity we have stated this result for general n but that the case $n = 2$ suffices. □

We would like to integrate by parts in the terms I and II using (4.193). Unfortunately the r.v. η_k is not Gaussian, it is a random sign. We now describe the technique, called "approximate integration by parts", that is a substitute of integration by parts for such variables.

The basic fact is that if v is a three times differentiable function on $\mathbb{R}$, then

$$v(1) - v(-1) = v'(1) + v'(-1) + \frac{1}{2}\int_{-1}^{1}(x^2 - 1)v'''(x)\mathrm{d}x \,. \tag{4.198}$$

This is proved by integrating the last term by parts,

$$\int_{-1}^{1} \frac{1}{2}(x^2-1)v'''(x)\mathrm{d}x = \frac{1}{2}(x^2-1)v''(x)\Big|_{x=-1}^{x=1} - \int_{-1}^{1} xv''(x)\mathrm{d}x \; ,$$

$$\begin{aligned}\int_{-1}^{1} xv''(x)\mathrm{d}x &= xv'(x)\Big|_{x=-1}^{x=1} - \int_{-1}^{1} v'(x)\mathrm{d}x \\ &= v'(1) + v'(-1) - (v(1) - v(-1)) \; .\end{aligned}$$

If η is a r.v. such that $\mathsf{P}(\eta = \pm 1) = 1/2$, then (4.198) implies

$$\mathsf{E}\,\eta v(\eta) = \mathsf{E}\,v'(\eta) + \frac{1}{4}\int_{-1}^{1} (x^2-1)v'''(x)\mathrm{d}x \; . \tag{4.199}$$

We will call $\mathsf{E}\,v'(\eta)$ the *main term* and the last term the *error term*. This term will have a tendency to be small because v will depend little on η. Typically every occurrence of η in v is multiplied by a small factor (e.g. $1/\sqrt{N}$). We will always bound the error term through the crude inequality

$$\left|\frac{1}{4}\int_{-1}^{1} (x^2-1)v'''(x)\mathrm{d}x\right| \le \sup_{|x|\le 1} |v'''(x)| \; . \tag{4.200}$$

The contribution of the main term is what we would get if the r.v. η had been Gaussian.

We start to apply approximate integration by parts to (4.194). We take care of the main terms first. These terms are the same as if we were integrating by parts for Gaussian r.v.s, and we have learned how to make this calculation in Chapter 3. Let us set

$$S_{\ell,\ell'} = \sum_{2\le k\le M} n_k^{\ell} n_k^{\ell'} \; ,$$

so that (being careful to distinguish between $g_t^{\ell'}$ and $g_t^{\ell'}$, where the position of the $\prime$ is not the same) the relations

$$\ell \ne \ell' \Rightarrow \mathsf{E}\,g_t^{\ell'} g_t^{\ell'} = S_{\ell,\ell'} - r$$

$$\mathsf{E}\,g_t^{\ell'} g_t^{\ell} = S_{\ell,\ell} - \rho$$

hold and integration by parts brings out factors $S_{\ell,\ell'} - r$ and $S_{\ell,\ell} - \rho$. The dependence on g_t^{ℓ} is through the Hamiltonian and f_t. It then should be clear that the contribution of the main terms to the integration by parts in II is bounded by

$$\begin{aligned}\mathrm{IV} = K\Bigg(&\sum_{\ell\ne\ell'\le n+2} \nu_t\bigg(\Big(|f_t| + \Big|\frac{\partial f_t}{\partial x_\ell}\Big|\Big)|S_{\ell,\ell'} - r|\bigg) \\ &+ \sum_{\ell\le n+1} \nu_t\bigg(\Big(|f_t| + \Big|\frac{\partial f_t}{\partial x_\ell}\Big|\Big)|S_{\ell,\ell} - \rho|\bigg)\Bigg) \; .\end{aligned} \tag{4.201}$$

Here, as well as in the rest of the section the quantity K is permitted to depend on n. In this bound we would like to have ν rather than ν_t. Since $|f_t|$ depends on g_t^ℓ, $\ell \le n$, we cannot readily use differential inequalities to relate ν and ν_t. The next page or so will take care of that technical problem. We will then show how to control the error terms in the approximate integration by parts, which is not trivial.

Sophistication is not needed to prove that we can replace ν_t by ν in (4.201), but the details are tedious.

Lemma 4.6.5. *Consider a function $f^* \ge 0$ of n replicas $\boldsymbol{\rho}^1, \ldots, \boldsymbol{\rho}^n$, $n \le 3$, that might also depend on η_k, z, ξ^ℓ for $\ell \le n$ and $k \le M$. Then*

$$\nu_t(f^*) \le K\nu\big((\mathsf{E}_0 f^{*2})^{1/2} \exp KT^-\big) \tag{4.202}$$

where

$$T^- = \sup_{\boldsymbol{\rho}} \sum_{2 \le k \le M} n_k^2(\boldsymbol{\rho}) \,, \tag{4.203}$$

and where E_0 denotes expectation in the r.v.s η_k, z and ξ^ℓ. Moreover

$$\nu_t(f^*) \le K\nu((\mathsf{E}_0 f^{*2})^{1/2}) + \exp(-N)\nu(\mathsf{E}_0 f^{*2})^{1/2} \,. \tag{4.204}$$

The restriction $n \le 3$ is simply so that K does not depend on n.

Proof. We write the definition of $\nu_t(f^*)$ as in (4.181). Let

$$Y_{t,\ell} = \beta(g_t^\ell + tn_1^\ell + (1-t)\mu) + h$$

so that $\mathrm{Av}_{\varepsilon_1,\ldots,\varepsilon_n} \mathcal{E}_t = \prod_{\ell \le n} \mathrm{ch} Y_{t,\ell} \ge 1$. Since f^* is a function of $\boldsymbol{\rho}^1, \ldots, \boldsymbol{\rho}^n$, equality (4.181) implies

$$\nu_t(f^*) = \mathsf{E} \frac{\langle f^* \prod_{\ell \le n} \mathrm{ch} Y_{t,\ell} \rangle_-}{\mathsf{E}_\xi \langle \prod_{\ell \le n} \mathrm{ch} Y_{t,\ell} \rangle_-} \le \mathsf{E} \Big\langle f^* \prod_{\ell \le n} \mathrm{ch} Y_{t,\ell} \Big\rangle_- \,.$$

Taking first expectation E_0 inside the bracket and using the Cauchy-Schwarz inequality we get

$$\nu_t(f^*) \le \mathsf{E} \Big\langle (\mathsf{E}_0 f^{*2})^{1/2} \Big(\mathsf{E}_0 \prod_{\ell \le n} \mathrm{ch}^2 Y_{t,\ell}\Big)^{1/2} \Big\rangle_- \,. \tag{4.205}$$

We claim that

$$\mathsf{E}_0 \prod_{\ell \le n} \mathrm{ch}^2 Y_{\ell,t} \le K \exp KT^- \,. \tag{4.206}$$

Using that $n \le 3$ and Hölder's inequality, it suffices to show that

$$\mathsf{E}_0 \mathrm{ch}^6 Y_{t,\ell} \le K \exp KT^- \,.$$

First we observe that $\mathrm{ch}^6 x \le L(\exp 6x + \exp(-6x))$. As in the proof of (A.21), for numbers a_k we have

$$\mathsf{E}\exp\sum_{k\le M}\eta_k a_k = \prod_{k\le M}\mathrm{ch} a_k = \exp\sum_{k\le M}\log\mathrm{ch} a_k \le \exp\sum_{k\le M} a_k^2/2 ,$$

and recalling the definition (4.180) of g_t^ℓ, using (A.6) and independence, we see that indeed

$$\mathsf{E}_0\exp(\pm 6Y_{t,\ell}) \le K\exp KT^- .$$

Combining (4.206) and (4.205) we get

$$\nu_t(f^*) \le K\mathsf{E}\big(\langle(\mathsf{E}_0 f^{*2})^{1/2}\rangle_- \exp K_1 T^-\big) . \tag{4.207}$$

For a function $f^\sim \ge 0$ that depends only on $\boldsymbol{\rho}^1,\ldots,\boldsymbol{\rho}^n$,

$$\begin{aligned}\mathsf{E}_0\langle f^\sim\rangle &= \mathsf{E}_0\frac{\langle f^\sim\prod_{\ell\le n}\mathrm{ch}Y_{1,\ell}\rangle_-}{\langle\prod_{\ell\le n}\mathrm{ch}Y_{1,\ell}\rangle_-}\\ &\ge \langle f^\sim\rangle_-\mathsf{E}_0\frac{1}{\langle\mathrm{ch}Y_{1,1}\rangle_-^n}\\ &\ge \langle f^\sim\rangle_-\exp(-K_2T^-) ,\end{aligned}$$

using that $\mathsf{E}_0(1/X) \ge 1/\mathsf{E}_0X$ for $X = \langle\mathrm{ch}Y_{1,1}\rangle^n$, and using (4.206) for $t = 1$. We write this inequality for $f^\sim = (\mathsf{E}_0 f^{*2})^{1/2}$ (that depends only on $\boldsymbol{\rho}^1,\ldots,\boldsymbol{\rho}^n$), we multiply by $\exp((K_1+K_2)T^-)$ and we take expectation to get

$$\mathsf{E}\big(\langle(\mathsf{E}_0 f^{*2})^{1/2}\rangle_- \exp K_1T^-\big) \le \nu\big((\mathsf{E}_0 f^{*2})^{1/2}\exp KT^-\big) .$$

Combining with (4.207) this proves (4.202). The point of (4.204) is that T^- is not bounded, so we write

$$\begin{aligned}\nu\big((\mathsf{E}_0 f^{*2})^{1/2}\exp KT^-\big) \le{}& \exp KL\nu((\mathsf{E}_0 f^{*2})^{1/2})\\ &+ \nu\big(\mathbf{1}_{\{T^-\ge L\}}(\mathsf{E}_0 f^{*2})^{1/2}\exp KT^-\big) .\end{aligned}$$

The last term is

$$\mathsf{E}\big(\mathbf{1}_{\{T^-\ge L\}}\exp KT^-\langle(\mathsf{E}_0 f^{*2})^{1/2}\rangle\big)$$

and using Hölder's inequality we bound it by

$$\mathsf{P}(T^-\ge L)^{1/4}(\mathsf{E}\exp 4KT^-)^{1/4}\nu(\mathsf{E}_0 f^{*2})^{1/2} .$$

Using Corollary 4.5.8 for $N-1$ rather than N yields that $\mathsf{E}\exp 4KT^- \le K$. Using (4.150) for for $N-1$ rather than N we then obtain that if L is large enough we have $\mathsf{P}(T^-\ge L) \le \exp(-4N)$. □

Corollary 4.6.6. *If f is one of the functions of Lemma 4.6.3 then the term (4.201) satisfies*

$$\mathrm{IV} \le K\nu(|S_{1,2} - r| + |S_{1,1} - \rho|) + K \exp\left(-\frac{N}{K}\right) .$$

Proof. First we note that $\mathsf{E}_0 f_t^2 \le K$ and $\mathsf{E}_0(\partial f_t/\partial x_\ell)^2 \le K$. Let

$$f^* = \left(|f_t| + \left|\frac{\partial f_t}{\partial x_\ell}\right|\right)|S_{\ell,\ell'} - r| ,$$

so that

$$\mathsf{E}_0 f^{*2} \le K|S_{\ell,\ell'} - r|^2 . \tag{4.208}$$

Now, the Cauchy-Schwarz inequality implies $|\sum_{2\le k\le M} m_k(\boldsymbol{\sigma}^1)m_k(\boldsymbol{\sigma}^2)| \le T$, so that recalling (4.164) and (4.149) we have $|r| \le \mathsf{E}T \le K$. In a similar manner we get $\nu(S_{\ell,\ell'}^2) \le K$. Thus (4.208) proves that $\nu(\mathsf{E}_0 f^{*2}) \le K$, and (4.204) proves that

$$\nu_t(f^*) \le K\nu(|S_{\ell,\ell'} - r|) + K\exp(-N) .$$

Proceeding in the same manner for the other terms of (4.201) completes the proof. □

Next we deduce from Proposition 4.6.1 that the term IV is $\le KN^{-1/4}$. Using obvious notation,

$$\begin{aligned} S_{1,2} - \sum_{2\le k\le M} m_k^1 m_k^2 &= \sum_{2\le k\le M} n_k^1 n_k^2 - m_k^1 m_k^2 \\ &= \sum_{2\le k\le M} ((n_k^1 - m_k^1)m_k^2 + m_k^1(n_k^2 - m_k^2)) \\ &+ \sum_{2\le k\le M} (n_k^1 - m_k^1)(n_k^2 - m_k^2) \end{aligned}$$

and using that $|n_k^\ell - m_k^\ell| \le 1/N$, the Cauchy-Schwarz inequality and (4.149) yields

$$\nu\left(\left|S_{1,2} - \sum_{2\le k\le M} m_k^1 m_k^2\right|\right) \le \frac{K}{\sqrt{N}} .$$

Using (4.170) for $k = 1$ we then obtain that $\nu(|S_{1,2} - r|) \le KN^{-1/4}$. We then proceed similarly for the other terms.

In this manner we can control the main terms produced by approximate integration by parts in the term II of (4.196). The case of the term I of (4.196) is entirely similar, and the term III of (4.197) is immediate to control as in Corollary 4.6.6. We turn to the control of the error terms produced by approximate integration by parts. Let us fix $2 \le k \le M$, and consider approximate integration by parts in η_k using (4.199). Consider e.g. the case of the term

$$\nu_t(\eta_k n_k^\ell \varepsilon_\ell f_t) ,$$

on $n \le 3$ replicas. We have to consider the case of the four functions of Lemma 4.6.3. We consider only the case where $f(\boldsymbol{\sigma}_1, x_1) = \sigma_N^1 x_1$. The other cases are completely similar. In this case the term $\nu_t(\eta_k n_k^\ell \varepsilon_\ell f_t)$ is simply

$$\nu_t(\eta_k n_k^\ell \varepsilon_1 \varepsilon_\ell g_t^1) . \tag{4.209}$$

Let us define $g_{t,x}^\ell$ as g_t^ℓ in (4.180) except that we replace the term $\sqrt{t}\eta_k n_k^\ell$ by $\sqrt{t} x n_k^\ell$. Recalling (4.182) let us define $\mathcal{E}_{t,x}$ as $\mathcal{E}_t$ but using $g_{t,x}^\ell$ instead of g_t^ℓ, so that $\mathcal{E}_{t,x}$ does not depend on η_k. For a possible random function f^* of $t, x, \boldsymbol{\sigma}_1, \cdots, \boldsymbol{\sigma}_n$ let us define

$$\langle f^* \rangle_{t,x} = \frac{\mathsf{E}_\xi \langle \mathrm{Av}_{\varepsilon_1,\ldots,\varepsilon_n=\pm 1} f^* \mathcal{E}_{t,x} \rangle_-}{\mathsf{E}_\xi \langle \mathrm{Av}_{\varepsilon_1,\ldots,\varepsilon_n=\pm 1} \mathcal{E}_{t,x} \rangle_-} .$$

We consider the function

$$v(x) = \mathsf{E}\, \langle n_k^\ell \varepsilon_1 \varepsilon_2 g_{t,x}^1 \rangle_{t,x} .$$

In words, in the definition of the term (4.209), we replace every occurrence of η_k by x. We note that $\mathsf{E}\eta_k v(\eta_k)$ is the quantity (4.209), and that $\mathsf{E}\, v'(\eta_k)$ is the "main term" in the approximate integration by parts, that we have already taken into account.

Since there is a factor n_k^ℓ in front of the occurrence of x in $g_{t,x}^\ell$, differentiation of $v(x)$ in x brings out such a factor in each term. It should then be obvious using the inequality $|x_1 x_2 x_3 x_4| \le \sum_{\ell \le 4} x_\ell^4$ that

$$|v'''(x)| \le K\mathsf{E}\Big\langle \sum_{\ell \le n+3} (n_k^\ell)^4 (1 + |g_{t,x}^1|) \Big\rangle_{t,x} .$$

We then reproduce the argument of Lemma 4.6.5 to find that this quantity is bounded by

$$K\nu((n_k)^4) + K\exp(-N/K) . \tag{4.210}$$

The bound (4.200) implies that the error term created by the approximate integration by parts in the quantity $\nu_t(\eta_\ell n_k^\ell \varepsilon_\ell f_t)$ is bounded by the quantity (4.210). The sum over all values of k of these errors is bounded by

$$\nu\Big(\sum_{2 \le k \le M} (n_k)^4\Big) + K\exp\Big(-\frac{N}{K}\Big) .$$

Writing $x^4 = x \cdot x^3$ and using the Cauchy-Schwarz inequality yields

$$\sum_{2 \le k \le M} (n_k)^4 \le \Big(\sum_{2 \le k \le M} (n_k)^2\Big)^{1/2} \Big(\sum_{2 \le k \le M} (n_k)^6\Big)^{1/2} ,$$

and using the Cauchy-Schwarz inequality for ν we get

$$\nu\Big(\sum_{2\le k\le M}(n_k)^4\Big)\le\Big(\nu\Big(\sum_{2\le k\le M}(n_k)^2\Big)\Big)^{1/2}\Big(\nu\Big(\sum_{2\le k\le M}(n_k)^6\Big)\Big)^{1/2}.$$

Now we use (4.171) for $k=3$ to see that $\nu((n_k)^6)\le KN^{-3/2}$ and thus

$$\nu\Big(\sum_{2\le k\le M}(n_k)^6\Big)\le\frac{K}{N^{1/2}}.$$

Finally, recalling (4.203) we have $\sum_{2\le k\le M}(n_k)^2\le T^-$, so that, using (4.149) for $N-1$ rather than N, we get $\nu(\sum_{2\le k\le M}(n_k)^2)\le \mathsf{E}T^-\le L$. Therefore

$$\nu\Big(\sum_{2\le k\le M}(n_k)^4\Big)\le\frac{K}{N^{1/4}}.$$

This completes the proof that the equations (4.188) and (4.192) are satisfied with error terms $\le KN^{-1/4}$.

4.7 Notes and Comments

The Hopfield model was introduced in [118], but became popular only after Hopfield [79], [80] put it forward as a model of memory. For this aspect as a model of memory, it is the energy landscape, i.e. the function $\boldsymbol{\sigma}\mapsto\sum_{k\le M}m_k^2(\boldsymbol{\sigma})$ that matters. There are some rigorous results, [112], [97], [142], [132], [56] but they are based on ad hoc methods, none of which deserves to appear in a book. A detailed study of the model from the physicists' point of view appears in [3].

The first attempt at justifying the replica-symmetric equations can be found in [121]. The authors try to duplicate the results of [120] for the Hopfield model, i.e. to establish the replica-symmetric equations under the assumption that a certain quantity does not fluctuate with the disorder. This paper contains many interesting ideas, but one could of course wonder, among other things, how one could prove anything at all without addressing the question of uniqueness of the solutions of these equations. See also [122].

My notation differs from the traditional one as I call r what is traditionally called $r\alpha$. Thus, the replica-symmetric equations usually read

$$q=\mathsf{E}\,\mathrm{th}^2(\beta z\sqrt{r\alpha}+\beta\mu+h)$$

$$\mu=\mathsf{E}\,\mathrm{th}(\beta z\sqrt{r\alpha}+\beta\mu+h)$$

$$r=\frac{q}{(1-\beta(1-q))^2}.$$

This might be natural when one derives these equations from the "replica trick". The reason for not following this tradition is that the entire approach starts with studying the sequence $(m_k(\boldsymbol{\sigma}))_{k\leq M}$, and its global behavior (as in the Bovier-Gayrard localization theorem). Thus it is natural to incorporate the data about the length of this sequence (i.e. α) in the parameter r. (Maybe it is not such a good idea after all, but it is too late to change it anyway!)

The Bovier-Gayrard localization theorem is the culmination of a series of papers of these authors, sometimes with P. Picco. I am unsure as to whether the alternate approach I give here is better than the original one, but at least it is different. Bovier and Gayrard put forward the law of $(m_k(\boldsymbol{\sigma}))$ under Gibbs' measure as the central object. This greatly influenced the paper [142] where I first proved the validity of the replica-symmetric solution, using the cavity method. Very soon after seeing the paper [142], Bovier and Gayrard gave a simpler proof [28], based on convexity properties of the function ψ of (4.34) (which they proved) and on the Brascamp-Lieb inequalities. It is quite interesting that the convexity of ψ does not seem to hold in the whole region where there is replica-symmetry (the physicists' way to say that $R_{1,2} \simeq q$). Despite this the Bovier-Gayrard approach is of interest, as will become even clearer in Section 6.7. I have largely followed it here, rewriting of course some of the technicalities in the spirit of the rest of the book. In Volume II I will present my own approach, which is not really that much more difficult, although it yields much more accurate results.

In her paper [128], Shcherbina claims that her methods allow her to prove that the replica-symmetric solution holds on a large region. It would indeed be very nice to have a proof of the validity of the replica-symmetric solution that does not require to prove first something like the Bovier-Gayrard localization theorem. It is sad to see how some authors apparently do not care whether their ideas will be transmitted to the community or will be lost. More likely than not, in the present case they will be lost.

The paper [17] should not be missed. The interesting paper [20] is also related to the present chapter.

5. The V-statistics Model

5.1 Introduction

The model presented in this chapter was invented in an effort to discover natural Hamiltonians of mathematical interest. It illustrates well the power of the methods we have developed so far. It presents genuinely new features compared with the previous models, and these new features are the main motivations for studying it. The discovery of this model raises the question as to whether the models presented in this book represent well the main types of possible features of mean-field models, or whether genuinely new types remain to be discovered.

The model of the present chapter is related to the Perceptron model of Chapter 2 at the technical level, so we advise the reader to be comfortable with that chapter before reading the details of the proofs here. We consider independent standard normal r.v.s $(g_{i,k})_{i,k\geq 1}$ and for $\boldsymbol{\sigma} \in \Sigma_N$ we define as usual

$$S_k = S_k(\boldsymbol{\sigma}) = \frac{1}{\sqrt{N}} \sum_{i \leq N} g_{i,k} \sigma_i \,. \tag{5.1}$$

We consider a function $u : \mathbb{R}^2 \to \mathbb{R}$. We assume that it is symmetric,

$$u(x,y) = u(y,x) \tag{5.2}$$

and, given an integer M, we consider the Hamiltonian

$$-H_{N,M}(\boldsymbol{\sigma}) = \frac{1}{N} \sum_{1 \leq k_1 < k_2 \leq M} u(S_{k_1}, S_{k_2}) \,. \tag{5.3}$$

The name of the model is motivated by the fact that the right-hand side of (5.3) resembles an estimator known as a V-statistics. However no knowledge about these seems relevant for the present chapter. The case of interest is when M is a proportion of N. Then $H_{N,M}$ is of order N. As in Chapter 2, we will be interested only in the "algebraic" structure connected with the Hamiltonian (5.3), so we will decrease technicalities by making a strong assumption on u. We assume that for a certain number D,

$$u \text{ and all its partial derivatives of order} \leq 6 \text{ are bounded by } D. \tag{5.4}$$

M. Talagrand, *Mean Field Models for Spin Glasses*, Ergebnisse der Mathematik und ihrer Grenzgebiete. 3. Folge / A Series of Modern Surveys in Mathematics 54, DOI 10.1007/978-3-642-15202-3_5,

How, and to which extent, this condition can be relaxed is an open problem, although one could expect that techniques similar to those presented later in Chapter 9 should bear on this question. We assume

$$D \geq 1\,. \tag{5.5}$$

Let us first try to describe what happens at a global level. In the high-temperature regime, we expect to have the usual relation

$$R_{1,2} \simeq q \tag{5.6}$$

where $R_{1,2} = N^{-1}\sum_{i\leq N}\sigma_i^1\sigma_i^2$, and where the number q depends on the system.

Throughout the chapter we use the notation

$$w = \frac{\partial u}{\partial x}\,. \tag{5.7}$$

Thus the symmetry condition (5.2) implies

$$w(x,y) = \frac{\partial u}{\partial x}(x,y) = \frac{\partial u}{\partial y}(y,x)\,. \tag{5.8}$$

The relation (5.6) has to be complemented by the relation

$$\frac{1}{4N^3}\sum_{k_1,k_2,k_3\leq M} w(S_{k_1}^1,S_{k_2}^1)w(S_{k_1}^2,S_{k_3}^2) \simeq r\,, \tag{5.9}$$

where as usual $S_k^\ell = S_k(\boldsymbol{\sigma}^\ell)$. The new and unexpected feature is that the computation of r seems to require the use of an auxiliary function $\gamma(x)$. Intuitively this function γ satisfies

$$\gamma(x) \simeq \mathsf{E}\Big\langle \frac{1}{N}\sum_{k\leq M} u(x,S_k)\Big\rangle = \frac{M}{N}\mathsf{E}\langle u(x,S_M)\rangle\,,$$

where of course the bracket denotes an average for the Gibbs measure with Hamiltonian (5.3). The reason behind the occurrence of this function is the "cavity in M" argument. Going from $M-1$ to M we add the term

$$\frac{1}{N}\sum_{k<M} u(S_k,S_M)$$

to the Hamiltonian, and we will prove that in fact this term acts somewhat as $\gamma(S_M)$. The function γ will be determined through a self-consistency equation and will in turn allow the computation of r. In the present model, the "replica symmetric equations" are a system of there equations with three unknowns, one of which is the function γ.

5.2 The Smart Path

We use the same interpolation as in Chapter 2. We consider independent standard Gaussian r.v.s $(\xi_k)_{k\le M}$ and, as in (2.15), we consider

$$S_{k,t} = \frac{1}{\sqrt{N}}\sum_{i<N} g_{i,k}\sigma_i + \sqrt{\frac{t}{N}}g_{N,k}\sigma_N + \sqrt{\frac{1-t}{N}}\xi_k \ , \qquad (5.10)$$

and the Hamiltonian

$$-H_{N,M,t} = \frac{1}{N}\sum_{1\le k_1<k_2\le M} u(S_{k_1,t},S_{k_2,t}) + \sigma_N\sqrt{1-t}Y \ , \qquad (5.11)$$

where Y is a Gaussian r.v. independent of any other randomness, and where

$$r = \mathsf{E}Y^2 \qquad (5.12)$$

will be determined later.

Consider independent copies $\boldsymbol{\xi}^\ell$ of $\boldsymbol{\xi} = (\xi_k)_{k\le M}$. We recall that as usual, E_ξ denotes expectation in all the r.v.s ξ_k^ℓ. For a function

$$f = f(\boldsymbol{\sigma}^1,\ldots,\boldsymbol{\sigma}^n,\boldsymbol{\xi}^1,\ldots,\boldsymbol{\xi}^n)$$

we define $\langle f\rangle_t$ by the formula (2.19), i.e.

$$\langle f\rangle_t = \frac{1}{Z_t^n}\mathsf{E}_\xi\sum_{\boldsymbol{\sigma}^1,\ldots,\boldsymbol{\sigma}^n} f(\boldsymbol{\sigma}^1,\ldots,\boldsymbol{\sigma}^n,\boldsymbol{\xi}^1,\ldots,\boldsymbol{\xi}^n)\exp\Bigl(-\sum_{\ell\le n}H_t^\ell\Bigr) \ , \qquad (5.13)$$

where $Z_t = \mathsf{E}_\xi\sum_{\boldsymbol{\sigma}}\exp(-H_t(\boldsymbol{\sigma}))$ and $H_t^\ell = H_{t,N,M}(\boldsymbol{\sigma}^\ell,\boldsymbol{\xi}^\ell)$. We write

$$\nu_t(f) = \mathsf{E}\langle f\rangle_t \quad ; \quad \nu_t'(f) = \frac{\mathrm{d}}{\mathrm{d}t}\nu_t(f) \ ,$$

and, as usual, $\varepsilon_\ell = \sigma_N^\ell$. We also recall that $\nu = \nu_1$.

This interpolation is designed to decouple the last spin. The following is proved as Lemma 1.6.2.

Lemma 5.2.1. *For a function f^- on Σ_{N-1}^n, and a subset I of $\{1,\ldots,n\}$ we have*

$$\nu_0\Bigl(f^-\prod_{\ell\in I}\varepsilon_\ell\Bigr) = \mathsf{E}(\mathrm{th}Y)^{\mathrm{card}I}\nu_0(f^-) = \nu_0\Bigl(\prod_{\ell\in I}\varepsilon_\ell\Bigr)\nu_0(f^-) \ .$$

Throughout this chapter, we write $\alpha = M/N$. We recall that $r = \mathsf{E}Y^2$, and that $w = \partial u/\partial x$. As usual we use the notation $\varepsilon_\ell = \sigma_N^\ell$.

Proposition 5.2.2. *For a function f on Σ_N^n, we have*

$$\nu_t'(f) = \mathrm{I} + \mathrm{II} \,, \tag{5.14}$$

where, defining

$$A_{\ell,\ell'} = \frac{1}{N^3} \sum \nu_t\big(\varepsilon_\ell \varepsilon_{\ell'} w(S^\ell_{k_1,t}, S^\ell_{k_2,t}) w(S^{\ell'}_{k_1,t}, S^{\ell'}_{k_3,t}) f\big) \tag{5.15}$$

for a summation over $k_1, k_2, k_3 \le M$, $k_2 \ne k_1$, $k_3 \ne k_1$, we have

$$\mathrm{I} = \sum_{1 \le \ell < \ell' \le n} A_{\ell,\ell'} - n \sum_{\ell \le n} A_{\ell,n+1} + \frac{n(n+1)}{2} A_{n+1,n+2} \tag{5.16}$$

and

$$\mathrm{II} = -r \bigg(\sum_{1 \le \ell < \ell' \le n} \nu_t(\varepsilon_\ell \varepsilon_{\ell'} f) - n \sum_{\ell \le n} \nu_t(\varepsilon_\ell \varepsilon_{n+1} f) + \frac{n(n+1)}{2} \nu_t(\varepsilon_{n+1} \varepsilon_{n+2} f) \bigg) \,. \tag{5.17}$$

Proof. Let us write

$$S^{\ell\prime}_{k,t} = \frac{\partial}{\partial t} S^\ell_{k,t} = \frac{1}{2\sqrt{tN}} g_{N,k} \varepsilon_\ell - \frac{1}{2\sqrt{(1-t)N}} \xi^\ell_k \,.$$

The reader should carefully distinguish between $S^{\ell\prime}_{k,t}$ and $S^{\ell'}_{k,t}$ (the position of the $\prime$ is not the same). From (5.8), we get

$$\frac{\partial u}{\partial y}(x,y) = w(y,x) \,,$$

and the relations (5.10) and (5.11) imply

$$\begin{aligned} \frac{\mathrm{d}}{\mathrm{d}t}(-H_{N,M,t}(\boldsymbol{\sigma}^\ell)) &= \frac{1}{N} \sum_{1 \le k_1 < k_2 \le M} (S^{\ell\prime}_{k_1,t} w(S^\ell_{k_1,t}, S^\ell_{k_2,t}) + S^{\ell\prime}_{k_2,t} w(S^\ell_{k_2,t}, S^\ell_{k_1,t})) \\ &\quad - \frac{1}{2\sqrt{1-t}} \varepsilon_\ell Y \\ &= \frac{1}{N} \sum_{k_1 \ne k_2} S^{\ell\prime}_{k_1,t} w(S^\ell_{k_1,t}, S^\ell_{k_2,t}) - \frac{1}{2\sqrt{1-t}} \varepsilon_\ell Y \,. \end{aligned}$$

Therefore, differentiation of the formula (5.13) yields

$$\nu_t'(f) = \mathrm{III} + \mathrm{IV} \,, \tag{5.18}$$

where

$$\mathrm{III} = \sum_{\ell \le n} \nu_t(D_\ell f) - n \nu_t(D_{n+1} f)$$

for

$$D_\ell = \frac{1}{N} \sum_{k_1 \neq k_2} S^{\ell\prime}_{k_1,t} w(S^\ell_{k_1,t}, S^\ell_{k_2,t})$$

and where

$$\mathrm{IV} = -\frac{1}{2\sqrt{1-t}} \left(\sum_{\ell \le n} \nu_t(\varepsilon_\ell Y f) - n\nu_t(\varepsilon_{n+1} Y f) \right) .$$

It remains of course to integrate by parts. The integration by parts in Y in the term IV has been done many times and IV = II. Concerning the term III, we have explained in great detail a similar case in Chapter 2. We think of the r.v.s $S^{\ell\prime}_{k,t}$, $S^\ell_{k,t}$ as ℓ and k vary as a jointly Gaussian family of r.v.s. The relations

$$\mathsf{E} S^{\ell\prime}_{k,t} S^\ell_{k',t} = 0$$

imply that, when integrating by parts in the r.v.s $S^{\ell\prime}_{k,t}$, only the dependence of the Hamiltonian on the randomness creates terms (but not the randomness of $w(S^\ell_{k_1,t}, S^\ell_{k_2,t})$). For $\ell' \neq \ell$, the relations

$$\mathsf{E} S^{\ell\prime}_{k,t} S^{\ell'}_{k,t} = \frac{1}{2N} \varepsilon_\ell \varepsilon_{\ell'} \; ; \qquad \mathsf{E} S^{\ell\prime}_{k,t} S^{\ell'}_{k',t} = 0 \;\; \text{if } k' \neq k$$

hold and (with the usual abuse of notation)

$$\begin{aligned} \frac{\partial H_{N,M,t}}{\partial S_{k,t}} &= \frac{1}{N} \left(\sum_{k<k_2} w(S_{k,t}, S_{k_2,t}) + \sum_{k_1<k} w(S_{k,t}, S_{k_1,t}) \right) \\ &= \frac{1}{N} \sum_{k' \neq k} w(S_{k,t}, S_{k',t}) \,. \end{aligned}$$

Then the result follows by carrying out the computation as in Chapter 2, following the method outlined in Exercise 2.3.3. □

We recall that $\alpha = M/N$.

Corollary 5.2.3. *Assume that $D^2\alpha^3 \le 1$ and $|r| \le 1$. Then for any function $f \ge 0$ on Σ_N^n we have*

$$\nu_t(f) \le L^{n^2} \nu(f) \,. \tag{5.19}$$

Of course the conditions $D^2\alpha^3 \le 1$ and $|r| \le 1$ are simply convenient choices and do not have any intrinsic meaning.

Proof. It follows from (5.14) that $|\nu'_t(f)| \le 2n^2 \nu_t(f)$, and we integrate. □

5.3 Cavity in M

We would like, with the appropriate choice of r (i.e. if (5.9) holds), that the terms I and II of (5.14) nearly cancel out. So we need to make sense of the term $A_{\ell,\ell'}$. To lighten notation we assume $\ell = 1$, $\ell' = 2$.

In the summation (5.15) there are at most M^2 terms for which $k_2 = k_3$. Defining

$$A'_{1,2} = \frac{1}{N^3} \sum_{k_1,k_2,k_3 \text{ all different}} \nu_t\big(\varepsilon_1\varepsilon_2 w(S^1_{k_1,t}, S^1_{k_3,t}) w(S^2_{k_1,t}, S^2_{k_2,t}) f\big)$$

(and keeping the dependence on t implicit) we get

$$|A_{1,2} - A'_{1,2}| \le \frac{KM^2}{N^3}\nu_t(|f|) = \frac{K}{N}\alpha^2\nu_t(|f|) \le \frac{K'}{N}\nu_t(|f|) , \tag{5.20}$$

where K is a number depending only on D and α. Each triplet (k_1, k_2, k_3) brings the same contribution to $A'_{1,2}$, so that

$$A'_{1,2} = \frac{M(M-1)(M-2)}{N^3}\nu_t\big(\varepsilon_1\varepsilon_2 w(S^1_{M,t}, S^1_{M-1,t}) w(S^2_{M,t}, S^2_{M-2,t}) f\big) .$$

Therefore, defining

$$C_{1,2} = \nu_t\big(\varepsilon_1\varepsilon_2 w(S^1_{M,t}, S^1_{M-1,t}) w(S^2_{M,t}, S^2_{M-2,t}) f\big) , \tag{5.21}$$

we have

$$A'_{1,2} = \frac{M(M-1)(M-2)}{N^3} C_{1,2} ,$$

so that

$$|A'_{1,2} - \alpha^3 C_{1,2}| \le \frac{K}{N} .$$

Combining with (5.20) we reach that

$$|A_{1,2} - \alpha^3 C_{1,2}| \le \frac{K}{N}\nu_t(|f|) . \tag{5.22}$$

To estimate $C_{1,2}$ it seems a good idea to make explicit the dependence of the Hamiltonian on $S_{M,t}, S_{M-1,t}$ and $S_{M-2,t}$. Defining

$$-H_{N,M-3,t} = \frac{1}{N} \sum_{1 \le k_1 < k_2 \le M-3} u(S_{k_1,t}, S_{k_2,t}) + \sqrt{1-t}\,\sigma_N Y , \tag{5.23}$$

it holds that

$$-H_{N,M,t} = -H_{N,M-3,t} - H , \tag{5.24}$$

where

$$-H = \frac{1}{N} \sum_{1 \le k_1 < k_2 \le M, k_2 \ge M-2} u(S_{k_1,t}, S_{k_2,t}) \,. \tag{5.25}$$

When in this formula we replace $\boldsymbol{\sigma}$ by $\boldsymbol{\sigma}^\ell$ and $\boldsymbol{\xi}$ by $\boldsymbol{\xi}^\ell$ we denote the result by $-H^\ell$, and when we do the same for $H_{N,M-3,t}$ we denote the result by $H^\ell_{N,M-3,t}$.

Let us denote by $\langle \cdot \rangle_{t,\sim}$ an average for the Hamiltonian $H_{N,M-3,t}$, in the sense of (5.13). That is, for a function $f = f(\boldsymbol{\sigma}^1, \ldots, \boldsymbol{\sigma}^n, \boldsymbol{\xi}^1, \ldots, \boldsymbol{\xi}^n)$ we define $\langle f \rangle_{t,\sim}$ by the formula

$$\langle f \rangle_{t,\sim} = \frac{1}{Z^n_{t,\sim}} \mathsf{E}^-_\xi \sum_{\boldsymbol{\sigma}^1, \ldots, \boldsymbol{\sigma}^n} f(\boldsymbol{\sigma}^1, \ldots, \boldsymbol{\sigma}^n, \boldsymbol{\xi}^1, \ldots, \boldsymbol{\xi}^n) \exp\Bigl(-\sum_{\ell \le n} H^\ell_{N,M-3,t}\Bigr) \,, \tag{5.26}$$

where $Z_{t,\sim}$ is the normalization factor, $Z_{t,\sim} = \mathsf{E}^-_\xi \sum_{\boldsymbol{\sigma}} \exp(-H_{N,M-3,t}(\boldsymbol{\sigma}))$ and where E^-_ξ denotes expectation in the r.v.s ξ^ℓ_k for $\ell \ge 1$ and $k \le M-3$. Let us then define

$$\mathcal{E} = \exp\Bigl(\sum_{\ell \le n} -H^\ell\Bigr) \,. \tag{5.27}$$

Then for a function h of $\boldsymbol{\sigma}^1, \ldots, \boldsymbol{\sigma}^n$ and of $S^\ell_{M-j,t}$ for $j = 0, 1, 2$ and $\ell \le n$, the identity

$$\langle h \rangle_t = \frac{\mathsf{E}_\xi \langle h\mathcal{E} \rangle_{t,\sim}}{\mathsf{E}_\xi \langle \mathcal{E} \rangle_{t,\sim}} \tag{5.28}$$

holds, where, as usual, E_ξ denotes expectation in all the r.v.s "labeled ξ". Here $\langle h\mathcal{E} \rangle_{t,\sim}$ and $\langle \mathcal{E} \rangle_{t,\sim}$ depend only on the r.v.s ξ^ℓ_k for $k = M-1, M-2, M-3$.

Exercise 5.3.1. Rather than (5.26), let us define

$$\langle f \rangle_{t,\sim} = \frac{1}{Z^n_{t,\sim}} \mathsf{E}_\xi \sum_{\boldsymbol{\sigma}^1, \ldots, \boldsymbol{\sigma}^n} f(\boldsymbol{\sigma}^1, \ldots, \boldsymbol{\sigma}^n, \boldsymbol{\xi}^1, \ldots, \boldsymbol{\xi}^n) \exp\Bigl(-\sum_{\ell \le n} H^\ell_{N,M-3,t}\Bigr) \,, \tag{5.29}$$

where E_ξ denotes expectation in *all* the r.v.s ξ^ℓ_k. Show that then that rather than (5.27) we have

$$\langle h \rangle_t = \frac{\langle h\mathcal{E} \rangle_{t,\sim}}{\langle \mathcal{E} \rangle_{t,\sim}} \,. \tag{5.30}$$

Of course, (5.28) and (5.30) are simply two different manners to write the same identity.

The convention used in (5.29) (i.e. that E_ξ stands for expectation in *all* the r.v.s. ξ) was used in Chapter 2. It is somewhat more natural than the convention used in (5.28). As in Chapter 3 we shall not use it here, to avoid having to constantly remind the reader of it.

Our best guess is that the quantities $S^\ell_{k,t}$, $\ell \le n$, $k = M, M-1, M-2$ will have a jointly Gaussian behavior when seen as functions on the system with Hamiltonian (5.23). For different values of k they will be independent, and for the same value of k their pairwise correlation will be a new parameter q. So we fix $0 \le q \le 1$ (which will be determined later) and for $j = 0,1,2$, $\ell \le n$, we consider independent standard Gaussian r.v.s z_j and $\hat{\xi}_j^\ell$ (that are independent of all the other sources of randomness) and we set

$$\theta_j^\ell = z_j\sqrt{q} + \hat{\xi}_j^\ell\sqrt{1-q}\,.$$

For $0 \le v \le 1$ we define

$$S_{j,v}^\ell = \sqrt{v}S_{M-j,t}^\ell + \sqrt{1-v}\theta_j^\ell\,, \tag{5.31}$$

keeping the dependence of $S_{j,v}^\ell$ on t implicit. (The reader will observe that, despite the similarity of notation, it is in practice impossible to confuse the quantity $S_{k,t}$ with the quantity $S_{j,v}$. Here again we choose a bit of informality over heavy notation.) Let us denote by

$\mathcal{E}_v$ the quantity (5.27) when one replaces each occurrence of $S_{M-j,t}^\ell$ by $S_{j,v}^\ell$ for $\ell \le n$ and $j = 0,1,2$. (5.32)

For any function h of $\boldsymbol{\sigma}^1,\dots,\boldsymbol{\sigma}^n$ and of $S_{j,v}^\ell$ for $j = 0,1,2$ and $\ell \le n$, we define $\langle h\rangle_{t,\sim}$ by (5.26) and

$$\nu_{t,v}(h) = \mathsf{E}\frac{\langle h\mathcal{E}_v\rangle_{t,\sim}}{\mathsf{E}_\xi\langle\mathcal{E}_v\rangle_{t,\sim}}\,, \tag{5.33}$$

where (following our usual convention) E_ξ now denotes expectation in the variables ξ_k^ℓ **and the r.v.s** $\hat{\xi}_j^\ell$. Therefore, if h depends on $\boldsymbol{\sigma}^1,\dots,\boldsymbol{\sigma}^n$ only, taking expectation in (5.28) yields

$$\nu_{t,1}(h) = \nu_t(h)\,. \tag{5.34}$$

Lemma 5.3.2. *Consider a function f depending on $\boldsymbol{\sigma}^1,\dots,\boldsymbol{\sigma}^n$ only. Then we have*

$$\left|\frac{\mathrm{d}}{\mathrm{d}v}\nu_{t,v}(f)\right| \le Ln^2\alpha^2D^2\nu_{t,v}(f^2)^{1/2}\nu_{t,v}\big((R_{1,2}-q)^2\big)^{1/2} + \frac{K}{N}\nu_{t,v}(|f|)\,. \tag{5.35}$$

Moreover, if $\alpha D \le 1$ and

$$B_v = w(S_{0,v}^1, S_{1,v}^1)w(S_{0,v}^2, S_{2,v}^2)\,, \tag{5.36}$$

we have

$$\left|\frac{\mathrm{d}}{\mathrm{d}v}\nu_{t,v}(B_vf)\right| \le Ln^2D^2\nu_{t,v}(f^2)^{1/2}\nu_{t,v}\big((R_{1,2}-q)^2\big)^{1/2} + \frac{K}{N}\nu_{t,v}(|f|)\,. \tag{5.37}$$

Proof. It is as in Lemma 2.3.2. We compute the derivatives of the function $v \mapsto \nu_{t,v}(f)$ (resp. $v \mapsto \nu_{t,v}(B_v f)$) and we integrate by parts. Defining $S^{\ell\prime}_{j,v} = \mathrm{d}S^{\ell}_{j,v}/\mathrm{d}v$, we observe the relations (where the reader will carefully distinguish $S^{\ell\prime}_{j,v}$ from $S^{\ell'}_{j,v}$)

$$
\begin{aligned}
\mathsf{E} S^{\ell\prime}_{j,v} S^{\ell'}_{j',v} &= 0 \ \text{ if } j \neq j' \\
\mathsf{E} S^{\ell\prime}_{j,v} S^{\ell}_{j,v} &= 0 \\
\ell \neq \ell' \Rightarrow \mathsf{E} S^{\ell\prime}_{j,v} S^{\ell'}_{j,v} &= \frac{1}{2}\left(\frac{1}{N}\sum_{i<N} \sigma^{\ell}_i \sigma^{\ell'}_i + \frac{t}{N}\sigma^{\ell}_N \sigma^{\ell'}_N - q\right) = \frac{1}{2}(R^t_{\ell,\ell'} - q) \,.
\end{aligned}
$$

So, integration by parts "creates a factor $R^t_{\ell,\ell'} - q$ in each term". There are $3M - 6 \leq 3M$ terms in the expression (5.25). Each has a factor $1/N$. Before integration by parts, the expression for the derivative of the function $v \mapsto \nu_{t,v}(f)$ contains $\leq LnM$ terms. Each of these terms uses $n+1$ replicas, and its integration by parts creates $\leq LnM$ terms. All told, there are at most Ln^2M^2 terms in the expression for $\mathrm{d}\nu_{t,v}(f)/\mathrm{d}v$, and each of them is bounded by a term of the type

$$
\frac{D^2}{N^2}\nu_{t,v}(|f||R^t_{\ell,\ell'} - q|)
$$

for certain values of ℓ and ℓ', $\ell \neq \ell'$. So the bound (5.35) simply follows from the Cauchy-Schwarz inequality. We proceed similarly in the case of (5.36). The reason why we cannot get a factor α in (5.37) is that when computing $\mathrm{d}\nu_{t,v}(B_v f)/\mathrm{d}v$ we find the term $\nu_{t,v}(S^{1\prime}_{0,v} w'(S^1_{0,v}, S^1_{1,v})) w(S^2_{0,v}, S^2_{2,v}) f)$ where $w'(x,y) = \partial w(x,y)/\partial x = \partial^2 u(x,y)/\partial x^2$. When integrating by parts, this creates a term

$$
\nu_{t,v}((R^t_{1,2} - q) w'(S^1_{0,v}, S^1_{1,v})) w'(S^2_{0,v}, S^2_{2,v}) f) \,,
$$

and the best we can do is to bound this term by $D^2 \nu_{t,v}(|f||R^t_{1,2} - q|)$. □

The factor α^2 in (5.35) is not really needed for the rest of the proof. There is a lot of room in the arguments. However it occurs so effortlessly that we see no reason to omit it. This might puzzle the reader.

Lemma 5.3.3. *If $\alpha D \leq 1$, we have*

$$
\nu_{t,v}(|h|) \leq L^n \nu_t(|h|) \ ; \ \mathsf{E}\langle |h| \rangle_{t,\sim} \leq L^n \nu_t(|h|) \,. \tag{5.38}
$$

Proof. The quantity $-H$ of (5.25) satisfies $|-H| \leq 3\alpha D \leq 3$ (bounding each term $u(S_{k_1,t}, S_{k_2,t})$ by D) so that the quantity $\mathcal{E}_v$ of (5.36) satisfies $L^{-n} \leq \mathcal{E}_v \leq L^n$. Thus (5.33) (used for $|h|$ rather than h) implies $\nu_{t,v}(|h|) \leq L^n \mathsf{E}\langle |h\mathcal{E}_v| \rangle_{t,\sim} \leq L^{2n} \mathsf{E}\langle |h| \rangle_{t,\sim}$. Using again (5.33) in the case $v = 1$ we get $\mathsf{E}\langle |h| \rangle_{t,\sim} \leq L^n \nu_{t,1}(|h|) = L^n \nu_t(|h|)$, using (5.34) in the equality. □

Exercise 5.3.4. Prove the first part of (5.38) using a differential inequality.

5.4 The New Equation

The purpose of the interpolation of the previous section is that we expect that $\nu_{t,0}(B_0 f)$ will be easier to understand than $\nu_{t,1}(B_1 f) = \nu_t(B_1 f)$. This is the case, but the (un)pleasant surprise is that it will still require significant work to understand $\nu_{t,0}(B_0 f)$. Let us consider the random function

$$\gamma_\ell(x) = \frac{1}{N} \sum_{k \le M-3} u(S_{k,t}^\ell, x) ,$$

where the dependence on t is kept implicit in the left-hand side. Then, for $v = 0$, the quantity $\mathcal{E}_v$ of (5.32) is equal to

$$\mathcal{E}_0 = \mathcal{E}'\mathcal{E}'' \tag{5.39}$$

for

$$\mathcal{E}' = \exp \sum_{\ell \le n} \sum_{j=0,1,2} \gamma_\ell(\theta_j^\ell) \tag{5.40}$$

$$\mathcal{E}'' = \exp \frac{1}{N} \sum_{\ell \le n} (u(\theta_0^\ell, \theta_2^\ell) + u(\theta_0^\ell, \theta_1^\ell) + u(\theta_1^\ell, \theta_2^\ell)) . \tag{5.41}$$

This is seen by separating in the sum (5.25) the terms for which $k_1 = M - 2$ or $k_1 = M - 1$ (these create $\mathcal{E}''$).

Of course the influence of $\mathcal{E}''$ will be very small; but to understand the influence of $\mathcal{E}'$, we must understand the function γ_ℓ. We explain first the heuristics.

We hope that the quantities $(S_{k,t}^\ell)_{k \le M-3}$ behave roughly like independent r.v.s under the averages $\langle \cdot \rangle_{t,\sim}$, so that by the law of large numbers, we should have that for each ℓ,

$$\gamma_\ell(x) \simeq \frac{M-3}{N} \mathsf{E}\langle u(S_{1,t}, x)\rangle_{t,\sim} \simeq \alpha \mathsf{E}\langle u(S_{1,t}, x)\rangle_{t,\sim} . \tag{5.42}$$

This shows that (in the limit $N \to \infty$) γ_ℓ does not depend on ℓ and is not random. We denote by γ this non-random function, and we now look for the relation it should satisfy. It seems very plausible that

$$\mathsf{E}\langle u(S_{1,t}, x)\rangle_{t,\sim} \simeq \nu_t(u(S_{1,t}, x)) = \nu_t(u(S_{M,t}, x)) \tag{5.43}$$

by symmetry. We expect that (5.37) still holds, with a similar proof, if we define now $B_v = u(S_v, x)$. Assuming that as expected $R_{1,2} \simeq q$, we should have $\nu_t(B_1) \simeq \nu_{t,0}(B_0)$ i.e. (with obvious notation: since there is only one replica, we no longer need replica indices)

$$\begin{aligned} \nu_t(u(S_{M,t}, x)) &\simeq \nu_{t,0}(u(\theta_0, x)) \\ &\simeq \mathsf{E} \frac{u(\theta_0, x) \exp \sum_{0 \le j \le 2} \gamma(\theta_j)}{\mathsf{E}_\xi \exp \sum_{0 \le j \le 2} \gamma(\theta_j)} . \end{aligned} \tag{5.44}$$

Now, using independence

$$\mathsf{E}_\xi u(\theta_0,x)\exp\sum_{0\le j\le 2}\gamma(\theta_j)=\mathsf{E}_\xi u(\theta_0,x)\exp\gamma(\theta_0)\prod_{j=1,2}\mathsf{E}_\xi\exp\gamma(\theta_j)$$

$$\mathsf{E}_\xi\exp\sum_{0\le j\le 2}\gamma(\theta_j)=\prod_{j=0,1,2}\mathsf{E}_\xi\exp\gamma(\theta_j)$$

so that from (5.44) we get

$$\nu_t(u(S_{M,t},x))\simeq\mathsf{E}\frac{u(\theta_0,x)\exp\gamma(\theta_0)}{\mathsf{E}_\xi\exp\gamma(\theta_0)}\ . \tag{5.45}$$

Exercise 5.4.1. Find a more economical interpolation to reach (5.45). (Hint: in (5.25) replace $M-2$ by M.)

Let us now write

$$\theta=\sqrt{q}z+\sqrt{1-q}\xi\ ,$$

where z and q are independent standard Gaussian r.v.s, and repeat that E_ξ denotes expectation in ξ only. Combining the previous chain of equations (5.42) to (5.45) we reach that the non-random function γ should satisfy the relation

$$\gamma(x)=\alpha\mathsf{E}\frac{u(\theta,x)\exp\gamma(\theta)}{\mathsf{E}_\xi\exp\gamma(\theta)}\ . \tag{5.46}$$

The first task is to prove that this functional equation has a solution.

Lemma 5.4.2. *If $LD\alpha\le 1$, given any value of q then there exists a unique function $\gamma=\gamma_{\alpha,q}$ from $\mathbb{R}$ to $[0,1]$ that satisfies (5.46). Moreover, given any other function γ_* from $\mathbb{R}$ to $[0,1]$ we have*

$$\sup_x|\gamma(x)-\gamma_*(x)|\le 2\sup_y\left|\gamma(y)-\alpha\mathsf{E}\frac{u(\theta,y)\exp\gamma_*(\theta)}{\mathsf{E}_\xi\exp\gamma_*(\theta)}\right|\ . \tag{5.47}$$

We remind the reader that throughout the book a statement such as "If $LD\alpha\le 1$..." is a short-hand for "There exists a universal constant L with the following property. If $LD\alpha\le 1$..."

Proof. This is of course a "contraction argument". Consider the supremum norm $\|\cdot\|_\infty$ on the space $\mathcal{C}$ of functions from $\mathbb{R}$ to $[-1,1]$. Consider the operator U that associates to a function $\psi\in\mathcal{C}$ the function $U(\psi)$ given by

$$U(\psi)(x)=\alpha\mathsf{E}\frac{u(\theta,x)\exp\psi(\theta)}{\mathsf{E}_\xi\exp\psi(\theta)}\ .$$

Since $1/e\le\exp\psi(\theta)\le e$ and $|u(\theta,x)|\le D$ we have $|U(\psi)(x)|\le\alpha De^2$, so if $\alpha De^2\le 1$ we have $U(\psi)\in\mathcal{C}$. Consider $\psi_1,\psi_2\in\mathcal{C}$ and $\varphi(t)=U(t\psi_1+(1-t)\psi_2)\in\mathcal{C}$, so that, writing $\mathcal{E}_t=\exp(t\psi_1(\theta)+(1-t)\psi_2(\theta))$, we get

$$\varphi'(t) = \alpha \mathsf{E}\frac{u(\theta,x)(\psi_1(\theta)-\psi_2(\theta))\mathcal{E}_t}{\mathsf{E}_\xi \mathcal{E}_t} - \alpha \mathsf{E}\frac{u(\theta,x)\mathcal{E}_t \mathsf{E}_\xi(\psi_1(\theta)-\psi_2(\theta))\mathcal{E}_t}{(\mathsf{E}_\xi \mathcal{E}_t)^2} ,$$

and since $\|t\psi_1 + (1-t)\psi_2\|_\infty \le 1$, we have $1/e \le \mathcal{E}_t \le e$ and $\|\varphi'(t)\|_\infty \le L_0\alpha D\|\psi_1 - \psi_2\|_\infty$. Therefore $\|\varphi(1) - \varphi(0)\|_\infty \le L_0\alpha D\|\psi_1 - \psi_2\|_\infty$, i.e.

$$\|U(\psi_1) - U(\psi_2)\|_\infty \le L_0\alpha D\|\psi_1 - \psi_2\|_\infty . \tag{5.48}$$

Thus for $2L_0\alpha D \le 1$, the map U is a contraction of $\mathcal{C}$ and thus it has a unique fixed point γ.

We turn to the proof of (5.47). We write that, since $\gamma = U(\gamma)$, for any function $\gamma^* \in \mathcal{C}$, we have, when $2L_0\alpha D \le 1$, and using (5.48),

$$\begin{aligned}\|\gamma_* - \gamma\|_\infty &\le \|\gamma_* - U(\gamma_*)\|_\infty + \|U(\gamma_*) - U(\gamma)\|_\infty \\ &\le \|\gamma_* - U(\gamma_*)\|_\infty + \frac{1}{2}\|\gamma_* - \gamma\|_\infty .\end{aligned}$$

Therefore

$$\frac{1}{2}\|\gamma_* - \gamma\|_\infty \le \|U(\gamma_*) - U(\gamma)\|_\infty ,$$

which is (5.47). □

Theorem 5.4.3. *Consider any value of $0 \le q \le 1$, and γ as provided by Lemma 5.4.2. Then assuming*

$$L\alpha D \le 1 \tag{5.49}$$

we have

$$\forall x ,\ \mathsf{E}\left\langle\left(\frac{1}{N}\sum_{k\le M-3} u(S_{k,t},x) - \gamma(x)\right)^2\right\rangle_{t,\sim} \le L(\alpha D)^2 \mathsf{E}\langle (R_{1,2}-q)^2\rangle_{t,\sim} + \frac{K}{N} . \tag{5.50}$$

Proof. Let us define

$$A(x) = \frac{1}{N}\sum_{k\le M-3} u(S_{k,t},x) - \gamma(x) \ ; \quad A_*(x) = \frac{1}{N}\sum_{k\le M-4} u(S_{k,t},x) - \gamma(x) , \tag{5.51}$$

so that

$$|A(x) - A_*(x)| \le \frac{K}{N} . \tag{5.52}$$

Using symmetry between the values of k in the first line and (5.52) in the second line yields

$$\mathsf{E}\langle A(x)^2\rangle_{t,\sim} = \mathsf{E}\left\langle\left(\frac{M-3}{N}u(S_{M-3,t},x)-\gamma(x)\right)A(x)\right\rangle_{t,\sim}$$
$$\leq \frac{K}{N} + \mathsf{E}\langle(\alpha u(S_{M-3,t},x)-\gamma(x))A_*(x)\rangle_{t,\sim} \,. \tag{5.53}$$

We are again in a "cavity in M" situation. We need to make explicit the influence of the term $S_{M-3,t}$ in the Hamiltonian. So we introduce the Hamiltonian $H_{N,M-4,t}$ as in (5.23) and we denote by $\langle\cdot\rangle_*$ an average for this Hamiltonian, keeping the dependence on t implicit. Thus, for a function h of $\boldsymbol{\sigma}$ and of the quantities $S_{k,t}$, $k \leq M-3$, the formula

$$\mathsf{E}\langle h\rangle_{t,\sim} = \mathsf{E}\frac{\langle h\mathcal{E}_*\rangle_*}{\mathsf{E}_\xi\langle\mathcal{E}_*\rangle_*} \tag{5.54}$$

holds, where

$$\mathcal{E}_* = \exp\frac{1}{N}\sum_{k\leq M-4} u(S_{k,t},S_{M-3,t}) \,, \tag{5.55}$$

and where E_ξ denotes expectation in the variables ξ_k^ℓ. Again, we must devise a cavity argument with the underlying belief that $S_{M-3,t}$ will have a Gaussian behavior. So, considering independent standard Gaussian r.v.s z and $\hat{\xi}$, and defining

$$\theta = z\sqrt{q} + \hat{\xi}\sqrt{1-q} \,,$$

for $0 \leq v \leq 1$ we set $S_v = \sqrt{v}S_{M-3,t} + \sqrt{1-v}\theta$. We consider the function

$$\psi(v) = \mathsf{E}\frac{\big\langle(\alpha u(S_v,x)-\gamma(x))A_*(x)\exp\frac{1}{N}\sum_{k\leq M-4}u(S_{k,t},S_v)\big\rangle_*}{\mathsf{E}_\xi\big\langle\exp\frac{1}{N}\sum_{k\leq M-4}u(S_{k,t},S_v)\big\rangle_*} \,, \tag{5.56}$$

where E_ξ denotes expectation in the r.v.s ξ_k^ℓ **and** $\hat{\xi}$. This is exactly the same procedure we used in (5.33). Thus the relations (5.54) and (5.55) imply

$$\psi(1) = \mathsf{E}\langle(\alpha u(S_{M-3,t},x)-\gamma(x))A_*(x)\rangle_{t,\sim} \,. \tag{5.57}$$

We will bound $|\psi'(v)|$ (as in Lemma 5.3.2) but let us first look at $\psi(0)$. Defining

$$B(x) = \frac{1}{N}\sum_{k\leq M-4} u(S_{k,t},x) = A_*(x) + \gamma(x) \,, \tag{5.58}$$

we have

$$\psi(0) = \mathsf{E}\frac{\langle(\alpha u(\theta,x)-\gamma(x))A_*(x)\exp B(\theta)\rangle_*}{\mathsf{E}_\xi\langle\exp B(\theta)\rangle_*} \,.$$

Since we are following the pattern of Section 5.3, it should not come as a surprise that the value of $\psi(0)$ is not completely trivial to estimate; but a last interpolation will suffice. For $0 \leq s \leq 1$ we consider

$$\psi_*(s) = \mathsf{E}\frac{\big\langle(\alpha u(\theta,x) - \gamma(x))A_*(x)\exp(sB(\theta) + (1-s)\gamma(\theta))\big\rangle_*}{\mathsf{E}_\xi\langle\exp(sB(\theta) + (1-s)\gamma(\theta))\rangle_*} . \tag{5.59}$$

Thus $\psi_*(1) = \psi(0)$. Using independence and recalling (5.46) yields

$$\begin{aligned}\psi_*(0) &= \mathsf{E}\frac{\big\langle(\alpha u(\theta,x) - \gamma(x))A_*(x)\exp\gamma(\theta)\big\rangle_*}{\mathsf{E}_\xi\langle\exp\gamma(\theta)\rangle_*} \\ &= \mathsf{E}\langle A_*(x)\rangle_*\mathsf{E}\frac{(\alpha u(\theta,x) - \gamma(x))\exp\gamma(t)}{\mathsf{E}_\xi\exp\gamma(x)} = 0 .\end{aligned}$$

We compute $\psi_*'(s)$ in a straightforward manner, observing that

$$\frac{\mathrm{d}}{\mathrm{d}s}(sB(\theta) + (1-s)\gamma(\theta)) = B(\theta) - \gamma(\theta) = A_*(\theta) .$$

Writing $\mathcal{E}_s = \exp(sB(\theta) + (1-s)\gamma(\theta))$, we find

$$\begin{aligned}\psi_*'(s) &= \mathsf{E}\frac{\langle(\alpha u(\theta,x) - \gamma(x))A_*(x)A_*(\theta)\mathcal{E}_s\rangle_*}{\mathsf{E}_\xi\langle\mathcal{E}_s\rangle_*} \\ &\quad - \mathsf{E}\frac{\langle(\alpha u(\theta,x) - \gamma(x))A_*(x)\mathcal{E}_s\rangle_*\mathsf{E}_\xi\langle A_*(\theta)\mathcal{E}_s\rangle_*}{(\mathsf{E}_\xi\langle\mathcal{E}_s\rangle_*)^2} .\end{aligned}$$

To bound $|\psi_*'(s)|$, believe it or not, no integration by parts is required! First we observe that since $|B(x)|, |\gamma(x)| \le L\alpha D$, we have $1/L \le \mathcal{E}_s \le L$. Also,

$$|\alpha u(\theta,x) - \gamma(x)| \le L\alpha D . \tag{5.60}$$

Using the Cauchy-Schwarz inequality it is then straightforward to get the bound

$$|\psi_*'(s)| \le L\alpha D\mathsf{E}\langle A_*(x)^2\rangle_*^{1/2}\mathsf{E}\langle A_*(\theta)^2\rangle_*^{1/2} . \tag{5.61}$$

Since $\psi_*(0) = 0$ it follows that

$$\psi_*(1) = \psi(0) \le L\alpha D\mathsf{E}\langle A_*(x)^2\rangle_*^{1/2}\mathsf{E}\langle A_*(\theta)^2\rangle_*^{1/2} . \tag{5.62}$$

To bound $|\psi'(v)|$ we proceed as in Lemma 5.3.2. We compute $\psi'(v)$ through differentiation and integration by parts, and this integration by parts "creates a factor $R^t_{\ell,\ell'}$ in each term". Using the Cauchy-Schwarz inequality and (5.60) we then get

$$|\psi'(v)| \le L\alpha D\mathsf{E}\langle A_*(x)^2\rangle_*^{1/2}\mathsf{E}\langle(R_{1,2} - q)^2\rangle_*^{1/2} + \frac{K}{N} ,$$

so that

$$\begin{aligned}\psi(1) &\le L\alpha D\mathsf{E}\langle A_*(x)^2\rangle_*^{1/2}\mathsf{E}\langle A_*(\theta)^2\rangle_*^{1/2} \\ &\quad + L\alpha D\mathsf{E}\langle A_*(x)^2\rangle_*^{1/2}\mathsf{E}\langle(R_{1,2} - q)^2\rangle_*^{1/2} + \frac{K}{N} .\end{aligned} \tag{5.63}$$

For $L'\alpha D \le 1$ we have

$$L\alpha D \mathsf{E}\langle A_*(x)^2\rangle_*^{1/2} \mathsf{E}\langle A_*(\theta)^2\rangle_*^{1/2} \le \frac{1}{16}\mathsf{E}\langle A_*(x)^2\rangle_* + \frac{1}{16}\mathsf{E}\langle A_*(\theta)^2\rangle_* \,,$$

and the inequality $ab \le a^2/t + tb^2$ for $t = L\alpha D$ implies

$$L\alpha D \mathsf{E}\langle A_*(x)^2\rangle_*^{1/2} \mathsf{E}\langle (R_{1,2}-q)^2\rangle_*^{1/2} \le \frac{1}{16}\mathsf{E}\langle A_*(x)^2\rangle_* + L(\alpha D)^2 \mathsf{E}\langle (R_{1,2}-q)^2\rangle_* \,.$$

Combining with (5.63) we get

$$\psi(1) \le \frac{K}{N} + \frac{1}{8}\mathsf{E}\langle A_*(x)^2\rangle_* + \frac{1}{16}\mathsf{E}\langle A_*(\theta)^2\rangle_* + L(\alpha D)^2 \mathsf{E}\langle (R_{1,2}-q)^2\rangle_* \,.$$

Combining with (5.53) and (5.57) we then obtain

$$\begin{aligned} \mathsf{E}\langle A(x)^2\rangle_{t,\sim} \le \frac{K}{N} &+ \frac{1}{8}\mathsf{E}\langle A_*(x)^2\rangle_* + \frac{1}{16}\mathsf{E}\langle A_*(\theta)^2\rangle_* \\ &+ L(\alpha D)^2 \mathsf{E}\langle (R_{1,2}-q)^2\rangle_* \,. \end{aligned} \tag{5.64}$$

Now since $|A(x) - A_*(x)| \le K/N$ we have $A_*(x)^2 \le A(x)^2 + K/N$ and thus

$$\mathsf{E}\langle A_*(\theta)^2\rangle_* \le \mathsf{E}\langle A(\theta)^2\rangle_* + \frac{K}{N} \,.$$

In the quantity $\mathsf{E}\langle A(\theta)^2\rangle_*$, the r.v. θ is independent of the randomness of $\langle\cdot\rangle_*$. So, denoting by E_* expectation in the randomness of this bracket only, we have

$$\mathsf{E}_*\langle A(\theta)^2\rangle_* \le \sup_y \mathsf{E}_*\langle A(y)^2\rangle_* = \sup_y \mathsf{E}\langle A(y)^2\rangle_* \,,$$

and thus, taking expectation,

$$\mathsf{E}\langle A(\theta)^2\rangle_* \le \sup_y \mathsf{E}\langle A(y)^2\rangle_* \,. \tag{5.65}$$

Moreover, as in Lemma 5.3.3, if $L\alpha D \le 1$ we have $\mathsf{E}\langle |h|\rangle_* \le 2\mathsf{E}\langle |h|\rangle_{t,\sim}$. Combining these relations we then get from (5.64) that for any x,

$$\begin{aligned} \mathsf{E}\langle A(x)^2\rangle_{t,\sim} \le \frac{K}{N} &+ \frac{1}{4}\mathsf{E}\langle A(x)^2\rangle_{t,\sim} + \frac{1}{4}\sup_y \mathsf{E}\langle A(y)^2\rangle_{t,\sim} \\ &+ L(\alpha D)^2 \mathsf{E}\langle (R_{1,2}-q)^2\rangle_{t,\sim} \end{aligned}$$

and thus

$$\frac{3}{4}\sup_x \mathsf{E}\langle A(x)^2\rangle_{t,\sim} \le \frac{1}{4}\sup_y \mathsf{E}\langle A(y)^2\rangle_{t,\sim} + \frac{K}{N} + L(\alpha D)^2 \mathsf{E}\langle (R_{1,2}-q)^2\rangle_{t,\sim} \,.$$

Therefore we get

$$\sup_x \mathsf{E}\langle A(x)^2\rangle_{t,\sim} \le L(\alpha D)^2 \mathsf{E}\langle (R_{1,2}-q)^2\rangle_{t,\sim} + \frac{K}{N} \,, \tag{5.66}$$

and recalling the definition (5.51) of $A(x)$ this is exactly (5.50). □

5.5 The Replica-Symmetric Solution

Now that we have proved Theorem 5.4.3, we can go back to the study of the quantity $\nu_{t,0}(B_0 f)$ of Section 5.3. Given $0 \le q \le 1$, and the function γ provided by Lemma 5.4.2, we define

$$r^* = \mathsf{E}\left(\frac{\mathsf{E}_\xi \gamma'(\theta) \exp \gamma(\theta)}{\mathsf{E}_\xi \exp \gamma(\theta)}\right)^2 . \tag{5.67}$$

Differentiating in x the relation (5.46) yields

$$\gamma'(x) = \alpha \mathsf{E} \frac{w(x,\theta) \exp \gamma(\theta)}{\mathsf{E}_\xi \exp \gamma(\theta)} , \tag{5.68}$$

and thus $|\gamma'(x)| \le L\alpha D$ so that $|r^*| \le L$ when $\alpha D \le 1$.

Proposition 5.5.1. *Assume $L\alpha D \le 1$. Then with the notation of Lemma 5.3.2, we have*

$$|\nu_{t,0}(f) - \langle f \rangle_{t,\sim}| \le L^n \alpha D (\mathsf{E}\langle f^2 \rangle_{t,\sim})^{1/2} (\mathsf{E}\langle (R_{1,2} - q)^2 \rangle_{t,\sim})^{1/2} + \frac{K}{N} (\mathsf{E}\langle f^2 \rangle_{t,\sim})^{1/2} . \tag{5.69}$$

When B_v is given by (5.36) we have

$$\begin{aligned} &|\alpha^2 \nu_{t,0}(B_0 f) - r^* \langle f \rangle_{t,\sim}| \\ &\le L^n \alpha D (\mathsf{E}\langle f^2 \rangle_{t,\sim})^{1/2} (\mathsf{E}\langle (R_{1,2} - q)^2 \rangle_{t,\sim})^{1/2} + \frac{K}{N} (\mathsf{E}\langle f^2 \rangle_{t,\sim})^{1/2} . \end{aligned} \tag{5.70}$$

Proof. We prove (5.70). For $0 \le s \le 1$ we define, recalling the notation (5.32),

$$\begin{aligned} \mathcal{E}(s) &= \mathcal{E}_0^s \exp(1-s)\Big(\sum_{j=0,1,2,\, \ell \le n} \gamma(\theta_j^\ell)\Big) \\ &= \exp\Bigg(\sum_{j=0,1,2,\, \ell \le n} \gamma(\theta_j^\ell) + s \sum_{\ell \le n} \sum_{j=0,1,2} \Big(\frac{1}{N} \sum_{k \le M-3} u(S_{k,t}^\ell, \theta_j^\ell) - \gamma(\theta_j^\ell)\Big) \\ &\quad + \frac{s}{N} \sum_{\ell \le n} \big(u(\theta_0^\ell, \theta_2^\ell) + u(\theta_0^\ell, \theta_1^\ell) + u(\theta_1^\ell, \theta_2^\ell)\big)\Bigg) , \end{aligned} \tag{5.71}$$

and we consider

$$\psi(s) = \alpha^2 \mathsf{E} \frac{\langle B_0 f \mathcal{E}(s) \rangle_{t,\sim}}{\mathsf{E}_\xi \langle \mathcal{E}(s) \rangle_{t,\sim}} . \tag{5.72}$$

The fundamental formula (5.33) shows that $\psi(1) = \alpha^2 \nu_{t,0}(B_0 f)$. As expected we will compute $\psi(0)$ and bound $\psi'(s)$. Using that $B_0 = w(\theta_0^1, \theta_1^1) w(\theta_0^2, \theta_2^2)$ we get, by independence of θ_j^ℓ and of the randomness $\langle \cdot \rangle_{t,\sim}$

$$\psi(0) = \alpha^2 \mathsf{E} \frac{w(\theta_0^1, \theta_1^1) w(\theta_0^2, \theta_2^2) \exp \sum_{j=0,1,2\ \ell \le n} \gamma(\theta_j^\ell)}{\mathsf{E}_\xi \exp \sum_{j=0,1,2,\ \ell \le n} \gamma(\theta_j^\ell)} \mathsf{E}\langle f \rangle_{t,\sim} . \quad (5.73)$$

Using independence between the r.v.s $\hat{\xi}_j^\ell$ we obtain

$$\mathsf{E}_\xi \exp \sum_{j=0,1,2,\ \ell \le n} \gamma(\theta_j^\ell) = \prod_{j=0,1,2,\ \ell \le n} \mathsf{E}_\xi \exp \gamma(\theta_j^\ell)$$

and

$$\begin{aligned} &\mathsf{E}_\xi w(\theta_0^1, \theta_1^1) w(\theta_0^2, \theta_2^2) \exp \sum_{j=0,1,2,\ \ell \le n} \gamma(\theta_j^\ell) \\ &= \mathsf{E}_\xi w(\theta_0^1, \theta_1^1) \exp(\gamma(\theta_0^1) + \gamma(\theta_1^1)) \mathsf{E}_\xi w(\theta_0^2, \theta_2^2) \exp(\gamma(\theta_0^2) + \gamma(\theta_2^2)) \\ &\quad \times \mathsf{E}_\xi \exp \gamma(\theta_2^1) \mathsf{E}_\xi \exp \gamma(\theta_1^2) \prod_{3 \le \ell \le n} \prod_{j=0,1,2} \mathsf{E}_\xi \exp \gamma(\theta_j^\ell) . \end{aligned}$$

Therefore

$$\psi(0) = \mathsf{E} U_1 U_2 \mathsf{E}\langle f \rangle_{t,\sim} , \quad (5.74)$$

where

$$\begin{aligned} U_1 &= \alpha \frac{\mathsf{E}_\xi w(\theta_0^1, \theta_1^1) \exp(\gamma(\theta_0^1) + \gamma(\theta_1^1))}{\mathsf{E}_\xi \exp \gamma(\theta_0^1) \mathsf{E}_\xi \exp \gamma(\theta_1^1)} \\ U_2 &= \alpha \frac{\mathsf{E}_\xi w(\theta_0^2, \theta_2^2) \exp(\gamma(\theta_0^2) + \gamma(\theta_2^2))}{\mathsf{E}_\xi \exp \gamma(\theta_0^2) \mathsf{E}_\xi \exp \gamma(\theta_2^2)} . \end{aligned}$$

Let us now recall that $\theta_j^\ell = z_j \sqrt{q} + \hat{\xi}_j^\ell \sqrt{1-q}$, where the Gaussian r.v.s $z_j, \hat{\xi}_j^\ell$ are all independent of each other. Let us denote by E_j expectation in z_j only, and $\mathsf{E}_{\ell,j}$ expectation in $\hat{\xi}_j^\ell$ only. Then

$$\mathsf{E} U_1 U_2 = \mathsf{E}(\mathsf{E}_1 U_1)(\mathsf{E}_2 U_2)$$

and

$$\begin{aligned} \mathsf{E}_1 U_1 &= \alpha \mathsf{E}_1 \frac{\mathsf{E}_{1,0} \mathsf{E}_{1,1} w(\theta_0^1, \theta_1^1) \exp \gamma(\theta_0^1) \exp \gamma(\theta_1^1)}{\mathsf{E}_{1,0} \exp \gamma(\theta_0^1) \mathsf{E}_{1,1} \exp(\theta_1^1)} \\ &= \alpha \mathsf{E}_1 \left(\mathsf{E}_{1,0} \frac{\exp \gamma(\theta_0^1)}{\mathsf{E}_{1,0} \exp \gamma(\theta_0^1)} \mathsf{E}_{1,1} \frac{w(\theta_0^1, \theta_1^1) \exp \gamma(\theta_1^1)}{\mathsf{E}_{1,1} \exp(\theta_1^1)} \right) \\ &= \alpha \mathsf{E}_{1,0} \left(\frac{\exp \gamma(\theta_0^1)}{\mathsf{E}_{1,0} \exp \gamma(\theta_0^1)} \mathsf{E}_1 \mathsf{E}_{1,1} \frac{w(\theta_0^1, \theta_1^1) \exp \gamma(\theta_1^1)}{\mathsf{E}_{1,1} \exp \gamma(\theta_1^1)} \right) . \end{aligned}$$

Now, using (5.68)

$$\alpha \mathsf{E}_1 \mathsf{E}_{1,1} \frac{w(\theta_0^1, \theta_1^1) \exp \gamma(\theta_1^1)}{\mathsf{E}_{1,1} \exp \gamma(\theta_1^1)} = \gamma'(\theta_0^1) ,$$

so that

$$\mathsf{E}_1 U_1 = \frac{\mathsf{E}_{1,0}\gamma'(\theta_0^1)\exp\gamma(\theta_0^1)}{\mathsf{E}_{1,0}\exp\gamma(\theta_0^1)} \, .$$

In a similar manner,

$$\mathsf{E}_2 U_2 = \frac{\mathsf{E}_{2,0}\gamma'(\theta_0^2)\exp\gamma(\theta_0^2)}{\mathsf{E}_{2,0}\exp\gamma(\theta_0^2)} \, ,$$

so that

$$\mathsf{E}_1 U_1 = \mathsf{E}_2 U_2 = \frac{\mathsf{E}_\xi\gamma'(\theta)\exp\gamma(\theta)}{\mathsf{E}_\xi\exp\gamma(\theta)} \, ,$$

and thus $\mathsf{E}U_1U_2 = \mathsf{E}(\mathsf{E}_1U_1)^2 = r^*$ by (5.67). Thus we have shown that $\psi(0) = r^*\mathsf{E}\langle f\rangle_{t,\sim}$.

To bound $\psi'(s)$, we proceed very much as in the proof of (5.61). We define

$$A^\ell(x) = \frac{1}{N}\sum_{k\le M-3} u(S_{k,t}^\ell, x) - \gamma(x) \, .$$

Comparing with (5.51) yields

$$\mathsf{E}\langle A^\ell(\theta_j^\ell)^2\rangle_{t,\sim} = \mathsf{E}\langle A(\theta)^2\rangle_{t,\sim} \, . \tag{5.75}$$

We observe the relation

$$\mathcal{E}'(s) = \Big(\sum_{\ell\le n}\sum_{j=0,1,2} A^\ell(\theta_j^\ell) + C\Big)\mathcal{E}(s) \, ,$$

where $C = N^{-1}\sum_{\ell\le n}\big(u(\theta_0^\ell,\theta_2^\ell) + u(\theta_0^\ell,\theta_1^\ell) + u(\theta_1^\ell,\theta_2^\ell)\big)$, so that $|C| \le K/N$. We observe that $\exp(-Ln) \le \mathcal{E}(s) \le \exp(Ln)$, since $\alpha D \le 1$. We differentiate the formula (5.72), we use the Cauchy-Schwarz inequality and that $\alpha^2|B_0| \le 1$ (since $|B_0| \le D$ and $\alpha D \le 1$) to obtain, using (5.75):

$$|\psi'(s)| \le L^n(\mathsf{E}\langle f^2\rangle_{t,\sim})^{1/2}(\mathsf{E}\langle A(\theta)^2\rangle_{t,\sim})^{1/2} + \frac{K}{N}\mathsf{E}\langle |f|\rangle_{t,\sim} \, . \tag{5.76}$$

The random variable θ is independent of the randomness of $\langle\cdot\rangle_{t,\sim}$, and therefore as in (5.65) we have

$$\mathsf{E}\langle A(\theta)^2\rangle_{t,\sim} \le \sup_y \mathsf{E}\langle A(y)^2\rangle_{t,\sim} \, .$$

Combining with (5.76) and (5.66) we get

$$|\psi'(s)| \le L^n\alpha D(\mathsf{E}\langle f^2\rangle_{t,\sim})^{1/2}(\mathsf{E}\langle (R_{1,2}-q)^2\rangle_{t,\sim})^{1/2} + \frac{K}{N}(\mathsf{E}\langle f^2\rangle_{t,\sim})^{1/2}$$

and this proves (5.70). The proof of (5.69) is similar but much simpler. □

We are finally ready to control the terms (5.15) in Proposition 5.2.2. We consider only the case $n = 2$ for simplicity.

Proposition 5.5.2. *Assume that $L\alpha D \le 1$. Then we have*

$$|A_{1,2} - \alpha r^* \nu_t(\varepsilon_1\varepsilon_2 f)| \le L\alpha^2 D \nu_t(f^2)^{1/2}\nu_t\big((R_{1,2}-q)^2\big)^{1/2} + \frac{K}{N}\nu_t(f^2)^{1/2} . \tag{5.77}$$

Proof. From (5.69) and (5.70) we get that, using Lemma 5.3.3 in the second inequality,

$$\begin{aligned}|\alpha^2\nu_{t,0}(B_0 f) - r^*\nu_{t,0}(f)| &\le L\alpha D(\mathsf{E}\langle f^2\rangle_{t,\sim})^{1/2}(\mathsf{E}\langle (R_{1,2}-q)^2\rangle_{t,\sim})^{1/2} \\ &\quad + \frac{K}{N}\mathsf{E}\langle f^2\rangle_{t,\sim}^2 \\ &\le L\alpha D\nu_t(f^2)^{1/2}\nu_t\big((R_{1,2}-q)^2\big)^{1/2} \\ &\quad + \frac{K}{N}\nu_t(f^2)^{1/2} .\end{aligned} \tag{5.78}$$

It follows from (5.37) and Lemma 5.3.3 again that

$$|\alpha^2\nu_{t,0}(B_0 f) - \alpha^2\nu_t(B_1 f)| \le L\alpha^2 D^2 \nu_t(f^2)^{1/2}\nu_t\big((R_{1,2}-q)^2\big)^{1/2} + \frac{K}{N}\nu_t(f^2)^{1/2} .$$

Moreover from (5.35) we see that the quantity $r^*|\nu_{t,0}(f) - \nu_t(f)|$ satisfies the same bound. Combining with (5.78) we obtain

$$|\alpha^2\nu_t(B_1 f) - r^*\nu_t(f)| \le L\alpha D\nu_t(f^2)\nu_t\big((R_{1,2}-q)^2\big)^{1/2} + \frac{K}{N}\nu_t(f^2)^{1/2} .$$

Replacing f by $\varepsilon_1\varepsilon_2 f$, and since $\nu_t(B_1\varepsilon_1\varepsilon_2 f) = C_{1,2}$ by (5.21), the result follows from (5.22). □

Corollary 5.5.3. *If f is a function on Σ_N^2, if $L\alpha D \le 1$, and if*

$$r = \alpha r^* \tag{5.79}$$

we have

$$|\nu_t'(f)| \le L\alpha^2 D\nu(f^2)^{1/2}\nu\big((R_{1,2}-q)^2\big)^{1/2} + \frac{K}{N}\nu(f^2)^{1/2} . \tag{5.80}$$

Proof. We combine (5.14) and (5.77). □

Theorem 5.5.4. *If $L\alpha D \le 1$, $\alpha \le 1$, writing as usual $\theta = z\sqrt{q} + \xi\sqrt{1-q}$, the system of three equations (5.46),*

$$r = \alpha\mathsf{E}\left(\frac{\mathsf{E}_\xi\gamma'(\theta)\exp\gamma(\theta)}{\mathsf{E}_\xi\exp\gamma(\theta)}\right)^2 \tag{5.81}$$

$$q = \mathsf{E}\mathrm{th}^2 z\sqrt{r} \tag{5.82}$$

with unknown (q, r, γ) has a unique solution and

$$\nu\big((R_{1,2}-q)^2\big) \le \frac{K}{N} . \tag{5.83}$$

Proof. First we will show that (5.46) and (5.81) define r as a continuous function $r(q)$ of q. Thinking of α as fixed once and for all, we denote by γ_q the solution of (5.46). We will first show that the map $q \mapsto \gamma_q \in \mathcal{C}$ is continuous when $\mathcal{C}$ is provided with the topology induced by the supremum norm $\|\cdot\|$. Let us write $\theta = \theta_q$ to make explicit the dependence on q. Let us fix q_0 and let us consider the function $q \mapsto \psi(q) \in \mathcal{C}$ given by

$$\psi(q)(y) = \alpha \mathsf{E} \frac{u(\theta_q, y) \exp \gamma_{q_0}(\theta_q)}{\mathsf{E}\xi \exp \gamma_{q_0}(\theta_q)} .$$

It is straightforward to show that the function ψ is continuous, and by (5.81) we have $\psi(q_0) = \gamma_{q_0}$. It then follows from (5.47) used for $\gamma = \gamma_q$ and $\gamma_* = \gamma_{q_0}$ that

$$\|\gamma_q - \gamma_{q_0}\| \le 2\|\gamma_q - \psi(q_0)\| ,$$

and this shows that the function $q \mapsto \gamma_q$ is continuous at $q = q_0$, and hence everywhere. It follows from (5.68) that the map $q \mapsto \gamma_q'$ is continuous, and this shows that r is a continuous function of q.

Therefore the map $q \mapsto \mathsf{E}\mathrm{th}^2 z\sqrt{r(q)}$ is continuous from $[0,1]$ to itself and has a fixed point. This proves the existence of a solution to these equations, and this solution is unique by (5.83). The rest of the proof follows from (5.80) through our standard scheme of proof. Namely, we write

$$\begin{aligned} \nu\big((R_{1,2} - q)^2\big) &= \nu((\varepsilon_1\varepsilon_2 - q)(R_{1,2} - q)) \\ &\le \frac{2}{N} + \nu((\varepsilon_1\varepsilon_2 - q)f) , \end{aligned} \tag{5.84}$$

where $f = R_{1,2}^- - q$. By Lemma 5.2.1 and since $q = \mathsf{E}\mathrm{th}^2 z\sqrt{r} = \mathsf{E}\mathrm{th}^2 Y$ we have

$$\nu_0((\varepsilon_1\varepsilon_2 - q)f) = (\mathsf{E}\mathrm{th}^2 Y - q)\nu_0(f) = 0 ,$$

and using (5.80) for $(\varepsilon_1\varepsilon_2 - q)f$ we obtain

$$\begin{aligned} \nu\big((\varepsilon_1\varepsilon_2 - q)f\big) &\le \frac{K}{N} + L\alpha^2 D\nu(f^2)^{1/2}\nu\big((R_{1,2} - q)^2\big)^{1/2} \\ &\le \frac{K}{N} + L\alpha^2 D\nu\big((R_{1,2} - q)^2\big) , \end{aligned}$$

so comparing with (5.84) yields

$$\nu\big((R_{1,2} - q)^2\big) \le \frac{K}{N} + L\alpha^2 D\nu\big((R_{1,2} - q)^2\big) , \tag{5.85}$$

and this finishes the proof. □

The last result of this chapter deals with the computation of

$$p_{N,M} = \frac{1}{N}\mathsf{E}\log\sum_{\boldsymbol{\sigma}} \exp(-H_{N,M}(\boldsymbol{\sigma})) .$$

We will follow the method of the first proof of Theorem 2.4.2. We consider q and r as in Theorem 5.5.4. We consider independent standard Gaussian r.v.s $z, (z_k)_{k \le M}, (z'_i)_{i \le N}, (\xi_k)_{k \le M}$, we write

$$\theta_k = z_k\sqrt{q} + \xi_k\sqrt{1-q} \; ; \; S_{k,s} = \sqrt{s}S_k + \sqrt{1-s}\theta_k , \tag{5.86}$$

and we consider the following interpolating Hamiltonian for $0 \le s \le 1$:

$$-H_{N,M,s} = \frac{1}{N} \sum_{1 \le k_1 < k_2 \le M} u(S_{k_1,s}, S_{k_2,s}) - \sum_{i \le N} \sigma_i \sqrt{1-s} z'_i \sqrt{r} \, . \tag{5.87}$$

We define

$$p_{N,M,s} = \frac{1}{N} \mathsf{E} \log \mathsf{E}_\xi \sum_\sigma \exp(-H_{N,M,s}) \, .$$

An in the case of Theorem 2.4.2, this interpolation is designed to preserve the replica-symmetric equations along the interpolation. The interesting twist is that the computation of $p_{N,M,0}$ is no longer trivial. It should be obvious that

$$p_{N,M,0} = \log 2 + \mathsf{E} \log \mathrm{ch}(z\sqrt{r}) + p^*_{N,M} \, , \tag{5.88}$$

where

$$p^*_{N,M} = \frac{1}{N} \mathsf{E} \log \mathsf{E}_\xi \exp\left(\frac{1}{N} \sum_{1 \le k_1 < k_2 \le M} u(\theta_{k_1}, \theta_{k_2}) \right) , \tag{5.89}$$

but how should one compute $p^*_{N,M}$?

Research Problem 5.5.5. (Level unknown) Consider $q, \alpha > 0$, and the function u. Recall that z_k, ξ_k denote independent standard Gaussian r.v.s, that $\theta_k = z_k\sqrt{q} + \xi_k\sqrt{1-q}$, and that E_ξ denotes expectation in the r.v.s ξ_k only. Recalling (5.89), compute

$$\lim_{N \to \infty, M/N \to \alpha} p^*_{N,M} \, .$$

We do not assume in Problem 5.5.5 that α is small. When we write (5.89), we think of the quantities ξ_k as "spins", so there is no telling how difficult this problem might be (although it could well be an exercise for an expert in large deviation theory). In the present case however, we are concerned only with the case $L\alpha D \le 1$, and the result in this case is described as follows.

Proposition 5.5.6. *There is a number L with the following property. Assume that $D \ge 1$. For $\alpha \le 1/LD$ and $q \in [0,1]$, denote by $\gamma_{\alpha,q}$ the function obtained by solving (5.46), and define*

$$W(\alpha, q) = \int_0^\alpha \mathsf{E} \log \mathsf{E}_\xi \exp \gamma_{x,q} \mathrm{d}x \, , \tag{5.90}$$

where as usual $\theta = z\sqrt{q} + \xi\sqrt{1-q}$, z and ξ are independent standard Gaussian r.v.s, and E_ξ denotes expectation in ξ only. Then if $LDM \le N$ and $0 \le q \le 1$ we have

$$\left| p^*_{N,M} - W\left(\frac{M}{N}, q\right) \right| \le \frac{K}{\sqrt{N}} . \tag{5.91}$$

The function W satisfies $W(0,q) = 0$ and

$$\frac{\partial W}{\partial \alpha}(\alpha, q) = \mathsf{E} \log \mathsf{E}_\xi \exp \gamma_{\alpha,q}(\theta) . \tag{5.92}$$

Moreover

$$\frac{\partial W}{\partial q}(\alpha, q) = -\frac{r(\alpha, q)}{2} \tag{5.93}$$

where $r(\alpha, q)$ is given by (5.79) and (5.67) for $\gamma = \gamma_{\alpha,q}$.

The following question is called an exercise rather than a Research Problem, because the solution might not be publishable; but the author does not know this solution.

Exercise 5.5.7. Consider the function W defined by (5.90). Find a direct proof that W satisfies (5.93).

The obstacle here is that it is not clear how to use condition (5.46). Comparing (5.92) and (5.93) we get the relation

$$\frac{\partial}{\partial q}(\mathsf{E} \log \mathsf{E}_\xi \exp \gamma_{\alpha,q}(\theta)) = \frac{\partial}{\partial \alpha}\left(-\frac{r(\alpha, q)}{2}\right) . \tag{5.94}$$

A direct proof of this mysterious relation would provide a solution to the exercise. The difficulty is of course that $\gamma_{\alpha,q}$ depends on q and α.

Proof of Proposition 5.5.6. From now on until the end of the chapter, the arguments will be complete but sketchy, as they will rely on simplified versions of techniques we have already used in this chapter. We define the function $W(\alpha, q)$ by $W(0,q) = 0$ and (5.92).

Since the very definition of $p^*_{N,M}$ involves thinking of the variables ξ_k as spins, we will approach the problem by the methods we have developed to study spin systems. We write the identity

$$N(p^*_{N,M+1} - p^*_{N,M}) = \mathsf{E} \log \mathsf{E}_\xi \left\langle \exp \frac{1}{N} \sum_{1 \le k \le M} u(\theta_k, \theta_{M+1}) \right\rangle \tag{5.95}$$

where, for a function $h(\theta_1, \ldots, \theta_M)$ we define

$$\langle h \rangle = \frac{1}{Z} \mathsf{E}_\xi h \exp(-H'_{N,M}) , \tag{5.96}$$

where $Z = \mathsf{E}_\xi \exp(-H'_{N,M})$ and where

$$-H'_{N,M} = \frac{1}{N} \sum_{1 \le k_1 < k_2 \le M} u(\theta_{k_1}, \theta_{k_2}) \,. \tag{5.97}$$

The next step is to prove that (recalling that $\alpha = M/N$),

$$\mathsf{E}\left\langle \left(\frac{1}{N} \sum_{1 \le k \le M} u(\theta_k, x) - \gamma_{\alpha,q}(x) \right)^2 \right\rangle \le \frac{K}{N} \,. \tag{5.98}$$

The argument is as in Theorem 5.4.3 but much simpler. We define

$$A(x) = \frac{1}{N} \sum_{1 \le k \le M} u(\theta_k, x) - \gamma_{\alpha,q}(x) \; ; \; A_*(x) = \frac{1}{N} \sum_{1 \le k < M} u(\theta_k, x) - \gamma_{\alpha,q}(x) \,,$$

so that as in (5.53) we have

$$\mathsf{E}\langle A(x)^2 \rangle \le \mathsf{E}\langle (\alpha u(\theta_M, x) - \gamma_{\alpha,q}(x)) A_*(x) \rangle + \frac{K}{N} \,. \tag{5.99}$$

Let us denote by $\langle \cdot \rangle_*$ an average as in (5.96) but for the Hamiltonian $H'_{N,M-1}$. Let

$$B(x) = \frac{1}{N} \sum_{1 \le k < M} u(\theta_k, \theta_M) = A_*(x) + \gamma_{\alpha,q}(x) \,,$$

so that

$$\langle (\alpha u(\theta_M, x) - \gamma_{\alpha,q}(x)) A_*(x) \rangle = \frac{\langle (\alpha u(\theta_M, x) - \gamma_{\alpha,q}(x)) A_*(x) \exp B(\theta_M) \rangle_*}{\mathsf{E}_\xi \langle \exp B(\theta_M) \rangle_*} \,.$$

Let us then define $\psi_*(s)$ by the formula (5.59). Proceeding as in (5.62) we obtain

$$\psi_*(1) \le L\alpha D \mathsf{E}\langle A_*(x)^2 \rangle_*^{1/2} \mathsf{E}\langle A_*(\theta)^2 \rangle_*^{1/2} \,,$$

and combining with (5.99) we get

$$\mathsf{E}\langle A(x)^2 \rangle \le L\alpha D \mathsf{E}\langle A_*(x)^2 \rangle_*^{1/2} \mathsf{E}\langle A_*(\theta)^2 \rangle_*^{1/2} + \frac{K}{N} \,.$$

Also, we have $\mathsf{E}\langle h \rangle_* \le L\langle h \rangle$ when h is a positive function, so that

$$\mathsf{E}\langle A(x)^2 \rangle \le L\alpha D \mathsf{E}\langle A_*(x)^2 \rangle^{1/2} \mathsf{E}\langle A_*(\theta)^2 \rangle^{1/2} + \frac{K}{N} \,,$$

after which we conclude the proof of (5.98) as in the few lines of the proof of Theorem 5.4.3 that follow (5.64).

Combining (5.98) and (5.95) yields

$$|N(p^*_{N,M+1} - p^*_{N,M}) - \mathsf{E} \log \mathsf{E}_\xi \exp \gamma_{\alpha,q}(\theta)| \le \frac{K}{\sqrt{N}} \,. \tag{5.100}$$

The right-hand side of (5.92) is a function $f(\alpha)$ of α (since γ is a function of α), and (5.92) implies

$$W(\alpha+1/N,q) - W(\alpha,q) = \int_\alpha^{\alpha+1/N} \mathsf{E}\log \mathsf{E}_\xi \exp \gamma_{x,q}(\theta)\mathrm{d}x \,,$$

so that

$$\left| W(\alpha+1/N,q) - W(\alpha,q) - \frac{1}{N}\mathsf{E}\log \mathsf{E}_\xi \exp \gamma_{\alpha,q}(\theta) \right| \le \frac{K}{N^2}\,,$$

i.e.

$$\left| N\left(W\left(\frac{M+1}{N},q\right) - W\left(\frac{M}{N},q\right)\right) - \mathsf{E}\log \mathsf{E}_\xi \exp \gamma_{\alpha,q}(\theta) \right| \le \frac{K}{N}\,.$$

Comparing with (5.100) and summing over M yields (5.91).

It remains only to prove the elusive relation (5.93). For this we compute

$$\frac{\partial}{\partial q} p^*_{N,M} = \frac{1}{N^2}\mathsf{E}\Big\langle \sum_{1\le k_1<k_2\le M} (\theta'_{k_1} w(\theta_{k_1},\theta_{k_2}) + \theta'_{k_2} w(\theta_{k_2},\theta_{k_1}))\Big\rangle$$

where

$$\theta'_k = \frac{1}{2\sqrt{q}} z_k - \frac{1}{2\sqrt{1-q}}\xi_k$$

and where the bracket $\langle\cdot\rangle$ is as in (5.96). Thus

$$\begin{aligned}\frac{\partial}{\partial q} p^*_{N,M} &= \frac{1}{N^2}\mathsf{E}\Big\langle \sum_{k_1\ne k_2} \theta'_{k_1} w(\theta_{k_1},\theta_{k_2})\Big\rangle \\ &= \frac{1}{N^2}\mathsf{E}\frac{\sum_{k_1\ne k_2}\theta'_{k_1} w(\theta_{k_1},\theta_{k_2})\exp(-H'_{N,M})}{\mathsf{E}_\xi \exp(-H'_{N,M})}\,.\end{aligned}$$

We then need to integrate by parts in the r.v.s θ'_{k_1} i.e. to compute

$$\mathsf{E}\,\theta'_{k_1}\frac{w(\theta_{k_1},\theta_{k_2})\exp(-H'_{N,M})}{\mathsf{E}_\xi \exp(-H'_{N,M})}\,.$$

The straightforward method is to replace θ'_{k_1} by its value and to integrate by parts in the r.v.s z_k and ξ_k. One can also obtain the formula by using the heuristic principle (2.58), although of course to really prove the formula one has to perform the calculations again. Here (2.58) means that we can pretend to perform the computation that the denominator is a function of the quantities $\theta^\sim_k = \sqrt{q}z_k + \sqrt{1-q}\xi^\sim_k$, where $\xi^\sim_k$ are independent copies of the r.v.s ξ_k. Since $\mathsf{E}\theta'_{k_1}\theta_{k_2} = 0$ and $\mathsf{E}\theta'_{k_1}\theta^\sim_k = 0$ if $k \ne k_1$ and $= 1/2$ if $k = k_1$, one then gets that the only terms occurring are created by the denominator, and this gives

$$\mathsf{E}\theta'_{k_1} \frac{w(\theta_{k_1}, \theta_{k_2}) \exp(-H'_{N,M})}{\mathsf{E}_\xi \exp(-H'_{N,M})} = -\frac{1}{2N} \mathsf{E}\Big\langle \sum_{k_3 \neq k_1} w(\theta_{k_1}, \theta_{k_2}) w(\theta_{k_1}, \theta_{k_3}) \Big\rangle ,$$

so that finally

$$\frac{\partial}{\partial q} p^*_{N,M} = -\frac{1}{2N^3} \mathsf{E}\Big\langle \sum_{k_1 \neq k_2, k_1 \neq k_3} w(\theta_{k_1}, \theta_{k_2}) w(\theta_{k_1}, \theta_{k_3}) \Big\rangle . \tag{5.101}$$

Symmetry between the values of k yields

$$\left| \frac{\partial}{\partial q} p^*_{N,M} + \frac{\alpha^3}{2} \mathsf{E}\langle w(\theta_M, \theta_{M-1}) w(\theta_M, \theta_{M-2}) \rangle \right| \leq \frac{K}{N} .$$

Using the familiar "cavity in M argument" of (5.100) for $M-3$ rather than M and reproducing the computation following (5.73) we then get

$$\left| \frac{\partial}{\partial q} p^*_{N,M} + \frac{r(\alpha, q)}{2} \right| \leq \frac{K}{\sqrt{N}} . \tag{5.102}$$

For $q = 1$, we have $\theta_k = z_k$ and (5.89) yields

$$p^*_{N,M}\Big|_{q=1} = \frac{M(M-1)}{2N^2} \mathsf{E}u(z, z) , \tag{5.103}$$

and combining with (5.103) gives

$$\left| \left(\frac{\alpha^2}{2} \mathsf{E}u(z, z) - p^*_{N,M} \right) + \frac{1}{2} \int_q^1 r(\alpha, x) \mathrm{d}x \right| \leq \frac{K}{\sqrt{N}} .$$

Comparison with (5.91) yields (taking $N \to \infty$ and $M/N \to \alpha$)

$$W(\alpha, q) = \frac{\alpha^2}{2} \mathsf{E}u(z, z) + \frac{1}{2} \int_q^1 r(\alpha, x) \mathrm{d}x$$

and this proves that

$$\frac{\partial W}{\partial q}(\alpha, q) = -\frac{r(\alpha, q)}{2} .$$

□

Theorem 5.5.8. *Recalling the function W of Proposition 5.5.6 let*

$$\mathrm{RS}(\alpha) = W(\alpha, q) - \frac{r}{2}(1 - q) + \mathsf{E} \log \operatorname{ch}(z\sqrt{q}) + \log 2 ,$$

where γ, q and r are as in Theorem 5.5.4. Then, if $L\alpha D \leq 1$ snd $\alpha = M/N$, we have

$$|p_{N,M} - \mathrm{RS}(\alpha)| \leq \frac{K}{\sqrt{N}} . \tag{5.104}$$

Proof. Since

$$p_{N,M} = p_{N,M,1} = p_{N,M,0} + \int_0^1 \frac{\partial}{\partial s} p_{N,M,s} \mathrm{d}s ,$$

combining with (5.88) and (5.91) it suffices to prove that

$$\left| \frac{\partial}{\partial s} p_{N,M,s} + \frac{r}{2}(1-q) \right| \le \frac{K}{\sqrt{N}} . \tag{5.105}$$

First we compute $\partial p_{N,M,s}/\partial s$ using straightforward differentiation. Denoting by ν_s the average corresponding to the Hamiltonian (5.87) and defining

$$S'_{k,s} = \frac{1}{2\sqrt{s}} S_k - \frac{1}{2\sqrt{1-s}} \theta_k$$

we get

$$\frac{\partial}{\partial s} p_{N,M,s} = \mathrm{I} + \mathrm{II} ,$$

where

$$\mathrm{I} = \frac{1}{N^2} \sum_{k_1 \ne k_2} \nu_s(S'_{k_1,s} w(S_{k_1,s}, S_{k_2,s}))$$

and

$$\mathrm{II} = -\frac{1}{2\sqrt{1-s}} \nu_s \left(\sum_{i \le N} \sigma_i z'_i \sqrt{r} \right) .$$

We then integrate by parts. This is similar to the integration by parts in (2.81). This is easy for the term II. We will explain the result of the computation for the term I using the heuristic principle (2.58). The relation $\mathsf{E} S'_{k_1,s} S_{k_2,s} = 0$ shows that as in the derivation of (5.101) "the only terms created come from the denominator in the expression of ν_s". Moreover, the action of the expectation E_ξ in the denominator amount to "shift the quantities $S_{k,s}$ to a new replica." As in the case of (2.81) the definition of replicas here involves replacing ξ_k by an independent copy ξ_k^ℓ. That is, defining S_k^ℓ in the obvious manner, we set

$$S_{k,s}^\ell = \sqrt{s} S_k^\ell + \sqrt{1-s}(\sqrt{q} z_k + \sqrt{1-q} \xi_k^\ell)$$
$$S_{k,s}^{\ell\prime} = \frac{1}{2\sqrt{s}} S_k^\ell - \frac{1}{2\sqrt{1-s}} (\sqrt{q} z_k + \sqrt{1-q} \xi_k^\ell) .$$

We observe the relation $\mathsf{E} S_{k,s}^{1\prime} S_{k,s}^2 = R_{1,2} - q$, so that in the terms arising from the denominator we get the factor $R_{1,2} - q$. Therefore we get

$$\mathrm{I} = -\frac{1}{2} \nu_s \left((R_{1,2} - q) \frac{1}{N^3} \sum_{k_1 \ne k_2 , k_1 \ne k_3} w(S_{k_1,s}^1, S_{k_2,s}^1) w(S_{k_1,s}^2, S_{k_2,s}^2) \right) ,$$

and as usual we have

$$\mathrm{II} = -\frac{r}{2}(1 - \nu(R_{1,2})) \,.$$

Finally we have obtained the relation

$$\begin{aligned}
&\frac{\partial}{\partial s} p_{N,M,s} = \mathrm{I} + \mathrm{II} \\
&= -\frac{1}{2}\nu_s\left((R_{1,2} - q)\left(\frac{1}{N^3}\sum_{k_1 \neq k_2\,, k_1 \neq k_3} w(S^1_{k_1,s}, S^1_{k_2,s}) w(S^2_{k_1,s}, S^2_{k_2,s}) - r\right)\right) \\
&- \frac{r}{2}(1 - q) \,.
\end{aligned}$$

One then extends (5.83) to the interpolating system to obtain (5.105) through the Cauchy-Schwarz inequality. □

Exercise 5.5.9. Improve the rate of (5.104) into the usual rate K/N. (This requires very significant work.)

6. The Diluted SK Model and the K-Sat Problem

6.1 Introduction

In the SK model, each individual (or spin) interacts with every other individual. For large N, this does not make physical sense. Rather, we would like that, as $N \to \infty$, a given individual typically interacts only with a bounded number of other individuals. This motivates the introduction of the *diluted SK model*. In this model, the Hamiltonian is given by

$$-H_N(\boldsymbol{\sigma}) = \beta \sum_{i<j} g_{ij}\gamma_{ij}\sigma_i\sigma_j \;. \tag{6.1}$$

As usual, $(g_{ij})_{i<j}$ are i.i.d. standard Gaussian r.v.s. The quantities $\gamma_{ij} \in \{0,1\}$ determine which of the interaction terms are actually present in the Hamiltonian. There is an interaction term between σ_i and σ_j only when $\gamma_{ij} = 1$. The natural choice for these quantities is to consider a parameter $\gamma > 0$ (that does not depend on N) indicating "how diluted is the interaction", and to decide that the quantities γ_{ij} are i.i.d. r.v.s with $\mathsf{P}(\gamma_{ij} = 1) = \gamma/N$, $\mathsf{P}(\gamma_{ij} = 0) = 1 - \gamma/N$, and are independent from the r.v.s g_{ij}. Thus, the expected number of terms in (6.1) is

$$\frac{\gamma}{N}\frac{N(N-1)}{2} = \frac{\gamma(N-1)}{2} \,,$$

and the expected number of terms that contain σ_i is about $\gamma/2$. That is, the average number of spins that interact with one given spin is about $\gamma/2$. One should observe that the usual normalizing factor $1/\sqrt{N}$ does not occur in (6.1).

If we draw an edge between i and j when $\gamma_{ij} = 1$, the resulting random graph is well understood [12]. When $\gamma < 1$, this graph has only small connected components, so there is no "global interaction" and the situation is not so interesting. In order to get a challenging model we must certainly allow the case where γ takes any positive value.

In an apparently unrelated direction, let us remind the reader that the motivation of Chapter 2 is the problem as to whether certain random subsets of $\{-1,1\}^N$ have a non-empty intersection. In Chapter 2, we considered "random half-spaces". These somehow "depend on all coordinates". What would

M. Talagrand, *Mean Field Models for Spin Glasses*, Ergebnisse der Mathematik und ihrer Grenzgebiete. 3. Folge / A Series of Modern Surveys in Mathematics 54, DOI 10.1007/978-3-642-15202-3_6, © Springer-Verlag Berlin Heidelberg 2011

happen if instead we considered sets depending only on a given number p of coordinates? For example sets of the type

$$\{\boldsymbol{\sigma}\,;\,(\sigma_{i_1},\ldots,\sigma_{i_p})\neq(\eta_1,\ldots,\eta_p)\}\tag{6.2}$$

where $1\leq i_1<i_2<\ldots<i_p\leq N$, and $\eta_1,\ldots,\eta_p=\pm1$?

The question of knowing whether M random independent sets of the type (6.2) have a non-empty intersection is known in theoretical computer science as the random K-sat problem, and is of considerable interest. (There K is just another notation for what we call p. "Sat" stands for "satisfiability", as the problem is presented under the equivalent form of whether one can assign values to N Boolean variables in order to satisfy a collection of M random logical clauses of a certain type.) By a random subset of the type (6.2), we of course mean a subset that is chosen uniformly at random among all possible such subsets. This motivates the introduction of the Hamiltonian

$$-\,H_N(\boldsymbol{\sigma})=-\beta\sum_{k\leq M}W_k(\boldsymbol{\sigma})\tag{6.3}$$

where $W_k(\boldsymbol{\sigma})=0$ if $(\sigma_{i(k,1)},\ldots,\sigma_{i(k,p)})\neq(\eta_{k,1},\ldots,\eta_{k,p})$, and $W_k(\boldsymbol{\sigma})=1$ otherwise. The indices $1\leq i(k,1)<i(k,2)<\ldots<i(k,p)\leq N$ and the numbers $\eta_{k,i}=\pm1$ are chosen randomly uniformly over all possible choices. The interesting case is when M is proportional to N.

In a beautiful paper, S. Franz and S. Leone [60] observed that many technicalities disappear (and that one obtains a similar model) if rather than insisting that the Hamiltonian contains exactly a given number of terms, this number of terms is a Poisson r.v. M (independent of the other sources of randomness). Since we are interested in the case where M is proportional to N we will assume that $\mathsf{E}M$ is proportional to N, i.e. $\mathsf{E}M=\alpha N$, where of course α does not depend on N.

To cover simultaneously the cases of (6.1) and (6.3), we consider a random real-valued function θ on $\{-1,1\}^p$, i.i.d. copies $(\theta_k)_{k\geq1}$ of θ, and the Hamiltonian

$$-\,H_N(\boldsymbol{\sigma})=\sum_{k\leq M}\theta_k(\sigma_{i(k,1)},\ldots,\sigma_{i(k,p)})\,.\tag{6.4}$$

Here, M is a Poisson r.v. of expectation αN, $1\leq i(k,1)<\ldots<i(k,p)\leq N$, the sets $\{i(k,1),\ldots,i(k,p)\}$ for $k\geq1$ are independent and uniformly distributed, and the three sources of randomness (these sets, M, and the θ_k) are independent of each other. There is no longer a coefficient β, since this coefficient can be thought of as a part of θ. For example, a situation very similar to (6.1) is obtained for $p=2$ and $\theta(\sigma_1,\sigma_2)=\beta g\sigma_1\sigma_2$ where g is standard Gaussian. It would require no extra work to allow an external field in the formula (6.4). We do not do this for simplicity, but we stress that our approach does not require any special symmetry property. (On the other hand, precise specific results such as those of [78] seem to rely on such properties.)

It turns out that the mean number of terms of the Hamiltonian that depend on a given spin is of particular relevance. This number is $\gamma = \alpha p$ (where α is such that $\mathsf{E}M = \alpha N$), and for simplicity of notation this will be our main parameter rather than α.

The purpose of this chapter is to describe the behavior of the system governed by the Hamiltonian (6.4) under a "high-temperature condition" asserting in some sense that this Hamiltonian is small enough. This condition will involve the r.v. S given by

$$S = \sup |\theta(\sigma_1, \ldots, \sigma_p)| \; , \tag{6.5}$$

where the supremum is of course over all values of $\sigma_1, \sigma_2, \ldots, \sigma_p = \pm 1$, and has the following property: if γ (and p) are given, then the high-temperature condition is satisfied when S is small enough.

Generally speaking, the determination of exactly under which conditions there is high-temperature behavior is a formidable problem. The best that our methods can possibly achieve is to reach qualitatively optimal conditions, that capture "a fixed proportion of the high-temperature region". This seems to be the case of the following condition:

$$16 p \gamma \mathsf{E}\, S \exp 4S \le 1 \; . \tag{6.6}$$

Since the mean number of spins interacting with a given spin remains bounded independently of N, the central limit theorem does not apply, and the ubiquitous Gaussian behavior of the previous chapters is now absent. Despite this fundamental difference, and even though this is hard to express explicitly, there are many striking similarities.

We now outline the organization of this chapter. A feature of our approach is that, in contrast with what happened for the previous models, we do not know how to gain control of the model "in one step". Rather, we will first prove in Section 6.2 that for large N a small collection of spins are approximately independent under a condition like (6.6). This is the main content of Theorem 6.2.2. The next main step takes place in Section 6.4, where in Theorem 6.4.1 we prove that under a condition like (6.6), a few quantities $\langle\sigma_1\rangle, \ldots, \langle\sigma_k\rangle$ are approximately independent with law μ_γ where μ_γ is a probability measure on $[0,1]$, that is described in Section 6.3 as the fixed point of a (complicated) operator. This result is then used in the last part of Section 6.4 to compute $\lim_{N\to\infty} p_N(\gamma)$, where $p_N(\gamma) = N^{-1}\mathsf{E}\log\sum\exp(-H_N(\boldsymbol{\sigma}))$, still under a "high-temperature" condition of the type (6.6). In Section 6.5 we prove under certain conditions an upper bound for $p_N(\gamma)$, that is true for *all values of* γ and that asymptotically coincides with the limit previously computed under a condition of the type (6.6). In Section 6.6 we investigate the case of continuous spins, and in Section 6.7 we demonstrate the very strong consequences of a suitable concavity hypothesis on the Hamiltonian, and we point out a number of rather interesting open problems.

6.2 Pure State

The purpose of this section is to show that under (6.6) "the system is in a pure state", that is, the spin correlations vanish. In fact we will prove that

$$\mathsf{E}\,|\langle\sigma_1\sigma_2\rangle - \langle\sigma_1\rangle\langle\sigma_2\rangle| \le \frac{K}{N} \tag{6.7}$$

where K depends only on p and γ. The proof, by induction over N, is similar in spirit to the argument occurring at the end of Section 1.3. In order to make the induction work, it is necessary to carry a suitable induction hypothesis, that will prove a stronger statement than (6.7). This stronger statement will be useful later in its own right.

Given $k \ge 1$ we say that two functions f, f' on Σ_N^n *depend on k coordinates* if we can find indices $1 \le i_1 < \ldots < i_k \le N$ and functions $\overline{f}, \overline{f}'$ from $\{-1,1\}^{kn}$ to $\mathbb{R}$ such that

$$f(\boldsymbol{\sigma}^1, \ldots, \boldsymbol{\sigma}^n) = \overline{f}(\sigma_{i_1}^1, \ldots, \sigma_{i_k}^1, \sigma_{i_1}^2, \ldots, \sigma_{i_k}^2, \ldots, \sigma_{i_1}^n, \ldots, \sigma_{i_k}^n)$$

and similarly for f'. The reason we define this for two functions is to stress that both functions depend on the same set of k coordinates.

For $i \le N$, consider the transformation T_i of Σ_N^n that, for a point $(\boldsymbol{\sigma}^1, \ldots, \boldsymbol{\sigma}^n)$ of Σ_N^n, exchanges the i-th coordinates of $\boldsymbol{\sigma}^1$ and $\boldsymbol{\sigma}^2$, and leaves all the other coordinates unchanged.

The following technical condition should be interpreted as an "approximate independence condition".

Definition 6.2.1. *Given three numbers $\gamma_0 > 0$, $B > 0$ and $B^* > 0$, we say that Property $C(N, \gamma_0, B, B^*)$ holds if the following is true. Consider two functions f, f' on Σ_N^n, and assume that they depend on k coordinates. Assume that $f \ge 0$, that for a certain $i \le N$ we have*

$$f' \circ T_i = -f' , \tag{6.8}$$

and that for a certain number Q we have $|f'| \le Qf$ at each point of Σ_N^n. Then if $\gamma \le \gamma_0$ we have

$$\mathsf{E}\left|\frac{\langle f'\rangle}{\langle f\rangle}\right| \le \frac{(kB + B^*)Q}{N} . \tag{6.9}$$

Condition $C(N, \gamma_0, B, B^*)$ is not immediately intuitive. It is an "approximate independence condition" because if the spins were really independent, the condition $f' \circ T_i = -f'$ would imply that $\langle f'\rangle = \langle f' \circ T_i\rangle = \langle -f'\rangle$ so that $\langle f'\rangle = 0$.

To gain intuition, let us relate condition $C(N, \gamma_0, B, B^*)$ with (6.7). We take $n = 2$, $f = 1$,

$$f'(\boldsymbol{\sigma}^1, \boldsymbol{\sigma}^2) = \sigma_1^1(\sigma_2^1 - \sigma_2^2) ,$$

so that (6.8) holds for $i = 2$, $k = 2$ and $|f'| \le 2f$. Thus under condition $C(N, \gamma_0, B, B^*)$ we get by (6.9) that

$$\mathsf{E}\,|\langle \sigma_1^1(\sigma_2^1 - \sigma_2^2)\rangle| \le \frac{2B + B^*}{N}$$

i.e.

$$\mathsf{E}\,|\langle \sigma_1\sigma_2\rangle - \langle \sigma_1\rangle\langle \sigma_2\rangle| \le \frac{2B + B^*}{N} ,$$

which is (6.7). More generally, basically the same argument shows that when condition $C(N, \gamma_0, B, B^*)$ holds (for each N and numbers B and B^* that do not depend on N), to compute Gibbs averages of functions that depend only on a number of spin that remains bounded independently of N, one can pretend that these spins are independent under Gibbs' measure. We will return to this important idea later.

Theorem 6.2.2. *There exists a number $K_0 = K_0(p, \gamma_0)$ such that if $\gamma \le \gamma_0$ and*

$$16\gamma_0 p \mathsf{E}\, S \exp 4S \le 1, \tag{6.10}$$

then Property $C(N, \gamma_0, K_0, K_0)$ holds for each N.

When property $C(N, \gamma_0, K_0, K_0)$ holds, for two functions f, f' on Σ_N^n, that depend on k coordinates, and with $f \ge 0$, $|f'| \le Qf$, then under (6.8), and if $\gamma \le \gamma_0$, we have

$$\mathsf{E}\left|\frac{\langle f'\rangle}{\langle f\rangle}\right| \le \frac{(kK_0 + K_0)Q}{N} \le \frac{2kK_0Q}{N} . \tag{6.11}$$

The point of distinguishing in the definition of $C(N, \gamma_0, B, B^*)$ the values B and B^* will become apparent during the proofs.

To prove Theorem 6.2.2, we will proceed by induction over N. The smallest value of N for which the model is defined is $N = p$. We first observe that $|\langle f'\rangle| \le Q\langle f\rangle$, so that $C(p, \gamma_0, K_1, K_1^*)$ is true if $K_1 \ge p$. We will show that if K_1 and K_1^* are suitably chosen, then under (6.10) we have

$$C(N-1, \gamma_0, K_1, K_1^*) \Rightarrow C(N, \gamma_0, K_1, K_1^*) . \tag{6.12}$$

This will prove Theorem 6.2.2.

The main idea to prove (6.12) is to relate the N-spin system with an $(N-1)$-spin system through the cavity method, and we first need to set up this method. We write $-H_N(\boldsymbol{\sigma}) = -H_{N-1}(\boldsymbol{\sigma}) - H(\boldsymbol{\sigma})$, where

$$-\,H_{N-1}(\boldsymbol{\sigma}) = \sum \theta_k(\sigma_{i(k,1)}, \dots, \sigma_{i(k,p)}) , \tag{6.13}$$

where the sum is over those $k \le M$ for which $i(k,p) \le N-1$, and where $H(\boldsymbol{\sigma})$ is the sum of the other terms of (6.4), those for which $i(k,p) = N$.

Since the set $\{i(k,1),\ldots,i(k,p)\}$ is uniformly distributed over the subsets of $\{1,\ldots,N\}$ of cardinality p, the probability that $i(k,p)=N$ is exactly p/N.

A remarkable property of Poisson r.v.s is as follows: when M is a Poisson r.v., if $(X_k)_{k\geq 1}$ are i.i.d. $\{0,1\}$-valued r.v.s then $\sum_{k\leq M} X_k$ and $\sum_{k\leq M}(1-X_k)$ are independent Poisson r.v.s with mean respectively $\mathsf{E}M\mathsf{E}X_k$ and $\mathsf{E}M\mathsf{E}(1-X_k)$. The simple proof is given in Lemma A.10.1. Using this for $X_k=1$ if $i(k,p)=N$ and $X_k=0$ otherwise implies that the numbers of terms in $H(\boldsymbol{\sigma})$ and $H_{N-1}(\boldsymbol{\sigma})$ are independent Poisson r.v.s of mean respectively $(p/N)\alpha N=\gamma$ and $\alpha N-\gamma$. Thus the pair $(-H_{N-1}(\boldsymbol{\sigma}),-H(\boldsymbol{\sigma}))$ has the same distribution as the pair

$$\left(\sum_{k\leq M'}\theta'_k(\sigma_{i'(k,1)},\ldots,\sigma_{i'(k,p)}),\sum_{j\leq r}\theta_j(\sigma_{i(j,1)},\ldots,\sigma_{i(j,p-1)},\sigma_N)\right). \tag{6.14}$$

Here M' and r are Poisson r.v.s of mean respectively $\alpha N-\gamma$ and γ; θ'_k and θ_j are independent copies of θ; $i'(k,1)<\ldots<i'(k,p)$ and the set $\{i'(k,1),\ldots,i'(k,p)\}$ is uniformly distributed over the subsets of $\{1,\ldots,N-1\}$ of cardinality p; $i(j,1)<\ldots<i(j,p-1)\leq N-1$ and the set $I_j=\{i(j,1),\ldots,i(j,p-1)\}$ is uniformly distributed over the subsets of $\{1,\ldots,N-1\}$ of cardinality $p-1$; all these random variables are globally independent.

The following exercise describes another way to think of the Hamiltonian H_N, which provides a different intuition for the fact that the pair $(-H_{N-1}(\boldsymbol{\sigma}),-H(\boldsymbol{\sigma}))$ has the same distribution as the pair (6.14).

Exercise 6.2.3. For each p-tuple $\mathbf{i}=(i_1,\ldots,i_p)$ with $1\leq i_1<\ldots<i_p\leq N$, and each $j\geq 1$ let us consider independent copies $\theta_{\mathbf{i},j}$ of θ, and define

$$-H_{\mathbf{i}}(\boldsymbol{\sigma})=\sum_{j\leq r_{\mathbf{i}}}\theta_{\mathbf{i},j}(\sigma_{i_1},\ldots,\sigma_{i_p})\,,$$

where $r_{\mathbf{i}}$ are i.i.d. Poisson r.v.s (independent of all other sources of randomness) with

$$\mathsf{E}r_{\mathbf{i}}=\frac{\alpha M}{\binom{M}{p}}\,.$$

Prove that the Hamiltonian H_N has the same distribution as the Hamiltonian $\sum_{\mathbf{i}} H_{\mathbf{i}}$.

Since the properties of the system governed by the Hamiltonian H_N depend only of the distribution of this Hamiltonian, from now on in this section we will assume that, using the same notation as in (6.14),

$$-H_N(\boldsymbol{\sigma})=-H_{N-1}(\boldsymbol{\sigma})-H(\boldsymbol{\sigma})\,, \tag{6.15}$$

where

$$-H_{N-1}(\boldsymbol{\sigma})=\sum_{k\leq M'}\theta'_k(\sigma_{i'(k,1)},\ldots,\sigma_{i'(k,p)})\,, \tag{6.16}$$

and

$$-H(\boldsymbol{\sigma}) = \sum_{j \le r} \theta_j(\sigma_{i(j,1)}, \ldots, \sigma_{i(j,p-1)}, \sigma_N) \; . \tag{6.17}$$

Let us stress that in this section and in the next, the letter r will stand for the number of terms in the summation (6.17), which is a Poisson r.v. of expectation γ.

We observe from (6.16) that if we write $\boldsymbol{\rho} = (\sigma_1, \ldots, \sigma_{N-1})$ when $\boldsymbol{\sigma} = (\sigma_1, \ldots, \sigma_N)$, $-H_{N-1}(\boldsymbol{\sigma}) = -H_{N-1}(\boldsymbol{\rho})$ is the Hamiltonian of a $(N-1)$-spin system, except that we have replaced γ by a different value γ_-. To compute γ_- we recall that the mean number of terms of the Hamiltonian H_{N-1} is $\alpha N - \gamma$, so that the mean number γ_- of terms that contain a given spin is

$$\gamma_- = \frac{p}{N-1}(\alpha N - \gamma) = \gamma \frac{N-p}{N-1} \; , \tag{6.18}$$

since $p\alpha = \gamma$. We note that $\gamma_- \le \gamma$, so that

$$\gamma < \gamma_0 \Rightarrow \gamma_- \le \gamma_0 \; , \tag{6.19}$$

a fact that will help the induction.

Given a function f on Σ_N^n, the algebraic identity

$$\langle f \rangle = \frac{\langle \mathrm{Av}\, f\mathcal{E} \rangle_-}{\langle \mathrm{Av}\, \mathcal{E} \rangle_-} \tag{6.20}$$

holds. Here,

$$\mathcal{E} = \mathcal{E}(\boldsymbol{\sigma}^1, \ldots, \boldsymbol{\sigma}^n) = \exp\Bigl(\sum_{\ell \le n} -H(\boldsymbol{\sigma}^\ell)\Bigr) \; , \tag{6.21}$$

and as usual Av means average over $\sigma_N^1, \ldots, \sigma_N^n = \pm 1$. Thus $\mathrm{Av}\, f\mathcal{E}$ is a function of $(\boldsymbol{\rho}^1, \ldots, \boldsymbol{\rho}^n)$ only, and $\langle \mathrm{Av}\, f\mathcal{E} \rangle_-$ means that it is then averaged for Gibbs' measure relative to the Hamiltonian (6.13).

In the right-hand side of (6.20), we have two distinct sources of randomness: the randomness in $\langle \cdot \rangle_-$ and the randomness in $\mathcal{E}$. It will be essential that these sources of randomness are probabilistically independent. In the previous chapters we were taking expectation given $\langle \cdot \rangle_-$. We find it more convenient to now take expectation given $\mathcal{E}$. This expectation is denoted by E_-, so that, according to (6.20) we have

$$\mathsf{E}\left|\frac{\langle f' \rangle}{\langle f \rangle}\right| = \mathsf{E}\left|\frac{\langle \mathrm{Av}\, f'\mathcal{E} \rangle_-}{\langle \mathrm{Av}\, f\mathcal{E} \rangle_-}\right| = \mathsf{E}\mathsf{E}_-\left|\frac{\langle \mathrm{Av}\, f'\mathcal{E} \rangle_-}{\langle \mathrm{Av}\, f\mathcal{E} \rangle_-}\right| \; . \tag{6.22}$$

After these preparations we describe the structure of the proof. Let us consider a pair (f', f) as in Definition 6.2.1. The plan is to write

$$\mathrm{Av}\, f'\mathcal{E} = \frac{1}{2}\sum_s f'_s$$

for some functions f'_s on Σ^n_{N-1}, such that the number of terms does not depend on N, and that all pairs $(f'_s, \mathrm{Av}\, f\mathcal{E})$ have the property of the pair (f', f), but in the $(N-1)$-spin system. Since

$$\frac{\mathrm{Av} f'\mathcal{E}}{\mathrm{Av}\, f\mathcal{E}} = \frac{1}{2}\sum_s \frac{f'_s}{\mathrm{Av}\, f\mathcal{E}} ,$$

we can now apply the induction hypothesis to each term to get a bound for the sum and hence for

$$\mathsf{E}_- \left| \frac{\langle \mathrm{Av}\, f'\mathcal{E}\rangle_-}{\langle \mathrm{Av}\, f\mathcal{E}\rangle_-} \right| ,$$

and finally (6.22) completes the induction step.

We now start the proof. We consider a pair (f', f) as in Definition 6.2.1, that is $|f'| \le Qf$, $f' \circ T_i = -f'$ for some $i \le N$, and f, f' depend on k coordinates. We want to bound $\mathsf{E}|\langle f'\rangle/\langle f\rangle|$, and for this we study the last term of (6.22). Without loss of generality, we assume that $i = N$ and that f and f' depend on the coordinates $1, \ldots, k-1, N$. First, we observe that, since we assume $|f'| \le Qf$, we have $|f'\mathcal{E}| \le Qf\mathcal{E}$, so that $|\mathrm{Av}\, f'\mathcal{E}| \le \mathrm{Av}\, |f'\mathcal{E}| \le Q\mathrm{Av}\, f\mathcal{E}$, and thus

$$\mathsf{E}_- \left| \frac{\langle \mathrm{Av}\, f'\mathcal{E}\rangle_-}{\langle \mathrm{Av}\, f\mathcal{E}\rangle_-} \right| \le Q . \tag{6.23}$$

We recall (6.21) and (6.17), and in particular that r is the number of terms in the summation (6.17) and is a Poisson r.v. of expectation γ. We want to apply the induction hypothesis to compute the left-hand side of (6.23). The expectation E_- is expectation given $\mathcal{E}$, and it helps to apply the induction hypothesis if the functions $\mathrm{Av} f'\mathcal{E}$ and $\mathrm{Av} f\mathcal{E}$ are not too complicated. To ensure this it will be desirable that all the points $i(j,\ell)$ for $j \le r$ and $\ell \le p-1$ are different and $\ge k$. In the rare event Ω (we recall that Ω denotes an event, and not the entire probability space) where this not the case, we will simply use the crude bound (6.23) rather than the induction hypothesis. Recalling that $i(j,1) < \ldots < i(j,p-1)$, to prove that Ω is a rare event we write $\Omega = \Omega_1 \cup \Omega_2$ where

$$\Omega_1 = \Big\{ \exists j \le r ,\ i(j,1) \le k-1 \Big\}$$

$$\Omega_2 = \Big\{ \exists j, j' \le r ,\ j \ne j' ,\ \exists \ell, \ell' \le p-1 ,\ i(j,\ell) = i(j',\ell') \Big\} .$$

These two events depend only on the randomness of $\mathcal{E}$. Let us recall that for $j \le r$ the sets

$$I_j = \{ i(j,1), \ldots, i(j,p-1) \} \tag{6.24}$$

are independent and uniformly distributed over the subsets of $\{1, \ldots, N-1\}$ of cardinality $p-1$. The probability that any given $i \le N-1$ belongs to I_j is therefore $(p-1)/(N-1)$. Thus the probability that $i(j,1) \le k-1$, i.e. the probability that there exists $\ell \leq k-1$ that belongs to I_j is at most $(p-1)(k-1)/(N-1)$. Therefore

$$\mathsf{P}(\Omega_1) \leq \frac{(p-1)(k-1)}{N-1} \mathsf{E}r \leq \frac{kp\gamma}{N} .$$

Here and below, we do not try to get sharp bounds. There is no point in doing this, as anyway our methods cannot reach the best possible bounds. Rather, we aim at writing explicit bounds that are not too cumbersome. For $j < j' \leq r$, the probability that a given point $i \leq N-1$ belongs to both sets I_j and $I_{j'}$ is $((p-1)/(N-1))^2$. Thus the random number U of points $i \leq N-1$ that belong to two different sets I_j for $j \leq r$ satisfies

$$\mathsf{E}U = (N-1)\left(\frac{p-1}{N-1}\right)^2 \mathsf{E}\frac{r(r-1)}{2} \leq \frac{p^2\gamma^2}{2N} ,$$

using that $\mathsf{E}r(r-1) = (\mathsf{E}r)^2$ since r is a Poisson r.v., see (A.64). Since U is integer valued, we have $\mathsf{P}(\{U \neq 0\}) \leq \mathsf{E}U$ and since $\Omega_2 = \{U \neq 0\}$ we get

$$\mathsf{P}(\Omega_2) \leq \frac{p^2\gamma^2}{2N} ,$$

so that finally, since $\Omega = \Omega_1 \cup \Omega_2$, we obtain

$$\mathsf{P}(\Omega) \leq \frac{kp\gamma + p^2\gamma^2}{N} . \tag{6.25}$$

Using (6.22), (6.23) and (6.25), we have

$$\begin{aligned} \mathsf{E}\left|\frac{\langle f'\rangle}{\langle f\rangle}\right| &= \mathsf{E}\left(1_\Omega \mathsf{E}_-\left|\frac{\langle \mathrm{Av}\, f'\mathcal{E}\rangle_-}{\langle \mathrm{Av}\, f\mathcal{E}\rangle_-}\right|\right) + \mathsf{E}\left(1_{\Omega^c} \mathsf{E}_-\left|\frac{\langle \mathrm{Av}\, f'\mathcal{E}\rangle_-}{\langle \mathrm{Av}\, f\mathcal{E}\rangle_-}\right|\right) \\ &\leq \frac{kp\gamma + p^2\gamma^2}{N} Q + \mathsf{E}\left(1_{\Omega^c} \mathsf{E}_-\left|\frac{\langle \mathrm{Av}\, f'\mathcal{E}\rangle_-}{\langle \mathrm{Av}\, f\mathcal{E}\rangle_-}\right|\right) . \end{aligned} \tag{6.26}$$

The next task is to use the induction hypothesis to study the last term above. When Ω does not occur (i.e. on Ω^c), all the points $i(j,\ell)$, $j \leq r$, $\ell \leq p-1$ are different and are $\geq k$. Recalling the notation (6.24) we have

$$J = \{i(j,\ell);\ j \leq r,\ \ell \leq p-1\} = \bigcup_{j \leq r} I_j ,$$

so that $\operatorname{card} J = r(p-1)$ and

$$J \cap \{1, \ldots, k-1, N\} = \emptyset . \tag{6.27}$$

For $i \leq N-1$ let us denote by U_i the transformation of Σ_{N-1}^n that exchanges the coordinates σ_i^1 and σ_i^2 of a point $(\boldsymbol{\rho}^1, \ldots, \boldsymbol{\rho}^n)$ of Σ_{N-1}^n, and that leaves all the other coordinates unchanged. That is, U_i is to $N-1$ what T_i is to N.

Lemma 6.2.4. *Assume that f' satisfies (6.8) for $i = N$, i.e. $f' \circ T_N = -f'$ and depends only on the coordinates in $\{1, \ldots, k-1, N\}$. Then when Ω does not occur (i.e. on Ω^c) we have*

$$(\mathrm{Av}\, f'\mathcal{E}) \circ \prod_{i \in J} U_i = -\mathrm{Av}\, f'\mathcal{E} \,. \tag{6.28}$$

Here $\prod_{i \in J} U_i$ denotes the composition of the transformations U_i for $i \in J$ (which does not depend on the order in which this composition is performed). This (crucial...) lemma means that something of the special symmetry of f' (as in (6.8)) is preserved when one replaces f' by $\mathrm{Av}\, f'\mathcal{E}$.

Proof. Let us write $T = \prod_{i \in J} T_i$. We observe first that

$$f' \circ T = f'$$

because f' depends only on the coordinates in $\{1, \ldots, k-1, N\}$, a set disjoint from J. Thus

$$f' \circ T \circ T_N = f' \circ T_N = -f' \tag{6.29}$$

since $f' \circ T_N = -f'$. We observe now that $T \circ T_N$ exchanges σ_i^1 and σ_i^2 for all $i \in J \cup \{N\}$. These values of i are precisely the coordinates of which $\mathcal{E}$ depends, so that

$$\mathcal{E} \circ T \circ T_N(\boldsymbol{\sigma}^1, \boldsymbol{\sigma}^2, \ldots, \boldsymbol{\sigma}^n) = \mathcal{E}(\boldsymbol{\sigma}^2, \boldsymbol{\sigma}^1, \ldots, \boldsymbol{\sigma}^n) = \mathcal{E}(\boldsymbol{\sigma}^1, \boldsymbol{\sigma}^2, \ldots, \boldsymbol{\sigma}^n) \,,$$

and hence

$$\mathcal{E} \circ T \circ T_N = \mathcal{E} \,.$$

Combining with (6.29) we get

$$(f'\mathcal{E}) \circ T \circ T_N = (f' \circ T \circ T_N)(\mathcal{E} \circ T \circ T_N) = -f'\mathcal{E}$$

so that, since T_N^2 is the identity,

$$(f'\mathcal{E}) \circ T = -(f'\mathcal{E}) \circ T_N \,. \tag{6.30}$$

Now, for any function f we have $\mathrm{Av}(f \circ T_N) = \mathrm{Av} f$ and $\mathrm{Av}(f \circ T) = (\mathrm{Av} f) \circ \prod_{i \in J} U_i$. Therefore we obtain

$$\mathrm{Av}\,((f'\mathcal{E}) \circ T_N) = \mathrm{Av}\, f'\mathcal{E}$$

$$\mathrm{Av}((f'\mathcal{E}) \circ T) = (\mathrm{Av}\, f'\mathcal{E}) \circ \prod_{i \in J} U_i \,,$$

so that applying Av to (6.30) proves (6.28). □

Let us set $k' = r(p-1) = \mathrm{card}\, J$, and let us enumerate as $i_1, \ldots, i_{k'}$ the points of J. Now (6.28) implies

$$\operatorname{Av} f'\mathcal{E} = \frac{1}{2}\left(\operatorname{Av} f'\mathcal{E} - (\operatorname{Av} f'\mathcal{E}) \circ \prod_{s \le k'} U_{i_s}\right) = \frac{1}{2} \sum_{1 \le s \le k'} f'_s , \tag{6.31}$$

where

$$f'_s = (\operatorname{Av} f'\mathcal{E}) \circ \prod_{u \le s-1} U_{i_u} - (\operatorname{Av} f'\mathcal{E}) \circ \prod_{u \le s} U_{i_u} . \tag{6.32}$$

Since U_i^2 is the identity, we have

$$f'_s \circ U_{i_s} = -f'_s . \tag{6.33}$$

In words, (6.31) decomposes $\operatorname{Av} f'\mathcal{E}$ as a sum of $k' = r(p-1)$ pieces that possess the symmetry property required to use the induction hypothesis. In order to apply this induction hypothesis, it remains to establish the property that will play for the pairs $(f'_s, \operatorname{Av} f\mathcal{E})$ the role the inequality $|f'| \le Qf$ plays for the pair (f', f). This is the purpose of the next lemma. For $j \le r$ we set

$$S_j = \sup |\theta_j(\varepsilon_1, \varepsilon_2, \ldots, \varepsilon_p)| ,$$

where the supremum is over all values of $\varepsilon_1, \varepsilon_2, \ldots, \varepsilon_p = \pm 1$. We recall the notation (6.24).

Lemma 6.2.5. *Assume that Ω does not occur and that $i_s \in I_v$ for a certain (unique) $v \le r$. Then*

$$|f'_s| \le 4QS_v \exp\left(4 \sum_{u \le r} S_u\right) \operatorname{Av} f\mathcal{E} . \tag{6.34}$$

A crucial feature of this bound is that it does not depend on the number n of replicas.

Proof. Let us write

$$\mathcal{E}' = \exp\left(\sum_{3 \le \ell \le n} -H(\boldsymbol{\sigma}^\ell)\right) ; \; \mathcal{E}'' = \exp\left(\sum_{\ell=1,2} -H(\boldsymbol{\sigma}^\ell)\right) ,$$

so that $\mathcal{E} = \mathcal{E}'\mathcal{E}''$. Since $|H(\boldsymbol{\sigma})| \le \sum_{j \le r} S_j$, we have

$$\mathcal{E}'' \ge \exp\left(-2 \sum_{j \le r} S_j\right) ,$$

and therefore

$$\mathcal{E} \ge \mathcal{E}' \exp\left(-2 \sum_{j \le r} S_j\right) . \tag{6.35}$$

This implies

$$\operatorname{Av} f\mathcal{E} \ge (\operatorname{Av} f\mathcal{E}') \exp\left(-2 \sum_{j \le r} S_j\right) . \tag{6.36}$$

Next,

$$\begin{aligned} f'_s &= (\mathrm{Av}\, f'\mathcal{E}) \circ \prod_{u\le s-1} U_{i_u} - (\mathrm{Av}\, f'\mathcal{E}) \circ \prod_{u\le s} U_{i_u} \\ &= \mathrm{Av}\Bigl((f'\mathcal{E}) \circ \prod_{u\le s-1} T_{i_u} - (f'\mathcal{E}) \circ \prod_{u\le s} T_{i_u} \Bigr) \\ &= \mathrm{Av}\Bigl(f' \Bigl(\mathcal{E} \circ \prod_{u\le s-1} T_{i_u} - \mathcal{E} \circ \prod_{u\le s} T_{i_u} \Bigr)\Bigr) , \end{aligned} \tag{6.37}$$

using in the last line that $f' \circ T_{i_u} = f'$ for each u, since f' depends only on the coordinates $1, \ldots, k-1, N$. Recalling that $\mathcal{E} = \mathcal{E}''\mathcal{E}'$, and observing that for each i, we have $\mathcal{E}' \circ T_i = \mathcal{E}'$, we get

$$\mathcal{E} \circ \prod_{u\le s-1} T_{i_u} - \mathcal{E} \circ \prod_{u\le s} T_{i_u} = \mathcal{E}' \Bigl(\mathcal{E}'' \circ \prod_{u\le s-1} T_{i_u} - \mathcal{E}'' \circ \prod_{u\le s} T_{i_u} \Bigr) ,$$

and, if we set

$$\Delta = \sup \Bigl| \mathcal{E}'' \circ \prod_{u\le s-1} T_{i_u} - \mathcal{E}'' \circ \prod_{u\le s} T_{i_u} \Bigr| = \sup |\mathcal{E}'' - \mathcal{E}'' \circ T_{i_s}| ,$$

we get from (6.37) that, using that $|f'| \le Qf$ in the first inequality and (6.35) in the second one,

$$|f'_s| \le \Delta \mathrm{Av}\,(|f'|\mathcal{E}') \le Q\Delta \mathrm{Av}(f\mathcal{E}') \le Q\Delta \mathrm{Av}(f\mathcal{E}) \exp\Bigl(2\sum_{j\le r} S_j\Bigr) . \tag{6.38}$$

To bound Δ, we write $\mathcal{E}'' = \prod_{j\le r} \mathcal{E}_j$, where

$$\mathcal{E}_j = \exp \sum_{\ell=1,2} \theta_j(\sigma^\ell_{i(j,1)}, \ldots, \sigma^\ell_{i(j,p-1)}, \sigma^\ell_N) .$$

We note that $\mathcal{E}_j \circ T_{i_s} = \mathcal{E}_j$ if $j \ne v$, because then $\mathcal{E}_j$ depends only on the coordinates in I_j, and $i_s \notin I_j$ if $j \ne v$, since $i_s \in I_v$ and $I_j \cap I_v = \emptyset$. Thus

$$\mathcal{E}'' - \mathcal{E}'' \circ T_{i_s} = (\mathcal{E}_v - \mathcal{E}_v \circ T_{i_s}) \prod_{j\ne v} \mathcal{E}_j .$$

Now, using the inequality $|e^x - e^y| \le |x-y|e^a \le 2ae^a$ for $|x|, |y| \le a$ and $a = 2S_v$, we get

$$|\mathcal{E}_v - \mathcal{E}_v \circ T_{i_s}| \le 4S_v \exp 2S_v .$$

Since for all j we have $\mathcal{E}_j \le \exp 2S_j$, we get $\Delta \le 4S_v \exp 2\sum_{j\le r} S_j$. Combining with (6.38) completes the proof. □

Proposition 6.2.6. *Assume that $N \geq p+1$ and that condition $C(N-1, \gamma_0, B, B^*)$ holds. Consider f' and f as in Definition 6.2.1, and assume that $\gamma \leq \gamma_0$. Then*

$$\mathsf{E}\left|\frac{\langle f'\rangle}{\langle f\rangle}\right| \leq \frac{Qp}{N}\Big(k(\gamma+4BD\exp 4D)+4pBU\mathsf{E}r^2V^{r-1}+p\gamma^2+4B^*D\exp 4D\Big)\,, \tag{6.39}$$

where

$$D = \gamma \mathsf{E}\, S\exp 4S\,.$$

Proof. We keep the notation of Lemmas 6.2.4 and 6.2.5. Since $\gamma_- \leq \gamma$, we can use $C(N-1, \gamma_0, B, B^*)$ to conclude from (6.33) and (6.34) that, since f'_s and $\mathrm{Av}\,\mathcal{E}f$ depend on $k-1+r(p-1) \leq k+rp$ coordinates, and since $1/(N-1) \leq 2/N$ because $N \geq 2$, on Ω^c we have

$$\mathsf{E}_-\left|\frac{\langle f'_s\rangle_-}{\langle \mathrm{Av}\, f\mathcal{E}\rangle_-}\right| \leq \frac{8Q}{N}((k+rp)B+B^*)S_v\exp\Big(4\sum_{j\leq r} S_j\Big)\,.$$

Let us denote by E_θ expectation in the r.v.s $\theta_1, \ldots, \theta_r$ only. Then we get

$$\mathsf{E}_\theta\mathsf{E}_-\left|\frac{\langle f'_s\rangle_-}{\langle \mathrm{Av}\, f\mathcal{E}\rangle_-}\right| \leq \frac{8Q}{N}((k+rp)B+B^*)UV^{r-1}\,,$$

where

$$U = \mathsf{E}\, S\exp 4S\,; \quad V = \mathsf{E}\exp 4S\,.$$

Combining with (6.31), and since there are $k' = r(p-1) \leq rp$ terms we get

$$\mathsf{E}_\theta\, \mathsf{E}_-\left|\frac{\langle \mathrm{Av}\, f'\mathcal{E}\rangle_-}{\langle \mathrm{Av}\, f\mathcal{E}\rangle_-}\right| \leq \frac{4Qp}{N}((kr+r^2p)B+rB^*)UV^{r-1}\,.$$

This bound assumes that Ω does not occur; but combining with (6.26) we obtain the bound

$$\mathsf{E}\left|\frac{\langle f'\rangle}{\langle f\rangle}\right| \leq \frac{Qp}{N}\Big(k\gamma+p\gamma^2+4B\big(kU\mathsf{E}rV^{r-1}+pU\mathsf{E}r^2V^{r-1}\big)+4B^*U\mathsf{E}rV^{r-1}\Big)\,.$$

Since r is a Poisson r.v. of expectation γ a straightforward calculation shows that $\mathsf{E}rV^{r-1} = \gamma\exp\gamma(V-1)$. Since $e^x \leq 1+xe^x$ for all $x \geq 0$ (as is trivial using power series expansion) we have $V \leq 1+4U$, so $\exp\gamma(V-1) \leq \exp 4\gamma U$ and $U\mathsf{E}rV^{r-1} \leq D\exp 4D$. The result follows. □

Proof of Theorem 6.2.2. If

$$D_0 = \gamma_0\mathsf{E}\, S\exp 4S$$

is small enough that $16pD_0 \leq 1$ then

$$4pD_0\exp 4D_0 \leq 1/2\,, \tag{6.40}$$

and (6.39) implies

$$\mathsf{E}\left|\frac{\langle f'\rangle}{\langle f\rangle}\right| \le \frac{Q}{N}\left(k\left(p\gamma_0 + \frac{B}{2}\right) + 4p^2 BU\mathsf{E}r^2V^{r-1} + p^2\gamma^2 + \frac{B^*}{2}\right) .$$

Thus condition

$$C(N,\gamma_0, p\gamma_0 + B/2, 4p^2 BU\mathsf{E}r^2V^{r-1} + p^2\gamma_0^2 + B^*/2)$$

holds. That is, we have proved under (6.40) that

$$C(N-1,\gamma_0,B,B^*) \Rightarrow C(N,\gamma_0,p\gamma_0 + B/2, 4p^2BU\mathsf{E}r^2V^{r-1} + p^2\gamma_0^2 + B^*/2)) . \tag{6.41}$$

Now, we observe that $U\mathsf{E}r^2V^{r-1} \le K^{\sim}$ and that if $K_1 = 2p\gamma_0$ and $K_1^* = 8p^2K_1K^{\sim} + 2p^2\gamma_0^2$, (6.41) shows that (6.12) holds, and we have completed the induction. □

Probably at this point it is good to stop for a while and to wonder what is the nature of the previous argument. In essence this is "contraction argument". The operation of "adding one spin" essentially acts as a type of contraction, as is witnessed by the factor 1/2 in front of B and B^* in the right-hand side of (6.41). As it turns out, almost every single argument used in this work to control a model under a "high-temperature condition" is of the same type, whether this is rather apparent, as in Section 1.6, or in a more disguised form as here. (The one exception being Latala's argument of Section 1.4.)

We explained at length in Section 1.4 that we expect that at high temperature, as long as one considers a number of spins that remains bounded independently of N, Gibbs' measure is nearly a product measure. For the present model, this property follows from Theorem 6.2.2 and we now give quantitative estimates to that effect, in the setting we need for future uses.

Let us consider the product measure μ on Σ_{N-1} such that

$$\forall i \le N-1\,, \int \sigma_i \,\mathrm{d}\mu(\boldsymbol{\rho}) = \langle \sigma_i \rangle_- \,,$$

and let us denote by $\langle\cdot\rangle_\bullet$ an average with respect to μ. Equivalently, for a function f on Σ_{N-1}, we have

$$\langle f\rangle_\bullet = \langle f(\sigma_1^1,\ldots,\sigma_{N-1}^{N-1})\rangle_- \,, \tag{6.42}$$

where σ_i^i is the i-th coordinate of the i-th replica $\boldsymbol{\rho}^i$. The following consequence of property $C(N,\gamma_0,K_0,K_0)$ will be used in Section 6.4. It expresses, in a form that is particularly adapted to the use of the cavity method the fact that under property $C(N,\gamma_0,K_0,K_0)$, a given number of spins (independent of N) become nearly independent for large N.

Proposition 6.2.7. *If property* $C(N,\gamma_0,K_0,K_0)$ *holds for each* N, *and if* $\gamma \le \gamma_0$, *the following occurs. Consider for* $j \le r$ *sets* $I_j \subset \{1,\dots,N\}$ *with* $\operatorname{card} I_j = p$, $N \in I_j$, *and such that* $j \ne j' \Rightarrow I_j \cap I_{j'} = \{N\}$. *For* $j \le r$ *consider functions* W_j *on* Σ_N *depending only on the coordinates in* I_j *and let* $S_j = \sup|W_j(\boldsymbol{\sigma})|$. *Let*

$$\mathcal{E} = \exp\sum_{j\le r} W_j(\boldsymbol{\sigma}) \,.$$

Then, recalling the definition (6.42), we have

$$\mathsf{E}_-\left|\frac{\langle \operatorname{Av}\sigma_N\mathcal{E}\rangle_-}{\langle \operatorname{Av}\mathcal{E}\rangle_-} - \frac{\langle \operatorname{Av}\sigma_N\mathcal{E}\rangle_\bullet}{\langle \operatorname{Av}\mathcal{E}\rangle_\bullet}\right| \le \frac{8r(p-1)^2K_0}{N-1}\sum_{j\le r}\exp 2S_j \,. \tag{6.43}$$

This is a powerful principle, since it is very much easier to work with the averages $\langle\cdot\rangle_\bullet$ than with the Gibbs averages $\langle\cdot\rangle_-$. We will use this result when r is as usual the number of terms in (6.17) but since in (6.43) the expectation E_- is only in the randomness of $\langle\cdot\rangle_-$ we can, in the proof, think of the quantities r and W_j as being non-random.

Proof. Let $f' = \operatorname{Av}\sigma_N\mathcal{E}$ and $f = \operatorname{Av}\mathcal{E}$. For $0 \le i \le N-1$, let us define

$$f_i = f_i(\boldsymbol{\rho}^1,\dots,\boldsymbol{\rho}^{N-1}) = f(\sigma_1^1,\sigma_2^2,\dots,\sigma_i^i,\sigma_{i+1}^1,\dots,\sigma_{N-1}^1)$$

and f_i' similarly. The idea is simply that "we make the spins independent one at a time". Thus

$$\frac{\langle \operatorname{Av}\sigma_N\mathcal{E}\rangle_-}{\langle \operatorname{Av}\mathcal{E}\rangle_-} = \frac{\langle f_1'\rangle_-}{\langle f_1\rangle_-}\ ;\ \frac{\langle \operatorname{Av}\sigma_N\mathcal{E}\rangle_\bullet}{\langle \operatorname{Av}\mathcal{E}\rangle_\bullet} = \frac{\langle f_{N-1}'\rangle_-}{\langle f_{N-1}\rangle_-} \,, \tag{6.44}$$

and the left-hand side of (6.43) is bounded by

$$\sum_{2\le i\le N-1}\mathsf{E}_-\left|\frac{\langle f_{i-1}'\rangle_-}{\langle f_{i-1}\rangle_-} - \frac{\langle f_i'\rangle_-}{\langle f_i\rangle_-}\right| \,.$$

The terms in the summation are zero unless i belongs to the union of the sets I_j, $j \le r$, for otherwise f' and f do not depend on the i-th coordinate and $f_i = f_{i-1}$, $f_i' = f_{i-1}'$. We then try to bound the terms in the summation when $i \in I_j$ for a certain $j \le r$. Since $|f_i'| \le f_i$ we have

$$\begin{aligned}\left|\frac{\langle f_{i-1}'\rangle_-}{\langle f_{i-1}\rangle_-} - \frac{\langle f_i'\rangle_-}{\langle f_i\rangle_-}\right| &\le \left|\frac{\langle f_{i-1}' - f_i'\rangle_-}{\langle f_{i-1}\rangle_-}\right| + \left|\frac{\langle f_i'\rangle_-\langle f_{i-1} - f_i\rangle_-}{\langle f_{i-1}\rangle_-\langle f_i\rangle_-}\right| \\ &\le \left|\frac{\langle f_{i-1}' - f_i'\rangle_-}{\langle f_{i-1}\rangle_-}\right| + \left|\frac{\langle f_{i-1} - f_i\rangle_-}{\langle f_{i-1}\rangle_-}\right|\end{aligned}$$

so that, taking expectation in the previous inequality we get

$$\mathsf{E}_- \left| \frac{\langle f'_{i-1} \rangle_-}{\langle f_{i-1} \rangle_-} - \frac{\langle f'_i \rangle_-}{\langle f_i \rangle_-} \right| \le \mathsf{E}_- \left| \frac{\langle f'_{i-1} - f'_i \rangle_-}{\langle f_{i-1} \rangle_-} \right| + \mathsf{E}_- \left| \frac{\langle f_{i-1} - f_i \rangle_-}{\langle f_{i-1} \rangle_-} \right| . \quad (6.45)$$

We will use $C(N-1, \gamma_0, K_0, K_0)$ to bound these terms. First, we observe that the function $f'_{i-1} - f'_i$ changes sign if we exchange σ_i^1 and σ_i^i. Next, we observe that since W_u does not depend on σ_i for $u \ne j$ (where j is defined by $i \in I_j$) we have

$$\begin{aligned} \mathcal{E}' &:= \exp \sum_{u \ne j} W_u(\sigma_1^1, \sigma_2^2, \ldots, \sigma_i^i, \sigma_{i+1}^1, \ldots, \sigma_N^1) \\ &= \exp \sum_{u \ne j} W_u(\sigma_1^1, \sigma_2^2, \ldots, \sigma_{i-1}^{i-1}, \sigma_i^1, \sigma_{i+1}^1, \ldots, \sigma_N^1) . \end{aligned}$$

Then

$$f_{i-1} = \mathrm{Av}\, \mathcal{E}(\sigma_1^1, \ldots, \sigma_{i-1}^{i-1}, \sigma_i^1, \ldots, \sigma_N^1) \ge \exp(-S_j) \mathrm{Av}\, \mathcal{E}' ,$$

where Av denotes average over $\sigma_N^1 = \pm 1$. In a similar fashion, we get $|f'_{i-1}| \le \exp S_j \mathrm{Av}\, \mathcal{E}'$, $|f'_i| \le \exp S_j \mathrm{Av}\, \mathcal{E}'$, and thus

$$|f'_{i-1} - f'_i| \le (2 \exp 2S_j) f_{i-1} ,$$

so that using (6.11) property $C(N-1, \gamma_0, K_0, K_0)$ implies

$$\mathsf{E}_- \left| \frac{\langle f'_{i-1} - f'_i \rangle_-}{\langle f_{i-1} \rangle_-} \right| \le \frac{4K_0}{N-1} r(p-1) \exp 2S_j , \quad (6.46)$$

because these functions depend on $r(p-1)$ coordinates. We proceed similarly to handle the last term on the right-hand side of (6.45). We then perform the summation over $i \le N-1$. A new factor $p-1$ occurs because each set I_j contains $p-1$ such values of i. □

6.3 The Functional Order Parameter

As happened in the previous models, we expect that if we fix a number n and take N very large, at a given disorder, n spins $(\sigma_1, \ldots, \sigma_n)$ will asymptotically be independent, and that the r.v.s $\langle \sigma_1 \rangle, \ldots, \langle \sigma_n \rangle$ will asymptotically be independent. In the case of the SK model, the limiting law of $\langle \sigma_i \rangle$ was the law of $\mathrm{th}(\beta z \sqrt{q} + h)$ where z is a standard Gaussian r.v. and thus this law depended only on the single parameter q.

The most striking feature of the present model is that the limiting law is now a complicated object, that no longer depends simply on a few parameters. It is therefore reasonable to think of this limiting law μ as being itself a kind of parameter (the correct value of which has to be found). This is what the physicists mean when they say "that the order parameter of the model is a function" because they identify a probability distribution μ on $\mathbb{R}$ with the tail function $t \mapsto \mu([t,\infty))$.

The purpose of the present section is to find the correct value of this parameter. As is the case of the SK model this value will be given as the solution of a certain equation. The idea of the construction we will perform is very simple. While using the cavity method in the previous section, we have seen in (6.34) (used for $n = 1$ and $f(\boldsymbol{\sigma}) = \sigma_N$) that

$$\langle \sigma_N \rangle = \frac{\langle \mathrm{Av} \sigma_N \mathcal{E} \rangle_-}{\langle \mathrm{Av} \mathcal{E} \rangle_-} \,, \tag{6.47}$$

where

$$\mathcal{E} = \exp \sum_{j \le r} \theta_j(\sigma_{i(j,1)}, \ldots, \sigma_{i(j,p-1)}, \sigma_N) \,. \tag{6.48}$$

In the limit $N \to \infty$ the sets $I_j = \{i(j,1), \ldots, i(j,p-1)\}$ are disjoint. The quantity $\mathcal{E}$ depends on a number of spins that in essence does not depend on N. If we know the asymptotic behavior of any fixed number (i.e. of any number that does not depend on N) of the spins $(\sigma_i)_{i<N}$, we can then compute the behavior of the spin σ_N. This behavior has to be the same as the behavior of the spins σ_i for $i < N$, and this gives rise to a "self-consistency equation".

To define formally this equation, consider a Poisson r.v. r with $\mathsf{E}r = \gamma$, and independent of the r.v.s θ_j. For $\boldsymbol{\sigma} \in \{-1,1\}^{\mathbb{N}}$ and $\varepsilon \in \{-1,1\}$ we define

$$\mathcal{E}_r = \mathcal{E}_r(\boldsymbol{\sigma}, \varepsilon) = \exp \sum_{1 \le j \le r} \theta_j(\sigma_{(j-1)(p-1)+1}, \ldots, \sigma_{j(p-1)}, \varepsilon) \,. \tag{6.49}$$

This definition will be used **many times** in the sequel. We note that $\mathcal{E}_r$ depends on $\boldsymbol{\sigma}$ only through the coordinates of rank $\le r(p-1)$.

Given a sequence $\mathbf{x} = (x_i)_{i \ge 1}$ with $|x_i| \le 1$ we denote by $\lambda_{\mathbf{x}}$ the probability on $\{-1,1\}^{\mathbb{N}}$ that "has a density $\prod_i (1 + x_i \sigma_i)$ with respect to the uniform measure". More formally, $\lambda_{\mathbf{x}}$ is the product measure such that $\int \sigma_i \mathrm{d}\lambda_{\mathbf{x}}(\boldsymbol{\sigma}) = x_i$ for each i. We denote by $\langle \cdot \rangle_{\mathbf{x}}$ an average for $\lambda_{\mathbf{x}}$.

Similarly, if $\mathbf{x} = (x_i)_{i \le M}$ we also denote by $\lambda_{\mathbf{x}}$ the probability measure on $\Sigma_M = \{-1,1\}^M$ such that $\int \sigma_i \mathrm{d}\lambda_{\mathbf{x}}(\boldsymbol{\sigma}) = x_i$ and we denote by $\langle \cdot \rangle_{\mathbf{x}}$ an average for $\lambda_{\mathbf{x}}$, so that we have

$$\langle f \rangle_{\mathbf{x}} = \int \prod_{i \le M} (1 + x_i \sigma_i) f(\boldsymbol{\sigma}) \mathrm{d}\boldsymbol{\sigma} \,,$$

where $\mathrm{d}\boldsymbol{\sigma}$ means average for the uniform measure on Σ_M.

These definitions are also of **central importance** in this chapter. The idea underlying these definitions has already been used implicitly in (6.42) since for a function f on Σ_{N-1} we have

$$\langle f \rangle_\bullet = \langle f \rangle_{\mathbf{Y}} \,, \tag{6.50}$$

where $\mathbf{Y} = (\langle\sigma_1\rangle_-, \ldots, \langle\sigma_{N-1}\rangle_-)$.

Consider a probability measure μ on $[-1,1]$, and an i.i.d. sequence $\mathbf{X} = (X_i)_{i\geq 1}$ such that X_i is of law μ. We define $T(\mu)$ as the law of the r.v.

$$\frac{\langle \mathrm{Av}\, \varepsilon \mathcal{E}_r\rangle_{\mathbf{X}}}{\langle \mathrm{Av}\, \mathcal{E}_r\rangle_{\mathbf{X}}}\,, \tag{6.51}$$

where Av denotes the average over $\varepsilon = \pm 1$. We note that $\mathcal{E}$ depends on $\boldsymbol{\sigma}$ and ε, so that $\mathrm{Av}\,\varepsilon\mathcal{E}_r$ and $\mathrm{Av}\,\mathcal{E}_r$ depend on $\boldsymbol{\sigma}$ only and (6.51) makes sense. The intuition is that if μ is the law of $\langle\sigma_i\rangle$ for $i < N$, then $T(\mu)$ is the law of $\langle\sigma_N\rangle$. This is simply because if the spins "decorrelate" as we expect, and if in the limit any fixed number of the averages $\langle\sigma_i\rangle_i$ are i.i.d. of law μ, then the right-hand side of (6.47) will in the limit have the same distribution as the quantity (6.51).

Theorem 6.3.1. *Assume that*

$$4\gamma p \mathsf{E}(S \exp 2S) \leq 1\,. \tag{6.52}$$

Then there exists a unique probability measure μ on $[-1,1]$ such that

$$\mu = T(\mu)\,.$$

The proof will consist of showing that T is a contraction for the Monge-Kantorovich transportation-cost distance d defined in (A.66) on the set of probability measures on $[-1,1]$ provided with the usual distance. In the present case, this distance is simply given by the formula

$$d(\mu_1,\mu_2) = \inf \mathsf{E}|X - Y|\,,$$

where the infimum is taken over all pairs of r.v.s (X,Y) such that the law of X is μ_1 and the law of Y is μ_2. The very definition of d shows that to bound $d(\mu_1,\mu_2)$ there is no other method than to produce a pair (X,Y) as above such that $\mathsf{E}|X-Y|$ is appropriately small. Such a pair will informally be called a coupling of the r.v.s X and Y.

Lemma 6.3.2. *For a function f on $\{-1,1\}^{\mathbb{N}}$, we have*

$$\frac{\partial}{\partial x_i}\langle f\rangle_{\mathbf{x}} = \langle \Delta_i f\rangle_{\mathbf{x}} \tag{6.53}$$

where $\Delta_i f(\boldsymbol{\eta}) = (f(\boldsymbol{\eta}_i^+) - f(\boldsymbol{\eta}_i^-))/2$, and where $\boldsymbol{\eta}_i^+$ (resp. $\boldsymbol{\eta}_i^-$) is obtained by replacing the i-th coordinate of $\boldsymbol{\eta}$ by 1 (resp. -1).

Proof. The measure λ_x on $\{-1,1\}$ such that $\int \eta\, \mathrm{d}\lambda_x(\eta) = x$ gives mass $(1+x)/2$ to 1 and mass $(1-x)/2$ to -1, so that for a function f on $\{-1,1\}$ we have

$$\langle f \rangle_x = \int f(\eta)\, \mathrm{d}\lambda_x(\eta) = \frac{1}{2}(f(1) + f(-1)) + \frac{x}{2}(f(1) - f(-1))\,.$$

Thus, using in the second inequality the trivial fact that $a = \langle a \rangle_x$ for any number a implies

$$\frac{\mathrm{d}}{\mathrm{d}x}\langle f \rangle_x = \frac{1}{2}(f(1) - f(-1)) = \left\langle \frac{1}{2}(f(1) - f(-1)) \right\rangle_x\,. \tag{6.54}$$

Since $\lambda_{\mathbf{x}}$ is a product measure, using (6.54) given all the coordinates different from i, and then Fubini's theorem, we obtain (6.53). □

Lemma 6.3.3. *If $\mathcal{E}_r$ is as in (6.49), if $1 \le j \le r$ and if $(j-1)(p-1) < i \le j(p-1)$, then*

$$\left| \frac{\partial}{\partial x_i} \frac{\langle \mathrm{Av}\, \varepsilon \mathcal{E}_r \rangle_{\mathbf{x}}}{\langle \mathrm{Av}\, \mathcal{E}_r \rangle_{\mathbf{x}}} \right| \le 2S_j \exp 2S_j$$

where $S_j = \sup |\theta_j|$. For the other values of i the left-hand side of the previous inequality is 0.

Proof. Lemma 6.3.2 implies:

$$\frac{\partial}{\partial x_i} \frac{\langle \mathrm{Av}\, \varepsilon \mathcal{E}_r \rangle_{\mathbf{x}}}{\langle \mathrm{Av}\, \mathcal{E}_r \rangle_{\mathbf{x}}} = \frac{\langle \Delta_i(\mathrm{Av}\, \varepsilon \mathcal{E}_r) \rangle_{\mathbf{x}}}{\langle \mathrm{Av}\, \mathcal{E}_r \rangle_{\mathbf{x}}} - \frac{\langle \mathrm{Av}\, \varepsilon \mathcal{E}_r \rangle_{\mathbf{x}} \langle \Delta_i \mathrm{Av}\, \mathcal{E}_r \rangle_{\mathbf{x}}}{\langle \mathrm{Av}\, \mathcal{E}_r \rangle_{\mathbf{x}}^2}\,. \tag{6.55}$$

Now

$$|\Delta_i(\mathrm{Av}\, \varepsilon \mathcal{E}_r)| = |\mathrm{Av}\,(\varepsilon \Delta_i \mathcal{E}_r)| \le \mathrm{Av}\, |\Delta_i \mathcal{E}_r|\,.$$

We write $\mathcal{E}_r = \mathcal{E}'\mathcal{E}''$, where $\mathcal{E}' = \exp \theta_j(\sigma_{(j-1)(p-1)+1}, \dots, \sigma_{j(p-1)}, \varepsilon)$, and where $\mathcal{E}''$ does not depend on σ_i. Thus, using that $|e^x - e^y| \le |x-y|e^a \le 2ae^a$ for $|x|, |y| \le a$, we get (keeping in mind the factor 1/2 in the definition of Δ_i, that offsets the factor 2 above) that $\Delta_i \mathcal{E}' \le S_j \exp S_j$, and since $\mathcal{E}'' \le \mathcal{E}_r \exp S_j$ we get

$$|\Delta_i \mathcal{E}_r| = |\mathcal{E}'' \Delta_i \mathcal{E}'| \le (S_j \exp S_j)\mathcal{E}'' \le (S_j \exp 2S_j)\mathcal{E}_r$$

and thus

$$\left| \frac{\langle \Delta_i(\mathrm{Av}\, \varepsilon \mathcal{E}_r) \rangle_{\mathbf{x}}}{\langle \mathrm{Av}\, \mathcal{E}_r \rangle_{\mathbf{x}}} \right| \le S_j \exp 2S_j\,.$$

The last term of (6.55) is bounded similarly. □

Proof of Theorem 6.3.1. This is a fixed point argument. It suffices to prove that under (6.52), for any two probability measures μ_1 and μ_2 on $[-1,1]$, we have

$$d(T(\mu_1), T(\mu_2)) \le \frac{1}{2} d(\mu_1, \mu_2)\,. \tag{6.56}$$

First, Lemma 6.3.3 yields that given $\mathbf{x}, \mathbf{y} \in [-1,1]^{\mathbb{N}}$ it holds:

$$\left|\frac{\langle \mathrm{Av}\, \varepsilon \mathcal{E}_r \rangle_{\mathbf{x}}}{\langle \mathrm{Av}\, \mathcal{E}_r \rangle_{\mathbf{x}}} - \frac{\langle \mathrm{Av}\, \varepsilon \mathcal{E}_r \rangle_{\mathbf{y}}}{\langle \mathrm{Av}\, \mathcal{E}_r \rangle_{\mathbf{y}}}\right| \le 2 \sum_{j \le r} S_j \exp 2S_j \sum_{(j-1)(p-1) < i \le j(p-1)} |x_i - y_i| \,. \tag{6.57}$$

Consider a pair (X, Y) of r.v.s and independent copies $(X_i, Y_i)_{i \ge 1}$ of this pair. Let $\mathbf{X} = (X_i)_{i \ge 1}$, $\mathbf{Y} = (Y_i)_{i \ge 1}$, so that from (6.57) we have

$$\left|\frac{\langle \mathrm{Av}\, \varepsilon \mathcal{E}_r \rangle_{\mathbf{X}}}{\langle \mathrm{Av}\, \mathcal{E}_r \rangle_{\mathbf{X}}} - \frac{\langle \mathrm{Av}\, \varepsilon \mathcal{E}_r \rangle_{\mathbf{Y}}}{\langle \mathrm{Av}\, \mathcal{E}_r \rangle_{\mathbf{Y}}}\right| \le 2 \sum_{j \le r} S_j \exp 2S_j \sum_{(j-1)(p-1) < i \le j(p-1)} |X_i - Y_i| \,. \tag{6.58}$$

Let us assume that the randomness of the pairs (X_i, Y_i) is independent of the other sources of randomness in (6.58). Taking expectations in (6.58) we get

$$\mathsf{E}\left|\frac{\langle \mathrm{Av}\, \varepsilon \mathcal{E}_r \rangle_{\mathbf{X}}}{\langle \mathrm{Av}\, \mathcal{E}_r \rangle_{\mathbf{X}}} - \frac{\langle \mathrm{Av}\, \varepsilon \mathcal{E}_r \rangle_{\mathbf{Y}}}{\langle \mathrm{Av}\, \mathcal{E}_r \rangle_{\mathbf{Y}}}\right| \le 2\gamma(p-1)(\mathsf{E} S \exp 2S)\mathsf{E}|X - Y| \,. \tag{6.59}$$

If X and Y have laws μ_1 and μ_2 respectively, then

$$\frac{\langle \mathrm{Av}\, \varepsilon \mathcal{E}_r \rangle_{\mathbf{X}}}{\langle \mathrm{Av}\, \mathcal{E}_r \rangle_{\mathbf{X}}} \quad \text{and} \quad \frac{\langle \mathrm{Av}\, \varepsilon \mathcal{E}_r \rangle_{\mathbf{Y}}}{\langle \mathrm{Av}\, \mathcal{E}_r \rangle_{\mathbf{Y}}}$$

have laws $T(\mu_1)$ and $T(\mu_2)$ respectively, so that (6.59) implies

$$d(T(\mu_1), T(\mu_2)) \le 2\gamma(p-1)(\mathsf{E} S \exp 2S)\mathsf{E}|X - Y| \,.$$

Taking the infimum over all possible choices of X and Y yields

$$d(T(\mu_1), T(\mu_2)) \le 2\gamma(p-1) d(\mu_1, \mu_2) \mathsf{E} S \exp 2S \,,$$

so that (6.52) implies (6.56). □

Let us denote by T_γ the operator T when we want to insist on the dependence on γ. The unique solution of the equation $\mu = T_\gamma(\mu)$ depends on γ, and we denote it by μ_γ when we want to emphasize this dependence.

Lemma 6.3.4. *If γ and γ' satisfy (6.52) we have*

$$d(\mu_\gamma, \mu_{\gamma'}) \le 4|\gamma - \gamma'| \,.$$

Proof. Without loss of generality we can assume that $\gamma \le \gamma'$. Since $\mu_\gamma = T_\gamma(\mu_\gamma)$ and $\mu_{\gamma'} = T_{\gamma'}(\mu_{\gamma'})$, we have

$$\begin{aligned} d(\mu_\gamma, \mu_{\gamma'}) &\le d(T_\gamma(\mu_\gamma), T_\gamma(\mu_{\gamma'})) + d(T_\gamma(\mu_{\gamma'}), T_{\gamma'}(\mu_{\gamma'})) \\ &\le \frac{1}{2} d(\mu_\gamma, \mu_{\gamma'}) + d(T_\gamma(\mu_{\gamma'}), T_{\gamma'}(\mu_{\gamma'})) \,, \end{aligned} \tag{6.60}$$

using (6.56). To compare $T_\gamma(\mu)$ and $T_{\gamma'}(\mu)$ the basic idea is that there is natural coupling between a Poisson r.v. of expectation γ and another Poisson r.v. of expectation γ' (an idea that will be used again in the next section).

Namely if r'' is a Poisson r.v. with $\mathsf{E}r'' = \gamma'' := \gamma' - \gamma$, and r'' is independent of the Poisson r.v. r such that $\mathsf{E}r = \gamma$ then $r + r''$ is a Poisson r.v. of expectation γ'. Consider $\mathcal{E}_r$ as in (6.49) and, with the same notation,

$$\mathcal{E}' = \exp \sum_{r<j\leq r+r''} \theta_j(\sigma_{(j-1)(p-1)+1}, \ldots, \sigma_{j(p-1)}, \varepsilon) ,$$

so that $\mathcal{E}_r \mathcal{E}' = \mathcal{E}_{r+r''}$. Consider an i.i.d. sequence $\mathbf{X} = (X_i)_{i\geq 1}$ of common law μ. Then the r.v.s

$$\frac{\langle \mathrm{Av}\, \varepsilon \mathcal{E}_r \rangle_{\mathbf{X}}}{\langle \mathrm{Av}\, \mathcal{E}_r \rangle_{\mathbf{X}}} \quad \text{and} \quad \frac{\langle \mathrm{Av}\, \varepsilon \mathcal{E}_r \mathcal{E}' \rangle_{\mathbf{X}}}{\langle \mathrm{Av}\, \mathcal{E}_r \mathcal{E}' \rangle_{\mathbf{X}}}$$

have respectively laws $T_\gamma(\mu)$ and $T_{\gamma'}(\mu)$. Thus

$$\begin{aligned} d(T_\gamma(\mu), T_{\gamma'}(\mu)) &\leq \mathsf{E} \left| \frac{\langle \mathrm{Av}\, \varepsilon \mathcal{E}_r \rangle_{\mathbf{X}}}{\langle \mathrm{Av}\, \mathcal{E}_r \rangle_{\mathbf{X}}} - \frac{\langle \mathrm{Av}\, \varepsilon \mathcal{E}_r \mathcal{E}' \rangle_{\mathbf{X}}}{\langle \mathrm{Av}\, \mathcal{E}_r \mathcal{E}' \rangle_{\mathbf{X}}} \right| \qquad (6.61) \\ &\leq 2\mathsf{P}(r'' \neq 0) = 2(1 - e^{-(\gamma' - \gamma)}) \leq 2(\gamma' - \gamma) , \end{aligned}$$

so that (6.60) implies that $d(\mu_\gamma, \mu_{\gamma'}) \leq d(\mu_\gamma, \mu_{\gamma'})/2 + 2(\gamma' - \gamma)$, hence the desired result. □

Exercise 6.3.5. Consider three functions U, V, W on Σ_N^n. Assume that $V \geq 0$, that for a certain number Q, we have $|U| \leq QV$, and let $S^* = \sup_{\boldsymbol{\sigma}^1, \ldots, \boldsymbol{\sigma}^n} |W|$. Prove that for any Gibbs measure $\langle \cdot \rangle$ we have

$$\left| \frac{\langle U \exp W \rangle}{\langle V \exp W \rangle} - \frac{\langle U \rangle}{\langle V \rangle} \right| \leq 2QS^* \exp 2S^* .$$

Exercise 6.3.6. Use the idea of Exercise 6.3.5 to control the influence of $\mathcal{E}'$ in (6.61) and to show that if γ and γ' satisfy (6.52) then $d(\mu_\gamma, \mu_{\gamma'}) \leq 4|\gamma - \gamma'|\mathsf{E}S \exp 2S$.

6.4 The Replica-Symmetric Solution

In this section we will first prove that asymptotically as $N \to \infty$ any fixed number of the quantities $\langle \sigma_i \rangle$ are i.i.d. of law μ_γ, where μ_γ was defined in the last section. We will then compute the quantity $\lim_{N\to\infty} p_N(\gamma) = \lim_{N\to\infty} N^{-1} \mathsf{E} \log Z_N(\gamma)$.

Theorem 6.4.1. *Assume that*

$$16p\gamma_0 \mathsf{E}S \exp 4S \leq 1 . \qquad (6.62)$$

Then there exists a number $K_2(p, \gamma_0)$ such that if we define for $n \geq 0$ the numbers $A(n)$ as follows:

$$A(0) = K_2(p, \gamma_0)\mathsf{E}\exp 2S \;, \tag{6.63}$$

$$A(n+1) = A(0) + \big(40p^3(\gamma_0 + \gamma_0^3)\mathsf{E}S\exp 2S\big)A(n) \;, \tag{6.64}$$

then the following holds. If $\gamma \le \gamma_0$, given any integers $k \le N$ and n we can find i.i.d. r.v.s $z_1, \ldots, z_k$ of law μ_γ such that

$$\mathsf{E}\sum_{i\le k} |\langle\sigma_i\rangle - z_i| \le 2^{1-n}k + \frac{k^3 A(n)}{N} \;. \tag{6.65}$$

In particular when

$$80p^3(\gamma_0 + \gamma_0^3)\mathsf{E}S\exp 2S \le 1 \;, \tag{6.66}$$

we can replace (6.65) by

$$\mathsf{E}\sum_{i\le k} |\langle\sigma_i\rangle - z_i| \le \frac{2k^3 K_2(\gamma_0, p)}{N}\mathsf{E}\exp 2S \;. \tag{6.67}$$

The last statement of the Theorem simply follows from the fact that under (6.66) we have $A(n) \le 2A(0)$, so that we can take n very large in (6.90). When (6.66) need not hold, optimisation over n in (6.65) yields a bound $\le KkN^{-\alpha}$ for some $\alpha > 0$ depending only on γ_0, p and S.

The next problem need not be difficult. This issue came at the very time where the book was ready to be sent to the publisher, and it did not seem appropriate to either delay the publication or to try to make significant changes in a rush.

Research Problem 6.4.2. (level 1-) Is it true that (6.67) follows from (6.62)? More specifically, when $\gamma_0 \gg 1$, and when S is constant, does (6.67) follow from a condition of the type $K(p)\gamma_0 S \le 1$?

Probably the solution of this problem will not require essentially new ideas. Rather, it should require technical work and improvement of the estimates from Lemma 6.4.3 to Lemma 6.4.7, trying in particular to bring out more "small factors" such as $\mathsf{E}S\exp 2S$, in the spirit of Exercice 6.3.6. It seems however that it will also be necessary to proceed to a finer study of what happens on the set Ω defined page 349.

It follows from Theorem 6.2.2 that we can assume throughout the proof that property $C(\gamma_0, N, K_0, K_0)$ holds for every N. It will be useful to consider the metric space $[-1,1]^k$, provided with the distance d given by

$$d((x_i)_{i\le k}, (y_i)_{i\le k}) = \sum_{i\le k} |x_i - y_i| \;. \tag{6.68}$$

The Monge-Kantorovich transportation-cost distance on the space of probability measures on $[-1,1]^k$ that is induced by (6.68) will also be denoted by d. We define

$$D(N,k,\gamma_0)=\sup_{\gamma\le\gamma_0} d\big(\mathcal{L}(\langle\sigma_1\rangle,\ldots,\langle\sigma_k\rangle),\mu_\gamma^{\otimes k}\big) \tag{6.69}$$

where $\mathcal{L}(X_1,\ldots,X_k)$ denotes the law of the random vector $(X_1,\ldots,X_k)$.

By definition of the transportation-cost distance in the right-hand side of (6.69), the content of Theorem 6.4.1 is that if γ_0 satisfies (6.62) we have $D(N,k,\gamma_0)\le 2^{1-n}k+k^3A(n)/N$ for each $k\le N$ and each n. This inequality will be proved by obtaining a suitable induction relation between the quantities $D(N,k,\gamma_0)$. The overall idea of the proof is to use the cavity method to express $\langle\sigma_1\rangle,\ldots,\langle\sigma_k\rangle$ as functions of a smaller spin system, and to use Proposition 6.2.7 and the induction hypothesis to perform estimates on the smaller spin system.

We start by a simple observation. Since $\sum_{i\le k}|x_i-y_i|\le 2k$ for $x_i,y_i\in[-1,1]$, we have $D(N,k,\gamma_0)\le 2k$. Assuming, as we may, that $K_2(p,\gamma_0)\ge 4p$, we see that there is nothing to prove unless $N\ge 2pk^2$ so in particular $N\ge p+k$ and $N\ge 2k$. We will always assume below that this is the case. We also observe that, by symmetry,

$$\mathcal{L}(\langle\sigma_1\rangle,\ldots,\langle\sigma_k\rangle)=\mathcal{L}(\langle\sigma_{N-k+1}\rangle,\ldots,\langle\sigma_N\rangle)\,.$$

The starting point of the proof of Theorem 6.4.1 is a formula similar to (6.20), but where we remove the last k coordinates rather than the last one. Writing now $\boldsymbol{\rho}=(\sigma_1,\ldots,\sigma_{N-k})$, we consider the Hamiltonian

$$-H_{N-k}(\boldsymbol{\rho})=\sum_s \theta_s(\sigma_{i(s,1)},\ldots,\sigma_{i(s,p)})\,, \tag{6.70}$$

where the summation is restricted to those $s\le M$ for which $i(s,p)\le N-k$. This is the Hamiltonian of an $(N-k)$-spin system, except that we have replaced γ by a different value γ_-. To compute γ_- we observe that since the set $\{i(s,1),\ldots,i(s,p)\}$ is uniformly distributed among the subsets of $\{1,\ldots,N\}$ of cardinality p, the probability that $i(s,p)\le N-k$, i.e. the probability that this set is a subset of $\{1,\ldots,N-k\}$ is exactly

$$\tau=\frac{\binom{N-k}{p}}{\binom{N}{p}}\,,$$

so that the mean number of terms of this Hamiltonian is $N\alpha\tau$, and

$$\gamma_-(N-k)=pN\alpha\tau=\gamma N\tau\,,$$

and thus

$$\gamma_-=\gamma\frac{(N-k-1)\cdots(N-k-p+1)}{(N-1)\cdots(N-p+1)}\,. \tag{6.71}$$

In particular $\gamma_-\le\gamma_0$ whenever $\gamma\le\gamma_0$. Let us denote again by $\langle\cdot\rangle_-$ an average for the Gibbs measure with Hamiltonian (6.70). (The value of k will be clear from the context.) Given a function f on Σ_N, we then have

$$\langle f\rangle = \frac{\langle \mathrm{Av}\, f\mathcal{E}\rangle_-}{\langle \mathrm{Av}\,\mathcal{E}\rangle_-}\ , \tag{6.72}$$

where Av means average over $\sigma_{N-k+1},\ldots,\sigma_N = \pm 1$, and where

$$\mathcal{E} = \exp\sum \theta_s(\sigma_{i(s,1)},\ldots,\sigma_{i(s,p)})\ ,$$

for a sum over those values of $s \le M$ for which $i(s,p) \ge N-k+1$. As before, in distribution,

$$\mathcal{E} = \exp\sum_{j\le r} \theta_j(\sigma_{i(j,1)},\ldots,\sigma_{i(j,p)})\ , \tag{6.73}$$

where now the sets $\{i(j,1),\ldots,i(j,p)\}$ are uniformly distributed over the subsets of $\{1,\ldots,N\}$ of cardinality p that intersect $\{N-k+1,\ldots,N\}$, and where r is a Poisson r.v. The expected value of r is the mean number of terms in the Hamiltonian $-H_N$ that are not included in the summation (6.70), so that

$$\mathsf{E}r = \alpha N\left(1 - \frac{\binom{N-k}{p}}{\binom{N}{p}}\right) = \frac{\gamma N}{p}\left(1 - \frac{(N-k)\cdots(N-k-p+1)}{N\cdots(N-p+1)}\right)\ . \tag{6.74}$$

The quantity r will keep this meaning until the end of the proof of Theorem 6.4.1, and the quantity $\mathcal{E}$ will keep the meaning of (6.73). It is good to note that, since $N \ge 2kp$, for $\ell \le p$ we have

$$\frac{N-k-\ell}{N-\ell} = 1 - \frac{k}{N-\ell} \ge 1 - \frac{2k}{N}\ .$$

Therefore

$$\frac{(N-k)\cdots(N-k-p-1)}{N\cdots(N-p+1)} \ge \left(1-\frac{2k}{N}\right)^p \ge 1 - \frac{2kp}{N}\ , \tag{6.75}$$

and thus

$$\mathsf{E}r \le 2k\gamma\ . \tag{6.76}$$

We observe the identity

$$\mathcal{L}(\langle\sigma_{N-k+1}\rangle,\ldots,\langle\sigma_N\rangle) = \mathcal{L}\left(\frac{\langle \mathrm{Av}\,\sigma_{N-k+1}\mathcal{E}\rangle_-}{\langle \mathrm{Av}\,\mathcal{E}\rangle_-},\ldots,\frac{\langle \mathrm{Av}\,\sigma_N\mathcal{E}\rangle_-}{\langle \mathrm{Av}\,\mathcal{E}\rangle_-}\right)\ . \tag{6.77}$$

The task is now to use the induction hypothesis to approximate the right-hand side of (6.77); this will yield the desired induction relation. There are three sources of randomness on the right-hand of (6.77). There is the randomness associated with the $(N-k)$-spin system of Hamiltonian (6.70); the randomness associated to r and the sets $\{i(j,1),\ldots,i(j,p)\}$; and the randomness associated to the functions θ_s, $s \le r$. These three sources of randomness are independent of each other.

To use the induction hypothesis, it will be desirable that for $j \le r$ the sets

$$I_j = \{i(j,1), \ldots, i(j,p-1)\} \tag{6.78}$$

are disjoint subsets of $\{1, \ldots, N-k\}$, so we first control the size of the rare event Ω where this is not the case. We have $\Omega = \Omega_1 \cup \Omega_2$, where

$$\Omega_1 = \Big\{\exists j \le r,\ i(j,p-2) \ge N-k+1\Big\}$$

$$\Omega_2 = \Big\{\exists j, j' \le r,\ j \ne j'\ ,\ \exists \ell, \ell' \le p-1\ , i(j,\ell) = i(j',\ell')\Big\}\ .$$

Proceeding as in the proof of (6.25) we easily reach the crude bound

$$\mathsf{P}(\Omega) \le \frac{4k^2}{N}(\gamma p + \gamma^2 p^2)\ . \tag{6.79}$$

We recall that, as defined page 341, given a sequence $\mathbf{x} = (x_1, \ldots, x_{N-k})$ with $|x_i| \le 1$ and a function f on Σ_{N-k}, we denote by $\langle f \rangle_{\mathbf{x}}$ the average of f with respect to the product measure $\lambda_{\mathbf{x}}$ on Σ_{N-k} such that $\int \sigma_i \, \mathrm{d}\lambda_{\mathbf{x}}(\boldsymbol{\rho}) = x_i$ for $1 \le i \le N-k$.

We now start a sequence of lemmas that aim at deducing from (6.77) the desired induction relations among the quantities $D(N,k,\gamma_0)$. There will be four steps in the proof. In the first step below, in each of the brackets in the right-hand side of (6.77) we replace the Gibbs measure $\langle\cdot\rangle_-$ by $\langle\cdot\rangle_{\mathbf{Y}}$ where $\mathbf{Y} = (\langle\sigma_1\rangle_-, \ldots, \langle\sigma_{N-k}\rangle_-)$. The basic reason why this creates only a small error is that $C(N,\gamma_0,K_0,K_0)$ holds true for each N, a property which is used as in Proposition 6.2.7.

Lemma 6.4.3. *Consider the sequence*

$$\mathbf{Y} = (\langle\sigma_1\rangle_-, \ldots, \langle\sigma_{N-k}\rangle_-)\ .$$

Set

$$u_\ell = \langle\sigma_{N-k+\ell}\rangle = \frac{\langle \mathrm{Av}\, \sigma_{N-k+\ell}\mathcal{E}\rangle_-}{\langle \mathrm{Av}\, \mathcal{E}\rangle_-}\ ;\ v_\ell = \frac{\langle \mathrm{Av}\, \sigma_{N-k+\ell}\mathcal{E}\rangle_{\mathbf{Y}}}{\langle \mathrm{Av}\, \mathcal{E}\rangle_{\mathbf{Y}}}\ .$$

Then we have

$$d(\mathcal{L}(u_1, \ldots, u_k), \mathcal{L}(v_1, \ldots, v_k)) \le \frac{k^3}{N} K(p,\gamma_0)\mathsf{E}\exp 2S\ . \tag{6.80}$$

Proof. From now on E_- denotes expectation in the randomness of the $N-k$ spin system only. When Ω does not occur, there is nothing to change to the proof of Proposition 6.2.7 to obtain that

$$\mathsf{E}_-|u_\ell - v_\ell| \le \frac{8r(p-1)^2 K_0}{N-k} \sum_{j \le r} \exp 2S_j\ ,$$

where we recall that r denotes the number of terms in the summation in (6.73), and is a Poisson r.v. which satisfies $\mathsf{E}r \le 2k\gamma$. We always have $\mathsf{E}_-|u_\ell - v_\ell| \le 2$, so that

$$\mathsf{E}_-|u_\ell - v_\ell| \le \frac{8r(p-1)^2 K_0}{N-k} \sum_{j \le r} \exp 2S_j + 2 1_\Omega \,. \tag{6.81}$$

Taking expectation in (6.81) then yields

$$\begin{aligned}\mathsf{E}|u_\ell - v_\ell| &\le \frac{8(p-1)^2 K_0}{N-k} \mathsf{E} \exp 2S \, \mathsf{E}r^2 + 2\mathsf{P}(\Omega) \\ &\le \frac{k^2 K(p,\gamma_0)}{N} \mathsf{E} \exp 2S \,,\end{aligned}$$

using (6.79), that $N - k \ge N/2$ and that $\mathsf{E}r^2 = \mathsf{E}r + (\mathsf{E}r)^2 \le 2\gamma k + 4\gamma^2 k^2$. Since the left-hand side of (6.80) is bounded by $\sum_{\ell \le k} \mathsf{E}|u_\ell - v_\ell|$, the result follows. □

In the second step, we replace the sequence $\mathbf{Y}$ by an appropriate i.i.d. sequence of law μ_{γ_-}. The basic reason this creates only a small error is the "induction hypothesis" i.e. the control of the quantities $D(N-k, m, \gamma_0)$.

Proposition 6.4.4. *Consider an independent sequence* $\mathbf{X} = (X_1, \ldots, X_{N-k})$ *where each* X_i *has law* $\mu_- := \mu_{\gamma_-}$. *We set*

$$w_\ell = \frac{\langle \mathrm{Av}\, \sigma_{N-k+\ell} \mathcal{E} \rangle_{\mathbf{X}}}{\langle \mathrm{Av} \mathcal{E} \rangle_{\mathbf{X}}} \,, \tag{6.82}$$

and we recall the quantities v_ℓ *of the previous lemma. Then we have*

$$\begin{aligned} d(\mathcal{L}(v_1, \ldots, v_k), \mathcal{L}(w_1, \ldots, w_k)) \le \frac{k^3}{N} K(p, \gamma_0) \\ + 4\mathsf{E}S \exp 2S \mathsf{E} D(N-k, r(p-1), \gamma_0) \,, \end{aligned} \tag{6.83}$$

where the last expectation is taken with respect to the Poisson r.v. r.

The proof will rely on the following lemma.

Lemma 6.4.5. *Assume that* Ω *does not occur. Consider* $\ell \le k$ *and*

$$\mathcal{E}_\ell = \exp \sum \theta_j(\sigma_{i(j,1)}, \ldots, \sigma_{i(j,p-1)}, \sigma_{N-k+\ell}) \,, \tag{6.84}$$

where the summation is over those $j \le r$ *for which* $i(j,p) = N - k + \ell$. *Then for any sequence* $\mathbf{x}$ *we have*

$$\frac{\langle \mathrm{Av}\, \sigma_{N-k+\ell} \mathcal{E} \rangle_{\mathbf{x}}}{\langle \mathrm{Av}\, \mathcal{E} \rangle_{\mathbf{x}}} = \frac{\langle \mathrm{Av}\, \sigma_{N-k+\ell} \mathcal{E}_\ell \rangle_{\mathbf{x}}}{\langle \mathrm{Av}\, \mathcal{E}_\ell \rangle_{\mathbf{x}}} \,. \tag{6.85}$$

Consequently

$$\frac{\partial}{\partial x_i} \frac{\langle \mathrm{Av}\, \sigma_{N-k+\ell}\mathcal{E}\rangle_{\mathbf{x}}}{\langle \mathrm{Av}\, \mathcal{E}\rangle_{\mathbf{x}}} = 0 \tag{6.86}$$

unless $i \in I_j$ *for some* j *with* $i(j,p) = N-k+\ell$. *In that case we have moreover*

$$\left| \frac{\partial}{\partial x_i} \frac{\langle \mathrm{Av}\, \sigma_{N-k+\ell}\mathcal{E}\rangle_{\mathbf{x}}}{\langle \mathrm{Av}\, \mathcal{E}\rangle_{\mathbf{x}}} \right| \le 4S_j \exp 2S_j\, . \tag{6.87}$$

Proof. Define $\mathcal{E}'_\ell$ by $\mathcal{E} = \mathcal{E}_\ell \mathcal{E}'_\ell$. Since Ω does not occur, the quantities $\sigma_{N-k+\ell}\mathcal{E}_\ell$ and $\mathcal{E}_{\ell'}$ depend on disjoint sets of coordinates. Consequently

$$\mathrm{Av}\, \sigma_{N-k+\ell}\mathcal{E} = (\mathrm{Av}\, \sigma_{N-k+\ell}\mathcal{E}_\ell)(\mathrm{Av}\, \mathcal{E}'_\ell) \tag{6.88}$$

$$\mathrm{Av}\, \mathcal{E} = (\mathrm{Av}\, \mathcal{E}_\ell)(\mathrm{Av}\, \mathcal{E}'_\ell)\, . \tag{6.89}$$

In both (6.88) and (6.89) the two factors on the right depend on disjoint sets of coordinates. Since $\langle\cdot\rangle_{\mathbf{x}}$ is a product measure, we get

$$\langle \mathrm{Av}\, \sigma_{N-k+\ell}\mathcal{E}\rangle_{\mathbf{x}} = \langle \mathrm{Av}\, \sigma_{N-k+\ell}\mathcal{E}_\ell\rangle_{\mathbf{x}} \langle \mathrm{Av}\, \mathcal{E}'_\ell\rangle_{\mathbf{x}}$$

and similarly with (6.89), so that (6.85) follows, of which (6.86) is an obvious consequence. As for (6.87), it is proved exactly as in Lemma 6.3.3. □

Proof of Proposition 6.4.4. The strategy is to construct a specific realization of $\mathbf{X}$ for which the quantity $\mathsf{E}\sum_{\ell\le N-k}|v_\ell - w_\ell|$ is small. Consider the set $J = \bigcup_{j\le r} I_j$ (so that $\mathrm{card} J \le (p-1)r$). The construction takes place given the set J. By definition of $D(N-k, r(p-1), \gamma_0)$, given J we can construct an i.i.d. sequence $(X_i)_{i\le N-k}$ distributed like μ_- that satisfies

$$\mathsf{E}_- \sum_{i\in J} |X_i - \langle\sigma_i\rangle_-| \le 2D(N-k, r(p-1), \gamma_0)\, . \tag{6.90}$$

We can moreover assume that the sequence $(\theta_j)_{j\ge 1}$ is independent of the randomness generated by J and the variables X_i. The sequence $(X_i)_{i\le N-k}$ is our specific realization. It is i.i.d. distributed like μ_-.

It follows from Lemma 6.4.5 that if Ω does not occur,

$$\begin{aligned} |w_\ell - v_\ell| &= \left| \frac{\langle \mathrm{Av}\, \sigma_{N-k+\ell}\mathcal{E}\rangle_{\mathbf{X}}}{\langle \mathrm{Av}\, \mathcal{E}\rangle_{\mathbf{X}}} - \frac{\langle \mathrm{Av}\, \sigma_{N-k+\ell}\, \mathcal{E}\rangle_{\mathbf{Y}}}{\langle \mathrm{Av}\, \mathcal{E}\rangle_{\mathbf{Y}}} \right| \\ &\le \sum \Big(\sum_{i\in I_j} |X_i - \langle\sigma_i\rangle_-| \Big) 2S_j \exp 2S_j\, , \end{aligned}$$

where the first sum is over those $j \le r$ for which $i(j,p) = N-k+\ell$. By summation over $\ell \le k$, we get that when Ω does not occur,

$$\sum_{\ell\le k} |w_\ell - v_\ell| \le 2\sum_{i\in J} |X_i - \langle\sigma_i\rangle_-| S_{j(i)} \exp 2S_{j(i)}\, ,$$

where $j(i)$ is the unique $j \le r$ with $i \in I_j$. Denoting by E_θ expectation in the r.v.s $(\theta_j)_{j\ge 1}$ and using independence we get

$$\mathsf{E}_\theta \sum_{\ell\le k} |w_\ell - v_\ell| \le 2\sum_{i\in J} |X_i - \langle\sigma_i\rangle_-| \mathsf{E}S\exp 2S\ .$$

Taking expectation E_- and using (6.90) implies that when Ω does not occur,

$$\mathsf{E}_\theta\mathsf{E}_- \sum_{\ell\le k} |w_\ell - v_\ell| \le 4(\mathsf{E}S\exp 2S)D(N-k, r(p-1), \gamma_0)\ ,$$

i.e.

$$\mathbf{1}_{\Omega^c}\mathsf{E}_\theta\mathsf{E}_- \sum_{\ell\le k} |w_\ell - v_\ell| \le 4(\mathsf{E}S\exp 2S)D(N-k, r(p-1), \gamma_0)\ . \tag{6.91}$$

On the other hand, on Ω we have trivially

$$\mathsf{E}_\theta\mathsf{E}_- \sum_{\ell\le k} |w_\ell - v_\ell| \le 2k\ ,$$

and combining with (6.91) we see that

$$\mathsf{E}_\theta\mathsf{E}_- \sum_{\ell\le k} |w_\ell - v_\ell| \le 4(\mathsf{E}S\exp 2S)D(N-k, r(p-1), \gamma_0) + 2k\mathbf{1}_\Omega\ .$$

Taking expectation and using (6.79) again yields

$$\mathsf{E}\sum_{\ell\le k} |w_\ell - v_\ell| \le \frac{k^3K(p,\gamma_0)}{N} + 4(\mathsf{E}S\exp 2S)\mathsf{E}D(N-k, r(p-1), \gamma_0)\ ,$$

and this implies (6.83). □

Now comes the key step: by definition of the operator T of (6.51) the r.v.s w_ℓ of (6.82) are nearly independent with law $T(\mu_-)$.

Proposition 6.4.6. *We have*

$$d(\mathcal{L}(w_1,\ldots,w_k), T(\mu_-)\otimes\cdots\otimes T(\mu_-)) \le \frac{k^2}{N}K(p,\gamma_0)\ . \tag{6.92}$$

Proof. Let us define, for $\ell \le k$

$$r(\ell) = \operatorname{card}\{j \le r;\ i(j,p-1) \le N-k,\ i(j,p) = N-k+\ell\}\ , \tag{6.93}$$

so that when Ω does not occur, $r(\ell)$ is the number of terms in the summation of (6.84), and moreover for different values of ℓ, the sets of indices occurring in (6.84) are disjoint. The sequence $(r(\ell))_{\ell\le k}$ is an i.i.d. sequence of Poisson r.v.s. (and their common mean will soon be calculated).

For $\ell \geq 1$ and $j \geq 1$ let us consider independent copies $\theta_{\ell,j}$ of θ and for $m \geq 1$ let us define, for $\boldsymbol{\sigma} \in \mathbb{R}^N$,

$$\mathcal{E}_{\ell,m} = \mathcal{E}_{\ell,m}(\boldsymbol{\sigma}, \varepsilon) = \exp \sum_{1 \leq j \leq m} \theta_{\ell,j}(\sigma_{(j-1)(p-1)+1}, \ldots, \sigma_{j(p-1)}, \varepsilon) ,$$

a formula that should be compared to (6.49).

For $\ell \leq k$, let us consider sequences $\mathbf{X}_\ell = (X_{i,\ell})_{i \geq 1}$, where the r.v.s $X_{i,\ell}$ are all independent of law μ_-. Let us define $w'_\ell = w_\ell$ when Ω occurs, and otherwise

$$w'_\ell = \frac{\langle \mathrm{Av}\, \varepsilon \mathcal{E}_{\ell,r(\ell)} \rangle_{\mathbf{X}_\ell}}{\langle \mathrm{Av}\, \mathcal{E}_{\ell,r(\ell)} \rangle_{\mathbf{X}_\ell}} . \tag{6.94}$$

The basic fact is that the sequences $(w_\ell)_{\ell \leq k}$ and $(w'_\ell)_{\ell \leq k}$ have the same law. This is because they have the same law given the r.v. r and the numbers $i(j,1), \ldots, i(j,p)$ for $j \leq r$. This is obvious when Ω occurs, since then $w'_\ell = w_\ell$. When Ω does not occur we simply observe from (6.85) and the definition of w_ℓ that

$$w_\ell = \frac{\langle \mathrm{Av}\, \sigma_{N-k+\ell} \mathcal{E}_\ell \rangle_{\mathbf{X}}}{\langle \mathrm{Av}\, \mathcal{E}_\ell \rangle_{\mathbf{X}}} .$$

We then compare with (6.94), keeping in mind that there are $r(\ell)$ terms in the summation (6.84), and then using symmetry.

Therefore we have shown that

$$\mathcal{L}(w_1, \ldots, w_k) = \mathcal{L}(w'_1, \ldots, w'_k) . \tag{6.95}$$

Since the sequence $(r(\ell))_{\ell \leq k}$ is an i.i.d. sequence of Poisson r.v.s, the sequence $(w'_\ell)_{\ell \leq k}$ is i.i.d. It has almost law $T(\mu_-)$, but not exactly because the Poisson r.v.s $r(\ell)$ do not have the correct mean. This mean $\gamma' = \mathsf{E}r(\ell)$ is given by

$$\gamma' = \frac{N\gamma}{p} \frac{\binom{N-k}{p-1}}{\binom{N}{p}} = \gamma \frac{(N-k)\cdots(N-k-p+2)}{(N-1)\cdots(N-p+1)} \leq \gamma .$$

To bound the small error created by the difference between γ and γ' we proceed as in the proof of Lemma 6.3.4. We consider independent Poisson r.v.s $(r''(\ell))_{\ell \leq k}$ of mean $\gamma - \gamma'$, so that $s(\ell) = r(\ell) + r''(\ell)$ is an independent sequence of Poisson r.v.s of mean γ. Let

$$w''_\ell = \frac{\langle \mathrm{Av}\, \varepsilon \mathcal{E}_{\ell,s(\ell)} \rangle_{\mathbf{X}_\ell}}{\langle \mathrm{Av}\, \mathcal{E}_{\ell,s(\ell)} \rangle_{\mathbf{X}_\ell}} .$$

The sequence $(w''_\ell)_{\ell \leq k}$ is i.i.d. and the law of w''_ℓ is $T(\mu_-)$. Thus (6.95) implies:

$$\begin{aligned} d(\mathcal{L}(w_1, \ldots, w_k), T(\mu_-) \otimes \cdots \otimes T(\mu_-)) &= d(\mathcal{L}(w'_1, \ldots, w'_k), \mathcal{L}(w''_1, \ldots, w''_k)) \\ &\leq \sum_{\ell \leq k} \mathsf{E}|w'_\ell - w''_\ell| . \end{aligned}$$

Now, since $w''_\ell = w'_\ell$ unless Ω occurs or $s(\ell) \neq r(\ell)$, we have

$$\mathsf{E}|w'_\ell - w''_\ell| \leq 2\big(\mathsf{P}(s(\ell) \neq r(\ell)) + \mathsf{P}(\Omega)\big)$$

and

$$\mathsf{P}(s(\ell) \neq r(\ell)) = \mathsf{P}(r''(\ell) \neq 0) \leq \gamma - \gamma' \,.$$

Moreover from (6.75) we see that $\gamma - \gamma' \leq 2\gamma kp/N$. The result follows. □

The next lemma is the last step. It quantifies the fact that $T(\mu_-)$ is nearly μ.

Lemma 6.4.7. *We have*

$$d(T(\mu_-)^{\otimes k}, \mu^{\otimes k}) \leq \frac{4\gamma k^2 p}{N} \,. \tag{6.96}$$

Proof. The left-hand side is bounded by

$$kd(T(\mu_-), \mu) = kd(T(\mu_-), T(\mu)) \leq \frac{k}{2} d(\mu, \mu_-) \leq 2k(\gamma - \gamma_-) \,,$$

using Lemma 6.3.4. The result follows since by (6.75) we have $\gamma - \gamma_- \leq 2kp\gamma/N$. □

Proof of Theorem 6.4.1. We set $B = 4\mathsf{E}S \exp 2S$. Using the triangle inequality for the transportation-cost distance and the previous estimates, we have shown that for a suitable value of $K_2(\gamma_0, p)$ we have (recalling the definition (6.63) of $A(0)$),

$$d\big(\mathcal{L}(\langle\sigma_{N-k+1}\rangle, \ldots, \langle\sigma_N\rangle), \mu^{\otimes k}\big) \leq \frac{k^3 A(0)}{N} + B\mathsf{E}D(N-k, r(p-1), \gamma_0) \,. \tag{6.97}$$

Given an integer n we say that property $C^*(N, \gamma_0, n)$ holds if

$$\forall p \leq N' \leq N \,,\ \forall k \leq N' \,,\ D(N', k, \gamma_0) \leq 2^{1-n} k + \frac{k^3 A(n)}{N'} \,. \tag{6.98}$$

Since $D(N', k, \gamma_0) \leq 2k$, $C^*(N, \gamma_0, 0)$ holds for each N. And since $A(n) \geq A(0)$, $C^*(p, \gamma_0, n)$ holds as soon as $K_2(\gamma_0, p) \geq 2p$, since then $D(p, k, \gamma_0) \leq 2k \leq k^3 A(0)/p \leq k^3 A(n)/p$. We will prove that

$$C^*(N-1, \gamma_0, n) \Rightarrow C^*(N, \gamma_0, n+1) \,, \tag{6.99}$$

thereby proving that $C^*(N, \gamma_0, n)$ holds for each N and n, which is the content of the theorem.

To prove (6.99), we assume that $C^*(N-1, \gamma_0, n)$ holds and we consider $k \leq N/2$. It follows from (6.98) used for $N' = N - k \leq N - 1$ and $r(p-1)$ instead of k that since $k \leq N/2$ we have

$$D(N-k, r(p-1), \gamma_0) \leq 2^{1-n} rp + \frac{p^3 r^3 A(n)}{N-k} \leq 2^{1-n} rp + \frac{2p^3 r^3 A(n)}{N} \,, \tag{6.100}$$

and going back to (6.97),

$$d(\mathcal{L}(\langle\sigma_{N-k+1}\rangle,\ldots,\langle\sigma_N\rangle),\mu^{\otimes k}) \le 2^{1-n}pB\mathsf{E}r + \frac{k^3A(0)}{N} + \frac{2p^3A(n)}{N}B\mathsf{E}(r^3)\,. \tag{6.101}$$

Since r is a Poisson r.v., (A.64) shows that $\mathsf{E}r^3 = (\mathsf{E}r)^3 + 3(\mathsf{E}r)^2 + \mathsf{E}r$, so that since $\mathsf{E}r \le 2k\gamma$ we have crudely

$$\mathsf{E}r^3 \le 20(\gamma+\gamma^3)k^3\,, \tag{6.102}$$

using that $\gamma^2 \le \gamma + \gamma^3$. Since $pB\mathsf{E}r = 2pBk\gamma \le k/2$ by (6.62), using (6.102) to bound the last term of (6.101) we get

$$d(\mathcal{L}(\langle\sigma_{N-k+1}\rangle,\ldots,\langle\sigma_N\rangle),\mu^{\otimes k}) \le 2^{-n}k + \frac{k^3}{N}(A(0) + 40p^3(\gamma+\gamma^3)BA(n))\,,$$

and since this holds for each $\gamma \le \gamma_0$, the definition of $D(N,k,\gamma_0)$ shows that

$$D(N,k,\gamma_0) \le 2^{-n}k + \frac{k^3}{N}(A(0)+40p^3(\gamma_0+\gamma_0^3)BA(n)) = 2^{-n}k + \frac{k^3A(n+1)}{N}\,. \tag{6.103}$$

We have assumed $k \le N/2$, but since $D(N,k,\gamma_0) \le 2k$ and $A(n+1) \ge A(0)$, (6.103) holds for $k \ge N/2$ provided $K_2(\gamma_0,p) \ge 8$. This proves $C^*(N,\gamma_0,n+1)$ and concludes the proof. □

We now turn to the computation of

$$p_N(\gamma) = \frac{1}{N}\mathsf{E}\log\sum_{\boldsymbol{\sigma}}\exp(-H_N(\boldsymbol{\sigma}))\,. \tag{6.104}$$

We will only consider the situation where (6.66) holds, leaving it to the reader to investigate what kind of rates of convergence she can obtain when assuming only (6.62). We consider i.i.d. copies $(\theta_j)_{j\ge1}$ of the r.v. θ, that are independent of θ, and we recall the notation (6.49). Consider an i.i.d. sequence $\mathbf{X} = (X_i)_{i\ge1}$, where X_i is of law μ_γ (given by Theorem 6.3.1). Recalling the definition (6.49) of $\mathcal{E}_r$ we define

$$p(\gamma) = \log 2 - \frac{\gamma(p-1)}{p}\mathsf{E}\log\langle\exp\theta(\sigma_1,\ldots,\sigma_p)\rangle_{\mathbf{X}} + \mathsf{E}\log\langle\mathrm{Av}\,\mathcal{E}_r\rangle_{\mathbf{X}}\,. \tag{6.105}$$

Here as usual Av means average over $\varepsilon = \pm1$, the notation $\langle\cdot\rangle_{\mathbf{X}}$ is as in e.g. (6.51), and r is a Poisson r.v. with $\mathsf{E}r = \gamma$.

Theorem 6.4.8. *Under (6.62) and (6.66), for $N \ge 2$, and if $\gamma \le \gamma_0$ we have*

$$|p_N(\gamma) - p(\gamma)| \le \frac{K\log N}{N}\,, \tag{6.106}$$

where K does not depend on N or γ.

As we shall see later, the factor $\log N$ above is parasitic and can be removed.

Let $\gamma_- = \gamma(N-p)/(N-1)$ as in (6.18). Theorem 6.4.8 will be a consequence of the following two lemmas, that use the notation (6.104), and where K does not depend on N or γ.

Lemma 6.4.9. *We have*

$$|Np_N(\gamma) - (N-1)p_{N-1}(\gamma_-) - \log 2 - \mathsf{E}\log\langle \mathrm{Av}\,\mathcal{E}_r\rangle_{\mathbf{X}}| \le \frac{K}{N}\,. \tag{6.107}$$

Lemma 6.4.10. *We have*

$$\Big|(N-1)p_{N-1}(\gamma) - (N-1)p_{N-1}(\gamma_-) \\ - \gamma\frac{p-1}{p}\mathsf{E}\log\langle\exp\theta(\sigma_1,\ldots,\sigma_p)\rangle_{\mathbf{X}}\Big| \le \frac{K}{N}\,. \tag{6.108}$$

Proof of Theorem 6.4.8. Combining the two previous relations we get

$$|Np_N(\gamma) - (N-1)p_{N-1}(\gamma) - p(\gamma)| \le \frac{K}{N}\,,$$

and by summation over N that

$$N|p_N(\gamma) - p(\gamma)| \le K\log N\,.$$

□

The following prepares for the proof of Lemma 6.4.10.

Lemma 6.4.11. *We have*

$$p'_N(\gamma) = \frac{1}{p}\mathsf{E}\log\langle\exp\theta(\sigma_1,\ldots,\sigma_p)\rangle\,. \tag{6.109}$$

Proof. As N is fixed, it is obvious that $p'_N(\gamma)$ exists. A pretty proof of (6.109) is as follows. Consider $\delta > 0$, i.i.d. copies $(\theta_j)_{j\ge1}$ of θ, sets $\{i(j,1),\ldots,i(j,p)\}$ that are independent uniformly distributed over the subsets of $\{1,\ldots,N\}$ of cardinality p, and define

$$-H_N^\delta(\boldsymbol{\sigma}) = \sum_{j\le u}\theta_j(\sigma_{i(j,1)},\ldots,\sigma_{i(j,p)})\,, \tag{6.110}$$

where u is a Poisson r.v. of mean $N\delta/p$. All the sources of randomness in this formula are independent of each other and of the randomness in H_N. In distribution, $H_N(\boldsymbol{\sigma}) + H_N^\delta(\boldsymbol{\sigma})$ is the Hamiltonian of an N-spin system with parameter $\gamma+\delta$, so that

$$\frac{p_N(\gamma+\delta) - p_N(\delta)}{\delta} = \frac{1}{N\delta}\mathsf{E}\log\big\langle\exp(-H_N^\delta(\boldsymbol{\sigma}))\big\rangle\,. \tag{6.111}$$

When $u = 0$, we have $H_N^\delta \equiv 0$ so that $\log \langle \exp(-H_N^\delta(\boldsymbol{\sigma})) \rangle = 0$. For very small δ, the probability that $u = 1$ is at the first order in δ equal to $N\delta/p$. The contribution of this case to the right-hand side of (6.111) is, by symmetry among sites,

$$\frac{1}{p}\mathsf{E}\log\left\langle \exp\theta_1(\sigma_{i(1,1)},\ldots,\sigma_{i(1,p)})\right\rangle = \frac{1}{p}\mathsf{E}\log\langle\exp\theta(\sigma_1,\ldots,\sigma_p)\rangle \,.$$

The contribution of the case $u > 1$ is of second order in δ, so that taking the limit in (6.111) as $\delta \to 0$ yields (6.109). □

Lemma 6.4.12. *Recalling that* $\mathbf{X} = (X_i)_{i\geq 1}$ *where* X_i *are i.i.d. of law* μ_γ *we have*

$$\left| p_N'(\gamma) - \frac{1}{p}\mathsf{E}\log\langle\exp\theta(\sigma_1,\ldots,\sigma_p)\rangle_{\mathbf{X}}\right| \leq \frac{K}{N} \,. \tag{6.112}$$

Proof. From Lemma 6.4.11 we see that it suffices to prove that

$$\left|\mathsf{E}\log\langle\exp\theta(\sigma_1,\ldots,\sigma_p)\rangle - \mathsf{E}\log\langle\exp\theta(\sigma_1,\ldots,\sigma_p)\rangle_{\mathbf{X}}\right| \leq \frac{K}{N} \,. \tag{6.113}$$

Let us denote by E_0 expectation in the randomness of $\langle\cdot\rangle$ (but not in θ), and let $S = \sup|\theta|$. It follows from Theorem 6.2.2 (used as in Proposition 6.2.7) that

$$\mathsf{E}_0\left|\langle\exp\theta(\sigma_1,\ldots,\sigma_p)\rangle - \left\langle\exp\theta(\sigma_1^1,\ldots,\sigma_p^p)\right\rangle\right| \leq \frac{K}{N}\exp S \,.$$

Here and below, the number K depends only on p and γ_0, but not on S or N. Now

$$\left\langle\exp\theta(\sigma_1^1,\ldots,\sigma_p^p)\right\rangle = \langle\exp\theta(\sigma_1,\ldots,\sigma_p)\rangle_{\mathbf{Y}} \,,$$

where $\mathbf{Y} = (\langle\sigma_1\rangle,\ldots,\langle\sigma_p\rangle)$. Next, since

$$\left|\frac{\partial}{\partial x_i}\langle\exp\theta(\sigma_1,\ldots,\sigma_p)\rangle_{\mathbf{x}}\right| \leq \exp S \,,$$

considering (as provided by Theorem 6.4.1) a joint realization of the sequences $(\mathbf{X},\mathbf{Y})$ with $\mathsf{E}_0|X_\ell - \langle\sigma_\ell\rangle| \leq K/N$ for $\ell \leq p$, we obtain as in Section 6.3 that

$$\mathsf{E}_0\left|\langle\exp\theta(\sigma_1,\ldots,\sigma_p)\rangle_{\mathbf{X}} - \langle\exp\theta(\sigma_1,\ldots,\sigma_p)\rangle_{\mathbf{Y}}\right| \leq \frac{K}{N}\exp S \,.$$

Combining the previous estimates yields

$$\mathsf{E}_0\left|\langle\exp\theta(\sigma_1,\ldots,\sigma_p)\rangle - \langle\exp\theta(\sigma_1,\ldots,\sigma_p)\rangle_{\mathbf{X}}\right| \leq \frac{K}{N}\exp S \,.$$

Finally for $x, y > 0$ we have

$$|\log x - \log y| \leq \frac{|x-y|}{\min(x,y)}$$

so that

$$\mathsf{E}_0 \left|\log \langle \exp \theta(\sigma_1, \ldots, \sigma_p) \rangle - \log \langle \exp \theta(\sigma_1, \ldots, \sigma_p) \rangle_{\mathbf{X}} \right| \le \frac{K}{N} \exp 2S \,,$$

and (6.113) by taking expectation in the randomness of θ. □

Proof of Lemma 6.4.10. We observe that

$$p_{N-1}(\gamma) - p_{N-1}(\gamma_-) = \int_{\gamma-}^{\gamma} p'_{N-1}(t)dt \,.$$

Combining with Lemma 6.4.12 and Lemma 6.3.4 implies

$$\gamma_- \le t \le \gamma \quad \Rightarrow \quad \left| p'_{N-1}(t) - \frac{1}{p} \log \langle \exp \theta(\sigma_1, \ldots, \sigma_p) \rangle_{\mathbf{X}} \right| \le \frac{K}{N}$$

and we conclude using that

$$\gamma - \gamma_- = \gamma \left(1 - \frac{N-p}{N-1}\right) = \gamma \left(\frac{p-1}{N-1}\right) \,.$$

□

Proof of Lemma 6.4.9. Let us denote by $\langle \cdot \rangle_-$ an average for the Gibbs measure of an $(N-1)$-spin system with Hamiltonian (6.13). We recall that we can write in distribution

$$-H_N(\boldsymbol{\sigma}) \stackrel{\mathcal{D}}{=} -H_{N-1}(\boldsymbol{\rho}) + \sum_{j \le r} \theta_j(\sigma_{i(j,1)}, \ldots, \sigma_{i(j,p-1)}, \sigma_N) \,,$$

where $(\theta_j)_{j \ge 1}$ are independent distributed like θ, where r is a Poisson r.v. of expectation γ and where the sets $\{i(j,1), \ldots, i(j,p-1)\}$ are uniformly distributed over the subsets of $\{1, \ldots, N-1\}$ of cardinality $p-1$. All these randomnesses, as well as the randomness of H_{N-1} are globally independent. Thus the identity

$$\begin{aligned} \mathsf{E} \log \sum_{\boldsymbol{\sigma}} \exp(-H_N(\boldsymbol{\sigma})) &= \mathsf{E} \log \sum_{\boldsymbol{\rho}} \exp(-H_{N-1}(\boldsymbol{\rho})) \\ &\quad + \log 2 + \mathsf{E} \log \langle \mathrm{Av}\, \mathcal{E} \rangle_- \end{aligned} \tag{6.114}$$

holds, where

$$\mathcal{E} = \mathcal{E}(\boldsymbol{\rho}, \varepsilon) = \exp \sum_{j \le r} \theta_j(\sigma_{i(j,1)}, \ldots, \sigma_{i(j,p-1)}, \varepsilon) \,.$$

The term $\log 2$ occurs from the identity $a(1) + a(-1) = 2\mathrm{Av}\, a(\varepsilon)$. Moreover (6.114) implies the equality

$$N p_N(\gamma) - (N-1) p_{N-1}(\gamma_-) = \log 2 + \mathsf{E} \log \langle \mathrm{Av}\, \mathcal{E} \rangle_- \,.$$

Thus (6.107) boils down to the fact that

$$\left|\mathsf{E}\log\langle\mathrm{Av}\,\mathcal{E}\rangle_- - \mathsf{E}\log\langle\mathrm{Av}\,\mathcal{E}_r\rangle_{\mathbf{X}}\right| \le \frac{K}{N}\,. \tag{6.115}$$

The reason why the left-hand side is small should be obvious, and the arguments have already been used in the proof of Lemma 6.4.12. Indeed, it follows from Theorems 6.2.2 and 6.4.1 that if F is a function on the $(N-1)$-spin system that depends only on k spins, the law of the r.v. $\langle F\rangle_-$ is nearly that of $\langle F\rangle_{\mathbf{Y}}$ where Y_i are i.i.d. r.v.s of law $\mu_- = \mu_{\gamma_-}$ (which is nearly μ_γ). The work consists in showing that the bound in (6.115) is actually in K/N. Writing the full details is a bit tedious, but completely straightforward. We do not give these details, since the exact rate in (6.107) will never be used. As we shall soon see, all we need in (6.106) is a bound that goes to 0 as $N \to \infty$. □

Theorem 6.4.13. *Under (6.62) and (6.66)we have in fact*

$$|p_N(\gamma) - p(\gamma)| \le \frac{K}{N}\,. \tag{6.116}$$

Proof. It follows from (6.112) that the functions $p'_N(\gamma)$ converge uniformly over the interval $[0,\gamma_0]$. On the other hand, Theorem 6.4.8 shows that $p(\gamma) = \lim p_N(\gamma)$. Thus $p(\gamma)$ has a derivative $p'(\gamma) = \lim_{N\to\infty} p'_N(\gamma)$, so that (6.112) means that $|p'_N(\gamma) - p'(\gamma)| \le K/N$, from which (6.116) follows by integration. □

Comment. In this argument we have used (6.106) only to prove that

$$p'(\gamma) = \frac{1}{p}\mathsf{E}\log\langle\exp\theta(\sigma_1,\ldots,\sigma_p)\rangle_{\mathbf{X}}\,.$$

One would certainly wish to find a simple direct proof of this fact from the definition of (6.105). A complicated proof can be found in [56], Proposition 7.4.9.

6.5 The Franz-Leone Bound

In the previous section we showed that, under (6.62), the value of $p_N(\gamma)$ is nearly given by the value (6.105). In the present section we prove a remarkable fact. If the function θ is nice enough, one can bound $p_N(\gamma)$ by a quantity similar to (6.105) for *all* values of γ. Hopefully this bound can be considered as a first step towards the very difficult problem of understanding the present model without a high-temperature condition. It is in essence a version of Guerra's replica-symmetric bound of Theorem 1.3.7 adapted to the present setting.

We make the following assumptions on the random function θ. We assume that there exists a random function $f : \{-1,1\} \to \mathbb{R}$ such that

$$\exp\theta(\sigma_1,\ldots,\sigma_p) = a(1 + bf_1(\sigma_1)\cdots f_p(\sigma_p))\,, \tag{6.117}$$

where $f_1, \ldots, f_p$ are independent copies of f, b is a r.v. independent of $f_1, \ldots, f_p$ that satisfies the condition

$$\forall n \geq 1, \quad \mathsf{E}(-b)^n \geq 0\,, \tag{6.118}$$

and a is any r.v. Of course (6.118) is equivalent to saying that $\mathsf{E}b^{2k+1} \leq 0$ for $k \geq 0$. We also assume two further conditions:

$$|bf_1(\sigma_1) \cdots f_p(\sigma_p)| \leq 1 \quad \text{a.e.}, \tag{6.119}$$

and

$$\text{either } \; f \geq 0 \; \text{ or } \; p \; \text{ is even.} \tag{6.120}$$

Let us consider two examples where these conditions are satisfied. First, let

$$\theta(\sigma_1, \ldots, \sigma_p) = \beta J \sigma_1 \cdots \sigma_p\,,$$

where J is a symmetric r.v. Then (6.117) holds for $a = \mathrm{ch}(\beta J)$, $b = \mathrm{th}(\beta J)$, $f(\sigma) = \sigma$, (6.118) holds by symmetry and (6.120) holds when p is even.

Second, let

$$\theta(\sigma_1, \ldots, \sigma_p) = -\beta \prod_{j \leq p} \frac{(1 + \eta_j \sigma_j)}{2}\,,$$

where η_i are independent random signs. This is exactly the Hamiltonian relevant to the K-sat problem (6.2). We observe that for $x \in \{0,1\}$ we have the identity $\exp(-\beta x) = 1 + (e^{-\beta} - 1)x$. Let us set $f_j(\sigma) = (1 + \eta_j \sigma)/2 \in \{0,1\}$. Since $\theta(\sigma_1, \ldots, \sigma_p) = -\beta x$ for $x = f_1(\sigma_1) \cdots f_p(\sigma_p) \in \{0,1\}$ we see that (6.117) holds for $a = 1$, $b = e^{-\beta} - 1$ and $f_j(\sigma) = (1 + \eta_j \sigma)/2$; (6.118) holds since $b < 0$, and (6.120) holds since $f \geq 0$.

Given a probability measure μ on $[-1,1]$, consider an i.i.d. sequence $\mathbf{X}$ distributed like μ, and let us denote by $p(\gamma, \mu)$ the right-hand side of (6.105). (Thus, under (6.62), μ_γ is well defined and $p(\gamma) = p(\gamma, \mu_\gamma)$).

Theorem 6.5.1. *Conditions (6.117) to (6.119) imply*

$$\forall \gamma,\, \forall \mu,\; p_N(\gamma) = \frac{1}{N} \mathsf{E} \log \sum_{\boldsymbol{\sigma}} \exp(-H_N(\boldsymbol{\sigma})) \leq p(\gamma, \mu) + \frac{K\gamma}{N}\,, \tag{6.121}$$

where K does not depend on N or γ.

Let us introduce for $\varepsilon = \pm 1$ the r.v.

$$U(\varepsilon) = \log \langle \exp \theta(\sigma_1, \ldots, \sigma_{p-1}, \varepsilon) \rangle_{\mathbf{X}}\,,$$

and let us consider independent copies $(U_{i,s}(1), U_{i,s}(-1))_{i,s \geq 1}$ of the pair $(U(1), U(-1))$.

Exercise 6.5.2. As a motivation for the introduction of the quantity U prove that if we consider the 1-spin system with Hamiltonian $-\sum_{s\le r} U_{i,s}(\varepsilon)$, the average of ε for this Hamiltonian is equal, in distribution, to the quantity (6.51). (Hence, it is distributed like $T(\mu)$.)

For $0 \le t \le 1$ we consider a Poisson r.v. M_t of mean $\alpha tN = \gamma tN/p$, and independent Poisson r.v.s $r_{i,t}$ of mean $\gamma(1-t)$, independent of M_t. We consider the Hamiltonian

$$-H_{N,t}(\boldsymbol{\sigma}) = \sum_{k\le M_t} \theta_k(\sigma_{i(k,1)},\ldots,\sigma_{i(k,p)}) + \sum_{i\le N}\sum_{s\le r_{i,t}} U_{i,s}(\sigma_i)\;, \tag{6.122}$$

where as usual the different sources of randomness are independent of each other, and we set

$$\varphi(t) = \frac{1}{N}\mathsf{E}\log\sum_{\boldsymbol{\sigma}}\exp(-H_{N,t}(\boldsymbol{\sigma}))\;.$$

Proposition 6.5.3. *We have*

$$\varphi'(t) \le -\frac{\gamma(p-1)}{p}\,\mathsf{E}\log\langle\exp\theta(\sigma_1,\ldots,\sigma_p)\rangle_{\mathbf{X}} + \frac{K\gamma}{N}\;. \tag{6.123}$$

This is of course the key fact.

Proof of Theorem 6.5.1. We deduce from (6.5.3) that

$$p_N(\gamma) = \varphi(1) \le \varphi(0) - \frac{\gamma(p-1)}{p}\,\mathsf{E}\log\langle\exp\theta(\sigma_1,\ldots,\sigma_p)\rangle_{\mathbf{X}} + \frac{K\gamma}{N}\;.$$

Therefore to prove Theorem 6.5.1 it suffices to show that $\varphi(0) = \log 2 + \mathsf{E}\log\langle\mathrm{Av}\,\mathcal{E}_r\rangle_{\mathbf{X}}$. For $t=0$ the spins are decoupled, so this reduces to the case $N=1$. Since $r_{1,0}$ has the same distribution as r, we simply observe that if $(\mathbf{X}_s)_{s\le r}$ are independent copies of $\mathbf{X}$, the quantity

$$\prod_{s\le r}\langle\exp\theta_s(\sigma_1,\ldots,\sigma_{p-1},\varepsilon)\rangle_{\mathbf{X}_s}$$

has the same distribution as the quantity $\langle\mathrm{Av}\,\mathcal{E}_r\rangle_{\mathbf{X}}$. Therefore,

$$\begin{aligned}\mathsf{E}\log\sum_{\varepsilon=\pm1}\exp\sum_{s\le r}U_{1,s}(\varepsilon) &= \mathsf{E}\log\sum_{\varepsilon=\pm1}\prod_{s\le r}\langle\exp\theta_s(\sigma_1,\ldots,\sigma_{p-1},\varepsilon)\rangle_{\mathbf{X}_s}\\ &= \log 2 + \mathsf{E}\log\langle\mathrm{Av}\,\mathcal{E}_r\rangle_{\mathbf{X}}\;,\end{aligned}$$

and this completes the proof of Theorem 6.5.1. □

We now prepare for the proof of (6.5.3).

Lemma 6.5.4. *We have*

$$\varphi'(t) \le \frac{\gamma}{p}\Bigg(\frac{1}{N^p}\sum_{i_1,\dots,i_p=1}^{N} \mathsf{E}\log\big\langle\exp\theta(\sigma_{i_1},\dots,\sigma_{i_p})\big\rangle \\ -\frac{p}{N}\sum_{i\le N}\mathsf{E}\log\langle\exp U(\sigma_i)\rangle\Bigg)+\frac{K\gamma}{N}\,. \tag{6.124}$$

Here, as in the rest of the section, we denote by $\langle\cdot\rangle$ an average for the Gibbs measure with Hamiltonian (6.122), keeping the dependence on t implicit. On the other hand, the number K in (6.124) is of course independent of t.

Proof. In $\varphi'(t)$ there are terms coming from the dependence on t of M_t and terms coming from the dependence on t of $r_{i,t}$.

As shown by Lemma 6.4.11, the term created by the dependence of M_t on t is

$$\frac{\gamma}{p}\mathsf{E}\log\langle\exp\theta(\sigma_1,\dots,\sigma_p)\rangle \le \frac{\gamma}{pN^p}\sum_{i_1,\dots,i_p=1}^{N}\mathsf{E}\log\big\langle\exp\theta(\sigma_{i_1},\dots,\sigma_{i_p})\big\rangle+\frac{\gamma K}{N}\,,$$

because all the terms where the indices $i_1,\dots,i_p$ are distinct are equal. The same argument as in Lemma 6.4.11 shows that the term created by the dependence of $r_{i,t}$ on t is $-(\gamma/N)\mathsf{E}\log\langle\exp U(\sigma_i)\rangle$. □

Thus, we have reduced the proof of Proposition 6.5.3 (hence, of Theorem 6.5.1) to the following:

Lemma 6.5.5. *We have*

$$\sum_{i_1,\dots,i_p=1}^{N}\frac{1}{N^p}\mathsf{E}\log\big\langle\exp\theta(\sigma_{i_1},\dots,\sigma_{i_p})\big\rangle-\frac{p}{N}\sum_{i\le N}\mathsf{E}\log\langle\exp U(\sigma_i)\rangle \\ +(p-1)\mathsf{E}\log\langle\exp\theta(\sigma_1,\dots,\sigma_p)\rangle_{\mathbf{X}}\le 0\,. \tag{6.125}$$

The proof is not really difficult, but it must have been quite another matter when Franz and Leone discovered it.

Proof. We will get rid of the annoying logarithms by power expansion,

$$\log(1+x)=-\sum_{n\ge1}(-1)^n\frac{x^n}{n}$$

for $|x|<1$. Let us denote by E_0 the expectation in the randomness of $\mathbf{X}$ and of the functions f_j of (6.117) only. Let us define

$$C_n=\mathsf{E}_0\langle f(\sigma_1)\rangle_{\mathbf{X}}^n \tag{6.126}$$

$$A_{j,n}=A_{j,n}(\boldsymbol{\sigma}^1,\dots,\boldsymbol{\sigma}^n)=\frac{1}{N}\sum_{i\le N}\prod_{\ell\le n}f_j(\sigma_i^\ell) \tag{6.127}$$

$$B_n = B_n(\boldsymbol{\sigma}^1, \ldots, \boldsymbol{\sigma}^n) = \mathsf{E}_0 A_{j,n} . \tag{6.128}$$

We will prove that the left-hand side quantity (6.125) is equal to

$$-\sum_{n=1}^{\infty} \frac{\mathsf{E}(-b)^n}{n} \mathsf{E}\left\langle B_n^p - pB_n C_n^{p-1} + (p-1)C_n^p \right\rangle . \tag{6.129}$$

The function $x \mapsto x^p$ is convex on $\mathbb{R}^+$, and when p is even it is convex on $\mathbb{R}$. Therefore $x^p - pxy^{p-1} + (p-1)y^p \geq 0$ for all $x, y \in \mathbb{R}^+$, and when p is even this is true for all $x, y \in \mathbb{R}$. Now (6.120) shows that either $B_n \geq 0$ and $C_n \geq 0$ or p is even, and thus it holds that $B_n^p - pB_n C_n^{p-1} + (p-1)C_n^p \geq 0$. Consequently the right-hand side of (6.129) is ≤ 0 because $\mathsf{E}(-b)^n \geq 0$ by (6.118).

By (6.117) we have

$$\exp\theta(\sigma_1, \ldots, \sigma_p) = a(1 + b\prod_{j\leq p} f_j(\sigma_j)) , \tag{6.130}$$

so that, taking the average $\langle\cdot\rangle_{\mathbf{X}}$ and logarithm, and using (6.119) to allow the power expansion in the second line,

$$\begin{aligned} \log\langle\exp\theta(\sigma_1, \ldots, \sigma_p)\rangle_{\mathbf{X}} &= \log a + \log\left(1 + b\left\langle\prod_{j\leq p} f_j(\sigma_j)\right\rangle_{\mathbf{X}}\right) \\ &= \log a - \sum_{n=1}^{\infty} \frac{(-b)^n}{n} \left\langle\prod_{j\leq p} f_j(\sigma_j)\right\rangle_{\mathbf{X}}^n . \end{aligned} \tag{6.131}$$

Now, by independence

$$\mathsf{E}_0\left\langle\prod_{j\leq p} f_j(\sigma_j)\right\rangle_{\mathbf{X}}^n = \mathsf{E}_0 \prod_{j\leq p} \langle f_j(\sigma_j)\rangle_{\mathbf{X}}^n = C_n^p$$

so that

$$\mathsf{E}_0 \log\langle\exp\theta(\sigma_1, \ldots, \sigma_p)\rangle_{\mathbf{X}} = \mathsf{E}_0 \log a - \sum_{n=1}^{\infty} \frac{(-b)^n}{n} C_n^p .$$

As in (6.131),

$$\begin{aligned} &\frac{1}{N^p} \sum_{i_1,\ldots,i_p}^{N} \log\left\langle\exp\theta(\sigma_{i_1}, \ldots, \sigma_{i_p})\right\rangle \\ &= \log a - \sum_{n=1}^{\infty} \frac{(-b)^n}{n} \left(\frac{1}{N^p} \sum_{i_1,\ldots,i_p=1}^{N} \left\langle\prod_{j\leq p} f_j(\sigma_{i_j})\right\rangle^n\right) . \end{aligned}$$

Using replicas, we get

$$\left\langle \prod_{j\le p} f_j(\sigma_{i_j}) \right\rangle^n = \left\langle \prod_{\ell\le n} \prod_{j\le p} f_j(\sigma^\ell_{i_j}) \right\rangle ,$$

so that, using (6.127) in the second line yields

$$\begin{aligned} \frac{1}{N^p} \sum_{i_1,\dots,i_p=1}^{N} \left\langle \prod_{j\le p} f_j(\sigma_{i_j}) \right\rangle^n &= \left\langle \frac{1}{N^p} \sum_{i_1,\dots,i_p=1}^{N} \prod_{\ell\le n} \prod_{j\le p} f_j(\sigma^\ell_{i_j}) \right\rangle^n \\ &= \left\langle \prod_{j\le p} A_{j,n} \right\rangle . \end{aligned}$$

Now from (6.128) and independence we get $E_0 \prod_{j\le p} A_{j,n} = B_n^p$, so that

$$\mathsf{E}_0 \frac{1}{N^p} \sum_{i_1,\dots,i_p=1}^{N} \log \left\langle \exp \theta(\sigma_{i_1},\dots,\sigma_{i_p}) \right\rangle = \mathsf{E}_0 \log a - \sum_{n=1}^{\infty} \frac{(-b)^n}{n} \langle B_n^p \rangle .$$

In a similar manner, recalling the definition of U, one shows that

$$\mathsf{E}_0 \frac{1}{N} \sum_{i\le n} \log \langle \exp U(\sigma_i) \rangle = \mathsf{E}_0 \log a - \sum_{n=1}^{\infty} \frac{(-b)^n}{n} \langle B_n C_n^{p-1} \rangle ,$$

and this concludes the proof of Lemma 6.5.5. □

6.6 Continuous Spins

In this section we consider the situation of the Hamiltonian (6.4) when the spins are real numbers. There are two motivations for this. First, the "main parameter" of the system is no longer "a function" but rather "a random function". This is both a completely natural and fun situation. Second, this will let us demonstrate in the next section the power of the convexity tools we developed in Chapters 3 and 4. We consider a (Borel) function θ on $\mathbb{R}^p$, i.i.d. copies $(\theta_k)_{k\ge1}$ of θ, and for $\boldsymbol{\sigma} \in \mathbb{R}^N$ the quantity $H_N(\boldsymbol{\sigma})$ given by (6.4). We consider a given probability measure η on $\mathbb{R}$, and we lighten notation by writing η_N for $\eta^{\otimes N}$, the corresponding product measure on $\mathbb{R}^N$. The Gibbs measure is now defined as the random probability measure on $\mathbb{R}^N$ which has a density with respect to η_N that is proportional to $\exp(-H_N(\boldsymbol{\sigma}))$. Let us fix an integer k and, for large N, let us try to guess the law of $(\sigma_1,\dots,\sigma_k)$ under Gibbs' measure. This is a random probability measure on $\mathbb{R}^k$. We expect that it has a density $Y_{k,N}$ with respect to $\eta_k = \eta^{\otimes k}$. What is the simplest possible structure? It would be nice if we had

$$Y_{k,N}(\sigma_1,\dots,\sigma_k) \simeq X_1(\sigma_1)\cdots X_k(\sigma_k) ,$$

where $X_1, \ldots, X_k \in L^1(\eta)$ are random elements of $L^1(\eta)$, which are probability densities, i.e. $X_i \in \mathcal{D}$, where

$$\mathcal{D} = \left\{ X \in L^1(\eta) \; ; \; X \geq 0 \; ; \int X \mathrm{d}\eta = 1 \right\} . \tag{6.132}$$

The nicest possible probabilistic structure would be that these random elements $X_1, \ldots, X_k$ be i.i.d, with a common law μ, a probability measure on the metric space $\mathcal{D}$. This law μ is the central object, the "main parameter". (If we wish, we can equivalently think of μ as the law of a random element of $\mathcal{D}$.) The case of Ising spins is simply the situation where $\eta(\{1\}) = \eta(\{-1\}) = 1/2$, in which case

$$\mathcal{D} = \{(x(-1), x(1)) \; ; \; x(1), x(-1) \geq 0 \, , \; x(1) + x(-1) = 2\}$$

and

$$\begin{gathered} \mathcal{D} \text{ can be identified with the interval } [-1, 1] \\ \text{by the map } (x(-1), x(1)) \mapsto (x(1) - x(-1))/2 \, . \end{gathered} \tag{6.133}$$

Thus, in that case, as we have seen, the main parameter is a probability measure on the interval $[-1, 1]$.

We will assume in this section that θ is uniformly bounded, i.e.

$$S = \sup_{\sigma_1, \ldots, \sigma_p \in \mathbb{R}} |\theta(\sigma_1, \ldots, \sigma_p)| < \infty \tag{6.134}$$

for a certain r.v. S. Of course $(S_k)_{k \geq 1}$ denote i.i.d. copies of S with $S_k = \sup |\theta_k(\sigma_1, \ldots, \sigma_p)|$. Whether or how this boundedness condition can be weakened remains to be investigated. Overall, once one gets used to the higher level of abstraction necessary compared with the case of Ising spins, the proofs are really not more difficult in the continuous case. In the present section we will control the model under a high-temperature condition and the extension of the methods of the previous sections to this setting is really an exercise. The real point of this exercise is that in the next section, we will succeed to partly control the model *without* assuming a high-temperature condition but assuming instead the concavity of θ, a result very much in the spirit of Section 3.1.

Our first task is to construct the "order parameter" $\mu = \mu_\gamma$. We keep the notation (6.49), that is we write

$$\mathcal{E}_r = \mathcal{E}_r(\boldsymbol{\sigma}, \varepsilon) = \exp \sum_{1 \leq j \leq r} \theta_j(\sigma_{(j-1)(p-1)+1}, \ldots, \sigma_{j(p-1)}, \varepsilon) \, ,$$

where now σ_i and ε are real numbers.

Given a sequence $\mathbf{X} = (X_i)_{i \geq 1}$ of elements of $\mathcal{D}$, for a function f of $\sigma_1, \ldots, \sigma_N$, we define (and this will be fundamental)

$$\langle f \rangle_{\mathbf{X}} = \int f(\sigma_1, \ldots, \sigma_N) X_1(\sigma_1) \cdots X_N(\sigma_N) \mathrm{d}\eta(\sigma_1) \cdots \mathrm{d}\eta(\sigma_N) , \tag{6.135}$$

that is, we integrate the generic k-th coordinate with respect to η after making the change of density X_k.

For consistency with the notation of the previous section, for a function $h(\varepsilon)$ we write

$$\mathrm{Av} h = \int h(\varepsilon) \mathrm{d}\eta(\varepsilon) . \tag{6.136}$$

Thus

$$\mathrm{Av}\mathcal{E}_r = \int \mathcal{E}_r(\boldsymbol{\sigma}, \varepsilon) \mathrm{d}\eta(\varepsilon)$$

is a function of $\boldsymbol{\sigma}$ only, and $\langle \mathrm{Av}\mathcal{E}_r \rangle_{\mathbf{X}}$ means that we integrate in $\sigma_1, \ldots, \sigma_N$, as in (6.135). We will also need the quantity $\langle \mathcal{E}_r \rangle_{\mathbf{X}}$, where we integrate in $\sigma_1, \ldots, \sigma_N$ as in (6.135), but we do *not* integrate this factor in ε. Thus $\langle \mathcal{E}_r \rangle_{\mathbf{X}}$ is a function of ε only, and by Fubini's theorem we have $\mathrm{Av}\langle \mathcal{E}_r \rangle_{\mathbf{X}} = \langle \mathrm{Av}\mathcal{E}_r \rangle_{\mathbf{X}}$. In particular, the function f of ε given by

$$\frac{\langle \mathcal{E}_r \rangle_{\mathbf{X}}}{\langle \mathrm{Av}\mathcal{E}_r \rangle_{\mathbf{X}}} \tag{6.137}$$

is such that $f \geq 0$ and $\mathrm{Av} f = 1$, i.e. $f \in \mathcal{D}$.

Consider a probability measure μ on $\mathcal{D}$, and $(X_i)_{i \geq 1}$ a sequence of elements of $\mathcal{D}$ that is i.i.d. of law μ. We denote by $T(\mu)$ the law (in $\mathcal{D}$) of the random element (6.137) when $\mathbf{X} = (X_i)_{i \geq 1}$. When the spins take only the values ± 1, and provided we then perform the identification (6.133), this coincides with the definition (6.51).

Theorem 6.6.1. *Assuming (6.52), i.e.* $4\gamma p(\mathsf{E} S \exp 2S) \leq 1$, *there exists a unique probability measure* μ *on* $\mathcal{D}$ *such that* $\mu = T(\mu)$.

On $\mathcal{D}$, the natural distance is induced by the L_1 norm relative to η, i.e. for $x, y \in \mathcal{D}$

$$d(x, y) = \|x - y\|_1 = \int |x(\varepsilon) - y(\varepsilon)| \mathrm{d}\eta(\varepsilon) . \tag{6.138}$$

The key to prove Theorem 6.6.1 is the following estimate, where we consider a pair (X, Y) of random elements of $\mathcal{D}$, and independent copies $(X_i, Y_i)_{i \geq 1}$ of this pair. Let $\mathbf{X} = (X_i)_{i \geq 1}$ and $\mathbf{Y} = (Y_i)_{i \geq 1}$.

Lemma 6.6.2. *We have*

$$\left\| \frac{\langle \mathcal{E}_r \rangle_{\mathbf{X}}}{\langle \mathrm{Av}\mathcal{E}_r \rangle_{\mathbf{X}}} - \frac{\langle \mathcal{E}_r \rangle_{\mathbf{Y}}}{\langle \mathrm{Av}\mathcal{E}_r \rangle_{\mathbf{Y}}} \right\|_1 \leq 2 \sum_{j \leq r} S_j \exp 2S_j \sum_{(j-1)(p-1) < i \leq j(p-1)} \|X_i - Y_i\|_1 . \tag{6.139}$$

Once this estimate has been obtained we proceed exactly as in the proof of Theorem 6.3.1. Namely, if μ and μ' are the laws of X and Y respectively, and since the law of the quantity (6.137) is $T(\mu)$, the expected value of the left-hand side of (6.139) is an upper bound for the transportation-cost distance $d(T(\mu), T(\mu'))$ associated to the distance d of (6.138) (by the very definition of the transportation-cost distance). Thus taking expectation in (6.139) implies that

$$d(T(\mu), T(\mu')) \le 2\gamma p(\mathsf{E} S \exp 2S)\mathsf{E}\|X - Y\|_1 .$$

Since this is true for any choice of X and Y with laws μ and μ' respectively, we obtained that

$$d(T(\mu), T(\mu')) \le 2\gamma p(\mathsf{E} S \exp 2S) d(\mu, \mu') ,$$

so that under (6.52) the map T is a contraction for the transportation-cost distance. This completes the proof of Theorem 6.6.1, modulo the fact that the set of probability measures on a complete metric space is itself a complete metric space when provided with the transportation-cost distance.

Proof of Lemma 6.6.2. It is essentially identical to the proof of (6.57), although we find it convenient to write it a bit differently "replacing Y_j by X_j one at a time". Let

$$\mathbf{X}(i) = (X_1, \ldots, X_i, Y_{i+1}, Y_{i+2} \ldots) .$$

To ease notation we write

$$\langle \cdot \rangle_i = \langle \cdot \rangle_{\mathbf{X}(i)} ,$$

so that

$$\frac{\langle \mathcal{E}_r \rangle_{\mathbf{X}}}{\langle \mathrm{Av}\mathcal{E}_r \rangle_{\mathbf{X}}} = \frac{\langle \mathcal{E}_r \rangle_{r(p-1)}}{\langle \mathrm{Av}\mathcal{E}_r \rangle_{r(p-1)}} ; \qquad \frac{\langle \mathcal{E}_r \rangle_{\mathbf{Y}}}{\langle \mathrm{Av}\mathcal{E}_r \rangle_{\mathbf{Y}}} = \frac{\langle \mathcal{E}_r \rangle_0}{\langle \mathrm{Av}\mathcal{E}_r \rangle_0} ,$$

and to prove (6.136) it suffices to show that if $(j-1)(p-1) < i \le j(p-1)$ we have

$$\left\| \frac{\langle \mathcal{E}_r \rangle_i}{\langle \mathrm{Av}\mathcal{E}_r \rangle_i} - \frac{\langle \mathcal{E}_r \rangle_{i-1}}{\langle \mathrm{Av}\mathcal{E}_r \rangle_{i-1}} \right\|_1 \le (2S_j \exp 2S_j)\|X_i - Y_i\|_1 . \tag{6.140}$$

We bound the left-hand side by $\mathrm{I} + \mathrm{II}$, where

$$\mathrm{I} = \left\| \frac{\langle \mathcal{E}_r \rangle_i - \langle \mathcal{E}_r \rangle_{i-1}}{\langle \mathrm{Av}\mathcal{E}_r \rangle_i} \right\|_1 \tag{6.141}$$

$$\mathrm{II} = \left\| \frac{\langle \mathcal{E}_r \rangle_{i-1}(\langle \mathrm{Av}\mathcal{E}_r \rangle_i - \langle \mathrm{Av}\mathcal{E}_r \rangle_{i-1})}{\langle \mathrm{Av}\mathcal{E}_r \rangle_i \langle \mathrm{Av}\mathcal{E}_r \rangle_{i-1}} \right\|_1 . \tag{6.142}$$

Now we observe that to bound both terms by $S_j \exp 2S_j \|X_i - Y_i\|_1$ it suffices to prove that

$$|\langle \mathcal{E}_r \rangle_i - \langle \mathcal{E}_r \rangle_{i-1}| \le S_j \exp 2S_j \|X_i - Y_i\|_1 \langle \mathcal{E}_r \rangle_i \ , \tag{6.143}$$

(where both sides are functions of ε). Indeed to bound the term I using (6.143) we observe that

$$\left\| \frac{\langle \mathcal{E}_r \rangle_i}{\langle \mathrm{Av}\mathcal{E}_r \rangle_i} \right\|_1 = \mathrm{Av} \frac{\langle \mathcal{E}_r \rangle_i}{\langle \mathrm{Av}\mathcal{E}_r \rangle_i} = 1 \tag{6.144}$$

and to bound the term II using (6.143) we observe that

$$\begin{aligned} |\langle \mathrm{Av}\mathcal{E}_r \rangle_i - \langle \mathrm{Av}\mathcal{E}_r \rangle_{i-1}| &\le \mathrm{Av}|\langle \mathcal{E}_r \rangle_i - \langle \mathcal{E}_r \rangle_{i-1}| \\ &\le S_j \exp 2S_j \|X_i - Y_i\|_1 \mathrm{Av}\langle \mathcal{E}_r \rangle_i \end{aligned}$$

and we use (6.144) again (for $i-1$ rather than i).

Thus it suffices to prove (6.143). For this we write $\mathcal{E}_r = \mathcal{E}'\mathcal{E}''$, where

$$\mathcal{E}' = \exp \theta_j(\sigma_{(j-1)(p-1)+1}, \dots, \sigma_{j(p-1)}, \varepsilon) \ ,$$

and where $\mathcal{E}''$ does not depend on σ_i. Therefore

$$A := \langle \mathcal{E}'' \rangle_i = \langle \mathcal{E}'' \rangle_{i-1} \ .$$

Since $\mathcal{E}'$ and $\mathcal{E}''$ depend on different sets of coordinates, we have

$$\langle \mathcal{E}_r \rangle_i = \langle \mathcal{E}' \rangle_i \langle \mathcal{E}'' \rangle_i = A \langle \mathcal{E}' \rangle_i \ ; \ \langle \mathcal{E}_r \rangle_{i-1} = \langle \mathcal{E}' \rangle_{i-1} \langle \mathcal{E}'' \rangle_{i-1} = A \langle \mathcal{E}' \rangle_{i-1} \ .$$

Let us define $B = B(\sigma_i, \varepsilon)$ the quantity obtained by integrating $\mathcal{E}'$ in each spin σ_k, $k < i$, with respect to η, and change of density X_k and each spin σ_k, $k > i$ with respect to η with change of density Y_k. Integrating first in the σ_k for $k \ne i$ we obtain

$$\langle \mathcal{E}' \rangle_i = \int B X_i(\sigma_i) \mathrm{d}\eta(\sigma_i) \ \ ; \ \ \langle \mathcal{E}' \rangle_{i-1} = \int B Y_i(\sigma_i) \mathrm{d}\eta(\sigma_i) \ ,$$

and therefore

$$\langle \mathcal{E}_r \rangle_i = A \int B X_i(\sigma_i) \mathrm{d}\eta(\sigma_i) \ \ ; \ \ \langle \mathcal{E}_r \rangle_{i-1} = A \int B Y_i(\sigma_i) \mathrm{d}\eta(\sigma_i) \ . \tag{6.145}$$

Consequently,

$$\begin{aligned} \langle \mathcal{E}_r \rangle_i - \langle \mathcal{E}_r \rangle_{i-1} &= A \int B(X_i(\sigma_i) - Y_i(\sigma_i)) \mathrm{d}\eta(\sigma_i) \\ &= A \int (B-1)(X_i(\sigma_i) - Y_i(\sigma_i)) \mathrm{d}\eta(\sigma_i) \end{aligned} \tag{6.146}$$

because $\int X_i \mathrm{d}\eta = \int Y_i \mathrm{d}\eta = 1$. Now, since $|\theta_j| \le S_j$, Jensen's inequality shows that $|\log B| \le S_j$. Using that $|\exp x - 1| \le |x| \exp |x|$ for $x = \log B$ we obtain that $|B - 1| \le S_j \exp S_j$. Therefore (6.146) implies

$$|\langle\mathcal{E}_r\rangle_i - \langle\mathcal{E}_r\rangle_{i-1}| \le AS_j \exp S_j \|X_i - Y_i\|_1 . \tag{6.147}$$

Finally since $\exp(-S_j) \le \mathcal{E}'$ we have $\exp(-S_j) \le B$, so that $\exp(-S_j) \le \int BX_i(\sigma_i)\mathrm{d}\eta(\sigma_i)$. The first part of (6.145) then implies that $A\exp(-S_j) \le \langle\mathcal{E}_r\rangle_i$ and combining with (6.147) this finishes the proof of (6.143) and Lemma 6.6.2. □

A suitable extension of Theorem 6.2.2 will be crucial to the study of the present model. As in the case of Theorem 6.6.1, once we have found the proper setting, the proof is not any harder than in the case of Ising spins.

Let us consider a probability space $(\mathcal{X}, \lambda)$, an integer n, and a family $(f'_\omega)_{\omega\in\mathcal{X}}$ of functions on $(\mathbb{R}^N)^n$. We assume that there exists $i \le N$ such that for each ω we have

$$f'_\omega \circ T_i = -f'_\omega , \tag{6.148}$$

where T_i is defined as in Section 6.2 i.e. T_i exchanges the ith components σ_i^1 and σ_i^2 of the first two replicas and leaves all the other components unchanged. Consider another function $f \ge 0$ on $(\mathbb{R}^N)^n$. We assume that f and the functions f'_ω depend on k coordinates (of course what we mean here is that they depend on the same k coordinates whatever the choice of ω). We assume that for a certain number Q, we have

$$\int |f'_\omega|\mathrm{d}\lambda(\omega) \le Qf . \tag{6.149}$$

Theorem 6.6.3. *Under (6.10), and provided that $\gamma \le \gamma_0$, with the previous notation we have*

$$\mathsf{E}\frac{\int |\langle f'_\omega\rangle|\mathrm{d}\lambda(\omega)}{\langle f\rangle} \le \frac{K_0 kQ}{N} . \tag{6.150}$$

Proof. The fundamental identity (6.20):

$$\langle f\rangle = \frac{\langle \mathrm{Av}\, f\mathcal{E}\rangle_-}{\langle \mathrm{Av}\,\mathcal{E}\rangle_-}$$

remains true if we define Av as in (6.136). We then copy the proof of Theorem 6.2.2 "by replacing everywhere f' by the average of f'_ω in ω" as follows. First, we define the property $C(N, \gamma_0, B, B^*)$ by requiring that under the conditions of Theorem 6.6.3, rather than (6.9):

$$\mathsf{E}\left|\frac{\langle f'\rangle}{\langle f\rangle}\right| \le \frac{Q(kB + B^*)}{N} ,$$

we get instead

$$\mathsf{E}\frac{\int |\langle f'_\omega\rangle|\mathrm{d}\lambda(\omega)}{\langle f\rangle} \le \frac{Q(kB + B^*)}{N} .$$

Rather than (6.32) we now define

$$f'_{\omega,s} = (\mathrm{Av}\, f'_\omega \mathcal{E}) \circ \prod_{u \le s-1} U_{i_u} - (\mathrm{Av}\, f'_\omega \mathcal{E}) \circ \prod_{u \le s} U_{i_u} \,.$$

We replace (6.34) by

$$\int |f'_{\omega,s}| \mathrm{d}\lambda(\omega) \le 4QS_v \exp\Bigl(4 \sum_{u \le r} S_u\Bigr) \mathrm{Av}\, f\mathcal{E} \,,$$

and again the left-hand side of (6.39) by

$$\mathsf{E} \frac{\int |\langle f'_\omega \rangle| \mathrm{d}\lambda(\omega)}{\langle f \rangle} \,. \qquad \square$$

We now describe the structure of the Gibbs measure, under a "high-temperature" condition.

Theorem 6.6.4. *There exists a number $K_1(p)$ such that whenever*

$$K_1(p)(\gamma_0 + \gamma_0^3)\mathsf{E}S \exp 4S \le 1 \,, \tag{6.151}$$

if $\gamma \le \gamma_0$, given any integer k, we can find i.i.d. random elements $X_1, \dots, X_k$ in $\mathcal{D}$ of law μ such that

$$\mathsf{E} \int |Y_{N,k}(\sigma_1, \dots, \sigma_k) - X_1(\sigma_1) \cdots X_k(\sigma_k)| \mathrm{d}\eta(\sigma_1) \cdots \mathrm{d}\eta(\sigma_k)$$
$$\le \frac{k^3 K(p, \gamma_0)}{N} \mathsf{E} \exp 2S \,, \tag{6.152}$$

where $Y_{N,k}$ denotes the density with respect to $\eta_k = \eta^{\otimes k}$ of the law of $\sigma_1, \dots, \sigma_k$ under Gibbs' measure, and μ is as in Theorem 6.6.1 and where $K(p, \gamma_0)$ depends only on p and γ_0.

It is convenient to denote by $\bigotimes_{\ell \le k} X_\ell$ the function

$$(\sigma_1, \dots, \sigma_k) \mapsto \prod_{\ell \le k} X_\ell(\sigma_\ell) \,,$$

so that the left-hand side of (6.152) is simply $\mathsf{E}\|Y_{N,k} - \bigotimes_{\ell \le k} X_\ell\|_1$.

Overall the principle of the proof is very similar to that of the proof of Theorem 6.4.1, but the induction hypothesis will not be based on (6.152). The starting point of the proof is the fundamental cavity formula (6.72), where Av now means that $\sigma_{N-k+1}, \dots, \sigma_N$ are averaged independently with respect to η. When f is a function of k variables, this formula implies that

$$\begin{aligned} \langle f(\sigma_{N-k+1}, \dots, \sigma_N) \rangle &= \frac{\langle \mathrm{Av} f(\sigma_{N-k+1}, \dots, \sigma_N) \mathcal{E} \rangle_-}{\langle \mathrm{Av} \mathcal{E} \rangle_-} \\ &= \mathrm{Av}\Bigl(f(\sigma_{N-k+1}, \dots, \sigma_N) \frac{\langle \mathcal{E} \rangle_-}{\langle \mathrm{Av}\mathcal{E} \rangle_-} \Bigr) \,. \end{aligned} \tag{6.153}$$

The quantity $\langle\mathcal{E}\rangle_-$ is a function of $\sigma_{N-k+1},\ldots,\sigma_N$ only since $(\sigma_1,\ldots,\sigma_{N-k})$ is averaged for $\langle\cdot\rangle_-$, and (6.153) means that the density with respect to η_k of the law of $\sigma_{N-k+1},\ldots,\sigma_N$ under Gibbs' measure is precisely the function

$$\frac{\langle\mathcal{E}\rangle_-}{\langle\mathrm{Av}\mathcal{E}\rangle_-} . \tag{6.154}$$

Before deciding how to start the proof of Theorem 6.6.4, we will first take full advantage of Theorem 6.6.3. For a function f on $\mathbb{R}^{N-k}$ we denote

$$\langle f\rangle_\bullet = \langle f(\sigma_1^1,\sigma_2^2,\ldots,\sigma_{N-k}^{N-k})\rangle_- ,$$

that is, we average every coordinate in a different replica. We recall the set Ω of (6.79).

Proposition 6.6.5. *We have*

$$\mathsf{E}1_{\Omega^c}\mathrm{Av}\left|\frac{\langle\mathcal{E}\rangle_-}{\langle\mathrm{Av}\mathcal{E}\rangle_-}-\frac{\langle\mathcal{E}\rangle_\bullet}{\langle\mathrm{Av}\mathcal{E}\rangle_\bullet}\right|\le\frac{k^2K}{N}\mathsf{E}\exp 2S . \tag{6.155}$$

Here and in the sequel, K denotes a constant that depends on p and γ_0 only. This statement approximates the true density (6.154) by a quantity which will be much simpler to work with, since it is defined via integration for the product measure $\langle\cdot\rangle_\bullet$.

The proof of Proposition 6.6.5 greatly resembles the proof of Proposition 6.2.7. Let us state the basic principle behind this proof. It will reveal the purpose of condition (6.149), that might have remained a little bit mysterious.

Lemma 6.6.6. *For $j\le r$ consider sets $I_j\subset\{1,\ldots,N\}$ with $\mathrm{card}I_j=p$, $\mathrm{card}I_j\cap\{N-k+1,\ldots,N\}=1$, and assume that*

$$j\ne j' \;\Rightarrow\; I_j\cap I_{j'}\subset\{N-k+1,\ldots,N\} ,$$

or, equivalently, that the sets $I_j\backslash\{1,\ldots,N-k\}$ for $j\le r$ are all disjoint. Consider functions $W_j(\boldsymbol{\sigma})$, depending only on the coordinates in I_j, and assume that $\sup_{\boldsymbol{\sigma}}|W_j(\boldsymbol{\sigma})|\le S_j$. Consider

$$\mathcal{E}=\exp\sum_{j\le r}W_j(\boldsymbol{\sigma}) .$$

Then we have

$$\mathsf{E}_-\mathrm{Av}\left|\frac{\langle\mathcal{E}\rangle_-}{\langle\mathrm{Av}\mathcal{E}\rangle_-}-\frac{\langle\mathcal{E}\rangle_\bullet}{\langle\mathrm{Av}\mathcal{E}\rangle_\bullet}\right|\le\frac{4K_0k(p-1)}{N-k}\sum_{j\le r}\exp 2S_j . \tag{6.156}$$

Here E_- means expectation in the randomness of $\langle\cdot\rangle_-$ only.

Proof. We "decouple the spins one at a time" for $i \le N-k$, that is, we write

$$\mathcal{E}_i = \mathcal{E}(\sigma_1^1, \sigma_2^2, \dots, \sigma_i^i, \sigma_{i+1}^1, \dots, \sigma_N^1)\,,$$

so that

$$\frac{\langle \mathcal{E}\rangle_-}{\langle \mathrm{Av}\mathcal{E}\rangle_-} = \frac{\langle \mathcal{E}_1\rangle_-}{\langle \mathrm{Av}\mathcal{E}_1\rangle_-}\,;\qquad \frac{\langle \mathcal{E}\rangle_\bullet}{\langle \mathrm{Av}\mathcal{E}\rangle_\bullet} = \frac{\langle \mathcal{E}_{N-k+1}\rangle_-}{\langle \mathrm{Av}\mathcal{E}_{N-k+1}\rangle_-}\,.$$

We bound

$$\mathsf{E}_-\mathrm{Av}\left|\frac{\langle \mathcal{E}_i\rangle_-}{\langle \mathrm{Av}\mathcal{E}_i\rangle_-} - \frac{\langle \mathcal{E}_{i-1}\rangle_-}{\langle \mathrm{Av}\mathcal{E}_{i-1}\rangle_-}\right|\,. \tag{6.157}$$

When i belongs to no set I_j this is zero because then $\mathcal{E}_i = \mathcal{E}_{i-1}$. Suppose otherwise that $i \in I_j$ for a certain $j \le r$. The term (6.157) is bounded by $\mathrm{I} + \mathrm{II}$, where

$$\mathrm{I} = \mathsf{E}_-\mathrm{Av}\left|\frac{\langle \mathcal{E}_i - \mathcal{E}_{i-1}\rangle_-}{\langle \mathrm{Av}\mathcal{E}_i\rangle_-}\right|\,;\qquad \mathrm{II} = \mathsf{E}_-\mathrm{Av}\left|\frac{\langle \mathcal{E}_i\rangle_-\langle \mathrm{Av}(\mathcal{E}_i - \mathcal{E}_{i-1})\rangle_-}{\langle \mathrm{Av}\mathcal{E}_i\rangle_-\langle \mathrm{Av}\mathcal{E}_{i-1}\rangle_-}\right|\,.$$

We first bound the term II. We introduce a "replicated copy" $\mathcal{E}'_i$ of $\mathcal{E}_i$ defined by

$$\mathcal{E}'_i = \mathcal{E}(\sigma_1^{N+1}, \sigma_2^{N+2}, \dots, \sigma_i^{N+i}, \sigma_{i+1}^{N+1}, \dots, \sigma_N^{N+1})$$

and we write

$$\langle \mathcal{E}_i\rangle_-\langle \mathrm{Av}(\mathcal{E}_i - \mathcal{E}_{i-1})\rangle_- = \langle \mathcal{E}'_i \mathrm{Av}(\mathcal{E}_i - \mathcal{E}_{i-1})\rangle_-\,.$$

Exchanging the variables σ_i^i and σ_i^1 exchanges $\mathcal{E}_i$ and $\mathcal{E}_{i-1}$ and changes the sign of the function $f' = \mathcal{E}'_i\mathrm{Av}(\mathcal{E}_i - \mathcal{E}_{i-1})$. Next we prove the inequality

$$|\mathcal{E}_i - \mathcal{E}_{i-1}| \le (2\exp 2S_j)\mathcal{E}_{i-1}\,.$$

To prove this we observe that $\mathcal{E}$ is of the form AB where A does not depend on the ith coordinate and $\exp(-S_j) \le B \le \exp S_j$. Thus with obvious notation $|B_i - B_{i-1}| \le 2\exp S_j \le 2\exp 2S_j B_{i-1}$ and since A does not depend on the ith coordinate we have $A_i = A_{i-1}$ and thus

$$\begin{aligned}|\mathcal{E}_i - \mathcal{E}_{i-1}| &= |A_iB_i - A_{i-1}B_{i-1}| = A_{i-1}|B_i - B_{i-1}|\\ &\le 2\exp 2S_j A_{i-1}B_{i-1} = (2\exp 2S_j)\mathcal{E}_{i-1}\,.\end{aligned}$$

Therefore

$$|\mathrm{Av}(\mathcal{E}_i - \mathcal{E}_{i-1})| \le (2\exp 2S_j)\mathrm{Av}\mathcal{E}_{i-1}$$

and

$$\mathrm{Av}|f'| \le (2\exp 2S_j)\mathrm{Av}\mathcal{E}'_i\mathrm{Av}\mathcal{E}_{i-1}\,. \tag{6.158}$$

Thinking of Av in the left-hand side as averaging over the parameter $\omega = (\sigma_i^\ell)_{N-k<i\le N, \ell \le N+1}$, we see that (6.158) is (6.149) when $A = 2\exp 2S_j$ and $f = \mathrm{Av}\mathcal{E}'_i\mathrm{Av}\mathcal{E}_{i-1}$. Applying (6.150) to the $(N-k)$-spin system we then obtain

$$\mathrm{II} \le (2\exp 2S_j)\frac{K_0 k}{N-k} \; .$$

Proceeding similarly we get the same bound for the term I (in a somewhat simpler manner) and this completes the proof of (6.156). □

Proof of Proposition 6.6.5. We take expected values in (6.156), and we remember as in the Ising case (i.e. when $\sigma_i = \pm 1$) that it suffices to consider the case $N \ge 2k$. □

It will be useful to introduce the following random elements $V_1, \ldots, V_k$ of $\mathcal{D}$. (These depend also on N, but the dependence is kept implicit.) The function V_ℓ is the density with respect to η of the law of $\sigma_{N-k+\ell}$ under Gibbs' measure. Let us denote by Y_k^* the function (6.154) of $\sigma_{N-k+1}, \ldots, \sigma_N$, which, as already noted, is the density with respect to η_k of the law of $\sigma_{N-k+1}, \ldots, \sigma_N$ under Gibbs' measure. Thus V_ℓ is the ℓth -marginal of Y_k^*, that is, it is obtained by averaging Y_k^* over all σ_{N-k+j} for $j \ne \ell$ with respect to η.

Proposition 6.6.7. *We have*

$$\mathsf{E}\Big\| Y_k^* - \bigotimes_{\ell\le k} V_\ell \Big\|_1 \le \frac{Kk^3}{N}\mathsf{E}\exp 2S \; . \tag{6.159}$$

Moreover, if $\mathcal{E}_\ell$ is defined as in (6.84), then

$$\forall \ell \le k\,, \;\; \mathsf{E}\Big\| V_\ell - \frac{\langle \mathcal{E}_\ell\rangle_\bullet}{\langle \mathrm{Av}\mathcal{E}_\ell\rangle_\bullet}\Big\|_1 \le \frac{Kk^2}{N}\mathsf{E}\exp 2S \; . \tag{6.160}$$

The L_1 norm is computed in $L^1(\eta_k)$ in (6.159) and in $L^1(\eta)$ in (6.160). The function $\bigotimes_{\ell\le k} V_\ell$ in (6.159) is of course given by

$$\Big(\bigotimes_{\ell\le k} V_\ell\Big)(\sigma_{N-k+1}, \ldots, \sigma_N) = \prod_{1\le \ell\le k} V_\ell(\sigma_{N-k+\ell}) \; .$$

Proof. Consider the event Ω as in (6.79). Using the L_1-norm notation as in (6.159), (6.155) means that

$$\mathsf{E}1_{\Omega^c}\Big\| \frac{\langle \mathcal{E}\rangle_-}{\langle \mathrm{Av}\mathcal{E}\rangle_-} - \frac{\langle \mathcal{E}\rangle_\bullet}{\langle \mathrm{Av}\mathcal{E}\rangle_\bullet}\Big\|_1 \le \frac{Kk^2}{N}\mathsf{E}\exp 2S \; . \tag{6.161}$$

When Ω does not occur, we have $\mathcal{E} = \prod_{\ell\le k}\mathcal{E}_\ell$, and the quantities $\mathcal{E}_\ell$ depend on different coordinates, so that

$$\langle \mathcal{E}\rangle_\bullet = \prod_{\ell\le k}\langle \mathcal{E}_\ell\rangle_\bullet \; .$$

Also, $\langle \mathcal{E}_\ell\rangle_\bullet$ depends on $\sigma_{N-k+\ell}$ but not on $\sigma_{N-k+\ell'}$ for $\ell \ne \ell'$ and thus

$$\mathrm{Av}\prod_{\ell\le k}\langle\mathcal{E}_\ell\rangle_\bullet=\prod_{\ell\le k}\mathrm{Av}\langle\mathcal{E}_\ell\rangle_\bullet\,.$$

Therefore

$$\frac{\langle\mathcal{E}\rangle_\bullet}{\langle\mathrm{Av}\mathcal{E}\rangle_\bullet}=\prod_{\ell\le k}U_\ell\,, \tag{6.162}$$

where

$$U_\ell=\frac{\langle\mathcal{E}_\ell\rangle_\bullet}{\langle\mathrm{Av}\mathcal{E}_\ell\rangle_\bullet}\,.$$

Let us think of U_ℓ as a function of $\sigma_{N-k+\ell}$ only, so we can write for consistency of notation $\prod_{\ell\le k}U_\ell=\bigotimes_{\ell\le k}U_\ell$. Thus (6.161) means

$$\mathsf{E}1_{\Omega^c}\Big\|Y_k^*-\bigotimes_{\ell\le k}U_\ell\Big\|_1\le\frac{Kk^2}{N}\mathsf{E}\exp 2S\,.$$

Now $\|Y_k^*-\bigotimes_{\ell\le k}U_\ell\|_1\le 2$, and combining with (6.79) we get

$$\mathsf{E}\Big\|Y_k^*-\bigotimes_{\ell\le k}U_\ell\Big\|_1\le\frac{Kk^2}{N}\mathsf{E}\exp 2S\,. \tag{6.163}$$

Now, we have

$$\|V_\ell-U_\ell\|_1\le\Big\|Y_k^*-\bigotimes_{\ell\le k}U_\ell\Big\|_1\,,$$

because the right-hand side is the average over $\sigma_{N-k+1},\ldots,\sigma_N$ of the quantity $|Y_k^*-\bigotimes_{\ell\le k}U_\ell|$, and if one averages over $\sigma_{N-k+\ell'}$ for $\ell'\ne\ell$ inside the absolute value rather than outside one gets the left-hand side. Thus (6.160) follows from (6.163). To deduce (6.159) from (6.163) it suffices to prove that

$$\Big\|\bigotimes_{\ell\le k}V_\ell-\bigotimes_{\ell\le k}U_\ell\Big\|_1\le\sum_{\ell\le k}\|V_\ell-U_\ell\|_1\,. \tag{6.164}$$

This inequality holds whenever $V_\ell,U_\ell\in\mathcal{D}$, and is obvious if "one replaces V_ℓ by U_ℓ one at a time" because

$$\begin{aligned}&\|V_1\otimes\cdots\otimes V_\ell\otimes U_{\ell+1}\otimes\cdots\otimes U_k-V_1\otimes\cdots\otimes V_{\ell-1}\otimes U_\ell\otimes\cdots\otimes U_k\|_1\\&=\|V_1\otimes\cdots\otimes V_{\ell-1}\otimes(V_\ell-U_\ell)\otimes U_{\ell+1}\otimes\cdots\otimes U_k\|_1=\|V_\ell-U_\ell\|_1\end{aligned}$$

since $V_{\ell'},U_{\ell'}\in\mathcal{D}$ for $\ell'\le k$. □

We recall that $Y_{N,k}$ denotes the density with respect to η_k of the law of $\sigma_1,\ldots,\sigma_k$ under Gibbs' measure. Let us denote by Y_ℓ the density with respect to η of the law of σ_ℓ under Gibbs' measure. We observe that $Y_{N,k}$ corresponds to Y_k^* if we use the coordinates $\sigma_1,\ldots,\sigma_k$ rather than $\sigma_{N-k+1},\ldots,\sigma_N$, and similarly $Y_1,\ldots,Y_k$ correspond to $V_1,\ldots,V_k$. Thus (6.159) implies

$$\mathsf{E}\|Y_{N,k} - \bigotimes Y_\ell\|_1 \le \frac{Kk^3}{N}\mathsf{E}\exp 2S\ .$$

Using as in (6.164) that

$$\Big\|\bigotimes_{\ell\le k} Y_\ell - \bigotimes_{\ell\le k} X_\ell\Big\|_1 \le \sum_{\ell\le k}\|Y_\ell - X_\ell\|_1\ ,$$

then (6.159) shows that to prove Theorem 6.6.4, the following estimates suffices.

Theorem 6.6.8. *Assuming (6.151), if $\gamma \le \gamma_0$, given any integer k, we can find i.i.d. random elements $X_1, \ldots, X_k$ in $\mathcal{D}$ with law μ such that*

$$\mathsf{E}\sum_{\ell\le k}\|Y_\ell - X_\ell\|_1 \le \frac{k^3 K(p,\gamma_0)}{N}\mathsf{E}\exp 2S\ . \tag{6.165}$$

We will prove that statement by induction on N. Denoting by $D(N,\gamma_0,k)$ the quantity

$$\sup_{\gamma\le\gamma_0}\ \inf_{X_1,\ldots,X_k}\ \mathsf{E}\sum_{\ell\le k}\|Y_\ell - X_\ell\|_1\ ,$$

one wishes to prove that

$$D(N,\gamma_0,k) \le \frac{k^3 K}{N}\mathsf{E}\exp 2S\ .$$

For this we relate the N-spin system with the $(N-k)$-spin system. For this purpose, the crucial equation is (6.162). The sequence $V_1, \ldots, V_k$ is distributed as $(Y_1, \ldots, Y_k)$. Moreover, if for $i \le N-k$ we denote by Y_i^- the density with respect to η of the law of σ_i under the Gibbs measure of the $(N-k)$-spin system, we have, recalling the notation (6.135)

$$\langle \mathcal{E}_\ell\rangle_\bullet = \langle \mathcal{E}_\ell\rangle_{\mathbf{Y}}\ ,$$

where $\mathbf{Y} = (Y_1^-, \ldots, Y_{N-k}^-)$, so that (6.160) implies

$$\sum_{\ell\le k}\mathsf{E}\mathbf{1}_{\Omega^c}\left\|V_\ell - \frac{\langle\mathcal{E}_\ell\rangle_{\mathbf{Y}}}{\langle \mathrm{Av}\mathcal{E}_\ell\rangle_{\mathbf{Y}}}\right\|_1 \le \frac{Kk^3}{N}\mathsf{E}\exp 2S\ . \tag{6.166}$$

We can then complete the proof of Theorem 6.6.8 along the same lines as in Theorem 6.4.1. The functions $(\mathcal{E}_\ell)_{\ell\le k}$ do not depend on too many spins. We can use the induction hypothesis and Lemma 6.6.2 to show that we can find a sequence $\mathbf{X} = (X_1, \ldots, X_{N-k+1})$ of identically distributed random elements of $\mathcal{D}$, of law μ_- $(= \mu_{\gamma_-}$, where γ_- is given by (6.74)), so that

$$\mathsf{E}\sum_{\ell\le k}\mathbf{1}_{\Omega^c}\left\|V_\ell-\frac{\langle\mathcal{E}_\ell\rangle_{\mathbf{X}}}{\langle\mathrm{Av}\mathcal{E}_\ell\rangle_{\mathbf{X}}}\right\|_1$$

is not too large. Then the sequence $(\langle\mathcal{E}_\ell\rangle_{\mathbf{X}}/\langle\mathrm{Av}\mathcal{E}_\ell\rangle_{\mathbf{X}})_{\ell\le k}$ is nearly i.i.d. with law $T(\mu_-)$, and hence nearly i.i.d. with law μ. Completing the argument really amounts to copy the proof of Theorem 6.4.1, so this is best left as an easy exercise for the motivated reader. There is nothing else to change either to the proof of Theorem 6.4.13.

We end this section by a challenging technical question. The relevance of this question might not yet be obvious to the reader, but it will become clearer in Chapter 8, after we learn how to approach the "spherical model" through the "Gaussian model". Let us consider the sphere

$$\mathbb{S}_N=\{\boldsymbol{\sigma}\in\mathbb{R}^N\ ;\ \|\boldsymbol{\sigma}\|=\sqrt{N}\}\tag{6.167}$$

and the uniform probability λ_N on $\mathbb{S}_N$.

Research Problem 6.6.9. Assume that the random function θ is Borel measurable, but not necessarily continuous. Investigate the regularity properties of the function

$$t\mapsto\psi(t)=\frac{1}{N}\mathsf{E}\log\int\exp\sum_{k\le M}\theta_k(t\sigma_{i(k,1)},\dots,t\sigma_{i(k,p)})\mathrm{d}\lambda_N(\boldsymbol{\sigma})\ .$$

In particular, if M is a proportion on N, $M=\alpha N$, is it true that for large N the difference $\psi(t)-\psi(1)$ becomes small whenever $|t-1|\le 1/\sqrt{N}$?

The situation here is that, even though each of the individual functions $t\mapsto\theta(t\sigma_{i(k,1)},\dots,t\sigma_{i(k,p)})$ can be wildly discontinuous, these discontinuities should be smoothed out by the integration for λ_N. Even the case θ is not random and $p=1$ does not seem obvious.

6.7 The Power of Convexity

Consider a random convex set V of $\mathbb{R}^p$, and $(V_k)_{k\ge1}$ an i.i.d. sequence of random convex sets distributed like V. Consider random integers $i(k,1)<\dots<i(k,p)$ such that the sets $I_k=\{i(k,1),\dots,i(k,p)\}$ are i.i.d. uniformly distributed over the subsets of $\{1,\dots,N\}$ of cardinality p. Consider the i.i.d. sequence of random convex subsets U_k of $\mathbb{R}^N$ given by

$$\boldsymbol{\sigma}\in U_k\quad\Leftrightarrow\quad(\sigma_{i(k,1)},\dots,\sigma_{i(k,p)})\in V_k\ .$$

We recall that λ_N is the uniform probability measure on the sphere $\mathbb{S}_N$, and that M is a Poisson r.v. of expectation αN.

Research Problem 6.7.1. (Level 3) Prove that, given p, V and α, there is a number a^* such that for N large

$$\frac{1}{N}\log\lambda_N\Big(\mathbb{S}_N\cap\bigcap_{k\le M}U_k\Big)\simeq a^* \tag{6.168}$$

with overwhelming probability, and compute a^*.

The value $a^* = -\infty$ is permitted; in that case we expect that given any number $a > 0$, for N large we have $\lambda_N(\mathbb{S}_N\cap\bigcap_{k\le M}U_k)\le\exp(-aN)$ with overwhelming probability. Problem 6.7.1 makes sense even if the random set V is not convex, but we fear that this case could be considerably more difficult.

Consider a number $\kappa > 0$, and the probability measure η $(=\eta_\kappa)$ on $\mathbb{R}$ of density $\sqrt{\kappa/\pi}\exp(-\kappa x^2)$ with respect to Lebesgue measure. After reading Chapter 8, the reader will be convinced that a good idea to approach Problem 6.7.1 is to first study the following, which in any case is every bit as natural and appealing as Problem 6.7.1.

Research Problem 6.7.2. (Level 3) Prove that, given p, V, α and κ there is a number a^* such for large N we have

$$\frac{1}{N}\log\eta^{\otimes N}\Big(\bigcap_{k\le M}U_k\Big)\simeq a^* \tag{6.169}$$

with overwhelming probability, and compute a^*.

Here again, the value $a^* = -\infty$ is permitted.

Consider a random concave function $\theta\le 0$ on $\mathbb{R}^p$ and assume that

$$V=\{\theta=0\}\,.$$

Then, denoting by $\theta_1,\ldots,\theta_M$ i.i.d. copies of θ, we have

$$\eta^{\otimes N}\Big(\bigcap_{k\le M}U_k\Big)=\lim_{\beta\to\infty}\int\exp\Big(\beta\sum_{k\le M}\theta_k(\sigma_{i(k,1)},\ldots,\sigma_{i(k,p)})\Big)\mathrm{d}\eta^{\otimes N}(\boldsymbol{\sigma})\,. \tag{6.170}$$

Therefore, to prove (6.169) it should be relevant to consider Hamiltonians of the type

$$-H_N(\boldsymbol{\sigma})=\sum_{k\le M}\theta_k(\sigma_{i(k,1)},\ldots,\sigma_{i(k,p)})\,, \tag{6.171}$$

where $\theta_1,\ldots,\theta_k$ are i.i.d. copies of a random concave function $\theta\le 0$. These Hamiltonians never satisfy a condition $\sup_{\sigma_1,\ldots,\sigma_p\in\mathbb{R}}|\theta(\sigma_1,\ldots,\sigma_p)|<\infty$ such as (6.134) unless $\theta\equiv 0$, and we cannot use the results of the previous sections. The purpose of the present section is to show that certain methods we have already used in Chapter 4 allow a significant step in the study of the Hamiltonians (6.171). In particular we will "prove in the limit the fundamental self-consistency equation $\mu = T(\mu)$". We remind the reader that we assume

$$\theta \text{ is concave, } \theta \le 0\,. \tag{6.172}$$

We will also assume that there exists a non random number A (possibly very large) such that θ satisfies the following Lipschitz condition:

$$\forall \sigma_1,\ldots,\sigma_p,\sigma'_1,\ldots,\sigma'_p\,,\quad |\theta(\sigma_1,\ldots,\sigma_p)-\theta(\sigma'_1,\ldots,\sigma'_p)| \le A\sum_{j\le p}|\sigma_j-\sigma'_j|\,. \tag{6.173}$$

The Gibbs measure is defined as usual as the probability measure on $\mathbb{R}^N$ with density with respect to $\eta^{\otimes N}$ that is proportional to $\exp(-H_N(\boldsymbol{\sigma}))$, and $\langle\cdot\rangle$ denotes an average for this Gibbs measure.

Lemma 6.7.3. *There exists a number K (depending on p, A, α and κ) such that we have*

$$\mathsf{E}\left\langle \exp\frac{|\sigma_1|}{K}\right\rangle \le K\,. \tag{6.174}$$

Of course it would be nice if we could improve (6.174) into $\mathsf{E}\langle\exp(\sigma_1^2/K)\rangle \le K$.

Lemma 6.7.4. *The density Y with respect to η of the law of σ_1 under Gibbs' measure satisfies*

$$\forall x,y\in\mathbb{R}\,,\quad Y(y)\le Y(x)\exp rA|y-x| \tag{6.175}$$

where $r=\operatorname{card}\{k\le M\,;\ i(k,1)=1\}$.

This lemma is purely deterministic, and is true for any realization of the disorder. It is good however to observe right away that r is a Poisson r.v. with $\mathsf{E}r=\gamma$, where as usual $\gamma=\alpha p$ and $\mathsf{E}M=\alpha N$.

Proof. Since the density of Gibbs' measure with respect to $\eta^{\otimes N}$ is proportional to $\exp(-H_N(\boldsymbol{\sigma}))$, the function $Y(\sigma_1)$ is proportional to

$$f(\sigma_1)=\int \exp(-H_N(\boldsymbol{\sigma}))\mathrm{d}\eta(\sigma_2)\cdots\mathrm{d}\eta(\sigma_N)\,.$$

We observe now that the Hamiltonian H_N depends on σ_1 only through the terms $\theta_k(\sigma_{i(k,1)},\ldots,\sigma_{i(k,p)})$ for which $i(k,1)=1$ so (6.173) implies that $f(\sigma'_1)\le f(\sigma_1)\exp rA|\sigma'_1-\sigma_1|$ and this in turn implies (6.175). □

Proof of Lemma 6.7.3. We use (6.175) to obtain

$$Y(0)\exp(-rA|x|)\le Y(x)\le Y(0)\exp rA|x|\,. \tag{6.176}$$

Thus, using Jensen's inequality:

$$
\begin{aligned}
1 = \int Y \mathrm{d}\eta \geq Y(0) \int \exp(-rA|x|)\mathrm{d}\eta(x) &\geq Y(0) \exp\left(-rA \int |x| \mathrm{d}\eta(x)\right) \\
&\geq Y(0) \exp\left(-\frac{LrA}{\sqrt{\kappa}}\right) \\
&\geq Y(0) \exp(-rK) ,
\end{aligned}
$$

where, throughout the proof K denotes a number depending on A, κ and p only, that may vary from time to time. Also,

$$
\begin{aligned}
\left\langle \exp \frac{\kappa}{2}\sigma_1^2 \right\rangle &= \int \exp \frac{\kappa}{2} x^2 Y(x) \mathrm{d}\eta(x) \\
&\leq Y(0) \int \exp \frac{\kappa x^2}{2} \exp rA|x| \mathrm{d}\eta(x) \\
&= Y(0)\sqrt{\kappa\pi} \int \exp\left(-\frac{\kappa x^2}{2}\right) \exp rA|x| \mathrm{d}x \\
&\leq KY(0) \exp Kr^2
\end{aligned}
$$

by a standard computation, or simply using that $-\kappa x^2/2 + rA|x| \leq -\kappa x^2/4 + Kx^2$. Combining with (6.176) yields

$$
\left\langle \exp \frac{\kappa}{2}\sigma_1^2 \right\rangle \leq K \exp Kr^2 \tag{6.177}
$$

so that Markov's inequality implies

$$
\langle \mathbf{1}_{\{|\sigma_1| \geq y\}} \rangle \leq K \exp\left(Kr^2 - \frac{\kappa y^2}{2}\right) .
$$

Using this for $y = K'x$, we obtain

$$
r \leq x \quad \Rightarrow \quad \langle \mathbf{1}_{\{|\sigma_1| \geq Kx\}} \rangle \leq K \exp(-x^2) .
$$

Now, since r is a Poisson r.v. with $\mathsf{E}r = \alpha p$ we have $\mathsf{E} \exp r \leq K$, and thus

$$
\mathsf{E}\langle \mathbf{1}_{\{|\sigma_1| \geq Kx\}} \rangle \leq K \exp(-x^2) + \mathsf{P}(r > x) \leq K \exp(-x) ,
$$

from which (6.174) follows. □

The essential fact, to which we turn now, is a considerable generalization of the statement of Theorem 3.1.11 that "the overlap is essentially constant". Throughout the rest of the section, we also assume the following mild condition:

$$
\mathsf{E}\theta^2(0, \ldots, 0) < \infty . \tag{6.178}
$$

Proposition 6.7.5. *Consider functions $f_1, \ldots, f_n$ on $\mathbb{R}$, and assume that for a certain number D we have*

$$|f_k^{(\ell)}(x)| \le D \tag{6.179}$$

for $\ell = 0, 1, 2$ and $k \le n$. Then the function

$$R = R(\boldsymbol{\sigma}^1, \ldots, \boldsymbol{\sigma}^n) = \frac{1}{N} \sum_{i \le N} f_1(\sigma_i^1) \cdots f_n(\sigma_i^n) \tag{6.180}$$

satisfies

$$\mathsf{E}\langle (R - \mathsf{E}\langle R \rangle)^2 \rangle \le \frac{K}{\sqrt{N}} , \tag{6.181}$$

where K depends only on κ, n, D and on the quantity (6.178).

The power of this statement might not be intuitive, but soon we will show that it has remarkable consequences. Throughout the proof, K denotes a number depending only on κ, n, A, D and on the quantity (6.178).

Lemma 6.7.6. *The conditions of Proposition 6.7.5 imply:*

$$\langle (R - \langle R \rangle)^2 \rangle \le \frac{K}{\sqrt{N}} . \tag{6.182}$$

Proof. The Gibbs' measure on $\mathbb{R}^{Nn}$ has a density proportional to

$$\exp\Big(- \sum_{\ell \le n} H_N(\boldsymbol{\sigma}^\ell) - \kappa \sum_{\ell \le n} \|\boldsymbol{\sigma}^\ell\|^2 \Big)$$

with respect to Lebesgue's measure. It is straightforward that the gradient of R at every point has a norm $\le K/\sqrt{N}$, so that

$$R \text{ has a Lipschitz constant } \le \frac{K}{N} . \tag{6.183}$$

Consequently (6.182) follows from (3.17) used for $k = 1$. □

To complete the proof of Proposition 6.7.5 it suffices to show the following.

Lemma 6.7.7. *We have*

$$\mathsf{E}(\langle R \rangle - \mathsf{E}\langle R \rangle)^2 \le \frac{K}{\sqrt{N}} . \tag{6.184}$$

Proof. This proof mimics the Bovier-Gayrard argument of Section 4.5. Writing $\eta_N = \eta^{\otimes N}$, we consider the random convex function

$$\varphi(\lambda) = \frac{1}{N} \log \int \exp\Big(- \sum_{\ell \le n} H_N(\boldsymbol{\sigma}^\ell) - \kappa \sum_{\ell \le n} \|\boldsymbol{\sigma}^\ell\|^2 + \lambda N R \Big) \mathrm{d}\boldsymbol{\sigma}^1 \cdots \mathrm{d}\boldsymbol{\sigma}^n ,$$

so that

$$\varphi'(0) = \langle R \rangle .$$

We will deduce (6.184) from Lemma 4.5.2 used for $k = 1$ and $\delta = 0$, $\lambda_0 = 1/K$, $C_0 = K$, $C_1 = K$, $C_2 = K/N$, and much of the work consists in checking conditions (4.135) to (4.138) of this lemma. Denoting by $\langle \cdot \rangle_\lambda$ an average for the Gibbs' measure with density with respect to Lebesgue's measure proportional to

$$\exp\Big(-\sum_{\ell \le n} H_N(\boldsymbol{\sigma}^\ell) - \kappa \sum_{\ell \le n} \|\boldsymbol{\sigma}^\ell\|^2 + \lambda N R \Big) , \tag{6.185}$$

we have $\varphi'(\lambda) = \langle R \rangle_\lambda$, so $|\varphi'(\lambda)| \le K$ and (4.135) holds for $C_0 = K$. We now prove the key fact that for $\lambda \le \lambda_0 = 1/K$, the function

$$-\sum_{\ell \le n} H_N(\boldsymbol{\sigma}^\ell) - \frac{\kappa}{2} \sum_{\ell \le n} \|\boldsymbol{\sigma}^\ell\|^2 + \lambda N R \tag{6.186}$$

is concave. We observe that (6.179) implies

$$\left| \frac{\partial^2 R}{\partial \sigma_i^\ell \partial \sigma_j^{\ell'}} \right| \le \frac{K}{N} ,$$

and that the left-hand side is zero unless $i = j$. This implies in turn that at every point the second differential D of R satisfies $|D(\mathbf{x}, \mathbf{y})| \le K \|\mathbf{x}\| \|\mathbf{y}\| / N$ for every $\mathbf{x}, \mathbf{y}$ in $\mathbb{R}^{Nn}$. On the other hand, the second differential $D^\sim$ of the function $-\kappa \sum_{\ell \le n} \|\boldsymbol{\sigma}^\ell\|^2/2$ satisfies at every point $D^\sim(\mathbf{x}, \mathbf{x}) = -\kappa \|\mathbf{x}\|^2$ for every $\mathbf{x}$ in $\mathbb{R}^{Nn}$. Therefore if $K\lambda \le \kappa$, at every point the second differential D^* of the function (6.186) satisfies $D^*(\mathbf{x}, \mathbf{x}) \le 0$ for every $\mathbf{x}$ in $\mathbb{R}^{Nn}$, and consequently this function is concave. Then the quantity (6.185) is of the type

$$\exp\Big(U - \frac{\kappa}{2} \sum_{\ell \le n} \|\boldsymbol{\sigma}^\ell\|^2 \Big)$$

where U is concave; we can then use (6.183) and (3.17) to conclude that

$$\varphi''(\lambda) = N \langle (R - \langle R \rangle_\lambda)^2 \rangle_\lambda \le K ,$$

and this proves (4.137) with $\delta = 0$ and hence also (4.136). It remains to prove (4.138). For $j \le N$ let us define

$$-H'_j = \sum_{k \le M, i(k,p) = j} \theta_k(\sigma_{i(k,1)}, \ldots, \sigma_{i(k,p)}) .$$

The r.v.s H'_j are independent, as is made obvious by the representation of H_N given in Exercise 6.2.3. For $m \le N$ we denote by Ξ_m the σ-algebra generated by the r.v.s H'_j for $j \le m$, and we denote by E_m the conditional expectation given Ξ_m, so that we have the identity

$$\mathsf{E}(\varphi(\lambda) - \mathsf{E}\varphi(\lambda))^2 = \sum_{0 \le m < N} \mathsf{E}(\mathsf{E}_{m+1}\varphi(\lambda) - \mathsf{E}_m\varphi(\lambda))^2 .$$

To prove (4.138), it suffices to prove that for any given value of m we have

$$\mathsf{E}(\mathsf{E}_{m+1}\varphi(\lambda) - \mathsf{E}_m\varphi(\lambda))^2 \le \frac{K}{N^2} .$$

Consider the Hamiltonian

$$-H^\sim(\boldsymbol{\sigma}) = -\sum_{j \ne m+1} H'_j \tag{6.187}$$

and

$$\varphi^\sim(\lambda) = \frac{1}{N} \log \int \exp\Big(\sum_{\ell \le n} H^\sim(\boldsymbol{\sigma}^\ell) - \kappa \sum_{\ell \le n} \|\boldsymbol{\sigma}^\ell\|^2 + \lambda N R\Big) \mathrm{d}\boldsymbol{\sigma}^1 \cdots \mathrm{d}\boldsymbol{\sigma}^n .$$

It should be obvious that (since we have omitted the term H'_{m+1} in (6.187))

$$\mathsf{E}_m \varphi^\sim(\lambda) = \mathsf{E}_{m+1}\varphi^\sim(\lambda) ,$$

so that

$$\begin{aligned}
\mathsf{E}(\mathsf{E}_{m+1}\varphi(\lambda) - \mathsf{E}_m\varphi(\lambda))^2 &= \mathsf{E}(\mathsf{E}_{m+1}(\varphi(\lambda) - \varphi^\sim(\lambda)) - \mathsf{E}_m(\varphi(\lambda) - \varphi^\sim(\lambda)))^2 \\
&\le 2\mathsf{E}\big(\mathsf{E}_{m+1}(\varphi(\lambda) - \varphi^\sim(\lambda))\big)^2 \\
&\quad + 2\mathsf{E}\big(\mathsf{E}_m(\varphi(\lambda) - \varphi^\sim(\lambda))\big)^2 \\
&\le 4\mathsf{E}(\varphi(\lambda) - \varphi^\sim(\lambda))^2 .
\end{aligned}$$

Therefore, it suffices to prove that

$$\mathsf{E}(\varphi(\lambda) - \varphi^\sim(\lambda))^2 \le \frac{K}{N^2} . \tag{6.188}$$

Thinking of λ as fixed, let us denote by $\langle \cdot \rangle_\sim$ an average on $\mathbb{R}^{Nn}$ with respect to the probability measure on $\mathbb{R}^{Nn}$ of density proportional to

$$\exp\Big(-\sum_{\ell \le n} H^\sim(\boldsymbol{\sigma}^\ell) - \kappa \sum_{\ell \le n} \|\boldsymbol{\sigma}^\ell\|^2 + \lambda N R\Big) .$$

We observe the identity

$$\varphi(\lambda) - \varphi^\sim(\lambda) = \frac{1}{N} \log \Big\langle \exp\Big(-\sum_{\ell \le n} (H_N(\boldsymbol{\sigma}^\ell) - H^\sim(\boldsymbol{\sigma}^\ell))\Big)\Big\rangle_\sim .$$

Now $H_N = H^{\sim} + H'_{m+1}$ and therefore

$$\varphi(\lambda) - \varphi^{\sim}(\lambda) = \frac{1}{N} \log \left\langle \exp\Bigl(-\sum_{\ell \le n} H'_{m+1}(\boldsymbol{\sigma}^\ell)\Bigr)\right\rangle_{\sim} . \tag{6.189}$$

Since $-H'_{m+1} \le 0$ we have $\varphi(\lambda) - \varphi^{\sim}(\lambda) \le 0$. Let us define

$$r = \operatorname{card}\{k \le M \; ; \; i(k,p) = m+1\} ,$$

the number of terms in H'_{m+1}, so that r is a Poisson r.v. with

$$\mathsf{E} r = \alpha N \frac{\binom{m}{p-1}}{\binom{N}{p}} \le \alpha p .$$

From (6.189) and Jensen's inequality it follows that

$$0 \ge \varphi(\lambda) - \varphi^{\sim}(\lambda) \ge \frac{1}{N}\Bigl\langle -\sum_{\ell \le n} H'_{m+1}(\boldsymbol{\sigma}^\ell)\Bigr\rangle_{\sim} , \tag{6.190}$$

and thus

$$(\varphi(\lambda) - \varphi^{\sim}(\lambda))^2 \le \frac{1}{N^2}\Bigl\langle -\sum_{\ell \le n} H'_{m+1}(\boldsymbol{\sigma}^\ell)\Bigr\rangle_{\sim}^2 \le \frac{1}{N^2}\Bigl\langle \Bigl(\sum_{\ell \le n} H'_{m+1}(\boldsymbol{\sigma}^\ell)\Bigr)^2\Bigr\rangle_{\sim} .$$

Therefore it suffices to prove that for $\ell \le n$ we have

$$\mathsf{E}\langle H'_{m+1}(\boldsymbol{\sigma}^\ell)^2\rangle_{\sim} \le K . \tag{6.191}$$

Writing $a_k = |\theta_k(0,\dots,0)|$ and using (6.173) we obtain

$$|\theta_k(\sigma^\ell_{i_1},\dots,\sigma^\ell_{i_k})| \le a_k + A\sum_{s \le p} |\sigma^\ell_{i_s}| , \tag{6.192}$$

and therefore

$$|H'_{m+1}(\boldsymbol{\sigma}^\ell)| \le \sum_{k \in I} a_k + A \sum_{i \le N} n_i |\sigma^\ell_i| ,$$

where $n_i \in \mathbb{N}$ and $\sum n_i = rp$, because each of the r terms in H'_{m+1} creates at most p terms in the right-hand side. The randomness of H'_{m+1} is independent of the randomness of $\langle\cdot\rangle_{\sim}$, and since $\mathsf{E} r^2 \le K$ and $\mathsf{E} a_k^2 < \infty$, by (6.178) it suffices to prove that if $i \le N$ then $\mathsf{E}\langle(\sigma^\ell_i)^2\rangle_{\sim} \le K$. This is done by basically copying the proof of Lemma 6.7.3. Using (6.183) the density Y with respect to η of the law of σ^ℓ_i under Gibbs' measure satisfies

$$\forall x, y \in \mathbb{R} \; , \; Y(x) \le Y(y) \exp\bigl((r_i A + K_0/N)|x-y|\bigr) ,$$

where $r_i = \operatorname{card}\{k \le M \; ; \; \exists s \le p, i(k,s) = i\}$. The rest is as in Lemma 6.7.3. □

The remarkable consequence of Proposition 6.7.5 we promised can be roughly stated as follows: to make any computation for the Gibbs measure involving only a finite number of spins, we can assume that different spins are independent, both for the Gibbs measure and probabilistically. To make this idea precise, let us recall the notation $\mathcal{D}$ of (6.132) (where now η has density proportional to $\exp(-\kappa x^2)$). Keeping the dependence on N implicit, let us denote by μ $(= \mu_N)$ the law in $\mathcal{D}$ of the density X with respect to η of the law of σ_1 under Gibbs' measure. Let us denote by $\mathbf{X} = (X_1, \dots, X_N)$ an i.i.d. sequence of random elements of law μ and recall the notation $\langle \cdot \rangle_{\mathbf{X}}$ of (6.135).

Theorem 6.7.8. *Consider two integers n, k. Consider continuous bounded functions $U_1, \dots, U_k$ from $\mathbb{R}^n$ to $\mathbb{R}$, and a continuous function $V : \mathbb{R}^k \to \mathbb{R}$. Then*

$$\lim_{N\to\infty} |\mathsf{E}V(\langle U_1(\sigma_1, \dots, \sigma_n)\rangle, \langle U_2(\sigma_1, \dots, \sigma_n)\rangle, \dots, \langle U_k(\sigma_1, \dots, \sigma_n)\rangle)$$
$$-\mathsf{E}V(\langle U_1(\sigma_1, \dots, \sigma_n)\rangle_{\mathbf{X}}, \dots, \langle U_k(\sigma_1, \dots, \sigma_n)\rangle_{\mathbf{X}})| = 0\,. \tag{6.193}$$

We leave to the reader to formulate and prove an even more general statement involving functions on several replicas.

Proof. Since $U_1, \dots, U_k$ are bounded, on their range we can uniformly approximate V by a polynomial, so that it suffices to consider the case where V is a monomial,

$$V(x_1, \dots, x_k) = x_1^{m_1} \cdots x_k^{m_k}\,. \tag{6.194}$$

The next step is to show that we can assume that for each $j \le k$ we have

$$\lim_{(\sigma_1,\dots,\sigma_n)\to\infty} U_j(\sigma_1, \dots, \sigma_n) = 0\,. \tag{6.195}$$

To see this, we first note that without loss of generality we can assume that $|U_j| \le 1$ for each j. Consider for each $j \le k$ a function $U_j^\sim$ with $|U_j^\sim| \le 1$ and assume that for some number S we have

$$\forall i \le n\,,\ |\sigma_i| \le S \Rightarrow U_j^\sim(\sigma_1, \dots, \sigma_n) = U_j(\sigma_1, \dots, \sigma_n)\,. \tag{6.196}$$

Then

$$|U_j(\sigma_1, \dots, \sigma_n) - U_j^\sim(\sigma_1, \dots, \sigma_n)| \le \sum_{s\le n} \mathbf{1}_{\{\sigma_s \ge S\}}\,,$$

and therefore

$$|\langle U_j(\sigma_1, \dots, \sigma_n)\rangle - \langle U_j^\sim(\sigma_1, \dots, \sigma_n)\rangle| \le \sum_{s\le n} \langle \mathbf{1}_{\{\sigma_s \ge S\}}\rangle\,.$$

We note that for numbers $x_1, \dots, x_k$ and $y_1, \dots, y_k$, all bounded by 1, we have the elementary inequality

$$|x_1^{m_1}\cdots x_k^{m_k} - y_1^{m_1}\cdots y_k^{m_k}| \le \sum_{j\le k} m_j|x_j - y_j| \,. \tag{6.197}$$

It then follows that if we set

$$C = \langle U_1(\sigma_1,\ldots,\sigma_n)\rangle^{m_1}\cdots\langle U_k(\sigma_1,\ldots,\sigma_n)\rangle^{m_k}$$
$$C^\sim = \langle U_1^\sim(\sigma_1,\ldots,\sigma_n)\rangle^{m_1}\cdots\langle U_k^\sim(\sigma_1,\ldots,\sigma_n)\rangle^{m_k}$$

then

$$|C - C^\sim| \le \sum_{j\le k} m_j \sum_{s\le n}\langle \mathbf{1}_{\{\sigma_s\ge S\}}\rangle \,,$$

and therefore

$$|\mathsf{E}C - \mathsf{E}C^\sim| \le \sum_{j\le k} m_j \sum_{s\le n}\mathsf{E}\langle \mathbf{1}_{\{\sigma_s\ge S\}}\rangle = n\sum_{j\le k} m_j\mathsf{E}\langle \mathbf{1}_{\{\sigma_1\ge S\}}\rangle \,.$$

By Lemma 6.7.3, the right-hand side can be made small for S large, and since we can choose the functions U_j that satisfy (6.196) and $U_j(\sigma_1,\ldots,\sigma_n) = 0$ if one of the numbers $|\sigma_s|$ is $\ge 2S$, this indeed shows that we can assume (6.195).

A function U_j that satisfies (6.195) can be uniformly approximated by a finite sum of functions of the type

$$f_1(\sigma_1)\cdots f_n(\sigma_n) \,,$$

where $|f_s^{(\ell)}|$ is bounded for $s \le n$ and $\ell = 0,1,2$. By expansion we then reduce to the case where

$$U_j(\sigma_1,\ldots,\sigma_n) = f_{1,j}(\sigma_1)\cdots f_{n,j}(\sigma_n) \tag{6.198}$$

and we can furthermore assume that $|f_{s,j}^{(\ell)}|$ is bounded for $\ell = 0,1,2$ and $s \le n$. Assuming (6.194) and (6.198) we have

$$\begin{aligned} B &:= \mathsf{E}V(\langle U_1(\sigma_1,\ldots,\sigma_n)\rangle,\ldots,\langle U_k(\sigma_1,\ldots,\sigma_n)\rangle) \\ &= \mathsf{E}\langle f_{1,1}(\sigma_1)\cdots f_{n,1}(\sigma_n)\rangle^{m_1}\cdots\langle f_{1,k}(\sigma_1)\cdots f_{n,k}(\sigma_n)\rangle^{m_k} \,. \end{aligned}$$

We will write this expression using replicas. Let $m = m_1 + \cdots + m_k$. Let us write $\{1,\ldots,m\}$ as the disjoint union of sets $I_1,\ldots,I_k$ with $\mathrm{card} I_j = m_j$; and for $\ell \in I_j$ and $s \le n$ let us set

$$g_{s,\ell} = f_{s,j} \,,$$

so that in particular for $\ell \in I_j$ we have $\prod_{s\le n} g_{s,\ell}(\sigma_s) = \prod_{s\le n} f_{s,j}(\sigma_s)$. Then, using independence of replicas in the first equality, we get

$$\begin{aligned} \Big\langle \prod_{\ell\le m}\prod_{s\le n} g_{s,\ell}(\sigma_s^\ell)\Big\rangle &= \prod_{\ell\le m}\Big\langle \prod_{s\le n} g_{s,\ell}(\sigma_s)\Big\rangle \\ &= \Big\langle \prod_{s\le n} f_{s,1}(\sigma_s)\Big\rangle^{m_1}\cdots\Big\langle \prod_{s\le n} f_{s,k}(\sigma_s)\Big\rangle^{m_k} \,, \end{aligned}$$

and therefore

$$B = \mathsf{E}\Big\langle \prod_{\ell\le m}\prod_{s\le n} g_{s,\ell}(\sigma_s^\ell)\Big\rangle = \mathsf{E}\Big\langle \prod_{s\le n}\prod_{\ell\le m} g_{s,\ell}(\sigma_s^\ell)\Big\rangle\,.$$

By symmetry among sites, for any indexes $i_1,\ldots,i_n \le N$, all different, we have

$$B = \mathsf{E}\Big\langle \prod_{s\le n}\prod_{\ell\le m} g_{s,\ell}(\sigma_{i_s}^\ell)\Big\rangle\,. \tag{6.199}$$

Therefore, for a number K that does not depend on N, we have

$$\Big| B - \mathsf{E}\frac{1}{N^n}\sum_{i_1,\ldots,i_n}\Big\langle \prod_{s\le n}\prod_{\ell\le m} g_{s,\ell}(\sigma_{i_s}^\ell)\Big\rangle\Big| \le \frac{K}{N}\,, \tag{6.200}$$

where the summation is over all values of $i_1,\ldots,i_n$. This is seen by using (6.199) for the terms of the summation where all the indices are different and by observing that there are at most KN^{n-1} other terms. Now

$$\frac{1}{N^n}\sum_{i_1,\ldots,i_n}\prod_{s\le n}\prod_{\ell\le m} g_{s,\ell}(\sigma_{i_s}^\ell) = \prod_{s\le n}\Big(\frac{1}{N}\sum_{i\le N}\prod_{\ell\le m} g_{s,\ell}(\sigma_i^\ell)\Big)\,.$$

Defining

$$R_s = \frac{1}{N}\sum_{i\le N}\prod_{\ell\le m} g_{s,\ell}(\sigma_i^\ell)\,,$$

we obtain from (6.200) that

$$\Big| B - \mathsf{E}\Big\langle \prod_{s\le n} R_s\Big\rangle\Big| \le \frac{K}{N}\,.$$

Proposition 6.7.5 shows that for each s we have $\mathsf{E}\langle |R_s - \mathsf{E}R_s|\rangle \le KN^{-1/4}$, so that, replacing in turn each R_s by $\mathsf{E}\langle R_s\rangle$ one at a time,

$$\Big|\mathsf{E}\Big\langle \prod_{s\le n} R_s\Big\rangle - \prod_{s\le n}\mathsf{E}\langle R_s\rangle\Big| \le \frac{K}{N^{1/4}}\,,$$

and therefore

$$\Big| B - \prod_{s\le n}\mathsf{E}\langle R_s\rangle\Big| \le \frac{K}{N^{1/4}}\,. \tag{6.201}$$

Now, using symmetry among sites in the first equality,

$$\mathsf{E}\langle R_s\rangle = \mathsf{E}\Big\langle \prod_{\ell\le m} g_{s,\ell}(\sigma_s^\ell)\Big\rangle = \mathsf{E}\prod_{\ell\le m}\langle g_{s,\ell}(\sigma_s)\rangle = \mathsf{E}\prod_{j\le k}\langle f_{s,j}(\sigma_s)\rangle^{m_j}\,,$$

and we have shown that

$$\lim_{N\to\infty}\left|B-\prod_{s\le n}\mathsf{E}\prod_{j\le k}\langle f_{s,j}(\sigma_s)\rangle^{m_j}\right|=0\,. \tag{6.202}$$

In the special case where V is given by (6.194) and U_j is given by (6.198), we have

$$\mathsf{E}V(\langle U_1(\sigma_1,\ldots,\sigma_n)\rangle_{\mathbf{X}},\ldots,\langle U_k(\sigma_1,\ldots,\sigma_n)\rangle_{\mathbf{X}})=\prod_{s\le n}\mathsf{E}\prod_{j\le k}\langle f_{s,j}(\sigma_s)\rangle^{m_j}\,,$$

so that (6.202) is exactly (6.193) in this special case. As we have shown, this special case implies the general one. □

Given n, k, and a number C, inspection of the previous argument shows that the convergence is uniform over the families of functions $U_1,\ldots,U_k$ that satisfy $|U_1|,\ldots,|U_k|\le C$.

We turn to the main result of this section, the proof that "in the limit $\mu=T(\mu)$". We recall the definition of $\mathcal{E}_r$ as in (6.49), and that r is a Poisson r.v. of expectation αp. Let us denote by $\mathbf{X}=(X_i)_{i\ge1}$ an i.i.d. sequence, where $X_i\in\mathcal{D}$ is a random element of law $\mu=\mu_N$ (the law of the density with respect to η of the law of σ_1 under Gibbs' measure), and let us define $T(\mu)$ as follows: if

$$Y=\frac{\langle\mathcal{E}_r\rangle_{\mathbf{X}}}{\langle\mathrm{Av}\mathcal{E}_r\rangle_{\mathbf{X}}}\in\mathcal{D}\,,$$

then $T(\mu)$ is the law of Y in $\mathcal{D}$. The following asserts in a weak sense that in the limit $T(\mu_N)=\mu_N$.

Theorem 6.7.9. *Consider an integer n, and continuous bounded functions $f_1,\ldots,f_n$ on $\mathbb{R}$. Then*

$$\lim_{N\to\infty}\left|\mathsf{E}\langle f_1(\sigma_1)\rangle\cdots\langle f_n(\sigma_1)\rangle-\mathsf{E}\frac{\langle\mathrm{Av}f_1(\varepsilon)\mathcal{E}_r\rangle_{\mathbf{X}}}{\langle\mathrm{Av}\mathcal{E}_r\rangle_{\mathbf{X}}}\cdots\frac{\langle\mathrm{Av}f_n(\varepsilon)\mathcal{E}_r\rangle_{\mathbf{X}}}{\langle\mathrm{Av}\mathcal{E}_r\rangle_{\mathbf{X}}}\right|=0\,. \tag{6.203}$$

To relate (6.203) with the statement that "$T(\mu)=\mu$", we note that

$$\frac{\langle\mathrm{Av}f_s(\varepsilon)\mathcal{E}_r\rangle_{\mathbf{X}}}{\langle\mathrm{Av}\mathcal{E}_r\rangle_{\mathbf{X}}}=\int Yf_s\mathrm{d}\eta\,,$$

so that writing $X=X_1$, (6.203) means that

$$\lim_{N\to\infty}\left|\mathsf{E}\int f_1X\mathrm{d}\eta\cdots\int f_nX\mathrm{d}\eta-\mathsf{E}\int f_1Y\mathrm{d}\eta\cdots\int f_nY\mathrm{d}\eta\right|=0\,. \tag{6.204}$$

In a weak sense this asserts that in the limit the laws of X (i.e μ) and Y (i.e. $T(\mu)$) coincide.

While we do not know how to prove this directly, in a second stage we will deduce from Theorem 6.7.9 that, as expected,

$$\lim_{N\to\infty} d(\mu_N, T(\mu_N)) = 0 \ , \tag{6.205}$$

where d is the transportation-cost distance.

Let us now explain the strategy to prove (6.203). The basic idea is to combine Theorem 6.7.8 with the cavity method. We find convenient to use the cavity method between an N-spin and an $(N+1)$-spin system. Let us define α' by

$$\alpha'(N+1)\frac{\binom{N}{p}}{\binom{N+1}{p}} = \alpha N \ , \tag{6.206}$$

and let us consider a Poisson r.v. r' with $\mathsf{E}r' = \alpha' p$. The letter r' keeps this meaning until the end of this chapter. For $j \geq 1$, let us consider independent copies θ_j of θ, and sets $\{i(j,1), \ldots, i(j,p-1)\}$ that are uniformly distributed among the subsets of $\{1, \ldots, N\}$ of cardinality $p-1$. Of course we assume that all the randomness there is independent of the randomness of $\langle\cdot\rangle$. Let us define

$$-H(\boldsymbol{\sigma}, \varepsilon) = \sum_{j\leq r'} \theta_j(\sigma_{i(j,1)}, \ldots, \sigma_{i(j,p-1)}, \varepsilon)$$

and $\mathcal{E} = \mathcal{E}(\boldsymbol{\sigma}, \varepsilon) = \exp(-H(\boldsymbol{\sigma}, \varepsilon))$. Recalling the Hamiltonian (6.171), the Hamiltonian $-H' = -H_N - H$ is the Hamiltonian of an $(N+1)$-spin system, where the value of α has been replaced by α' given by (6.206). Let us denote by $\langle\cdot\rangle'$ an average for the Gibbs measure relative to H'. Writing $\varepsilon = \sigma_{N+1}$, symmetry between sites implies

$$\mathsf{E}\langle f_1(\sigma_1)\rangle' \cdots \langle f_n(\sigma_1)\rangle' = \mathsf{E}\langle f_1(\varepsilon)\rangle' \cdots \langle f_n(\varepsilon)\rangle' \ . \tag{6.207}$$

Now, for a function $f = f(\boldsymbol{\sigma}, \varepsilon)$, the cavity formula

$$\langle f\rangle' = \frac{\langle \mathrm{Av} f\mathcal{E}\rangle}{\langle \mathrm{Av}\mathcal{E}\rangle}$$

holds, where Av means integration in ε with respect to η, and where $\mathcal{E} = \mathcal{E}(\boldsymbol{\sigma}, \varepsilon) = \exp(-H(\boldsymbol{\sigma}, \varepsilon))$. We rewrite (6.207) as

$$\mathsf{E}\frac{\langle f_1(\sigma_1)\mathrm{Av}\mathcal{E}\rangle}{\langle \mathrm{Av}\mathcal{E}\rangle} \cdots \frac{\langle f_n(\sigma_1)\mathrm{Av}\mathcal{E}\rangle}{\langle \mathrm{Av}\mathcal{E}\rangle} = \mathsf{E}\frac{\langle \mathrm{Av} f_1(\varepsilon)\mathcal{E}\rangle}{\langle \mathrm{Av}\mathcal{E}\rangle} \cdots \frac{\langle \mathrm{Av} f_n(\varepsilon)\mathcal{E}\rangle}{\langle \mathrm{Av}\mathcal{E}\rangle} \ . \tag{6.208}$$

We will then use Theorem 6.7.8 to approximately compute both sides of (6.208) to obtain (6.203). However an obstacle is that the denominators can be very small, or, in other words, that the function x/y is not continuous at $y = 0$. To solve this problem we consider $\delta > 0$ and we will replace these denominators by $\delta + \langle \mathrm{Av}\mathcal{E}\rangle$.

We will need to take limits as $\delta \to 0$, and in order to be able to exchange these limits with the limits as $N \to \infty$ we need the following.

Lemma 6.7.10. *Assume that $f = f(\boldsymbol{\sigma}, \varepsilon)$ is bounded. Then*

$$\limsup_{\delta \to 0} \sup_N \mathsf{E}\left| \frac{\langle \mathrm{Av} f \mathcal{E} \rangle}{\langle \mathrm{Av} \mathcal{E} \rangle} - \frac{\langle \mathrm{Av} f \mathcal{E} \rangle}{\delta + \langle \mathrm{Av} \mathcal{E} \rangle} \right| = 0 \,.$$

Proof. First, if $|f| \leq C$, we have

$$\left| \frac{\langle \mathrm{Av} f \mathcal{E} \rangle}{\langle \mathrm{Av} \mathcal{E} \rangle} - \frac{\langle \mathrm{Av} f \mathcal{E} \rangle}{\delta + \langle \mathrm{Av} \mathcal{E} \rangle} \right| = \frac{\delta |\langle \mathrm{Av} f \mathcal{E} \rangle|}{\langle \mathrm{Av} \mathcal{E} \rangle (\delta + \langle \mathrm{Av} \mathcal{E} \rangle)} \leq \frac{C\delta}{\delta + \langle \mathrm{Av} \mathcal{E} \rangle} \,.$$

Next, we have

$$\mathsf{E} \frac{\delta}{\delta + \langle \mathrm{Av} \mathcal{E} \rangle} \leq \sqrt{\delta} + \mathsf{P}(\langle \mathrm{Av} \mathcal{E} \rangle \leq \sqrt{\delta}) \,,$$

and, writing $H = H(\boldsymbol{\sigma}, \varepsilon)$,

$$\langle \mathrm{Av} \mathcal{E} \rangle = \langle \mathrm{Av} \exp(-H) \rangle \geq \exp \langle -\mathrm{Av} H \rangle \,,$$

so that

$$\mathsf{P}\big(\langle \mathrm{Av} \mathcal{E} \rangle \leq \sqrt{\delta}\big) \leq \mathsf{P}\left(\langle -\mathrm{Av} H \rangle \geq \log \frac{1}{\sqrt{\delta}} \right) \leq \frac{\mathsf{E} \langle \mathrm{Av} |H| \rangle}{\log(1/\sqrt{\delta})} \,.$$

It follows from (6.173) that

$$|H(\boldsymbol{\sigma}, \varepsilon)| \leq \sum_{j \leq r'} \bigg(|\theta_j(0, \ldots, 0)| + A\Big(\sum_{s \leq p-1} |\sigma_{i(j,s)}| + |\varepsilon| \Big) \bigg) \,,$$

so that (6.178) and Lemma 6.7.3 imply that $\sup_N \mathsf{E} \langle \mathrm{Av} |H| \rangle < \infty$ and the lemma is proved. □

Lemma 6.7.11. *We have*

$$\lim_{N \to \infty} \left| \mathsf{E} \frac{\langle f_1(\sigma_1) \mathrm{Av} \mathcal{E} \rangle}{\langle \mathrm{Av} \mathcal{E} \rangle} \cdots \frac{\langle f_n(\sigma_1) \mathrm{Av} \mathcal{E} \rangle}{\langle \mathrm{Av} \mathcal{E} \rangle} - \mathsf{E} \langle f_1(\sigma_1) \rangle \cdots \langle f_n(\sigma_1) \rangle \right| = 0 \,. \quad (6.209)$$

Proof. Consider the event $\Omega = \Omega_1 \cup \Omega_2 \cap \Omega_3$, where

$$\Omega_1 = \{\exists j \leq r' \,,\, i(j,1) = 1\}$$
$$\Omega_2 = \{\exists j, j' \leq r' \,,\, j \neq j' \,,\, \exists \ell, \ell' \leq p-1 \,,\, i(j,\ell) = i(j',\ell')\} \quad (6.210)$$
$$\Omega_3 = \{(p-1)(r'+1) \leq N\} \,, \quad (6.211)$$

so that as we have used many times we have

$$\mathsf{P}(\Omega) \leq \frac{K}{N} \,. \quad (6.212)$$

Let us now define

$$U = \mathrm{Av} \exp \sum_{1 \le j \le r'} \theta_j(\sigma_{j(p-1)+1}, \ldots, \sigma_{(j+1)(p-1)}, \varepsilon) \tag{6.213}$$

when $(r'+1)(p-1) \le N$ and $U = 1$ otherwise. The reader observes that U depends only on the spins σ_i for $i \ge p$. On Ω^c we have $i(j,1) > 1$ for all $j < r$, and the indexes $i(j,\ell)$ are all different. Thus symmetry between sites implies that for any $\delta > 0$,

$$\begin{aligned} &\mathsf{E}\left(\mathbf{1}_{\Omega^c} \frac{\langle f_1(\sigma_1)\mathrm{Av}\mathcal{E}\rangle}{\delta + \langle \mathrm{Av}\mathcal{E}\rangle} \cdots \frac{\langle f_n(\sigma_1)\mathrm{Av}\mathcal{E}\rangle}{\delta + \langle \mathrm{Av}\mathcal{E}\rangle}\right) \\ &= \mathsf{E}\left(\mathbf{1}_{\Omega^c} \frac{\langle f_1(\sigma_1)U\rangle}{\delta + \langle U\rangle} \cdots \frac{\langle f_n(\sigma_1)U\rangle}{\delta + \langle U\rangle}\right) . \end{aligned} \tag{6.214}$$

We claim that

$$\begin{aligned} \lim_{N\to\infty} \Bigg| &\mathsf{E}\frac{\langle f_1(\sigma_1)U\rangle}{\delta + \langle U\rangle} \cdots \frac{\langle f_n(\sigma_1)U\rangle}{\delta + \langle U\rangle} \\ &- \mathsf{E}\frac{\langle f_1(\sigma_1)\rangle_{\mathbf{x}}\langle U\rangle_{\mathbf{x}}}{\delta + \langle U\rangle_{\mathbf{x}}} \cdots \frac{\langle f_n(\sigma_1)\rangle_{\mathbf{x}}\langle U\rangle_{\mathbf{x}}}{\delta + \langle U\rangle_{\mathbf{x}}}\Bigg| = 0 . \end{aligned} \tag{6.215}$$

To see this we simply use Theorem 6.7.8 given r' and the functions θ_j, $j \le r'$. Since by (6.212) the influence of Ω vanishes in the limit, we get from (6.214) that

$$\begin{aligned} \lim_{N\to\infty} \Bigg| &\mathsf{E}\frac{\langle f_1(\sigma_1)\mathrm{Av}\mathcal{E}\rangle}{\delta + \langle \mathrm{Av}\mathcal{E}\rangle} \cdots \frac{\langle f_n(\sigma_1)\mathrm{Av}\mathcal{E}\rangle}{\delta + \langle \mathrm{Av}\mathcal{E}\rangle} \\ &- \mathsf{E}\frac{\langle f_1(\sigma_1)\rangle_{\mathbf{x}}\langle U\rangle_{\mathbf{x}}}{\delta + \langle U\rangle_{\mathbf{x}}} \cdots \frac{\langle f_n(\sigma_1)\rangle_{\mathbf{x}}\langle U\rangle_{\mathbf{x}}}{\delta + \langle U\rangle_{\mathbf{x}}}\Bigg| = 0 . \end{aligned} \tag{6.216}$$

Without loss of generality we can assume that $|f_s| \le 1$ for each s. The inequality (6.197) and Lemma 6.7.10 yield

$$\begin{aligned} &\limsup_{\delta\to 0}\sup_N \left|\mathsf{E}\frac{\langle f_1(\sigma_1)\mathrm{Av}\mathcal{E}\rangle}{\delta + \langle \mathrm{Av}\mathcal{E}\rangle} \cdots \frac{\langle f_n(\sigma_1)\mathrm{Av}\mathcal{E}\rangle}{\delta + \langle \mathrm{Av}\mathcal{E}\rangle} - \mathsf{E}\frac{\langle f_1(\sigma_1)\mathrm{Av}\mathcal{E}\rangle}{\langle \mathrm{Av}\mathcal{E}\rangle} \cdots \frac{\langle f_n(\sigma_1)\mathrm{Av}\mathcal{E}\rangle}{\langle \mathrm{Av}\mathcal{E}\rangle}\right| \\ &= 0 . \end{aligned} \tag{6.217}$$

Proceeding as in Lemma 6.7.10, we get

$$\limsup_{\delta\to 0}\sup_N \mathsf{E}\left|\frac{\langle U\rangle_{\mathbf{x}}}{\delta + \langle U\rangle_{\mathbf{x}}} - 1\right| = 0 , \tag{6.218}$$

and proceeding as in (6.217) we obtain

$$\limsup_{\delta\to 0}\sup_N \left|\mathsf{E}\langle f_1(\sigma_1)\rangle_{\mathbf{x}} \cdots \langle f_n(\sigma_1)\rangle_{\mathbf{x}} - \mathsf{E}\frac{\langle f_1(\sigma_1)\rangle_{\mathbf{x}}\langle U\rangle_{\mathbf{x}}}{\delta + \langle U\rangle_{\mathbf{x}}} \cdots \frac{\langle f_n(\sigma_1)\rangle_{\mathbf{x}}\langle U\rangle_{\mathbf{x}}}{\delta + \langle U\rangle_{\mathbf{x}}}\right| = 0.$$

Combining with (6.217) and (6.216) proves (6.209) since $\langle f_s(\sigma_1)\rangle = \langle f_s(\sigma_1)\rangle_{\mathbf{x}}$. □

To complete the proof of Theorem 6.7.9, we show the following, where we lighten notation by writing $f_s = f_s(\varepsilon)$.

Lemma 6.7.12. *We have*

$$\lim_{N\to\infty}\left|\mathsf{E}\frac{\langle \mathrm{Av} f_1\mathcal{E}\rangle}{\langle \mathrm{Av}\mathcal{E}\rangle}\cdots\frac{\langle \mathrm{Av} f_n\mathcal{E}\rangle}{\langle \mathrm{Av}\mathcal{E}\rangle} - \mathsf{E}\frac{\langle \mathrm{Av} f_1\mathcal{E}_r\rangle_{\mathbf{x}}}{\langle \mathrm{Av}\mathcal{E}_r\rangle_{\mathbf{x}}}\cdots\frac{\langle \mathrm{Av} f_n\mathcal{E}_r\rangle_{\mathbf{x}}}{\langle \mathrm{Av}\mathcal{E}_r\rangle_{\mathbf{x}}}\right| = 0\,.$$

Proof. We follow the method of Lemma 6.7.11, keeping its notation. For $s \le n$ we define

$$U_s = \mathrm{Av} f_s(\varepsilon)\exp\sum_{1\le j\le r'}\theta(\sigma_{j(p-1)+1},\ldots,\sigma_{(j+1)(p-1)},\varepsilon)$$

when $(r'+1)(p-1)\le N$ and $U_s = 1$ otherwise. Consider $\delta > 0$. Recalling (6.211) and (6.213), symmetry between sites yields

$$\begin{aligned}&\mathsf{E}\left(\mathbf{1}_{\Omega^c}\frac{\langle \mathrm{Av} f_1\mathcal{E}\rangle}{\delta+\langle \mathrm{Av}\mathcal{E}\rangle}\cdots\frac{\langle \mathrm{Av} f_n\mathcal{E}\rangle}{\delta+\langle \mathrm{Av}\mathcal{E}\rangle}\right)\\ &=\mathsf{E}\left(\mathbf{1}_{\Omega^c}\frac{\langle U_1\rangle}{\delta+\langle U\rangle}\cdots\frac{\langle U_n\rangle}{\delta+\langle U\rangle}\right)\,. \qquad (6.219)\end{aligned}$$

Moreover Theorem 6.7.8 implies

$$\lim_{N\to\infty}\left|\mathsf{E}\frac{\langle U_1\rangle}{\delta+\langle U\rangle}\cdots\frac{\langle U_n\rangle}{\delta+\langle U\rangle} - \mathsf{E}\frac{\langle U_1\rangle_{\mathbf{x}}}{\delta+\langle U\rangle_{\mathbf{x}}}\cdots\frac{\langle U_n\rangle_{\mathbf{x}}}{\delta+\langle U\rangle_{\mathbf{x}}}\right| = 0\,.$$

Since the influence of Ω vanishes in the limit, and exchanging again the limits $N\to\infty$ and $\delta\to 0$ as permitted by Lemma 6.7.10 (and a similar argument for the terms $\mathsf{E}\langle U_s\rangle_{\mathbf{x}}/(\delta+\langle U\rangle_{\mathbf{x}})$), we obtain

$$\lim_{N\to\infty}\left|\mathsf{E}\frac{\langle \mathrm{Av} f_1\mathcal{E}\rangle}{\langle \mathrm{Av}\mathcal{E}\rangle}\cdots\frac{\langle \mathrm{Av} f_n\mathcal{E}\rangle}{\langle \mathrm{Av}\mathcal{E}\rangle} - \mathsf{E}\frac{\langle U_1\rangle_{\mathbf{x}}}{\langle U\rangle_{\mathbf{x}}}\cdots\frac{\langle U_n\rangle_{\mathbf{x}}}{\langle U\rangle_{\mathbf{x}}}\right| = 0\,.$$

It then remains only to show that

$$\lim_{N\to\infty}\left|\mathsf{E}\frac{\langle U_1\rangle_{\mathbf{x}}}{\langle U\rangle_{\mathbf{x}}}\cdots\frac{\langle U_n\rangle_{\mathbf{x}}}{\langle U\rangle_{\mathbf{x}}} - \mathsf{E}\frac{\langle \mathrm{Av} f_1\mathcal{E}_r\rangle_{\mathbf{x}}}{\langle \mathrm{Av}\mathcal{E}_r\rangle_{\mathbf{x}}}\cdots\frac{\langle \mathrm{Av} f_n\mathcal{E}_r\rangle_{\mathbf{x}}}{\langle \mathrm{Av}\mathcal{E}_r\rangle_{\mathbf{x}}}\right| = 0\,,$$

which should be obvious by the definitions of U, $\mathcal{E}_r$ and U_s and since r' is a Poisson r.v. and, as $N\to\infty$, $\mathsf{E}r' = \alpha' p \to \alpha p = \mathsf{E}r$. □

We now state the desired strengthening of Theorem 6.7.9.

Theorem 6.7.13. *If d denotes the transportation-cost distance associated to the L^1 norm in $\mathcal{D}$, we have*

$$\lim_{N\to\infty} d(\mu_N, T(\mu_N)) = 0 . \tag{6.220}$$

As we shall see, the sequence $\mu = \mu_N$ is tight, and (6.220) implies that any cluster point of this sequence is a solution of the equation $\mu = T(\mu)$. If we knew that this equation has a unique solution, we would conclude that the sequence (μ_N) converges to this solution, and we could pursue the study of the model and in particular we could compute

$$\lim_{N\to\infty} \frac{1}{N} \mathsf{E} \log \int \exp(-H_N(\boldsymbol{\sigma}) - \kappa\|\boldsymbol{\sigma}\|^2) \mathrm{d}\boldsymbol{\sigma} .$$

Thus, further results seem to depend on the following.

Research Problem 6.7.14. (Level 2) Prove that the equation $\mu = T(\mu)$ has a unique solution.

One really wonders what kind of methods could be used to approach this question. Even if this can be solved, the challenge remains to find situations where in the relation (see (6.170))

$$\begin{aligned} &\mathsf{E}\frac{1}{N} \log \eta^{\otimes N}\Big(\bigcap_{k\le M} U_k\Big) \\ &= \lim_{\beta\to\infty} \frac{1}{N} \mathsf{E} \log \int \exp \beta \sum_{k\le M} \theta_k(\sigma_{i(k,1)}, \dots, \sigma_{i(k,p)}) \mathrm{d}\eta^{\otimes N}(\boldsymbol{\sigma}) \end{aligned}$$

one can exchange the limits $N \to \infty$ and $\beta \to \infty$. A similar problem in a different context will be solved in Chapter 8.

We turn to the technicalities required to prove Theorem 6.7.13. They are not difficult, although it is hard to believe that these measure-theoretic considerations are really relevant to spin glasses. For this reason it seems that the only potential readers for these arguments will be well versed in measure theory. Consequently the proofs (that use a few basic facts of analysis, which can be found in any textbook) will be a bit sketchy.

Lemma 6.7.15. *Consider a number B and*

$$\mathcal{D}(B) = \{f \in \mathcal{D} \; ; \; \forall x, y \; , \; f(y) \le f(x) \exp B|y - x|\} .$$

Then $\mathcal{D}(B)$ is norm-compact in $L^1(\eta)$.

Proof. A function f in $\mathcal{D}(B)$ satisfies

$$f(0) \exp(-B|x|) \le f(x) \le f(0) \exp B|x| ,$$

so that since $\int f(x)\mathrm{d}\eta(x) = 1$, we have $K^{-1} \le f(0) \le K$ where K depends on B and κ only. Moreover $\mathcal{D}(B)$ is equi-continuous on every interval, so a sequence (f_n) in $\mathcal{D}(B)$ has a subsequence that converges uniformly in any interval; since, given any $\varepsilon > 0$, there exists a number x_0 for which

$$f \in \mathcal{D}(B) \ \Rightarrow \int_{|x|\ge x_0} |f(x)|\mathrm{d}\eta(x) \le \varepsilon ,$$

it follows that this subsequence converges in $L^1(\eta)$. □

We recall the number A of (6.173).

Lemma 6.7.16. *For each N and each k we have*

$$\mu(\mathcal{D}(kA)) \ge \mathsf{P}(r \le k) , \tag{6.221}$$

where r is a Poisson r.v. of mean αp.

Proof. This is a reformulation of Lemma 6.7.4 since (6.175) means that $Y \in \mathcal{D}(rA)$. □

Proof of Theorem 6.7.13. The set of probability measures μ on $\mathcal{D}$ that satisfy (6.221) for each k is tight (and consequently is compact for the transportation-cost distance). Assuming if possible that (6.220) fails, we can find $\varepsilon > 0$ and a converging subsequence $(\mu_{N(k)})_{k\ge 1}$ of the sequence (μ_N) such that

$$\forall k , \quad d(\mu_{N(k)}, T(\mu_{N(k)})) \ge \varepsilon .$$

We defined $T(\nu)$ for $\nu = \mu_N$. We leave it to the reader to define (in the same manner) $T(\nu)$ for any probability measure ν on $\mathcal{D}$ and to show that the operator T is continuous for d. So that if we define $\nu = \lim_k \mu_{N(k)}$, then $T(\nu) = \lim_k T(\mu_{N(k)})$ and therefore $d(\nu, T(\nu)) \ge \varepsilon$. In particular we have $\nu \ne T(\nu)$. On the other hand, given continuous bounded functions $f_1, \ldots, f_n$ on $\mathbb{R}$, since μ_N is the law of Y (the density with respect to η of the law of σ_1 under Gibbs's measure) in $\mathcal{D}$ we have

$$\begin{aligned}\mathsf{E}\langle f_1(\sigma_1)\rangle \cdots \langle f_n(\sigma_1)\rangle &= \mathsf{E}\Big(\int f_1 Y \mathrm{d}\eta \cdots \int f_n Y \mathrm{d}\eta\Big) \\ &= \int \Big(\int f_1 Y \mathrm{d}\eta \cdots \int f_n Y \mathrm{d}\eta\Big)\, \mathrm{d}\mu_N(Y) . \end{aligned} \tag{6.222}$$

The map

$$\nu \mapsto \psi(\nu) := \int \Big(\int f_1 Y \mathrm{d}\eta \cdots \int f_n Y \mathrm{d}\eta\Big)\, \mathrm{d}\nu(Y)$$

is continuous for the transportation-cost distance; in fact if $|f_s| \le 1$ for each s, one can easily show that $|\psi(\nu) - \psi(\nu')| \le n d(\nu, \nu')$. Therefore the limit of the right-hand side of (6.222) along the sequence $(N(k))$ is

$$\int \left(\int f_1 Y \mathrm{d}\eta \cdots \int f_n Y \mathrm{d}\eta \right) \mathrm{d}\nu(Y) \, .$$

Also, the definition of $T(\mu_N)$ implies

$$\begin{aligned} &\mathsf{E} \frac{\langle \mathrm{Av} f_1 \mathcal{E}_r \rangle_{\mathbf{X}}}{\langle \mathrm{Av} \mathcal{E}_r \rangle_{\mathbf{X}}} \cdots \frac{\langle \mathrm{Av} f_n \mathcal{E}_r \rangle_{\mathbf{X}}}{\langle \mathrm{Av} \mathcal{E}_r \rangle_{\mathbf{X}}} \\ &= \int \left(\int f_1 Y \mathrm{d}\eta \cdots \int f_n Y \mathrm{d}\eta \right) \mathrm{d}T(\nu_N)(Y) \end{aligned} \qquad (6.223)$$

and the limit of the previous quantity along the sequence $(N(k))$ is

$$\int \left(\int f_1 Y \mathrm{d}\eta \cdots \int f_n Y \mathrm{d}\eta \right) \mathrm{d}T(\nu)(Y) \, .$$

Using (6.203) we get

$$\begin{aligned} &\int \left(\int f_1 Y \mathrm{d}\eta \cdots \int f_n Y \mathrm{d}\eta \right) \mathrm{d}\nu(Y) \\ &= \int \left(\int f_1 Y \mathrm{d}\eta \cdots \int f_n Y \mathrm{d}\eta \right) \mathrm{d}T(\nu)(Y) \, . \end{aligned} \qquad (6.224)$$

We will now show that this identity implies $\nu = T(\nu)$, a contradiction which completes the proof of the theorem. Approximating a function on a bounded set by a polynomial yields that if F is a continuous function of n variables, then

$$\begin{aligned} &\int F \left(\int f_1 Y \mathrm{d}\eta, \ldots, \int f_n Y \mathrm{d}\eta \right) \mathrm{d}\nu(Y) \\ &= \int F \left(\int f_1 Y \mathrm{d}\eta, \ldots, \int f_n Y \mathrm{d}\eta \right) \mathrm{d}T(\nu)(Y) \, . \end{aligned}$$

Consequently,

$$\int \varphi(Y) \mathrm{d}\nu(Y) = \int \varphi(Y) \mathrm{d}T(\nu)(Y) \, , \qquad (6.225)$$

whenever $\varphi(Y)$ is a pointwise limit of a sequence of uniformly bounded functions of the type

$$Y \mapsto F \left(\int f_1 Y \mathrm{d}\eta, \ldots, \int f_n Y \mathrm{d}\eta \right) \, .$$

These include the functions of the type

$$\varphi(Y) = \min \left(1, \min_{k \le k_1} (a_k + \|Y - Y_k\|_1) \right) \, , \qquad (6.226)$$

where a_k are ≥ 0 numbers. This is because

$$\varphi(Y) = \min\left(1, \min_{k \le k_1}\left(a_k + \max\left|\int fY\mathrm{d}\eta - \int fY_k\mathrm{d}\eta\right|\right)\right),$$

where the maximum is over $|f| \le 1$, f continuous. Any $[0,1]$-valued, 1-Lipschitz function φ on $\mathcal{D}$ is the pointwise limit of a sequence of functions of the type (6.226). It then follows that (6.225) implies that $\nu = T(\nu)$. □

6.8 Notes and Comments

The first paper "solving" a comparable model at high temperature is [153].

A version of Theorem 6.5.1 "with replica symmetry breaking" is presented in [115], where the proof of Theorem 6.5.1 given here can be found. This proof is arguably identical to the original proof of [60], but the computations are much simpler. This is permitted by the identification of which property of θ is really used (i.e. (6.117)). Another relevant paper is [78], but it deals only with a very special model.

An interesting feature of the present chapter is that we gain control of the model "in two steps", the first of which is Theorem 6.2.2. It would be esthetically pleasing to find a proof "in one step" of a statement including both Theorems 6.2.2 and 6.4.1.

There is currently intense interest in specific models of the type considered in this chapter, see e.g. [51] and [102].

7. An Assignment Problem

7.1 Introduction

Given positive numbers $c(i,j)$, $i,j \le N$, the assignment problem is to find

$$\min_\sigma \sum_{i\le N} c(i,\sigma(i)) \,, \tag{7.1}$$

where σ ranges over all permutations of $\{1,\ldots,N\}$. In words, if $c(i,j)$ represents the cost of assigning job j to worker i, we want to minimize the total cost when exactly one job is assigned to each worker.

We shall be interested in the random version of the problem, where the numbers $c(i,j)$ are independent and uniformly distributed over $[0,1]$.

Mézard and Parisi [103], [104] studied (7.1) by introducing a suitable Hamiltonian, and conjectured that

$$\lim_{N\to\infty} \mathsf{E} \min_\sigma \sum_{i\le N} c(i,\sigma(i)) = \frac{\pi^2}{6} \,. \tag{7.2}$$

This was proved by D. Aldous [2]. Aldous takes advantage of a feature of the present model, that makes it rather special among the various models we studied: the existence of a "limiting object" (which he discovered [1]).

In a related direction, G. Parisi conjectured the following remarkable identity. If the r.v.s $c(i,j)$ are independent exponential i.e. they satisfy $\mathsf{P}(c(i,j)\ge x) = e^{-x}$ for $x \ge 0$, then we have

$$\mathsf{E} \min_\sigma \sum_{i\le N} c(i,\sigma(i)) = 1 + \frac{1}{2^2} + \cdots + \frac{1}{N^2} \,. \tag{7.3}$$

The link with (7.2) is that it can be shown that if the r.v.s $c(i,j)$ are i.i.d., and their common distribution has a density f on $\mathbb{R}^+$ with respect to Lebesgue measure, then if f is continuous in a neighborhood of 0, the limit in (7.2) depends only on $f(0)$. (The intuition for this is simply that all the numbers $c(i,\sigma(i))$ relevant in the computation of the minimum in (7.2) should be very small for large N, so that only the part of the distribution of $c(i,j)$ close to 0 matters.) Thus it makes no difference to assume that $c(i,j)$ is uniform over $[0,1]$ or is exponential of mean 1.

M. Talagrand, *Mean Field Models for Spin Glasses*, Ergebnisse der Mathematik und ihrer Grenzgebiete. 3. Folge / A Series of Modern Surveys in Mathematics 54, DOI 10.1007/978-3-642-15202-3_7,

Vast generalizations of Parisi's conjecture have been recently proved [109], [96]. Yet the disordered system introduced by Mézard and Parisi remains of interest. This model is obviously akin to the other models we consider; yet it is rather different. In the author's opinion, this model demonstrates well the far-reaching nature of the ideas underlying the theory of mean field models for spin glasses.

It is a great technical challenge to prove rigorously anything at all concerning the original model of Mézard and Parisi. This challenge has yet to be met. We will consider a slightly different model, that turns out to be easier, but still of considerable interest. In this model, we consider two integers M, N, $M \geq N$. We consider independent r.v.s $(c(i,j))_{i \leq N, j \leq M}$ that are uniform over $[0,1]$. The configuration space is the set $\Sigma_{N,M}$ of all one-to-one maps σ from $\{1, \ldots, N\}$ to $\{1, \ldots, M\}$. On this space we consider the Hamiltonian

$$H_{N,M}(\sigma) = \beta N \sum_{i \leq N} c(i, \sigma(i)) , \tag{7.4}$$

where β is a parameter. The reader observes that there is no minus sign in this formula, that is, the Boltzmann factor is

$$\exp\Bigl(-\beta N \sum_{i \leq N} c(i, \sigma(i))\Bigr) .$$

Given a number $\alpha > 0$, we will study the system for $N \to \infty$, $M = \lfloor N(1+\alpha) \rfloor$, and our results will hold for $\beta \leq \beta(\alpha)$, where, unfortunately, $\lim_{\alpha \to 0} \beta(\alpha) = 0$. The original model of Mézard and Parisi is the case $M = N$, i.e. $\alpha = 0$. A step towards understanding this model would be the following.

Research Problem 7.1.1. (Level 2) Extend the results of the present chapter to the case $\beta \leq \beta_0$ where β_0 is independent of α.

Even in the domain $\beta \leq \beta(\alpha)$ our results are in a sense weaker than those of the previous chapters. We do not study the model for given large values of N and M, but only in the limit $N \to \infty$ and $M/N \to \alpha$, and we do not obtain a rate for several of the convergence results.

One of the challenges of the present situation is that it is not obvious how to formulate the correct questions. We expect (under our condition that β is small) that "the spins at two different sites are nearly independent". Here this should mean that when $i_1 \neq i_2$, under Gibbs' measure the variables $\sigma \mapsto \sigma(i_1)$ and $\sigma \mapsto \sigma(i_2)$ are nearly independent. But how could one quantify this phenomenon in a way suitable for a proof by induction?

We consider the partition function

$$Z_{N,M} = \sum_{\sigma} \exp(-H_{N,M}(\sigma)) , \tag{7.5}$$

where the summation is over all possible values of σ in $\Sigma_{N,M}$. Throughout the chapter we write

$$a(i,j) = \exp(-\beta N c(i,j)) , \tag{7.6}$$

so that

$$Z_{N,M} = \sum_{\sigma} \prod_{i \le N} a(i, \sigma(i)) .$$

The cavity method will require removing elements from $\{1,\ldots,N\}$ and $\{1,\ldots,M\}$. Given a set $A \subset \{1,\ldots,N\}$ and a set $B \subset \{1,\ldots,M\}$ such that $N - \operatorname{card} A \le M - \operatorname{card} B$, we write

$$Z_{N,M}(A;B) = \sum_{\sigma} \prod a(i,\sigma(i)) .$$

The product is taken over $i \in \{1,\ldots,N\} \backslash A$ and the sum is taken over the one-to-one maps σ from $\{1,\ldots,N\} \backslash A$ to $\{1,\ldots,M\} \backslash B$. Thus $Z_{N,M} = Z_{N,M}(\emptyset;\emptyset)$. When $A = \{i_1, i_2, \ldots\}$ and $B = \{j_1, j_2, \ldots\}$ we write

$$Z_{N,M}(A,B) = Z_{N,M}(i_1, i_2, \ldots; j_1, j_2, \ldots) .$$

Rather than working directly with Gibbs' measure, we will prove that

$$\frac{Z_{N,M}(i;j)}{Z_{N,M}} \simeq \frac{Z_{N,M}(\emptyset;j)}{Z_{N,M}} \frac{Z_{N,M}(i;\emptyset)}{Z_{N,M}} . \tag{7.7}$$

It should be obvious that this is a very strong property, and that it deals with independence. One can also get convinced that it deals with Gibbs' measure by observing that

$$G(\{\sigma(i) = j\}) = a(i,j) \frac{Z_{N,M}(i,j)}{Z_{N,M}} .$$

We consider the quantities

$$u_{N,M}(j) = \frac{Z_{N,M}(\emptyset;j)}{Z_{N,M}} \, ; \; w_{N,M}(i) = \frac{Z_{N,M}(i;\emptyset)}{Z_{N,M}} . \tag{7.8}$$

These quantities occur in the right-hand side of (7.7). The number $u_{N,M}(j)$ is the Gibbs probability that j does not belong to the image of $\{1,\ldots,N\}$ under the map σ. In particular we have $0 \le u_{N,M}(j) \le 1$. (On the other hand we only know that $w_{N,M}(i) > 0$.)

Having understood that these quantities are important, we would like to know something about the family $(u_{N,M}(j))_{j \le M}$ (or $(w_{N,M}(i))_{i \le N}$). An optimistic thought is that this family looks like an i.i.d. sequence drawn out of a certain distribution, that we would like to describe, probably as a fixed point of a certain operator. Analyzing the problem, it is not very difficult to guess what the operator should be; the unpleasant surprise is that it does not seem obvious that this operator has a fixed point, and this contributes significantly to the difficulty of the problem. In order to state our main result, let us describe this operator. Of course, the motivation behind this definition will become clear only gradually.

Consider a standard Poisson point process on $\mathbb{R}^+$ (that is, its intensity measure is Lebesgue's measure) and denote by $(\xi_i)_{i\geq 1}$ an increasing enumeration of the points it produces. Consider a probability measure η on $\mathbb{R}^+$, and i.i.d. r.v.s $(Y_i)_{i\geq 1}$ distributed according to η, which are independent of the r.v.s ξ_i. We define

$$A(\eta) = \mathcal{L}\Big(\frac{1}{\sum_{i\geq 1} Y_i \exp\left(-\beta\xi_i/(1+\alpha)\right)}\Big) \tag{7.9}$$

$$B(\eta) = \mathcal{L}\Big(\frac{1}{1+\sum_{i\geq 1} Y_i \exp(-\beta\xi_i)}\Big) , \tag{7.10}$$

where of course $\mathcal{L}(X)$ is the law of the r.v. X. The dependence on β and α is kept implicit.

Theorem 7.1.2. *Given $\alpha > 0$, there exists $\beta(\alpha) > 0$ such that for $\beta \leq \beta(\alpha)$ there exists a unique pair μ, ν where μ is a probability measure on $[0,1]$ and ν is a probability measure on $\mathbb{R}^+$ such that*

$$\int x \mathrm{d}\mu(x) = \frac{\alpha}{1+\alpha} \ ; \ \mu = B(\nu) \ ; \ \nu = A(\mu) \ . \tag{7.11}$$

Moreover if $M = \lfloor N(1+\alpha) \rfloor$, we have

$$\mu = \lim_{N\to\infty} \mathcal{L}(u_{N,M}(M)) \ ; \ \nu = \lim_{N\to\infty} \mathcal{L}(w_{N,M}(N)) \ . \tag{7.12}$$

Research Problem 7.1.3. (Level 2) Find a direct proof of the existence of the pair (μ, ν) as in (7.11).

One intrinsic difficulty is that there exists such a pair for each value of α (not too small); so one cannot expect that the operator $B \circ A$ is a contraction for a certain distance. The way we will prove (7.11) is by showing that a cluster point of the sequence $(\mathcal{L}(u_{N,M}(M)), \mathcal{L}(w_{N,M}(N)))$ is a solution of these equations.

While it is not entirely obvious what are the relevant questions one should ask about the system, the following shows that the objects of Theorem 7.1.2 are of central importance.

Theorem 7.1.4. *Given α, for $\beta \leq \beta(\alpha)$ we have*

$$\lim_{N\to\infty} \frac{1}{N} \mathsf{E} \log Z_{N,M} = -(1+\alpha) \int \log x \, \mathrm{d}\mu(x) - \int \log x \, \mathrm{d}\nu(x) \ . \tag{7.13}$$

7.2 Overview of the Proof

In this section we try to describe the overall strategy. The following fundamental identities are proved in Lemma 7.3.4 below

$$u_{N,M}(M) = \frac{1}{1 + \sum_{k \le N} a(k,M)\, w_{N,M-1}(k)} \tag{7.14}$$

$$w_{N,M}(N) = \frac{1}{\sum_{\ell \le M} a(N,\ell)\, u_{N-1,M}(\ell)} \,. \tag{7.15}$$

Observe that in the right-hand side of (7.14) the r.v.s $a(k, M)$ are independent of the numbers $w_{N,M-1}(k)$, and similarly in (7.15). We shall prove that

$$w_{N,M}(k) \simeq w_{N,M-1}(k) \simeq w_{N,M-2}(k) \,. \tag{7.16}$$

This fact is not easy. It is intimately connected to equation (7.7), and is rigorously established in Theorem 7.4.7 below.

Once we have (7.16) we see from (7.14) that

$$u_{N,M}(M) \simeq \frac{1}{1 + \sum_{k \le N} a(k,M)\, w_{N,M-2}(k)} \,, \tag{7.17}$$

and by symmetry between M and $M-1$ that

$$u_{N,M}(M-1) \simeq \frac{1}{1 + \sum_{k \le N} a(k,M-1)\, w_{N,M-2}(k)} \,. \tag{7.18}$$

As a consequence, given the numbers $w_{N,M-2}(k)$, the r.v.s $u_{N,M}(M)$ and $u_{N,M}(M-1)$ are nearly independent. Their common law depends only on the empirical measure

$$\frac{1}{N} \sum_{i \le N} \delta_{w_{N,M-2}(i)} \,,$$

which, by (7.16), is nearly

$$\nu_N = \frac{1}{N} \sum_{i \le N} \delta_{w_{N,M}(i)} \,. \tag{7.19}$$

We consider an independent sequence of r.v.s $(X_k)_{k \ge 1}$ uniformly distributed on $[0,1]$, independent of all the other sources of randomness, and we set

$$a(k) = \exp(-\beta N X_k) \,. \tag{7.20}$$

The reason this sequence is of fundamental importance for the present model is that, given j, the sequence $(a(k,j))_k$ of r.v.s has the same distribution as the sequence $(a(k))_k$, and, given i, this is also the case of the sequence $(a(i,k))_k$.

Consider the random measure $\overline{\mu}_N$ on $[0,1]$ given by

$$\overline{\mu}_N = \mathcal{L}_a\left(\frac{1}{1+\sum_{k\leq N} a(k) w_{N,M}(k)}\right),$$

where $\mathcal{L}_a$ denotes the law in the randomness of the variables $a(k)$, when all the other sources of randomness are fixed.

Thus, given the numbers $w_{N,M}(k)$, the r.v.s $u_{N,M}(M)$ and $u_{N,M}(M-1)$ are nearly independent with common law $\overline{\mu}_N$. By symmetry this is true for each pair of r.v.s $u_{N,M}(j)$ and $u_{N,M}(k)$.

Therefore we expect that the empirical measure

$$\mu_N = \frac{1}{M}\sum_{j\leq M} \delta_{u_{N,M}(j)}$$

is nearly $\overline{\mu}_N$.

Since $\overline{\mu}_N$ is a continuous function of ν_N, it follows that if ν_N is concentrated (in the sense that it is nearly non-random), then such is the case of $\overline{\mu}_N$, that is nearly concentrated around its mean μ'_N, and therefore μ_N itself is concentrated around μ'_N.

We can argue similarly that if μ_N is concentrated around μ'_N, then ν_N must be concentrated around a certain measure ν'_N that can be calculated from μ_N. The hard part of the proof is to get quantitative estimates showing that if β is sufficiently small, then these cross-referential statements can be combined to show that both μ_N and ν_N are concentrated around μ'_N and ν'_N respectively. Now, the way μ'_N is obtained from ν'_N means in the limit that $\mu'_N \simeq B(\nu'_N)$. Similarly, $\nu'_N \simeq A(\mu'_N)$. Also, $\mu'_N = \mathcal{L}(u_{N,M}(M))$ and $\nu'_M = \mathcal{L}(w_{N,M}(N))$, so $\mu = \lim_N \mathcal{L}(u_{N,M}(M))$ and $\nu = \lim_N \mathcal{L}(w_{N,M}(N))$ satisfy $\mu = B(\nu)$ and $\nu = A(\mu)$.

7.3 The Cavity Method

We first collect some simple facts.

Lemma 7.3.1. *If $i \notin A$, we have*

$$Z_{N,M}(A;B) = \sum_{\ell\notin B} a(i,\ell)\, Z_{N,M}(A\cup\{i\};\ B\cup\{\ell\})\,. \tag{7.21}$$

If $j \notin B$, we have

$$Z_{N,M}(A;B) = Z_{N,M}(A;B\cup\{j\}) + \sum_{k\notin A} a(k,j)\, Z_{N,M}(A\cup\{k\};\ B\cup\{j\})\,. \tag{7.22}$$

Proof. One replaces each occurrence of $Z_{N,M}(\cdot;\cdot)$ by its value and one checks that the same terms occur in the left-hand and right-hand sides. □

The following deserves no proof.

Lemma 7.3.2. *If $M \notin B$, we have*

$$Z_{N,M}(A; B \cup \{M\}) = Z_{N,M-1}(A; B) . \tag{7.23}$$

If $N \notin A$, we have

$$Z_{N,M}(A \cup \{N\}; B) = Z_{N-1,M}(A; B) . \tag{7.24}$$

In (7.24), and in similar situations below, we make the convention that $Z_{N-1,M}(\cdot;\cdot)$ is considered for a parameter β' such that $\beta'(N-1) = \beta N$.

The following is also obvious from the definitions, yet it is fundamental.

Lemma 7.3.3. *We have*

$$\sum_{\ell \le M} Z_{N,M}(\emptyset; \ell) = (M - N) Z_{N,M} \tag{7.25}$$

and thus

$$\sum_{\ell \le M} u_{N,M}(\ell) = M - N . \tag{7.26}$$

To prove (7.26) we can also observe that $u_{N,M}(\ell)$ is the Gibbs probability that ℓ does not belong to the image under σ of $\{1, \cdots, N\}$, so that the left-hand side of (7.26) is the expected number of integers that do not belong to this image, i.e. $M - N$. In particular (7.26) implies by symmetry between the values of ℓ that $\mathsf{E} u_{N,M}(M) = (M-N)/M \simeq \alpha/(1+\alpha)$, so that any cluster point μ of the sequence $\mathcal{L}(u_{N,M}(M))$ satisfies $\int x \mathrm{d}\mu(x) = \alpha/(1+\alpha)$.

Lemma 7.3.4. *We have*

$$u_{N,M}(M) = \frac{Z_{N,M-1}}{Z_{N,M}} = \frac{1}{1 + \sum_{k \le N} a(k, M)\, w_{N,M-1}(k)} \tag{7.27}$$

$$w_{N,M}(N) = \frac{Z_{N-1,M}}{Z_{N,M}} = \frac{1}{\sum_{\ell \le M} a(N, \ell)\, u_{N-1,M}(\ell)} . \tag{7.28}$$

Proof. We use (7.22) with $A = B = \emptyset$ and $j = M$ to obtain

$$Z_{N,M} = Z_{N,M}(\emptyset; M) + \sum_{k \le N} a(k, M)\, Z_{N,M}(k; M) .$$

Using (7.23) with $A = \emptyset$ or $A = \{k\}$ and $B = \emptyset$ we get

$$Z_{N,M} = Z_{N,M-1} + \sum_{k\le N} a(k,M)\, Z_{N,M-1}(k;\emptyset)$$
$$= Z_{N,M-1}\Bigl(1 + \sum_{k\le N} a(k,M)\, w_{N,M-1}(k)\Bigr)\,. \tag{7.29}$$

This proves (7.27). The proof of (7.28) is similar, using now (7.21) and (7.24). □

It will be essential to consider the following quantity, where $i \le N$:

$$L_{N,M}(i) = \frac{Z_{N,M}\, Z_{N,M-1}(i;\emptyset) - Z_{N,M}(i;\emptyset)\, Z_{N,M-1}}{Z_{N,M}^2}\,. \tag{7.30}$$

The idea is that (7.7) used for $j = M$ implies that $\mathsf{E}L_{N,M}(i)^2$ is small. (This expectation does not depend on i.) Conversely, if $\mathsf{E}L_{N,M}(i)^2$ is small this implies (7.7) for $j = M$ and hence for all values of j by symmetry.

We will also use the quantity

$$R_{N,M}(j) = \frac{Z_{N,M}\, Z_{N,M-1}(\emptyset;j) - Z_{N,M}(\emptyset;j)\, Z_{N,M-1}}{Z_{N,M}^2}\,. \tag{7.31}$$

It is good to notice that $|R_{N,M}(j)| \le 2$. This follows from (7.23) and the fact that the quantity $Z_{N,M}(A,B)$ decreases as B increases.

The reason for introducing the quantity $R_{N,M}(j)$ is that it occurs naturally when one tries to express $L_{M,N}(i)$ as a function of a smaller system (as the next lemma shows).

Lemma 7.3.5. *We have*

$$L_{N,M}(N) = -\frac{\sum_{\ell\le M-1} a(N,\ell)\, R_{N-1,M}(\ell) - a(N,M)\, u_{N-1,M}(M)^2}{\bigl(\sum_{\ell\le M} a(N,\ell)\, u_{N-1,M}(\ell)\bigr)^2} \tag{7.32}$$

$$R_{N,M}(M-1) = -\frac{\sum_{k\le N} a(k,M)\, L_{N,M-1}(k)}{\bigl(1+\sum_{k\le N} a(k,M)\, w_{N,M-1}(k)\bigr)^2}\,. \tag{7.33}$$

Proof. Using the definition (7.31) of $R_{N,M}(j)$ with $j = M-1$, we have

$$R_{N,M}(M-1) = \frac{Z_{N,M}\, Z_{N,M-1}(\emptyset;M-1) - Z_{N,M}(\emptyset;M-1)\, Z_{N,M-1}}{Z_{N,M}^2}\,. \tag{7.34}$$

As in (7.29), but using now (7.22) with $B = \{M-1\}$ and $j = M$ we obtain:

$$Z_{N,M}(\emptyset;M-1) = Z_{N,M-1}(\emptyset;M-1) + \sum_{k\le N} a(k,M)\, Z_{N,M-1}(k;M-1)\,. \tag{7.35}$$

Using this and (7.29) in the numerator of (7.34), and (7.29) in the denominator, and gathering the terms yields (7.33). The proof of (7.32) is similar. □

We end this section by a technical but essential fact.

Lemma 7.3.6. *We have*

$$\sum_{j\le M-1} R_{N,M}(j) = -u_{N,M}(M) + u_{N,M}(M)^2 \,. \tag{7.36}$$

Proof. From (7.25) we have

$$\sum_{j\le M-1} Z_{N,M}(\emptyset;j) = (M-N)\, Z_{N,M} - Z_{N,M}(\emptyset;M) = (M-N)\, Z_{N,M} - Z_{N,M-1},$$

and changing M into $M-1$ in (7.25) we get

$$\sum_{j\le M-1} Z_{N,M-1}(\emptyset,j) = (M-1-N)\, Z_{N,M-1} \,.$$

These two relations imply (7.36) in a straightforward manner. □

7.4 Decoupling

In this section, we prove (7.7) and, more precisely, the following.

Theorem 7.4.1. *Given $\alpha > 0$, there exists $\beta(\alpha) > 0$ such that if $\beta \le \beta(\alpha)$ and $M = \lfloor N(1+\alpha) \rfloor$, then for $\beta N \ge 1$*

$$\mathsf{E}\, L_{N,M}(N)^2 \le \frac{K(\alpha)}{N} \tag{7.37}$$

$$\mathsf{E}\, R_{N,M}(M-1)^2 \le \frac{K(\alpha)}{N} \,. \tag{7.38}$$

The method of proof consists of using Lemma 7.3.5 to relate $\mathsf{E}R_{N,M}(M-1)^2$ with $\mathsf{E}L_{N,M-1}(N)^2$ and $\mathsf{E}L_{N,M}(N)^2$ with $\mathsf{E}R_{N-1,M}(M-1)^2$, and to iterate these relations. In the right-hand sides of (7.32) and (7.33), we will first take expectation in the quantities $a(N,\ell)$ and $a(k,M)$, that are probabilistically independent of the other quantities (an essential fact). Our first task is to learn how to do this.

We recall the random sequence $a(k) = \exp(-\beta N X_k)$ of (7.20), where (X_k) are i.i.d., uniform over $[0,1]$, and independent of the other sources of randomness. The following lemma is obvious.

Lemma 7.4.2. *We have*

$$\mathsf{E}\, a(k)^p = \frac{1}{\beta p N}(1-\exp(-\beta p N)) \le \frac{1}{\beta p N} \,. \tag{7.39}$$

Lemma 7.4.3. *Consider numbers $(x_k)_{k\le N}$. Then we have*

$$\mathsf{E}\Big(\sum_{k\le N} a(k)\, x_k\Big)^2 \le \Big(\frac{1}{2\beta^2 N} + \frac{1}{2\beta N}\Big) \sum_{k\le N} x_k^2 \,. \tag{7.40}$$

Proof. Using (7.39) we have

$$\mathsf{E}\Big(\sum_{k\le N} a(k)\,x_k\Big)^2 = \sum_{k\le N} x_k^2\,\mathsf{E}\,a(k)^2 + \sum_{k\ne \ell} x_k\,x_\ell\,\mathsf{E}\,a(k)\,\mathsf{E}\,a(\ell)$$
$$\le \frac{1}{2\beta N}\sum_{k\le N} x_k^2 + \Big(\frac{1}{\beta N}\Big)^2 \sum_{k\ne\ell} |x_k|\,|x_\ell|\,.$$

Now, the Cauchy-Schwarz inequality implies:

$$\sum_{k\ne\ell} |x_k|\,|x_\ell| \le \frac{1}{2}\Big(\sum_{k\le N}|x_k|\Big)^2 \le \frac{N}{2}\sum_{k\le N} x_k^2\,.$$ □

Corollary 7.4.4. *If $\beta\le 1$ we have*

$$\mathsf{E}\,R_{N,M}(M-1)^2 \le \frac{1}{\beta^2}\,\mathsf{E}\,L_{N,M-1}(N)^2\,.$$

Proof. From (7.33) we have

$$R_{N,M}(M-1)^2 \le \Big(\sum_{k\le N} a(k,M)\,L_{N,M-1}(k)\Big)^2\,.$$

The sequence $(a(k,M))_{k\le N}$ has the same distribution as the sequence $(a(k))_{k\le N}$, so that taking expectation first in this sequence and using (7.40) we get, assuming without loss of generality that $\beta\le 1$,

$$\mathsf{E}\,R_{N,M}(M-1)^2 \le \frac{1}{\beta^2 N}\sum_{k\le N}\mathsf{E}\,L_{N,M-1}(k)^2 = \frac{1}{\beta^2}\,\mathsf{E}\,L_{N,M-1}(N)^2$$

by symmetry between the values of k. □

This is very crude because in (7.33) the denominator is not of order 1, but seems to be typically much larger. In order however to prove this, we need to know that a proportion of the numbers $(w_{N,M-1}(k))_{k\le M}$ are large. We will prove that this is indeed the case if $\beta\le\beta(\alpha)$, but we do not know it yet. To improve on the present approach it seems that we would need to have this information now. We could not overcome this technical difficulty, that seems related to Research Problem 7.1.1.

We next turn to the task of taking expectation in (7.32). The relation (7.26) is crucial here. Since $0\le u_{N,M}(\ell)\le 1$ and $M-N\simeq N\alpha$, this relation implies that at least a constant proportion of the numbers $(u(\ell))_{\ell\le M} = (u_{N,M}(\ell))_{\ell\le M}$ is not small. To understand what happens, consider an independent sequence X_ℓ uniformly distributed over $[0,1]$ and note that if we reorder the numbers $(NX_\ell)_{\ell\le M}$ by increasing order, they look like the sequence $(\xi_i/(1+\alpha))$ (where $(\xi_i)_{i\ge1}$ is an enumeration of the points of a Poisson point process on $\mathbb{R}^+$). The sum $\sum_{\ell\le N} a(\ell)u(\ell)$ then looks like the sum $\sum_{\ell\le N}\exp(-\beta\xi_\ell/(1+\alpha))u(\sigma(\ell))$ where σ is a random permutation, and it is easy to get convinced that typically it cannot be too small. The precise technical result we need is as follows.

Proposition 7.4.5. *Consider numbers $0\le u(\ell), u'(\ell)\le 1$, for $\ell\le M$. Assume that $\sum_{\ell\le M} u(\ell)\ge 4$ and $\sum_{\ell\le M} u'(\ell)\ge 4$. Consider b with $Nb\le \sum_{\ell\le M} u(\ell)$ and $Nb\le\sum_{\ell\le M} u'(\ell)$. Then if $\beta N\ge 1$ and if $\beta\le b/40$, for any numbers $(y(\ell))_{\ell\le M}$ we have*

$$\mathsf{E}\frac{\left(\sum_{\ell\le M} a(\ell)y(\ell)\right)^2}{\left(\sum_{\ell\le M} a(\ell)u(\ell)\right)^2\left(\sum_{\ell\le M} a(\ell)u'(\ell)\right)^2}\le \frac{L\beta^2}{b^4}\left(\frac{1}{N}\sum_{\ell\le M} y(\ell)\right)^2 + \frac{L\beta^3}{b^6 N}\sum_{\ell\le M} y(\ell)^2\,, \tag{7.41}$$

where $a(\ell)=\exp(-\beta N X_\ell)$ and L denotes a universal constant.

As will be apparent later, an essential feature is that the second term of this bound has a coefficient β^3 (rather than β^2).

Corollary 7.4.6. *If $\beta\le\alpha/80$, $\beta N\ge 1$, $M\ge\lfloor N(1+\alpha)\rfloor$, $M\le 3N$, we have*

$$\mathsf{E}L_{N,M}(N)^2\le\frac{L\beta^3}{\alpha^6}\mathsf{E}R_{N-1,M}(M-1)^2+\frac{K(\alpha)}{N}\,. \tag{7.42}$$

Proof. For $\ell\le M$, let $u(\ell)=u_{N-1,M}(\ell)$, and $a(\ell)=a(N,\ell)$. For $\ell\le M-1$ let $y(\ell)=R_{N-1,M}(\ell)$, and let $y(M)=-u_{N-1,M}(M)^2$. By (7.32) we have

$$L_{N,M}(N)^2=\frac{\left(\sum_{\ell\le M} a(\ell)y(\ell)\right)^2}{\left(\sum_{\ell\le M} a(\ell)u(\ell)\right)^4}\,.$$

We check first that $\sum_{\ell\le M} u(\ell)\ge 4$. Then (7.26) implies

$$\sum_{\ell\le M} u(\ell)=M-(N-1)\ge\lfloor N(1+\alpha)\rfloor-N=\lfloor N\alpha\rfloor\,,$$

and if $\beta\le\alpha/80$ and $\beta N\ge 1$, then $N\alpha\ge 80$ and this is certainly ≥ 4. Also

$$b:=\frac{1}{N}\sum_{\ell\le M} u(\ell)\ge\frac{\lfloor N\alpha\rfloor}{N}\ge\frac{\alpha}{2}$$

if $N\alpha\ge 2$ and in particular if $N\beta\ge 1$ and $\beta\le\alpha/80$. We then have $\beta\le b/40$. Taking expectation in the r.v.s $a(\ell)$, we can now use (7.41) with $u'(\ell)=u(\ell)$ to obtain

$$\mathsf{E}_a L_{N,M}(N)^2 \le \frac{L\beta^2}{\alpha^4}\Big(\frac{1}{N}\sum_{\ell\le M} y(\ell)\Big)^2 + \frac{L\beta^3}{\alpha^6 N}\sum_{\ell\le M} y(\ell)^2\,, \tag{7.43}$$

where E_a denotes expectation in the r.v.s $a(\ell)$ only. By (7.36) we have

$$\Big|\sum_{\ell\le M} y(\ell)\Big| = |u_{N-1,M}(M)| \le 1$$

and $y(M)^2 = u_{N-1,M}(M)^4 \le 1$. Thus (7.43) implies

$$\mathsf{E}_a L_{N,M}(N)^2 \le \frac{K(\alpha)}{N} + \frac{L\beta^3}{\alpha^6 N}\sum_{\ell\le M-1} y(\ell)^2\,. \tag{7.44}$$

To prove (7.42) we simply take expectation in (7.44), using that $M \le 3N$ and observing that $\mathsf{E}y(\ell)^2 = \mathsf{E}R_{N-1,M}(M-1)^2$ for $\ell \le M-1$. □

Proof of Theorem 7.4.1. To avoid trivial complications, we assume $\alpha \le 1$. Let us fix N, let us assume $M = \lfloor N(1+\alpha)\rfloor$, and, for $k \le N$ let us define

$$V(k) = \mathsf{E}\,R_{N-k,M-k}(M-k-1)^2\,.$$

In this definition we assume that the values of $Z_{N-k,M'}$ that are relevant for the computation of $R_{N-k,M-k}$ have been computed with the parameter β replaced by the value β' such that $\beta'(N-k) = \beta N$. We observe that $M-k = \lfloor N(1+\alpha)-k\rfloor \ge \lfloor (N-k)(1+\alpha)\rfloor$ and $M-k \le 3(N-k)$. Combining Corollaries 7.4.6 and 7.4.4, implies that if $\beta'(N-k) = \beta N \ge 1$ and $\beta' \le \alpha/80$ we have

$$V(k) \le \frac{L\beta}{\alpha^6}V(k+1) + \frac{K(\alpha)}{N}\,. \tag{7.45}$$

Let us assume that $k \le N/2$, so that $b' \le 2b$. Then (7.45) holds whenever $\beta \le \alpha/160$. Thus if $L\beta/\alpha^6 \le 1/2$, $k \le N/2$ and $\beta N \ge 1$, we obtain

$$V(k) \le \frac{1}{2}V(k+1) + \frac{K(\alpha)}{N}\,.$$

Combining these relations yields

$$V(0) \le 2^{-k}V(k) + \frac{K(\alpha)}{N} \le 2^{-k+2} + \frac{K(\alpha)}{N}$$

since $V(k) \le 4$. Taking $k \simeq \log N$ proves (7.38), and (7.37) follows by (7.42). □

Theorem 7.4.7. *Under the conditions of Theorem 7.4.1, for $j \le M-1$, $i \le N-1$ we have*

$$\mathsf{E}\left(u_{N,M}(j)-u_{N,M-1}(j)\right)^2 \le \frac{K(\alpha)}{N} \tag{7.46}$$

$$\mathsf{E}\left(u_{N,M}(j)-u_{N-1,M}(j)\right)^2 \le \frac{K(\alpha)}{N} \tag{7.47}$$

$$\mathsf{E}\left(w_{N,M}(i)-w_{N,M-1}(i)\right)^2 \le \frac{K(\alpha)}{N} \tag{7.48}$$

$$\mathsf{E}\left(w_{N,M}(i)-w_{N-1,M}(i)\right)^2 \le \frac{K(\alpha)}{N}\,. \tag{7.49}$$

Proof. The proofs are similar, so we prove only (7.46). We can assume $j = M-1$. Using (7.29) and (7.35) we get

$$\begin{aligned} u_{N,M}(M-1) &= \frac{Z_{N,M}(\emptyset; M-1)}{Z_{N,M}} \\ &= \frac{Z_{N,M-2}}{Z_{N,M-1}}\left(\frac{1+\sum_{k\le N} a(k,M)\,w_{N,M-2}(k)}{1+\sum_{k\le N} a(k,M)\,w_{N,M-1}(k)}\right)\,. \end{aligned}$$

We observe the identity

$$L_{N,M}(i) = \frac{Z_{N,M-1}}{Z_{N,M}}\left(w_{N,M-1}(i) - w_{N,M}(i)\right)\,,$$

which is obvious from (7.30). Using this identity for $M-1$ rather than M, we obtain

$$\begin{aligned} &u_{N,M}(M-1) - u_{N,M-1}(M-1) \\ &= \frac{Z_{N,M-2}}{Z_{N,M-1}}\left(\frac{1+\sum_{k\le N} a(k,N)\,w_{N,M-2}(k)}{1+\sum_{k\le N} a(k,N)\,w_{N,M-1}(k)} - 1\right) \\ &= \frac{\sum_{k\le N} a(k,N)\,L_{N,M-1}(k)}{1+\sum_{k\le N} a(k,N)\,w_{N,M-1}(k)}\,. \end{aligned}$$

Thus (7.47) follows from (7.37) and Lemma 7.4.3. □

We turn to the proof of Proposition 7.4.5, which occupies the rest of this section. It relies on the following probabilistic estimate.

Lemma 7.4.8. *Consider numbers $0 \le u(\ell) \le 1$, and let $b = N^{-1}\sum_{\ell\le M} u(\ell)$. Then if $\beta N \ge 1$ and $\beta \le b/20$ we have for $k \le 8$ that*

$$\mathsf{E}\left(\sum_{\ell\le M} a(\ell)u(\ell)\right)^{-k} \le \frac{L\beta^k}{b^k}\,, \tag{7.50}$$

where $a(\ell)$ is as in (7.20).

There is of course nothing magic about the number 8, this result is true for any other number (with a different condition on β). As the proof is tedious, it is postponed to the end of this section.

Proof of Proposition 7.4.5. First we reduce to the case $u(\ell) = u'(\ell)$ by using that $2cc' \leq c^2 + c'^2$ for

$$c = \Big(\sum_{\ell\leq M} a(\ell)u(\ell)\Big)^{-2} \quad ; \quad c' = \Big(\sum_{\ell\leq M} a(\ell)u'(\ell)\Big)^{-2} .$$

Next, let $\dot{a}(\ell) = a(\ell) - \mathsf{E}a(\ell) = a(\ell) - \mathsf{E}a(1)$, so that

$$\sum_{\ell\leq M} a(\ell)y(\ell) = \mathsf{E}a(1)\Big(\sum_{\ell\leq M} y(\ell)\Big) + \sum_{\ell\leq M} \dot{a}(\ell)y(\ell)$$

and since $\mathsf{E}a(1) \leq 1/(\beta N)$,

$$\Big(\sum_{\ell\leq M} a(\ell)y(\ell)\Big)^2 \leq \frac{2}{\beta^2}\Big(\frac{1}{N}\sum_{\ell\leq M} y(\ell)\Big)^2 + 2\Big(\sum_{\ell\leq M} \dot{a}(\ell)y(\ell)\Big)^2 .$$

Using (7.50) for $k = 4$, it suffices to prove that

$$\mathsf{E}\frac{\big(\sum_{\ell\leq M} \dot{a}(\ell)y(\ell)\big)^2}{\big(\sum_{\ell\leq M} a(\ell)u(\ell)\big)^4} \leq \frac{L\beta^3}{b^6 N}\sum_{\ell\leq M} y(\ell)^2 . \tag{7.51}$$

Expending the square in the numerator of the left-hand side, we see that it equals $\mathrm{I} + \mathrm{II}$, where

$$\mathrm{I} = \sum_{\ell'\leq M} y(\ell')^2 \mathsf{E}\frac{\dot{a}(\ell')^2}{\big(\sum_{\ell\leq M} a(\ell)u(\ell)\big)^4} \tag{7.52}$$

$$\mathrm{II} = \sum_{\ell_1\neq\ell_2} y(\ell_1)y(\ell_2)\mathsf{E}\frac{\dot{a}(\ell_1)\dot{a}(\ell_2)}{\big(\sum_{\ell\leq M} a(\ell)u(\ell)\big)^4} .$$

To bound the terms of I, let us set $S_{\ell'} = \sum_{\ell\neq\ell'} a(\ell)u(\ell)$, so

$$\mathsf{E}\frac{\dot{a}(\ell')^2}{\big(\sum_{\ell\leq M} a(\ell)u(\ell)\big)^4} \leq \mathsf{E}\frac{\dot{a}(\ell')^2}{S_{\ell'}^4} = \mathsf{E}\dot{a}(\ell')^2\mathsf{E}\frac{1}{S_{\ell'}^4}$$

by independence. Now since $\sum_{\ell\leq M} u(\ell) \geq 4$ and $u(\ell') \leq 1$, we have

$$\sum_{\ell\neq\ell'} u(\ell) \geq \frac{3}{4}\sum_{\ell\leq M} u(\ell) \geq \frac{3}{4}b , \tag{7.53}$$

so using (7.50) for $M - 1$ rather than M and $3b/4$ rather than b we get $\mathsf{E}S_{\ell'}^{-4} \leq L\beta^4/b^4$; since $\mathsf{E}\dot{a}(\ell')^2 \leq \mathsf{E}a(\ell')^2 \leq 1/\beta N$, we have proved that, using that $b \leq 1$ in the second inequality

$$\mathrm{I} \le \frac{L\beta^3}{Nb^4} \sum_{\ell \le M} y(\ell)^2 \le \frac{L\beta^3}{Nb^6} \sum_{\ell \le M} y(\ell)^2 \,.$$

To control the term II, let us set

$$S(\ell_1, \ell_2) = \sum_{\ell \ne \ell_1, \ell_2} a(\ell) u(\ell)$$

and

$$U = a(\ell_1)u(\ell_1) + a(\ell_2)u(\ell_2) \ge 0 \,.$$

Thus $\sum_{\ell \le M} a(\ell)u(\ell) = S(\ell_1, \ell_2) + U$. Since $U \ge 0$, a Taylor expansion yields

$$\frac{1}{\left(\sum_{\ell \le M} a(\ell)u(\ell)\right)^4} = \frac{1}{(S(\ell_1,\ell_2))^4} - \frac{4U}{S(\ell_1,\ell_2)^5} + \frac{\mathcal{R}}{S(\ell_1,\ell_2)^6} \tag{7.54}$$

where $|\mathcal{R}| \le 15U^2$. Since $S(\ell_1, \ell_2)$ is independent of $a(\ell_1)$ and $a(\ell_2)$, and since $\mathsf{E}\dot{a}(\ell_1)\dot{a}(\ell_2)U = 0$, multiplying (7.54) by $\dot{a}(\ell_1)\dot{a}(\ell_2)$ and taking expectation we get

$$\begin{aligned} \left| \mathsf{E} \frac{\dot{a}(\ell_1)\dot{a}(\ell_2)}{\left(\sum_{\ell \le M} a(\ell)u(\ell)\right)^4} \right| &\le \mathsf{E} \frac{15|\dot{a}(\ell_1)\dot{a}(\ell_2)|U^2}{S(\ell_1,\ell_2)^6} \\ &= 15\mathsf{E}(|\dot{a}(\ell_1)\dot{a}(\ell_2)|U^2)\mathsf{E}\frac{1}{S(\ell_1,\ell_2)^6} \,. \end{aligned}$$

Since $U^2 \le 2(a(\ell_1)^2 + a(\ell_2)^2)$ and $|\dot{a}(\ell_2)| \le 1$, independence implies

$$\mathsf{E}(|\dot{a}(\ell_1)\dot{a}(\ell_2)|U^2) \le 4\mathsf{E}(|\dot{a}(\ell_1)||\dot{a}(\ell_2)|a(\ell_2)^2) \le 4\mathsf{E}(|\dot{a}(\ell_1)|)\mathsf{E}a(\ell_2)^2 \,.$$

Now, $\mathsf{E}a(\ell)^2 \le 1/(2\beta N)$ and $\mathsf{E}|\dot{a}(\ell)| \le 2\mathsf{E}a(\ell) \le 2/(\beta N)$. Therefore we have

$$\mathsf{E}(|\dot{a}(\ell_1)\dot{a}(\ell_2)|U^2) \le \frac{L}{(\beta N)^2} \,.$$

We also have that $\mathsf{E}S(\ell_1,\ell_2)^{-6} \le L\beta^6/b^6$ by (7.50) (used for $k = 6$ and $M-2$ rather than M, and proceeding as in (7.53)). Thus

$$\begin{aligned} \mathrm{II} &\le \frac{L\beta^4}{b^6 N^2} \sum_{\ell_1 \ne \ell_2} |y(\ell_1)y(\ell_2)| \le \frac{L\beta^4}{b^6N^2}\Big(\sum_{\ell \le M} |y(\ell)|\Big)^2 \\ &\le \frac{L\beta^4}{b^6 N} \sum_{\ell \le M} y(\ell)^2 \,, \end{aligned}$$

and we conclude using that $\beta \le 1$. □

The following prepares the proof of Lemma 7.4.8.

Lemma 7.4.9. *If $\beta N \geq 1$ and $\lambda \geq 1$ we have*

$$\mathsf{E}\exp(-\lambda a(1)) \leq \exp\left(-\frac{\log\lambda}{2\beta N}\right) .$$

Proof. Assume first $\lambda \leq \exp\beta N$, so that $\log\lambda \leq \beta N$ and

$$\mathsf{P}(\lambda a(1) \geq 1) = \mathsf{P}(\exp\beta N X_1 \leq \lambda) = \mathsf{P}\left(X_1 \leq \frac{\log\lambda}{\beta N}\right) = \frac{\log\lambda}{\beta N} .$$

Thus, since $\exp(-x) \leq 1/2$ for $x \geq 1$, we have

$$\begin{aligned}\mathsf{E}\exp(-\lambda a(1)) &\leq 1 - \frac{1}{2}\mathsf{P}(\lambda a(1) \geq 1)\\ &\leq \exp\Big(-\frac{1}{2}\mathsf{P}(\exp(\beta N X_1) \leq \lambda)\Big)\\ &= \exp\Big(-\frac{\log\lambda}{2\beta N}\Big) .\end{aligned}$$

Consider next the case $\lambda \geq \exp\beta N$. Observe first that the function $\theta(x) = x/\log x$ increases for $x \geq e$ so that $\theta(\lambda) \geq \theta(\exp\beta N)$, i.e. $\lambda/\log(\lambda) \geq (\exp\beta N)/\beta N$, that is $\lambda\exp(-\beta N) \geq \log\lambda/\beta N$. Now, since $a(1) \geq \exp(-\beta N)$ we have

$$\mathsf{E}\exp(-\lambda a(1)) \leq \mathsf{E}\exp(-\lambda\exp(-\beta N)) \leq \exp\Big(-\frac{\log\lambda}{\beta N}\Big) . \qquad \square$$

Proof of Lemma 7.4.8. We use the inequality (A.8):

$$\mathsf{P}(Y \leq t) \leq (\exp\lambda t)\mathsf{E}\exp(-\lambda Y) \tag{7.55}$$

for $Y = \sum_{\ell\leq M} a(\ell)u(\ell)$ and any $\lambda \geq 0$. We have

$$\mathsf{E}\exp(-\lambda Y) = \mathsf{E}\exp\left(-\lambda\sum_{\ell\leq M} a(\ell)u(\ell)\right) = \prod_{\ell\leq M}\mathsf{E}\exp(-\lambda u_\ell a(\ell)) .$$

Since $u(\ell) \leq 1$, Hölder's inequality implies

$$\mathsf{E}\exp(-\lambda u_\ell a(\ell)) \leq \big(\mathsf{E}\exp(-\lambda a(\ell))\big)^{u(\ell)} = \big(\mathsf{E}\exp(-\lambda a(1))\big)^{u(\ell)} .$$

Therefore, assuming $\lambda \geq 1$, and using Lemma 7.4.9 in the second line,

$$\begin{aligned}\mathsf{E}\exp(-\lambda Y) &\leq \big(\mathsf{E}\exp(-\lambda a(1))\big)^{\sum_{\ell\leq M} u(\ell)}\\ &\leq \exp\Big(-\Big(\sum_{\ell\leq M} u(\ell)\Big)\frac{\log\lambda}{2\beta N}\Big)\\ &= \exp\left(-\frac{b\log\lambda}{2\beta}\right) ,\end{aligned} \tag{7.56}$$

using that $bN = \sum_{\ell \le M} u(\ell)$. Thus from (7.55) we get

$$\mathsf{P}\left(Y \le \frac{tb}{2e\beta}\right) \le \exp\left(-\frac{b}{2\beta}\left(\log\lambda - \frac{\lambda t}{e}\right)\right) . \tag{7.57}$$

For $t \le 1$, taking $\lambda = e/t$, and since then $\log\lambda - \lambda t/e = \log e/t - 1 = -\log t$, we get

$$\mathsf{P}\left(Y \le \frac{tb}{2e\beta}\right) \le t^{b/2\beta} .$$

Therefore whenever $t \ge 1$, the r.v. $X = 1/Y$ satisfies

$$\mathsf{P}\left(X \ge \frac{2te\beta}{b}\right) \le t^{-b/2\beta} . \tag{7.58}$$

Now we use (A.33) with $F(x) = x^k$ to get, making a change of variable in the second line,

$$\begin{aligned} \mathsf{E}X^k &= \int_0^\infty kt^{k-1}\mathsf{P}(X \ge t)\mathrm{d}t \\ &= \left(\frac{2e\beta}{b}\right)^k \int_0^\infty kt^{k-1}\mathsf{P}\left(X \ge \frac{2e\beta t}{b}\right)\mathrm{d}t . \end{aligned}$$

We bound $\mathsf{P}(X \ge 2e\beta t/b)$ by 1 for $t \le 1$ and using (7.58) for $t \ge 1$ to get

$$\mathsf{E}X^k \le \left(\frac{2e\beta}{b}\right)^k \left(1 + k\int_1^\infty t^{-b/(2\beta)+k-1}\mathrm{d}t\right) = \left(\frac{2e\beta}{b}\right)^k \left(1 + \frac{k}{b/(2\beta) - k}\right) ,$$

from which (7.50) follows since $k \le 8$ and $b/(2\beta) \ge 10$. □

Exercise 7.4.10. Prove that for a r.v. $Y \ge 0$ one has the formula

$$\mathsf{E}Y^{-k} = \frac{1}{(k-1)!}\int_0^\infty t^{k-1}\mathsf{E}\exp(-tY)\mathrm{d}t ,$$

and use it to obtain the previous bound on $\mathsf{E}X^k = \mathsf{E}Y^{-k}$ directly from (7.56).

7.5 Empirical Measures

Throughout the rest of this section, we assume the conditions of Theorem 7.4.1, that is, $\beta N \ge 1$, $M = \lfloor N(1+\alpha)\rfloor$ and $\beta \le \beta(\alpha)$.

Let us pursue our intuition that the sequence $(u_{N,M}(j))_{j \le M}$ looks like it is i.i.d. drawn out of a certain distribution. How do we find this distribution? The obvious candidate is the empirical measure

$$\mu_N = \frac{1}{M} \sum_{j \le M} \delta_{u_{N,M}(j)} \,. \tag{7.59}$$

We will also consider

$$\nu_N = \frac{1}{N} \sum_{i \le N} \delta_{w_{N,M}(i)} \,. \tag{7.60}$$

We recall the sequence $a(k) = \exp(-\beta N X_k)$, where (X_k) are i.i.d., uniform over $[0,1]$ and independent of the other sources of randomness. Consider the random measure $\overline{\mu}_N$ on $[0,1]$ given by

$$\overline{\mu}_N = \mathcal{L}_a\left(\frac{1}{1+\sum_{k\le N} a(k) w_{N,M}(k)}\right) ,$$

where $\mathcal{L}_a$ denotes the law in the randomness of the variables $a(k)$ with all the other sources of randomness fixed. Thus, for a continuous function f on $[0,1]$ we have

$$\int f \mathrm{d}\overline{\mu}_N = \mathsf{E}_a f\left(\frac{1}{1+\sum_{k\le N} a(k) w_{N,M}(k)}\right) ,$$

where E_a denotes expectation in the r.v.s $a(k)$ only. Consider the (non-random) measure $\mu'_N = \mathsf{E}\overline{\mu}_N$, so that

$$\int f \mathrm{d}\mu'_N = \mathsf{E} f\left(\frac{1}{1+\sum_{k\le N} a(k) w_{N,M}(k)}\right) .$$

In this section we shall show that $\mu_N \simeq \mu'_N$, and that, similarly, $\nu_N \simeq \nu'_N$ where

$$\int f \mathrm{d}\nu'_N = \mathsf{E} f\left(\frac{1}{\sum_{\ell\le M} a(\ell) u_{N,M}(\ell)}\right) .$$

In the next section we shall make precise the intuition that "ν'_N determines μ'_N" and "μ'_N determines ν'_N" to conclude the proof of Theorem 7.1.2.

It is helpful to consider an appropriate distance for probability measures. Given two probability measures μ, ν on $\mathbb{R}$, we consider the quantity

$$\Delta(\mu,\nu) = \inf \mathsf{E}(X-Y)^2 ,$$

where the infimum is over the pairs (X,Y) of r.v.s such that X has law μ and Y has law ν. The quantity $\Delta^{1/2}(\mu,\nu)$ is a distance. This statement is not obvious, but is proved in Section A.11, where the reader may find more information. This distance is called Wasserstein's distance between μ and ν. It is of course related to the transportation-cost distance considered in Chapter 6, but is more convenient here. Let us observe that since $\mathsf{E}(X-Y)^2 \ge (\mathsf{E}X - \mathsf{E}Y)^2$ we have

$$\left(\int x \mathrm{d}\mu(x) - \int x \mathrm{d}\nu(x)\right)^2 \le \Delta(\mu,\nu) \,. \tag{7.61}$$

Theorem 7.5.1. *The conditions of Theorem 7.4.1 imply*

$$\lim_{N\to\infty} \mathsf{E}\Delta(\mu_N,\mu'_N) = 0\ ;\ \lim_{N\to\infty} \mathsf{E}\Delta(\nu_N,\nu'_N) = 0\ . \tag{7.62}$$

We first collect some simple facts about Δ.

Lemma 7.5.2. *We have*

$$\Delta\Bigl(\frac{1}{N}\sum_{i\le N}\delta_{x_i}\,,\ \frac{1}{N}\sum_{i\le N}\delta_{y_i}\Bigr) = \inf_\sigma \frac{1}{N}\sum_{i\le N}(x_i - y_{\sigma(i)})^2\ , \tag{7.63}$$

where the infimum is over all permutations σ of $\{1,\ldots,N\}$.

We will use this lemma when $x_i = w_{N,M}(i)$, and almost surely any two of these points are distinct. For this reason, we will give the proof only in the (easier) case where any two of the points x_i (resp. y_i) are distinct.

Proof. The inequality $\le$ should be obvious. To prove the converse inequality, we observe that if X has law $N^{-1}\sum_{i\le N}\delta_{x_i}$ and Y has law $N^{-1}\sum_{i\le N}\delta_{y_i}$, then

$$\mathsf{E}\,(X-Y)^2 = \sum_{i,j\le N}\mathsf{P}\,(X=x_i, Y=y_j)(x_i-y_j)^2\ .$$

We observe that the bistochastic matrices are exactly the matrices $a_{ij} = N\mathsf{P}(X=x_i, Y=y_j)$. Thus the left-hand side of (7.63) is

$$\frac{1}{N}\inf\sum_{i,j\le N} a_{ij}(x_i-y_j)^2\ ,$$

where the infimum is over all bistochastic matrices (a_{ij}). The infimum is attained at an extreme point, and it is a classical result ("Birkhoff's theorem") that this extreme point is a permutation matrix. □

Lemma 7.5.3. *Given numbers $w(k), w'(k) \ge 0$ we have*

$$\mathsf{E}\left(\frac{1}{1+\sum_{k\le N}a(k)w(k)} - \frac{1}{1+\sum_{k\le N}a(k)w'(k)}\right)^2 \le \frac{2}{\beta^2 N}\sum_{k\le N}(w(k)-w'(k))^2\ . \tag{7.64}$$

Consequently

$$\Delta\left(\mathcal{L}\Bigl(\frac{1}{1+\sum_{k\le N}a(k)w(k)}\Bigr), \mathcal{L}\Bigl(\frac{1}{1+\sum_{k\le N}a(k)w'(k)}\Bigr)\right) \le \frac{2}{\beta^2 N}\sum_{k\le N}(w(k)-w'(k))^2\ . \tag{7.65}$$

Proof. We use Lemma 7.4.3 together with the inequality

$$\left(\frac{1}{1+\sum_{k\le N} a(k)w(k)} - \frac{1}{1+\sum_{k\le M} a(k)w'(k)}\right)^2$$
$$\le \left(\sum_{k\le N} a(k)(w(k)-w'(k))\right)^2 . \qquad \square$$

The following fact is crucial.

Lemma 7.5.4. *For any continuous function f we have*

$$\lim_{N\to\infty} \mathsf{E}\left(f(u_{N,M}(M)) - \int f \mathrm{d}\overline{\mu}_N\right)\left(f(u_{N,M}(M-1)) - \int f \mathrm{d}\overline{\mu}_N\right) = 0 . \tag{7.66}$$

Proof. Recalling the numbers $a(k,\ell)$ of (7.6), let us consider

$$u = \frac{1}{1+\sum_{k\le N} a(k,M) w_{N,M-2}(k)} .$$

Using (7.27), (7.64) and (7.48) (with $M-1$ instead of M) we obtain

$$\mathsf{E}(u_{N,M}(M)-u)^2 \le \frac{K}{N} .$$

Exchanging the rôles of M and $M-1$ shows that if

$$u' = \frac{1}{1+\sum_{k\le N} a(k,M-1) w_{N,M-2}(k)}$$

we have

$$\mathsf{E}(u_{N,M}(M-1)-u')^2 = \mathsf{E}(u_{N,M}(M)-u)^2 \le \frac{K}{N} .$$

Therefore to prove (7.66) it suffices to prove that

$$\lim_{N\to\infty} \mathsf{E}\left(f(u) - \int f \mathrm{d}\overline{\mu}_N\right)\left(f(u') - \int f \mathrm{d}\overline{\mu}_N\right) = 0 . \tag{7.67}$$

Now by definition of $\overline{\mu}_N$ we have

$$\mathsf{E}\left(f(u) - \int f \mathrm{d}\overline{\mu}_N\right)\left(f(u') - \int f \mathrm{d}\overline{\mu}_N\right) = \mathsf{E}(f(u)-f(u_1))(f(u')-f(u'_1)) ,$$

where

$$u_1 = \frac{1}{1+\sum_{k\le N} a(k) w_{N,M}(k)} \;;\; u'_1 = \frac{1}{1+\sum_{k\le N} a'(k) w_{N,M}(k)} ,$$

and where $a(k) = \exp(-\beta N X_k)$ and $a'(k) = \exp(-\beta N X'_k)$ are independent of all the other r.v.s involved. Let

$$u_2 = \frac{1}{1 + \sum_{k \le N} a(k) w_{N,M-2}(k)} \ ; \ u'_2 = \frac{1}{1 + \sum_{k \le M} a'(k) w_{N,M-2}(k)} \ .$$

Using again (7.64) and (7.48) we get

$$\mathsf{E}(u_1 - u_2)^2 \le \frac{K}{N} \ ; \ \mathsf{E}(u'_1 - u'_2)^2 \le \frac{K}{N} \ .$$

Therefore, to prove (7.67) it suffices to show that

$$\lim_{N \to \infty} \mathsf{E}(f(u) - f(u_2))(f(u') - f(u'_2)) = 0 \ .$$

Let us denote by E_a expectation only in the r.v.s $a(k), a'(k), a(k, M)$ and $a(k, M-1)$, which are probabilistically independent of the r.v.s $w_{N,M-2}(k)$. Then, by independence,

$$\mathsf{E}_a(f(u) - f(u_2))(f(u') - f(u'_2)) = (\mathsf{E}_a f(u) - \mathsf{E}_a f(u_2))(\mathsf{E}_a f(u') - \mathsf{E}_a f(u'_2)).$$

This is 0 because $\mathsf{E}_a f(u) = \mathsf{E}_a f(u_2)$, as is obvious from the definitions. □

Corollary 7.5.5. *For any continuous function f we have*

$$\lim_{N \to \infty} \mathsf{E}\left(\int f \mathrm{d}\mu_N - \int f \mathrm{d}\overline{\mu}_N\right)^2 = 0 \ . \tag{7.68}$$

Proof. We have

$$\int f \mathrm{d}\mu_N = \frac{1}{M} \sum_{\ell \le M} f(u_{N,M}(\ell))$$

so that, expanding the square and by symmetry

$$\mathsf{E}\left(\int f \mathrm{d}\mu_N - \int f \mathrm{d}\overline{\mu}_N\right)^2 = \frac{1}{M} \mathsf{E}\left(f(u_{N,M}(M)) - \int f \mathrm{d}\overline{\mu}_N\right)^2$$
$$+ \frac{M-1}{M} \mathsf{E}\left(f(u_{N,M}(M)) - \int f \mathrm{d}\overline{\mu}_N\right)\left(f(u_{N,M}(M-1)) - \int f \mathrm{d}\overline{\mu}_N\right) .$$

We conclude with Lemma 7.5.4. □

It is explained in Section A.11 why Wasserstein distance defines the weak topology on the set of probability measures on a compact space. Using (A.73) we see that (7.68) implies the following.

Corollary 7.5.6. *We have*

$$\lim_{N \to \infty} \mathsf{E}\Delta(\mu_N, \overline{\mu}_N) = 0 \ . \tag{7.69}$$

Lemma 7.5.7. *Consider an independent copy* $\widehat{\mu}_N$ *of the random measure* $\overline{\mu}_N$*. Then, recalling that* $\mu'_N = \mathsf{E}\overline{\mu}_N$*, we have*

$$\mathsf{E}\Delta(\overline{\mu}_N, \mu'_N) \le \mathsf{E}\Delta(\overline{\mu}_N, \widehat{\mu}_N) . \tag{7.70}$$

Proof. Let $\mathcal{C}$ be the class of pairs f, g of continuous functions such that

$$\forall x, y \ , \ f(x) + g(y) \le (x-y)^2 \ ,$$

so that by the duality formula (A.74) and since $\mu'_N = \mathsf{E}\overline{\mu}_N = \mathsf{E}\widehat{\mu}_N$,

$$\begin{aligned}\mathsf{E}\Delta(\overline{\mu}_N, \mu'_N) &= \mathsf{E} \sup_{(f,g)\in\mathcal{C}} \left(\int f \mathrm{d}\overline{\mu}_N + \mathsf{E} \int g \mathrm{d}\widehat{\mu}_N \right) \\ &\le \mathsf{E} \sup_{(f,g)\in\mathcal{C}} \left(\int f \mathrm{d}\overline{\mu}_N + \int g \mathrm{d}\widehat{\mu}_N \right) = \mathsf{E}\Delta(\overline{\mu}_N, \widehat{\mu}_N) \ ,\end{aligned}$$

using Jensen's inequality. □

Lemma 7.5.8. *Consider an independent copy* $\nu_{\widetilde{N}}$ *of the random measure* ν_N *defined in (7.60). Then we have*

$$\mathsf{E}\Delta(\overline{\mu}_N, \widehat{\mu}_N) \le \frac{2}{\beta^2} \mathsf{E}\Delta(\nu_N, \nu_{\widetilde{N}}) .$$

Proof. Let $\nu_{\widetilde{N}} = N^{-1} \sum_{k\le N} \delta_{w_{\widetilde{N},M}(k)}$, where $(w_{\widetilde{N},M}(k))_{k\le N}$ is an independent copy of the family $(w_{N,M}(k))_{k\le N}$. By Lemma 7.5.2 we can find a permutation σ with

$$\frac{1}{N} \sum_{k\le N} \left(w_{N,M}(k) - w_{\widetilde{N},M}(\sigma(k)) \right)^2 = \Delta(\nu_N, \nu_{\widetilde{N}})$$

and by Lemma 7.5.3 we get

$$\Delta(\overline{\mu}_N, \widehat{\mu}_N) \le \frac{2}{\beta^2} \Delta(\nu_N, \nu_{\widetilde{N}}) \tag{7.71}$$

where

$$\widehat{\mu}_N = \mathcal{L}_a \left(\frac{1}{1 + \sum_{k\le N} a(k) w_{\widetilde{N},M}(\sigma(k))} \right) = \mathcal{L}_a \left(\frac{1}{1 + \sum_{k\le N} a(k) w_{\widetilde{N},M}(k)} \right) .$$

Taking expectation in (7.71) concludes the proof, since $\widehat{\mu}_N$ is an independent copy of $\widetilde{\mu}_N$. □

Let us observe the inequality

$$\Delta(\mu_1, \mu_2) \le 2(\Delta(\mu_1, \mu_3) + \Delta(\mu_3, \mu_2)) \ , \tag{7.72}$$

which is a consequence of the fact that $\Delta^{1/2}$ is a distance.

Proposition 7.5.9. *We have*

$$\limsup_{N\to\infty} \mathsf{E}\Delta(\mu_N,\mu'_N) \le \frac{4}{\beta^2}\limsup_{N\to\infty} \mathsf{E}\Delta(\nu_N,\nu_N^{\sim})\,. \tag{7.73}$$

Consequently, if $\mu_N^{\sim}$ *denotes an independent copy of the random measure* μ_N*, we have*

$$\limsup_{N\to\infty} \mathsf{E}\Delta(\mu_N,\mu_N^{\sim}) \le \frac{16}{\beta^2}\limsup_{N\to\infty} \mathsf{E}\Delta(\nu_N,\nu_N^{\sim})\,. \tag{7.74}$$

Proof. Inequality (7.72) implies

$$\Delta(\mu_N,\mu'_N) \le 2\Delta(\mu_N,\overline{\mu}_N) + 2\Delta(\overline{\mu}_N,\mu'_N)\,.$$

Therefore (7.69) yields

$$\limsup_{N\to\infty} \mathsf{E}\Delta(\mu_N,\mu'_N) \le 2\limsup_{N\to\infty} \mathsf{E}\Delta(\overline{\mu}_N,\mu'_N)\,.$$

By (7.70) and Lemma 7.5.8 this proves (7.73). To prove (7.74) we simply use (7.72) to write that

$$\Delta(\mu_N,\mu_N^{\sim}) \le 2\Delta(\mu_N,\mu'_N) + 2\Delta(\mu'_N,\mu_N^{\sim})\,,$$

and we note that $\mathsf{E}\Delta(\mu'_N,\mu_N^{\sim}) = \mathsf{E}\Delta(\mu_N,\mu'_N)$. □

At this point we have done half of the work required to prove Theorem 7.5.1. The other half is as follows.

Proposition 7.5.10. *We have*

$$\limsup_{N\to\infty} \mathsf{E}\Delta(\nu_N,\nu'_N) \le \frac{L\beta^3}{\alpha^6}\limsup_{N\to\infty} \mathsf{E}\Delta(\mu_N,\mu_N^{\sim}) \tag{7.75}$$

and

$$\limsup_{N\to\infty} \mathsf{E}\Delta(\nu_N,\nu_N^{\sim}) \le \frac{L\beta^3}{\alpha^6}\limsup_{N\to\infty} \mathsf{E}\Delta(\mu_N,\mu_N^{\sim})\,. \tag{7.76}$$

It is essential there to have a coefficient β^3 rather than β^2. Combining (7.76) and (7.74) shows that

$$\begin{aligned}\limsup_{N\to\infty} \mathsf{E}\Delta(\nu_N,\nu_N^{\sim}) &\le \frac{L\beta^3}{\alpha^6}\limsup_{N\to\infty} \mathsf{E}\Delta(\mu_N,\mu_N^{\sim})\\ &\le \frac{L\beta^3}{\alpha^6}\frac{16}{\beta^2}\limsup_{N\to\infty} \mathsf{E}\Delta(\nu_N,\nu_N^{\sim})\,,\end{aligned}$$

so that if $16L\beta/\alpha^6 < 1$ then

$$\limsup_{N\to\infty} \mathsf{E}\Delta(\nu_N,\nu_N^{\sim}) = \limsup_{N\to\infty} \mathsf{E}\Delta(\mu_N,\mu_N^{\sim}) = 0$$

and (7.73) and (7.75) prove Theorem 7.5.1.

The proof of Proposition 7.5.10 is similar to the proof of Proposition 7.5.9, using (7.41) rather than (7.40). Let us first explain the occurrence of the all important factor β^3 in (7.76).

Lemma 7.5.11. *Consider numbers $u(\ell), u'(\ell) \geq 0$ for $\ell \leq M$ and assume that $\sum_{\ell\leq M} u(\ell) = \sum_{\ell\leq M} u'(\ell) \geq N\alpha/2$. Then we have*

$$\mathsf{E}\left(\frac{1}{\sum_{\ell\leq M} a(\ell)u(\ell)} - \frac{1}{\sum_{\ell\leq M} a(\ell)u'(\ell)}\right)^2 \leq \frac{L\beta^3}{\alpha^6 N}\sum_{\ell\leq M}(u(\ell)-u'(\ell))^2 . \tag{7.77}$$

Consequently we have

$$\Delta\left(\mathcal{L}\left(\frac{1}{\sum_{\ell\leq M} a(\ell)u(\ell)}\right), \mathcal{L}\left(\frac{1}{\sum_{\ell\leq M} a(\ell)u'(\ell)}\right)\right) \leq \frac{L\beta^3}{\alpha^6 N}\sum_{\ell\leq M}(u(\ell)-u'(\ell))^2 . \tag{7.78}$$

Proof. We write

$$\left(\frac{1}{\sum_{\ell\leq M} a(\ell)u(\ell)} - \frac{1}{\sum_{\ell\leq M} a(\ell)u'(\ell)}\right)^2 \leq \frac{\left(\sum_{\ell\leq M}(u(\ell)-u'(\ell))a(\ell)\right)^2}{\left(\sum_{\ell\leq M} u(\ell)a(\ell)\right)^2\left(\sum_{\ell\leq M} u'(\ell)a(\ell)\right)^2} ,$$

and we use (7.41) with $y(\ell) = u(\ell) - u'(\ell)$, so that $\sum_{\ell\leq M} y(\ell) = 0$. □

Consider the random measure $\overline{\nu}_N$ on $\mathbb{R}^+$ given by

$$\overline{\nu}_N = \mathcal{L}_a\left(\frac{1}{\sum_{\ell\leq N} a(\ell)u_{N,M}(\ell)}\right) ,$$

so that $\nu'_N = \mathsf{E}\overline{\nu}_N$. We denote by $\widehat{\nu}_N$ an independent copy of $\overline{\nu}_N$. We recall that $\mu_{\widetilde{N}}$ denotes an independent copy of μ_N.

Lemma 7.5.12. *We have*

$$\mathsf{E}\Delta(\overline{\nu}_N, \widehat{\nu}_N) \leq \frac{L\beta^3}{\alpha^6}\mathsf{E}\Delta(\mu_N, \mu_{\widetilde{N}}) .$$

Proof. Let $\mu_{\widetilde{N}} = M^{-1}\sum_{\ell\leq M}\delta_{u_{\widetilde{N},M}(\ell)}$, where $(u_{\widetilde{N},M}(\ell))_{\ell\leq M}$ is an independent copy of the family $(u_{N,M}(\ell))_{\ell\leq M}$. By Lemma 7.5.2 we can find a permutation σ with

$$\frac{1}{M}\sum_{\ell\leq M}\left(u_{N,M}(\ell) - u_{\widetilde{N},M}(\sigma(\ell))\right)^2 = \Delta(\mu_N, \mu_{\widetilde{N}}) .$$

The essential point now is that (7.26) yields

$$\sum_{\ell\leq M} u_{N,M}(\ell) = \sum_{\ell\leq M} u_{\widetilde{N},M}(\sigma(\ell)) = M - N \geq \alpha N/2 ,$$

so that we can use Lemma 7.5.11 to get

$$\Delta(\overline{\nu}_N, \widehat{\nu}_N) \le \frac{L\beta^3}{\alpha^6} \Delta(\mu_N, \mu_N^\sim) \tag{7.79}$$

where

$$\widehat{\nu}_N = \mathcal{L}_a\left(\frac{1}{\sum_{\ell \le M} a(\ell) u_{N,M}^\sim(\sigma(\ell))}\right) = \mathcal{L}_a\left(\frac{1}{\sum_{\ell \le M} a(\ell) u_{N,M}^\sim(\ell)}\right).$$

Taking expectation in (7.79) concludes the proof, since $\widehat{\nu}_N$ is an independent copy of $\widetilde{\nu}_N$. □

The rest of the arguments in the proof of Proposition 7.5.10 is very similar to the arguments of Proposition 7.5.9. One extra difficulty is that the distributions ν_N (etc.) no longer have compact support. This is bypassed by a truncation argument. Indeed, it follows from (7.28) and (7.50) that

$$\mathsf{E} w_{N,M}^4(i) \le K(\alpha) .$$

If $b \ge 0$ is a truncation level, the quantities $w_{N,M,b}(i) := \min(w_{N,M}(i), b)$ satisfy

$$\mathsf{E}(w_{N,M}(i) - w_{N,M,b}(i))^2 \le \mathsf{E}\big(w_{N,M}^2(i) \mathbf{1}_{\{w_{N,M}(i) \ge b\}}\big) \le \frac{K(\alpha)}{b^2} .$$

If we define $\nu_{N,b} = N^{-1} \sum_{i \le N} \delta_{w_{N,M,b}(i)}$, then

$$\Delta(\nu_{N,b}, \nu_N) \le \frac{1}{N} \sum_{i \le N} (w_{N,M}(i) - w_{N,M,b}(i))^2$$

so that

$$\mathsf{E}\Delta(\nu_{N,b}, \nu_N) \le \frac{K(\alpha)}{b^2} , \tag{7.80}$$

and using such a uniformity, rather than (7.75) it suffices to prove for each b the corresponding result when in the left-hand side "everything is truncated at level b". More specifically, defining $\nu'_{N,b}$ by

$$\int f \mathrm{d}\nu'_{N,b} = \mathsf{E} f\left(\min\left(b, \frac{1}{\sum_{\ell \le M} a(\ell) u_{N,M}(\ell)}\right)\right),$$

one proves that

$$\limsup_{N \to \infty} \mathsf{E}\Delta(\nu_{N,b}, \nu'_{N,b}) \le \frac{L\beta^3}{\alpha^6} \limsup_{N \to \infty} \mathsf{E}\Delta(\mu_N, \mu_N^\sim) ,$$

and one uses that (7.80) implies

$$\limsup_{N \to \infty} \mathsf{E}\Delta(\nu_N, \nu'_N) \le \limsup_{N \to \infty} \mathsf{E}\Delta(\nu_{N,b}, \nu'_{N,b}) + \frac{K(\alpha)}{b^2} .$$

The details are straightforward.

7.6 Operators

The definition of the operators A and B given in (7.9) and (7.10) is pretty, but it does not reflect the property we need. The fundamental property of the operator A is that if the measure $M^{-1}\sum_{\ell\le M}\delta_{u(\ell)}$ approaches the measure μ, the law of $\left(\sum_{\ell\le M} a_N(\ell)u(\ell)\right)^{-1}$ approaches $A(\mu)$, where $a_N(\ell) = \exp(-N\beta X_\ell)$, $M/N \simeq 1+\alpha$, and where of course the r.v.s $(X_\ell)_{\ell\ge 1}$ are i.i.d. uniform over $[0,1]$. Since the description of A given in (7.9) will not be needed, its (non-trivial) equivalence with the definition we will give below in Proposition 7.6.2 will be left to the reader.

In order to prove the existence of the operator A, we must prove that if two measures

$$\frac{1}{M}\sum_{\ell\le M}\delta_{u(\ell)} \quad\text{and}\quad \frac{1}{M'}\sum_{\ell\le M'}\delta_{u'(\ell)}$$

both approach μ, and if $M/N \simeq M'/N'$, then

$$\mathcal{L}\left(\frac{1}{\sum_{\ell\le M} a_N(\ell)u(\ell)}\right) \simeq \mathcal{L}\left(\frac{1}{\sum_{\ell\le M'} a_{N'}(\ell)u'(\ell)}\right) .$$

This technical fact is contained in the following estimate.

Proposition 7.6.1. *Consider a number $\alpha > 0$. Consider integers M, N, M', N' with $N\le M\le 2N$, $N'\le M'\le 2N'$ and numbers $0\le u(\ell)\le 1$ for $\ell\le M$, numbers $0\le u'(\ell)\le 1$ for $\ell\le M'$. Let*

$$\eta = \frac{1}{M}\sum_{\ell\le M}\delta_{u(\ell)}\ ;\ \eta' = \frac{1}{M'}\sum_{\ell\le M'}\delta_{u'(\ell)} .$$

Assume that $\int x\mathrm{d}\eta(x)\ge\alpha/4$ and $\int x\mathrm{d}\eta'(x)\ge\alpha/4$. Assume that $\beta N\ge 1, \beta N'\ge 1$ and $\beta\le\alpha/80$. Then, with $a_N(\ell)=\exp(-\beta N X_\ell)$ as above, we have

$$\begin{aligned}&\Delta\left(\mathcal{L}\left(\frac{1}{\sum_{\ell\le M} a_N(\ell)u(\ell)}\right),\mathcal{L}\left(\frac{1}{\sum_{\ell\le M'} a_{N'}(\ell)u'(\ell)}\right)\right)\\ &\le K(\alpha)\left(\frac{1}{N}+\frac{1}{N'}+\left|\frac{M}{N}-\frac{M'}{N'}\right|\right)\\ &+\frac{L\beta^3}{\alpha^6}\Delta(\eta,\eta')+\frac{L\beta^2}{\alpha^4}\left(\int x\mathrm{d}\eta(x)-\int x\mathrm{d}\eta'(x)\right)^2 . \end{aligned} \tag{7.81}$$

Let us state an important consequence.

Proposition 7.6.2. *Given a number $\alpha>0$ there exists a number $\beta(\alpha)>0$ with the following property. If $\beta\le\beta(\alpha)$ and if μ is a probability measure on*

$[0,1]$ with $\int x\mathrm{d}\mu(x) \geq \alpha/4$, there exists a unique probability measure $A(\mu)$ on $\mathbb{R}^+$ with the following property. Consider numbers $0 \leq u(\ell) \leq 1$ for $\ell \leq M$, and set

$$\eta = \frac{1}{M} \sum_{\ell \leq M} \delta_{u(\ell)} \,.$$

Then

$$\begin{aligned}\Delta\left(A(\mu), \mathcal{L}\left(\frac{1}{\sum_{\ell\leq M} a_N(\ell) u(\ell)}\right)\right) &\leq K(\alpha)\left(\frac{1}{N} + \left|\frac{M}{N} - (1+\alpha)\right|\right) \\ &+ \frac{L\beta^2}{\alpha^4}\left(\int x\mathrm{d}\mu(x) - \int x\mathrm{d}\eta(x)\right)^2 \\ &+ \frac{L\beta^3}{\alpha^6}\Delta(\mu,\eta)\,. \end{aligned} \tag{7.82}$$

Moreover, if μ' is another probability measure and if $\int x\mathrm{d}\mu'(x) \geq \alpha/4$, we have

$$\Delta(A(\mu), A(\mu')) \leq \frac{L\beta^2}{\alpha^4}\left(\int x\mathrm{d}\mu(x) - \int x\mathrm{d}\mu'(x)\right)^2 + \frac{L\beta^3}{\alpha^6}\Delta(\mu,\mu')\,. \tag{7.83}$$

A little bit of measure-theoretic technique is required again here, because we are dealing with probability measures that are not supported by a compact interval. In the forthcoming lemma, there is really nothing specific about the power 4.

Lemma 7.6.3. *Given a number C, consider the set $D(C)$ of probability measures θ on $\mathbb{R}^+$ that satisfy $\int_0^\infty x^4 \mathrm{d}\theta(x) \leq C$. Then $D(C)$ is a compact metric space for the distance Δ.*

Proof. The proof uses a truncation argument similar to the one given at the end of the proof of Proposition 7.5.10. Given a number $b > 0$ and a probability measure θ in $D(C)$ we define the truncation θ^b as the image of θ under the map $x \mapsto \min(x,b)$. In words, all the mass that θ gives to the half-line $[b,\infty[$ is pushed to the point b. Then we have

$$\Delta(\theta,\theta^b) \leq \int_0^\infty (x - \min(x,b))^2 \mathrm{d}\theta(x) \leq \int_b^\infty x^2 \mathrm{d}\theta(x) \leq \frac{C}{b^2}\,. \tag{7.84}$$

Consider now a sequence $(\theta_n)_{n\geq 1}$ in $D(C)$. We want to prove that it has a subsequence that converges for the distance Δ. Since for each b the set of probability measures on the interval $[0,b]$ is compact for the distance Δ (as is explained in Section A.11), we assume, by taking a subsequence if necessary, that for each integer m the sequence $(\theta_n^m)_{n\geq 1}$ converges for Δ to a certain probability measure λ_m. Next we show that there exists a probability measure λ in $D(C)$ such that $\lambda_m = \lambda^m$ for each m. This is simply because if $m' < m$

then $\lambda_m^{m'} = \lambda_{m'}$ (the "pieces fit together") and because $\int_0^\infty x^4 \mathrm{d}\lambda_m(x) \le C$ for each m. Now, for each m we have $\lim_{n\to\infty} \Delta(\theta_n^m, \lambda^m) = 0$, and (7.84) and the triangle inequality imply that $\lim_{n\to\infty} \Delta(\theta_n, \lambda) = 0$. □

Proof of Proposition 7.6.2. The basic idea is to define $A(\mu)$ "as the limit" of the law λ of $\left(\sum_{\ell\le M} a_N(\ell)u(\ell)\right)^{-1}$ as $M^{-1}\sum_{\ell\le M}\delta_{u(\ell)} \to \mu$, $M,N \to \infty$, $M/N \to (1+\alpha)$. We note that by (7.50) used for $k = 8$, whenever $\sum_{\ell\le M} u(\ell) \ge \alpha N/8$, (and $\beta < \beta(\alpha)$) we have $\int x^4 \mathrm{d}\lambda(x) \le L$. Thus, recalling the notation of Lemma 7.6.3, we have $\lambda \in D(L)$, a compact set, and therefore the family of these measures has a cluster point $A(\mu)$, and (7.82) holds by continuity. Moreover (7.83) is a consequence of (7.82) and continuity (and shows that the cluster point $A(\mu)$ is in fact unique). □

We recall the probability measures $\mu_N, \nu_N, \nu'_N, \mu'_N$ of Section 7.5.

Proposition 7.6.4. *We have*

$$\lim_{N\to\infty} \Delta(\nu'_N, A(\mu'_N)) = 0 \,. \tag{7.85}$$

Proof. First we recall that by (7.26) we have

$$\int x \mathrm{d}\mu_N(x) = \frac{1}{M}\sum_{\ell\le M} u_{N,M}(\ell) = \frac{M-N}{M} \ge \frac{\alpha}{2}$$

for $M = \lfloor N(1+\alpha)\rfloor$ and N large. Since Theorem 7.5.1 asserts that $\mathsf{E}\Delta(\mu'_N, \mu_N) \to 0$, (7.61) implies that

$$\mathsf{E}\left(\int x\mathrm{d}\mu_N(x) - \int x \mathrm{d}\mu'_N(x)\right)^2 \to 0$$

and thus $\int x\mathrm{d}\mu'_N(x) \ge \alpha/4$ for N large. Therefore we can use (7.82) for $\mu = \mu'_N$ and $\eta = \mu_N$ to get (using (7.61) again)

$$\begin{aligned}\Delta&\left(A(\mu'_N), \mathcal{L}_a\left(\frac{1}{\sum_{\ell\le M} a_N(\ell)u_{N,M}(\ell)}\right)\right) \\ &\le \frac{K(\alpha)}{N} + L\left(\frac{\beta^2}{\alpha^4} + \frac{\beta^3}{\alpha^6}\right)\Delta(\mu'_N, \mu_N) \,. \end{aligned} \tag{7.86}$$

The expectation of the right-hand side goes to zero as $N \to \infty$ by Theorem 7.5.1. Since by definition

$$\nu'_N = \mathsf{E}\mathcal{L}_a\left(\frac{1}{\sum_{\ell\le M} a_N(\ell)u_{N,M}(\ell)}\right),$$

taking expectation in (7.86) and using Jensen's inequality as in (7.70) completes the proof. □

Proposition 7.6.4 is of course only half of the work because we also have to define the operators B. These operators B have the following defining property.

Proposition 7.6.5. *To each probability measure ν on $\mathbb{R}^+$ we can attach a probability measure $B(\nu)$ on $[0,1]$ with the following property. Consider numbers $w(k) \geq 0$ for $k \leq N$, and let*

$$\eta = \frac{1}{N} \sum_{k \leq N} \delta_{w(k)} \ .$$

Then

$$\Delta\left(B(\nu), \mathcal{L}\left(\frac{1}{1+\sum_{k\leq N} a_N(k) w(k)}\right)\right) \leq \frac{K}{N} + \frac{L}{\beta^2} \Delta(\nu, \eta) \ . \tag{7.87}$$

Moreover

$$\Delta(B(\nu), B(\nu')) \leq \frac{L}{\beta^2} \Delta(\nu, \nu') \ . \tag{7.88}$$

Proof. Similar, but simpler than the proof of Proposition 7.6.2. □

Proposition 7.6.6. *We have*

$$\lim_{N\to\infty} \Delta(\mu'_N, B(\nu'_N)) = 0 \ . \tag{7.89}$$

Proof. Similar (but simpler) than the proof of (7.85). □

Proof of Theorem 7.1.2. It follows from the definition of ν'_N and (7.50) that $\int x^4 \mathrm{d}\nu'_N(x) \leq L$, so that, recalling the set $D(L)$ of Lemma 7.6.3, we have $\nu'_N \in D(L)$. Since μ'_N lives on $[0,1]$, we can find a subsequence of the sequence (μ'_N, ν'_N) that converges for Δ to a pair (μ, ν). Using (7.85) and (7.89) we see that this pair satisfies the relations (7.11):

$$\int x \mathrm{d}\mu(x) = \frac{\alpha}{1+\alpha} \ ; \ \mu = B(\nu) \ , \ \nu = A(\mu) \ . \tag{7.90}$$

The equations (7.90) have a unique solution. Indeed, if (μ', ν') is another solution (7.83) implies

$$\Delta(\nu, \nu') \leq \frac{L\beta^3}{\alpha^6} \Delta(\mu', \mu)$$

and by (7.88) we have

$$\Delta(\mu, \mu') \leq \frac{L}{\beta^2} \Delta(\nu, \nu')$$

so that

$$\Delta(\mu, \mu') \leq \frac{L\beta}{\alpha^6} \Delta(\mu, \mu')$$

and $\Delta(\mu, \mu') = 0$ if $L\beta/\alpha^6 < 1$. Let us stress the miracle here. The condition (7.26) forces the relation $\int x \mathrm{d}\mu(x) = \alpha/(1+\alpha)$, and this neutralizes the first term on the right-hand side of (7.83). This term is otherwise devastating, because the coefficient $L\beta^2/\alpha^4$ does not compensate the coefficient L/β^2 of (7.88).

Since the pair (μ,ν) of (7.90) is unique, we have in fact that $\mu = \lim \mu'_N$, $\nu = \lim \nu'_N$. On the other hand, by definition of μ_N we have $\mathsf{E}\mu_N = \mathcal{L}(u_{N,M}(M))$, so Jensen's inequality implies as in (7.70) that

$$\Delta(\mathcal{L}(u_{N,M}(M)), \mu'_N) \le \mathsf{E}\Delta(\mu_N, \mu'_N) ,$$

so $\lim_{N\to\infty} \mathcal{L}(u_{N,M}(M)) = \mu$ by (7.62). Similarly $\lim_{N\to\infty} \mathcal{L}(w_{N,M}(N)) = \nu$. □

We turn to the proof of Proposition 7.6.1. Let us start by a simple observation.

Proposition 7.6.7. *The bound (7.81) holds when $M = M'$.*

Proof. Without loss of generality we assume that $N' \le N$. Let $S = \sum_{\ell\le M} a_N(\ell)u(\ell)$ and $S' = \sum_{\ell\le M} a_{N'}(\ell)u'(\ell)$. Then

$$\Delta\left(\mathcal{L}\left(\frac{1}{S}\right), \mathcal{L}\left(\frac{1}{S'}\right)\right) \le \mathsf{E}\left(\frac{1}{S} - \frac{1}{S'}\right)^2 = \mathsf{E}\frac{(S-S')^2}{S^2S'^2} \le \mathrm{I} + \mathrm{II} \tag{7.91}$$

where

$$\mathrm{I} = 2\mathsf{E}\,\frac{\left(\sum_{\ell\le M}(a_N(\ell) - a_{N'}(\ell))u'(\ell)\right)^2}{S^2S'^2} ;$$

$$\mathrm{II} = 2\mathsf{E}\,\frac{\left(\sum_{\ell\le M} a_N(\ell)(u(\ell) - u'(\ell))\right)^2}{S^2S'^2} .$$

We observe since $N' \le N$ that $a'_N(\ell) \ge a_N(\ell)$, so that

$$S' \ge S^\sim := \sum_{\ell\le M} a_N(\ell)u'(\ell) ,$$

and

$$\mathrm{II} \le 2\mathsf{E}\,\frac{\left(\sum_{\ell\le M} a_N(\ell)(u(\ell) - u'(\ell))\right)^2}{S^2S^{\sim 2}} .$$

To bound this quantity we will use the estimate (7.41). The relations $\int x\mathrm{d}\eta(x) \ge \alpha/4$ and $\int x\mathrm{d}\eta'(x) \ge \alpha/4$ mean that $\sum_{\ell\le M} u(\ell) \ge \alpha M/4 \ge \alpha N/4$ and $\sum_{\ell\le M} u'(\ell) \ge \alpha M/4 \ge \alpha N/4$. Thus in (7.41) we can take $b = \alpha/4$. This estimate then yields

$$\mathrm{II} \le \frac{L\beta^2}{\alpha^4}\left(\frac{M}{N}\right)^2\left(\int x\mathrm{d}\eta(x) - \int x\mathrm{d}\eta'(x)\right)^2 + \frac{L\beta^3}{\alpha^6}\frac{1}{N}\sum_{\ell\le M}(u(\ell) - u'(\ell))^2 . \tag{7.92}$$

We can assume from Lemma 7.5.2 that we have reordered the terms $u'(\ell)$ so that $M^{-1}\sum_{\ell\le M}(u(\ell)-u'(\ell))^2 \le \Delta(\eta,\eta')$, and then the bound (7.92) is as desired, since $M \le 2N$.

To control the term I, we first note that $0 \le a_{N'}(\ell) - a_N(\ell) \le 1$ since $N' \le N$; and $\sum_{\ell\le M}(a_{N'}(\ell)-a_N(\ell))u(\ell) \le M$ since $0 \le u'(\ell) \le 1$. Therefore

$$\mathrm{I} \le 2M \sum_{\ell\le M} \mathsf{E}\frac{a_{N'}(\ell)-a_N(\ell)}{S^2 S'^2}\,.$$

We control this term with the same method that we used to control the term (7.52). Namely, we define $S_\ell = \sum_{\ell'\ne\ell} a_N(\ell')u(\ell')$ and S'_ℓ similarly, and we write, using independence and the Cauchy-Schwarz inequality that

$$\begin{aligned}\mathsf{E}\frac{a_{N'}(\ell)-a_N(\ell)}{S^2S'^2} &\le \mathsf{E}\frac{a_{N'}(\ell)-a_N(\ell)}{S_\ell^2S_\ell'^2}\\ &\le \mathsf{E}(a_{N'}(\ell)-a_N(\ell))\left(\mathsf{E}\frac{1}{S_\ell^4}\right)^{1/2}\left(\mathsf{E}\frac{1}{S_\ell'^4}\right)^{1/2}.\end{aligned}$$

Using (7.50), and since $\sum_{\ell\ne\ell'} u(\ell) \ge N\alpha/4-1 \ge N\alpha/8$ because $N\beta \ge 1$ and $\beta \le \alpha/80$, we get

$$\left(\mathsf{E}\frac{1}{S_\ell^4}\right)^{1/2} \le K(\alpha)\beta^2\,,$$

and similarly for S'_ℓ. Using (7.39) for $p=1$, we obtain

$$\mathsf{E}(a_{N'}(\ell)-a_N(\ell)) \le \frac{L}{\beta}\left(\frac{1}{N'}-\frac{1}{N}\right).$$

The result follows. □

The main difficulty in the proof of Proposition 7.6.1 is to find how to relate the different values M and M'. Given a sequence $(u(\ell))_{\ell\le M}$ and an integer M', consider the sequence $(u^\sim(\ell))_{\ell\le MM'}$ that is obtained by repeating each term $u(\ell)$ exactly M' times.

Proposition 7.6.8. *We have*

$$\Delta\left(\mathcal{L}\left(\frac{1}{\sum_{\ell\le M} a_N(\ell)u(\ell)}\right),\mathcal{L}\left(\frac{1}{\sum_{\ell\le MM'} a_{NM'}(\ell)u^\sim(\ell)}\right)\right) \le \frac{K}{N}\,. \tag{7.93}$$

Proof of Proposition 7.6.1. The meaning of (7.93) is that within a small error (as in (7.81)) we can replace M by MM' and N by NM'. Similarly, we replace M' by MM' and N' by $N'M$, so we have reduced the proof to the case $M = M'$ of Proposition 7.6.7 (using that $\Delta^{1/2}$ is a distance). □

The proof of Proposition 7.6.8 relies on the following.

Lemma 7.6.9. *Consider independent r.v.s X_ℓ, X, uniform over $[0,1]$. Consider an integer $R \ge 1$, a number $\gamma \ge 2$ and the r.v.s*

$$a = \exp(-\gamma X)\,;\; a' = \sum_{\ell \le R} \exp(-\gamma R X_\ell)\,.$$

Then we can find a pair of r.v.s (Y, Y') such that Y has the same law as the r.v. a and Y' has the same law as the r.v. a' with

$$\mathsf{E}|Y - Y'| \le \frac{L}{\gamma^2}\,,\; \mathsf{E}(Y - Y')^2 \le \frac{L}{\gamma^2}\,. \tag{7.94}$$

Proof of Proposition 7.6.8. We use Lemma 7.6.9 for $\gamma = \beta N$, $R = M'$. Consider independent copies (Y_ℓ, Y'_ℓ) of the pair (Y, Y'). It should be obvious from the definition of the sequence $u^\sim(\ell)$ that $S' := \sum_{\ell \le M} Y'_\ell u(\ell)$ equals $\sum_{\ell \le MM'} a_{MM'}(\ell) u^\sim(\ell)$ in distribution. Writing $S = \sum_{\ell \le M} Y_\ell u(\ell)$, the left-hand side of (7.93) is

$$\begin{aligned} \Delta\left(\mathcal{L}\left(\frac{1}{S}\right), \mathcal{L}\left(\frac{1}{S'}\right)\right) \le \mathsf{E}\left(\frac{1}{S} - \frac{1}{S'}\right)^2 &= \mathsf{E}\,\frac{\left(\sum_{\ell \le M}(Y_\ell - Y'_\ell)u(\ell)\right)^2}{S^2 S'^2}\,, \\ &\le \mathsf{E}\,\frac{\left(\sum_{\ell \le M}|Y_\ell - Y'_\ell|\right)^2}{S^2 S'^2}\,. \end{aligned}$$

We expand the square, and we use (7.94) for $\gamma = \beta N$ and one more time the method used to control (7.52) to find that this is $\le K(\alpha)/N$. □

Proof of Lemma 7.6.9. Given any two r.v.s $a, a' \ge 0$, there is a canonical way to construct a coupling of them. Consider the function Y on $[0,1]$ given by

$$Y(x) = \inf\{t\,;\, \mathsf{P}(a \ge t) \le x\}\,.$$

The law of Y under Lebesgue's measure is the law of a. Indeed the definition of $Y(x)$ shows that

$$\begin{aligned} \mathsf{P}(a \ge y) > x &\Rightarrow Y(x) > y \\ \mathsf{P}(a \ge y) < x &\Rightarrow Y(x) < y\,, \end{aligned}$$

so that if λ denotes Lebesgue measure, we have $\lambda(\{Y(x) \ge y\}) = \mathsf{P}(a \ge y)$. Moreover "the graph of Y is basically obtained from the graph of the function $t \mapsto \mathsf{P}(a \ge t)$ by making a symmetry around the diagonal". Define Y' similarly. The pair (Y, Y') is the pair we look for, although it will require some work to prove this. First we note that

$$\mathsf{E}|Y - Y'| = \int_0^1 |Y(x) - Y'(x)|\mathrm{d}x\,.$$

This is the area between the graphs of Y of Y', and also the area between the graphs of the functions $t \mapsto \mathsf{P}(a \ge t)$ and $t \mapsto \mathsf{P}(a' \ge t)$ because these two areas are exchanged by symmetry around the diagonal (except maybe for their boundary). Therefore

$$\mathsf{E}|Y - Y'| = \int_0^\infty |\mathsf{P}(a \ge t) - \mathsf{P}(a' \ge t)| \mathrm{d}t .$$

The rest of the proof consists in elementary (and very tedious) estimates of this quantity when a and a' are as in Lemma 7.6.9. For $t \le 1$ we have

$$\mathsf{P}(a \ge t) = \mathsf{P}(\exp(-\gamma X) \ge t) = \mathsf{P}\Big(X \le \frac{1}{\gamma} \log \frac{1}{t} \Big) = \min \Big(1, \frac{1}{\gamma} \log \frac{1}{t} \Big) ,$$

and similarly

$$\mathsf{P}(\exp(-\gamma R X_\ell) \ge t) = \min \Big(1, \frac{1}{\gamma R} \log \frac{1}{t} \Big) .$$

Since $a' \ge t$ as soon as one of the summands $\exp(-\gamma R X_\ell)$ exceeds t, independence implies

$$\mathsf{P}(a' \ge t) \ge 1 - \Big(1 - \min \Big(1, \frac{1}{\gamma R} \log \frac{1}{t} \Big) \Big)^R := \psi(t) .$$

Since $(1-x)^R \ge 1 - Rx$ for $x \ge 0$, we have

$$\psi(t) \le R \min \Big(1, \frac{1}{\gamma R} \log \frac{1}{t} \Big) = \min \Big(R, \frac{1}{\gamma} \log \frac{1}{t} \Big) ,$$

and since $\psi(t) \le 1$, we have in fact

$$\psi(t) \le \min \Big(1, \frac{1}{\gamma} \log \frac{1}{t} \Big) = \mathsf{P}(a \ge t) .$$

We note that

$$x \ge 0 \Rightarrow (1-x)^R \le e^{-Rx} \le 1 - Rx + \frac{R^2 x^2}{2} .$$

Using this for

$$x = \min \Big(1, \frac{1}{R\gamma} \log \frac{1}{t} \Big)$$

this yields that

$$\psi(t) = 1 - (1-x)^R \ge Rx - \frac{R^2 x^2}{2}$$

and

$$0 \le \mathsf{P}(a \ge t) - \psi(t) \le \min \Big(1, \frac{1}{\gamma} \log \frac{1}{t} \Big) - Rx + \frac{R^2 x^2}{2} .$$

Since

$$\min\left(1, \frac{1}{\gamma}\log\frac{1}{t}\right) \le Rx \le \frac{1}{\gamma}\log\frac{1}{t} ,$$

we have proved that

$$0 \le \mathsf{P}(a \ge t) - \psi(t) \le \frac{1}{2}\left(\frac{1}{\gamma}\log\frac{1}{t}\right)^2 . \tag{7.95}$$

For a real number y we write $y^+ = \max(y, 0)$, so that $|y| = -y + 2y^+$. We use this relation for $y = \mathsf{P}(a \ge t) - \mathsf{P}(a' \ge t)$, so that since $\mathsf{P}(a' \ge t) \ge \psi(t)$ we obtain

$$y^+ \le (\mathsf{P}(a \ge t) - \psi(t))^+ = \mathsf{P}(a \ge t) - \psi(t) ,$$

and

$$|\mathsf{P}(a \ge t) - \mathsf{P}(a' \ge t)| \le \mathsf{P}(a' \ge t) - \mathsf{P}(a \ge t) + 2(\mathsf{P}(a \ge t) - \psi(t)) . \tag{7.96}$$

Since $a \le 1$, for $t > 1$ we then have

$$|\mathsf{P}(a \ge t) - \mathsf{P}(a' \ge t)| = \mathsf{P}(a' \ge t) = \mathsf{P}(a' \ge t) - \mathsf{P}(a \ge t) . \tag{7.97}$$

Using (7.96) for $t \le 1$ and (7.97) for $t > 1$ we obtain, using (7.95) in the second inequality,

$$\begin{aligned}\int_0^\infty |\mathsf{P}(a \ge t) - \mathsf{P}(a' \ge t)|\,\mathrm{d}t &\le 2\int_0^1 (\mathsf{P}(a \ge t) - \psi(t))\,\mathrm{d}t \\ &\quad + \int_0^\infty \mathsf{P}(a' \ge t)\,\mathrm{d}t - \int_0^\infty \mathsf{P}(a \ge t)\,\mathrm{d}t \\ &\le \frac{L}{\gamma^2} + \mathsf{E}a' - \mathsf{E}a .\end{aligned}$$

Finally we use that by (7.39) we have $|\mathsf{E}a - \mathsf{E}a'| \le L/\gamma^2$, and this concludes the proof that $\mathsf{E}\,|Y - Y'| \le L/\gamma^2$.

We turn to the control of $\mathsf{E}(Y - Y')^2$. First, we observe that

$$\mathsf{E}(Y - Y')^2 \le 2\mathsf{E}(Y - \min(Y', 2))^2 + 2\mathsf{E}(\min(Y', 2) - Y')^2 .$$

Now, since $Y \le 1$, we have

$$\begin{aligned}\mathsf{E}(Y - \min(Y', 2))^2 &= \mathsf{E}(\min(Y, 2) - \min(Y', 2))^2 \\ &\le 2\mathsf{E}|\min(Y, 2) - \min(Y', 2)| \\ &\le 2\mathsf{E}|Y - Y'| \le \frac{L}{\gamma^2} .\end{aligned}$$

The r.v. $A = Y' - \min(Y', 2)$ satisfies

$$A > 0 \Rightarrow A = Y' - 2 ,$$

so that if $t > 0$ we have $\mathsf{P}(A \geq t) = \mathsf{P}(Y' \geq t+2)$. Since Y' and a' have the same distribution, it holds:

$$\begin{aligned}\mathsf{E}(\min(Y',2) - Y')^2 = \mathsf{E}A^2 &= 2\int_0^\infty t\mathsf{P}(Y' \geq t+2)\mathrm{d}t \\ &= 2\int_0^\infty t\mathsf{P}(a' \geq t+2)\mathrm{d}t\,.\end{aligned}$$

To estimate $\mathsf{P}(a' \geq t)$, we write, for $\lambda > 0$

$$\begin{aligned}\mathsf{P}(a' \geq t) &\leq \exp(-\lambda t)\,\mathsf{E}\exp\lambda\, a' \\ &= \exp(-\lambda t)\big(\mathsf{E}\exp(\lambda\exp(-\gamma\, R\, X))\big)^R\end{aligned}$$

and, using (7.39) in the second inequality, and a power expansion of e^λ to obtain the third inequality, we get

$$\begin{aligned}\mathsf{E}\exp(\lambda\exp(-\gamma\, R\, X)) &= \sum_{p\geq 0}\frac{\lambda^p}{p!}\,\mathsf{E}\exp(-\gamma\, R\, p\, X) \\ &\leq 1 + \sum_{p\geq 1}\frac{\lambda^p}{p!\,p\,\gamma\, R} \leq 1 + \frac{e^\lambda}{\gamma\, R} \\ &\leq \exp\left(\frac{e^\lambda}{\gamma\, R}\right)\end{aligned}$$

so that

$$\mathsf{P}(a' \geq t) \leq \exp\left(\frac{e^\lambda}{\gamma} - \lambda\, t\right)\,.$$

Taking $\lambda = \log\gamma > 0$, we get

$$\mathsf{P}(a' \geq t) \leq L\,\gamma^{-t}$$

so that since $\gamma \geq 2$ we obtain

$$\int_0^\infty t\,\mathsf{P}(a' \geq t+2)\,\mathrm{d}t \leq \frac{L}{\gamma^2}\,.$$

□

Research Problem 7.6.10. (Level 2) Is it true that given an integer n, there exists a constant $K(\alpha, n)$, and independent r.v.s $U_1, \ldots, U_n$ of law μ with

$$\mathsf{E}\sum_{i\leq n}(u_{N,M}(i) - U_i)^2 \leq \frac{K(\alpha,n)}{N}\ ? \tag{7.98}$$

Proof of Theorem 7.1.2. We will stay somewhat informal in this proof. We write $A_{N,M} = \mathsf{E}\log Z_{N,M}$, so that

$$A_{N,M} - A_{N,M-1} = \mathsf{E} \log \frac{Z_{N,M}}{Z_{N,M-1}} = -\mathsf{E} \log u_{N,M}(M-1)$$

$$A_{N,M} - A_{N-1,M} = \mathsf{E} \log \frac{Z_{N,M}}{Z_{N-1,M}} = -\mathsf{E} \log w_{N,M}(N) \,.$$

By Theorem 7.1.2, these quantities have limits $-\int \log x \, d\mu(x)$ and $-\int \log x \, d\nu(x)$ respectively. (To obtain the required tightness, we observe that from (7.27), (7.28) and Markov's inequality we have $\mathsf{P}(u_{N,M}(M-1) < t) \le Kt$ and $\mathsf{P}(w_{N,M}(N) < t) \le Kt$.) Setting $M(R) = \lfloor R(1+\alpha) \rfloor$, we write

$$A_{N,M} - A_{1,1} = \mathrm{I} + \mathrm{II} \,,$$

where

$$\mathrm{I} = \sum_{2 \le R \le M} A_{R,M(R)} - A_{R-1,M(R)}$$

$$\mathrm{II} = \sum_{2 \le R \le M} A_{R-1,M(R)} - A_{R-1,M(R-1)} \,.$$

For large R we have

$$A_{R,M(R)} - A_{R-1,M(R)} \simeq -\int \log x \, d\nu(x) \,,$$

and since $M(R) - 2 \le M(R-1) \le M(R) - 1$, we also have

$$A_{R-1,M(R)} - A_{R-1,M(R-1)} \simeq -(M(R) - M(R-1)) \int \log x \, d\mu(x) \,.$$

The result follows. □

A direction that should be pursued is the detailed study of Gibbs' measure; the principal difficulty might be to discover fruitful formulations. If G denotes Gibbs' measure, we should note the relation

$$G(\{\sigma(i) = j\}) = a(i,j) \frac{Z_{N,M}(i;j)}{Z_{N,M}} \simeq a(i,j) w_{N,M}(i) u_{N,M}(j) \,. \tag{7.99}$$

Also, if $i_1 \neq i_2$ and $j_1 \neq j_2$, we have

$$G(\{\sigma(i_1) = j_1 \,;\, \sigma(i_2) = j_2\}) = a(i_1,i_2) a(j_1,j_2) \frac{Z_{N,M}(i_1,i_2;j_1,j_2)}{Z_{N,M}} \,. \tag{7.100}$$

One can generalize (7.7) to show that

$$\frac{Z_{N,M}(i_1,i_2;j_1,j_2)}{Z_{N,M}} \simeq w_{N,M}(i_1) w_{N,M}(i_2) u_{N,M}(j_1) u_{N,M}(j_2)$$

so comparing (7.99) and (7.100) we get

$$G(\{\sigma(i_1) = j_1 \,;\, \sigma(i_2) = j_2\}) \simeq G(\{\sigma(i_1) = j_1\}) G(\{\sigma(i_2) = j_2\}) \,.$$

The problem however to find a nice formulation is that the previous relation holds for most values of j_1 and j_2 simply because both sides are nearly zero!

7.7 Notes and Comments

A recent paper [169] suggests that it could be of interest to investigate the following model. The configuration space consists of all pairs (A, σ) where A is a subset of $\{1, \ldots, N\}$, and where σ is a one to one map from A to $\{1, \ldots, N\}$. The Hamiltonian is then given by

$$H_N((A, \sigma)) = -C \operatorname{card} A + \beta N \sum_{i \in A} c(i, \sigma(i)), \tag{7.101}$$

where C is a constant and $c(i, j)$ are as previously. The idea of the Hamiltonian is that the term $-C \operatorname{card} A$ favors the pairs (A, σ) for which $\operatorname{card} A$ is large. It seems likely that, given C, results of the same nature as those we proved can be obtained for this model when $\beta \leq \beta(C)$, but that it will be difficult to prove the existence of a number β_0 such than these results hold for $\beta \leq \beta_0$, independently of the value of C, and even more difficult to prove that (as the results of [169] seem to indicate) they will hold for any value of C and of β.

A. Appendix: Elements of Probability Theory

A.1 How to Use this Appendix

This appendix lists some well-known and some less well-known facts about probability theory. The author does not have the energy to give a reference in the printed literature for the well known facts, for the simple reason that he has not opened a single textbook over the last three decades. However all the statements that come without proof should be in standard textbooks, two of which are [10] and [161]. Of course the less well-known facts are proved in detail.

The appendix is not designed to be read from the first line. Rather one should refer to each section as the need arises. If you do not follow this advice, you might run into difficulties, such as meeting the notation L before having learned that this always stands for a universal constant (= a number).

A.2 Differentiation Inside an Expectation

For the purpose of derivation inside an integral sign, or, equivalently, inside an expectation, the following result will suffice. It follows from Lebesgue's dominated convergence theorem. If that is too fancy, much more basic versions of the same principle suffice, and can be found in Wikipedia.

Proposition A.2.1. *Consider a random function $\psi(t)$ defined on an interval J of $\mathbb{R}$, and assume that $\mathsf{E}|\psi(t)| < \infty$ for each $t \in J$. Assume that the function $\psi(t)$ is always continuously differentiable, and that for each compact subinterval I of J one has*

$$\mathsf{E} \sup_{t \in I} |\psi'(t)| < \infty \,. \tag{A.1}$$

Then the function $\varphi(t) = \mathsf{E}\psi(t)$ is continuously differentiable and

$$\varphi'(t) = \mathsf{E}\psi'(t) \,. \tag{A.2}$$

As an illustration we give a proof of (1.41).

M. Talagrand, *Mean Field Models for Spin Glasses*, Ergebnisse der Mathematik und ihrer Grenzgebiete. 3. Folge / A Series of Modern Surveys in Mathematics 54, DOI 10.1007/978-3-642-15202-3, © Springer-Verlag Berlin Heidelberg 2011

Proposition A.2.2. *Consider an infinitely differentiable function F on $\mathbb{R}^M$, such that all its partial derivatives are of "moderate growth" in the sense of (A.18). Consider two independent centered jointly Gaussian families $\mathbf{u} = (u_i)_{i\le M}$, $\mathbf{v} = (v_i)_{i\le M}$, and let $u_i(t) = \sqrt{t}u_i + \sqrt{1-t}v_i$, $\mathbf{u}(t) = (u_i(t))_{i\le M}$. Consider the function*

$$\varphi(t) = \mathsf{E}F(\mathbf{u}(t)) \,. \tag{A.3}$$

Let

$$u_i'(t) = \frac{\mathrm{d}}{\mathrm{d}t}u_i(t) = \frac{1}{2\sqrt{t}}u_i - \frac{1}{2\sqrt{1-t}}v_i \,.$$

Then

$$\varphi'(t) = \mathsf{E}\sum_{i\le M} u_i'(t)\frac{\partial F}{\partial x_i}(\mathbf{u}(t)) \,. \tag{A.4}$$

Proof. We prove (A.1) with $\psi(t) = F(\mathbf{u}(t))$. We write first that for a compact subinterval I of $]0,1[$ we have

$$\sup_{t\in I}\left|u_i'(t)\frac{\partial F}{\partial x_i}(\mathbf{u}(t))\right| \le \sup_{t\in I}|u_i'(t)|\sup_{t\in I}\left|\frac{\partial F}{\partial x_i}(\mathbf{u}(t))\right| \,.$$

Using the Cauchy-Schwarz inequality, to prove (A.1) it suffices to prove that

$$\mathsf{E}\left(\sup_{t\in I}|u_i'(t)|\right)^2 < \infty$$

and

$$\mathsf{E}\left(\sup_{t\in I}\left|\frac{\partial F}{\partial x_i}(\mathbf{u}(t))\right|\right)^2 < \infty \,. \tag{A.5}$$

We prove only the second inequality, since the first one is rather immediate. Using that $\partial F/\partial x_i$ is of moderate growth (as in (A.18)), given any $a > 0$ we first see that there is a constant A such that

$$\left|\frac{\partial F}{\partial x_i}(\mathbf{x})\right| \le A\exp a\|\mathbf{x}\|^2 \,,$$

and since

$$\|\mathbf{u}(t)\| \le \sqrt{t}\|\mathbf{u}\| + \sqrt{1-t}\|\mathbf{v}\| \le \sqrt{2}\max(\|\mathbf{u}\|,\|\mathbf{v}\|)$$

we obtain

$$\begin{aligned}\sup_{t\in I}\left|\frac{\partial F}{\partial x_i}(\mathbf{u}(t))\right| &\le A\max(\exp 2a\|\mathbf{u}\|^2, \exp 2a\|\mathbf{v}\|^2)\\ &\le A\exp 2a\sum_{i\le M}(u_i^2+v_i^2) \,,\end{aligned}$$

so that (A.5) follows from Hölder's inequality and the integrability properties of Gaussian r.v.s, namely the fact that if g is a Gaussian r.v. then $\mathsf{E}\exp ag^2 < \infty$ for $a\mathsf{E}g^2 < 1/2$ as follows from (A.11) below. □

A.3 Gaussian Random Variables

A (centered) Gaussian r.v. g has a density of the type

$$\frac{1}{\sqrt{2\pi}\tau}\exp\left(-\frac{t^2}{2\tau^2}\right)$$

so that $\mathsf{E}\,g^2 = \tau^2$. When $\tau = 1$, g is called standard Gaussian. We hardly ever use non-centered Gaussian r.v., so that the expression "consider a Gaussian r.v. z" means "consider a centered Gaussian r.v. z". A fundamental fact is that

$$\mathsf{E}\exp ag = \exp\frac{a^2\tau^2}{2}\,. \tag{A.6}$$

Indeed,

$$\begin{aligned}\mathsf{E}\exp ag &= \frac{1}{\sqrt{2\pi}\tau}\int_{-\infty}^{\infty}\exp\left(at-\frac{t^2}{2\tau^2}\right)\mathrm{d}t\\ &= \left(\exp\frac{a^2\tau^2}{2}\right)\frac{1}{\sqrt{2\pi}\tau}\int_{-\infty}^{\infty}\exp\left(-\frac{(t-a\tau^2)^2}{2\tau^2}\right)\mathrm{d}t\\ &= \exp\frac{a^2\tau^2}{2}\,.\end{aligned}$$

For a r.v. $Y \geq 0$ and $s > 0$ we have Markov's inequality

$$\mathsf{P}(Y \geq s) \leq \frac{1}{s}\mathsf{E}Y\,. \tag{A.7}$$

Using this for $Y = \exp(\lambda X)$, where X is any r.v., we obtain for any $\lambda \geq 0$ the following fundamental inequality:

$$\mathsf{P}(X \geq t) = \mathsf{P}(\exp(\lambda X) \geq e^{\lambda t}) \leq e^{-\lambda t}\mathsf{E}\exp(\lambda X)\,. \tag{A.8}$$

Changing X into $-X$ and t into $-t$, we get the following equally useful fact:

$$\mathsf{P}(X \leq t) \leq e^{\lambda t}\mathsf{E}\exp(-\lambda X)\,.$$

Combining (A.6) with (A.8) we get for any $t \geq 0$ that

$$\mathsf{P}(g \geq t) \leq \exp\left(-\lambda t + \frac{\lambda^2\tau^2}{2}\right),$$

and taking $\lambda = t/\tau^2$

$$\mathsf{P}(g \geq t) \leq \exp\left(-\frac{t^2}{2\tau^2}\right)\,. \tag{A.9}$$

Elementary estimates (to be found in any probability book worth its price) show that for $t > 0$ we have, for some number L,

$$\mathsf{P}(g \ge t) \ge \frac{1}{L(1+t/\tau)} \exp\left(-\frac{t^2}{2\tau^2}\right) . \tag{A.10}$$

This is actually proved in (3.137) page 229, a way to show that this book is worth what you paid for. There is of course a more precise understanding of the tails of g than (A.9) and (A.10); but (A.9) and (A.10) will mostly suffice here. Another fundamental formula is that when $\mathsf{E}g^2 = \tau^2$ then for $2a\tau^2 < 1$ and any b we have

$$\mathsf{E}\exp(ag^2 + bg) = \frac{1}{\sqrt{1-2a\tau^2}} \exp \frac{\tau^2 b^2}{2(1-2a\tau^2)} . \tag{A.11}$$

Indeed,

$$\mathsf{E}\exp(ag^2 + bg) = \frac{1}{\sqrt{2\pi}\tau} \int_{-\infty}^{\infty} \exp\left(at^2 - \frac{t^2}{2\tau^2} + bt\right) dt .$$

We then complete the squares by writing

$$at^2 - \frac{t^2}{2\tau^2} + bt = -\frac{1-2a\tau^2}{2\tau^2}\left(t - \frac{b\tau^2}{1-2a\tau^2}\right)^2 - \frac{b\tau^2}{2(1-2a\tau^2)}$$

and conclude by making the change of variable

$$t = \frac{b\tau^2}{1-2a\tau^2} + u\frac{\tau}{\sqrt{1-2a\tau^2}} .$$

The following is also important.

Lemma A.3.1. *Consider M Gaussian r.v.s $(g_i)_{i\le M}$ with $\mathsf{E}g_i^2 \le \tau$ for each $i \le N$. We do NOT assume that they are independent. Then we have*

$$\mathsf{E}\max_{i\le M} g_i \le \tau\sqrt{2\log M} . \tag{A.12}$$

Proof. Consider $\beta > 0$. Using Jensen's inequality (1.23) as in (1.24) and (A.6) we have

$$\begin{aligned}\mathsf{E}\log\Big(\sum_{i\le M} \exp \beta g_i\Big) &\le \log\Big(\mathsf{E}\sum_{i\le M} \exp\beta g_i\Big) \\ &\le \log\left(M\exp\left(\frac{1}{2}\beta^2\tau^2\right)\right) \\ &= \frac{\beta^2\tau^2}{2} + \log M . \end{aligned} \tag{A.13}$$

Now

$$\beta \max_{i \le M} g_i \le \log\Big(\sum_{i \le M} \exp \beta g_i\Big) ,$$

so that, using (A.13),

$$\beta \mathsf{E} \max_{i \le M} g_i \le \mathsf{E} \log\Big(\sum_{i \le M} \exp \beta g_i\Big) \le \frac{\beta^2 \tau^2}{2} + \log M .$$

Taking $\beta = \sqrt{2 \log M}/\tau$ yields (A.12). □

An important fact is that when the r.v.s g_i are independent, the inequality (A.12) can essentially be reversed,

$$\mathsf{E} \max_{i \le M} \ge \frac{\tau}{L} \sqrt{\log M} .$$

We do not provide the simple proof, since we will not use this statement.

Given independent standard Gaussian r.v.s $g_1, \ldots, g_M$, their joint law has density $(2\pi)^{-M/2} \exp(-\|\mathbf{x}\|^2/2)$, where $\|\mathbf{x}\|^2 = \sum_{i \le M} x_i^2$. This density is invariant by rotation, and, as a consequence, the law of every linear combination $z = \sum_{i \le M} a_i g_i$ is Gaussian. The set $\mathcal{G}$ of these linear combinations is a vector space, each element of which is a Gaussian r.v. Such a space is often called a Gaussian space. It has a natural dot product, given with obvious notation by $\mathsf{E}\, z z' = \sum_{k \le M} a_k a'_k$. Given two linear subspaces F_1, F_2 of F, if these spaces are orthogonal, i.e. $\mathsf{E}\, z_1 z_2 = 0$ whenever $z_1 \in F_1$, $z_2 \in F_2$, they are probabilistically independent. This is obvious from rotational invariance, since after a suitable rotation these spaces are spanned by two disjoint subsets of $g_1, \ldots, g_M$.

We say that a family $z_1, \ldots, z_N$ of r.v.s is jointly Gaussian if the law of every linear combination $\sum_{k \le N} a_k z_k$ is Gaussian. If $z_1, \ldots, z_N$ belong to a Gaussian space $\mathcal{G}$ as above, then obviously the family $z_1, \ldots, z_N$ is jointly Gaussian. All the jointly Gaussian families considered in this book will obviously be of this type, since they are defined by explicit formulas such as $z_k = \sum_{i \le M} a_{k,i} g_i$ where $g_1, \ldots, g_M$ are independent standard Gaussian r.v.s, a formula that we abbreviate by $z_k = \mathbf{g} \cdot \mathbf{a}_k$ where $\mathbf{g} = (g_1, \ldots, g_M)$, $\mathbf{a}_k = (a_{k,1}, \ldots, a_{k,M})$ and $\cdot$ denotes the dot product in $\mathbb{R}^M$. For the beauty of it, let us mention that, in distribution, any jointly Gaussian family $z_1, \ldots, z_N$ can be represented as above as $z_k = \mathbf{a}_k \cdot \mathbf{g}$ (with $M = N$). This is simply because the joint law of a jointly Gaussian family $z_1, \ldots, z_k$ is determined by the numbers $\mathsf{E} z_k z_\ell$, so that it suffices to find the vectors $\mathbf{a}_k$ in such a manner that $\mathsf{E} z_k z_\ell = \mathbf{a}_k \cdot \mathbf{a}_\ell$. If we think of the linear span of the r.v.s $z_1, \ldots, z_N$ provided with the dot product $z \cdot z' = \mathsf{E} z z'$ as an Euclidean space, and of $z_1, \ldots, z_N$ as points in this space, they provide exactly such a family of vectors.

Another interesting fact is the following. If $(q_{u,v})_{u,v \le n}$ is a symmetric positive definite matrix, there exists jointly Gaussian r.v.s $(Y_u)_{u \le n}$ such that $\mathsf{E}\, Y_u Y_v = q_{u,v}$. This is obvious when the matrix $(q_{u,v})$ is diagonal; the general case follows from the fact that a symmetric matrix diagonalizes in an orthogonal basis.

A.4 Gaussian Integration by Parts

Given a continuously differentiable function F on $\mathbb{R}$ (that satisfies the growth condition at infinity stated below in (A.15)) and a centered Gaussian r.v. g we have the integration by parts formula

$$\mathsf{E}\, gF(g) = \mathsf{E}\, g^2 \mathsf{E}\, F'(g)\ . \tag{A.14}$$

To see this, if $\mathsf{E}\, g^2 = \tau^2$, we have

$$\begin{aligned}\mathsf{E}\, gF(g) &= \frac{1}{\sqrt{2\pi}\tau}\int_{\mathbb{R}} t\exp\left(-\frac{t^2}{2\tau^2}\right)F(t)\mathrm{d}t\\ &= \frac{\tau^2}{\sqrt{2\pi}\tau}\int_{\mathbb{R}} \exp\left(-\frac{t^2}{2\tau^2}\right)F'(t)\mathrm{d}t\\ &= \mathsf{E}\, g^2\mathsf{E}\, F'(g)\end{aligned}$$

provided

$$\lim_{|t|\to\infty} F(t)\exp(-t^2/2\tau^2) = 0\ . \tag{A.15}$$

This formula is used over and over in this work. As a first application, if $\mathsf{E}g^2 = \tau^2$ and $2a\tau^2 < 1$ we have

$$\mathsf{E}g^2 \exp ag^2 = \mathsf{E}g(g\exp ag^2) = \tau^2(\mathsf{E}\exp ag^2 + \mathsf{E}\, 2ag\exp ag^2)\ , \tag{A.16}$$

so that

$$(1-2a\tau^2)\mathsf{E}g\exp ag^2 = \tau^2\mathsf{E}\exp ag^2 = \tau\frac{1}{\sqrt{1-2a\tau^2}}$$

by (A.11) and $\mathsf{E}g\exp ag^2 = \tau^2(1-2a\tau^2)^{-3/2}$. As another application, if $k\geq 2$

$$\mathsf{E}g^k = \mathsf{E}gg^{k-1} = \tau^2(k-1)\mathsf{E}g^{k-2}\ ,$$

so that in particular $\mathsf{E}g^4 = 3\tau^2$, and one can recursively compute all the moments of g. All kinds of Gaussian integrals can be computed effortlessly in this manner.

Condition (A.15) holds in particular if F is of moderate growth in the sense that $\lim_{|t|\to\infty} F(t)\exp(-at^2) = 0$ for each $a > 0$. A function F (with a regular behavior as will be the case of all the functions we consider) fails to be of moderate growth if "it grows as fast as $\exp(at^2)$ for some $a > 0$". The functions to which we will apply the integration by parts formula typically do not "grow faster than $\exp(At)$" for a certain number A (except in the case of certain very explicit functions such as in (A.16)).

Formula (A.14) generalizes as follows. Given $g, z_1, \ldots, z_n$ in a Gaussian space $\mathcal{G}$, and a function F of n variables (with a moderate behavior at infinity to be stated in (A.18) below), we have

$$\mathsf{E}gF(z_1,\ldots,z_n) = \sum_{\ell\leq n}\mathsf{E}(gz_\ell)\mathsf{E}\frac{\partial F}{\partial x_\ell}(z_1,\ldots,z_n)\ . \tag{A.17}$$

This is probably the single most important formula in this work. For a proof, consider the r.v.s

$$z'_\ell = z_\ell - g\frac{\mathsf{E}\, z_\ell g}{\mathsf{E}\, g^2} \,.$$

They satisfy $\mathsf{E}\, z'_\ell g = 0$; thus g is independent of the family $(z'_1, \ldots, z'_n)$. We then apply (A.14) at $(z'_\ell)_{\ell \le n}$ given. Since $z_\ell = z'_\ell + g\mathsf{E}\, g z_\ell / \mathsf{E}\, g^2$, (A.17) follows whenever the following is satisfied to make the use of (A.14) legitimate (and to allow the interchange of the expectation in z and in the family $(z'_1, \ldots, z'_n)$): for each number $a > 0$, we have

$$\lim_{\|\mathbf{x}\| \to \infty} |F(\mathbf{x})| \exp(-a\|\mathbf{x}\|^2) = 0 \,. \tag{A.18}$$

A.5 Tail Estimates

We recall that given any r.v. X and $\lambda > 0$, by (A.8) we have

$$\mathsf{P}(X \ge t) \le e^{-\lambda t} \mathsf{E} \exp \lambda X \,.$$

If $X = \sum_{i \le N} X_i$ where $(X_i)_{i \le N}$ are independent, then

$$\mathsf{E} \exp \lambda X = \prod_{i \le N} \mathsf{E} \exp \lambda X_i \,,$$

so that

$$\mathsf{P}(X \ge t) \le e^{-\lambda t} \prod_{i \le N} \mathsf{E} \exp \lambda X_i = \exp\Big(-\lambda t + \sum_{i \le N} \log \mathsf{E} \exp \lambda X_i\Big) \,. \tag{A.19}$$

If $(\eta_i)_{i \le N}$ are independent Bernoulli r.v.s, i.e. $\mathsf{P}(\eta_i = \pm 1) = 1/2$, then $\mathsf{E} \exp \lambda\, a_i \eta_i = \operatorname{ch} \lambda a_i$, and thus

$$\mathsf{P}\Big(\sum_{i \le N} a_i \eta_i \ge t\Big) \le \exp\Big(-\lambda t + \sum_{i \le N} \log \operatorname{ch} \lambda a_i\Big) \,. \tag{A.20}$$

It is obvious on power series expansions that $\operatorname{ch} t \le \exp(t^2/2)$, so that

$$\mathsf{P}\Big(\sum_{i \le N} a_i \eta_i \ge t\Big) \le \exp\Big(-\lambda t + \frac{\lambda^2}{2} \sum_{i \le N} a_i^2\Big) \,,$$

and by optimization over λ, for all $t \ge 0$,

$$\mathsf{P}\Big(\sum_{i \le N} a_i \eta_i \ge t\Big) \le \exp\Big(-\frac{t^2}{2 \sum_{i \le N} a_i^2}\Big) \,. \tag{A.21}$$

This inequality is often called the subgaussian inequality. By symmetry, $\mathsf{P}(\sum_{i\le N} a_i\eta_i \le -t)$ is bounded by the same expression, so that

$$\mathsf{P}\left(\left|\sum_{i\le N} a_i\eta_i\right| \ge t\right) \le 2\exp\left(-\frac{t^2}{2\sum_{i\le N} a_i^2}\right) . \tag{A.22}$$

As a consequence of (A.21) we have the following

$$\operatorname{card}\{(\boldsymbol{\sigma}^1, \boldsymbol{\sigma}^2) \in \Sigma_N^2 \; ; \; R_{1,2} \ge t\} \le 2^{2N} \exp\left(-\frac{Nt^2}{2}\right) . \tag{A.23}$$

This is seen by taking $a_i = 1/N$, by observing that for the uniform measure on Σ_N^2 the sequence $\eta_i = \sigma_i^1\sigma_i^2$ is an independent Bernoulli sequence and that $R_{1,2} = \sum_{i\le N} a_i\eta_i$. Related to (A.21) is the fact that

$$\mathsf{E}\exp\frac{1}{2}\left(\sum_{i\le N} a_i\eta_i\right)^2 \le \frac{1}{\sqrt{1-\sum_{i\le N} a_i^2}} . \tag{A.24}$$

Equivalently,

$$\sum \exp\frac{1}{2}\left(\sum_{i\le N} a_i\sigma_i\right)^2 \le \frac{2^N}{\sqrt{1-\sum_{i\le N} a_i^2}} ,$$

where the summation is over all sequences $(\sigma_i)_{i\le N}$ with $\sigma_i = \pm 1$. To prove (A.24) we consider a standard Gaussian r.v. g independent of the r.v.s η_i and, using (A.6), we have, denoting by E_g expectation in g only, and using again that $\log \operatorname{ch} t \le t^2/2$,

$$\begin{aligned}\mathsf{E}\exp\frac{1}{2}\left(\sum_{i\le N} a_i\eta_i\right)^2 &= \mathsf{E}\,\mathsf{E}_g \exp\sum_{i\le N} g a_i\eta_i \\ &= \mathsf{E}_g \exp\sum_{i\le N} \log \operatorname{ch} g a_i \\ &\le \mathsf{E}_g \exp\frac{g^2}{2}\sum_{i\le N} a_i^2 \\ &= \frac{1}{\sqrt{1-\sum_{i\le N} a_i^2}} .\end{aligned}$$

It follows from (A.24) that if $S = \sum_{i\le N} a_i^2$, then, if $b_i = a_i/\sqrt{2S}$, we have $\sum_{i\le N} b_i^2 = 1/2$ and

$$\mathsf{E}\exp\frac{1}{4S}\left(\sum_{i\le N} a_i\eta_i\right)^2 = \mathsf{E}\exp\frac{1}{2}\left(\sum_{i\le N} b_i\eta_i\right)^2 \le \frac{1}{\sqrt{1/2}} \le 2 .$$

Since $\exp x \geq x^n/n! \geq x^n/n^n$ for each n and $x \geq 0$ we see that

$$\mathsf{E}\Big(\sum_{i\leq N} a_i\eta_i\Big)^{2n} \leq 2(4n)^n S^n = 2(4n)^n \Big(\sum_{i\leq N} a_i^2\Big)^n , \tag{A.25}$$

a relation known as Khinchin's inequality.

Going back to (A.20), if $a_i = 1$ for each $i \leq N$, changing t into Nt, we get

$$\mathsf{P}\Big(\sum_{i\leq N} \eta_i \geq Nt\Big) \leq \exp N(-\lambda t + \log \operatorname{ch} \lambda) .$$

If $0 \leq t < 1$, the exponent is minimized for $\operatorname{th} \lambda = t$, i.e.

$$\frac{e^{\lambda} - e^{-\lambda}}{e^{\lambda} + e^{-\lambda}} = \frac{e^{2\lambda} - 1}{e^{2\lambda} + 1} = t ,$$

so that $e^{2\lambda} = (1+t)/(1-t)$ and

$$\lambda = \frac{1}{2}(\log(1+t) - \log(1-t)) .$$

Also, $\operatorname{ch}^{-2}\lambda = 1 - \operatorname{th}^2\lambda = 1 - t^2$, so that

$$\log \operatorname{ch} \lambda = -\frac{1}{2}\log(1-t^2) ,$$

and

$$\begin{aligned}\min_{\lambda}\,(-\lambda t + \log \operatorname{ch} \lambda) &= -\frac{1}{2}(t\log(1+t) - t\log(1-t)) \\ &\quad -\frac{1}{2}\log(1-t) - \frac{1}{2}\log(1+t) \\ &= -\mathcal{I}(t)\end{aligned} \tag{A.26}$$

where

$$\mathcal{I}(t) = \frac{1}{2}((1+t)\log(1+t) + (1-t)\log(1-t)) . \tag{A.27}$$

The function $\mathcal{I}(t)$ is probably better understood by noting that

$$\mathcal{I}(0) = \mathcal{I}'(0) = 0\,,\ \ \mathcal{I}''(t) = \frac{1}{1-t^2} . \tag{A.28}$$

It follows from (A.26) that

$$\mathsf{P}\Big(\sum_{i\leq N} \eta_i \geq Nt\Big) \leq \exp(-N\mathcal{I}(t)) ,$$

or, equivalently, that

$$\operatorname{card}\Big\{\boldsymbol{\sigma} \in \Sigma_N\,;\ \sum_{i\le N} \sigma_i \ge tN\Big\} \le 2^N \exp(-N\mathcal{I}(t))\,. \tag{A.29}$$

If k is an integer, then $\sum_{i\le N} \sigma_i = k$ exactly when the sequence $(\sigma_i)_{i\le N}$ contains $(N+k)/2$ times 1 and $(N-k)/2$ times -1. This is impossible when $N+k$ is odd. When $N+k$ is even, using Stirling's formula $n! \sim n^n e^{-n}\sqrt{2\pi n}$, we obtain

$$\begin{aligned}
\operatorname{card}\Big\{\boldsymbol{\sigma} \in \Sigma_N\,;\ \sum_{i\le N} \sigma_i = k\Big\} &= \binom{N}{\frac{N+k}{2}} = \frac{N!}{\left(\frac{N+k}{2}\right)!\left(\frac{N-k}{2}\right)!} \\
&\ge \frac{1}{L}\frac{\sqrt{N}}{\sqrt{(N-k)(N+k)}}\frac{N^N}{\left(\frac{N+k}{2}\right)^{(N+k)/2}\left(\frac{N-k}{2}\right)^{(N-k)/2}} \\
&\ge \frac{2^N}{L\sqrt{N}}\frac{1}{\left(1+\frac{k}{N}\right)^{(N+k)/2}\left(1-\frac{k}{N}\right)^{(N-k)/2}} \\
&= \frac{2^N}{L\sqrt{N}}\exp\left(-N\mathcal{I}\left(\frac{k}{N}\right)\right)\,.
\end{aligned} \tag{A.30}$$

This reverses the inequality (A.29) within the factor $L\sqrt{N}$.

Since by Lemma 4.3.5 the function $t \mapsto \log \operatorname{ch}\sqrt{t}$ is concave, it follows from (A.20) that

$$\mathsf{P}\Big(\sum_{i\le N} a_i\eta_i \ge t\sqrt{N}\Big) \le \exp N\Big(-\lambda t + \log\operatorname{ch}\lambda\sqrt{\sum_{i\le N} a_i^2}\Big)$$

and, using (A.26)

$$\mathsf{P}\Big(\sum_{i\le N} a_i\eta_i \ge t\sqrt{N}\Big) \le \exp\Big(-N\mathcal{I}\Big(\frac{t}{\sqrt{\sum_{i\le N} a_i^2}}\Big)\Big)\,. \tag{A.31}$$

A.6 How to Use Tail Estimates

It will often occur that for a r.v. X, we know an upper bound for the probabilities $\mathsf{P}(X \ge t)$, and that we want to deduce an upper bound for $\mathsf{E}F(X)$ for a certain function F. For example, if Y is a r.v., $Y \ge 0$, then

$$\mathsf{E}Y = \int_0^\infty \mathsf{P}(Y \ge t)\,dt\,, \tag{A.32}$$

using Fubini theorem to compute the "area under the graph of Y".

More generally, if $X \ge 0$ and F is a continuously differentiable nondecreasing function on $\mathbb{R}^+$ we have

$$F(X) = F(0) + \int_0^X F'(t)\mathrm{d}t = F(0) + \int_{\{t \le X\}} F'(t)\mathrm{d}t \,.$$

Taking expectation, and using Fubini's theorem to exchange the integral in t and the expectation, we get that

$$\mathsf{E}F(X) = F(0) + \int_0^\infty F'(t)\mathsf{P}(X \ge t)\mathrm{d}t \,. \tag{A.33}$$

For a typical application of (A.33) let us assume that X satisfies the following tail condition:

$$\forall t \ge 0 \,,\ \mathsf{P}(|X| \ge t) \le 2\exp\Big(-\frac{t^2}{2A^2}\Big) \,, \tag{A.34}$$

where A is a certain number. Then, using (A.33) for $F(x) = x^k$ and $|X|$ instead of X we get

$$\mathsf{E}|X|^k \le 2k \int_0^\infty t^{k-1} \exp\Big(-\frac{t^2}{2A^2}\Big)\mathrm{d}t \,.$$

The right-hand side can be recursively computed by integration by parts. If $k \ge 3$,

$$\int_0^\infty t^{k-1} \exp\Big(-\frac{t^2}{2A^2}\Big)\mathrm{d}t = (k-2)A^2 \int_0^\infty t^{k-3} \exp\Big(-\frac{t^2}{2A^2}\Big)\mathrm{d}t.$$

In this manner one obtains e.g.

$$\mathsf{E}X^{2k} \le 2^{k+1} k! A^{2k} \,.$$

This shows in particular that "the moments of order k of X grow at most like $\sqrt{k}$." Indeed, using the crude inequality $k! \le k^k$ we obtain

$$(\mathsf{E}|X|^k)^{1/k} \le (\mathsf{E}X^{2k})^{1/2k} \le 2A\sqrt{k} \,. \tag{A.35}$$

Suppose, conversely, that for a given r.v. X we know that for a certain number B and any $k \ge 1$ we have $\mathsf{E}X^{2k} \le B^{2k}k^k$ (i.e. an inequality of the type (A.35) for even moments). Then, using the power expansion $\exp x^2 = \sum_{k \ge 0} x^{2k}/k!$, for any number C we have

$$\mathsf{E}\exp\frac{X^2}{C^2} = \sum_{k \ge 0} \frac{\mathsf{E}X^{2k}}{C^{2k}k!} \le \sum_{k \ge 0} \frac{B^{2k}k^k}{C^{2k}k!} \,.$$

Now, by Stirling's formula, there is a constant L_0 such that $k^k \le L_0^k k!$, and therefore there is a number L (e.g. $L = 2L_0$) such that

$$\mathsf{E}\exp\frac{X^2}{LB^2} \le 2 \,.$$

This implies in turn that

$$\mathsf{P}(X \geq t) \leq 2\exp\left(-\frac{t^2}{LB^2}\right) .$$

Many r.v.s considered in this work satisfy the condition (A.34). The previous considerations explain why, when convenient, we control these r.v.s through their moments.

If F is a continuously differentiable non-decreasing function on $\mathbb{R}$, $F \geq 0$, $F(-\infty) = 0$, we have

$$F(X) = \int_{-\infty}^{X} F'(t)\mathrm{d}t = \int_{\{t \leq X\}} F'(t)\mathrm{d}t .$$

Taking expectation, and using again Fubini's theorem to exchange the integral in t and the expectation, we get now that

$$\mathsf{E}\, F(X) = \int_{-\infty}^{\infty} F'(t)\, \mathsf{P}(X \geq t)\, \mathrm{d}t . \tag{A.36}$$

This no longer assumes that $X \geq 0$. Considering now $a < b$ we have

$$\mathsf{E}(F(\min(X,b))\mathbf{1}_{\{X \geq a\}}) = F(a)\mathsf{P}(X \geq a) + \int_a^b F'(t)\mathsf{P}(X \geq t)\mathrm{d}t . \tag{A.37}$$

This is seen by using (A.36) for the conditional probability that $X \geq a$, and for the r.v. $\min(X,b)$ instead of X.

A.7 Bernstein's Inequality

Theorem A.7.1. *Consider a r.v.* X *with* $\mathsf{E}X = 0$ *and an independent sequence* $(X_i)_{i \leq N}$ *distributed like* X*. Assume that, for a certain number* A*, we have*

$$\mathsf{E}\exp\frac{|X|}{A} \leq 2 . \tag{A.38}$$

Then, for all $t > 0$ *we have*

$$\mathsf{P}\Big(\sum_{i \leq N} X_i \geq t\Big) \leq \exp\left(-\min\left(\frac{t^2}{4NA^2}, \frac{t}{2A}\right)\right) \tag{A.39}$$

$$\mathsf{P}\Big(\sum_{i \leq N} X_i \geq t\Big) \leq \exp\left(-\frac{t^2}{2N\mathsf{E}X^2}\left(1 - \frac{4A^3 t}{N(\mathsf{E}X^2)^2}\right)\right) . \tag{A.40}$$

Proof. From (A.19) we obtain

$$\mathsf{P}\Big(\sum_{i\le N} X_i \ge t\Big) \le \exp(-\lambda t + N \log \mathsf{E}\exp \lambda X)\,. \tag{A.41}$$

We have

$$\mathsf{E}\exp \lambda X = 1 + \mathsf{E}\,\varphi(\lambda X) \tag{A.42}$$

where $\varphi(x) = e^x - x - 1$. We observe that $\mathsf{E}\varphi(|X|/A) \le \mathsf{E}\exp(|X|/A) - 1 = 2 - 1 = 1$. Now power series expansion yields that $\varphi(x) \le \varphi(|x|)$ and that for $x > 0$, the function $\lambda \to \varphi(\lambda x)/\lambda^2$ increases. Thus, for $\lambda \le 1/A$, we have

$$\mathsf{E}\,\varphi(\lambda X) \le \lambda^2 A^2 \mathsf{E}\,\varphi(|X|/A) \le \lambda^2 A^2\,.$$

Combining (A.42) with the inequality $\log(1+x) \le x$, we obtain $\log \mathsf{E}\exp \lambda X \le \lambda^2 A^2$. Consequently (A.41) implies

$$\mathsf{P}\Big(\sum_{i\le N} X_i \ge t\Big) \le \exp(-\lambda t + N\lambda^2 A^2)\,.$$

We choose $\lambda = t/2NA^2$ if $t \le 2NA$ (so that $\lambda \le 1/A$). When $t \ge 2NA$, we choose $\lambda = 1/A$, and then

$$-\lambda t + N\lambda^2 A^2 = -\frac{t}{A} + N \le -\frac{t}{2A}\,.$$

This proves (A.39). To prove (A.40) we replace (A.42) by

$$\mathsf{E}\exp \lambda X = 1 + \frac{\lambda^2 \mathsf{E}X^2}{2} + \mathsf{E}\,\varphi_1(\lambda X)$$

where $\varphi_1(x) = e^x - x^2/2 - x - 1$. We observe that $\mathsf{E}\varphi_1(|X|/A) \le \mathsf{E}\varphi(|X|/A) \le 1$. Using again power series expansion yields $\varphi_1(x) \le \varphi_1(|x|)$ and that for $x > 0$ the function $\lambda \mapsto \varphi_1(\lambda x)/\lambda^3$ increases. Thus, if $\lambda \le 1/A$, we get

$$\mathsf{E}\,\varphi_1(\lambda X) \le \lambda^3 A^3 \mathsf{E}\,\varphi_1(|X|/A) \le \lambda^3 A^3$$

so that $\log \mathsf{E}\exp \lambda X \le \lambda^2 \mathsf{E}X^2/2 + \lambda^3 A^3$ and we choose $\lambda = t/N\mathsf{E}X^2$ to obtain (A.40) when $t \le N\mathsf{E}X^2/A$. When $t \ge N\mathsf{E}X^2/A$, then

$$\frac{4A^3 t}{N(\mathsf{E}X^2)^2} \ge \frac{4A^2}{\mathsf{E}X^2} \ge 1$$

because $\mathsf{E}X^2/2A^2 \le \mathsf{E}\exp|X|/A \le 2$. Thus (A.40) is automatically satisfied in that case since the right-hand side is ≥ 1. □

Another important version of Bernstein's inequality assumes that

$$|X| \le A\,. \tag{A.43}$$

In that case for $p \geq 2$ we have $\mathsf{E}X^p \leq A^{p-2}\mathsf{E}X^2$, so that when $\lambda \leq 1$, and since $\sum_{p\geq 2} 1/p! = e - 2 \leq 1$,

$$\mathsf{E}\,\varphi(\lambda X) = \sum_{p\geq 2} \frac{\lambda^p}{p!}\,\mathsf{E}\,X^p \leq \lambda^2\,\mathsf{E}\,X^2 \sum_{p\geq 2} \frac{(\lambda A)^{p-2}}{p!} \leq \lambda^2 \mathsf{E}\,X^2\,.$$

Proceeding as before, and taking now $\lambda = \min(t/\mathsf{E}\,X^2, 1/A)$, we get

$$\mathsf{P}\Big(\sum_{i\leq N} X_i \geq t\Big) \leq \exp\left(-\min\left(\frac{t^2}{4N\mathsf{E}\,X^2}, \frac{t}{2A}\right)\right)\,. \tag{A.44}$$

We will also need a version of (A.39) for martingale difference sequences. Assume that we are given an increasing sequence $(\Xi_i)_{0\leq i\leq N}$ of σ-algebras. A sequence $(X_i)_{1\leq i\leq N}$ is called a martingale difference sequence if X_i is Ξ_i-measurable and $\mathsf{E}_{i-1}(X_i) = 0$, where E_{i-1} denotes conditional expectation given Ξ_{i-1}. Let us assume that for a certain number A we have

$$\forall i \leq N\,,\ \mathsf{E}_{i-1} \exp\frac{|X_i|}{A} \leq 2\,. \tag{A.45}$$

Exactly as before, this implies that for $|\lambda|\,A \leq 1$ we have $\mathsf{E}_{i-1}\exp \lambda X_i \leq \exp \lambda^2 A^2$. Thus

$$\begin{aligned}\mathsf{E}_{k-1} \exp \lambda \sum_{i\leq k} X_i &= \exp\Big(\lambda \sum_{i\leq k-1} X_i\Big)\,\mathsf{E}_k \exp \lambda\, X_k \\ &\leq \exp\Big(\lambda \sum_{i\leq k-1} X_i + \lambda^2 A^2\Big)\,.\end{aligned}$$

By decreasing induction over k, this shows that for each k we have

$$\mathsf{E}_{k-1} \exp \lambda \sum_{i\leq N} X_i \leq \exp\Big(\lambda \sum_{i\leq k-1} X_i + (N-k+1)\,\lambda^2 A^2\Big)\,.$$

Using this for $k = 1$ and taking expectation yields $\mathsf{E}\exp\lambda\sum_{i\leq N} X_i \leq \exp N\lambda^2 A^2$. Use of Chebyshev inequality as before gives

$$\mathsf{P}\Big(\sum_{i\leq N} X_i \geq t\Big) \leq \exp\left(-\min\left(\frac{t^2}{4NA^2}, \frac{t}{2A}\right)\right)\,. \tag{A.46}$$

A.8 ε-Nets

A *ball* of $\mathbb{R}^M$ is a convex balanced set with non-empty interior. The convex hull of a set A is denoted by conv$A$.

Proposition A.8.1. *Given a ball B of $\mathbb{R}^M$, we can find a subset A of B such that*

$$\operatorname{card} A \leq \left(1+\frac{1}{\varepsilon}\right)^M \tag{A.47}$$

$$\forall x \in B\,,\ A \cap (x + 2\varepsilon B) \neq \emptyset \tag{A.48}$$

$$\operatorname{conv} A \supset (1-2\varepsilon)B\,. \tag{A.49}$$

Moreover, given a linear functional φ on $\mathbb{R}^M$, we have

$$\sup_{x\in A} \varphi(x) \geq (1-2\varepsilon) \sup_{x \in B} \varphi(x)\,. \tag{A.50}$$

As a corollary, we can find a subset A of $(1-2\varepsilon)^{-1}B$ such that $\operatorname{card} A \leq (1+\varepsilon^{-1})^M$ and $B \subset \operatorname{conv} A$. The case $\varepsilon = 1/4$ is of interest: $\operatorname{card} A \leq 5^M$ and $\sup_{x\in A}\varphi(x) \geq (1/2)\sup_{x\in B}\varphi(x)$.

Proof. We simply take for A a maximal subset of B such that the sets $x+\varepsilon B$ are disjoint for $x \in A$. These sets are of volume $\varepsilon^M \operatorname{Vol} B$, and are entirely contained in the set $(1+\varepsilon)B$, which is of volume $(1+\varepsilon)^M \operatorname{Vol} B$. This proves (A.47).

Given x in B, we can find y in A with $(x+\varepsilon B) \cap (y+\varepsilon B) \neq \emptyset$, for otherwise this would contradict the maximality of A. Thus $y \in (x + 2\varepsilon B) \cap A$. This proves (A.48).

Using (A.48), given x in B, we can find y_0 in A with $x - y_0 \in 2\varepsilon B$. Applying this to $(x-y_0)/2\varepsilon$, we find y_1 in A with $x - y_0 - 2\varepsilon y_1 \in (2\varepsilon)^2 B$, and in this manner we find a sequence (y_i) in A with

$$y = \sum_{i\geq 0} (2\varepsilon)^i y_i \in (1-2\varepsilon)^{-1} \operatorname{conv} A\,,$$

since A is finite. This proves (A.49), of which (A.50) is an immediate consequence. □

A.9 Random Matrices

In this section we get some control of the norm of certain random matrices. Much more detailed (and difficult) results are known.

Lemma A.9.1. *If $(g_{ij})_{1\leq i<j\leq N}$ are independent standard Gaussian r.v.s, then, with probability at least $1 - L\exp(-N)$ we have*

$$\forall\, (x_i)_{i\leq N}\,, \forall\, (y_i)_{i\leq N}\,,\ \Big|\sum_{i<j} g_{ij}\, x_i\, y_j\Big| \leq L\sqrt{N}\Big(\sum_{i\leq N} x_i^2 \sum_{i\leq N} y_i^2\Big)^{1/2}\,. \tag{A.51}$$

Proof. Let us denote by B the Euclidean ball of $\mathbb{R}^N$, and by A a subset of $2B$ with $\operatorname{card} A \le 5^N$ and $\operatorname{conv} A \supset B$, as provided by Proposition A.8.1. If $(x_i)_{i\le N}$ and $(y_i)_{i\le N}$ belong to A, then $\mathsf{E}\big(\sum_{i<j} g_{ij}\, x_i\, y_j\big)^2 \le \sum_{i\le N} x_i^2 \sum_{j\le N} y_j^2 \le 16$ and (A.9) implies

$$\mathsf{P}\Big(\Big|\sum_{i<j} g_{ij}\, x_i\, y_j\Big| \ge t\Big) \le 2\exp\Big(-\frac{t^2}{32}\Big)\,,$$

so that with probability at least $1 - 2(25)^N \exp(-64N)$ it holds that

$$\forall\,(x_i)_{i\le N}\,,\ \forall\,(y_i)_{i\le N} \in A\,,\ \Big|\sum_{i<j} g_{ij}\, x_i\, y_j\Big| \le 32\sqrt{N}\,,$$

and hence

$$\forall\,(x_i)_{i\le N}\,,\ \forall\,(y_i)_{i\le N} \in B\,,\ \Big|\sum_{i<j} g_{ij}\, x_i\, y_j\Big| \le 32\sqrt{N}\,,$$

and this implies (A.51). □

We consider independent Bernoulli r.v.s $(\eta_{i,k})_{i\le N, k\le M}$, that is, $\mathsf{P}(\eta_{i,k} = \pm 1) = 1/2$.

Lemma A.9.2. *Consider numbers $(\alpha_{k,k'})_{k,k'\le M}$ with $\sum \alpha_{k,k'}^2 \le 1$. Then, for $t > 0$ we have*

$$\mathsf{P}\Big(\sum_{k\ne k'}\sum_{i\le N} \alpha_{k,k'}\,\eta_{i,k}\,\eta_{i,k'} \ge t\Big) \le \exp\Big(-\min\Big(\frac{t^2}{NL},\frac{t}{L}\Big)\Big) \tag{A.52}$$

$$\mathsf{P}\Big(\sum_{k\ne k'}\sum_{i\le N} \alpha_{k,k'}\,\eta_{i,k}\,\eta_{i,k'} \ge t\Big) \le \exp\Big(-\frac{t^2}{2N}\Big(1-\frac{Lt}{N}\Big)\Big)\,. \tag{A.53}$$

Proof. The r.v.s $X_i = \sum_{k\ne k'} \alpha_{k,k'}\,\eta_{i,k}\,\eta_{i,k'}$ are i.i.d., and obviously $\mathsf{E}X_i = 0$, $\mathsf{E}X_i^2 = \sum \alpha_{k,k'}^2 \le 1$. An important result of C. Borell [14] implies that then $\mathsf{E}\exp(|X_i|/L) \le 2$ so that (A.52) is a consequence of (A.39) and (A.53) is a consequence of (A.40). □

Proposition A.9.3. *Consider a number $0 < a \le 1$ and $n \le M$. If $n\log(eM/n) \le Na^2$, the following event occurs with probability at least $1 - \exp(-a^2 N)$. Given any subset I of $\{1,\ldots,M\}$ with $\operatorname{card} I = n$, and any sequences $(x_k)_{k\le M}$, $(y_k)_{k\le M}$, we have*

$$\sum_{i\le N}\Big(\sum_{k\in I} x_k\,\eta_{i,k}\Big)\Big(\sum_{k\in I} y_k\,\eta_{i,k}\Big)$$
$$\le N\sum_{k\in I} x_k\, y_k + NLa\Big(\sum_{k\in I} x_k^2\Big)^{1/2}\Big(\sum_{k\in I} y_k^2\Big)^{1/2}\,. \tag{A.54}$$

Corollary A.9.4. *If $a \leq 1$ and $M \leq Na^2$, then with probability at least $1 - \exp(-a^2N/L)$, for any sequences $(x_k)_{k\leq M}$ and $(y_k)_{k\leq M}$ we have*

$$\sum_{i\leq N}\Big(\sum_{k\leq M} x_k\,\eta_{i,k}\Big)\Big(\sum_{k\leq M} y_k\,\eta_{i,k}\Big) \leq N\sum_{k\leq M} x_k\,y_k + NLa\Big(\sum_{k\leq M} x_k^2\Big)^{1/2}\Big(\sum_{k\leq M} y_k^2\Big)^{1/2}, \quad (A.55)$$

and

$$\sum_{i\leq N}\Big(\sum_{k\leq M} x_k\,\eta_{i,k}\Big)^2 \leq N(1+La)\Big(\sum_{k\leq M} x_k^2\Big). \quad (A.56)$$

Proof. The case $n = M$ of (A.54) is (A.55) and the case $y_k = x_k$ of (A.55) is (A.56). □

Proof of Proposition A.9.3. We rewrite (A.54) as

$$\sum_{i\leq N}\ \sum_{k\neq k',\,k,\,k'\in I} x_k\,y_{k'}\,\eta_{i,k}\,\eta_{i,k'} \leq LNa\Big(\sum_{k\in I} x_k^2\Big)^{1/2}\Big(\sum_{k\in I} y_k^2\Big)^{1/2}. \quad (A.57)$$

Consider a subset A of $\mathbb{R}^n$, with $\operatorname{card} A \leq 5^n$, $A \subset 2B$ and $\operatorname{conv} A \supset B$, where B is the Euclidean ball $\sum_{k\leq n} x_k^2 = 1$. To ensure (A.57) it suffices that

$$\sum_{i\leq N}\ \sum_{k\neq k',k,k'\in I} x_k\,y_{k'}\,\eta_{i,k}\,\eta_{i,k'} \leq LNa \quad (A.58)$$

whenever $(x_k)_{k\in I} \in A$ and $(y_k)_{k\in I} \in A$. Now, given any such sequences (A.52) implies

$$\mathsf{P}\Big(\sum_{i\leq N}\ \sum_{k\neq k',k,k'\in I} x_k\,y_{k'}\,\eta_{i,k}\,\eta_{i,k'} \geq Nu\Big) \leq \exp\Big(-\frac{N}{L}\min(u^2,u)\Big). \quad (A.59)$$

Since $n \leq M$ and $n\log(eM/n) \leq Na^2$ it holds that $n \leq Na^2$. We observe also that $25 \leq e^4$. Thus the number of possible choices for I and the sequences $(x_k)_{k\in I}$, $(y_k)_{k\in I}$ is at most

$$\binom{M}{n}(\operatorname{card} A)^2 \leq \Big(\frac{eM}{n}\Big)^n 25^n = 25^n\exp\Big(n\log\Big(\frac{eM}{n}\Big)\Big) \leq \exp 5Na^2$$

so that taking $u = L'a$ where L' large enough, all the events (A.58) simultaneously occur with a probability at least $1 - \exp(-Na^2)$. □

Our next result resembles Proposition A.9.3, but rather than restricting the range of k we now restrict the range of i.

Proposition A.9.5. *Consider a number $0 < a < 1$. Consider a number $N_0 \le N$ such that $N_0 \log(eN/N_0) \le a^2 N$, and assume that $M \le a^2 N$. Then the following event occurs with probability at least $1 - \exp(-a^2 N)$: Given any subset J of $\{1, \ldots, N\}$ with $\mathrm{card} J \le N_0$, and any sequence $(x_k)_{k \le M}$, we have*

$$\sum_{i \in J} \Big(\sum_{k \le M} x_k \eta_{i,k} \Big)^2 \le N_0 \sum_{k \le M} x_k^2 + L \max(Na^2, \sqrt{NN_0}a) \Big(\sum_{k \le M} x_k^2 \Big) . \tag{A.60}$$

Proof. The proof is very similar to the proof of Proposition A.9.3. It suffices to prove that for all choices of (x_k) and (y_k) we have

$$\sum_{i \in J} \sum_{k \ne k'} x_k y_{k'} \eta_{i,k} \eta_{i,k'} \le L \max(Na^2, \sqrt{NN_0}a) \Big(\sum_{k \le M} x_k^2 \Big)^{1/2} \Big(\sum_{k \le M} y_k^2 \Big)^{1/2} . \tag{A.61}$$

Consider a subset A of $\mathbb{R}^M$, with $\mathrm{card} A \le 5^M$, $A \subset 2B$, $B \subset \mathrm{conv} A$, where B is the Euclidean ball $\sum_{k \le M} x_k^2 \le 1$. To ensure (A.61) it suffices that

$$\sum_{i \in J} \sum_{k \ne k'} x_k y_{k'} \eta_{i,k} \eta_{i,k'} \le L \max(Na^2, \sqrt{NN_0}a)$$

whenever $\mathrm{card} J \le N_0$, $(x_k)_{k \le M}, (y_k)_{k \le M} \in A$. It follows from (A.52) that for $v > 0$,

$$\mathsf{P}\Big(\sum_{i \in J} \sum_{k \ne k'} x_k y_k \eta_{i,k} \eta_{i,k'} \ge v \mathrm{card} J \Big) \le \exp\Big(-\frac{\mathrm{card} J}{L} \min(v^2, v) \Big) ,$$

and using this for $v = uN_0 / \mathrm{card} J \ge u$ entails

$$\mathsf{P}\Big(\sum_{i \in J} \sum_{k \ne k'} x_k y_k \eta_{i,k} \eta_{i,k'} \ge N_0 u \Big) \le \exp\Big(-\frac{N_0}{L} \min(v^2, u) \Big) . \tag{A.62}$$

The number of possible choices for J and the sequences $(x_k)_{k \le M}$, $(y_k)_{k \le M}$ is at most

$$\sum_{n \le N_0} \binom{N}{n} (\mathrm{card} A)^2 \le \Big(\frac{eN}{N_0} \Big)^{N_0} 25^M \le \exp 5Na^2 ,$$

so that by taking $u = L' \max(a^2 N / N_0, a\sqrt{N/N_0})$ where L' is large enough, all the events (A.61) simultaneously occur with a probability at least $1 - \exp(-Na^2)$. □

Here is another nice consequence of Lemma A.9.2.

Lemma A.9.6. *If $\varepsilon > 0$ we have*

$$\mathsf{P}\Big(\sum_{1\le k<k'\le M} N R_{k,k'}^2 \ge (1-2\varepsilon)^{-2}\, u\Big) \le \Big(1+\frac{1}{\varepsilon}\Big)^{M^2} \exp\Big(-\frac{u}{2}\Big(1 - L\sqrt{\frac{u}{N}}\Big)\Big)$$

where $R_{k,k'} = N^{-1}\sum_{i\le N} \eta_{i,k}\,\eta_{i,k'}$.

Proof. We start the proof by observing that

$$\Big(\sum_{k<k'} R_{k,k'}^2\Big)^{1/2} = \sup \sum_{k<k'} \alpha_{k,k'}\, R_{k,k'}$$

where the supremum is taken over the subset B of $\mathbb{R}^{M(M-1)/2}$ of sequences $\alpha_{k,k'}$ with $\sum_{k<k'} \alpha_{k,k'}^2 \le 1$. We use Proposition A.8.1 to find a subset A of B with $\operatorname{card} A \le (1+\varepsilon^{-1})^{M^2}$ such that

$$\sup_A \sum \alpha_{k,k'}\, R_{k,k'} \ge (1-2\varepsilon)\Big(\sum_{k<k'} R_{k,k'}^2\Big)^{1/2} .$$

Thus

$$\begin{aligned}
&\mathsf{P}\Big(\sum_{k<k'} N R_{k,k'}^2 \ge (1-2\varepsilon)^{-2}\, u\Big)\\
&= \mathsf{P}\Big(\Big(\sum_{k<k'} R_{k,k'}^2\Big)^{1/2} \ge (1-2\varepsilon)^{-1}\sqrt{\frac{u}{N}}\Big)\\
&\le \mathsf{P}\Big(\sup_A \sum_{k<k'} \alpha_{k,k'}\, R_{k,k'} \ge \sqrt{\frac{u}{N}}\Big)\\
&\le \Big(1+\frac{1}{\varepsilon}\Big)^{M^2} \exp\Big(-\frac{u}{2}\Big(1 - L\sqrt{\frac{u}{N}}\Big)\Big) ,
\end{aligned}$$

where we use (A.53) for $t = \sqrt{uN}$ in the last line. □

Corollary A.9.7. *We have*

$$2^{-Nn}\operatorname{card}\Big\{(\boldsymbol{\sigma}^1,\ldots,\boldsymbol{\sigma}^n)\,;\ \sum_{1\le \ell<\ell'\le n} N R_{\ell,\ell'}^2 \ge (1-2\varepsilon)^{-2}\, u\Big\}$$

$$\le \Big(1+\frac{1}{\varepsilon}\Big)^{n^2} \exp\Big(-\frac{u}{2}\Big(1 - L\sqrt{\frac{u}{N}}\Big)\Big)$$

where $R_{\ell,\ell'} = N^{-1}\sum_{i\le N} \sigma_i^\ell\, \sigma_i^{\ell'}$.

Proof. This is another way to formulate Lemma A.9.6 when $M = n$. □

A.10 Poisson Random Variables and Point Processes

A Poisson random variable X of expectation a is an integer-valued r.v. such that, for $k = 0, 1, \ldots$

$$\mathsf{P}(X = k) = \frac{a^k}{k!} e^{-a}$$

so that

$$\mathsf{E} \exp \lambda X = \sum_{k \geq 0} \frac{a^k}{k!} e^{\lambda k - a} = \exp a(e^\lambda - 1) . \tag{A.63}$$

Differentiating 1, 2, or 3 times this relation in λ and setting $\lambda = 0$ we see that

$$\mathsf{E}X = a \; ; \; \mathsf{E}X^2 = a + a^2 \; ; \; \mathsf{E}X^3 = a + 3a^2 + a^3 . \tag{A.64}$$

Using from (A.8) that for $\lambda > 0$ and a r.v. Y we have $\mathsf{P}(Y \geq t) \leq e^{-\lambda t} \mathsf{E} \exp \lambda Y$ and $\mathsf{P}(Y \leq t) \leq e^{\lambda t} \mathsf{E} \exp(-\lambda Y)$, and optimizing over λ we get that for $t > 1$ we have

$$\mathsf{P}(X \geq at) \leq \exp(-a(t \log t - t - 1))$$

and

$$\mathsf{P}(X \leq a/t) \leq \exp(-a(t \log t - t - 1)) .$$

In particular we have

$$\mathsf{P}(|X - a| \geq a/2) \leq \exp\left(-\frac{a}{L}\right) . \tag{A.65}$$

Of course, such an inequality holds for any constant instead of 1/2.

If X_1, X_2 are independent Poisson r.v.s, $X_1 + X_2$ is a Poisson r.v. The following lemma prove a less known remarkable property of these variables.

Lemma A.10.1. *Consider a Poisson r.v X and i.i.d. r.v.s $(\delta_i)_{i \geq 1}$ such that $\mathsf{P}(\delta_i = 1) = \delta, \mathsf{P}(\delta_i = 0) = 1 - \delta$ for a certain number δ. Then the r.v.s*

$$X_1 = \sum_{i \leq X} \delta_i \; ; \; X_2 = \sum_{i \leq X} (1 - \delta_i)$$

are independent Poisson r.v.s, of expectation respectively $\delta \mathsf{E}X$ and $(1-\delta)\mathsf{E}X$.

In this lemma we "split X in two pieces". In a similar manner, we can split X in any number of pieces.

Proof. We compute

$$\begin{aligned}
\mathsf{E}\exp(\lambda X_1+\mu X_2) &= \mathsf{E}\exp\Big(\lambda\sum_{i\le X}\delta_i+\mu\sum_{i\le X}(1-\delta_i)\Big)\\
&=\sum_{k\ge 0}\frac{a^k}{k!}e^{-a}\mathsf{E}\exp\Big(\lambda\sum_{i\le k}\delta_i+\mu\sum_{i\le k}(1-\delta_i)\Big)\\
&=\sum_{k\ge 0}\frac{a^k}{k!}e^{-a}\big(\mathsf{E}\exp(\lambda\delta_i+\mu(1-\delta_1))\big)^k\\
&=\sum_{k\ge 0}\frac{a^k}{k!}e^{-a}(\delta e^{\lambda}+(1-\delta)e^{\mu})^k\\
&=\exp a(\delta e^{\lambda}+(1-\delta)e^{\mu}-1)\\
&=\exp a\delta(e^{-\lambda}-1)\exp a(1-\delta)(e^{-\mu}-1)\\
&=\mathsf{E}\exp(\lambda Y_1+\mu Y_2)\ ,
\end{aligned}$$

where Y_1 and Y_2 are independent Poisson r.v.s with expectation respectively δ and $1-\delta$. □

Consider a positive measure μ of finite total mass $|\mu|$ (say on $\mathbb{R}^3$), and assume for simplicity that μ has no atoms. A Poisson point process of intensity measure μ is a random finite subset $\Pi=\Pi_\mu$ with the following properties:

1. $\operatorname{card}\Pi$ is a Poisson r.v. of expectation $|\mu|$.
2. Given that $\operatorname{card}\Pi=k$, Π is distributed like the set $\{X_1,\ldots,X_k\}$ where $X_1,\ldots,X_k$ are i.i.d. r.v.s of law $\mu/|\mu|$.

(Some inessential complications occur when μ has atoms, and one has to count points of the Poisson point process "with their order of multiplicity".) We list without proof some of the main properties of Poisson point processes. (The proofs are all very easy.)

Given two disjoint Borel sets, A, B, $\Pi\cap A$ and $\Pi\cap B$ are independent Poisson point processes.

Given two finite measures μ_1,μ_2, if Π_{μ_1} and Π_{μ_2} are independent Poisson point processes of intensity measure μ_1 and μ_2 respectively, then $\Pi_{\mu_1}\cup\Pi_{\mu_2}$ is a Poisson point process of intensity measure $\mu_1+\mu_2$.

Given a (continuous) map φ, $\varphi(\Pi)$ is a Poisson point process of intensity measure $\varphi(\mu)$, the image measure of the intensity measure μ of Π by φ.

Consider a positive measure μ and a Poisson point process Π_μ of intensity measure μ. If ν is a probability (say on $\mathbb{R}^3$), and $(U_\alpha)_{\alpha\ge 1}$ are i.i.d. r.v.s of law ν, we can construct a Poisson point process of intensity measure $\mu\otimes\nu$ as follows. We number in a random order the points of Π as $x_1,\ldots,x_k$, and we consider the couples $(x_1,U_1),\ldots,(x_k,U_k)$.

Consider now a positive measure μ on $\mathbb{R}^+$. We do not assume that μ is finite, but we assume that $\mu([a,\infty))$ is finite for each $a\ge 0$. We denote by μ_0 the restriction of μ to $[1,\infty)$, by μ_k its restriction to $[2^{-k},2^{-k+1}[$, $k\ge 1$. Consider for $k\ge 0$ a Poisson point process Π_k of intensity measure μ_k, and assume that these are independent. We can define a Poisson point process of intensity measure μ as $\Pi = \cup_{k\ge 0}\Pi_k$. Then for each a, $\Pi \cap [a,\infty)$ is a Poisson point process, the intensity measure of which is the restriction of μ to $[a,\infty)$.

A.11 Distances Between Probability Measures

The set $\mathcal{M}_1(\mathcal{X})$ of probability measures on a compact metric space $(\mathcal{X}, d)$ is provided with a natural topology, the weakest topology that makes all the maps $\mu \mapsto \int f(x)\mathrm{d}\mu(x)$ continuous, where $f \in \mathcal{C}(\mathcal{X})$, the space of continuous functions on $\mathcal{X}$. For this topology the set $\mathcal{M}_1(\mathcal{X})$ is a compact metric space. The compactness is basically obvious if one knows the fundamental Riesz representation theorem. This theorem identifies $\mathcal{M}_1(\mathcal{X})$ with the set of positive linear functionals Φ on $\mathcal{C}(\mathcal{X})$ that have the property that $\Phi(\mathbf{1}) = 1$ where the function $\mathbf{1}$ is the function that takes the value 1 at every point.

The so-called Monge-Kantorovich transportation-cost distance on $\mathcal{M}_1(\mathcal{X})$ is particularly useful. Given a compact metric space $(\mathcal{X}, d)$, and two probability measures μ_1 and μ_2 on $\mathcal{X}$, their transportation-cost distance is defined as

$$d(\mu_1, \mu_2) = \inf \mathsf{E}\, d(X_1, X_2) \;, \tag{A.66}$$

where the infimum is taken over all pairs (X_1, X_2) of r.v.s such that the law of X_j is μ_j for $j = 1, 2$. Equivalently,

$$d(\mu_1, \mu_2) = \inf \int d(x_1, x_2)\mathrm{d}\theta(x_1, x_2) \;,$$

where the infimum is over all probability measures θ on $\mathcal{X}^2$ with marginals μ_1 and μ_2 respectively. It is not immediately clear that the formula (A.66) defines a distance. This is however obvious due to the (fundamental) "duality formula"

$$d(\mu_1, \mu_2) = \sup \left(\int f(x)\,\mathrm{d}\mu_1(x) - \int f(x)\,\mathrm{d}\mu_2(x) \right) \tag{A.67}$$

where the supremum is taken over all functions f from $\mathcal{X}$ to $\mathbb{R}$ with Lipschitz constant 1, i.e. that satisfy $|f(x) - f(y)| \le d(x, y)$ for all x, y in $\mathcal{X}$. The classical formula (A.67) is a simple consequence of the Hahn-Banach theorem. We will not use it in any essential way, so we refer the reader to Lemma A.11.1 below for the complete proof of a similar result.

Another proof that d is a distance uses the classical notion of disintegration of measures (or, equivalently of conditional probability), and we sketch it now. Consider a probability measure θ on $\mathcal{X}^2$ with marginals μ_1 and μ_2 respectively. Then there exists a (Borel measurable) family of probability measures θ_x on $\mathcal{X}$ such that for any continuous function h on $\mathcal{X}^2$ we have

$$\int h\mathrm{d}\theta = \int \left(\int h(x, y)\mathrm{d}\theta_x(y) \right) \mathrm{d}\mu_1(x) \,. \tag{A.68}$$

Consider another probability measure θ' on $\mathcal{X}^2$ with marginals μ_1 and μ_3 respectively, and a family of probability measures θ'_x on $\mathcal{X}$ such that for any continuous function h on $\mathcal{X}^2$ we have

$$\int h\mathrm{d}\theta' = \int\Big(\int h(x,y)\mathrm{d}\theta'_x(y)\Big)\mathrm{d}\mu_1(x)\,. \tag{A.69}$$

Consider then the probability measure θ'' on $\mathcal{X}^2$ such that for any continuous function we have

$$\int h\mathrm{d}\theta'' = \int\Big(\int h(y,z)\mathrm{d}\theta_x(y)\mathrm{d}\theta'_x(z)\Big)\mathrm{d}\mu_1(x)\,. \tag{A.70}$$

Using (A.70) in the case where $h(y,z) = f(y)$ in the first line and (A.68) in the third line we get that

$$\begin{aligned}\int f(y)\mathrm{d}\theta''(y,z) &= \int\Big(\int f(y)\mathrm{d}\theta_x(y)\mathrm{d}\theta'_x(z)\Big)\mathrm{d}\mu_1(x)\\ &= \int\Big(\int f(y)\mathrm{d}\theta_x(y)\Big)\mathrm{d}\mu_1(x)\\ &= \int f(y)\mathrm{d}\theta(x,y) = \int f(y)\mathrm{d}\mu_2(y)\,,\end{aligned}$$

using in the last inequality that μ_2 is the second marginal of θ. This proves that the first marginal of θ'' is μ_2, and similarly, its second marginal is μ_3. Using the triangle inequality

$$d(y,z) \le d(y,x) + d(x,z)\,,$$

and using (A.70), (A.69) and (A.68) we obtain

$$\int d(y,z)\mathrm{d}\theta''(y,z) \le \int d(x,y)\mathrm{d}\theta(x,y) + \int d(x,z)\mathrm{d}\theta'(x,z)\,,$$

and in this manner we can easily complete the proof that d is a distance on $\mathcal{M}_1(\mathcal{X})$.

The topology defined by the distance d is the weak topology on $\mathcal{M}_1(\mathcal{X})$. To see this we observe first that the weak topology is also the weakest topology that makes all the maps $\mu \mapsto \int f(x)\mathrm{d}\mu(x)$ where f is a Lipschitz function on $\mathcal{X}$ with Lipschitz constant ≤ 1. This is simply because the linear span of the classes of such functions is dense in $\mathcal{C}(\mathcal{X})$ for the uniform norm. Therefore the weak topology is weaker than the topology defined by d. To see that it is also stronger we note that in (A.67) we can also take the supremum on the class of Lipschitz functions that take the value 0 at a given point of $\mathcal{X}$. This class is compact for the supremum norm. Therefore given $\varepsilon > 0$ there is a finite class $\mathcal{F}$ of Lipschitz functions on $\mathcal{X}$ such that

$$d(\mu_1,\mu_2) \le \varepsilon + \sup_{\mathcal{F}}\left|\int f(x)\,\mathrm{d}\mu_1(x) - \int f(x)\,\mathrm{d}\mu_2(x)\right|\,. \tag{A.71}$$

Given two probability measures (μ, ν) on a compact metric space $(\mathcal{X}, d)$ we consider the quantity

$$\Delta(\mu, \nu) = \inf \mathsf{E}d(X, Y)^2 , \tag{A.72}$$

where the infimum is taken over all pairs of r.v.s (X, Y) with laws μ and ν respectively. The quantity $\Delta^{1/2}(\mu, \nu)$ is a distance, called Wasserstein's distance between μ and ν. This is not obvious from the definition, but can be proved following the scheme we outlined in the case of the Monge-Kantorovich transportation-cost distance (A.66). It also follows from the duality formula given in Lemma A.11.1 below. Of course Wasserstein's distance is a close cousin of the transportation-cost distance, simply we replace the "linear" measure of the "cost of transportation" by a "quadratic measure" of this cost.

Denoting by D the diameter of $\mathcal{X}$, i.e.

$$D = \sup\{d(x, y) \; ; \; x \; , \; y \in \mathcal{X}\} ,$$

for any two r.v.s X and Y we have the inequalities

$$(\mathsf{E}d(X, Y))^2 \le \mathsf{E}d(X, Y)^2 \le D\mathsf{E}d(X, Y) ,$$

so that

$$d(\mu, \nu)^2 \le \Delta(\mu, \nu) \le Dd(\nu, \mu) .$$

Consequently the topology induced by Wasserstein distance on $\mathcal{M}_1(\mathcal{X})$ also coincides with the weak topology. Let us note in particular from (A.71) that, given a number $\varepsilon > 0$ there exists a finite set $\mathcal{F}$ of continuous functions on $\mathcal{X}$ such that

$$\Delta(\mu_1, \mu_2) \le \varepsilon + \sup_{\mathcal{F}} \left| \int f(x) \, \mathrm{d}\mu_1(x) - \int f(x) \, \mathrm{d}\mu_2(x) \right| . \tag{A.73}$$

The following is the "duality formula" for Wasserstein's distance.

Lemma A.11.1. *If μ and ν are two probability measures on $\mathcal{X}$, then*

$$\Delta(\mu, \nu) = \sup \Big\{ \int f \, \mathrm{d}\mu + \int g \, \mathrm{d}\nu \, ; \; f, g \text{ continuous}, \\ \forall x, y \in \mathcal{X} \; , \; f(x) + g(y) \le d(x, y)^2 \Big\} . \tag{A.74}$$

Proof. If f and g are continuous functions such that

$$\forall x, y \in \mathcal{X}, \;\; f(x) + g(y) \le d(x, y)^2 ,$$

then for each pair (X, Y) of r.v.s valued in $\mathcal{X}$ we have $\mathsf{E}f(X) + \mathsf{E}g(Y) \le \mathsf{E}d(X, Y)^2$, so that if X has law μ and Y has law ν we have $\int f \, \mathrm{d}\mu + \int g \, \mathrm{d}\nu \le$

$\mathsf{E}d(X,Y)^2$. Taking the infimum over all choices of X and Y we see that $\int f\,\mathrm{d}\mu + \int g\,\mathrm{d}\nu \le \Delta(\mu,\nu)$. Therefore if a denotes the right-hand side of (A.74), we have proved that $a \le \Delta(\mu,\nu)$, and we turn to the proof of the converse. We consider the subset $\mathcal{S}$ of the set $\mathcal{C}(\mathcal{X}\times\mathcal{X})$ of continuous functions on $\mathcal{X}\times\mathcal{X}$ that consists of the functions $w(x,y)$ such that there exists continuous functions f and g on $\mathcal{X}$ for which

$$\int f\,\mathrm{d}\mu + \int g\,\mathrm{d}\nu = a \tag{A.75}$$

and

$$\forall\, x,y \in \mathcal{X}\ ,\ w(x,y) > f(x) + g(y) - d(x,y)^2\ . \tag{A.76}$$

It follows from the definition of a that for each function w in $\mathcal{S}$ there exist x and y with $w(x,y) > 0$. Since $\mathcal{S}$ is convex and open, the Hahn-Banach separation theorem asserts that we can find a linear functional Φ on $\mathcal{C}(\mathcal{X}\times\mathcal{X})$ such that $\Phi(w) > 0$ for each w in $\mathcal{S}$. If $w \in \mathcal{S}$ and $w' \ge 0$ it follows from the definition of $\mathcal{S}$ that $w + \lambda w' \in \mathcal{S}$, so that $\Phi(w+\lambda w') > 0$. Thus $\Phi(w') \ge 0$, i.e. Φ is positive, it is a positive measure on $\mathcal{X}\times\mathcal{X}$. Since it is a matter of normalization, we can assume that it is a probability, which we denote by θ. If f and g are as in (A.75), then for each $\varepsilon > 0$ we see by (A.76) that the function $w(x,y) = f(x) + g(y) - d(x-y)^2 + \varepsilon$ belongs to $\mathcal{S}$ and thus

$$\int d(x,y)^2\,\mathrm{d}\theta(x,y) \le \int (f(x)+g(y))\,\mathrm{d}\theta(x,y)\ . \tag{A.77}$$

Now this holds true if we replace f by $f + f'$ where $\int f'\,\mathrm{d}\mu = 0$. Thus this latter condition must imply that $\int f'(x)\,\mathrm{d}\theta(x,y) = 0$. It follows that if θ_1 is the first marginal of θ then $\int f'(x)\mathrm{d}\theta_1(x) = 0$ whenever $\int f'\,\mathrm{d}\mu = 0$. Using this for $f'(x) = f(x) - \int f\,\mathrm{d}\mu$ where f is any continuous function, we see that $\theta_1 = \mu$, i.e. μ is the first marginal of θ. Similarly, ν is the second marginal of θ so that

$$\int (f(x)+g(y))\,\mathrm{d}\theta(x,y) = \int f\,\mathrm{d}\mu + \int g\,\mathrm{d}\nu = a\ ,$$

and (A.77) then implies that $\int d(x,y)^2\,\mathrm{d}\theta(x,y) \le a$. A pair (X,Y) of r.v.s of joint law θ then witnesses that $\Delta(\mu,\nu) \le a$. □

The previous distances must not be confused with the total variation distance given by

$$\|\mu-\nu\| = \sup\left\{\left|\int f\mathrm{d}\mu(x) - \int f\mathrm{d}\nu(x)\right|\ ;\ |f| \le 1\right\}\ . \tag{A.78}$$

The total variation distance induces the weak topology on $\mathcal{M}_1(\mathcal{X})$ only when $\mathcal{X}$ is finite. When this is the case, we have

$$\|\mu-\nu\| = \sum_{x\in\mathcal{X}} |\mu(\{x\}) - \nu(\{x\})|\ . \tag{A.79}$$

Exercise A.11.2. Prove that $\|\mu - \nu\| = 2\Delta(\mu,\nu)$, where $\Delta^{1/2}(\mu,\nu)$ is Wasserstein's distance when $\mathcal{X}$ is provided with the distance d given by $d(x,x) = 0$ and $d(x,y) = 1$ when $x \neq y$.

When $\mathcal{X}$ is a metric space, that is not necessarily compact, the formulas (A.66) and (A.72) still make sense, although the infimum might be infinite. The corresponding "distances" still satisfy the triangle inequality.

A.12 The Paley-Zygmund Inequality

This simple (yet important) argument is also known as the second moment method. It goes back to the work of Paley and Zygmund on trigonometric series.

Proposition A.12.1. *Consider a r.v. $X \geq 0$. Then*

$$\mathsf{P}\left(X \geq \frac{1}{2}\mathsf{E}X\right) \geq \frac{1}{4}\frac{(\mathsf{E}X)^2}{\mathsf{E}X^2}. \tag{A.80}$$

Proof. If $A = \{X \geq \mathsf{E}X/2\}$, then, since $X \leq \mathsf{E}X/2$ on the complement A^c of A, we have

$$\mathsf{E}X = \mathsf{E}(X\mathbf{1}_A) + \mathsf{E}(X\mathbf{1}_{A^c}) \leq \mathsf{E}(X\mathbf{1}_A) + \frac{1}{2}\mathsf{E}X .$$

Thus, using the Cauchy-Schwarz inequality,

$$\frac{1}{2}\mathsf{E}X \leq \mathsf{E}(X\mathbf{1}_A) \leq (\mathsf{E}X^2)^{1/2}\,\mathsf{P}(A)^{1/2} . \qquad \square$$

A.13 Differential Inequalities

We will often meet simple differential inequalities, and it is worth to learn how to handle them. The following is a form of the classical Gronwall's lemma.

Lemma A.13.1. *If a function $\varphi \geq 0$ satisfies*

$$|\varphi'_r(t)| \leq c_1\varphi(t) + c_2$$

for $0 < t < 1$, where $c_1, c_2 \geq 0$ and where φ'_r is the right-derivative of φ, then

$$\varphi(t) \leq \exp(c_1 t)\left(\varphi(0) + \frac{c_2}{c_1}\right). \tag{A.81}$$

Proof. We note that

$$\left|\left(\varphi(t) + \frac{c_2}{c_1}\right)'_r\right| \leq c_1\left(\varphi(t) + \frac{c_2}{c_1}\right),$$

so that

$$\varphi(t) + \frac{c_2}{c_1} \leq \exp(c_1 t)\left(\varphi(0) + \frac{c_2}{c_1}\right). \qquad \square$$

A.14 The Latala-Guerra Lemma

In this section we prove Proposition 1.3.8. We present the proof due to F. Guerra. M. Yor noticed that essentially the same proof gives a more general result that is probably less mysterious.

Proposition A.14.1. *Consider an increasing bounded function* $\varphi(y)$, *that satisfies* $\varphi(-y) = -\varphi(y)$ *and* $\varphi''(y) < 0$ *for* $y > 0$. *Then the function* $\Psi(x) = \mathsf{E}\varphi(z\sqrt{x} + h)^2/x$ *is strictly decreasing on* $\mathbb{R}^+$ *and vanishes as* $x \to \infty$.

Proof. To prove that the function Ψ is strictly decreasing, working conditionally on h, we can assume that h is a number. We set $Y = z\sqrt{x} + h$. We have

$$\begin{aligned} x^2\Psi'(x) &= \mathsf{E}\left(z\sqrt{x}\,\varphi'(Y)\varphi(Y) - \varphi(Y)^2\right) \\ &= \mathsf{E}\big(\varphi(Y)(Y\varphi'(Y) - \varphi(Y))\big) - h\mathsf{E}\,\varphi'(Y)\varphi(Y)\,. \end{aligned}$$

The reader should observe here how tricky we have been: we resisted the temptation to use Gaussian integration by parts.

To study φ, we note first that $\varphi(0) = 0$ since φ is odd, so that since φ is increasing $\varphi(y) > 0$ for $y > 0$ and $\varphi(y) < 0$ for $y < 0$. The function $\psi(y) = y\varphi'(y) - \varphi(y)$ satisfies $\psi(0) = 0$ and $\psi'(y) = y\varphi''(y)$. Thus $\psi'(y) < 0$ for $y \neq 0$ and thus $\psi(y) < 0$ for $y > 0$ and $\psi(y) > 0$ for $y < 0$. Therefore $\varphi(y)(y\varphi'(y) - \varphi(y)) = \varphi(y)\psi(y) < 0$ for $y \neq 0$ and hence

$$\mathsf{E}\big(\varphi(Y)(Y\varphi'(Y) - \varphi(Y))\big) < 0\,.$$

Consequently all we have to prove is that $h\mathsf{E}\,\varphi(Y)\varphi'(Y) \geq 0$. We start by writing that

$$\mathsf{E}\,\varphi(Y)\varphi'(Y) = \frac{1}{\sqrt{2\pi}}\int \varphi(z\sqrt{x} + h)\varphi'(z\sqrt{x} + h)\, e^{-z^2/2}\,\mathrm{d}z\,. \tag{A.82}$$

Now comes the beautiful trick. We make the change of variable

$$z = \frac{y - h}{\sqrt{x}}\,,$$

so that $y = z\sqrt{x} + h$ and

$$\begin{aligned} &\frac{1}{\sqrt{2\pi}}\int \varphi(z\sqrt{x} + h)\varphi'(z\sqrt{x} + h)\, e^{-z^2/2}\mathrm{d}z \\ &= \frac{1}{\sqrt{2\pi x}}\int \varphi(y)\varphi'(y)\exp\left(-\frac{y^2}{2x} + \frac{hy}{x} - \frac{h^2}{2x}\right)\mathrm{d}y\,. \end{aligned} \tag{A.83}$$

Making the change of variable $y' = -y$ we get

$$\frac{1}{\sqrt{2\pi}}\int \varphi(z\sqrt{x}+h)\varphi'(z\sqrt{x}+h)\,e^{-z^2/2}\mathrm{d}z$$
$$= -\frac{1}{\sqrt{2\pi x}}\int \varphi(y)\varphi'(y)\exp\left(-\frac{y^2}{2x}-\frac{hy}{x}-\frac{h^2}{2x}\right)\mathrm{d}y\,. \tag{A.84}$$

Recalling (A.82) and adding (A.83) and (A.84) we get

$$h\mathsf{E}\,\varphi(Y)\varphi'(Y) = \frac{1}{\sqrt{2\pi x}}\int \varphi(y)\varphi'(y)h\mathrm{sh}(hy/x)\exp\left(-\frac{y^2}{2x}-\frac{h^2}{2x}\right)\mathrm{d}y \ge 0\,,$$

because $\varphi'(y) \ge 0$ and $h\varphi(y)\mathrm{sh}(hy/x) \ge 0$. This proves that Ψ is strictly decreasing. □

A.15 Proof of Theorem 3.1.4

Proof. We start with $N = 1$ and then perform induction on the dimension N. By homogeneity, we may assume that $\int U\mathrm{d}x = \int V\mathrm{d}x = 1$ and by approximation that U and V are continuous with strictly positive values. Define $x, y :]0,1[\to \mathbb{R}$ by

$$\int_{-\infty}^{x(t)} U(q)\mathrm{d}q = t\,, \quad \int_{-\infty}^{y(t)} V(q)\mathrm{d}q = t\,.$$

Therefore x and y are increasing and differentiable and

$$x'(t)U(x(t)) = y'(t)V(y(t)) = 1\,.$$

Set $z(t) = sx(t) + (1-s)y(t)$, $t \in]0,1[$. By the arithmetic-geometric mean inequality, for every t,

$$z'(t) = sx'(t) + (1-s)y'(t) \ge (x'(t))^s(y'(t))^{1-s}\,. \tag{A.85}$$

Now, since z is injective, by the hypothesis (3.11) on W and (A.85),

$$\begin{aligned}\int W\mathrm{d}x &\ge \int_0^1 W(z(t))z'(t)\mathrm{d}t \\ &\ge \int_0^1 U(x(t))^s V(y(t))^{1-s}(x'(t))^s(y'(t))^{1-s}\mathrm{d}t \\ &= \int_0^1 [U(x(t))x'(t)]^s[V(y(t))y'(t)]^{1-s}\mathrm{d}t \\ &= 1\,.\end{aligned}$$

This proves the case $N = 1$. It is then easy to deduce the general case by induction on N as follows. Suppose $N > 1$ and assume that the functional version of the Brunn-Minkowski theorem holds in $\mathbb{R}^{N-1}$. Let U, V, W be non-negative measurable functions on $\mathbb{R}^N$ satisfying (3.11) for some $s \in [0,1]$. Let $q \in \mathbb{R}$ be fixed and define $U_q : \mathbb{R}^{N-1} \to [0,\infty[$ by $U_q(x) = U(x,q)$ and similarly for V_q and W_q. Clearly, if $q = sq_0 + (1-s)q_1$, $q_0, q_1 \in \mathbb{R}$,

$$W_q(sx + (1-s)y) \geq U_{q_0}(x)^s V_{q_1}(y)^{1-s}$$

for all $x, y \in \mathbb{R}^{N-1}$. Therefore, by the induction hypothesis,

$$\int_{\mathbb{R}^{N-1}} W_q(x)\mathrm{d}x \geq \left(\int_{\mathbb{R}^{N-1}} U_{q_0}(x)\mathrm{d}x\right)^s \left(\int_{\mathbb{R}^{N-1}} V_{q_1}(x)\mathrm{d}x\right)^{1-s} . \tag{A.86}$$

Let us define $W^*(q) = \int_{\mathbb{R}^{N-1}} W_q(x)\mathrm{d}x$, and $U^*(q)$, $V^*(q)$ similarly. We see from (A.86) that

$$W^*(sq_0 + (1-s)q_1) \geq U^*(q_0)^s V^*(q_1)^{1-s} ,$$

so applying the one-dimensional case shows that

$$\int_{\mathbb{R}} W^*(q)\mathrm{d}q \geq \left(\int_{\mathbb{R}} U^*(q)\mathrm{d}q\right)^s \left(\int_{\mathbb{R}} V^*(q)\mathrm{d}q\right)^{1-s} .$$

Since

$$\int_{\mathbb{R}^N} W(x)\mathrm{d}x = \int_{\mathbb{R}} \left(\int_{\mathbb{R}^{N-1}} W_q(x)\mathrm{d}x\right) \mathrm{d}q = \int_{\mathbb{R}} W^*(q)\mathrm{d}q ,$$

and similarly for U^* and V^* this is the desired result. Theorem 3.1.4 is established. □

References

1. Aldous D. (1992) Asymptotics in the random assignment problem. *Probab. Theory Related Fields* **93**, no. 4, pp. 507–534.
2. Aldous D. (2001) The $\zeta(2)$ limit in the random assignment problem. *Random Structures and Algorithms* **18**, no. 4, pp. 381–418.
3. Amit D.J., Gutfreund H., Sompolinsky H. (1987) Statistical mechanics of neural networks near saturation. *Annals of Physics* **173**, pp. 30–67.
4. Aizenman M., Lebowitz J.L., Ruelle D. (1987) Some rigorous results on the Sherrington-Kirkpatrick model. *Comm. Math. Phys.* **112**, pp. 3–20.
5. Almeida J.R.L., Thouless D.J. (1978) Stability of the Sherrington-Kirkpatrick solution of a spin glass model. *J. Phys. A: Math. Gen.* **II**, pp. 983–990.
6. Albeverio S., Tirozzi B., Zegarlinski B. (1992) Rigorous results for the free energy in the Hopfield model. *Comm. Math. Phys.* **150**, no. 2, pp. 337–373.
7. Barbina X., Márquez-Carreras D., Rovira C., Tindel S. (2004) Higher order expansions for the overlap of the SK model, Seminar on Stochastic Analysis, Random Fields and Applications IV, pp. 21–43, Progr. Probab., 58, Birkhäuser, Basel, 2004.
8. Bayati M., Gamarnik D., Tetali P. (2009) A combinatorial approach to Guerra's interpolation method. Manuscript.
9. Bianchi A., Contucci P., Giardina C. (2003) Thermodynamic Limit for Mean Field Spin Models. *Math. Phys. El. Jour.* **9**, n. 6, pp. 1–15.
10. Billingsley, P. (1995) *Probability and Measure.* Second edition. Wiley Series in Probability and Mathematical Statistics: Probability and Mathematical Statistics. John Wiley & Sons, Inc., New York. xiv+622 pp.
11. Bobkov S.G., Ledoux M. (2000) From Brunn-Minkowski to Brascamp-Lieb and to logarithmic Sobolev inequalities. *Geom. Funct. Anal.* **10**, no. 5, pp. 1028–1052.
12. Bollobás B. (2001) *Random graphs.* Second edition. Cambridge Studies in Advanced Mathematics **73**, Cambridge University Press, Cambridge, 2001, xviii+498 pp.
13. Bolina O., Wreszinski W.F. (2004) A Self Averaging "Order Parameter" for the Sherrington-Kirkpatrick Spin Glass Model. *Journal Stat. Phys.* **116**, pp. 1389–1404.
14. Borell C. (1984) On polynomial chaos and integrability. *Probab. Math. Statist.* **3**, no. 2, pp. 191–203.
15. Bouten M. (1988) Replica symmetry instability in perceptron models. Comment on: "Optimal storage properties of neural network models", *J. Phys. A* **21**, no. 1, pp. 271–284, by E. Gardner and B. Derrida. With a reply by Derrida. *J. Phys. A* **27** (1994), no. 17, pp. 6021-6025.
16. Bovier A. (1994) Self-averaging in a class of generalized Hopfield models. *J. Phys. A* **27**, no. 21, pp. 7069–7077.

M. Talagrand, *Mean Field Models for Spin Glasses*, Ergebnisse der Mathematik und ihrer Grenzgebiete. 3. Folge / A Series of Modern Surveys in Mathematics 54, DOI 10.1007/978-3-642-15202-3,

17. Bovier A. (1997) Comment on “capacity of Hopfield model”. *J. Phys. A* **30**, pp. 7993–7996.
18. Bovier A. (1999) Sharp upper bounds on perfect retrieval in the Hopfield model. *J. Appl. Probab.* **36**, no. 3, pp. 941–950.
19. Bovier A. (2006) *Statistical mechanics of disordered systems. A mathematical perspective.* Cambridge Series in Statistical and Probabilistic Mathematics. Cambridge University Press, Cambridge. xiv+312 pp.
20. Bovier A., van Enter A.C.D., Niederhauser B. (1999) Stochastic symmetry-breaking in a Gaussian Hopfield model. *J. Statist. Phys.* **95**, no. 1-2, pp. 181–213.
21. Bovier A., Gayrard V. (1992) Rigorous bounds on the storage capacity of the dilute Hopfield model. *J. Statist. Phys.* **69**, no. 3-4, pp. 597–627.
22. Bovier A., Gayrard V. (1993) Rigorous results on the thermodynamics of the dilute Hopfield model. *J. Statist. Phys.* **72**, no. 1-2, pp. 79–112.
23. Bovier A., Gayrard V. (1993) Lower bounds on the memory capacity of the dilute Hopfield model. Cellular automata and cooperative systems (Les Houches, 1992), pp. 55–66. *NATO Adv. Sci. Inst. Ser. C Math. Phys. Sci.* **396**, Kluwer Acad. Publ., Dordrecht.
24. Bovier A., Gayrard V. (1993) The Hopfield model on random graphs: lower bounds on the storage capacity. Dynamics of complex and irregular systems (Bielefeld, 1991), pp. 277–288. *Bielefeld Encount. Math. Phys.* VIII, World Sci. Publishing, River Edge, NJ.
25. Bovier A., Gayrard V. (1996) An almost sure large deviation principle for the Hopfield model. *Ann. Probab.* **24**, no. 3, pp. 1444–1475.
26. Bovier A., Gayrard V. (1997) The retrieval phase of the Hopfield model: a rigorous analysis of the overlap distribution. *Probab. Theory Relat. Fields* **107**, no. 1, pp. 61–98.
27. Bovier A., Gayrard V. (1997) An almost sure central limit theorem for the Hopfield model. *Markov Process. Related Fields* **3**, no. 2, pp. 151–173.
28. Bovier A., Gayrard V. (1998) Hopfield models as generalized random mean field models. Mathematical aspects of spin glasses and neural networks. *Progr. Probab.* **41**, pp. 3–89. Birkhäuser Boston, Inc., Boston, MA.
29. Bovier A., Gayrard V. (1998) Metastates in the Hopfield model in the replica symmetric regime. *Math. Phys. Anal. Geom.* **1**, no. 2, pp. 107–144.
30. Bovier A., Mason D.M. (2001) Extreme value behavior in the Hopfield model. *Ann. Appl. Probab.* **11**, no. 1, pp. 91–120.
31. Bovier A., Niederhauser B. (2001) The spin-glass phase-transition in the Hopfield model with p-spin interactions. *Adv. Theor. Math. Phys.* **5**, no. 6, pp. 1001–1046.
32. Bovier A., Gayrard V., Picco P. (1994) Gibbs states for the Hopfield model in the regime of perfect memory. *Prob. Theory Relat. Fields* **100**, pp. 1329–363.
33. Bovier A., Gayrard V., Picco P. (1995) Gibbs states for the Hopfield model with extensively many patterns. *J. Stat. Phys.* **79**, pp. 395–414.
34. Bovier A., Gayrard V., Picco P. (1995) Gibbs states of the Hopfield model with extensively many patterns. *J. Statist. Phys.* **79**, no. 1-2, pp. 395–414.
35. Bovier A., Gayrard V., Picco P. (1995) Large deviation principles for the Hopfield model and the Kac-Hopfield model. *Probab. Theory Relat. Fields* **101**, no. 4, pp. 511–546.
36. Bovier A., Picco P. (editors) (1997) Mathematical aspects of spin glasses and Neural networks. *Progress in Probability* **41**, Birkhäuser, Boston.
37. Brascamp H., Lieb E. (1976) On extension of the Brunn-Minkowski and the Prékopa-Leindler Theorems, including inequalities for log-concave functions,

and with an application to the diffusion equation, *J. Funct. Anal.* **22**, pp. 366–389.
38. Canisius J., van Enter A.C.D., van Hemmen J.L. (1983) On a classical spin glass model. *Z. Phys. B* **50**, no. 4, pp. 311–336.
39. Carmona P., Hu Y. (2006) Universality in Sherrington-Kirkpatrick's spin glass model. *Ann. Inst. H. Poincaré Probab. Statist.* **42**, no. 2, pp. 215–222.
40. Carvalho S., Tindel S. (2007) On the multiple overlap function of the SK model. *Publicacions Matematiques* **51**, pp. 163–199.
41. Carvalho S., Tindel S. (2007) A central limit theorem for a localized version of the SK model. *Potential Analysis* **26**, pp. 323–343.
42. Catoni O. (1996) The Legendre transform of two replicas of the Sherrington-Kirkpatrick spin glass model. A free energy inequality. *Probab. Theory Relat. Fields* **105**, no. 3, pp. 369–392.
43. Cavagna A., Giardina I., Parisi G. (1997) Structure of metastable states in spin glasses by means of a three replica potential. *J. Phys. A: Math. and Gen.* **30**, no. 13, pp. 4449–4466.
44. Chatterjee S. (2007) Estimation in spin glasses: a first step. *Ann. Statist.* **35**, no. 5, pp. 1931–1946.
45. Chatterjee S. (2010) Spin glasses and Stein's method. *Probab. Theory Relat. Fields* **148**, pp. 567–600.
46. Chatterjee S., Crawford N. (2009) Central limit theorems for the energy density in the Sherrington-Kirkpatrick model. *J. Stat. Phys.* **137**, no. 4 pp. 639–666.
47. Comets F. (1996) A spherical bound for the Sherrington-Kirkpatrick model. Hommage à P.A. Meyer et J. Neveu. *Astérisque* **236**, pp. 103–108.
48. Comets F. (1998) The martingale method for mean-field disordered systems at high temperature. Mathematical aspects of spin glasses and neural networks. *Progr. Probab.* **41**, pp. 91–113. Birkhäuser Boston, Inc., Boston, MA.
49. Comets F., Neveu J. (1995) The Sherrington-Kirkpatrick model of spin glasses and stochastic calculus: the high temperature case. *Comm. Math. Phys.* **166**, no. 3, pp. 549–564.
50. Crawford N. (2008) The intereaction between multioverlap in the high temperature phase of the Sherrington-Kirkpatrick spin glass. *J. Math. Phys.* **49**, 125201 (24 pages).
51. Dembo A., Montanari A. (2010) Gibbs measures and phase transitions on sparse random graphs. *Braz. J. Probab. Stat.* **24**, no. 2, pp. 137–211.
52. Derrida B., Gardner E. (1988) Optimal storage properties of neural network models. *J. Phys. A* **21**, pp. 271–284.
53. Derrida B., Gardner E. (1989) Three unfinished works on the optimal storage capacity of networks. Special issue in memory of Elizabeth Gardner, pp. 1957-1988. *J. Phys. A* **22**, no. 12, pp. 1983–1994.
54. Ellis R.S. (1985) *Entropy, large Deviations and Statistical Mechanics* Grundlehren der Mathematischen Wissenschaften **271**. Springer-Verlag, New York. xiv+364 pp.
55. Feng J., Shcherbina M., Tirozzi B. (2001) On the critical capacity of the Hopfield model. *Comm. Math. Phys.* **216**, no. 1, pp. 139–177.
56. Feng J., Shcherbina M., Tirozzi B. (2001) On the critical capacity of the Hopfield model. *Comm. Math. Phys.* **216**, no. 1, pp. 139-177.
57. Feng J., Tirozzi B. (1995) The SLLN for the free-energy of a class of neural networks. *Helv. Phys. Acta* **68**, no. 4, pp. 365–379.
58. Feng J., Tirozzi B. (1997) Capacity of the Hopfield model. *J. Phys. A* **30**, no. 10, pp. 3383–3391.

59. Fischer K.H., Hertz J., (1991) *Spin glasses.* Cambridge Studies in Magnetism, 1. Cambridge University Press, Cambridge, 1991. x+408 pp.
60. Franz S., Leone M. (2003) Replica bounds for optimization problems and diluted spin systems. *J. Statist. Phys.* **111**, no. 3-4, pp. 535–564.
61. Fröhlich J., Zegarlinski B. (1987) Some comments on the Sherrington-Kirkpatrick model of spin glasses. *Comm. Math. Phys.* **112**, pp. 553–566.
62. Gamarnik, D. (2004) Linear phase transition in random linear constraint satisfaction problems. *Probab. Theory Relat. Fields* **129**, pp. 410–440.
63. Gardner E. (1988) The space of interactions in neural network models. *J. Phys. A* **21**, pp. 257–270.
64. Gentz B. (1996) An almost sure central limit theorem for the overlap parameters in the Hopfield model. *Stochastic Process. Appl.* **62**, no. 2, pp. 243–262.
65. Gentz B. (1996) A central limit theorem for the overlap in the Hopfield model. *Ann. Probab.* **62**, no. 4, pp. 1809–1841.
66. Gentz B. (1998) On the central limit theorem for the overlap in the Hopfield model. Mathematical aspects of spin glasses and neural networks *Progr. Probab.* **41**, pp. 115–149, Birkhäuser Boston, Boston, MA.
67. Gentz B., Löwe M. (1999) The fluctuations of the overlap in the Hopfield model with finitely many patterns at the critical temperature. *Probab. Theory Relat. Fields* **115**, no. 3, pp. 357–381.
68. Gentz B., Löwe M. (1999) Fluctuations in the Hopfield model at the critical temperature. *Markov Process. Related Fields* **5**, no. 4, pp. 423–449.
69. Guerra F. (1995) Fluctuations and thermodynamic variables in mean field spin glass models. *Stochastic Processes, Physics and Geometry*, S. Albeverio et al. editors, World Scientific, Singapore.
70. Guerra F. (1996) About the overlap distribution in mean field spin glass models. *International Journal of Modern Physics B* **10**, pp. 1675–1684.
71. Guerra F. (2001) Sum rules for the free energy in the mean field spin glass model. *Fields Institute Communications* **30**, pp. 161–170.
72. Guerra F. (2005) Mathematical aspects of mean field spin glass theory. European Congress of Mathematics, pp. 719–732, Eur. Math. Soc., Zürich, 2005.
73. Guerra F., Toninelli F.L. (2002) Quadratic replica coupling for the Sherrington-Kirkpatrick mean field spin glass model. *J. Math. Phys.* **43**, no. 7, pp. 3704–3716.
74. Guerra F., Toninelli F.L. (2002) Central limit theorem for fluctuations in the high temperature region of the Sherrington-Kirkpatrick spin glass model. *J. Math. Phys.* **43**, no. 12, pp. 6224–6237.
75. Guerra F., Toninelli F.L. (2002) The Thermodynamic Limit in Mean Field Spin Glass Models. *Commun. Math. Phys.* **230**, pp. 71–79.
76. Guerra F., Toninelli F.L. (2003) The Infinite Volume Limit in Generalized Mean Field Disordered Models. *Markov Process. Related Fields* **9**, no. 2, pp. 195–207.
77. Guerra F., Toninelli F.L. (2003) Infinite volume limit and spontaneous replica symmetry breaking in mean field spin glass models. *Ann. Henri Poincaré* **4**, suppl. 1, S441–S444.
78. Guerra F., Toninelli F.L. (2004) The high temperature region of the Viana-Bray diluted spin glass model. *J. Statist. Phys.* **115**, no. 1-2, pp. 531–555.
79. Hanen A. (2007) Un théorème limite pour les covariances des spins dans le modèle de Sherrington-Kirkpatrick avec champ externe. *Ann. Probab.* **35**, no. 1, pp. 141–179.
80. Hanen A. (2008) A limit theorem for mean magnetization in the Sherrington-Kirkpatrick model with an external field. To appear.

81. van Hemmen J.L., Palmer R.G. (1979) The replica method and a solvable spin glass system. *J. Phys. A* **12**, no. 4, pp. 563–580.
82. van Hemmen J.L., Palmer R.G. (1982) The thermodynamic limit and the replica method for short-range random systems. *J. Phys. A* **15**, no. 12, pp. 3881–3890.
83. Hertz J., Krogh A., Palmer R.C. (1991) *Introduction to the theory of neural computation.* Santa Fe Institute Studies in the Sciences of Complexity. Lecture Notes, I. Addison-Wesley Publishing Company, Advanced Book Program, Redwood City. xxii+327 pp.
84. Hopfield J.J. (1982) Neural networks and physical systems with emergent collective computational abilities. *Proc. Natl. Acad. Sci. USA* **79**, pp. 1554–2558.
85. Hopfield J.J. (1982) Neural networks and physical systems with emergent collective computational abilities. *Proc. Natl. Acad. Sci. USA* **81**, pp. 3088.
86. Ibragimov A., Sudakov V.N., Tsirelson B.S. (1976) Norms of Gaussian sample functions. Proceedings of the Third Japan USSR Symposium on Probability theory. *Lecture Notes in Math.* **550**, Springer Verlag, pp. 20–41.
87. Kahane J.-P. (1986) Une inégalité du type de Slepian et Gordon sur les processus gaussiens. *Israel J. Math.* **55**, no. 1, pp. 109–110.
88. Kösters H. (2006) Fluctuations of the free energy in the diluted SK-model. *Stochastic Process. Appl.* **116**, no. 9, pp. 1254-1268.
89. Kim J.H., Roche J.R. (1998) Covering cubes by random half cubes, with applications to binary neural networks: rigorous combinatorial approaches. Eight annual Workshop on Computational Learning Theory, Santa Cruz, 1995. *J. Comput. System Sci.* **56**, no. 2, pp. 223–252.
90. Krauth W., Mézard M. (1989) Storage capacity of memory network with binary couplings. *J. Phys.* **50**, pp. 3057-3066.
91. Kurkova, I. (2005) Fluctuations of the free energy and overlaps in the high-temperature p-spin SK and Hopfield models. *Markov Process. Related Fields* **11**, no. 1, pp. 55–80.
92. Latala R. (2002) Exponential inequalities for the SK model of spin glasses, extending Guerra's method. Manuscript.
93. Ledoux M. (2001) *The concentration of measure phenomenon.* Mathematical Surveys and Monographs **89**, American Mathematical Society, Providence, RI, x+181 pp.
94. Ledoux M. (2000) On the distribution of overlaps in the Sherrington-Kirkpatrick spin glass model, *J. Statist. Phys.* **100**, no. 5-6, pp. 871–892.
95. Ledoux M., Talagrand M. (1991) *Probability in Banach Spaces.* Springer-Verlag, Berlin.
96. Linusson S., Wästlund J. (2004) A proof of Parisi's conjecture on the random assignment problem. *Probab. Theory Relat. Fields* **128**, no. 3, pp. 419–440.
97. Loukianova D. (1997) Lower bounds on the restitution error of the Hopfield model. *Probab. Theory Relat. Fields* **107**, pp. 161–176.
98. Löwe M. (1998) On the storage capacity of Hopfield models with correlated patterns. *Ann. Appl. Probab.* **8**, no. 4, pp. 1216–1250.
99. Löwe M. (1999) The storage capacity of generalized Hopfield models with semantically correlated patterns. *Markov Process. Related Fields* **5**, no. 1, pp. 1–19.
100. Márquez-Carreras D., Rovira C., Tindel S. (2006) Asymptotic behavior of the magnetization for the perceptron model. *Ann. Inst. H. Poincaré Probab. Statist.* **42**, no. 3, pp. 327–342.
101. Mézard M. (1988) The space of interactions in neural networks: Gardner's computation with the cavity method. *J. Phys. A* **22**, pp. 2181–2190.

102. Mézard M., Montanari A. (2009) *Information, Physics, and Computation.* Oxford Graduate Texts. Oxford University Press, xiv+569 pp.
103. Mézard M., Parisi G. (1985) Replicas and optimization. *J. Physique Lett.* **46**, L771.
104. Mézard M., Parisi G. (1986) Mean field equations for the matching and the traveling salesman problem. *Europhys. Lett.* **916**.
105. Mézard M., Parisi G., Virasoro M. (1987) *Spin glass theory and beyond*, World Scientific, Singapore.
106. Milman V. (1988) The heritage of P. Lévy in functional analysis. Colloque P. Lévy sur les processus stochastiques, *Astérisque* **157-158**, pp. 273–301.
107. Milman V., Schechtman G. (1986) Asymptotic theory of finite dimensional normed spaces. *Lecture Notes in Math.* **1200**, Springer Verlag.
108. Monasson R., Zecchina R. (1997) Statistical mechanics of the random K-sat model. *Phys. Rev. E* pp. 1357–1370.
109. Nair C., Prabhakar B., Sharma, M. (2005) Proofs of the Parisi and Coppersmith-Sorkin random assignment conjectures. *Random Structures Algorithms* **27** (2005), no. 4, pp. 413–444.
110. Newman C. (1988) Memory capacity in neural network models: rigorous lower bounds. *Neural Networks* **I**, pp. 223–238.
111. Newman C., Stein D. (1993) Chaotic size dependence in spin glasses. Cellular automata and cooperative systems (Les Houches, 1992). *NATO Adv. Sci. Inst. Ser. C Math. Phys. Sci* **396**, Kluwer Acad. Publ., Dordrecht.
112. Nishimori, H. (2001) *Statistical physics of spin glasses and information processing. An introduction.* International Series of Monographs on Physics **111**, Oxford University Press, New York, xii+243 pp.
113. Newman C., Stein D. (1997) Spatial inhomogeneity and thermodynamic chaos. *Phys. Rev. Letters* **76**, pp. 4821–4824.
114. Panchenko, D. (2005) A central limit theorem for weighted averages of spins in the high temperature region of the Sherrington-Kirkpatrick model. *Electron. J. Probab.* **10**, no. 14, pp. 499–524.
115. Panchenko, D., Talagrand, M. (2004) Bounds for diluted mean-fields spin glass models. *Probab. Theory Relat. Fields* **130**, no. 3, pp. 319–336.
116. Parisi G. (1992) *Field theory, disorder, simulation.* World Scientific Lecture Notes in Physics **49**, World Scientific Publishing Co., Inc., River Edge, NJ, 1992. vi+503 pp.
117. Parisi G. (1992) Statistical physics and spectral theory of disordered systems: some recent developments. *Mathematical Physics* **X**, (Leipzig, 1991), pp. 70–86, Springer, Berlin.
118. Pastur L.A., Figotin A. (1977) Exactly soluble model of a spin glass. *Sov. J. Low Temp. Phys.* **3**, pp. 378–383.
119. Pastur L.A., Figotin A. (1978) On the theory of disordered spin systems. *Theor. Math. Phys.* **35**, pp. 403–414.
120. Pastur L.A., Shcherbina M.V. (1991) Absence of self-averaging of the order parameter in the Sherrington-Kirkpatrick model. *J. Statist. Phys.* **62**, no. 1-2, pp. 1–19.
121. Pastur L.A., Shcherbina M.V., Tirozzi B. (1994) The replica-symmetric solution without replica trick for the Hopfield model. *J. Statist. Phys.* **74**, no. 5-6, pp. 1161–1183.
122. Pastur L.A., Shcherbina M.V., Tirozzi B. (1999) On the replica-symmetric equations for the Hopfield model. *J. Math. Phys.* **40**, no. 8, pp. 3930–3947.
123. Petritris D. (1996) Equilibrium statistical mechanics of frustrated spin glasses: a survey of mathematical results. *Ann. Inst. H. Poincaré Phys. Théor.* **64**, pp. 255–288.

124. Pisier G. (1986) Probabilistic methods in the geometry of Banach spaces. Probability and analysis (Varenna, 1985), pp. 167–241, Lecture Notes in Math. **1206**, Springer, Berlin.
125. Ruelle, D. (1999) *Statistical Mechanics: Rigorous Results.* Reprint of the 1989 edition. World Scientific Publishing Co., Inc., River Edge, NJ, xvi+219 pp.
126. Ruelle, D. (2004) *Thermodynamic Formalism: The Mathematical Structure of Equilibrium Statistical Mechanics.* Second edition. Cambridge Mathematical Library. Cambridge University Press, Cambridge. xx+174 pp.
127. Shcherbina M.V. (1991) More about absence of self averaging of the order parameter in the Sherrington-Kirkpatrick model. *CARR Reports in Mathematical Physics* no 3/91, Department of Mathematics, University of Rome "la Sapienza".
128. Shcherbina M.V. (1997) On the replica-symmetric solution for the Sherrington-Kirkpatrick model. *Helv. Phys. Acta* **70**, pp. 838–853.
129. Shcherbina M.V. (1999) Some estimates for the critical temperature of the Sherrington-Kirkpatrick model with magnetic field. In: *Mathematical Results in Statistical Mechanics*, World Scientific, Singapore, pp. 455–474.
130. Shcherbina M.V., Tirozzi B. (1994) The free energy of a class of Hopfield models. *J. Statist. Phys.* **72**, no. 1-2, pp. 113–125.
131. Shcherbina M.V., Tirozzi B. (1995) A perturbative expansion for the Hopfield model. *Helv. Phys. Acta* **68**, no. 5, pp. 470–491.
132. Shcherbina M.V., Tirozzi B. (2001) On the critical capacity of the Hopfield model. *Comm. Math. Phys.* **216**, no. 1, pp. 139–177.
133. Shcherbina M.V., Tirozzi B. (2003) On the rigorous solution of Gardner's problem. *Comm. Math. Phys.* **234**, no. 3, pp. 383–422.
134. Shcherbina M.V., Tirozzi B. (2003) Central limit theorems for order parameters of the Gardner problem. *Markov Process. Related Fields* **9**, no. 4, pp. 803–828.
135. Shcherbina M.V., Tirozzi B. (2005) Central limit theorems for the free energy of the modified Gardner model. *Markov Process. Related Fields* **11**, no. 1, pp. 133–144.
136. Sherrington D., Kirkpatrick S. (1972) Solvable model of a spin glass. *Phys. Rev. Lett.* **35**, pp. 1792–1796.
137. Slepian D. (1962) The one-side barrier for Gaussian white noise. *Bell Systems Tech. J.* **41** pp. 463–501.
138. Talagrand M. (1987) Regularity of Gaussian processes. *Acta. Math.* **159**, pp. 99–149.
139. Talagrand M. (1995) Concentration of measure and isoperimetric inequalities in product spaces. *Publ. Math. I.H.E.S.* **81**, pp. 73–205.
140. Talagrand M. (1996) A new look at independence. *Ann. Probab.* **24**, pp. 1–34.
141. Talagrand M. (1998) The Sherrington-Kirkpatrick model: a challenge to mathematicians. *Probab. Theory Relat. Fields* **110**, pp. 109–176.
142. Talagrand M. (1998) Rigorous results for the Hopfield model with many patterns. *Probab. Theory Relat. Fields* **110**, pp. 177–276.
143. Talagrand M. (1998) Huge random structures and mean field models for spin glasses, Proceedings of the Berlin International Congress of mathematicians. *Documenta Math.*, Extra Vol. **I**, pp. 507–536.
144. Talagrand M. (1999) Intersecting random half cubes. *Random Structures and Algorithms* **15**, pp. 436–449.
145. Talagrand M. (1999) Self-averaging and the space of interactions in neural networks. *Random Structures and Algorithms* **14**, pp. 199–213.

146. Talagrand M. (2000) Verres de spin et optimisation combinatoire, Séminaire Bourbaki, Vol. 1998-99. *Astérisque* **15**, pp. 436–449.
147. Talagrand M. (2000) Rigorous low temperature results for the mean field p-spin interaction model. *Probab. Theor. Relat. Fields* **117**, pp. 303–360.
148. Talagrand M. (2000) Intersecting random half-spaces: towards the Derrida-Gardner formula. *Ann. Probab.* **28**, pp. 725–758.
149. Talagrand M. (2000) Exponential inequalities and replica-symmetry breaking for the Sherrington-Kirkpatrick model. *Ann. Probab.* **28**, pp. 1018–1062.
150. Talagrand M. (2000) Exponential inequalities and convergence of moments in the replica-symmetric phase of the Hopfield model. *Ann. Probab.* **28**, pp. 1393–1469.
151. Talagrand M. (2000) Large deviation principles and generalized Sherrington-Kirkpatrick models. *Ann. Fac. Sci. Toulouse Math.* **9**, pp. 203–244.
152. Talagrand M. (2000) Spin glasses: a new direction for probability theory? Mathematics towards the third millennium (Rome, 1999). *Atti Accad. Naz. Lincei Cl. Sci. Fis. Mat. Natur. Rend. Lincei* **9**, Mat. Appl. Special Issue, pp. 127–146.
153. Talagrand M. (2001) The high temperature phase of the random K-sat problem. *Probab. Theory Relat. Fields* **119**, pp. 187–212.
154. Talagrand M. (2001) The Hopfield model at the critical temperature. *Probab. Theory Relat. Fields* **121**, pp. 237–268.
155. Talagrand M. (2002) On the high temperature phase of the Sherrington-Kirkpatrick model. *Ann. Probab.* **30**, pp. 364–381.
156. Talagrand M. (2002) On the Gaussian perceptron at high temperature. *Math. Phys. Anal. Geom.* **5**, no. 1, pp. 77–99.
157. Talagrand M. (2003) *Spin glasses: a challenge for mathematicians. Cavity and mean field models.* Ergebnisse der Mathematik und ihrer Grenzgebiete. 3. Folge. A Series of Modern Surveys in Mathematics, **46**. Springer-Verlag, Berlin. x+586 pp.
158. Talagrand M. (2003) Mean field models for spin glasses: a first course, Lectures on probability theory and statistics (Saint-Flour, 2000), pp. 181–285, Lecture Notes in Math. **1816**, Springer, Berlin.
159. Talagrand M. (2007) Large Deviations, Guerra's and A.S.S. Schemes, and the Parisi Hypothesis. *J. Stat. Phys.* **126**, no. 4, pp. 837–894.
160. Thouless D.J., Anderson P.W., Palmer R.G. (1977) Solution of 'Solvable model of a spin glass' *Philosphical Magazine* **35**, no. 3, pp. 593–601.
161. Tijms H. (2007) *Understanding Probability. Chance rules in everyday life.* Second edition. Cambridge University Press, Cambridge. x+442 pp.
162. Tindel S. (2003) Quenched large deviation principle for the overlap of a p-spins system *J. Stat. Phys.* **110**, pp. 51–72.
163. Tindel S. (2005) On the stochastic calculus method for spins systems. *Ann. Probab.* **33**, no. 2, pp. 561–581.
164. Toubol A. (1998) High temperature regime for a multidimensional Sherrington-Kirkpatrick model of spin glass. *Probab. Theory Relat. Fields* **110**, no. 4, pp. 497–534.
165. Toubol A. (1999) Small random perturbation of a classical mean field model. *Stochastic Process. Appl.* **81**, no. 1, pp. 1–24.
166. Toulouse G. (1983) Frustration and disorder, new problems in statistical mechanics, spin glasses in a historical perspective. Heidelberg Colloquium on Spin Glasses, J.L. van Hemmen and I. Morgenstern eds. *Lecture Notes in Physics* **192**, Springer Verlag.
167. Wästlund J. (2010), The mean field traveling salesman and related problems, *Acta Mathematica* **204**, no. 1, pp. 91-150.

168. Wästlund J. (2009) An easy proof of the $\zeta(2)$ limit in the random assignment problem. *Electron. Commun. Probab.* **14**, pp. 261–269.
169. Wästlund J. Replica-symmetry and combinatorial optimization, Manuscript, 2009.

Index

M. Talagrand, *Mean Field Models for Spin Glasses*, Ergebnisse der Mathematik und ihrer Grenzgebiete. 3. Folge / A Series of Modern Surveys in Mathematics 54, DOI 10.1007/978-3-642-15202-3, © Springer-Verlag Berlin Heidelberg 2011

Glossary

G'	In the Hopfield model, the image of the Gibbs' measure under the map $\boldsymbol{\sigma} \mapsto \mathbf{m}(\boldsymbol{\sigma}) = (m_k(\boldsymbol{\sigma}))_{k\leq M}$, 244
G_N	Gibbs' measure on Σ_N, 5
K	A quantity that does not depend on N, although it might depend on other parameters of the model. Its value might not be the same at each occurrence, 30
L	A universal constant, i.e. a number, that does not depend on anything. Its value might not be the same at each occurrence, 30
$O(k)$	Any quantity A such that $\lvert A\rvert \leq KN^{-k/2}$ where K does not depend on N, 41
$R_{1,2}$	The overlap of configurations $\boldsymbol{\sigma}^1$ and $\boldsymbol{\sigma}^2$, that is the quantity $N^{-1}\sum_{1\leq i\leq N}\sigma_i^1\sigma_i^2$, 3
$R_{\ell,\ell'}$	The overlap between configurations $\boldsymbol{\sigma}^\ell$ and $\boldsymbol{\sigma}^{\ell'}$, that is $N^{-1}\sum_{1\leq i\leq N}\sigma_i^\ell\sigma_i^{\ell'}$, 35
$R^-_{\ell,\ell'}$	The quantity $N^{-1}\sum_{1\leq i<N}\sigma_i^\ell\sigma_i^{\ell'}$, 55
S_k	In the models of Chapter 2 and 3, this denotes the quantity $N^{-1/2}\sum_{i\leq N} g_{i,k}\sigma_i$. We can also denote this quantity by $S_k(\boldsymbol{\sigma})$. We then use the short-hand notation $S_k^\ell = S_k(\boldsymbol{\sigma}^\ell)$, 153
S_v^ℓ	The quantity corresponding to S_k when using the "cavity in M", where t is now fixed, and v is the interpolation parameter, 161
$S_{k,t}$	The quantity corresponding to S_k when using the cavity method (interpolation along the last spin), e.g. in the case of the perceptron model this quantity is given by (2.15). One then need the "replicated versions" $S_{k,t}^\ell$ of $S_{k,t}$ such as in (2.22), 155
$T_{\ell,\ell'}$; T_ℓ ; T	$T_{\ell,\ell'} = \frac{1}{N}\sum_{i\leq N}\dot\sigma_i^\ell\dot\sigma_i^{\ell'}$; $T_\ell = \frac{1}{N}\sum_{i\leq N}\dot\sigma_i^\ell\langle\sigma_i\rangle$; $T = \frac{1}{N}\sum_{i\leq N}\langle\sigma_i\rangle^2 - q$, 87

M. Talagrand, *Mean Field Models for Spin Glasses*, Ergebnisse der Mathematik und ihrer Grenzgebiete. 3. Folge / A Series of Modern Surveys in Mathematics 54,
DOI 10.1007/978-3-642-15202-3, © Springer-Verlag Berlin Heidelberg 2011

W	In the Hopfield model, $W = (N\beta/2\pi)^{M/2}$, 244
Y_N	Often $Y = \beta z\sqrt{q} + h$, 40
E	Mathematical expectation, VII
$\mathsf{E}Y^2$	Short-hand for $\mathsf{E}(Y^2)$, 3
E_ξ	Expectation only in the r.v. ξ, that is, for all the other r.v.s given. More generally, expectation only in the r.v.s "named from ξ" such as ξ^ℓ, 51
$\mathcal{E}$	This notation and its avatars such as $\mathcal{E}_\ell$, etc. is used throughout the book to denote an exponential term that occurs when using the cavity method, 79
Ω	This denotes an *event*, not the whole of the probability space, 247
RS	The typical name for a "replica-symmetric formula", i.e. and expression that gives the limiting value of p_N at high temperature. In the case of the SK model, this is denoted SK(β, h), VII
$\mathbb{S}_N$	The sphere of $\mathbb{R}^N$of center 0 and radius $\sqrt{N}$, VII
SK(β, h)	The expression giving the replica-symmetric formula in the case of the SK model, 24
Σ_N	$\{-1, 1\}^N$, IX
α	In a model with two parameters M and N, such as in Chapters 2 to 4, this often denotes the ratio M/N. This might also denote a number > 0, such as in the expression "$M/N \to \alpha$", 153
ch$x$	The hyperbolic cosine of x, 9
$\langle\cdot\rangle$	An average for the Gibbs measure or its products, 6
$\langle\cdot\rangle_t$	A Gibbs average for an interpolating Hamiltonian, when the interpolating parameter is equal to t, 20
log	The natural logarithm, VII
th$x$	The hyperbolic tangent of x, 9
$\nu_t'(f)$	A short-hand for $\mathrm{d}\nu_t(f)/\mathrm{d}t$, 31
$\nu(f)$	A short-hand for $\mathsf{E}\langle f\rangle$, 30
$\nu(f)^{1/2}$	A short-hand for $(\nu(f))^{1/2}$, 131
$\nu_t(f)$	A short-hand for $\mathsf{E}\langle f\rangle_t$, 31
$\nu_{t,v}$	This is the quantity that corresponds to ν_t when we interpolate "in the cavity in M" method along the parameter v, so $\nu_t = \nu_{t,1}$, and supposedly, $\nu_{t,0}$ is easier to compute than ν_t, 162

$\overline{G}$	In the Hopfield model, the convolution of G' with γ, the Gaussian measure of density $W\exp(-\beta N\\|\mathbf{z}\\|^2/2)$ with respect to Lebesgue measure on $\mathbb{R}^M$. It is a small perturbation of G', 244
$\psi(\mathbf{z})$	In the Hopfield model, the quantity $\psi(\mathbf{z}) = -N\beta\\|\mathbf{z}\\|^2/2 + \sum_{i\le N}\log\mathrm{ch}(\beta\boldsymbol{\eta}_i\cdot\mathbf{z}+h)$, 245
$\dot{\sigma}_i$	A short-hand for $\sigma_i - \langle\sigma_i\rangle$, 47
$\mathrm{sh}x$	The hyperbolic sine of x, 9
$\stackrel{\mathcal{D}}{=}$	Equality in distribution, 62
$\mathbf{b}$	The barycenter of μ, 51
ε_ℓ	A short-hand for σ_N^ℓ, the last spin of the ℓ-th replica, 55
$\boldsymbol{\nabla} F$	The gradient of F, 16
$\dot{\boldsymbol{\sigma}}$	The sequence $(\dot{\sigma}_i)_{1\le i\le N}$, 47
$\mathbf{g}$	A standard Gaussian vector, that is $\mathbf{g} = (g_1,\dots,g_M)$ where $g_1,\dots,g_M$ are i.i.d. standard Gaussian r.v.s, and where M should be clear from the context, 15
$\mathbf{m}(\boldsymbol{\sigma})$	In the Hopfield model, $\mathbf{m}(\boldsymbol{\sigma}) = (m_k(\boldsymbol{\sigma}))_{k\le M}$, 244
$\boldsymbol{\rho}$	When considering a configuration $\boldsymbol{\sigma} = (\sigma_1,\dots,\sigma_N)$ we denote by $\boldsymbol{\rho}$ the configuration $(\sigma_1,\dots,\sigma_{N-1})$ is the $(N-1)$-spin system, 53
$\boldsymbol{\sigma}^\ell$	The standard name for a configuration in the ℓ-th replica, 6
$\widehat{q}$	Most of the time, $\widehat{q} = \mathsf{E}\mathrm{th}^4 Y = \mathsf{E}\mathrm{th}^4(\beta z\sqrt{q}+h)$, 83
$a(k)$	The k-th moment of a standard Gaussian r.v., except in chapter 7 where the meaning is different, 40
a^*	In the Hopfield model, $a^* = 1-\beta(1-m^{*2})$, 256
b^*	In the Hopfield model, $b^* = \log\mathrm{ch}(\beta m^* + h) - \frac{\beta}{2}m^{*2}$, 242
m^*	In the Hopfield model, the solution of the equation $m^* = \mathrm{th}(\beta m^* + h)$, 241
m_k	In the Hopfield model, $m_k = m_k(\boldsymbol{\sigma}) = N^{-1}\sum_{i\le N}\eta_{i,k}\sigma_i$, 240
p_N	$p_N = N^{-1}\mathsf{E}\log Z_N$. This quantity is also denoted $p_N(\beta)$ or $p_N(\beta,h)$. In models where there are two parameters N and M, as in Chapters 2, 3, 4, might be denoted $p_{N,M}$, 6

$\mathcal{A}(x)$	The function $-\frac{\mathrm{d}}{\mathrm{d}x}\log \mathcal{N}(x) = \frac{1}{\sqrt{2\pi}}\frac{e^{-x^2/2}}{\mathcal{N}(x)}$, 228
$\mathcal{I}(t)$	The function $$\mathcal{I}(t) = \frac{1}{2}\big((1+t)\log(1+t) + (1-t)\log(1-t)\big),$$ which satisfies $\mathcal{I}(0) = \mathcal{I}'(0) = 0$ and $\mathcal{I}''(0) = 1/(1-t^2)$, see (A.29), 237
$\mathcal{N}(x)$	The probability that a standard Gaussian r.v. g is $\geq x$, VII
$\mathcal{R}$	Denotes a quantity which is a remainder, of "smaller order", such as in (1.217), 81
$\mathbf{1}_A$	The indicator function of the set A, 49
Av	Typically denotes the average over one or a few spins that take values ± 1, 53
approximate integration by parts	A central technique to handle situations where the randomness is generated by Bernoulli r.v.s rather than by Gaussian r.v.s. It relies on the identity (4.198), 289
AT line	For the SK model, the line of equation $\beta^2 \mathsf{E}\mathrm{ch}^{-4}(\beta z\sqrt{q} + h) = 1$, where q is the solution of (1.74), 80
Bernoulli r.v.	A r.v. η such that $\mathsf{P}(\eta = 1) = \mathsf{P}(\eta = -1) = 1/2$, 13
Boltzmann factor	At the configuration $\boldsymbol{\sigma}$ is has the value $\exp(-\beta H_N(\boldsymbol{\sigma}))$, IX
configuration	An element of the configuration space, which is most of the time either Σ_N are $\mathbb{S}_N$, IX
decorrelate	The spins σ_1 and σ_2 decorrelate when $\lim_{N\to\infty} \mathsf{E}\lvert\langle\sigma_1\sigma_2\rangle - \langle\sigma_1\rangle\langle\sigma_2\rangle\rvert = 0$. One expects this behavior at high temperature. One also expects that asymptotically the r.v.s $\langle\sigma_1\rangle$ and $\langle\sigma_2\rangle$ are independent, 47
disorder	The randomness of the Hamiltonian, X
energy	A number associated to each configuration, which is often random, IX

essentially supported A random measure G is essentially supported by a set A (depending on N and M) if the complement A^c of A is negligible, 254

external field A term $h\sum_{1\leq i\leq N}\sigma_i$ occurring in the Hamiltonian, 4

finite connectivity A situation where the average number of spins that interact with a given spin remains bounded as the size of the system increases, XII

Gibbs' measure The probability on the configuration space with density proportional to the Boltzmann factor, IX

Griffiths' lemma If a sequence φ_N of convex differentiable functions converges pointwise in an interval to a (necessarily convex) function φ, then $\lim_{N\to\infty}\varphi'_N(x) = \varphi'(x)$ at every point x for which $\varphi'(x)$ exists, 25

Hamiltonian The function that associates to each configuration its energy, IX

Hamming distance The Hamming distance of two sequences $\boldsymbol{\sigma}^1$ and $\boldsymbol{\sigma}^2$ of Σ_N is the proportion of coordinates where they differ, 3

high-temperature behavior The situation where the spins decorrelate, and where the limiting value of p_N is given by the replica-symmetric formula, 11

independent This word is always understood in the probabilistic sense, 1

interchange of limits A very sticky point, 8

Jensen's inequality For a convex function φ and a r.v. X, the fact that $\varphi(\mathsf{E}X)\leq\mathsf{E}\varphi(X)$, 8

negligible If G is a random measure, as set A (depending on N and M) is negligible if $\mathsf{E}G(A)\leq K\exp(-N/K)$ where K does not depend on N or M, 254

overlap	The overlap between two configurations $\boldsymbol{\sigma} = (\sigma_1, \ldots, \sigma_N)$ and $\boldsymbol{\tau} = (\tau_1, \ldots, \tau_N)$ is the quantity $N^{-1}\boldsymbol{\sigma} \cdot \boldsymbol{\tau} = N^{-1}\sum_{1\le i\le N} \sigma_i \tau_i$, 3
overwhelming probability	An event Ω (depending on M and N) occurs with overwhelming probability if $\mathsf{P}(\Omega) \ge 1 - K\exp(-N/K)$ for some number K that does not depend either on N and M, 247
partition function	The normalizing factor in Gibbs' measure, $Z_N = Z_N(\beta) = \sum_{\boldsymbol{\sigma}} \exp(-\beta H_N(\boldsymbol{\sigma}))$, IX
r.v.s	A short-hand for random variables, 1
random external field	A term $\sum_{1\le i\le N} h_i \sigma_i$ in the Hamiltonian, where (h_i) are i.i.d. r.v.s, 20
replica-symmetric	In physics' terminology, this describes "high-temperature behavior". The limiting value of p_N is then given by the *replica-symmetric formula.* This formula depends on some "free parameters" that are specified by the *replica-symmetric equations.* These equations always seem to express that the free parameters are a critical point of the replica-symmetric formula. A simple example of replica-symmetric formula is given by the right-hand side of (1.73), and the corresponding replica-symmetric equation is (1.74), 22
replicas	Configurations that are averaged independently for Gibbs' measure. They are typically denoted by $\boldsymbol{\sigma}^1, \ldots, \boldsymbol{\sigma}^\ell, \ldots$, 6
self-averaging	Informally, a random quantity X_N such that $\mathsf{E}\lvert X_N\rvert \ge 1/L$ but that the variance of X_N goes to 0 as $N \to \infty$. The value of $\mathsf{E}X_N$ then "gives all the first-order information about X_N", 7
site	An integer $1 \le i \le N$, 8
symmetry between replicas	A consequence of the fact that $(\boldsymbol{\sigma}^\ell)$ is an i.i.d. sequence for Gibbs' measure. A good place to learn about it is the beginning of the proof of Proposition 1.8.7, 32

symmetry between sites — A general principle that for many Hamiltonians, the sites "play the same rôle", 8

typical — A situation that occurs with probability near 1, as opposed to an exceptional situation, which occurs with probability near 0, VIII